Comprehensive Treatise of Electrochemistry

Volume 8
Experimental Methods
in Electrochemistry

COMPREHENSIVE TREATISE OF ELECTROCHEMISTRY

A Continuation Order Plan is available for this series. A continuation order will bring delivery of each new volume immediately upon publication. Volumes are billed only upon actual shipment. For further information please contact the publisher.

Comprehensive Treatise of Electrochemistry

Volume 8
Experimental Methods
in Electrochemistry

Edited by

Ralph E. White

Texas A&M University
College Station, Texas

J. O'M. Bockris

Texas A&M University
College Station, Texas

Brian E. Conway

University of Ottawa
Ottawa, Ontario, Canada

Ernest Yeager

Case Western Reserve University
Cleveland, Ohio

PLENUM PRESS ▪ NEW YORK AND LONDON

Library of Congress Cataloging in Publication Data

Main entry under title:

Experimental methods in electrochemistry.

(Comprehensive treatise of electrochemistry; v. 8)
Includes bibliographical references and index.
1. Electrochemistry—Laboratory manuals. I. White, Ralph E. II. Series.
QD552.C64   vol. 8 [QD557]   541.37s [541.37]   84-4826
ISBN 0-306-41448-1

©1984 Plenum Press, New York
A Division of Plenum Publishing Corporation
233 Spring Street, New York, N.Y. 10013

Printed in the United States of America

Contributors

Richard P. Buck • Department of Chemistry, University of North Carolina, Chapel Hill, North Carolina 27514

Martin Fleischmann • Department of Chemistry, University of Southampton, Southampton SO9 5NH, England

Robert Greef • Department of Chemistry, University of Southampton, Southampton SO9 5NH, England

J. S. Hammond • Physical Electronic Division, Perkin-Elmer Corp., 6509 Flying Cloud Drive, Eden Prairie, Minnesota 55344

I. R. Hill • Department of Chemistry, University of Southampton, Southampton SO9 5NH, England

D. J. Kampe • Union Carbide Corporation, Carbon Products Division, Parma Technical Center, P.O. Box 6116, Cleveland, Ohio 44101

B. Kastening • Institute für Physikalische Chemie der Universität Hamburg, Laufgraben 24, D-2000 Hamburg 13, West Germany

Jaroslav Kúta • The J. Heyrovsky Institute of Physical Chemistry and Electrochemistry, Czechoslovak Academy of Sciences, 11840 Prague 1, Czechoslovakia

Digby D. Macdonald • Fontana Corrosion Center, Department of Metallurgical Engineering, The Ohio State University, Columbus, Ohio 43210

Harry B. Mark, Jr. • Department of Chemistry, University of Cincinnati, Cincinnati, Ohio 45221

Michael C. H. McKubre • Materials Research Laboratory, SRI International, Menlo Park, California 94025

Leonard Nanis • Consultant, Electrochemical Engineering, 627 Georgia Ave., Palo Alto, California 94306

Thomas H. Ridgway • Department of Chemistry, University of Cincinnati, Cincinnati, Ohio 45221

V. K. Venkatesan • Central Electrochemical Research Institute, Karaikudi-6203006, Tamilnadu, India

N. Winograd • Department of Chemistry, The Pennsylvania State University, University Park, Pennsylvania 16802

v

Preface to Comprehensive Treatise of Electrochemistry

Electrochemistry is one of the oldest defined areas in physical science, and there was a time, less than 50 years ago, when one saw "Institute of Electrochemistry and Physical Chemistry" in the chemistry buildings of European universities. But, after early brilliant developments in electrode processes at the beginning of the twentieth century and in solution chemistry during the 1930s, electrochemistry fell into a period of decline which lasted for several decades. Electrochemical systems were too complex for the theoretical concepts of the quantum theory. They were too little understood at a phenomenological level to allow their ubiquity in application in so many fields to be comprehended.

However, a new growth began faintly in the late 1940s, and clearly in the 1950s. This growth was exemplified by the formation in 1949 of what is now called The International Society for Electrochemistry. The usefulness of electrochemistry as a basis for understanding conservation was the focal point in the founding of this Society. Another very important event was the choice by NASA in 1958 of fuel cells to provide the auxiliary power for space vehicles.

With the new era of diminishing usefulness of the fossil fuels upon us, the role of electrochemical technology is widened (energy storage, conversion, enhanced attention to conservation, direct use of electricity from nuclear–solar plants, finding materials which interface well with hydrogen). This strong new interest is not only in the technological applications of electrochemistry. Quantum chemists have taken great interest in redox processes. Organic chemists are interested in situations where the energy of electrons is as easily controlled as it is at electrodes. Some biological processes are now seen in electrodic terms, with electron transfer to and from materials which would earlier have been considered to be insulators.

It is now time for a comprehensive treatise to look at the whole field of electrochemistry.

The present treatise was conceived in 1974, and the earliest invitations to authors for contributions were made in 1975. The completion of the early volumes has been delayed by various factors.

There has been no attempt to make each article emphasize the most recent situation at the expense of an overall statement of the modern view. This treatise is not a collection of articles from *Recent Advances in Electrochemistry* or *Modern Aspects of Electrochemistry*. It is an attempt at making a mature statement about the present position in the vast area of what is best looked at as a new interdisciplinary field.

Texas A & M University   J. O'M. Bockris
University of Ottawa   B. E. Conway
Case Western Reserve University   Ernest Yeager
Texas A & M University   Ralph E. White

Preface to Volume 8

Experimental methods in electrochemistry are becoming more diverse. This volume describes many of the new techniques that are being used as well as some of the well-established techniques. It begins with two chapters (1 and 2) on electronic instrumentation and methods for utilization of microcomputers for experimental data acquisition and reduction. Next, two chapters (3 and 4) on classical methods of electrochemical analysis are presented: ion selective electrodes and polarography. Relatively new methods are then presented beginning with ellipsometry (Chapter 5), followed by Raman spectroscopy (Chapter 6), electron spin resonance (Chapter 7), electron spectroscopy (Chapter 9), and electron microscopy (Chapter 10). Finally, an extensive list of references is presented in Chapter 11 for both old and new experimental methods in electrochemistry.

Texas A&M University   Ralph E. White
Case Western Reserve University   Ernest Yeager
Texas A&M University   J. O'M. Bockris
University of Ottawa   B. E. Conway

Contents

1. Electronic Instrumentation for Electrochemical Studies

Michael C. H. McKubre and Digby D. Macdonald

4. Polarography

Jaroslav Kůta

5. Ellipsometry

Robert Greef

6. Raman Spectroscopy

M. Fleischmann and I. R. Hill

7. Electron Spin Resonance
Bertel Kastening

8. Electron Spectroscopy for Chemical Analysis (ESCA) and Electrode Surface Chemistry

J. S. Hammond and N. Winograd

9. Field Ion Microscopy

Leonard Nanis

10. Application of Electron Microscopy to Electrochemical Analysis

D. J. Kampe

11. Classified Bibliography of Electroanalytical Applications

V. K. Venkatesan

1

Electronic Instrumentation for Electrochemical Studies

MICHAEL C. H. McKUBRE
AND DIGBY D. MACDONALD

Auxiliary Notation

A	Op amp open loop gain; V/V
A_o	open loop gain at the limit of low frequency; V/V
A^0	signal amplitude; V
BN	normalized bandwidth of a bandpass filter; dimensionless
BW	bandwidth of a bandpass filter; Hz
CE	counter electrode
$CMRR$	common mode rejection ratio; db
e	output voltage of a potential control device; V
$e(j\omega)$	frequency domain potential; V
$E(t)$	time domain potential; V
e_a	potential at the noninverting input of an Op Amp; V
e_b	potential at the inverting input of an Op Amp; V
e_c	control potential at the input of a voltage comparator; V
e_{cm}	input common mode potential at a differential amplifier; V
e_i	generalized input potential in an Op Amp circuit; V
e_n	a-c noise potential; V

MICHAEL C. H. McKUBRE • SRI International, Materials Research Laboratory, Menlo Park, California 94025. **DIGBY D. MACDONALD** • Fontana Corrosion Center, Department of Metallurgical Engineering, The Ohio State University, Columbus, Ohio 43210.

e_o generalized output potential in an Op Amp circuit; V

e_o^0 initial condition of an integrator circuit; V

e_{os} d-c offset voltage of an Op Amp; V

e_r reference potential at the input of a voltage comparator; V

e_{ref} reference electrode potential in a three-terminal electrochemical cell; V

E_{WE} working electrode potential in a three-terminal electrochemical cell; V

$G(j\omega)$ open loop, frequency domain transfer function; dimensionless

G_{CA} open loop gain of the control amplifier in a simple potentiostat; V/V

$H(j\omega)$ Op Amp feedback characteristic; dimensionless

$\{H_E(s)\}$ potential control transfer function for a single-amplifier potentiostat; dimensionless

$\{H_I(s)\}$ current control transfer function for a single-amplifier potentiostat; dimensionless

$i(j\omega)$ frequency domain current; A

$I(t)$ time domain current; A

I_f feedback current; A

i_i input current; A

i_n noise current; A

i_n^+ error current in the noninverting input of an Op Amp; A

i_n^- error current in the inverting input of an Op Amp; A

$i_n^\pm$ differential error current; A

i_o output current; A

IR product of the current and the uncompensated electrolyte resistance for a three terminal electrochemical cell; V

j complex operator $= (-1)^{1/2}$

$L(j\omega)$ loop gain; dimensionless

n order of a high-pass or low-pass filter; dimensionless

NIC   negative impedance convertor

PSD   phase sensitive detector

Q   selectivity of a bandpass or band-reject filter; dimensionless

RE   reference electrode

R_f feedback resistance; Ω

R_i input resistance; Ω

R_n ratio of noise voltage to noise current; Ω

R_s resistance between the counter and reference electrodes in a three-terminal electrochemical cell; Ω

R_u uncompensated electrolyte resistance between the reference and working electrodes in a three-terminal electrochemical cell; Ω

s summing point

(s) Laplacian frequency operator

t_d transient time response of a potentiostat circuit

$T(j\omega)$ closed-loop gain; dimensionless

TWR   transient wave recorder

u voltage maintained by a potential control device; V

WE   working electrode

$Y(j\omega)$ test transfer function; dimensionless

Z_{cm} common mode input impedance; Ω

Z_d differential mode input impedance; Ω

Z_f feedback impedance; Ω

Z_i input impedance; Ω

α resistance ratio; dimensionless

β Butterworth filter damping ratio; dimensionless

β resistance ratio; dimensionless

γ resistance ratio; dimensionless

δ resistance ratio; dimensionless

$\phi_{(j\omega)}$ phase response of active filter circuit; radians

ρ the sum of all series resistance terms not associated with the interface being controlled, for a potential control device; Ω

ω frequency; Hz

ω_h cut-off frequency for a high-pass filter; Hz

ω_l cut-off frequency for a low-pass filter; Hz

ω_o^* center frequency for a bandpass or band-reject filter; Hz

ω_o turnover frequency for Op Amp gain rolloff; Hz

ω_p Op Amp gain-bandwidth product; Hz

1. Introduction

Electrochemistry has undergone a revolution over the past two decades that has completely transformed existing experimental techniques and has led to the introduction of a myriad of new sophisticated methods for studying charge transfer processes at electrode/solution interfaces. The reason for this revolution was the introduction of integrated circuit operational amplifiers, which now form the inexpensive "building blocks" of modern electrochemical instrumentation. Operational amplifiers having bandwidths up to the mega-hertz range enabled experimenters to "tailor make" control instruments for their needs, and also led to a whole range of commercial instrumentation becoming available for use by those not well-versed in basic electronics.

Over the last few years, a second revolution has been taking place. This revolution has resulted from the development and general availability of inexpensive mini- and micro-computers. These, in turn, have greatly increased experimental flexibility and our ability to handle the vast amounts of data that are now frequently produced in many electrochemical experiments. This has been made possible by handling data in the digital form rather than in the hitherto analog domain.

It is our belief that a knowledge of basic electronics and computational techniques is essential for any practicing electrochemist. Sadly, few university courses on electrochemistry include an adequate discussion of the subject, and those that do are often very superficial in their treatment. In this chapter we present a brief review of the essentials of this subject as we see them, for the purpose of providing a base on which the reader can develop a more sophisticated understanding of the role of electronics in modern electrochemistry.

2. Operational Amplifiers

2.1. Introduction

We have chosen to employ the "building block" or "black box" approach as a means to form a basis of the electronic fundamentals involved. This approach allows us to describe, with a minimum of complexity, the aspects of electronic circuit design that are important to an experimentalist.

The principal building block of "linear" circuitry unquestionably is the Operational Amplifier or "Op Amp." In this section we seek to provide a familiarity with Op Amp principles and limitations sufficient for scientists with no formal training in electronics to enjoy the consequent discussion.

2.2. Functional Description

The name Operational Amplifier is derived from the use of differential, high gain d-c amplifiers, to perform mathematical operations in analog computers. The precision and flexibility of the Op Amp is a direct result of the use of a negative feedback, which is achieved when a fraction of the amplified output signal is returned, out of phase to the input. It is important at the outset to distinguish between the "open-loop" characteristics, which are specified by the manufacturer, and the expected "closed-loop" characteristics obtained when employing a particular feedback combination.

The simplicity of Op Amp circuit design resides in the fact that, as a first-order effect, the functioning of a circuit is independent of the amplifier open-loop chracteristics, and may be predicted by a few simple rules. Operation thus is implementation, and not device dependent. However, while open-loop specifications do not define circuit operation, they frequently may limit it.

Figure 1a shows on Op Amp, represented schematically by a triangle, with no declared feedback path, and thus in "open-loop" operation. The inputs are classified as inverting ($-$) and noninverting ($+$), and the output is the apex of the triangle opposite the inputs. A common ground reference for inputs e_a and e_b and output e_o is implied by a dashed line. The amplifier shown is differential in that the output (ideally) is a function of the difference between e_a and e_b, and not of the magnitude of either. The input/output transfer

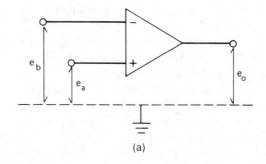

(a)

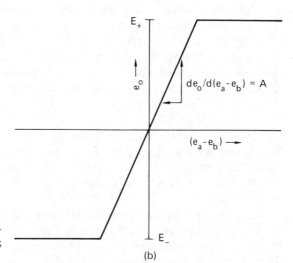

Figure 1. (a) Open-loop operational amplifier circuit symbol;
(b) Open-loop voltage response.

(b)

characteristics of a typical device are shown in Figure 1b. Because of the large
open-loop gain, typically $10^4 < A < 10^7$, the amplifier responds linearly only
within a narrow range of input potentials. The output saturates (*i.e.*, can no
longer change) when the product $A(e_a - e_b)$ exceeds the amplifier power
supply voltage.

While its limitations are obvious, the device in Figure 1a may be practically
employed as a voltage comparator, since the output will switch states when
e_a exceeds e_b by as little as $1\ \mu V$.

2.3. The Ideal Op Amp

In discussing closed-loop operation, it is customary to define the ideal
open-loop amplifier so that the open-loop characteristic can be ignored, at
least for the present. The following are a set of conditions sufficient for the

amplifier response to be determined only by the feedback path:

   • Infinite gain: $A = e_o/(e_a - e_b) = \infty$.

   • Infinite input impedance: The current drawn by either input from a voltage source referenced to ground (common mode) or to the other terminal (differential mode) is zero.

   • Zero output impedance: The amplifier can supply unlimited current to the load being driven, independent of e_o and without altering e_o.

   • Zero response time: The output occurs at the same time as the input. For a-c signals, the phase shift is 180° with respect to the inverting input, and 0° with respect to the noninverting input, independent of frequency.

   • Zero d-c offset: $e_o = 0$ when $e_a - e_b = 0$, independent of input voltage level.

   The ideal simplifier in closed-loop operation is shown in Figure 2. Since the input is supplied to the inverting terminal, the circuit is part of the general class of operational inverters. However, without specifying Z_i and Z_f, the input and feedback impedance, respectively, we cannot predict the phase relationship between e_i and e_o or whether feedback is positive, negative, or nonexistent. Nevertheless, a number of important consequences arise immediately by assuming ideal behavior as specified above for the amplifier within the dashed lines.

   1. The summing point constraint: Since no current may flow to either input, any current which flows from the source, i.e., to the point labeled s, must flow through Z_f. For reasons that will become apparent, the point labeled s at the intersection of the input and feedback circuits is known as the summing point.

   2. The virtual ground conditions: Due to the (assumed) infinite amplifier gain, under stable operating conditions, $e_b - e_a = 0$.

   Two points need clarification. The first concerns the feedback. It will be shown later that stable operating conditions require negative feedback, i.e., that the phase shift in the feedback loop be within the range $-\pi/2 \leq \Phi \leq \pi/2$ radians. The second point is that, under conditions of negative feedback, the summing point is a virtual ground. It is "virtual" in that a passive, low-resistance path to the actual circuit ground does not exist. The behavior of this point emulates a ground because of the feedback loop.

   Taken together with the (assumed) instantaneous response and zero output inmpedance of the control amplifier, these two "laws" or assertions provide a sufficient basis to characterize the operation of the vast majority of closed-loop amplifier circuits.

   For the circuit shown in Figure 2a, if we ignore the output impedance of the source, then the left-hand side of Z_i is held at e_i. By reason of assertion 2, $e_s = 0$, and by assertion 1 the current flowing through Z_i also flows in Z_f. The circuit thus has the form of an *impedance lever* as shown in Figure 2b.

$$\frac{e_o}{e_i} = \frac{-Z_f}{Z_i} \tag{1}$$

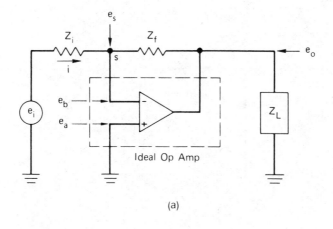

(a)

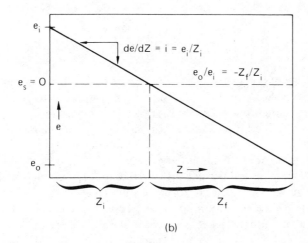

(b)

Figure 2. The closed-loop inverting amplifier (a) with gain represented as an impedance lever (b).

Notice that if Z_f and Z_i are pure resistances, then the circuit has conventional inverting amplifier properties with frequency independent gain.

Figure 3 demonstrates an important consequence of the fact that under conditions of negative feedback the summing point maintains virtual ground potential, and the reason this point is so named. Since $e_s \approx 0$, an impedance element connected to the inverting input meters a current proportional only to the source voltage. Thus, several impedances so connected do not affect each other, and such inputs are said to be decoupled. By assertion 1, and provided the inputs are decoupled (as a consequence of assertion 2), then the current in Z_f

$$i_f = \sum_{i=1}^{n} e_{i,n}/Z_{i,n} \tag{2}$$

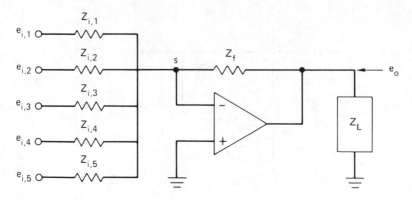

Figure 3. The summing amplifier circuit.

and

$$e_0 = -Z_f \sum_{i=1}^{n} e_{i,n}/Z_{i,n} \tag{3}$$

Two limitations of the inverting amplifier immediately are apparent. First, we frequently do not desire a 180° phase shift. More importantly, since $e_s = 0$, the impedance presented to e_i is simply Z_i; that is, the circuit input impedance also is determined by the loop configuration and not by the characteristics of the control amplifier. Due to noise pick-up and nonlinearity, as well as stray capacitance and inductance effects in large resistances, Z_i cannot practically exceed 1 $M\Omega$ in high precision circuits. Fortunately another option is available.

Figure 4a shows the general form of the closed-loop noninverting amplifier, and Figure 4b an important special case, the voltage follower. In each case, the circuit function becomes immediately apparent by invoking assertions 1 and 2 above. The voltage follower (Figure 4b) utilizes 100% negative feedback and thus the output maintains the source voltage perfectly. Since no current flows into the ideal amplifier, the input impedance of the voltage follower is infinite. The output impedance is just that of the ideal amplifier itself, i.e., zero. Note also that no current flows to the inverting input, so any arbitrary (but finite) impedance may be placed in the feedback loop without changing the properties of the ideal amplifier.

We will employ the concept of the impedance lever to analyze the circuit in Figure 4a. Under conditions of negative feedback $e_s = e_i$. Since the other end of Z_i is held at ground potential, and the current that flows in Z_i flows also in Z_f

$$e_o = e_i + \frac{e_i}{Z_i} Z_f = e_i(1 + Z_f/Z_i) \tag{4}$$

For an ideal Op Amp in an inverting amplifier configuration, with Z_i and Z_f

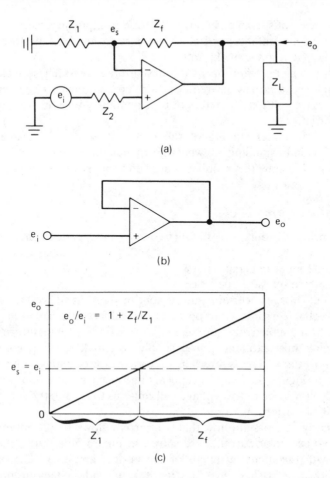

Figure 4. The noninverting amplifier (a) with gain; (b) unity gain voltage follower; and (c) impedance lever.

both resistors, the amplifier has infinite input impedance, zero output impedance, and gain >1, independent of frequency. It should be noted that negative feedback functions to keep the differential inputs at similar potentials. Thus, the summing point in a noninverting amplifier circuit is held at the input potential, and not at virtual ground.

2.4. Reality Regained

In the following two sections we shall critically examine the influences of amplifier nonideality on the accuracy of circuit operation. This analysis should point the way for device selection, but we should state that no golden rule is presented. In order for the thoughtful experimentalist to choose between necessary trade-offs (gain vs. bandwidth, noise vs. input impedance, differential operation vs. d-c stability, and of course cost vs. amplifier performance), he must have a clear grasp of the following:

1. A complete definition of the design objectives. These include signal levels employed, accuracy desired, bandwidth requirements, circuit impedance, and ambient conditions of temperature, humidity, time in use, and supply voltage variation.

2. A firm understanding of the pertinent electrical parameters. This includes an understanding of what the manufacturer means by a specified value, as well as how this number is applied to meet designed objectives.

2.5. Static Errors

The principal sources of static (frequency independent) error are shown in Figure 5a to be:
- Finite input impedance
- Nonzero output impedance
- Input voltage uncertainty due to spurious current and voltage sources.

The voltage error is a composite of several factors and will in general be a complex (and effectively unpredictable) function of temperature, semiconductor aging, mistreatment, power supply potential, etc. It is convenient to represent all the voltage sources as occurring in the noninverting input lead as shown in Figure 5b. There it appears and behaves as an unwanted signal source. The presence of this voltage will cause us to modify the summing point constraint (assertion 2).

Similarly, the assumption that no current flows to the inputs must be changed. Input error currents are shown in Figure 5c as current sources in parallel with the input terminals of the perfect amplifier. The operational circuit shown by dashed lines in Figure 5c provide a measurement of the inverting input error current. The error current of the noninverting input, i_n^+ is not observed because the input is grounded. However, i_n^- is transduced by Z_f into an erroneous output voltage.

The output impedance both of the voltage source and amplifier, and the amplifier gain all contribute to circuit function error. Figure 6 shows a voltage follower encumbered with five sources of static error: nonzero source impedance, nonzero output impedance, inverting input current error, noninverting input current error, input voltage error, and finite amplifier gain. The error analysis presented with Figure 6 is instructive in several points:
- e_o is reduced from e_i by the factor $(1 + 1/A)^{-1}$, irrespective of other sources of error.
- The error attributable to i_n^- and R_o is reduced by the factor $1/(1 + A)$.
- For $e_i = 0$

$$e_o = (e_n - i_n^+ R_i)/(1 + 1/A) - i_n^- R_o/(1 + A) \qquad (5)$$

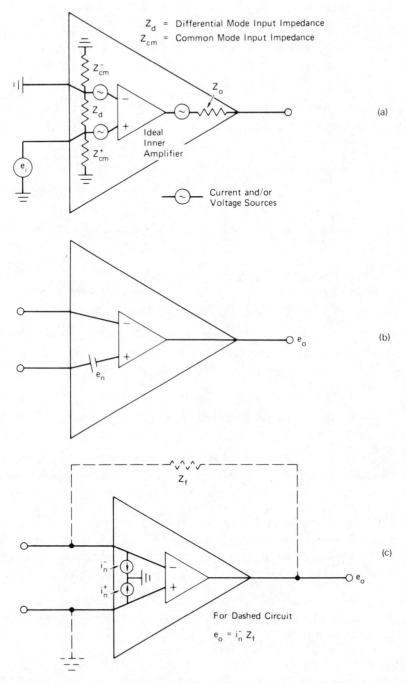

Figure 5. Sources of static error for a nonideal amplifier. (a) Current and voltage sources; (b) the effect of voltage error sources; (c) the effect of input error current sources.

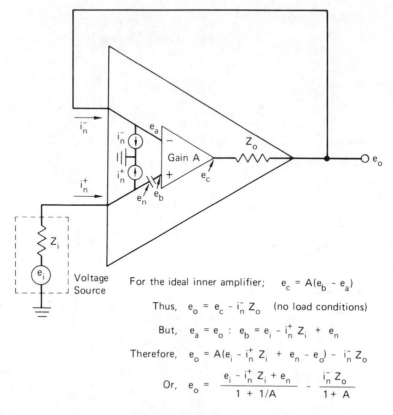

For the ideal inner amplifier; $e_c = A(e_b - e_a)$

Thus, $e_o = e_c - i_n^- Z_o$ (no load conditions)

But, $e_a = e_o : e_b = e_i - i_n^+ Z_i + e_n$

Therefore, $e_o = A(e_i - i_n^+ Z_i + e_n - e_o) - i_n^- Z_o$

Or, $e_o = \dfrac{e_i - i_n^+ Z_i + e_n}{1 + 1/A} - \dfrac{i_n^- Z_o}{1 + A}$

Figure 6. The effect of static errors on the voltage follower.

This provides a measure of the total circuit output voltage error, and is a function both of the input current and input voltage errors. Input voltage error is a direct measurement error about which nothing can be done except to use an amplifier having better specifications. Input current causes error by inducing error in the input, output, and feedback impedances.

The d-c offset voltage, e_{os}, is defined as the voltage required at the input, from a zero source impedance, to drive the output to zero. A manufacturer will specify the initial offset voltage for a stated supply voltage, and temperature (usually 25°C). In most amplifiers provisions are made to adjust initial offset voltage to zero with an external trim potentiometer. Nevertheless, drift of the offset voltage and bias current from their initial values, both with temperature and time, constitutes by far the most important source of error in most d-c and low-frequency circuit applications. The temperature coefficients of these parameters, $\Delta e_{os}/\Delta T$ and $\Delta i_n/\Delta T$, are not generally defined by manufacturers as the instantaneous derivative, but as the average slope over a specified temperature range. In general, however, drift is a nonlinear function of temperature with slopes greater at temperature extremes than at 25°C, which generally means that for small temperature excursions the errors experienced will be well within the specifications.

Typical values of the temperature coefficients are given in Table 1 for the low cost 741 and low drift 180 Op Amps.[1] The third column represents the source resistance value at which the temperature dependence of the input current error equals the voltage error (found by dividing the two coefficients).

It follows that, for source resistances exceeding a few hundred ohms, the current error drift dominates, even though the input impedance of the follower may exceed $10^9 \Omega$. The circuit in Figure 7 provides about a factor of 10 decrease in the current error drift for the voltage follower, provided Z_i is known and constant. For $A \to \infty$

$$e_o = e_i - i_n^+ Z_i + i_n^- Z_f + e_n \qquad (6)$$

The difference current $i_n^\pm$ is often referred to as the input offset current, and is defined as the absolute difference between the bias current at each input. The input circuitry of differential amplifiers is generally symmetric[1] so that i_n^+ and i_n^- tend to be equal and to track with changes in temperature and supply voltage such that $i_n^\pm \approx -0.1\, i_n^+ \approx 0.1\, i_n^-$. Thus, if $Z_i = Z_f$, then

$$e_o \approx e_i - 0.1\, i_n^+ Z_i + e_n \qquad (7)$$

It should be noted that, while this artifice serves well to reduce the influence of bias currents on the output voltage, it does not reduce the input current which may be intolerably large, as for example in the measurement of very large interfacial impedances. In this case the experimentalist must consider the input impedance in addition to the bias and offset currents. Differential input impedance, Z_d, is defined as the impedance between the two input terminals, at a specified temperature, with the offset error voltage nulled to zero. Common mode impedance, Z_{cm}, is defined as the impedance between each input and the power supply common potential. Z_{cm} in general is a nonlinear function both of temperature and of common mode voltage.

It has been assumed to this point that the output responds only to the difference between the input voltages (plus error terms), and produces no output for a common mode voltage; i.e., that $e_o = 0$ when $e_a = e_b = e_{cm}$. Due to slightly different inverting and noninverting input gains, however, common

Table 1

Device	$\Delta e_{os}/\Delta T$ (μV/°C)	$\Delta i_n \Delta T$ (nA/°C)	R_i^0 (Ω)	Z_{input} (Ω)
AD 741 J[1]	±20	±200	100	10^6
AD 180 K[1]	±0.5	±4	125	10^9

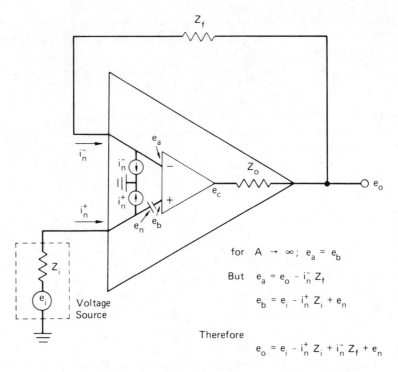

for $A \to \infty$; $e_a = e_b$

But $e_a = e_o - i_n^- Z_f$

$e_b = e_i - i_n^+ Z_i + e_n$

Therefore

$e_o = e_i - i_n^+ Z_i + i_n^- Z_f + e_n$

Figure 7. Voltage follower with improved immunity to current error drift.

mode voltages may not sum to zero at the output. The output error voltage generally is treated as deriving from a common mode error voltage, Δe_{cm}, at the inputs (i.e., a differential mode voltage), by dividing by the closed loop gain. The common mode rejection ratio, *CMRR*, is defined as the ratio of common mode to common mode error voltages, usually expressed in db

$$CMRR = 20 \log_{10}(e_{cm}/\Delta e_{cm}) \qquad (8)$$

Regrettably, *CMRR* is not well defined and, particularly for FET[1] input devices, may be a strong function not only of temperature, but of the magnitude and frequency of e_{cm}. *CMRR* is of particular significance for true differential and instrumentation amplifiers and will be discussed more fully in Section 3.

2.6. Dynamic Error and Stability

The sources of dynamic error are twofold. First, the amplifier cannot respond instantaneously to an imposed differential mode input voltage change. Second, the open-loop gain (which we had initially considered to be infinite) decreases with frequency, becoming less than unity at some finite frequency.

Dynamic stability is dependent both on the amplifier open-loop gain characteristics, and on the components of the closed loop, both intentional and unintentional.

The closed-loop circuits we have described to this point have been shown with generalized impedances as input and feedback elements, but we have described only "amplifiers," that is, with Z_i and Z_f both nonreactive. We implied that the circuits were stable. Not all feedback circuits produce stable conditions, however. Some are definitely unstable and oscillate freely, while others are only conditionally stable and will oscillate when provided with the right stimulus. The Nyquist stability criterion states that the closed loop must have gain less than unity at frequencies for which the phase shift exceeds π radians. We will show that the open-loop transfer function incorporates a phase shift of $\pi/2$ radians. Thus, instability may be induced either by the intentional use of reactive input or feedback devices, which impose a phase shift in the feedback path of magnitude greater than $\pi/2$ radians at some critical frequency, or by unintentional stray shunt capacitance or series inductance effects at high frequencies.

In describing why a phase shift of $\pi/2$ radians (or 90°) should be critical, it is convenient to adopt the formalism of linear systems analysis[2] and operator calculus. Figure 8a shows a generalized Op Amp with transfer function $G(j\omega)$ and feedback $H(j\omega)$. Tracing around the loops we see

$$[e_i - e_o G(j\omega)]H(j\omega) = e_o \tag{9}$$

$$e_o/e_i = G(j\omega)/[1 + G(j\omega)H(j\omega)] \tag{10}$$

The small signal open-loop transfer function of a typical Op Amp has the form

$$G(j\omega) = \frac{-A_o}{1 + j(\omega/\omega_0)} \approx jA_o\omega_o/\omega \tag{11}$$

where A_o is the gain at limiting low frequencies ("d-c") and ω_0 is the turnover frequency, characteristic of the device. The approximate form holds for $\omega/\omega_0 \gg 1$ (typically, for $\omega \geq 100\,\text{rad. s}^{-1}$), and implies that the open-loop phase response is a constant $\pi/2$ radians as shown in Figure 8b.

By rearranging Eq. (10), we obtain an expression for the closed-loop transfer function

$$T(j\omega) = [1/H(j\omega)]/[1 + 1/L(j\omega)] \tag{12}$$

where $L(j\omega) = G(j\omega)H(j\omega)$.

$L(j\omega)$ is frequently referred to as the loop gain. This parameter determines the most important characteristics of the closed loop system, including the function accuracy and the loop stability. From Eq. (12) it is clear that for $L(j\omega) \gg 1$, the closed-loop transfer function is determined only by the feedback characteristics, $H(j\omega)$.

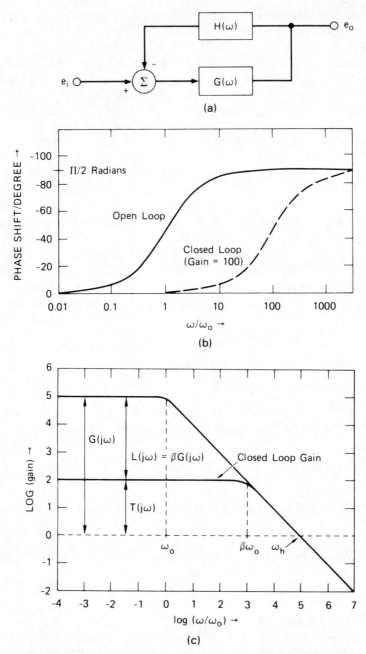

Figure 8. Dynamic response of an operational amplifier. (a) Linear system representation; (b) open and closed loop phase response with first-order rolloff, and open loop gain = 10^5; and (c) Bode plot with first-order rolloff, under open loop (DC gain = 10^5) and closed loop (DC gain = 10^2) conditions.

By way of example, for a system with open-loop gain given by Eq. (11), and feedback β, which is equivalent to a noninverting amplifier with $R_f = (\beta^{-1} - 1)R_i$

$$T(j\omega) = (1/\beta)/[1 - 1/\beta A_o - j(\omega/\beta A_o \omega_o)] \tag{13}$$

The closed-loop characteristics predicted by Eq. (13) are shown together with the open-loop response in Figure 8b and 8c. Application of feedback has extended the frequency below which the phase shift sensibly is zero, by the factor β. The Bode plot representation in Figure 8c serves to emphasize the importance of loop gain. With large loop gain, $\log L(j\omega) > 0$, the amplifier is said to possess a dynamic reserve sufficient to reinforce the function accuracy, and the closed-loop gain error is small. Because of the logarithmic form, $L(j\omega)$ can be calculated from the Bode plot simply by subtracting the log gain from the open-loop response. At frequencies greater than $\beta\omega_o$, the loop gain falls to zero, and circuit function becomes that of the open-loop device.

Gain and phase errors directly attributable to the form of $G(j\omega)$ constitute the major source of function error at frequencies ≥ 100 radians s^{-1}. Defining the error transfer function, $E_f(j\omega)$, as the fractional steady-state output due to $G(j\omega)$, referred to the input

$$E_f(j\omega) = [e_1 - e_o H(j\omega)]/e_i$$
$$E_f(j\omega) = [e_i - e_i T(j\omega) H(j\omega)]/e_i \tag{14}$$
$$E_f(j\omega) = 1/[1 + L(j\omega)]$$

Despite the errors indicated by Eq. (14), the form of $G(j\omega)$ is deliberately chosen by the manufacturer, and is necessary to ensure loop stability at all frequencies under conditions of reactive feedback, either deliberate or unintentional. The most frequent cause of additional phase lag that engenders oscillation in operational circuits is stray capacitance from the summing point to ground. This case is shown in Figure 9.

The addition of a capacitive shunt to the load of an Op Amp also may lead to instability at high frequencies by the introduction of a peak in the closed-loop transfer function. The effect is shown in Figure 10 for a typical amplifier, with progressively increased capacitive loading.

Instability may occur in ways other than those displayed in Figures 9 and 10. For example, the simple differentiator formed by inserting a capacitance as the input impedance of an inverting amplifier is *a priori* unstable, because the feedback network has only phase lag characteristics (see Section 3). In general, the Nyquist stability criterion may be used to predict the stability of circuit stability. Under this criterion, a closed-loop, linear circuit is stable if a plot of the loop gain, $L(j\omega)$, as a function of frequency, does not enclose the point $(-1, j0)$ in the complex plane. Relative stability may be gauged by the gain and phase margins. The gain margin is the reciprocal of $|L(j\omega)|$ at the frequency where the phase shift is 180°. The phase margin is 180° − the phase of $[L(j\omega)]$ at the frequency where $L(j\omega) = 1$.

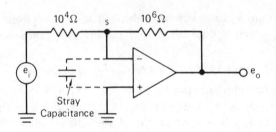

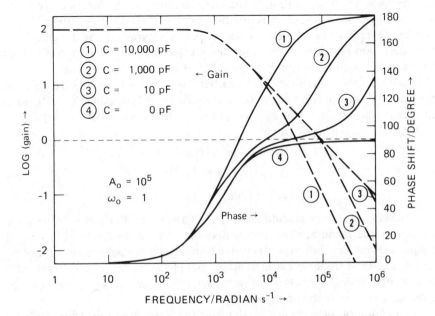

Figure 9. The influence of a stray capacitance between the summing point and ground on the dynamic response of an inverting amplifier.

It frequently is not possible to calculate the absolute and relative stability margins when the contributions to the feedback, $H(j\omega)$, due to the presence of stray capacitance and inductance effects is not known. A simple, but expensive, alternative is to calculate $L(j\omega)$ from a measurement of the system closed-loop transfer function, $T(j\omega)$, as a function of frequency. With reference to Eq. (12)

$$L(j\omega) = G(j\omega)/T(j\omega) - 1 \tag{15}$$

In this way $L(j\omega)$ can be calculated provided the form of the amplifier open-loop gain, $G(j\omega)$, is well known. Methods by which $T(j\omega)$ may be measured are described briefly in Section 4, and in detail elsewhere.[3]

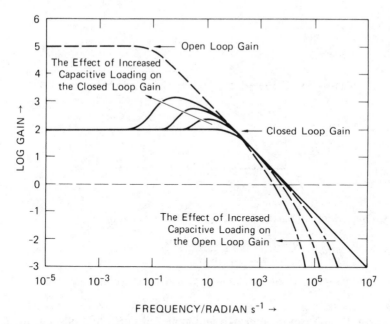

Figure 10. The effect of capacitive loading on the dynamic response of a closed-loop Amp circuit.

A more elegant method of measuring $L(j\omega)$, which does not require preknowledge of the form of $G(j\omega)$, may be accomplished by introducing a test signal to an additional summing mode, as shown in Figure 11. Defining the test transfer function $Y(j\omega)$, as the ratio $X(j\omega)/N(j\omega)$, then, with reference to Figure 11, and neglecting $e_i(j\omega)$

$$Y(j\omega) = G(j\omega)H(j\omega)/[1 + G(j\omega)H(j\omega)] \tag{16}$$

$$L(j\omega) = G(j\omega)H(j\omega) = Y(j\omega)[1 - Y(j\omega)] \tag{17}$$

Equation (17) demonstrates that the loop gain may be calculated from a measurement of $Y(j\omega)$ only. This measurement may be made with the circuit in operation ($e_i(j\omega) \neq 0$). It is possible to neglect $e_i(j\omega)$ in the derivation of Eq. (16) only when the test and input signals are not coherent; that is, when the detector is capable of discriminating between these two signals. It is common to employ random noise as the test signal,[3-5] since this correlates least with any other signal, and is likely to result in minimum perturbation of the output.

2.7. Noise

We use the term noise here to describe all spurious or unwanted signals that appear at the output, uncorrelated with the input signal. Input voltage

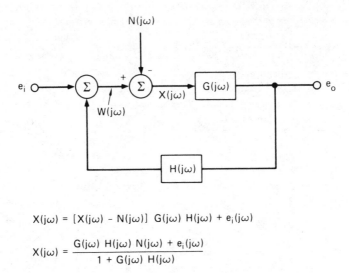

$$X(j\omega) = [X(j\omega) - N(j\omega)] \; G(j\omega) \; H(j\omega) + e_i(j\omega)$$

$$X(j\omega) = \frac{G(j\omega) \; H(j\omega) \; N(j\omega) + e_i(j\omega)}{1 + G(j\omega) \; H(j\omega)}$$

Figure 11. Method of measuring loop gain in closed-loop operation.

and current noise can be specified and analyzed much like offset voltage and bias currents. In fact, offset voltage drift constitutes the major source of noise in Op Amp circuits at frequencies below 0.1 Hz. Above this frequency it is convenient to divide the noise spectrum into several frequency ranges, each of which is dominated by certain noise components. Above 1 kHz, noise tends to be dominated by resistor (or thermal) noise in passive components, and shot noise in semiconductor devices. Both are readily predicted from simple theoretical considerations, and result in a noise power that is constant (or "white") per square root frequency.

The predominant noise source from 10 Hz to 1 kHz is pick-up from main frequency harmonics and electrical switching noise. This noise may propagate by radiation or via the power supply.

The noise spectrum from 0.1 to 10 Hz is dominated by $1/f$ or "flicker" noise, associated with structural imperfections in components, particularly transistors. For this noise source, the power is distributed uniformly with log frequency, resulting, as the frequency decreases, in a relatively higher amplitude for a constant frequency interval.

White noise and flicker noise are intrinsic to a device, and components must be selected such that the sum of the noise, including appropriate amplification factors, is within the designer's tolerance. In general a manufacturer will specify separately the narrow band ($1/f$) and broadband (white) noise components, by listing both low-frequency and high-frequency current and voltage noise figures.

The root mean square output voltage response to a noise current, i_n, injected at the summing point, and a noise voltage, e_n, in series with the input of an otherwise ideal inverting amplifier, as shown in Figure 2a, is

$$e_{rms} = |1/\beta|e_n[1 + |(1/R_n)/1/Z_i + 1/Z_f)|]^2 \tag{18}$$

where

$$R_n = e_n/i_n \tag{19}$$

and

$$1/\beta = (1 + Z_f/Z_i) \tag{20}$$

The factor $1/\beta$ is called the noise gain. In order to maximize the signal-to-noise ratio at the output, it is desirable that noise gain be as low as possible. In order to minimize the influence of i_n, the circuit should be designed to operate at the lowest possible impedance level (Z_i, Z_f). These requirements will be tempered by the requirements of circuit gain and input impedance.

Contributions due to pick-up obviously cannot be specified in advance, and will be a strong function of the proximity and type of noise sources, circuit type and layout, and of the adequacy of shielding. Noise may propagate both by radiation and via the device's power supply. With regard to the latter, additional smoothing can be achieved by installing large electrolytic capacitors across the d-c supplies. Since electrolytic capacitors do not function as capacitors (and even display inductive properties) at high frequencies, the electrolytic capacitors should be shunted by good polystyrene or ceramic capacitors. Also, for a variety of reasons, the power supplies to each Op Amp should be decoupled at the device itself with 0.01–0.1 μF nonelectrolytic capacitors.

Radiated noise can be reduced by adequate shielding, particularly for high-impedance signal sources, and at the summing point. Figure 12 shows a standard method of achieving a high degree of pick-up isolation for measurements requiring long cables with the following pertinent features:
- The isolated voltage source is contained within a grounded electrostatic shield.
- Input cables are "double coax." The outer coaxial sheath is grounded and acts as an electrostatic shield. The inner coaxial sheath is driven from a high-precision voltage follower to the same potential as the central conductors. This method serves to minimize capacitive coupling between the source and ground because there is no (or very little) potential difference between the central conductor and the inner coaxial sheath.
- The regulated dual voltage supply is built on a separate chassis to isolate a-c mains sources, and is decoupled within the instrument chassis.

Incorporation of these features has been estimated[6] to result in an order of magnitude noise reduction, between 1 Hz and 1 kHz, for a sample impedance of approximately 10 kΩ.

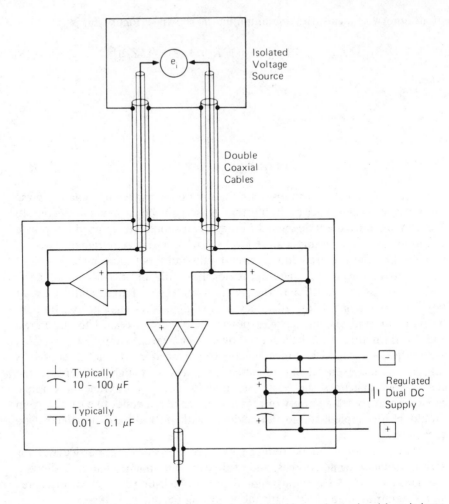

Figure 12. The use of double coaxial cable with a driven shield, to reduce noise pick-up in input cables.

3. Further Building Blocks

3.1. Introduction

Having described the fundamental parameters and characteristics of devices that incorporate negative feedback, we intend to discuss a few classes of the functional blocks that will be of particular importance to the electrochemist in the pursuit of measurement, computation, and control by analog means. Measurement is usually restricted to amplifiers of various types, and therefore to linear circuits of the type described in Section 2. More complex measurement circuits (linear, nonlinear, digital, and hybrid), are described in Sections 5 and 6.

Control devices measure, compute, and correct deviations from programmed behavior. These are exemplified in electrochemical circuitry by potentiostats, galvanostats, thermostats, feedback motor controllers, servo systems, and similar closed-loop devices. Control may be accomplished by switching at a threshold value as in an on/off temperature controller, or by linear feedback of an analog signal proportional to the deviation between the measured and preset control value, as in a conventional potentiostat or galvanostat. These latter devices are of particular electrochemical significance; their circuit function will be analyzed in detail in Section 4.

Feedback also may be applied by digital or digital-hybrid means. In the case of a d-c motor control operating in a closed loop with a tachometer pick-up, the frequency-to-voltage conversion may be made by a digital computer or by an analog device, for example a phase sensitive detector (see Section 5). Very often the advantages of speed and reliability, and (in the economics of today's technology) the relative costs, favor an analog approach.

The classical use of Op Amps as analog computation machines remains a basic function, as an adjunct to measurement or control. In general, a computation device refines data. This may be a linear function, as addition/subtraction, or may be nonlinear as for example multiplication/division, logarithm/exponentiation, integration/differentiation, level sensing, etc.

The remainder of this section will be concerned principally with computational circuits as components of more complex devices. It is not our intention to provide an exhaustive list circuit of ideas or "application notes." These tasks have been performed elsewhere.[1,7-10] The circuit functions discussed below have been chosen as representative of particular functional types, which are of particular significance to electrochemists, with the intention of facilitating a generalization of knowledge.

3.2. Threshold Detectors

It was stated in Section 2.2 that an Op Amp with no declared feedback, as shown in Figure 1a, functions as a voltage comparator. With the control signal connected to the noninverting input, and a reference signal to the inverting input, the output will switch state from negative to positive saturation when the control signal, e_c, exceeds the voltage reference value, e_r. This output may be used, for example, as an event marker or counter to drive a thermostat or subsequent digital circuitry.

The circuit in Figure 1a is accurate but slow, since it may take on the order of a millisecond for an amplifier to recover from voltage saturation. To increase speed by ensuring that amplifier output remains in the linear operating range, the output may be "bounded" by employing parallel diodes in the feedback loop. Since a diode represents a high resistance at low forward bias, and a low resistance at large forward bias, such a device will behave as a high-gain amplifier near the threshold point, but as a voltage follower for $|e_c - e_r| \geq 0.65$ V for a silicon diode.

Figure 13a shows a double-bounded, high-precision comparator that incorporates adjustable positive feedback to increase switching speed. This

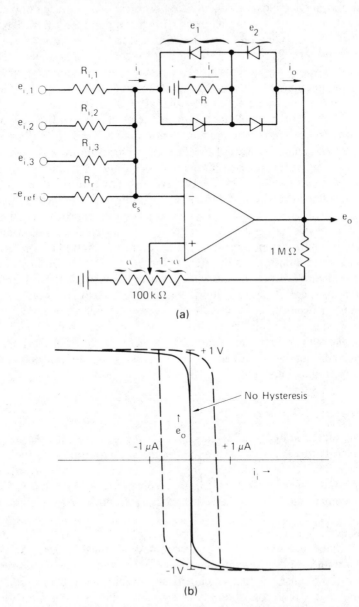

(a)

(b)

Figure 13. (a) Precision comparator circuit, and (b) transfer function.

circuit will switch when $i_i \approx 0$, that is, when the weighted sum of the input voltages is equal to $-e_{ref}$. It should be noted that, under conditions of positive feedback $(\alpha > 0)$, $e_s \approx 0.1\ \alpha e_o$, and the input voltages are not completely decoupled at the summing point.

The analysis of this circuit is instructive in several points. By inspection

$$e_o = e_s + e_1 + e_2 \tag{21}$$

$$i_o = i_i + i_r \tag{22}$$

$$i_r = (e_1 + e_s)/R \tag{23}$$

For small applied potentials the diodes have an exponential current/voltage response. Thus, for each parallel combination of reversed diodes

$$i_n = i_o[\exp(\beta e_n) - \exp(-\beta e_n)] \tag{24}$$

or

$$\beta e_n = \sinh^{-1}(i_n/2) \tag{25}$$

In the absence of positive feedback, $\alpha = 0$, and $e_s \approx 0$, thus

$$\beta e_0 = \sinh^{-1}(i_i/2i_0) + \sinh^{-1}[i_i/2i_o + \sinh^{-1}(i_i/2i_0)/2\beta R i_o] \tag{26}$$

Figure 13b shows the response of the comparator calculated from Eq. (26) for typical values of i_o and β. The influence of positive feedback on operation speed and hysteresis also is shown.

3.3. Instrumentation Amplifier

A very common requirement in measurement circuits is for a balanced differential input, dual high-impedance amplifier. Such an amplifier is necessary when the voltage to be measured is not referenced to circuit ground.

The simplest method of achieving this is shown in Figure 14. Dual voltage followers present a very high input impedance to the source, and drive a difference amplifier that produces a differential output with gain. In the general case, $R_1 \neq R_2$ and $\alpha \neq \beta$. We can analyze this circuit in terms of two voltage dividers, formed by R_1 and αR_1, R_2 and βR_2

$$e_r = e_2 - e_2/(1 + \beta) \tag{27}$$

$$e_s = e_1 - (e_1 - e_o)/(1 + \alpha) \tag{28}$$

Under ideal conditions (the "virtual ground" condition), $e_r = e_s$, and

$$e_o = e_2\alpha(1 + \beta)/(1 + \alpha) - e_1\beta \tag{29}$$

By equating α and β, Eq. (29) reduces to the more usual form for an instrumentation amplifier with gain

$$e_o = \alpha(e_2 - e_1) \tag{30}$$

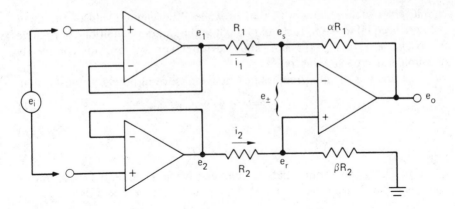

Figure 14. Simple instrumentation amplifier circuit.

The common mode rejection ratio (*CMRR*) for the circuit of Figure 14 may be determined by setting $e_2 = e_1 = e_{cm}$, in Eq. (29)

$$CMRR = e_o/e_{cm} = (\alpha - \beta)(1 + \alpha) \tag{31}$$

Thus a 10% mismatch between α and β will result in an output error voltage of magnitude 0.1 e_{cm} (*CMRR* = 20 db) for an amplifier of gain 1000, and 0.05 e_{cm} (*CMRR* = 26 db) for a gain of unity.

This analysis highlights the principal limitation of this circuit. In order to preserve a precise response function, the resistive ratios, α and β, must track closely. Changing the circuit gain necessarily involves adjusting and trimming two resistances.

It should be noted that the difference amplifier may have multiple inverting and noninverting inputs. If there are "*m*" inputs, e_i, to the negative terminal summing point, and "*n*" inputs, e_j, to the positive summing point, then the general expression for the response is

$$e_o = \sum_{j=1}^{n} \beta_j e_j - \sum_{i=1}^{m} \alpha_i e_i \tag{32}$$

provided that

$$\sum_{j=1}^{n} \beta_j = \sum_{i=1}^{m} \alpha_i \tag{33}$$

where α_i is the ratio of the feedback resistance to the inverting input resistance of the i'th input, and β_j is the ratio of the resistance to ground, to the noninverting input resistance of the j'th input.

The addition of three resistors results in a very significant refinement of the basic instrumentation amplifier. With reference to Figure 15

$$e_1' = e_2 + (\gamma + 1)(e_1 - e_2) \tag{34}$$

$$e_2' = e_1 + (\delta + 1)(e_2 - e_1) \tag{35}$$

$$(e_2' - e_1')/(e_2 - e_1) = \gamma + \delta + 1 \tag{36}$$

Thus the input stage provides a differential mode gain equal to $\gamma + \delta + 1$. This stage does not need to be balanced (i.e., γ need not equal δ), and the overall circuit gain can be varied with a single resistor (R_1). Even more significantly, the input stage affords some common mode rejection, in addition to that of the differential output stage. Setting $e_1 = e_2 = e_{cm}$ in Eq. (34) and (35), we see that $e_2' = e_1' = e_{cm}$. That is, common mode signals pass the input stage at unity gain, while differential mode signals are amplified by the factor $\gamma + \delta + 1$. This results in a *CMRR* improvement of an equivalent amount in the overall circuit.

3.4. Current Follower

One of the simplest Op Amp circuits performs the function of a current-to-voltage converter, presenting a low impedance to the current source, and providing an output voltage that is proportional to the current. Analysis of

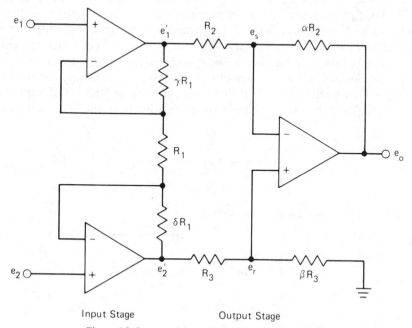

Input Stage   Output Stage

Figure 15. Improved instrumentation amplifier circuit.

the circuit shown in Figure 16 may be performed by inspection. For an ideal amplifier, the current flowing to the inverting input flows through the feedback resistor, and the output voltage

$$e_o = iR_f \tag{37}$$

The important characteristic of the current follower is the input impedance, which must be maintained as *low* as possible to decouple the current measurement circuit from the current source. We know that

$$e_o = e_i G(j\omega) \tag{38}$$

and combining Eqs. (37) and (38) we obtain the input impedance

$$Z_i = e_i/i = R_f/G(j\omega) \tag{39}$$

Equation (39) highlights a very common limitation of commercial potentiostats that employ current followers to provide an output voltage proportional to cell current. At frequencies for which $G(j\omega) \lesssim 1000$, appreciable errors are introduced in the measured current and in the potentiostatting voltage, since the current follower acts as an uncompensated, frequency-dependent resistance in series with the working electrode.

3.5. Voltage References

The inherent high input and low output impedances of Op Amps make precise voltage and current reference sources possible. The simplest of these employs a high-impedance noninverting amplifier in conjunction with a standard cell to provide a variable but stable output voltage from which currents of the order of 10 mA may be drawn without significant voltage errors. In this application, the parameters chiefly of interest in determining circuit precision are the amplifier d-c offset drift and output impedance. The latter we define as the drop in output voltage induced by a load current. With reference

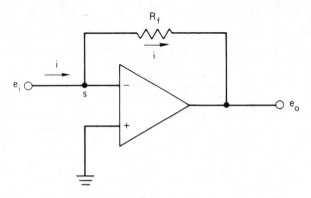

Figure 16. Current follower circuit.

to Figure 17a, $Z_{out} = [e_o^{no\,load} - e_o^{load}]/i_l$, or

$$Z_{out} \approx \frac{r_o/G(j\omega)}{1/G(j\omega) + [R_i + r_o/G(j\omega)]/(R_i + R_o)} \qquad (40)$$

where r_o is the intrinsic, internal Op Amp output impedance, which typically is of the order 1 k to 3 kΩ.

Under d-c operating conditions for which $G(j\omega) \gg 1$

$$Z_{out} \approx r_o/[G(j\omega)R_i/(r_i + R_f)] = r_o/[G(j\omega)H(j\omega)] \qquad (41)$$

That is, the intrinsic amplifier output impedance is divided by the loop gain [see Eq. (12)].

An alternate method of providing a precise, stable voltage reference is shown in Figure 17b. R_z is adjusted to set the current through a zener diode to minimize the voltage temperature coefficient. Noise and pick-up are attenuated by the input filter $R_1 C_1$, and the resultant voltage is amplified and buffered by the noninverting amplifier, as for Figure 17a. The additional capacitor, C_f, is utilized primarily to ensure dynamic stability. For the circuit shown

$$e_o = \left[1 + \frac{R_f}{R_i(1 + \omega R_f C_f)}\right]\left[\frac{e_z}{1 + \omega R_1 C_1}\right] \qquad (42)$$

3.6. Current References

The current flowing in the feedback loop of an inverting or noninverting amplifier is determined only by the requirement that the output voltage equal the product of the open-loop gain, and the potential difference between the two inputs. In each instance, therefore

$$e_o/G(j\omega) = e_i - i_f R_i \qquad (43)$$

where e_i is the input voltage and R_i is the inverting input resistance.

According to Eq. (43), a current $i_f = e_i/R_i$ may be driven through a load placed within the feedback loop. The precision of control of this current will be within the limits of error imposed by a finite $G(j\omega)$ and the amplifier leakage currents. These circuits are much used, but suffer the inconvenience that the load cannot be grounded. A circuit that eliminates this disadvantage by floating the voltage reference source is shown in Figure 18.

A feature common to the current sources described here is instability when driving inductive loads. In the circuit of Figure 18, the tendency to oscillate is reduced by the parallel RC filter comprised by R_1 and C_1. Analysis of the response of this circuit at low frequencies may be performed, with reference to Figure 18, by letting $G(j\omega)$ and the impedance of C_1 approach infinity. Under this condition

$$i_i = e_r/R_i \qquad (44)$$

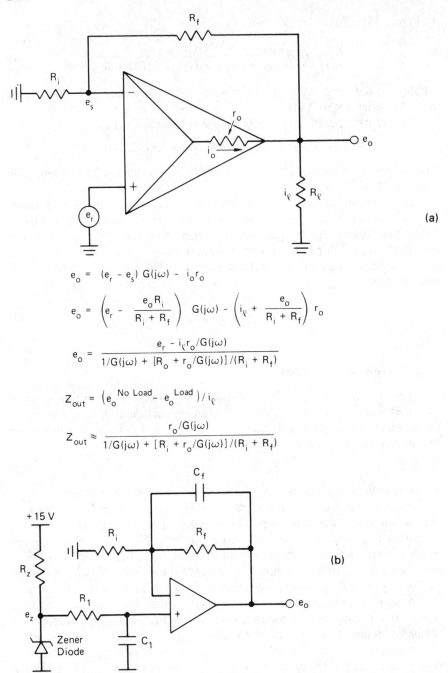

$$e_o = (e_r - e_s)\, G(j\omega) - i_o r_o$$

$$e_o = \left(e_r - \frac{e_o R_i}{R_i + R_f}\right) G(j\omega) - \left(i_\varrho + \frac{e_o}{R_i + R_f}\right) r_o$$

$$e_o = \frac{e_r - i_\varrho r_o / G(j\omega)}{1/G(j\omega) + [R_o + r_o/G(j\omega)]/(R_i + R_f)}$$

$$Z_{out} = \left(e_o^{\,No\ Load} - e_o^{\,Load}\right) / i_\varrho$$

$$Z_{out} \approx \frac{r_o / G(j\omega)}{1/G(j\omega) + [R_i + r_o/G(j\omega)]/(R_i + R_f)}$$

Figure 17. Voltage reference sources. (a) The influence of output resistance on the source voltage; (b) stable voltage reference circuit.

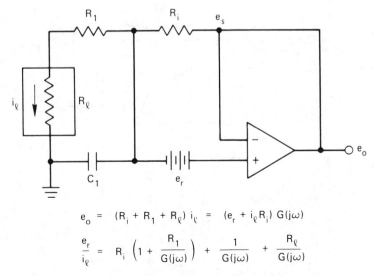

$$e_o = (R_i + R_1 + R_\varrho) i_\varrho = (e_r + i_\varrho R_i) G(j\omega)$$

$$\frac{e_r}{i_\varrho} = R_i \left(1 + \frac{R_1}{G(j\omega)}\right) + \frac{1}{G(j\omega)} + \frac{R_\varrho}{G(j\omega)}$$

Figure 18. Current source with a grounded load.

as for Eq. (43), and the load current may be adjusted up to the amplifier output voltage and current limits, by varying R_i.

3.7. Operational Integrator

Analog integrators have been largely superseded in analytical applications by their digital cousins. Nevertheless, the operational integrator deserves separate treatment here both for its historical importance, and for its striking parallels to inverting amplifier circuitry.

The simple integrator is obtained by replacing the feedback resistance of an inverting amplifier with a capacitor, as shown in Figure 19a. The response of this circuit clearly is frequency dependent. Reverting to the techniques of dynamic analysis described in Section 2.6, the fractional feedback in the simple integrator is given by

$$H(j\omega) = \frac{R}{R + j/\omega C} = \frac{1 - j/\omega CR}{1 + (\omega CR)^{-2}} \tag{45}$$

$$1/H(j\omega) = 1 + j/\omega CR \tag{46}$$

Figure 19b is the Bode plot of the integrator. The closed-loop response parallels the open-loop gain at intermediate and high frequencies, but the two are coincident at low frequencies. Thus the integrator runs out of loop gain, represented by the cross-hatched area in Figure 19b, at low frequencies when $1/H(j\omega) = G(j\omega)$ [refer to Eq. (23)], and at high frequencies when $H(j\omega) = 1$.

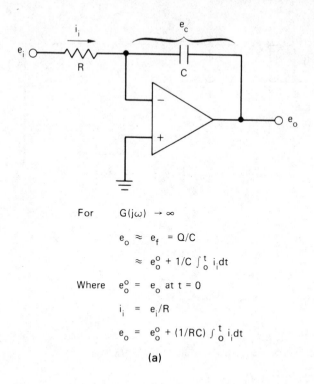

For $\quad G(j\omega) \rightarrow \infty$

$$e_o \approx e_f = Q/C$$

$$\approx e_o^o + 1/C \int_o^t i_i dt$$

Where $\quad e_o^o = e_o$ at $t = 0$

$$i_i = e_i/R$$

$$e_o = e_o^o + (1/RC) \int_o^t i_i dt$$

(a)

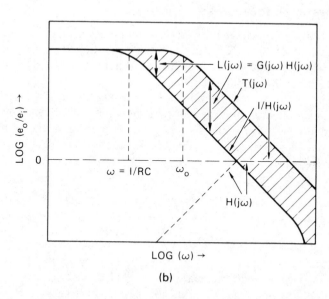

(b)

Figure 19. The operational integrator. (a) The simple integrator circuit; (b) Bode plot for the simple integrator.

The decrease in loop gain and thus decrease in circuit accuracy at low frequencies results in the integrator output decaying toward zero for the condition of no input signal ($e_i = 0$).

The integrator circuit of Figure 19a contains no provision to force the output to an arbitrary (initial) condition, e_o^0, at time zero, and a free running integrator will, in time, drift to one at its bounds due to the input leakage currents. This difficulty may be alleviated by periodically resetting the integrator by short-circuiting C. For nonzero values of e_o^0, a voltage source may be switched to charge C to the desired initial condition.

The integrator demonstrates further similarities with an inverting amplifier. A summing integrator results when the feedback resistance of the summing amplifier shown in Figure 3 is replaced with a capacitor, and a difference integrator that responds to the time integral of the differential input may be constructed by replacing the feedback resistor and the resistor between the noninverting input and ground of the instrumentation amplifiers in Figures 14 and 15, each with capacitors.

3.8. Operational Differentiator

As a general rule of operational amplifier circuitry, if one places within the feedback loop a circuit that performs a particular mathematical function, then the inverse function results. Thus, if the input resistor and feedback capacitor of the simple integrator shown in Figure 19a are interchanged, then we achieve the circuit of an "ideal" differentiator shown in Figure 20a. A similar interchange in Figures 3, 14, and 15 will result in "ideal" summing and differencing differentiators.

The term "ideal" is a testament to the caution one must use in applying rules such as that given in the previous paragraph. The circuit in Figure 20a comprises a first-order lead network, exactly opposite in character to the lag network provided for in the Op Amp by the manufacturer to ensure loop stability (see Section 2.6). Any small additional phase shift will result in oscillation; the "ideal" differentiator is not dependably stable at any frequency without the addition of further components.

A circuit for a practical differentiator appears in Figure 20b. The added resistor and capacitor modify the response at frequencies above the highest data frequency of interest. This circuit has characteristics of both a differentiator and an integrator. R_1 and C_1 serve to decrease the closed-loop gain, and provide a lag phase shift, at high frequencies. This is necessary both for dynamic stability, and to curtail the noise that would be accentuated at high frequencies by the "ideal" differentiator.

The Bode plot in Figure 20c presents schematically the closed-loop frequency response of the practical differentiator circuit, for the transfer function given in Figure 20b. To achieve well-damped stability, the turn-over frequency, $\omega_c (=1/\tau_c)$, should be chosen to be no greater than the geometric

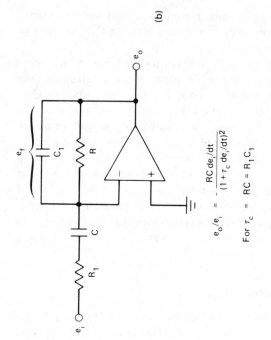

(a)

For $A \to \infty$; $e_o \approx -e_f \approx -i_i R$

$i_i = C \, de_i/dt$

$e_o = -RC \, de_i/dt$

(b)

$e_o/e_i = - \dfrac{RC \, de_i/dt}{(1 + \tau_c \, de_i/dt)^2}$

For $\tau_c = RC = R_1 C_1$

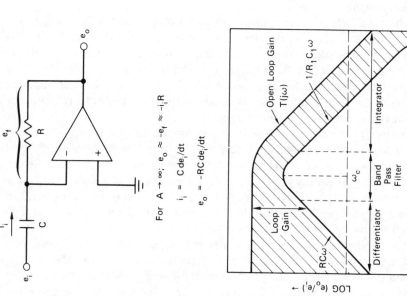

(c)

Figure 20. The operational differentiator. (a) The ideal differentiator circuit; (b) a practical differentiator circuit; (c) Bode plot for the practical differentiator.

mean between the differentiator characteristic frequency $1/RC$ and the amplifier's gain-bandwidth product, ω_p. That is

$$\tau_c > (RC/\omega_p)^{1/2} \text{ or } RC > 1/\omega_p \qquad (47)$$

Wisdom often dictates choosing a substantially larger value of τ_c in order to limit the sensitivity of the differentiator to high-frequency noise. The differentiator characteristic time, RC, is selected so that the maximum rate of change of input signal will produce a full-scale output

$$RC = |e_i|_{max}/|de_i/dt|_{max} \qquad (48)$$

3.9. Active Filter Design

The consideration of integrators and differentiators leads naturally into a discussion of filter design using operational amplifiers. In general, a filter is a device with a frequency selective transfer function. Thus, the simple integrator in Figure 19a can be considered as a low-pass filter since it passes low frequencies at an increased amplitude. Similarly, the ideal differentiator in Figure 20a might function as a high-pass filter, and the practical differentiator circuit in Figure 20b could be employed as a bandpass filter. In each case, however, the circuit gain and phase shift are not constant in the pass band, and such devices are far from ideal as filters for signal analysis.

Passive filters may be designed using passive LCR networks, but the use of active components can greatly simplify the design of high performance signal filters. The active element, in this case the operational amplifier, in active networks is necessary to permit the realization of complex left-hand plane poles in the Nyquist plot of the system transfer function, using only resistors and capacitors for the passive elements.[8] The operational amplifier eliminates the need for inductors and for impedance matching between cascaded filter stages. Filters designed to satisfy sophisticated mathematical criteria can be realized without resorting to "equalization" or trimming.

The simplest useful active circuit low-pass filter is a first-order lag circuit, as shown in Figure 21a. This system displays a single-pole transfer function with the following frequency domain magnitude and phase characteristics

$$|H(j\omega)| = \frac{e_o(j\omega)}{e_i(j\omega)} = \left[\frac{H_o^2 \omega_o^2}{\omega^2 + \omega_o^2} \right]^{1/2} \qquad (49)$$

$$\Phi(j\omega) = -\tan^{-1}(\omega/\omega_o) \qquad (50)$$

where $H_o = R/R_1$ and $\omega_o = 1/RC$.

This transfer function should be recognized as exactly that used for simple frequency compensation of the open-loop Op Amp gain (see Section 2.6). In addition, the frequency/gain characteristics of the closed-loop device shown in Figure 21a are identical in form to the open-loop gain characteristics shown in Figure 8b.

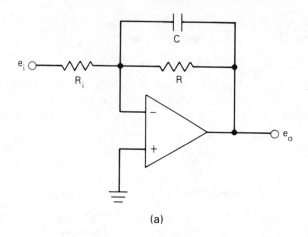

(a)

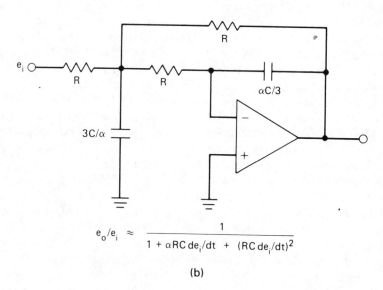

$$e_o/e_i \approx \frac{1}{1 + \alpha RC\, de_i/dt + (RC\, de_i/dt)^2}$$

(b)

Figure 21. Low-pass filters. (a) First-order lag circuit; (b) second-order Butterworth low-pass filter.

In practice, single-pole filters using a single Op Amp are rather uneconomical and seldom used. Efficient higher-order filters exhibit a response equation having no more than one real root, the other factors being damped quadratics. Figure 21b shows a typical second-order or complex conjugate pole pair filter section, with frequency domain transfer function

$$|H(j\omega)| = \left[\frac{\omega_o^4}{\omega^4 + \omega^2\omega_o^2(\alpha^2 - 2) + \omega_o^4} \right]^{1/2} \tag{51}$$

and phase

$$\Phi(j\omega) = -\tan^{-1}\left[\frac{2\omega}{\alpha\omega_o} + \left(\frac{4}{\alpha^2} - 1\right)^{1/2}\right] - \tan^{-1}\left[\frac{2\omega}{\alpha\omega_o} - \left(\frac{4}{\alpha^2} - 1\right)^{1/2}\right] \quad (52)$$

where $\omega_o = 1/RC$. The Q of a complex pole pair equals $1/\alpha$.

Several such sections may be cascaded to achieve a higher even-order filter characteristic.

The damping ratio, α, must be maintained at less than two or the poles will not be complex. Tailoring the damping ratios of successive filter stages can be used as an efficient way of achieving a filter with very high performance characteristics conforming to a chosen mathematical criterion. For example, even-order Butterworth (minimal phase shift) filters can be constructed by cascading quadratic filter sections of the types shown in Figure 21b, using the damping ratios presented in Table 2.

Table 2
Low-Pass Butterworth Filter Factors

Order, n	Damping ratio, α
2	1.414
4	1.845, 0.7654
6	1.932, 1.414, 0.4176
8	1.961, 1.663, 1.111, 0.3896
10	1.976, 1.893, 1.414, 0.9081, 0.3128

The frequency domain gain characteristics of these cascaded Butterworth filters have a form generalized from that of the single-pole filter presented in Eq. (49)

$$|H(j\omega)| = \left[\frac{1}{1 + (\omega/\omega_o)^{2n}}\right]^{1/2} \quad (53)$$

where n is the number of sections.

Not unexpectedly, the first-order lead circuit shown in Figure 22a constitutes a single-pole high-pass filter. The frequency domain gain and phase shift characteristics of this filter can be predicted by conformal transformation of the equivalent low-pass filter transfer function. By substituting $-\omega/\omega_o$ for ω_o/ω in Eq. (49) and (50) we obtain the single-pole high-pass filter characteristics

$$|H(j\omega)| = \left[\frac{H_o^2\omega^2}{\omega^2 + \omega_o^2}\right]^{1/2} \quad (54)$$

$$\Phi(j\omega) = \pi/2 - \tan^{-1}(\omega/\omega_o) \quad (55)$$

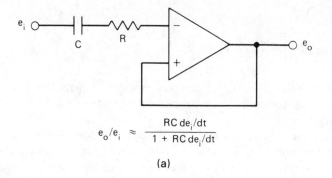

$$e_o/e_i \;\approx\; \frac{RC\,de_i/dt}{1 + RC\,de_i/dt}$$

(a)

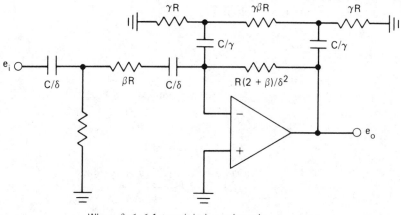

Where $\beta \ll 1$ to minimize noise gain

$$\gamma \;=\; 1 + X^2 / [X^2 + (2 - \alpha^2)X + 1]$$

$$\delta \;=\; \alpha[1 + X]$$

$$X \;=\; \beta/[2 + \beta]$$

$$e_o/e_i \;\approx\; \frac{(RC\,de_i/dt)^2}{1 + \alpha RC\,de_i/dt \;+\; (RC\,de_i/dt)^2}$$

(b)

Figure 22. High-pass filter. (a) First-order lead circuit; (b) second-order Butterworth high-pass filter.

Figure 22b shows an efficient second-order high-pass filter, which is capable of being used as the quadratic factor in higher-order cascade filters. Unlike the low-pass filter, the pass band characteristics of a high-pass filter ultimately become determined by the amplifier open-loop gain characteristics.

The frequency domain transfer function of the complex conjugate pole-pair high-pass filter presented in Figure 22b is given by

$$|H(j\omega)| = \left[\frac{\omega^4}{\omega^4 + \omega^2\omega_o(\alpha^2 - 2) + \omega_o^4}\right]^{1/2} \tag{56}$$

$$\Phi(j\omega) = \pi - \tan^{-1}\left[\frac{2\omega}{\alpha\omega_o} + \left(\frac{4}{\alpha^2} - 1\right)^{1/2}\right] - \tan^{-1}\left[\frac{2\omega}{\alpha\omega_o} - \left(\frac{4}{\alpha^2} - 1\right)^{1/2}\right] \tag{57}$$

The circuits in Figures 22a and 22b can be cascaded separately or jointly to produce odd- or even-order filters. High-order filters employing circuit 22b suffer a significant phase distortion in the vicinity of the cut-off frequency, ω_o. As with all filters of this type, the transient response deteriorates in fidelity as the number of sections, n, is increased.

The simplest way to view a bandpass filter is as a combination of low-pass and high-pass filters with cut-off frequencies that do not overlap. Thus, cascading filters of the types shown in Figures 21 and 22 will result in a filter having a pass bandwidth (BW)

$$BW = \omega_h - \omega_l \tag{58}$$

where ω_h and ω_l are the high- and low-pass filter cut-off frequencies.

Bandpass filters are frequently classified according to their "normalized" bandwidth

$$BN = BW/2\omega_o^* \tag{59}$$

where $\omega_o^* = (\omega_h\omega_l)$. Filters with $BN \geq 1$ are classified as wide-band, and those with $BN < 1$ as narrow-band.

A band-reject or notch filter can be realized by performing the operation $1 - H_{BP}(j\omega)$, where $H_{BP}(j\omega)$ is a bandpass transfer function. Notch filters are very commonly used to eliminate noise due to specific frequency sources associated particularly with mains frequency pick-up. In such applications, a very narrow-band, high Q filter is desirable, and regenerative (positive feedback) filters are often employed. Figure 23 shows a notch filter that is convenient for use in rejecting mains frequencies. In this circuit, voltage follower 1 is provided to prevent loading of the passive twin-T filter, while follower 2 provides a fractional positive feedback from the output. The variable resistance, R, is adjusted to fine tune the notch frequency, while the 5 kΩ potentiometer is used to adjust the circuit Q.

4. Potential and Current Control

4.1. Introduction

The two most commonly used control instruments in electrochemistry are potentiostats and galvanostats. A potentiostat is a device that controls the

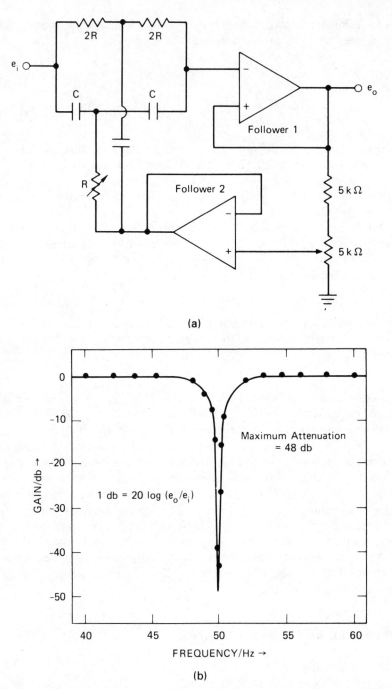

(a)

(b)

Figure 23. Notch filters. (a) Regenerative notch filter circuit; (b) frequency response with $R = 7170\,\Omega$, $C = 0.22\,\mu\mathrm{f}$.

potential difference between a working electrode (WE) and a reference electrode (RE) at some predetermined value in a three-electrode cell by continuously modifying the current between the working electrode and a counter electrode (CE). On the other hand, a galvanostat maintains a constant current between a working electrode and a counter electrode, irrespective of the impedance of the cell. The potential difference between the WE and the RE is the independent variable in the case of potentiostatic control, whereas cell current is the independent variable for galvanostatic control.

A discussion on the history of development of these devices is beyond the scope of this review. However, it should be noted that inexpensive and easily usable potentiostats became available only on advent of operational amplifiers (see Sections 2 and 3) in the early 1960s. Since that time, numerous designs for both potentiostats and galvanostats have been published, some of which are discussed later in this review.

4.2. Principles of Polarization Control

Probably the most fundamental, and also the most commonly ignored, question concerning the application of control devices in electrochemistry is: will the device maintain the intended control without itself distorting the true *electrochemical* response of the system? The answer to this very important question is not obvious, at least for potentiostats, since it depends on both the properties of the control device and the electrochemical characteristics of the system under control. Epelboin *et al.*[11] have considered this problem in terms of load line analysis, and certain aspects of their treatment are reproduced here.

Consider the general control device/electrochemical cell arrangement shown schematically in Figure 24. In this equivalent circuit, we have lumped together as a single resistance, ρ, all impedances not associated directly with the interfaces. These include the output impedance of the control device, the resistance of the solution in the cell, and any stray resistances (e.g., contact resistance) associated with external circuitry. The resistance R is included to represent specifically resistances that appear between the reference electrode and the ground connection of the working electrode. Irrespective of the properties of the electrochemical cell, the current is related to the voltage that appears across the cell and to the output voltage of the control device as follows

$$i = (e/\rho) - (u/\rho) \tag{60}$$

A plot of i vs. u therefore yields a straight line (the "load line") of slope $-1/\rho$ (Figure 24). However, we may also characterize the response of the *cell* in i, u space as shown schematically in Figure 25. The point of intersection, X, defines the current-voltage data pair that will be observed for a given output voltage (e) of the control device. *A stable system is characterized by a single point of intersection, X.*

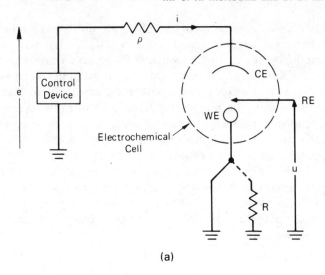

(a)

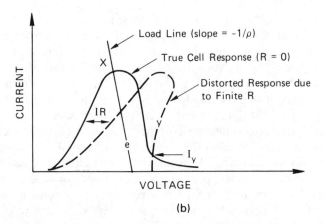

(b)

Figure 24. The influence of uncompensated resistance on the potentiostatic control of an electro-chemical cell. (a) Schematic of an electrochemical control device and cell; (b) schematic current–voltage curves showing the effects of IR distortion and the load line for the system.

A variety of causes exists for instability in control devices, and one of these is illustrated in Figure 25. In this case, the value of the resistance ρ is assumed to be large, resulting in a load line that is far from vertical. If an attempt is now made to measure the current–voltage curve for the cell by increasing the control device output, e, the correct curve is obtained over the region O–X. However, an additional increment in e results in three points of intersection (U, V, W) and the system becomes unstable. This results in a sharp transition in the current to point Y. Under these circumstances the observed response differs from the true response *because of limitations of the*

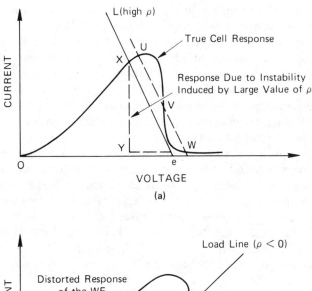

Figure 25. Schematic current–voltage curves illustrating the cause of instability $[p \gg 0$, plot (a)$]$ and correction by using negative impedance conversion $[p < 0$, plot (b)$]$.

control device (high p). Clearly this problem could have been avoided by maintaining a sufficiently low value of p to yield a load line that is close to vertical. This could have been achieved by insuring that the output impedance of the control device was small and that the resistance of the solution between the reference electrode and the counter electrode also was low. This then is the formal reason why well-designed electrochemical control devices should be characterized by low output impedances; that is, the control device should be capable of supplying whatever current is required to determine the true response of the cell.

A very common problem in experimental electrochemistry is the notorious "IR" drop that may arise from a finite solution resistance between WE and RE, or because of unwanted resistance between the WE and ground (e.g., due to the mercury capillary in classical polarography). The influence of an uncompensated electrolyte resistance on the performance of a control device is best illustrated by returning to Figure 24, in which the causes for the IR drop are lumped together as a single resistance R. Because the IR drop is included in the measured voltage between the RE and the connection to the WE (or ground), the cell response becomes distorted as shown by the broken line. Clearly, even with an ideal control device (zero output impedance) and no resistance between the CE and the RE (i.e., one having a vertical load line) the true response of the WE could be characterized only at a voltage as high as $u = Y$. An incrementally higher voltage would then result in instability and a transition in current down to i_y.

The above problem can be circumvented by using a control device whose load line is characterized by a sufficiently high *positive* slope. Since no assumptions were made in deriving Eq. (60) as to the nature of the control device, such a device must possess a sufficiently large *negative* output impedance to yield the desired positive slope. Such a device is called a "negative impedance converter" (NIC). Although NICs are well known in electrical engineering, their use in electrochemistry is less than a decade old.[11] The operation of NICs is discussed in detail in Section 4.6.

4.3. Potentiostat Design

Modern potentiostats are generally based on one of two designs: the single (or differential) amplifier type or the adder variety. These two designations refer to the configuration of the input and feedback loops around the control amplifier. Although both varieties are designed to accomplish essentially the same task, they differ significantly in versatility and in their dynamic responses. Not unexpectedly, these differences have led to various hybrid varieties, which will also be discussed where appropriate. No consideration is given to early servo-mechanical devices.

The simplest variety, the single amplifier or differential amplifier potentiostat, is shown schematically in Figure 26a. In this case, the reference electrode (RE) voltage is fed back to the inverting input of the control amplifier and the input voltage, e_i, is connected to the noninverting input. Because both inputs have very high input impedances (typically $10^{12} \, \Omega$ for FET devices), negligible current is drawn from the reference electrode or the voltage source; that is, they are not polarized. The control equation for this configuration is given as

$$e_{\text{ref}} + iR_u = e_i - e_o / G_{CA} \qquad (61)$$

where the quantity of the left-hand side is the voltage of the reference electrode with respect to ground, G_{CA} is the open-loop gain of the control amplifier, and e_o is the output voltage from the amplifier that is required to maintain the required current. For most commercially available potentiostats, the (boosted) output voltage is limited to less than 100 V and with $G_{CA} \approx 10^5$, the last term on the right-hand side of Eq. (61) is of the order of 1 mV at the

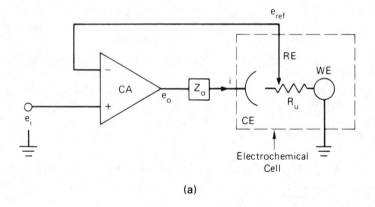

(a)

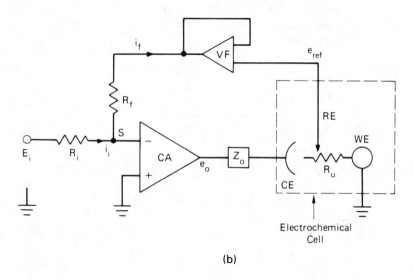

(b)

Figure 26. Schematics of (a) single amplifier potentiostat; (b) adder-type potentiostat.

most. Accordingly, stable operation ensures that the voltage applied between the RE and the WE (which is grounded) is very close to the input voltage, e_i.

The "adder" configuration is shown schematically in Figure 26b. In this case the voltage from the reference electrode is added to the input voltage at the summing point junction, s. Because the summing point has a low impedance with respect to the circuit ground, it is necessary to place a high-impedance buffer between the summing point, s, and RE in order to avoid polarizing the reference electrode. Only a single input is shown in Figure 26b. However, additional inputs are readily incorporated in an adder configuration (see Section 2.3).

As before, the control function for this configuration is readily derived

$$(e_{ref} + iR_u) = -(R_f/R_i)[(1 + G_{VF})/G_{VF}]e_i$$
$$-(e_{o,CA}/G_{CA})[(1 + G_{VF})/G_{VF}][(R_i + R_f)/R_i] \quad (62)$$

where subscripts CA and VF designate parameters for the control amplifier and voltage follower, respectively. Since G_{VF} and G_{CA} are typically 10^5 and customarily, $R_i \approx R_f$, the condition for control is

$$(e_{ref} + iR_u) = -(R_f/R_i)e_i \quad (63)$$

That is, the potential drop between the working and reference electrodes is again determined by the input voltage e_i and additionally, in this case, by the ratio R_f/R_i.

As noted above, hybrids of these two designs have appeared, the most common of which is shown schematically in Figure 27. The reader will recognize the second amplifier (A) as being in the voltage adder configuration which provides a voltage [see Eq. (3)]

$$e_o = -R_f \sum_{i=1}^{n} (e_{i,n}/R_{i,n}) \quad (64)$$

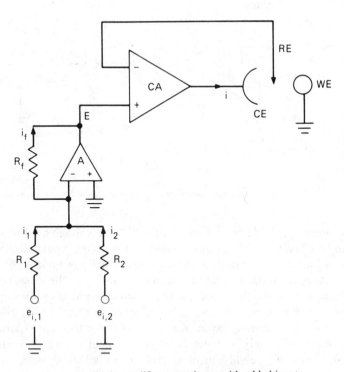

Figure 27. Single-amplifier potentiostat with added inputs.

for n inputs. The potentiostat control function is now modified to read

$$e_{\text{ref}} + iR_u = -R_f \sum_{i=1}^{n} (e_{i,n}/R_{i,n}) - e_o/G_{CA} \tag{65}$$

In this case, the voltage applied between the RE and WE may be adjusted by changing $e_{i,n}$ or $R_{i,n}$, or both.

Measurement of the current flowing through the cell is most easily accomplished by placing a resistor between the control amplifier and the counter electrode and then measuring the voltage drop using a high-impedance voltmeter having floating inputs. Note that this resistor should never be inserted between the working electrode and ground unless provisions are made to clamp the working electrode potential at virtual ground. A resistor in this position would cause distortion of the current–voltage curve and hence control problems of the type discussed in the previous section. As noted above, the exception to this rule is when the voltage of the WE is clamped at virtual ground using a current follower of the type shown in Figure 16, and discussed in Section 3.4. Applying the amplifier circuit equations

$$e_{\text{WE}} = e_o - iR_f \tag{66}$$

$$e_o = -Ae_{\text{WE}} \tag{67}$$

yields an expression for the voltage of the working electrode

$$e_{\text{WE}} = iR_f/(1 + A) \tag{68}$$

For d-c experiments, the open-loop gain $A(=A_o)$ is of the order of 10^5 so that e_{WE} is effectively zero, i.e., at ground potential. Because G decreases with increasing frequency, the clamping error increases dramatically in the frequency domain, and the working electrode may tend to float well above ground. Finally, the output voltage of this device may be used to give a direct measure of the current, since from Eq. (37) with $e_{\text{WE}} \approx 0$

$$i = -e_o/R_f \tag{69}$$

The output of the current follower or voltage clamp provides a particularly convenient signal for *IR* compensation, as discussed in the following section. However, it does so at a cost in performance under dynamic conditions (see Section 3.4).

4.4. IR Compensation

The distortion of the true current–voltage response by an uncompensated *IR* potential drop between the RE and the connection to the WE has been briefly alluded to in Section 4.2. However, the problem is much more serious than indicated, since in transient or relaxation studies *IR* potential losses also distort the form of the perturbation applied between the working and reference electrodes, and the degree of distortion may be sufficiently large so as to invalidate subsequent data analysis in terms of a preconceived model. Consequently a great deal of effort has been expended in developing techniques for automatically compensating for the *IR* drop at the input to the potentiostat. Britz[12] and others[13] have published excellent reviews of this topic to which the reader is referred. Only a brief outline of the subject will be given here.

Two commonly employed methods for *IR* compensation are shown schematically in Figure 28. Both circuits are based on the "adder" potentiostat design (similar configurations can be used with single amplifier poteniostats) and involve feeding back to the input a voltage that is proportional to the current flowing through the cell. The circuits differ only in the placement of the current-sensing resistance; in the first case it is placed in the counter electrode circuit, whereas in the second it is an integral part of the voltage clamp (current follower) at the working electrode. Note that in the first case, provided R_i is small compared with the cell resistance, the dynamic behavior is not affected significantly by *IR* compensation, whereas that of the second configuration is.[12] However, the first configuration requires the use of a differential amplifier (*DA*) for which both inputs may be operating at high potentials with respect to circuit ground. This may lead to common-mode error problems, as discussed in Section 3.3.

Application of the summing point constraint for an ideal control amplifier yields the following operating equation for both circuits

$$e_i(R_f/R_i) - \alpha i R_i(R_f/R_{fb}) = -e_{\text{ref}} - iR_u \qquad (70)$$

Complete *IR* compensation is achieved when the second terms on each side of Eq. (70) cancel. This condition is met when

$$\alpha = R_u R_{fb}/R_i R_f \qquad (71)$$

to yield

$$e_{\text{ref}} = -(R_f/R_i)e_i \qquad (72)$$

Equation (71) indicates that if R_u is known from independent measurement it should be possible to calculate the fraction of feedback α required to compensate exactly for the *IR* drop. In practice, this method is inconvenient, and in many cases impractical. Instead, the most common method of determining the correct amount of feedback is to vary α (e.g., by adjusting R_i or the gain of *DA*) until the response of the overall system indicates that the desired effect has been achieved. Not surprisingly, the response observed depends on the form of the input (e_i) used to perturb the system. Only three of many such perturbations are discussed below; for a discussion of others the reader is referred to the reviews by Britz[12] and Macdonald.[14]

One leading manufacturer of commercial potentiostats recommends that for a constant e_1 the fraction α will be increased until oscillation sets in. The amount of feedback (α) is then backed-off slightly to yield the "correct"

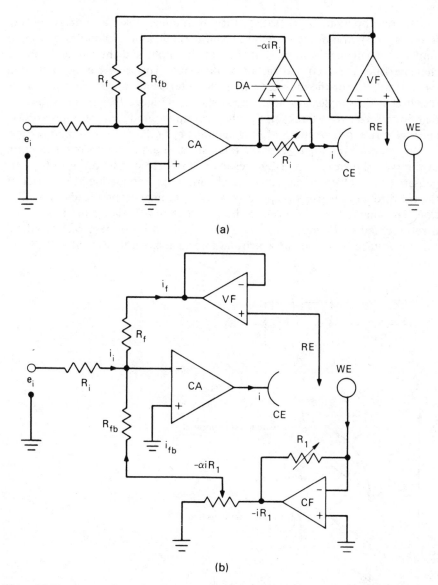

Figure 28. Two commonly employed methods for *IR* compensation using an adder-type potentiostat. (a) Instrumentation amplifier in series with counter electrode; (b) proportional feedback from current follower.

compensation. This technique is not very satisfactory for accurate work because, although excessive feedback *may* cause oscillation, it does not always do so and it is possible to overcompensate. This may be as detrimental as no compensation at all.

In our opinion, more satisfactory techniques take advantage of the frequency response of the system to some periodic input. In one case, a small amplitude triangular voltage is applied to the input and the current–voltage characteristics are recorded on an oscilloscope. This case has been treated theoretically[15] for the instance where the interfacial impedance can be represented by the simple equivalent circuit shown in Figure 29a. Theory shows[15] that finite (positive) values for R_u (the uncompensated solution-phase resistance) give rise to relaxations at the beginning of both the forward and reverse sweeps. If, however, R_u is zero, the relaxations disappear and a parallelogram of the type shown in Figure 29c results. This corresponds to the fully compensated case. Overcompensation results in "blips" appearing at the start of the forward and reverse sweeps; excessive overcompensation leads to oscillation. This method is valid even in the presence of a Faradaic process, which is represented by R_p in Figure 29a. Accordingly, it is not necessary to adjust the d-c voltage to a value at which the interface is purely capacitive.

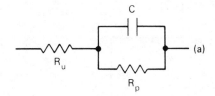

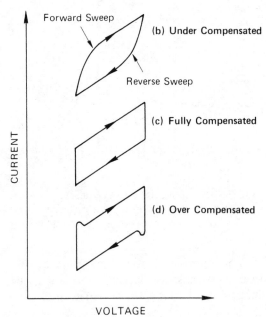

Figure 29. Electrical equivalent circuit for an electrochemical interface (a) and small amplitude cyclic voltammograms for varying degrees of *IR* compensation (b, c, d).

A closely related technique[16] makes use of a sinusoidal input and records the real (in-phase) and imaginary (quadrature) components of the impedance. The impedance of the interface can be represented as a complex number

$$Z = R - jX \tag{73}$$

where R and X are frequency-dependent parameters. At sufficiently high frequency, $X \to 0$ and $R \to R_s$, as shown in Figure 30a. Accordingly, the correct level of compensation can be achieved simply by increasing α until the intercept of the impedance locus on the real axis, at sufficiently high frequencies, occurs at the origin (Figure 30b). It is stressed, however, that the frequency should

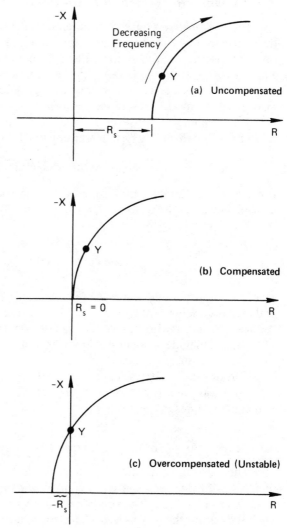

Figure 30. Schematic impedance loci illustrating the effect of *IR* compensation. Overcompensation can result if the amount of feedback required is determined at a frequency that is not sufficiently high that $X_y \neq 0$.

be sufficiently high that the imaginary component is negligibly small. If this is not so, then overcompensation may result, as shown schematically in Figure 30c. In this case, the real component at point Y would be nulled out, thereby resulting in a negative intercept on the real axis at infinite frequency.

The two frequency-domain methods described above are generally effective for the quantitative elimination of uncompensated resistance, provided that the precautions noted are recognized. They are by no means the only techniques nor are they necessarily the best. However, we have found them to be effective for a wide range of systems.

4.5. Dynamic Response of Potentiostats

The dynamic response of a potentiostat determines the time taken for the device to establish control after application of some specified perturbation. In manufacturers' specifications, the dynamic response is often erroneously equated (expressed or implied) with the slew rate of the control amplifier. As is shown below, the actual response rate depends also on the nature of the load that is being driven and on the form of the input.

The dynamic response of a system can be formulated using *transform analysis*[17] in which the system transfer function $H(s)$

$$H(s) = \overline{\text{Output}(s)}/\overline{\text{Input}(s)} \tag{74}$$

is expressed in terms of the properties of the components that make up the control device and the load. The superscript bars denote linear (Laplace) transforms involving the operator, s

$$f(s) = \int_{o}^{\infty} F(t) \, e^{-st} \, dt \tag{75}$$

Once the transfer function has been specified, Eq. (75) may be rearranged and then inversely transformed to yield the response in real time. In this review we describe the transform analysis for both single amplifier and adder potentiostats, and include an analysis of the effect of cell parameters on the response time.

The transform analyses are presented for a potential step input voltage. This case was chosen because it represents the most rapid passage between two control states.

4.5.1. Single Amplifier Potentiostats

The electrical equivalent circuit that will be used here for the transform analysis of an electrochemical cell controlled by a single amplifier potentiostat is shown in Figure 31. The components external to the potentiostat itself are enclosed within the box labeled "electrochemical cell."

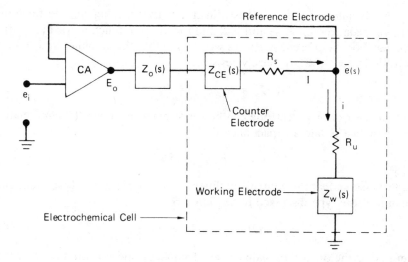

Figure 31. Electrical equivalent circuit for an electrochemical cell controlled with a single-amplifier potentiostat.

With reference to Figure 31, it is seen that the output voltage is given by

$$e_o = -G(s)[\bar{e}(s) - e_i] \tag{76}$$

where $G(s)$ is the open-loop gain of the amplifier. Similarly

$$i = \{e_o - \bar{e}(s)\}/\{Z_o + Z_{CE} + R_s\} \tag{77}$$

$$i = \bar{e}(s)/\{Z_w + R_4\} \tag{78}$$

where Z_0, Z_{CE}, and Z_w are the output, counter electrode, and working electrode impedances, respectively, R_s is the resistance between the counter electrode and reference electrode, and R_u is the resistance between the reference electrode and the working electrode. Manipulation of Eq. (76) to (78) yields the voltage applied at the RE point $[\bar{e}(s)]$ in terms of the input e_i

$$E = e_i\{G[R_u + Z_w]/[Z_o + Z_{CE} + R_s + (G+1)(R_u + Z_w)]\} = e_i\{H_E(s)\} \tag{79}$$

The quantity $\{H_E(s)\}$ is known as the *control transfer function*. Note that no assumptions have been made at this point concerning the performance characteristics of the control amplifier or the form of the impedance of the electrochemical cell. A second transfer function may be defined, which gives the current response of the system

$$i = -e_i\{G/[Z_o + Z_{CE} + R_s + (G+1)(R_u + Z_w)]\} = e_i\{H_I(s)\} \tag{80}$$

The two transfer functions $H_E(s)$ and $H_I(s)$ completely define the response of the system to any arbitrary input, $E_i(t)$.

The dynamic response is best illustrated by adopting specific functions for the impedances Z_o, Z_{CE}, and Z_W, and for the gain function for the control amplifier. For the present purposes we assume that the amplifier and counter electrode impedances are purely resistive

$$Z_o = R_o, Z_{CE} = R_{CE}$$

and that the working electrode can be represented by a parallel combination of a resistance R and a capacitance C

$$Z_W = R/(1 + sRC) \tag{81}$$

We also assume that the control amplifier is characterized by a first-order gain function of the type discussed in Section 2.6

$$G(s) = A_o/(1 + \tau s) \tag{82}$$

where A_o is the open-loop gain at zero frequency and τ is the time constant for the amplifier $(=1/\omega_o)$. Substitution of these relationships into Eq. (79) yields

$$e = e_i[(L + Ms)/(N + Os + Ps^2)] \tag{83}$$

where

$$L = A_o(R + R_u)$$

$$M = A_oRR_uC$$

$$N = (1 + A_o)(R + R_u) + R_o + R_{CE} + R_s$$

$$O = \tau(R_o + R_{CE} + R_s + R + R_u) + [R_o + R_{CE} + R_s + (A_o + 1)R_u]RC$$

$$P = \tau RC[R_o + R_{CE} + R_s + R_u]$$

In order to complete the analysis, we must obtain the Laplace transform of E_i (i.e., e_i) and then carry out the inverse transformation to obtain the response in real time.

For a step input voltage, e_i has the form

$$E_i = 0 \quad \text{for } t \leq 0$$
$$E_i = E_1^0 \quad \text{for } t > 0 \tag{84}$$

and hence

$$e_i = E_i/s \tag{85}$$

Substitution into Eq. (83) therefore yields

$$e = E_i(L/N)(1 + as)/s[1 + (2\zeta/\omega_1)s + (1/\omega_1^2)s^2] \tag{86}$$

which, on inverse transformation using standard tables,[18] gives the response in the time domain as

$$E = E_i \left(\frac{L}{N}\right)\left\{1 + \frac{1}{\sqrt{1-\zeta^2}}[1 - 2a\zeta\omega_1 + a^2\omega_1^2]^{1/2} \exp\left(-\zeta\omega_1 t\right)\cdot\right.$$

$$\left. \times \sin\left[\omega_1(1-\zeta^2)^{1/2}t + \psi\right]\right\} \tag{87}$$

where

$$a = M/L$$
$$\omega_1 = (N/P)^{1/2}$$
$$\zeta = O/2(NP)^{1/2}$$

and

$$\psi = \tan^{-1}\left[\frac{a\omega_w(1-\zeta^2)^{1/2}}{1 - a\zeta\omega_1}\right] - \tan^{-1}\left[\frac{(1-\zeta^2)^{1/2}}{-\zeta}\right]$$

Equation (87) demonstrates that the voltage applied between the reference and working electrodes is a damped oscillation in spite of the fact that the input was a perfect step, as shown in Figure 32. A convenient measure of the transient response time for the overall system is the time constant contained

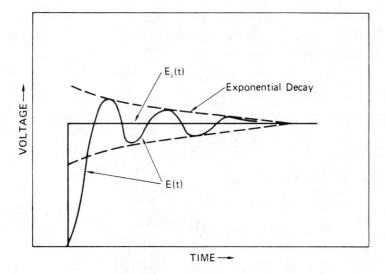

Figure 32. Voltage response $E(t)$ for a single amplifier potentiostat. The input voltage $E_i(t)$ is assumed to be a perfect step function.

within the exponential decay term in Eq. (87)

$$t_d = 1/\zeta\omega_1 = \frac{2\tau RC(R_o + R_{CE} + R_s + R_u)}{\tau(R_o + R_{CE} + R_s + R + R_u) + RC[R_o + R_{CE} + R_s + (A_oH)R_u}$$

(88)

Clearly the time constant, t_d, is a function of both the amplifier characteristics (A_o, τ, R_o) and of the properties of the external load (R, C, R_s, R_{CE}, R_u). For large values of RC, as sometimes exist for a mercury pool electrode in the absence of fast faradaic processes, Eq. (88) reduces to the approximate form

$$t_d \approx 2\tau(R_o + R_{CE} + R_s + R_u)/\{(1 - A_o)R_u\}$$

(89)

In this case, if an attempt is made to compensate fully for the solution resistance between the working and reference electrodes $(R_u \to 0)$, the system may oscillate since $t_d \to \infty$. Note that the response time is still a function of both the control amplifier and cell (but not the working electrode) parameters. At the other extreme, in cases for which RC becomes very small (fast faradaic reaction and low capacitance), the time constant tends toward $2RC$, provided that $(1 + A_o)RCR_u \ll \tau(R_o + R_{CE} + R_s + R + R_u)$. In this case, the response time is very short, and is independent of the properties of the control amplifier and the value of R_u.

We may carry out a similar analysis for an adder potentiostat (Figure 33). In this case, the control equations are written as

$$e_o = -G \cdot e_s$$

(90)

$$e_o - e = (Z_o Z_{CE} + R_s)i$$

(91)

$$= (R_u + R_w)i$$

(92)

in which it is assumed that the voltage follower in the feedback loop behaves ideally. Proceeding as before, we find that

$$e = \frac{-\{GR_f(R_u + Z_w)\}e_i}{[(R_i + R_f) + (Z_o + Z_{CE} + R_s + R_u + Z_z) + GR_i(R_u + Z_w)]}$$

(93)

which on substitution for the various impedance functions gives

$$e = -[(L' + M's)/(N' + O's + P's^2)]e_i$$

(94)

where

$$L' = A_o R_f(R + R_u)$$

$$M' = (R + R_u)/R_u RC$$

$$N' = (R_i + R_f)(R_o + R_{CE} + R_s) + (R + R_u)[R_f + (A_o + 1)R_i]$$

$$O' = (R_i + R_f)(R_o + R_{CE} + R_s)(RC + \tau)$$

$$+ (R_i + R_f)[\tau(R + R_u) + R_u RC] + A_o R_i R_u RC$$

$$P' = \tau RC(R_i + R_f)(R_o + R_{CE} + R_s + R_u)$$

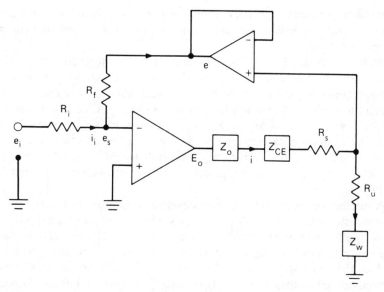

Figure 33. Schematic of an adder-type potentiostat showing lumped impedances as used for the transform analysis of the dynamic response.

Again, for a potential step, we obtain

$$e = -e_i(L' + M's)/[s(N' + O's + P's)] \tag{95}$$

which is of the same form as that for the single amplifier potentiostat [Eq. (83)] thereby resulting in damped oscillatory output. As before, the time constant for damping is given as

$$t_d' = 1/\zeta'\omega_1' = \frac{2\tau RC(R_o + R_{CE} + R_s + R_u)}{\tau(R_o + R_{CE} + R_s + R + R_u) + RC\left[R_o + R_{CE} + R_s + \left(1 + \frac{A_o R_i}{R_i + R_f}\right)R_u\right]} \tag{96}$$

which is identical to that for the single amplifier Eq. (88) except for the last term in the square brackets in the denominator. Because $R_i/(R_i + R_f) <$ 1, $t_d < t_d'$, and the damped oscillation will tend to die away more quickly for the single amplifier configuration than for the adder configuration. However, unless $R_i \ll R_f$, the difference is not large as shown by comparing Eqs. (89) and (96), and disappears completely for $R_u \to 0$.

The analyses presented above demonstrate that the dynamic response of a potentiostat depends not only on the properties of the external load, but also on the characteristics of the control amplifier. Accordingly, the trend among manufacturers of commercial instruments in quoting the slew rate of the amplifier as a measure of the dynamic response without defining the nature of the load is very misleading and cannot be taken to be reliable for most practical situations.

As far as the stability of potentiostat systems is concerned, Britz[12] notes that the most serious destabilizing effect is damping due to large resistance in the Luggin probe and stray capacitance to ground. The use of a current follower between the working electrode and ground is also recognized as a destabilizing influence, particularly at high frequencies. Purely capacitive working electrodes, for example a large mercury pool at a potential where no faradaic processes occur and for which the "uncompensated" resistance has been compensated for, are especially prone to stability problems as predicted by Eq. (96).

A number of methods have been devised to stabilize inherently unstable potentiostatic systems.[19-21] A particularly effective method, as reported by Brown, Smith, and Booman,[19] is to place a capacitor (C_1) between the output of the control amplifier and the reference electrode (Figure 34), which tends to reduce the phase shift in the cell transfer function. A capacitor (C_2) is also commonly added around the control amplifier, as shown in Figure 34. However, the work of Brown, Smith, and Booman[19] shows that this practice may accentuate instability, particularly if IR compensation is employed.

4.6. Galvanostats

The advent of operational amplifiers has also had a significant impact on the design of constant current sources, and some aspects of the operation of galvanostats are discussed below.

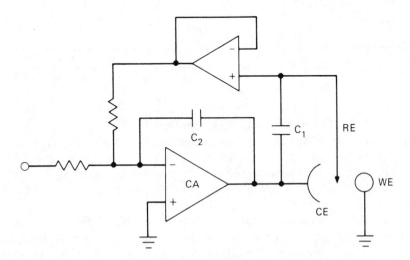

Figure 34. Use of capacitors C_1 and C_2 to stabilize an adder-type potentiostat.

The basic requirement of a galvanostat is that the current that is delivered should be independent of the impedance of the load (electrochemical cell). Two circuits that satisfy this requirement are shown schematically in Figure 35. Application of the control equations for the first case gives

$$(e_i - e_s)/R_i = (e_s - e_o)/Z_C = i_f \qquad (97)$$

Substituting $C_o = -Ge_s$ we obtain

$$i_f = e_i/[R_i + Z_C/(1 + G)] \qquad (98)$$

where G is the open-loop gain of the amplifier and Z_C is the impedance of the load. At low frequencies, $G \gtrsim 10^5$ for most amplifiers used, and the second

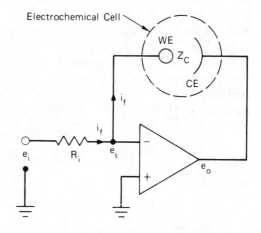

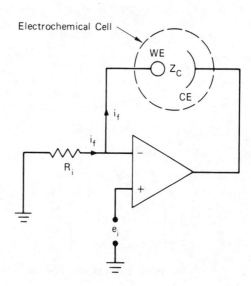

Figure 35. Commonly employed galvanostatic circuits for electrochemical studies.

term in the square brackets is negligible compared with $R_i(\sim 1\text{–}10\text{ k}\Omega)$, provided that the impedance of the cell is not too high. Accordingly, the current is independent of Z_C, as desired.

Application of the control equations to the second case shows that provided Z_C is not too large, the current is determined only by R_i and e_i. However, the two circuits differ in a number of important respects. In the first case, the working electrode is held at virtual ground during operation and the current is drawn from the voltage source. This latter problem is easily circumvented by placing a voltage follower between e_i and R_i. In the second case, the working electrode floats above ground at a potential e_i and no current is drawn from the voltage source. In this case, the potential of the WE with respect to a reference electrode must be measured with an instrumentation amplifier (see Section 3.3). Both circuits will continue to operate as galvanostats provided that the current and power output specifications of the amplifier are not exceeded.

Although the dynamic response of galvanostats is of less interest than that of potentiostats (because few galvanostatic transient studies are ever performed), it is nevertheless instructive to examine the time taken for a galvanostat to establish control. With reference to Figure 36 we can write

$$e_i - e_s = i_f R_i \tag{99}$$

$$e_s - e_o = i_f(Z_o + Z_W + R_s + Z_{CE}) \tag{100}$$

and

$$e_o = G \cdot e_s \tag{101}$$

where the impedances Z_W and Z_{CE} refer to the working electrode and counter electrode, respectively, and R_s is the solution resistance. As before, eliminating

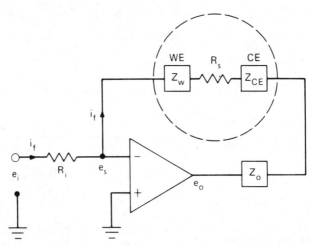

Figure 36. Schematic of the galvanostatic circuit used for transform analysis.

e_o and e_s, and assuming $Z_o = R_o$, $Z_{CE} = R_{CE}$, and $G = A_o/(1 - \tau s)$ yields

$$i_f = [(L + Ms + Ns^2)/(O + Ps + Qs^2)]e_i \qquad (102)$$

which has nearly the same functional form as the dynamic response function for a potentiostat. For a step input, $e_i = E_i/s$, and Eq. (102) becomes

$$i_f = e_i(L + Ms + Ns^2)/s(O + Ps + Qs^2) \qquad (103)$$

where

$$L = 1 + A_o$$

$$M = RC(1 + A_o) + \tau$$

$$N = \tau RC$$

$$O = (1 + A_o)R_i + R_o + R_s + R_{CE} + R$$

$$P = (1 + A_o)RR_iC + (R_o + R_s + R_{CE})(RC + \tau) + \tau(R + R_i)$$

$$Q = \tau RC[R_i + R_o + R_s + R_{CE}]$$

Inverse transformation is easily accomplished using standard tables to yield

$$i_f = (e_iL/O)\left\{1 + \frac{1}{(1 - \zeta^2)^{1/2}}[(1 - a\zeta\omega_1 - b\omega_1^2 + 2b\zeta^2\omega_1)^2 + \omega_1^2(1 - \zeta^2)\right.$$

$$\left. \times (a - 2b\zeta\omega_1)^2]^{1/2} \exp[-\zeta M\omega_1 t] \sin[\omega_1(1 - \zeta^2)^{1/2}t + \varphi]\right\} \qquad (104)$$

$$\Psi = \tan^{-1}\left\{\frac{(\omega_1 - \zeta^2)^{1/2}(a - 2b\zeta\omega_1)}{b\omega_1^2(2\zeta^2 - 1) + 1 - a\zeta\omega_1}\right\} - \tan^{-1}\left\{\frac{(1 - \zeta^2)^{1/2}}{-\zeta}\right\} \qquad (105)$$

with

$$\omega_1 = (O/Q)^{1/2}$$

$$\zeta = P/[2(OQ)^{1/2}]$$

$$a = M/L$$

$$b = N/L$$

As expected, the current response has the form of a damped oscillation of time constant

$$t_d = 2Q/P = \frac{2\tau RC(R_i + R_o + R_s + R_{CE})}{(1 + A_o)RR_iC + (R_o + R_s + R_{CE})(RC + \tau) + \tau(R + Ri)} \qquad (106)$$

As for the case of a potentiostat, the time taken for the galvanostat to establish control is a function of both the amplifier and cell characteristics. For most experimental systems, the first term in the denominator is considerably larger than the others so that Eq. (106) reduces to

$$t_d \approx 2\tau(R_i + R_o + R_s + R_{CE})/(1 + A_o)R_i \qquad (107)$$

substituting typical values of $A_o = 10^5$, $R_i = 1000\,\Omega$, $\tau = 0.1\,s$, $R_o = 10\,\Omega$, $R_s = 10\,\Omega$, and $R_{CE} = 10\,\Omega$, $t_d \sim 2\,\mu s$, and the decay time is relatively independent of the properties of the working electrode interface and cell $(R_i \gg R_o - R_s + R_{CE})$.

4.7. Negative Impedance Converters

In Section 4.2 we noted that a conventional potentiostat/electrochemical cell system is characterized by a load line having a negative slope. This property precludes control over electrodes that exhibit "Z"-shaped polarization curves. Over the past decade, Epelboin and co-workers,[11,22–24] and more recently Diard and LeGorrec,[25] have used negative impedance converters (NIC's) to explore the electrochemical properties of systems of this type.

The circuit used by Epelboin *et al.*[11] is shown schematically in Figure 37(a) and the equivalent circuit is given in Figure 37(b). Application of Kirchoff's Laws to the equivalent circuit yields

$$U' = R_1(i' + i_2) + Z_o(i' - i) + A(U' - U) \tag{108}$$

$$A(U' - U) = Z_o(i - i') + R_2(i + i_2) + U \tag{109}$$

$$O = Z_i i_2 + (i_2 + i')R_1 + (i_2 + i)R_2 \tag{110}$$

which for very large Z_i but very small Z_o, such that $R_1 + R_2 \ll Z_i$ and $R_2 \gg Z_o$, yields

$$u = Ei\rho \left(\frac{R_2}{R_1}\right) \tag{111}$$

Comparison of this expression with Eq. (60) demonstrates that the NIC is characterized by a load line that has a positive slope and hence is capable of fully characterizing current voltage curves of the type shown in Figure 25(b).

A significantly different approach to this problem was adopted by Diard and LeGorrec,[25] who devised a "load line variator" by manipulating the reference circuit of a classical potentiostat without modifying the main electrolysis circuit (cf. Epelboin *et al.*[11]). Their circuit is shown schematically in Figure 38. This device operates by detecting the potential difference across a resistor in the counter electrode circuit and then subtracting this quantity from the potential difference between the reference electrode and the working electrode (ground). The potentiostatic control function for this circuit is given as

$$e_{ref} = -i(R_u - R) + e_i - e_o/G_{CA} \tag{112}$$

where, as before, R_u is the uncompensated electrolyte resistance between the RE and ground. Normal potentiostatic regulation is obtained with $R = 0$ [cf. Eq. (60)], and hence the slope of the load line is negative. However, for

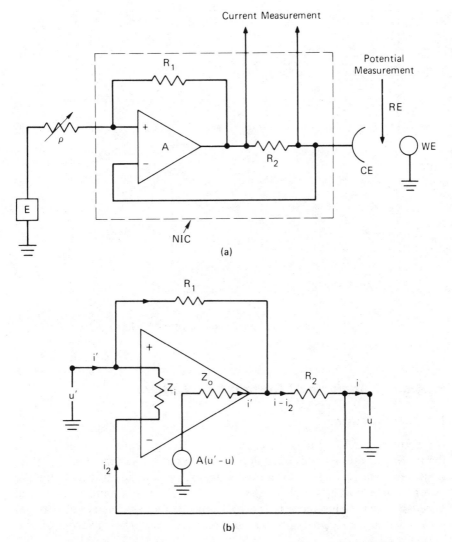

Figure 37. Negative impedance converter after Epelboin and coworkers (reference 11).

$R > R_u$, the slope of the load line reverses sign and control of "Z"-shaped polarization curves becomes feasible.

Epelboin *et al.*[11] and Diard and LeGorrec[25] have used these devices to study a number of electrochemical systems that under normal potentiostatic regulation exhibit sharp transitions from the active to the passive states. They were able to demonstrate that, in general, such discontinuities are not characteristic of the electrode under study but instead are artifacts due to the limitations of classical potentiostatic control.

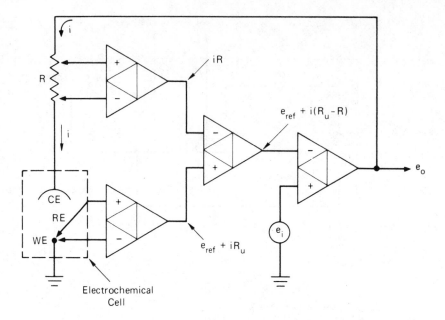

Figure 38. Negative impedance converter after Diard and LeGorrec (reference 25).

5. Analog Analysis of Current and Voltage

5.1. Introduction

Until the advent of digital computers, all electrochemical studies involved the processing and analysis of analog signals in either the time domain or the frequency domain. Typical examples of analog signal analysis include the use of a-c coupled bridges and of Lissajous figures for determining interfacial impedance. In both instances, the desired information (e.g., balance of a bridge) is obtained in purely analog format and no need exists for converting signals into digital form.

When describing analog instrumental methods, it is convenient to classify techniques according to the type of excitation function employed, particularly with respect to the independent variable. For example, frequency domain impedance measurements are carried out using a small-amplitude sinusoidal excitation with frequency as the independent variable. Alternatively, the perturbation and response may be recorded in the time domain with time as the independent variable, and the impedance as a function of frequency can then be extracted by time-to-frequency conversion techniques such as Laplace or Fourier transformation. However, these latter techniques invariably involve digital processing and, accordingly, they are discussed in Section 6.

5.2. Frequency Domain Measurements: General

The application of a sine wave excitation to a system under test often is the easiest method of determining the system transfer function. In previous sections we have been concerned with the transfer function of measurement and control devices. Here we are concerned with measuring or inferring a transfer function for an electrochemical cell, as a first step in determining reaction mechanistic and kinetic parameters.[3,14,16,27]

Once again the system transfer function is the output divided by the input

$$G(j\omega) = X_{out}(j\omega)/X_{in}(j\omega) \tag{113}$$

For the special case where the output signal is the system voltage and the input (or excitation function) is the current, the transfer function is the system impedance

$$G(j\omega) = E(j\omega)/I(j\omega) = Z(j\omega) \tag{114}$$

Since the output may be changed both in amplitude and phase with respect to the input, we must express the impedance as a complex number

$$Z(j\omega) = Z' - jZ'' \tag{115}$$

where primed and double primed variables refer to in-phase and quadrature components, respectively.

It is important to note that we are using the formalism of linear systems analysis. That is, Eq. (114) is considered to hold independently of the magnitude of the input perturbation. Electrochemical systems do not, in general, have linear current-voltage characteristics. However, since any continuous, differentiable function can be considered linear for limitingly small input perturbation amplitudes, this presents more of a practical problem than a theoretical one.

In the following sections we present a number of standard methods of measuring a system impedance, or a frequency domain transfer function. In an application of any of the methods described, it is necessary that the perturbation be sufficiently small that the response is linear. While this may be decided from theoretical considerations,[28-31] the most practical method is to increase the input signal to the maximum value at which the response is independent of the excitation function amplitude.

5.3. Audio Frequency Bridges

In the past, impedance measurements using reactively substituted Wheatstone bridges at audio frequencies have been the easiest to accomplish. Consequently, a great deal of emphasis has been placed historically on electrochemical processes having characteristic impedance spectra in the audio frequency range 20–20,000 Hz; namely double layer capacitive and moderately fast reaction kinetic effects at plane parallel electrodes.

The mathematics and methodology of such measurements are well understood.[32,33] However, considerable use still may be made of passive audio frequency bridge measurements in this age of active circuitry, principally in high-precision applications. Following a brief review of bridge circuits, we will restrict our discussion to the limitations imposed by the use of each type of bridge, since these will influence the point at which an experimentalist will select a more complex measuring device.

Figure 39 shows schematically the familiar representation of an audio frequency bridge adapted for use with an imposed d-c potential. The condition of balance for the bridge shown is

$$Z_x = (R_1/R_2)Z_s \qquad (116)$$

where subscripts x and s refer to unknown and standard impedances, respectively. A variety of RCL combinations is possible for Z_s. In the commonly employed Wien bridge,[33] Z_s takes the form of series variable resistance and capacitance standards, which are adjusted alternately until the real and imaginary components of the voltage at the null detector simultaneously are zero. For this null condition the real and imaginary components of the unknown impedance may be calculated as

$$Z_x' = (R_1/R_2)R_s \qquad (117)$$

$$Z_x'' = (R_1/R_2)/\omega C_s \qquad (118)$$

The form of Eqs. (117) and (118) has led to the widespread and unfortunate practice of tabulating and plotting measured impedance data in terms of the

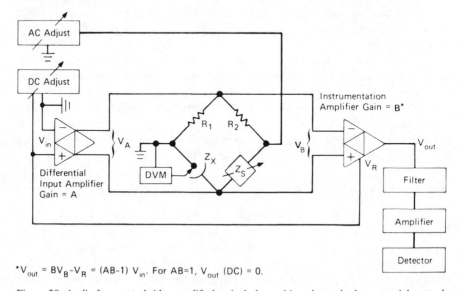

*$V_{out} = BV_B - V_R = (AB-1) V_{in}$. For AB=1, V_{out} (DC) = 0.

Figure 39. Audio frequency bridge modified to include working electrode d-c potential control.

complex pair $(R_s, j/\omega C_s)$ even when a Wien bridge has not been employed. The complex impedance notation $(Z', -jZ'')$ is significantly less ambiguous, and will be used here.

a. *High-Frequency Limitations.* The upper operating limit is imposed primarily by reactivity and nonlinearity of available resistive standards (chiefly inductive effects) and by the effects of stray capacitive shunts. By employing a Wagner earth,[33–35] the latter effect can often be reduced sufficiently to allow sensible measurements at frequencies up to 10^5 Hz. However, the importance of Wagner earthing varies greatly depending on the magnitude of the impedances being measured.[33] In general, elimination of stray capacitance is most important at high frequencies when measuring small capacitances or large resistances (i.e., for small area electrodes).

b. *Low-Frequency Limitations.* The null detection system traditionally employed with an audio frequency bridge consists of an amplifier, filter, and a-c voltmeter. This combination imposes three limitations at low frequencies:

• Null detection employing a magnitude voltmeter or oscilloscope is most sensitive when the resistive and reactive components of the unknown impedance are of the same magnitude, since the total bridge out-of-balance signal contains terms proportional to each. For an impedance bridge used to measure the electrical properties of electrochemical cells this imposes a limit on accuracy at low frequencies since the reactive terms, which are primarily capacitive, dominate the cell admittance with decreasing frequency. Increasing the gain in order to observe the resistive component more precisely, results in saturation of the detection system with the reactive out-of-balance signal.

• A significant source of noise at the detector may result from harmonic distortion originating in the oscillator or caused by nonlinearity in the system under test or in subsequent amplifiers. In such cases, the signal at balance consists mainly of the second harmonic. At high frequencies, this signal can be effectively removed by appropriate signal conditioning; specifically by the use of bandpass, low-pass, or notch filters (see Section 3.9). At low frequencies, however, analog filters of bandwidth less than 10 Hz are less easy to construct and control.

• The second major source of noise at low frequencies is mains pick-up. This may be of the order of hundreds of millivolts superimposed on the test signal unless major efforts are made at shielding and ground loop suppression. Usually, unless an adequate notch filter is used in addition, the experimentalist must be satisfied with reduced precision at frequencies below about 100 Hz.

These three effects can be reduced to a large extent by using a phase sensitive detector (PSD) to measure separately the real and imaginary components of the bridge out-of-balance signal. By separate amplification of the in-phase and quadrature components, differential sensitivities in excess of 100 : 1 can be accomplished. The advantages and limitations conferred by the use of PSD's are described in Section 5.6c.

In normal operation, a PSD is completely insensitive to the second harmonic, but most commercial instruments have the additional facility of being able to select a reference signal at twice the fundamental frequency. By this means the extent of second harmonic distortion can be measured. This often reflects not an error signal (i.e., noise) but an expected response induced by nonlinearity of the system under test.

In addition, and unlike traditional bandpass filters, a phase sensitive detector has a bandpass characteristic with bandwidth decreasing with decreasing frequency, and can frequently be used within ± 5 Hz of 60 Hz mains pick-up.

When phase sensitive null detection is employed, the practical low-frequency limit becomes a function of the particular form of bridge chosen. However, for the Wein bridge this limit is imposed by the selection of suitably large adjustable capacitance standards at frequencies below about 20 Hz.

c. *Limitations of Imposed Potential.* A considerable limitation on the use of this form of bridge is that it necessitates the use of a two-terminal cell. While it is often possible to construct a cell in which the working electrode impedance greatly exceeds that of the counter electrode, potentiostatted conditions cannot be established adequately with this type of bridge. Closely associated with this is the fact that in normal usage, the cell current and voltage are not explicit functions of the measured out-of-balance voltage, and vary with the settings of the resistive and reactive standards.

In electrochemical applications, these combined limitations may be severe. Figure 39 shows one of a variety of possible methods in which an imposed working electrode d-c potential can be adjusted to the desired value without influencing the detector circuit. The method shown can be used at frequencies less than the normal operating frequency limit of a-c coupled amplifiers.

5.4. Transformer Ratio Arm Bridges

The high-frequency limitation imposed by unavoidable stray capacitances on the operation of reactively substituted Wheatstone bridges prompted the development of the transformer ratio arm bridge.[36] By substituting a transformer for orthodox ratio arms, a bridge was produced for which the impedance ratio is proportional to the square of the number of turns, and which was capable of accepting heavy capacitive loads with virtually no effect on the voltage ratio.[36]

The operation of a transformer ratio arm bridge is shown schematically in Figure 40. Briefly, voltages 180° out of phase are fed from the secondary winding of the input "voltage" transformer to the cell or unknown impedance, and to resistance and capacitance standards. The "arms" of the bridge consist of a series of ratio taps of the primary windings of an output "current" transformer. The standard and unknown impedances are connected to the output transformer in such a way that a detector null is achieved when the

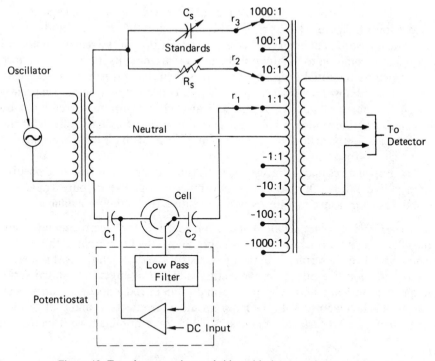

Figure 40. Transformer ratio arm bridge with d-c potentiostat control.

sum of the flux induced by the unknown and standard currents in the output transformer is zero. In this condition

$$\frac{r_1}{Z_x} = \frac{r_2}{R} + j\omega C r_3 \tag{119}$$

independent of V_{in}, where r_1, r_2, and r_3 are ratios (usually decade), separately selected.

The advantages that accrue from the use of this type of bridge are as follows:

• Error resulting from the impurity of standard variables can be virtually eliminated. Because ratios are selectable over a wide range (usually $1000:1$), standards can be small. Also, with decade spaced transformer ratios, standards need be variable only over a range of about $11:1$. Consequently, standards can be employed that closely approximate to ideality (e.g., the use of air gap capacitors and nonreactively wound metal resistances), and one standard can be used to measure a wide impedance range.

• By the use of precision transformers as ratio arms, highly accurate ratio values can be obtained that are essentially independent of frequency well into the megahertz range.

• The bridge is highly insensitive to the presence of stray capacitance. Inspection of Figure 41 will reveal the reason for this: C_1, C_2, and C_3 can cause no measurement error; C_1 because it merely produces a reactive potential drop that is common to the unknown and standard circuits, C_2 and C_3 because at balance no potential drop appears across them. C_u represents the capacity across the unknown terminals and its effect is cancelled by trimming capacitor C_t on the standard side. C_t is adjusted at each measurement frequency by disconnecting the standard and balancing the bridge. Similarly, effects of the stray capacities to earth virtually disappear if the neutral terminal is earthed[36] (see Figure 40).

• Impedances may be measured in all four quadrants by selecting positive or negative ratios. Of particular importance is the use of pure capacitive standards to measure unknowns with a positive (inductive) reactance.

a. *High-Frequency Limitations.* In normal use for electrochemical cells the effective upper operating limit is imposed by effects external to the bridge. These have been described in detail by Armstrong *et al.*[37] and consist primarily of transmission line effects in connecting cables, the effect of residual series inductance in leads and cell, and normally undesired impedance dispersion effects of solid electrodes. In the latter group, edge effects[26] and transmission line effects due to surface roughness[37–40] become dominant with increasing

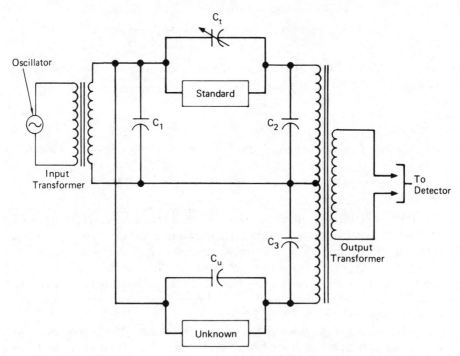

Figure 41. The effect of stray capacitances in the transformer ratio arm bridge.

frequency.[3,14,26] In electrochemical systems for which the interfacial impedance is the desired parameter, measurement precision becomes limited by the
dominance of the uncompensated electrolyte resistance in the total measured
impedance. This effect has prompted the use of very small electrodes for which
the ratio of uncompensated resistance to interfacial impedance is reduced.[41]

Series leakage inductances in the transformers within the bridge result in
an impedance measurement error that is proportional to frequency. This effect
has been examined by Calvert,[36] but is seldom likely to impose high-frequency
limitations in electrochemical applications.

b. *Low-Frequency Limitations.* The use of input and output transformers
results in cell current and voltage, and detector signals, which are dependent
on frequency. This effect becomes apparent only at low frequencies and imposes
a practical lower limit of the order of 100–200 Hz with commercial bridges.

c. *Limitations of Potential Control.* The limitations of potential control for
a transformer ratio arm bridge are similar to those imposed on classical bridge
measurement. That is, it is not possible to apply the a-c potential via a reference
electrode and potentiostat circuit, only to the interface of interest. The
measured impedance necessarily includes series terms associated with the lead
and electrolyte resistances and the counter electrode impedance.

D-C potentials can be applied to the interface of interest by using a circuit
of the form shown within the dashed lines in Figure 40, since at moderate
frequencies the low-pass filter will not observe the a-c component. However,
direct current must be excluded from the bridge windings by the use of blocking
capacitors C_1 and C_2. The impedance of these also will be included in the
measured "cell" impedance.

5.5. Berberian–Cole Bridge

An active null admittance measuring instrument that incorporates many
of the advantages of the transformer ratio arm technique, while obviating
many of the disadvantages of passive bridges, has been reported by Berberian
and Cole.[42] Figure 42 shows a form of this bridge modified to measure
impedance and to remove some of the limitations of the earlier instrument.[6,43]

The basic operation is as follows. The external variable decade standards
are R_1 and C, while R' and R'' are internal and fixed. With reference to
Figure 42

At all times,

$$i_1 + i_2 + i_3 = 0 \tag{120}$$

$$i_1 = A V_A / R_1 \qquad (V_A = IZ) \tag{121}$$

$$i_2 = A V_A (j\omega C) \qquad (V_A = IZ) \tag{122}$$

$$i_3 = B V_B / R' \qquad (V_B = -IR') \tag{123}$$

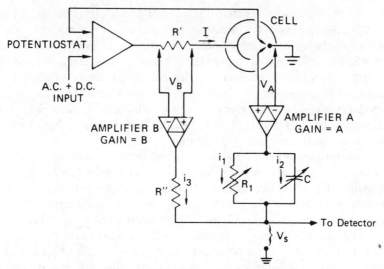

Figure 42. Modified Berberian–Cole bridge shown as a three-terminal interfacial impedance measuring system, with potentiostatic control of the working electrode.

where Z is the impedance between the working electrode and the reference electrode, and I is the current flowing through the cell. Therefore, for the condition of balance at the summing point

$$BIR'/R'' = AIZ/R_1 + AIZ(j\omega C) \tag{124}$$

Removing I and solving for the unknown impedance yields

$$Z = \left[\frac{BR'R_1}{AR''}\right]\left[\frac{(1 - j\omega R_1 C)}{1 + \omega^2 R_1^2 C^2}\right] \tag{125}$$

The advantages of this method apply principally at low (audio and sub-audio) frequencies, and arise due to the simultaneous measurement of current and voltages and because of the flexibility made possible by the use of active bridge elements. It is important to notice that the device shown schematically in Figure 42 is a bridge only in the sense that external variables are adjusted to produce an output null. Because of the use of buffer amplifiers, null adjustment does not vary the potential across (or current through) the unknown impedance, as is the case for classical and transformer ratio arm bridge measurement.

Other significant advantages are as follows:

• Measurements can be made on two-, three-, or four-terminal cells allowing the isolation of the impedance component of interest from the total cell impedance. This is not possible with a passive bridge and it is frequently infeasible to construct a cell for which the impedance of interest is much greater than all series terms. This is particularly true when measuring the impedance of an electrode of large area or in a highly resistive electrolyte, or when the impedance of interest is that of a highly conductive electrolyte.

• Measurements can be made effectively down to 0 Hz. The bridge shown in Figure 42 is direct coupled and thus the low frequency limits are those of the null detection system and the patience of the experimentalist.

• Measurements can be made in the presence of a d-c bias under potentiostatic control, without the use of blocking capacitors.

• Impedances can be measured over an extremely wide range from below 10^{-3} Ω to greater than 10^9 Ω.

• Error resulting from the impurity of standards can be virtually eliminated since standards can be selected according to their ideality, not magnitude.

• By employing differential gain for the real and reactive standards, a suitable range of measurement can be selected for each impedance component separately. This feature is incorporated in Figure 43.

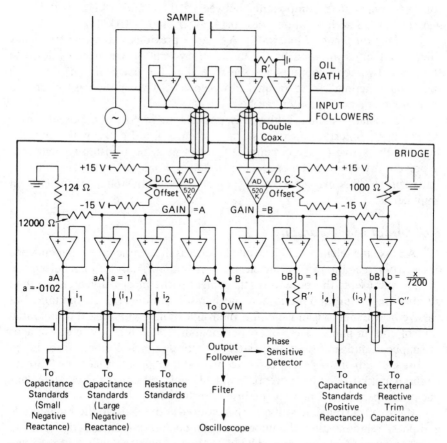

Figure 43. Schematic diagram of a working (modified) Berberian–Cole bridge shown as a four-terminal impedance measuring system.

• Impedances may be measured in all four quadrants ($RC, -RC, RL,$ $-RL$) using resistance and capacitance standards alone.

Because the gains of amplifiers A and B perform the same function as the ratios in a transformer ratio arm bridge, the two techniques have many features in common.

a. *High-Frequency Limitations.* Inaccuracies at high frequencies can occur due to errors in the gain functions A and B with decreasing amplifier open-loop gain as discussed in Section 2.6. Figure 43 shows schematically a practical bridge of this type. Gain errors in the voltage followers are negligible, and, since amplifiers A and B are identical devices, and their gains appear as a ratio in Eq. (125), inaccuracies in this term are partially compensated. Nevertheless, the upper operating frequency limit for the bridge shown in Figure 43 is about 10 kHz, depending somewhat on the magnitude of the unknown impedance. This device is capable of 0.01% measurement accuracy for both impedance components between 1.0 Hz and 1 kHz, with 0.1% accuracy in the peripheral decades (0.1–1.0 Hz, and 1–10 kHz).

b. *Low-Frequency Limitations.* As stated previously, the low-frequency operating limit is imposed by the detection system. At frequencies down to 0.5 Hz, a two-component phase sensitive detector performs an ideal null detection function.[43] At frequencies below 0.1 Hz, a low-pass filter and oscilloscope or picoammeter can be used.[42]

c. *Potential Control.* While it is possible to impose a-c potentiostatic control at the interface of interest, the presence of a d-c bias will result in a signal in the active bridge circuits. D-C offset must be adjusted to near zero to prevent overloading in subsequent gain stages. For a cell under d-c potentiostatic control, this may necessitate frequent offset adjustment of the current amplifier B.

5.6. Direct Measurement

An essential element of electrode kinetics is the characteristic dependence of electrode reaction rate on the electrode potential. Thus, for many electrode studies, the use of the potentiostatic control is the most convenient method of obtaining relevant kinetic and mechanistic parameters. A limitation of passive bridge methods in general is their inflexibility with regard to potential control and, in many cases, the experimentalist must forego the advantages of simplicity and sensitivity associated with bridge measurement in order to impose a-c and/or d-c potentiostatic control at a single interface. The "direct" methods permit effective potential control while retaining the relative simplicity of operation of may of the bridge techniques.

If the cell current and voltage are measured with regard to their magnitude and phase relations, then the impedance can be determined directly from Eq. (114). Figure 44 shows in simplified form a circuit that allows the direct measurement of impedance under potentiostatic control.

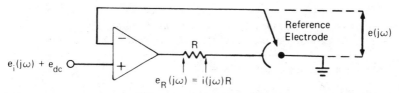

Figure 44. Direct measurement of interfacial impedance under a-c and d-c potentiostatic control.

It is necessary, at the outset, to separate phase shifts associated with the cell impedance from those attributable to the potentiostat control loop. Commercial potentiostats normally are optimized for fast step response, and their potentiostatting function becomes substantially in error for sinusoidal inputs with increasing frequency. As shown in Section 4.5 for both a single amplifier and adder-type potentiostat, the upper frequency for which $e(j\omega)$ can be considered equal to $e_i(j\omega)$ is a function of the cell impedance as well as the electrical parameters of the potentiostat.

The rather complicated result given by Eq. (87) indicates that the potential applied across the cell by a single amplifier potentiostat consists of two components: (1) a sinusoidal voltage that is phase shifted with respect to the input, and (2) a damped oscillation that decays with a time constant given by Eq. (88). In the frequency domain techniques, only the steady-state response is of direct interest as far as computation of the cell impedance is concerned. However, the damped oscillation term determines the minimum time that must be allowed prior to data collection for the transients to decay away. Failure to allow the system to attain a steady state can result in substantial errors in the measured impedance due to "tracking" of the current-voltage response.

Equations (83), (87), (88), (89), (94), (95), and (96) demonstrate that the system response depends on both the cell parameters as well as the properties of the control amplifier. The magnitude of the error in both amplitude and phase can be estimated by assuming typical values for the various parameters contained within these equations. Analyses of this type[3] have shown that the phase angle, φ_1, first deviates in the negative direction prior to shifting very rapidly to large positive values, which give rise to instability. The negative values for φ_1 imply a small inductive error in the potentiostat control at frequencies lower than that for the onset of instability. This inductive error is easily detected in commercial instruments of the single amplifier type.

Many commercial potentiostats are of the "adder" type, which permits feedback to be used to compensate for ohmic potential loss between the working and reference electrodes. Brown et al.[19,44] have carried out the transform analysis for this case and have shown that significant errors may result at relatively low frequencies depending on the cell resistance, interfacial capacitance, and extent of positive feedback. Some of their results are reproduced in Table 3. For the particular values shown for the amplifier and cell

Table 3
Summary of Upper Frequency Limits for Stable, Accurate Potentiostat Response
with Various Double-Layer Capacitance and Cell Resistance[a]

Cell and potentiostat parameters:[b] $R = 1.00 \times 10^3 \, \Omega$, $R_f = 1.00 \times 10^3 \, \Omega$, $C = 20 \, pF$
$C_f = 0.00 \, pF$, $\beta = 1.00$, $R_2 C_1 = 1.00 \times 10^{-6} \, s$, $R_1 = R_2$

Double layer Capacitance (μF) → Cell resistance R_2 (Ω)	2×10^{-6}	1×10^{-6}	5×10^{-7}	2×10^{-7}	1×10^{-7}
25	1900	3200	5000	10000	15000
100	1400	200	3200	4500	7000
200	1200	1500	2100	3300	5000
500	800	950	1300	2100	3200
1000	500	650	1000	1300	2100
2000	350	500	650	1000	1600

[a] After Brown *et al.*[19]
[b] Defined in reference 19.

parameters, it is clear that significant error ($>2\%$) occurs at frequencies as low as 350 Hz for a system having a high-cell resistance and high interfacial capacitance. The significant probability of potentiostat errors, even at quite modest frequencies, dictates that if accurate impedance measurements are to be made over a wide frequency range, it is imperative that measurements be made of both the cell current and the voltage applied between the working and reference electrodes (i.e., e not e_i, Figure 44). Ideally, these measurements should be made simultaneously and with identical measuring instruments, in order to eliminate systematic instrumental errors. This may be accomplished as follows.

a. *Two Channel Oscilloscope Measurement.* By recording $e(j\omega)$ and $i(j\omega)$ (as the voltage drop across a series resistance, R_s) using a twin beam oscilloscope, the magnitude of the impedance can be calculated from the ratio of the two peak-to-peak voltages, and the directly observed phase angle. Figure 45 shows the oscilloscopic traces for $e(j\omega)$ and $e_R(j\omega)$ that result from the imposition of a sine wave between the working and reference electrodes.

$$|Z| = R_S |e(j\omega)| / |e_R(j\omega)| \tag{126}$$

$$Z' = |Z| \cos(\phi) \tag{127}$$

$$Z'' = |Z| \sin(\phi) \tag{128}$$

The time base of available storage oscilloscopes limits low-frequency measurements to about 10^{-2} Hz. High-frequency limitations are imposed by effects external to the oscilloscope, principally stray capacitance and transmission line effects in the leads and cell. Measurements can often be made at frequencies above 10^5 Hz.

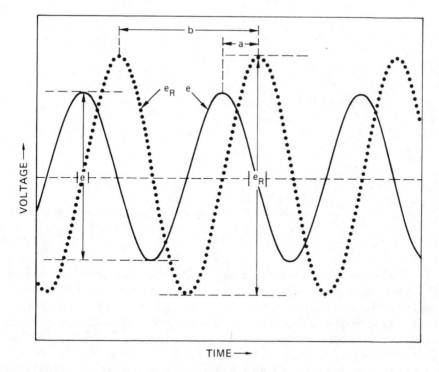

Figure 45. The direct measurement of impedance using a twin beam oscilloscope.

The primary limitation of this technique is precision. Oscilloscope linearity is seldom better than 1%, and it is difficult to measure phase angles directly with a precision of better than 2°. Measurements can usually be accomplished with an uncertainty in Z' and Z'' of ±3% of $|Z|$.

b. *Lissajous Figures.* Elimination of t between expressions for e and i of the form

$$e = e° \sin (\omega t) \tag{129}$$

$$i = i° \sin (\omega t + \phi) \tag{130}$$

leads to an equation of the form of an ellipse. When e and i are plotted orthogonally using an oscilloscope or "XY" recorder, this ellipse is traced out and the components of the impedance can be calculated from its dimensions. With reference to Figure 46

$$|Z| = \Delta e/\Delta i \tag{131}$$

$$\sin (\phi) = \Delta i'/\Delta i = \alpha\beta/(\Delta i \cdot \Delta e) \tag{132}$$

Z' and Z'' can be calculated from Eqs. (127) and (128).

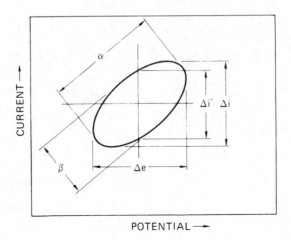

Figure 46. Lissajous figure for the evaluation of impedance.

Limitations of oscilloscopic recording are essentially those of precision described above for two-channel measurement. However, since time is not an explicit variable, time base limitations do not apply when recording Lissajous figures. Low-frequency limitations are imposed by electrochemical instabilities in the system under test, and electrical instabilities (particularly d-c offset drift) in the attendant circuitry. Electrochemical "XY" recording can be used to achieve a precision better than 1% of $|Z|$ at frequencies from 1 Hz to below 10^{-3} Hz.[33]

A good deal of caution is necessary when applying this last method. Electrochemical systems are susceptible to external noise pick-up. The use of high gain, without appropriate electrical filtering, to amplify low-level sine wave voltage and current perturbations may result in severe errors in the dimensions of the ellipse traced on an electromechanical XY plotter. This is because the mechanical damping of the plotter may disguise the fact that the input amplifiers are overloaded by the high frequency (≥ 10 Hz) noise envelope. This effect is shown schematically in Figure 47 for 50 or 60 Hz mains pick-up in the "Y" amplifier. Errors may, of course, occur in both channels. This phenomenon is often reflected as skewing or tracking of the recorded ellipse, but may result in a stable erroneous trace. In order to prevent errors in the calculated impedance values, it is imperative that appropriate electronic low-pass or notch (50 or 60 Hz mains pick-up) filtering be employed at an early stage of amplification.

c. *Phase Sensitive Detection.* The real and imaginary components of a voltage can be measured directly with respect to a reference signal using a phase sensitive detector (PSD). Because of the requirements for linearity (Section 5.1), small input signals must be used to measure electrochemical impedances, and noise problems often make it impractical to use either e or e_R (Figure 44) as a reference signal. Accordingly, e and e_R must be measured

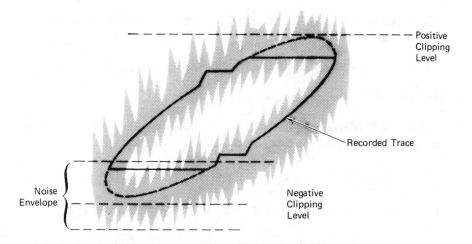

Figure 47. The errors in electromechanical Lissajous figure recording due to the presence of input noise.

alternately in terms of a coherent reference signal of arbitrary phase, and the impedance determined from the complex quotient

$$Z = \frac{e_R}{e_R} = \frac{(e' + je'')}{(e_R' + je_R'')} R \qquad (133)$$

In order to understand the advantages inherent to this method, it is appropriate to discuss briefly the detection technique.

Phase sensitive detection may be accomplished by the sequential operation of multiplexing and time averaging circuits. The multiplexer serves effectively to multiply the input sine wave, e_i with a reference square wave, e_{ref}. We can represent e_{ref} in terms of its Fourier components[45]

$$e_{ref} = \frac{4}{\pi} \{\sin(\omega_r t) + \tfrac{1}{3}\sin(3\omega_r t) + \tfrac{1}{5}\sin(5\omega_r t) + \cdots\} \qquad (134)$$

and the input sine wave can be written as $e_i = A^0 \sin(\omega_i t + \Phi)$ where A^0 is the input signal amplitude, ω is the angular frequency, and subscripts r and i refer to the reference and input signals.

The multiplexer output will be

$$e_{mpx} = e_{ref}\, e_i = \frac{2A^0}{\pi} \{\cos[(\omega_i - \omega_r)t + \Phi] - \cos[(\omega_i + \omega_r)t + \Phi]$$

$$+ \tfrac{1}{3}\cos[(\omega_i - 3\omega_r)t + \Phi] - \tfrac{1}{3}\cos[(\omega_i + 3\omega_r)t + \Phi] + \cdots\} \qquad (135)$$

In normal practice, ω_r and ω_i are derived from a common source (i.e., $\omega_r = \omega_i$), and the multiplexer output is

$$e_{mpx} = \frac{2A^0}{\pi} \{\cos(\Phi) - \cos(2\omega_r t + \Phi) + \tfrac{1}{3}\cos(-2\omega_r t + \phi)$$

$$- \tfrac{1}{3}\cos(4\omega_r t + \Phi) + \cdots\} \tag{136}$$

Only the first term in Eq. (136) is time independent and when applied to the time average circuit will result in a nonzero output

$$e_{out} = \frac{2A^0}{\pi} \cos(\Phi) \tag{137}$$

This is obviously a phase sensitive d-c output voltage, which is a maximum at $\Phi = 0$.

The PSD output is frequency selective since the time average of Eq. (135) for $\omega_r \neq \omega_i$ is zero. The important exception to this statement is for $\omega_i = 3\omega_r$, $5\omega_r$, $7\omega_r$, etc. That is, a phase sensitive detector responds to odd order harmonics of the input signal. This contribution diminishes with the order of the harmonic.

Time averaging may be accomplished by analog or digital means. In the vast majority of commercial instruments, an analog low-pass smoothing circuit is employed with a front panel adjustable time constant. This offers the advantage of simplicity and flexibility in high-frequency operation. The upper frequency limit is commonly[46,47] 10^5 Hz. The low-frequency limit of analog time averaging devices is imposed by the practical details of low-pass filter design,[48] smoothing capacitor ideality, current leakage in buffer amplifiers, and external asymmetric (nonrandom) noise effects. The low-frequency limit of commercial instruments is commonly[46,47] in the range 0.5–10 Hz. Impedance can usually be measured with 0.1% precision in both components over the specified frequency range.

By using digital integration methods, the low-frequency response can be extended to below 10^{-3} Hz. In this method, the average is taken digitally over an integral number of cycles.[49,50] At very low frequencies, information relating to e and i taken over a single cycle can be used to calculate the real and imaginary impedance components, with a precision of 0.1%.

In general, direct methods can be used to acquire impedance data significantly more rapidly than bridge methods. This is particularly true for digitally demodulated phase sensitive detectors, for which a single cycle only is required. Nevertheless, in systems that are unstable, such as rapidly corroding specimens, acquisition rate is an important consideration, and a major criticism of PSD methods is that these must be performed frequency-by-frequency. Fortunately, this is not a serious hindrance when such equipment is automated.[51] In the past decade, a number of experimentalists[51–55] have utilized the Solarton[56] 1170 Series Frequency Response Analyzers as automated, digitally demodulated, stepped frequency, impedance meters. Model 1174 is capable of 0.02% precision for each impedance component in the frequency range 10^{-4}–10^6 Hz. These instruments employ a sine wave reference and phase sensitive detection is accomplished by a correlation technique.[28,56]

6. Digital Analysis of Current and Voltage

6.1. Basic Principles

As noted previously, two methods of data processing are available to the experimenter: analog and digital. In the analog mode, a continuous signal (voltage) is acquired from the system and is processed using familiar analog recording devices, such as oscilloscopes and pen recorders. These devices are normally capable of yielding permanent records, for example, in the form of a photographic trace from a storage oscilloscope or as an ink trace on paper from strip-chart or "XY" recorders. Furthermore, the signals may be modified (e.g., amplified) using analog signal conditioners that are easily constructed using operational amplifiers as described in Sections 2 and 3. Indeed, the great majority of electrochemical instruments currently available operate purely in the analog mode.

However, with the advent of high-speed digital computers, a clear trend toward digital signal processing has become apparent. The advantage of digital over analog data processing is purely mathematical; a far wider range of mathematical computations can be performed in the digital mode than on analog signals. Digital signal processing using hard-wired devices has also expanded rapidly over the past decade and is likely to find even more application in the years to come.

In this section, we discuss briefly the techniques that are now available for processing and interconversion of analog and digital signals. The discussion is illustrated by reference to various commonly employed processing circuits.

6.2. Interconversion of Analog and Digital Signals

Analog-to-digital (A/D) and digital-to-analog (D/A) converters are employed extensively in modern electrochemical instrumentation, particularly for interfacing analog systems with computers. The essential operations in A/D and D/A conversion are shown schematically in Figures 48(a) and (b), respectively. In the first case (A/D), the analog signal is sampled at discrete (known) times and a digital record of amplitude vs. time is established. In the second case, a set of discrete amplitude vs. time data points are taken and an analog signal is devised by piecewise construction of a curve through them. Smoothing of the analog signal is accomplished by inserting a low-pass filter in series with the D/A output. The time constant of this filter should be of the order of Δt. The resultant data records may then be interfaced with appropriate signal processing devices, as shown schematically in Figure 49.

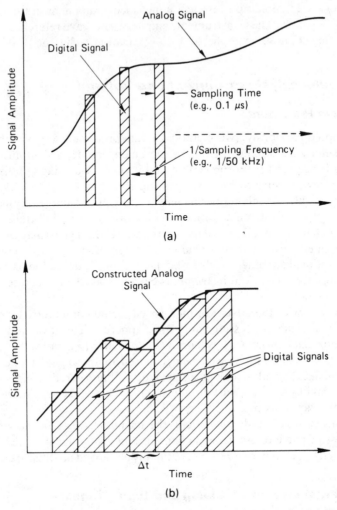

Figure 48. Basic operations in A/D and D/A conversions. (a) Schematic of A/D conversion operation; (b) schematic of D/A conversion operation.

In order to understand how D/A and A/D converters operate and are interfaced with other components, it is first necessary to consider the types of signals that are produced throughout the conversion train so that the information can be processed at a later stage. As noted above, the most important application of A/D and D/A converters is in the interfacing of analog and digital computer systems. However, digital computers employ binary code to represent real (including integer) numbers. This code is employed because any "bit" of information is represented by either an "on" ($\equiv 1$) or "off" ($\equiv 0$) state, as determined by the mechanism of information storage in the memory.

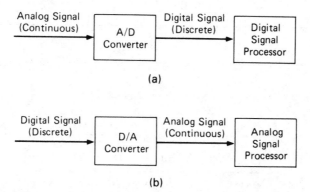

Figure 49. Idealized signal trains for the interconversion of analog and digital signals using A/D and D/A converters. (a) Analog signal with a digital processor; (b) digital signal with an analog processor.

Clearly, the amount of information stored by a single bit is extremely limited (two). However, by combining bits in sequence it is possible to represent a large single number (real or integer) by a single word or "byte", as shown in Table 4 for an integer, and represented mathematically below for a real or fixed-point number

$$\{A\}_n = A_{n-1}2^{n-1} + A_{n-2}2^{n-2} + A_{n-3}2^{n-3} + \cdots A_1 2^1$$
$$+ A_o 2^0 \cdot A_{-1}2^{-1} + A_{-2}2^{-2} + A_{-3}2^{-3} + \cdots A_{-n}2^{-n} \tag{138}$$

Because very large numbers are commonly encountered in data analysis and computation, the floating point representation is commonly employed. In this notation, the number is divided into two parts: a mantissa (number part) and an exponent (base). The relationship between this format and the fixed point format is shown in Figure 50. Note that in the floating point format two bits

Table 4
Binary Code Representation of Real (Integer) Numbers

Decimal	Binary	Decimal	Binary
1	1	11	1,011
2	10	13	1,100
3	11	12	1,101
4	100	14	1,110
5	101	15	1,110
6	110	16	10,000
7	111	17	10,001
8	1000	18	10,010
9	1001	19	10,011
10	1010	20	10,100

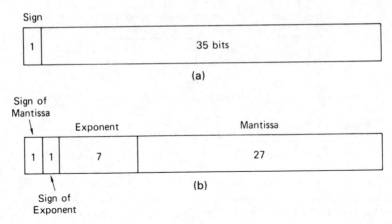

Figure 50. (a) Fixed point and (b) floating point formats for a 36-bit binary number.

are reserved for the sign: one for the sign of the mantissa and one for the exponent.

Other number systems are employed in computer systems, including the octal and hexadecimal codes. However, a full discussion of these systems is beyond the scope of this review, and the reader is referred to the excellent book by Soucek[57] for additional information.

An important consideration, when interfacing a digital processing/storage system to analog instrumentation, is the size of the memory since this will ultimately determine the size of the data base that can be handled. Various kinds of memory are available (see reference 57) including "random access" (RAM) and "read only" (ROM) in the computer itself, and passive storage in peripherals. A typical microcomputer, for example, can store up to 65,536 $(=2^{16} = 64$ "k" bytes) in random access memory (RAM), whereas much larger storage space is available in-core in mini and main-frame computers; the latter typically being in the megabyte range. Megabytes of data are also readily stored in peripherals such as magnetic tapes and disks (floppy and rigid). For our purposes, it is important to note only that the size of the word determines the precision of the information that can be stored and eventually manipulated. The relationship between the size of the word and the inherent precision is summarized in Table 5. For most electrochemical applications a minimum of a ten-bit word is necessary for reasonable precision (0.1%) but larger-sized words (14 or 16) are preferable if extensive ranges of data are to be recorded during a single transient (see below).

The maximum precision identified in Table 5 is not a good measure of the actual precision that can be achieved in a typical experiment, since the maximum word size is not normally utilized. For example, if the maximum input of an A/D converter is 10 V (typically), but the maximum analog input is 10 mV, then the precision is lowered by a factor of 1000. This potential

Table 5
Relationship between Size of Word and Precision for Fixed Point
Digitized Data

Size of word (bits)	Datum size	Maximum precision	Minimum error
8	2^8	1 : 256	0.4%
10	2^{10}	1 : 1024	0.1%
12	2^{10}	1 : 4096	0.25%
14	2^{14}	1 : 12394	0.01%
16	2^{16}	1 : 65536	0.0015%

loss in precision is easily corrected by scaling the input using appropriate amplifiers.

6.3. Analog-to-Digital Conversion

The conversion of analog signals into digital form and ultimately into binary-word representation is now a common practice in electrochemistry, particularly for interfacing analog instruments, such as potentiostats, with digital recording and processing equipment (e.g., computers). The essential operation desired is to convert the value of an analog signal into a binary word whose magnitude is proportional to the signal being sampled. This process involves two operations: sampling and quantization. The first involves momentarily "freezing" the analog signal in time in order to produce a discrete value (see Figure 48). This value is then converted to its binary representation during the second "quantization" step after which the cycle is repeated at some later time.

Sampling is normally achieved by using "sample and hold" amplifiers of the type shown schematically in Figure 51. In this circuit, a signal to the analog switch (e.g., 4066 CMOS) connects the analog input to the amplifier. Provided that the capacitance to ground is sufficiently small the capacitor will charge

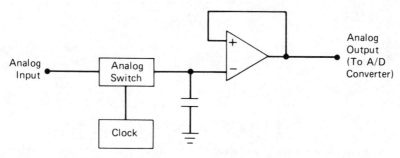

Figure 51. Schematic diagram of a sample-and-hold amplifier.

to the analog input with good fidelity. Removal of the control signal effectively disconnects the input from the capacitor so that the analog output assumes the value of the input at the instant that the switch is opened. The above cycle is then repeated with the sample rate being determined by the control signal from the clock. It is necessary in the case of A/D conversion, however, that the hold time be sufficiently long that the quantization step will generate an accurate digital representation of the analog input.

A number of quantization techniques are available, but only one, the successive approximation technique, will be discussed here. The principle of the technique is shown in Figure 52. In this particular example, we wish to convert the analog input voltage (10.3 V) into its floating point eight-bit binary form to the nearest 0.0625 V. The process involves eight successive steps, in which the field is divided into halves and a bit of 1 or 0 is assigned to each

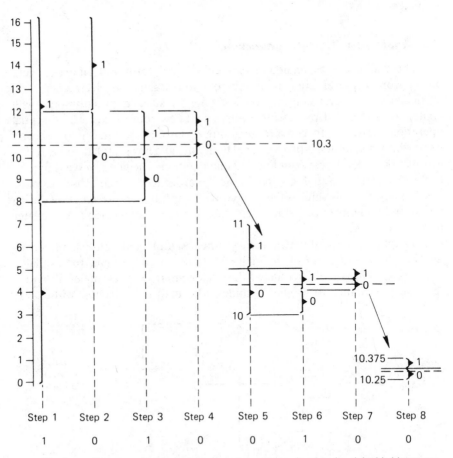

Figure 52. Successive approximation conversion of an analog signal to its eight-bit binary representation. Analog input = 10.3. Equivalent binary output = 1010·0100 ≡ 10.25.

step depending on where in each field the analog value lies. For example, in the first step 10.3 lies between 8 and 16 so that 1 is assigned as the most significant bit. However, in the second conversion step the analog input lies in the 8–12 field rather than in the upper half (12–16) so that 0 is assigned to the second most significant bit. This process is repeated until the desired precision is achieved. Clearly, only n steps are required to quantize an unknown voltage into its $\{A\}_n$ binary form.

A block diagram for a successive approximation A/D converter is shown in Figure 53. The circuit converts each successive approximation into an analog signal Y, which is equivalent to the center of each division shown in Figure 52. The analog input (A) is then compared with Y in the following cycle; if A is greater than Y the comparator swings to positive saturation and a "1" is loaded into the register.

The principal advantage of the successive approximation technique is high speed, and conversion rates in the megahertz range are possible. Although this is not the only quantization method available, it is one of the more popular andis frequently used in electrochemical instrumentation.

6.4. Digital-to-Analog Conversion

The process whereby a sequence of discrete numbers, each represented by a binary word, is converted into an analog signal is known as digital-to-analog (D/A) conversion. This process is commonly employed in electrochemistry when, for example, an analog potentiostat is operated under (digital) computer control. The process of D/A conversion is easily understood by noting that any binary number (integer) can be represented as

$$\{A\}_n = A_{n-1}2^{n-1} + A_{n-2}2^{n-2} + A_{n-3}2^{n-3} + \cdots + A_1 2^1 + A_o 2^0 \quad (139)$$

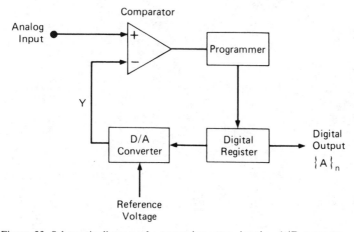

Figure 53. Schematic diagram of a successive approximation A/D converter.

where $A_o, A_1 \cdots A_{n-1}$ can assume one of two values: 0 or 1. Thus, the four bit binary representation of the integer number 7 is given by $A_4 = 0, A_3 = 1, A_2 = 1, A_o = 1$, that is {0111}. A circuit that converts this binary number into an analog voltage is shown schematically in Figure 54. The circuit consists essentially of a parallel network of resistors having values of $R/2^k$, $k = 0, 1, \ldots n - 1$, each in series with an electronic switch ("flip-flop"), which in turn is controlled by an external switch. This switch can adopt one of two states: a_1 closed $+a_2$ open or a_1 open $+a_2$ closed. In the first case, resistor $R/2^0$ is connected in series with the voltage source e_i so that a current flows through it. This state occurs when the "least significant" bit of the binary word

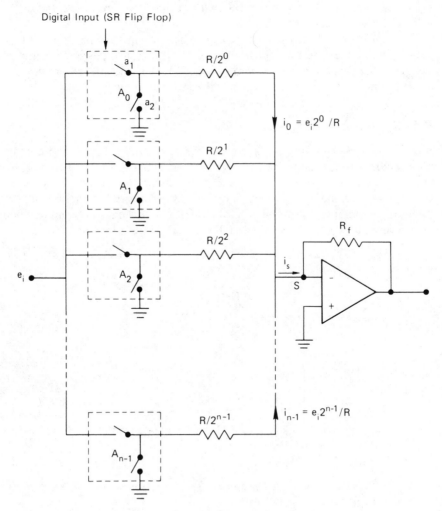

Figure 54. Basic circuit for a D/A converter.

is "1." On the other hand, the second state results in isolation of the resistor from e_i and subsequent grounding. Since the summing junction of the amplifier is held at virtual ground, no current flows through the resistor when the flip-flop is in this "0" state.

The reader will recognize this circuit as being a special example of a voltage adder (see Section 2.3). Therefore, the total current into the summing point is given by

$$i_s = i_i \sum_{k=0}^{n-1} (A_k 2^k / R) \tag{140}$$

and the voltage output of the amplifier

$$e_o = -e_i R_f \sum_{k=0}^{n-1} (A_k 2^k / R) \tag{141}$$

For the example given above, $\{A\}_4 = \{0111\}$, the voltage output is simply

$$e_0 = e_i R_f \left\{ \frac{1}{R} + \frac{2}{R} + \frac{4}{R} + 0 \right\}$$

$$= -7(e_i R_f / R) \tag{142}$$

Likewise, the binary word $\{A\}_4 = \{1100\}$ (i.e., "12") yields $e_o = -12 \, (e_i R_f / R)$. Clearly, the voltage output is proportional to the magnitude of the binary input, i.e., D/A conversion has been achieved.

6.5. Time-to-Digital Conversion

An accurate measure of time is frequently required in electrochemical studies (for example, in the analysis of transients), and it is sometimes necessary to store and manipulate the time interval data in binary form in a computer. A simple circuit for time-to-digital (T/D) conversion is shown in Figure 55. This circuit utilizes an AND gate to compare a sequence of pulses from a high-frequency clock with the analog input that is being sampled. Because the output of the AND gate is zero if *either* of the inputs is zero, then the counter stores only those pulses from the clock (continuous) that are sampled during the duration of the nonzero analog input. Clearly, the accuracy of T/D conversion is determined by the relative values of the clock frequency and the signal duration T.

6.6. Transient Waveform Recorders

In recent years, a class of digital recording instruments known as "transient waveform recorders" (TWR) has appeared in the marketplace. These recorders are now used extensively in electrochemical studies, particularly for the recording of transients.

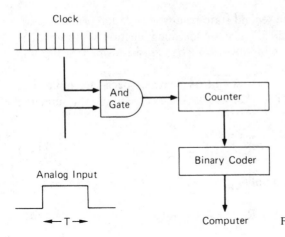

Figure 55. Time-to-signal conversion.

A block diagram for a typical TWR is shown in Figure 56. In essence, the instrument converts dual analog inputs into digital form using a single 10 bit A/D converter that is in turn controlled by a programmable clock. The digitized data are stored in a single 10 bit ×2048 word recirculating memory. In this particular instrument, the data from the two analog inputs are stored in the same memory in (for example) an alternating sequence (Ch. 1, Ch. 2, Ch. 1, Ch. 2,...) or as alternating sets ({Ch. 1, Ch. 1,.... }, {Ch. 2, Ch. 2,.... }). The digital (binary) data can be recovered using the memory control-driven demultiplexer, which selects the data from the storage memory in the sequence in which they were stored. These data can then be converted to analog form using standard D/A converters or can be accessed in their digital form for recording (e.g., on magnetic tape) or for direct manipulation off-line by a computer.

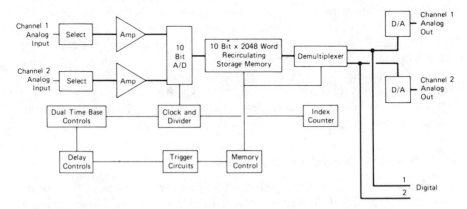

Figure 56. Block diagram for a typical transient waveform recorder (Physical Data, Inc., Model 513A). —— Data lines; —— control lines.

6.7. Computer Interfaces

The ability to perform experiments (as distinguished from calculations) under computer control, is largely determined by the flexibility of the interface structure between the computer and the experimental setup. In a typical case, the computer must issue control signals to initiate a measurement sequence, and receive the resultant data down the interface bus (IB). In this case, bus communication is bidirectional, but it frequently is necessary to coordinate multiple input/output (I/O) devices using a single bus.

The range and diversity of IB structures is enormous. Many have specialized applications such as printer or modem (i.e., telephone) communication and are limited in the number of devices handled, and speed of data transmission. We propose to discuss only one interface bus in this section, chosen for its widespread use and applicability in experiment control and data retrieval.

The IEEE standard 488–1978 interface† represents a highly flexible, moderate speed system that is well suited to laboratory use. The IEEE-488 IB consists of 16 signal lines. Eight lines are used for data, five for bus management, and three lines are used to establish a temporary communication link, or "handshake," between two devices that are properly attached to the bus. Because there are eight data lines, an eight-bit byte can be communicated in each handshake cycle. Thus 16, 24, etc., -bit words (either instructions or data) can be communicated with sequential handshake cycles. This is often referred to as "bit parallel, byte serial" transmission.[58]

A very large number of devices can be connected simultaneously to the IB. Each device is given a unique address, which is used in establishing a handshake. Handshake is used to ensure that data is transferred from a source to one or more designated acceptors. Figure 57 shows a portion of the bus structure. The three signals used for handshake are: DAta Valid (DAV), Not Ready For Data (NRFD), and No Data Accepted (NDAC). The DAV line is driven by the source, while the NRFD and NDAC lines are driven by the receiver. The handshake procedure ensures that all listeners are ready to receive data, that the data on the eight data lines are valid, and that the data have been accepted by all listeners. Data will be sent only as fast as they can be accepted by the slowest receiver.

The IEEE-488 IB is designed to interface with the four major types of devices, shown in Figure 57. A Master or Controller sends commands over the bus, using the bus control (uniline) and data lines (multiline). Normally one controller is present (e.g., a computer or microprocessor), but if more are present, only one may exercise control at any time. A controller issues a system initiation command InterFace Clear (IFC), and designates which devices are talkers and which are listeners. The controller has complete control of the

† Also known as Hewlett Packard Interface Bus (HPIB) and General Purpose Interface Bus (GPIB).

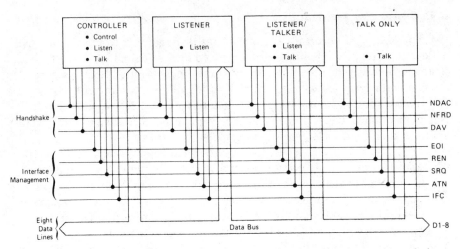

Figure 57. A section of an IEEE-488 interface bus, showing the major classes of user devices.

ATtentioN line (ATN). When ATN is "true," the controller is issuing messages or commands.

A listener receives data over the IB, following an acceptor handshake. Addressed listeners are capable of responding to controller commands; listen-only devices are intended for use in a circuit with no controller. An example of a listener might be a digitally controlled analog potentiostat.

A talker is a device capable of sending data over the IB, to a controller or listener. An unaddressed talk-only device, such as a digital voltmeter, may present a problem in a circuit with a controller, since this device may continue talking when the controller requires attention (ATN "true"), resulting in a garbled message.

Most commercial devices intended for use with an IEEE-488 IB are combined Talker/Listeners, capable of receiving instructions, setting the data collection mode and experimental conditions, and returning data to the controller. Details of the bidirectional communication between a source and one or more acceptors are shown as a flow chart in Figure 58.

6.8. On-Line Computer Data Analysis

Just as the operational amplifier revolutionized analog signal processing and analysis, on-line digital computers have greatly enhanced the electrochemist's ability to handle and process large amounts of experimental data that are frequently generated, for example in transient or perturbation studies. The emphasis in this discussion is on the use of computers in the on-line mode of data analysis as opposed to off-line applications, which are readily implemented using transient waveform recorders (for example) as discussed in Section 6.6.

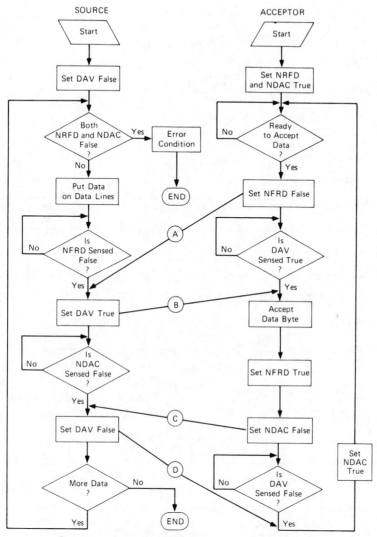

(A) NFRD signal line goes false only when all acceptors are ready.

(B) Data is now valid and may be accepted.

(C) NDAC signal line stays low until all acceptors have received data.

(D) Data now not valid.

Figure 58. IEEE-488 interface bus: the principle of operation.

A variety of configurations has been employed in on-line computer analysis of electrochemical data. Shown in Figure 59 is a particular configuration that has been used by Smith *et al.*[59] for impedance/admittance studies of electrochemical interfaces using the fast Fourier transform (FFT) algorithm. Since

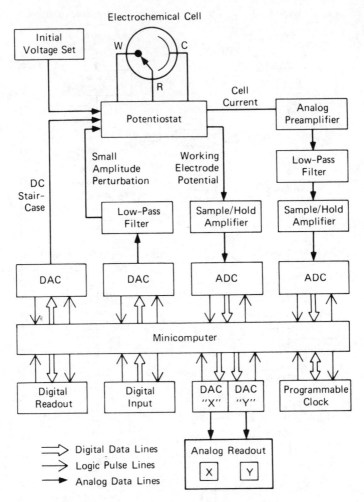

Figure 59. Schematic diagram of instrumentation for FFT electrochemical relaxation measurements. ADC = analog-to-digital converter; DAC = digital-to-analog converter; W = working electrode; R = reference electrode; C = counter electrode.

this particular application admirably illustrates the power of on-line computer analysis it is discussed in detail below.

The objective in the work described by Smith was to compute the interfacial admittance

$$A(\omega) = i(\omega)e^*(\omega)/e(\omega)e^*(\omega) \qquad (143)$$

from the response of the system [current $I(t)$] in the time domain to a time domain voltage perturbation $E(t)$. The frequency-domain current, $i(\omega)$, and voltage, $e(\omega)$, are computed from the corresponding digitized time-domain quantities using the FFT algorithm that was originally developed by Cooley and Tukey.[60] The complex conjugate of the transformed voltage $e^*(\omega)$ is then evaluated by substituting $-j$ for j in the complex number representation of $e(\omega)$. The interfacial admittance and hence impedance is then easily computed from Eq. (143). The actual data manipulations involved and the bandwidth limitations of the FFT are beyond the scope of this discussion and the reader is referred to the extensive publications by Smith and coworkers[59,61] for details.

The apparatus shown schematically in Figure 58 contains facilities for computer generation of the perturbation voltage, which is imposed between the working and reference electrodes by the potentiostat. In the particular case shown, the perturbation signal consists of a d-c ramp on which is superimposed a time-varying small amplitude excitation. The perturbation is generated in digital form and is then converted to analog form by two parallel D/A converters. Similarly, potentiostat output voltage and current are converted to binary form by parallel A/D converters for subsequent FFT analysis by the computer. The entire operation is controlled by a programmable clock, which permits the cell to be sampled at preselected times.

7. Future Trends

The various classes of instrumentation used in electrochemical systems may be loosely classified as:
- Input: voltage and current sources, function generators.
- Control: thermostats, galvanostats, potentiostats.
- Signal Conditioning: linear amplifiers, clipping amplifiers, filters.
- Computation: differentiators, integrators, digital computers.
- Output: oscilloscopes, electromechanical plotters, digital displays.

In the electrochemical studies, the system under test will invariably be analog. Thus input, control, and signal conditioning devices will necessarily contain linear circuit elements. Nevertheless, the clear trend is toward increasing use of digital devices, and an experimental electrochemist, who requires a complete understanding of the intrinsic advantages and potential limitations of laboratory instrumentation, must have a good grounding in the principles of analog and digital circuit design.

At the time of this writing, digital devices are being employed most heavily for data output, using digital plotters and digital voltmeters. In general these greatly outperform their analog ancestors, both in precision and speed, and little regard need be paid to possible circuit limitations. Synthesized digital inputs also are being intensively used to provide complex input functions, as well as traditional functions (e.g., sinewaves) at very low frequencies. In this way, input stimuli unrealizable by purely analog means can be constructed efficiently and inexpensively. However, some care must be taken in the direct substitution of digital for analog input devices. For example, whether a digital staircase can be substituted for a continuous voltage ramp depends largely on the characteristic response time function of the system under test, in combination with the control device, as discussed in Sections 4.5 and 4.6. Similarly, the spectrum of harmonic distortion associated with the piecewise generation of a sinewave is considerably different from that associated with the output of an analog oscillator.

Potentially, the greatest use of digital processing power is in computation and control devices. In addition to the obvious advantages of eliminating a machine-human-machine interface in the transcription of data, real-time computing power can be used directly to fit data to an expected response function, then pass information to the control device that moves the system under test into the desired operating range. Since it is rarely desirable (or even possible) to exceed 1 MHz in electrochemical applications, the present generation of high speed 16 and 32 bit processors allows an enormous flexibility in computation and control that will obviate many limitations of analog feedback devices.

When the computation requirements are severe, as for example in Fourier transformation, software limitations of available mini- or micro-computers make real time processing unfeasible, and dedicated, hard-wired, digital processors must be used. A number of such devices currently are available, performing mathematical operations useful to electrochemists (Fourier transform, autocorrelation, digital filtering, etc.), and we expect that considerable use will be made of these devices in committed laboratory instrumentation.

In the last class—signal conditioning—digital techniques have the greatest potential in signal filtering, particularly when processing is not performed in real time. Under these conditions, physically unrealizable filters can be emulated. For example a "brick-wall" filter can be "constructed" with no attenuation or phase shift in the passband, and many hundreds of db rejections in the stopband.

In conclusion it should be noted that this review is of the past and of the present. The devices and techniques discussed are available for use "off-the-shelf". There can be no doubt that experimental electrochemistry has a vigorous future. We foresee only two possible pitfalls: that scientists reach for the available electronic tools without an adequate grounding in circuit fundamentals, or that they do not reach at all.

Acknowledgments

The authors wish to thank SRI International and the Ohio State University for assistance during various stages in the preparation of this review. In particular, we would like to acknowledge the valuable contributions of Dr. Marc J. Madou and Mr. Robert D. Weaver.

References

1. Analog Devices Product Guide and Application Notes. Analog Devices Inc., P.O. Box 280, Norwood, Mass. (1975).
2. T. Kailath, *Linear Systems*, Prentice-Hall, Englewood Cliffs, N.J. (1980).
3. D. D. Macdonald and M. C. H. McKubre, In: *Modern Aspects of Electrochemistry*, J. O'M. Bockris, B. E. Conway, and R. E. White, eds., Vol. 14, Plenum, New York (1983).
4. Hewlett Packard Application Note No. 204.1, *Digital Signal Analysis*, Hewlett Packard, Palo Alto, Calif. (1980).
5. Hewlett Packard Model 5042A Digital Signal Analyzer Users Manual, Hewlett Packard, Palo Alto, Calif. (1980).
6. M. C. H. McKubre, Ph.D Thesis, Victoria University, Wellington, New Zealand (1976).
7. Analog Devices, *Data Acquisition Components and Subsystems Catalog*, Analog Devices, Inc., P.O. Box 280, Norwood, Mass. (1980).
8. J. G. Graeme, G. E. Tobey, and L. P. Huelsman, *Operational Amplifiers; Design and Application*, McGraw-Hill, New York (1971).
9. H. W. Fox, *Master Op Amp Applications Handbook*, Tab Book, Blue Ridge Summit, Penn. (1978).
10. J. I. Smith, *Modern Operational Circuit Design*, Wiley-Interscience, New York (1971).
11. I. Epelboin, C. Gabrielli, M. Keddam, J.-C. Lestrade, and H. Takenouti, *J. Electrochem. Soc.* **119**, 1632 (1972).
12. D. Britz, *J. Electroanal. Chem.* **88**, 309 (1978).
13. R. Bezman, *Anal. Chem.* **44**, 1781 (1972); D. Britz and W. A. Brocke, *J. Electroanal. Chem.* **58**, 301 (1975).
14. D. D. Macdonald, *Transient Techniques in Electrochemistry*, Plenum, New York (1977).
15. D. D. Macdonald, *J. Electrochem. Soc.* **125**, 1443 (1978).
16. D. D. Macdonald, unpublished data (1981).
17. S. Goldman, *Transformation Calculus and Electrical Transients*, Prentice-Hall, Inc., New York (1950).
18. G. E. Roberts and H. Kaufman, *Table of Laplace Transforms*, Saunders, Philadelphia (1966).
19. E. R. Brown, D. E. Smith, and G. L. Booman, *Anal. Chem.* **40**, 1411 (1968).
20. A. Bewick, M. Fleischmann, and M. Liler, *Electrochim. Acta* **1**, 83 (1959).
21. R. Koopmann, *Ber. Bunsenges. Phys. Chem.* **72**, 43 (1968).
22. D. Deroo, J. P. Diard, J. Guitton, and B. LeGorrec, *J. Electroanal. Chem.* **67**, 269 (1976).
23. J. P. Diard, Thesis, Grenoble, France (1977).
24. C. Gabrielli, Thesis, Paris, France (1973).
25. J. P. Diard and B. LeGorrec, *J. Electroanal. Chem.* **103**, 363 (1979).
26. M. Sluyters-Rehback and J H. Sluyters, In: *Electroanalytical Chemistry*, A. J. Bard, ed., Vol. 4, pp. 1–128, Marcel Dekker, New York (1970).
27. D. D. Macdonald and M. C. H. McKubre, *Electrochemical Impedance Techniques in Corrosion Science*, ASTM Special Technical Publication No. 717, p. 110 (1981).
28. M. C. H. McKubre, *The Measurement of Corrosion Rates in Cathodically Protected Systems*, Final report to the Electric Power Research Institute under contract number RP 1689-7-1 (1983).
29. M. C. H. McKubre, *The Use of Harmonic Analysis to Determine Corrosion Rates in Cathodically Protected Systems*, Paper No. 135 presented at the Denver Meeting of The Electrochemical Society, October 1981.
30. U. Bertocci, *Corrosion* **35**, 211 (1979).
31. G. P. Rao and A. K. Mishra, *J. Electroanal. Chem.* **77**, 121 (1977).
32. R. D. Armstrong, W. P. Race, and H. R. Thirsk, *Electrochim. Acta* **13**, 215 (1968).
33. B. Hague, *AC Bridge Methods*, Pitman, London (1957).
34. R. Parsons, *Trans. Faraday Soc.* **56**, 1340 (1960).

35. P. Delahay and I. Trachtenburg, *J. Am. Chem. Soc.* **79**, 2355 (1957).
36. R. Calvert, *Electron. Engin.* **20**, 28 (1948).
37. R. de Levie, *Electrochim. Acta* **10**, 113 (1965).
38. R. de Levie, *Electrochim. Acta* **8**, 751 (1963).
39. R. de Levie, *Electrochim. Acta* **9**, 1231 (1964).
40. R. de Levie, In: *Advances in Electrochemistry and Electrochemical Engineering*, P. Delahay, Ed., Vol. 6, p. 329 (1967).
41. T. Zeuthen, *Med. Biol. Eng. Comput.* **16**, 483 (1978).
42. J. G. Berberian and R. H. Cole, *Rev. Sci. Instrum.* **40**, 811 (1969).
43. M. C. H. McKubre and J. W. Tomlinson, *A Modified Berberian–Cole Bridge for High Precision, Low Frequency Impedance Measurement*, To be published.
44. E. R. Brown, *Electrochim. Acta* **13**, 319 (1968).
45. E. J. Crain, *Laplace and Fourier Transforms for Electrical Engineers*, Holt, Rinehart and Winston, New York (1970).
46. Brookdeal Electronics, *Phase Sensitive Detector Model 9412, Operation Manual*, Berkshire, England (1976).
47. Princeton Applied Research, *Lock-In Amplifier, Model 129A, Operation Manual*, Princeton, N.J. (1975).
48. R. P. Sallen and E. L. Key, *IRE Trans.* **CT-2**, 76 (1955).
49. K. B. Oldham, *Trans. Faraday Soc.* **53**, 80 (1957).
50. G. J. Hills and M. C. H. McKubre, Low Frequency Phase Sensitive Detection Employing Gated Digital Demodulation, To be published; L. E. A. Berlious, Ph.D. Thesis, The University, Southampton, England (1981).
51. R. D. Armstrong, M. F. Bell, and A. A. Metcalfe, *J. Electroanal. Chem.* **77**, 287 (1977).
52. C. Gabrielli and M. Keddam, *Electrochim. Acta* **19**, 355 (1974).
53. R. D. Armstrong M. F. Bell, and A. A. Metcalfe, *J. Electroanal. Chem.* **84**, 61 (1977); K. L. Bladen, *J. Appl. Electrochem.* **7**, 345 (1977); A. A. Metcalfe, *J. Electroanal. Chem.* **88**, 187 (1978).
54. D. Lelievre and V. Plichon, *Electrochim. Acta* **23**, 725 (1978).
55. P. Casson, N. A. Hampson, and M. J. Willars, *J. Electroanal. Chem.* **97**, 21 (1979).
56. Solartron, *1170 Series Frequency Response Analyzers, Operation Manual*, Farnborough, Hampshire, England (1977).
57. B. Soucek, *Minicomputers in Data Processing and Simulation*, Wiley-Interscience, New York (1971).
58. California Computing Systems, *Model 7490 GPIB Interface, Owners Manual*, California Computing Systems, Sunnyvale, Calif. (1980).
59. D. E. Smith, *CRC Crit. Rev. Anal. Chem.* 247 (September 1971).
60. J. W. Cooley and J. W. Tukey, *Math. Comp.* **19**, 297 (1965).
61. D. E. Smith, *Anal. Chem.* **48**, 221A (1976).
62. S. C. Creason and D. E. Smith, *J. Electroanal. Chem.* **47**, 9 (1973).
63. S. C. Creason and D. E. Smith, *J. Electroanal. Chem.* **36**, App. 1 (1972).

2

Computerization in Electroanalytical Chemistry

THOMAS H. RIDGWAY and HARRY B. MARK, Jr.

1. Introduction

Prior to the mid 1960s, the use of computers in electrochemistry was essentially limited to the simulation of response curves and the statistical treatment of manually entered data points. The watershed was the introduction of the first minicomputers by Digital Equipment Corporation (DEC), and the first models were followed shortly by the now almost universal DEC model PDP-8. Shortly after these came other computers that bore the then reasonable price of approximately $50,000.

The first extensive use of on-line laboratory computers was in the field of nuclear chemistry, with reviews of such applications first appearing in 1966.[1-3] By early 1967, the first applications to electrochemistry began to appear with Lauer and Osteryoung's article on a computerized chronocoulometric instrument.[4] Since that time, papers have appeared that describe computer-based data acquisition and control systems for cyclic voltammetry,[5-8] a-c polarography,[9-15] and stationary electrode voltammetry.[16-20] In fact, almost every facet of electrochemistry, including conductance measurements,[21] has received at least some attention. Although to date the vast majority of publications has come from academic laboratories, this is a very deceiving statistic as the industrial communities' extensive efforts in this area go largely unreported.

THOMAS H. RIDGWAY and HARRY B. MARK, Jr. • Department of Chemistry, University of Cincinnati, Cincinnati, Ohio 45221.

2. Architecture of a Laboratory Computer

A naked computer sitting in a laboratory is of little more utility to the chemist than a remote data entry to a large computer system. In order to be considered a laboratory computer, the system must be capable of monitoring and interacting with the instrumental environment. The general architecture of a laboratory computer is shown in Figure 1. The central processing unit (CPU) is the portion of the computer where the numeric and bookkeeping operations are performed. The data are stored in the memory, which can be thought of as a vast array of mail boxes, each with its own particular numeric address. The capacity of an individual memory location depends upon the nature of the computer.

As part of this writing, all currently available commercial computers operate upon binary numbers, that is, an N digit string, where the only possible values for each digit are 0 or 1. The four digit binary number 1010 is equivalent to the decimal number 10. In the terminology of computers, the digit capacity of a memory location is called the number of *bits* in the word. Thus, an eight-bit word is a memory location with a capacity of eight binary digits. Since there are 2^N possible states for an N bit binary number, the eight-bit word has 256 possible values, the 12-bit word has 4096 values, and the 16-bit word has 65,336 possible values, etc. These are normally treated by the arithmetic portion of the CPU as half positive and half negative numbers, i.e., a 16-bit word will range from $-32,768$ to $+32,767$. These word sizes are the most common in laboratory computers, although some machines, which are primarily meant for signal averaging, are available with a 24-bit word size. The use of binary numbers for the representation of the data is universal at

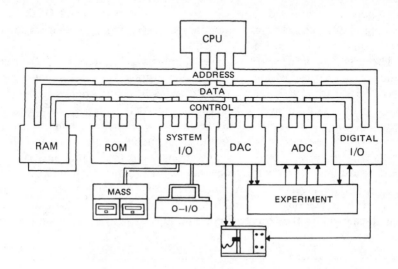

Figure 1. Architecture of a laboratory control computer.

present, although the earlier computers were decimal machines. The reason for the present universal use of the digital representation lies in the inherent noise immunity of circuits using binary representation. The internal circuitry only has to recognize two discrete states. Typically, the low or zero state is approximately from zero volts to one-third of the power supply voltage, while the high or "1" state is from two-thirds of the supply voltage to the full supply voltage. The outputs will normally produce a guaranteed "0," which is less than 10% of supply, and a "1," which is greater than 80% of supply. This allows fairly simple internal circuitry for the elemental input and output stages, while still maintaining high noise immunity, and has led to the mass production of inexpensive but highly complex integrated circuits. There is no particular philosophical advantage to the choice of base two numbers; any other base could be used, and in fact, there are theoretical advantages to higher bases, but manufacturing difficulties have prevented their commercial implementation. This may change in the near future for high-power machines. There have been a number of publications in the English, French, German, and Russian literature dealing with experimental circuits based on higher order bases, and some companies are now making limited production runs of multilevel logic elements.

2.1. Internal Communication Buss

Linking the CPU and the memory are three sets of signal lines or busses. These are the data buss, the address buss, and the control buss. The address buss is a monodirectional buss normally used by the CPU to tell the memory what particular storage location is to be selected for communication. The data buss is a bidirectional set of lines over which the data are sent to, and brought from, the memory. The control buss is a set of signal lines that identifies what is to happen on the address and data busses; for example, one line will be a read/write line, i.e., it will signal whether memory is to receive data from the data buss or present it to the buss. A second line in the control buss is normally used to synchronize transmission of the data. The number of bits in the address buss determines the maximum number of memory locations that can be accessed in a fully implemented computer, and, as in the case of the data buss, the resulting M-bit word is treated as a binary number. The address buss is considered as representing an unsigned or positive number, so that a 16-bit address buss means that the computer can directly access memory locations 0 through 65,535, inclusively.[22–25]

2.2. The Central Processor Unit

The memory is used both as a place to store data and as a source for the instructions to the CPU. The CPU itself has a number of special storage locations. One of these is the program counter, which is a register that has the same number of bits as the address buss, and points at the location of the next instruction to be executed. The second is the accumulator. This is the register where addition, subtraction, and logical operations take place. There is also usually a shift register associated with the accumulator so that the contents of the accumulator can be shifted left or right one or more places. A third register is known as an index register. This register usually has the same number of bits as the address buss, and is used for accessing data. The simplest use of the index register is to load the accumulator with the contents of the memory location specified by the index register, or to store the contents of the accumulator to the location specified by the index register. Many computers have a number of more complex and sophisticated ways of using the index register, but this simple use is essentially universal. Many modern minicomputers have more than one index register and some have several accumulators.

The CPU interprets the bits in an instruction word in a fixed way to determine what action is to be performed. For example, if the high-order or leftmost two bits are "1," this may signify that the accumulator is involved. If both are "0" it may signify that it is a memory reference instruction (a load or store), if the pattern is "01" the index register is involved, and if the pattern is "10" the program counter is involved. The next several bits would further specify the nature of the operation to be performed. For example, the next three bits might be used as "000," which means add data to the contents of the accumulator; "001" means subtract from the accumulator, "010" means *and* the data, "011" means *or* the data, "100" means shift contents right, "101" means shift contents left, and so forth.

In order to perform the mathematical operation $A = B + C$, the computer must first load the data residing in storage location B into the accumulator. This is done by executing a read from memory instruction with the address of location B on the address buss. Next, the contents of location C are added to the accumulator, and finally, the contents of the accumulator are stored to location A.

2.3. Memory

At one time the memory of a laboratory computer was invariably core memory, that is, small ferromagnetic toroids with wire windings, one toroid for each bit of memory. These toroids were organized in blocks; for example, a 4 k word block would occupy one structural element of the computer. In the mid to late 1960s, these memory blocks were physically large and expensive. A computer with 16 k words was considered a large system, both in terms of its computing power and its physical size. In 1968, a Raytheon 706 minicomputer 16 k word (16 bits/word) memory bay and its associated power and drive electronics occupied an entire 5 ft. × 4 ft. × 24 in. bay. Core memory technology has undergone significant evolutionary changes since then, and the

16 k (16 bits/word) memory for a DEC PDP-11 or DG Nova 2, etc., computer has been available on a single card (approximately 12 in. × 12 in.) since 1974. Core memory has one important property; if power is turned off, the toroids remain polarized and the data written in them are available to the computer after power is restored. This property is referred to as nonvolatility.

Most of today's new inexpensive laboratory computers use semiconductor memory, which is significantly smaller and requires less power than core memory. Numerous books have been written on the subject of semiconductor memory,[26] and we will only discuss a few features that are important to the user. Semiconductor memory can be broken down into two broad classes, random access memory (RAM) and read only memory (ROM). RAM is read/write memory just like core memory, with one major difference. When power is lost, the data disappear. This means that semiconductor memory is highly sensitive to short, inadvertent power losses. RAM can be further subdivided into two classes, static and dynamic. Static RAM will retain its contents as long as power is maintained, whereas dynamic RAM requires additional attention. Each bit in a dynamic RAM must be accessed every 1 or 2 msec, or it returns to a preferred state, either "1" or "0," depending on the manufacturing process used. To prevent this, memory systems employing dynamic RAM have additional hardware that generates a "refresh" cycle occasionally, thus guaranteeing that each bit is attended to in a time frame well within the critical limits. There is an important consequence to this requirement. In systems that do not use dynamic RAM, one can be certain that a given block of program codes will always take a fixed and known amount of time to execute, and the time interval between two instructions might be known to be 59 μsec. One can use this property to provide timing capability to a program segment. In dynamic memory systems, this is not possible, since one never knows when the refresh cycles will be inserted. One knows only that in total there will be a refresh cycle at fixed time intervals, for example, every 10 instructions, but where they will occur relative to execution of a given block of eight instructions is unknown; it may occur once or not at all. There are advantages to the use of dynamic RAMs: they require less power than a static RAM of identical bit size, they are, in general, internally simpler and they can operate significantly faster. This is not to say that all dynamic RAMs are faster and require less power than all static RAMs, but at any point in time, the state of the art dynamic RAMs will require lower power and be less expensive than corresponding static RAMs. Historically, the number of bits that can be put into a single dynamic RAM chip is greater than two to four times that of the static. Thus, in the spring of 1977, commercial 16 k but dynamic RAMs were available from a number of vendors, but only 4 k static RAMs are available in quantity.

ROM, by definition, cannot have its contents altered, so it is inherently nonvolatile and impervious to loss of power. Currently, 64 k ROMs are commercially available and 256 k versions are in limited production. The contents of a ROM are determined in the manufacturing process. Individual ROM chips, even 64 k versions, are relatively inexpensive, but the set-up charge for producing the first chip is typically $10–35 K. Two subsets of the class eliminate this cost problem. The first is the programmable ROM (PROM). This device can be programmed fairly easily by the user and makes possible small production runs without the set-up charge of the ROM, but the cost per chip is higher. The other version is the erasable PROM (EPROM). This chip is enclosed under a quartz lid and exposure to intense UV light will erase the contents, allowing one to reuse or reprogram the EPROM if an initial mistake was made, or if requirements change. The current state of the art is 64 K bits per EPROM chip.

There is one additional form of ROM, known as electrically erasable read only memory (EEROM). In this form the contents of ROM can actually be dynamically altered by the running system, yet the data remain after removal of power. The programming of these chips typically requires some unusual voltages not typically found in the computer, although at least one vendor is producing a chip that can be programmed by application of voltages that are normally available. Even for this part, however, the time required to write data into the memory is several orders of magnitude greater than that required for RAM so that the EEROM cannot be treated simply as read–write memory.

2.4. Operation Input and Output Devices

The computer must have some form of operator input/output (OI/O) capability whereby the CPU performs two-way communication with the operator. Input to the computer is usually accomplished by means of a full alphanumeric (A–Z, 0–9, and symbols) keyboard, although in a totally dedicated system, one occasionally finds a numeric key pad (0–9) and a few labelled pushbuttons instead of the full keyboard. The output to the operator is often both alphanumeric and graphic. The graphic output is often by means of an X–Y recorder and/or an oscilloscopic display. The alphanumeric output devices can be classified as to whether their records are permanent or volatile. The volatile devices are based on cathode ray tubes (CRTs), often merely an oscilloscope or television screen, although more sophisticated systems are available. The hard copy devices are either impact devices such as the familiar teletype (TTY), or more exotic methods such as electrostatic ink sprays, electrostatic discharge, thermal or photosensitive papers, etc.[27] Some of the earlier of the exotic printers were less than reliable, expensive to use, requiring either special paper or other supplies, and the output tended to fade with time. However, these problems seem to have been mostly overcome. The more exotic methods have the advantage of being quieter than impact methods, and usually much faster, which is an advantage for situations where large volumes of output are generated.

2.5. Experimental Input and Output Devices

The experimental I/O (EI/O) system in a laboratory computer is the portion that differentiates it from other computer systems. The electrochemist is normally interested in monitoring cell parameters such as current, potential, resistance, or capacitance, although measurement of optical properties such as absorbance, scattering, fluorescence, or phosphorescence is also becoming common. In order to produce these responses from the sample, we excite it with a current, potential, thermal, acoustical, magnetic, or photonic stimulus. The computer itself operates in binary space with all data being represented as an N bit string of "1"s and "0"s. These numbers are represented internally by one of two electrical potentials on the lines of the computer's buss. For most of the laboratory computers in existence today, these voltages are roughly 0.5 and 3.5 V. In order to establish two-way communications between the CPU and the experimental environment, it is necessary to convert the internal buss representation into a current, potential, magnetic field, flux of photons, etc., or vice versa.

2.5.1. Data Domains

The data domain concept has been used by Enke[28] to discuss generalized chemical instruments, and a modification of this approach will be used here. For our purposes it is convenient to define four basic signal domains as seen in Figure 2: the analog domain, which has subdomains such as current, voltage, charge, and power; the frequency domain, which has subdomains frequency and phase; the pulse domain with subdomains rate and temporal separation; and the numeric domain.

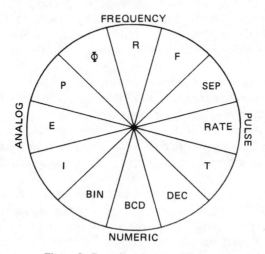

Figure 2. Data domain representation.

These classifications are somewhat arbitrary, and a single instrument can produce data in several of these defined domains. For example, consider a photomultiplier tube as a detector in an experiment. A photon striking the photocathode will have a fixed probability of causing the emission of an electron (the probability is a function of the energy of the photon and the surface material of the electrode). This emitted electron is attracted to the next electrode in the dynode chain, causing the release of further electrons, and so on down the chain. For each photon that caused the release of a single electron, there will be a pulse of electrons at the anode. Depending upon the timeframe of the observer and the incident photon flux, this will look like an impulse of N electrons, a gaussian pulse of electrons with a characteristic time functionality (dependent upon the characteristics of the detector), or at the shortest time scale one could observe that it is actually a series of electron bursts, with an overall gaussian envelope. When the photon flux is high enough the output will appear as a continuous current flow and the pulse character will be obscured, which is the type of response associated with normal spectroscopic measurements. However, if the photon flux is low, the flux rate is represented not by a current level but by the rate at which the individual pulses occur. At this level one must also watch for two complications. One complication is pulses that result from spontaneous emission of an electron from some electrode (the familiar dark current phenomenon), and the second is the simultaneous arrival (at the time resolution of the equipment) of the two photons resulting in a pulse of twice the height of a "normal" pulse. In both cases the amplitude of the pulse allows us to discriminate between "normal" pulses and abnormal pulses. Simultaneous pulses will be abnormally large; the vast majority of the spurious "dark current" pulses will be significantly lower in amplitude, as they will on the average originate from a lower dynode and, hence, undergo less amplification than a pulse originating at the target electrode. If the production of the photons is controlled either by pulsing the light beam or by electrochemical modulation, then one can also take advantage of the time correlation to eliminate or at least minimize contributions from spurious photons.

A careful examination of the nature of the types of excitation signals that are of interest reveals that almost all of them can be generated or controlled by means of either switching operations (for example, laser pulse initiation, coulostatic impulse generation by means of a relay, pressure jump experiments by fracture of a membrane with a solenoid, etc.), or by an analog voltage that can be further transduced into a current, charge, magnetic field, acoustical pressure, etc.

The switching operations are the simplest to implement because they are inherently binary. All that is required is to convert from the voltage or power level of the computer data buss to the levels required to drive the relay or solenoid or to trigger the laser, etc. There are commercial solid-state switches available that allow one to control several thousand watts of 110 V a-c power directly from the normal logic levels of most computers' data buss.

2.5.2. Digital to Analog Conversion[29]

The conversion from digital signals to analog voltage form as shown in Figure 3 is only somewhat more complex. The class of device that performs this operation is known as a digital to analog converter (DAC) and the simplest implementation shown in Figure 4 consists of a reference voltage source and a set of N parallel weighted resistors with a digitally controlled analog switch connecting one end of each resistor to the common buss. The resistors are weighted as $R, R/2, R/4, R/8, \ldots R/2^{N-1}$. When the first switch is closed, a current $E\text{ref}/R$ flows into the common buss. If the second switch is closed, a current $2E\text{ref}/R$ flows, etc.; thus the total current flow is directly proportional to the binary digital signal control inputs to these switches. This is a simple current DAC. To form a voltage DAC, one connects the common buss to the negative input of an operational amplifier with a suitable feedback resistor, normally $2R$ so that the output voltage becomes $-E\text{ref}(Dig/2^N)$ where Dig is the binary number formed by the N-bit digital input. The commercially available versions of this resistive ladder network are somewhat more complex, but the principle is the same. One feature of this type of DAC is that the reference voltage can be of either polarity, and in effect, one is performing a

ANALOG TO DIGITAL & DIGITAL TO ANALOG CONVERSION

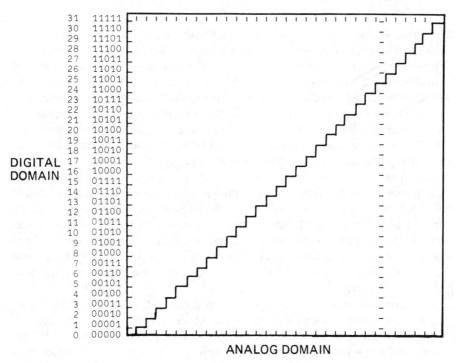

Figure 3. Digital–analog interconversion.

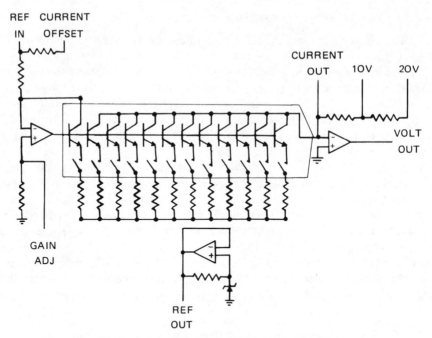

Figure 4. Structure of a digital to analog converter.

digital multiplication of an analog signal with an analog result. The second common form of DAC is a current-switching DAC. These DACs operate by turning on specially constructed transistors, which pass weighted amounts of current. The major difference between the two types of DACs is that the current-switching DAC can be made with higher accuracy, but is limited to monopolar reference voltages. This latter limitation can be circumvented by additional circuitry. There are several important parameters associated with DACs that determine their suitability. The resolution is a function of the number of bits in the input word. For example, 10-bit resolution means that the output will be parsed into 2^{10} or 1024 different levels. The linearity refers to the worst case difference between the actual output voltage and the straight line drawn through the outputs states. If the nonlinearity exceeds $\pm 1/2$ bit, then the output will not be monotonic, i.e., there will be one or more codes where the output signal does not increase as the code increases. A final parameter that can be important is the settling time, which refers to the amount of time required for the DAC to settle to its final value under some specified change in the digital inputs.

A consideration of the kinds of response signals that one might encounter in electrochemistry shows that the types of signals can be broken into two categories: those that are in essentially digital form to begin with, and those that can be transduced into an analog voltage. The intrinsically digital data types include pulse trains, which can be converted to binary form by either counting for a fixed time period or by using a timer to determine the time required to accumulate a fixed number of pulses. Other digital data types could be the conditions of control relays, solenoids, etc., which can be converted into digital voltages that are compatible with the data buss requirements.

2.5.3. Analog to Digital Conversion[30]

The conversion of an analog voltage to a digital representation is accomplished by an analog to digital converter (ADC). There are a number of schemes that are currently represented by commercial products. One of the simplest is the run-up converter. In this method a DAC is driven by a counter connected to an oscillator. The output of the DAC is compared against the input voltage. When the DAC output is greater than or equal to the input voltage, the oscillator is shut off and the counter output represents the analog input voltage. To initiate a new conversion, one initializes the counter to zero and enables the oscillator. This method of conversion requires a time equal to M/F_0 where M is the digital value of the input voltage and F_0 is the oscillator frequency. This form of ADC is rarely encountered in modern instrumentation, but a closely related form, the tracking ADC, is still fairly common. In this form the counter is bidirectional, that is, it can count up or down under the control of the comparison circuit. When the input voltage and the DAC output are equal, no counting occurs. As long as the (dE/dt) of the input voltage is less than $(E_m/2^N - 1)F_0$, where N is the number of bits in the converter and E_m is the full scale voltage, then the tracking ADC will continuously have a valid conversion result. The state of the conversion is monitored by observing the output of the comparison amplifier. Converters of this sort are very useful for continuously monitoring slowly changing signals. The successive approximation ADC is another modification of the run-up ADC. In this form a series of N comparisons are made for an N-bit conversion. The high-order or most significant bit is turned on first. If the DAC output is less than the input voltage, the high-order bit is left on; if not, it is turned off. Then the next most significant bit is turned on and the comparison is made again. This proceeds for all N bits. The conversion time is NT_c where T_c is the time required for each new bit addition and comparison. This type of ADC is more complex than the previous two, but can be quite fast. The fastest form of ADC is the simultaneous conversion type. In this approach, there are 2^N separate comparison amplifiers for an N-bit converter and the conversion is limited by the speed of the comparison amplifier and the decoding logic that follows it. These are, relatively speaking, quite expensive and the maximum number of bits available is far lower than for the previous types, as the complexity of the circuitry involved more than doubles for each additional bit of resolution. Converters of this type with resolution greater than 10 bits are quite rare.

It is important to recognize the limitations of these types of converters in terms of their response to changing signals and noise components. The run-up converter will produce a conversion result that will accurately represent the state of the input when the conversion complete signal comes true. However, this technique is highly sensitive to noise. In addition, it is fairly difficult to maintain time synchronization for rapidly changing signals with this technique due to the fact that the conversion time is proportional to the magnitude of the signal converted. This can be an important limitation for systems where accurately known potential–time coordinate pairs are required. In the case of the tracking ADC, if the (dE/dt) of the input (due to either the characteristics of the signal or superimposed noise) exceeds the time response of the converter, then the ADC will never accurately reflect the input signal. The simultaneous conversion ADC is extremely sensitive to noise or signal changes during the conversion period; fluctuations during the conversion time will invariably lead to erroneous results. For all three of these types of ADC, one will usually precede the ADC with a sample and hold amplifier (S&H), thus freezing the signal prior to initiation of the conversion. One also commonly finds low-pass filtering prior to the S&H. The conversion rate of a system will include not only the time required for the actual conversion itself, but also the time required for the S&H to acquire the new signal in the sample mode as well as the time for the S&H to go to the hold mode. Normally this latter time is small in comparison with the acquisition time. The simultaneous conversion ADC usually operates on a time scale of 25–1000 nsec; therefore, except for very rapidly changing signals one need not worry about the (dE/dt) or noise problems. If the signal contains very high-frequency components the usual approach is to place a sharp cut-off low-pass filter prior to the converter rather than to use an S&H, although there are now very fast amplifiers available commercially, so that the S&H approach is also a valid method of freezing the signal for this ADC. The actual behavior of a simultaneous conversion ADC working on a changing signal will depend upon the internal structure of the ADC and the nature of the signal, but one can assume that, in general, the resulting conversion will be erroneous.

There are several ADC techniques that are relatively insensitive to noise and signal changes; all of them are based on integration methods. Most digital multimeters are built using these techniques and the reader is directed to the manufacturers' literature for discussions of these methods. The one that will be discussed here is based upon the voltage-to-frequency converter approach. In these techniques the input voltage is integrated up to a reference potential, then disconnected and a reference current of opposite sign is integrated back to zero volts. The time required for the integration to the trip point and back to zero volts is directly proportional to the input voltage. If the ADC generates a high digital signal during the signal integration and a low digital signal during the reference reintegration, then a pulse train results whose frequency is directly proportional to the input voltage. If one counts the pulses for a fixed time period, the count accumulated is a digital representation of the input. ADCs based upon this approach are in a sense insensitive to (dE/dt) and noise effects in that they effectively integrate both signal and noise. The result in the absence of noise is effectively $\int_{t_1}^{t_2} F(t)\ dt$, where t_1 is time at initiation of digitization and t_2 is the time at the end of digitization. Noise components are also integratively cancelled. One particularly interesting feature of this approach is that noise components with the same period as the sampling period are effectively perfectly rejected. If one uses a sampling period of 1/60 or 1/50 s, the power line related components can be perfectly eliminated, as well as all higher harmonics of the power line frequency. A VFC type ADC with a maximum output of 100 kHz full scale can accumulate 1667 counts in 1/60 s, or 2000 counts in 1/50 s, so that the effective resolution of such an ADC is between 10 and 11 bits. Commercial VFCs with differential nonlinearities of less than 0.01% at 75 kHz are readily available and inexpensive.

The price and size of ADCs has shrunk remarkably in the past decade. In 1970, a 25-μsec 12-bit successive approximation ADC cost about \$1200 and was roughly the size of this volume. In early 1977, an identical performance integrated circuit ADC cost about \$75 and is smaller than a cigarette lighter, while a 10-bit VFC type ADC with a 2-msec period is even smaller and costs about \$25.

Even with these impressive cost and size decreases, one seldom finds more than one ADC in a given computer system even when several signal points are to be monitored. Instead, one precedes the S&H with an analog multiplexer. A typical moderate performance multiplexer (MPX) is an eight-channel version selling for between \$9 and \$18 each. This unit will have eight analog inputs, one analog output, three address lines [which are treated as a binary address for the eight channels (channels 0–7)], and an enable-disable line that turns all channels off and allows cascading to 16, 24, etc., channels systems. The channel settling time (time to settle from -10 V on one channel to $+10$ V on another to within 0.05%), is 1.5 μsec.

The overall conversion rate for a system consisting of an ADC, S&H, and MPX is controlled by the delays of all these components. For an 8-μsec ADC, 2-μsec S&H, and 1.5-μsec MPX, the total conversion time is 10 μsec: 8 for the ADC and 2 for the S&H–MPX combination, not 11.5. The reason for this apparent discrepancy is that the MPX can be switched to the new channel and allowed to settle while the ADC is converting. If the ADC conversion time is 1 μsec, the total conversion time will be 3 μsec and any further increase in the ADC conversion speed will produce little throughput gain. Often a significant increase can be obtained by the method known as "ping ponging," which requires the addition of a second S&H and a two-channel multiplexer. The second S&H is placed in parallel with the first and outputs of both go to the inputs of the new MPX. While one channel is being converted the second S&H is acquiring the data. As long as the settling time for the S&H-N channel MPX combination is less than the time for the ADC-2 channel MPX combination, these latter two components are the throughput limiting elements. Although complex, this is usually a very cost-effective solution to the problem.

2.6. Communication with I/O Devices

For peripheral devices such as ADCs, DACs, digital input/output (DI/O) ports, etc., there are two conventions in common use. The simplest structure conceptually is to treat peripheral devices as memory locations and to use the address lines to access each device. This is called memory-mapped I/O. Some computers using this architecture have a "wait line," which will cause the CPU to hold the transaction (input or output or both) until the peripheral is prepared to complete the transaction. The second architecture distinguishes between peripherals and memory, with separate instructions for input and output that are different from those involving memory. There are two different buss structure implementations of this approach. In the first, the address buss, or at least part of it (commonly only eight bits for 256 possible device addresses), is used to pass the address with an additional line used to signify whether memory or peripherals are being addressed. In the second realization of this approach, a separate peripheral address buss (normally eight bits wide) is used.

Generally, there is nothing to prevent one from using the memory-mapped I/O approach in a system that also supports separate I/O commands. The use of the memory-mapping technique does have the effect of diminishing the available memory for program or data by the number of peripheral device addresses implemented. However, since most minicomputers can now access 32 or 64 k memory locations and it is a rare system that uses more than 128 device addresses, the loss is usually not noticeable unless the memory block structure causes loss at large quantities of memory.

It is not uncommon for a single peripheral to require more than one device address. For example, if one must load a 14-bit DAC (resolution of one part in 16,384) with an 8- or 12-bit data buss, then two transactions are required; the first is to pass the low-order 8 (or 12) bits, and the second is to pass the remaining high-order bits. In a case like this, it is necessary to buffer digitally (or store) the low-order bits in the peripheral prior to the DAC itself and then transfer the high-order and low-order bits to the DAC simultaneously. Otherwise, the DAC will be in a spurious state representing neither the old nor the new value between the two transactions. Most laboratory computers will have more than one DAC, usually two for control of an X–Y recorder and one or more for experimental control purposes. There are several ways to implement multiple DAC systems. One method is to use one device address (two consecutive addresses, if multiple transactions are required for 8- or 12-bit data buss structures) and to include the binary address of the particular DAC chosen in the data word at either the least or the most significant end. For example, a 16-bit transaction utilizing 12-bit DACs has four unused bits, so a total of 16 DACs can be selected. This has the advantage of the optimal conservation of device addresses and has been popular in the past. It has the disadvantage of requiring a fairly complex section of program to generate the data word-device address combination. The second common method is to utilize two device addresses and two transactions (three for the case where the data buss is smaller than the input resolution of the DAC). The first transaction to the first device location selects the DAC for the second (and possibly the third) data transaction. This approach has the advantage of using only two device locations (or three), and permitting up to 2^N separate DACs, where N is the width of the data buss, but it does require a separate transaction to identify the particular DAC selected. The third approach is to assign each DAC an individual address (or adjacent pair of addresses for the case where the data buss is smaller than the width of the data to be sent). This method uses the most device addresses, but will usually require the fewest instructions to implement the desired operation. A hybrid of the last two methods is possible for the case where two data transactions are required to meet the DAC resolution. Here, one sends the low-order bits to a general address for buffering, which is common to all DACs, and sends the high-order bits to an address that is particular to the desired DAC. The low-order bits are then also piped to this selected DAC with a resultant conservation of device addresses.

A second case where multiple transactions are required is in servicing an ADC. Normally one must first start the ADC converting (unless the interface is such that the ADC is continuously converting), and then test the status of the ADC to see if valid data are available. The first is an output instruction, while the second is an input, so that the same device address could be used for both instructions. If a multiplexer is placed in front of the S&H amplifier, one may also use a separate address output to establish the MPX channel. Alternately, one could use one bit in the output to initiate conversion and the remaining bits to select the multiplexer channel. Similarly, the status check can be performed using the same address as that used to input the conversion results if there are more input bits than are provided by the precision of the ADC. (Note that two separate read operations are required to input 12 data bits on an eight-bit buss, or 14 to 16 data bits on a 12-bit data buss with extra unused bits available for the status in either case.) The use of the same address for status and data conserves device locations at the expense of extra program steps to remove the status from the data. This also has the effect of slowing down the minimum service loop for the ADC, and hence, can have the effect of decreasing the *system* data acquisition rate. One can often conserve device addresses and program steps at the expense of a more complex interface between the data buss. For example, the actual sequence of events in selecting an MPX channel and initiating conversion is: select the channel, allow time for settling, place the S&H in hold, initiate conversion. After completion of conversion one should place the S&H back into sample mode. A minimal interface would require separate commands for all of these steps, in addition to the status check and reading of the conversion results. A more complex interface could require only one command to select the MPX channel and would itself place the S&H in hold after a suitable settling delay and initiate conversion. At the end of the conversion, it would automatically place the S&H back into sample mode. Even more complex interfaces allow for automatic sequential scans of adjacent MPX channels by decreasing the channel address at the end of conversion and restarting the conversion after the settling delay. The latter type of interface requires that the previous conversion results be read before the new conversion is finished.

2.7. Experimental Timing

The electrochemist commonly requires data that have rigid timing requirements. For example, in stripping analysis, the potential at which a local current maxima occurs contains information about the nature of the material that was deposited into the electrode. As the stripping normally is produced by application of a linear potential–time program to the electrode system, the requirement for accurate potential information is translated into a requirement for accurate time control. There are four basic ways to provide accurate timing control in data acquisition and control programs. The first is to use the fact that for almost all computers there is a highly accurate crystal clock used to generate the internal timing for the computer and that one can accurately forecast the execution time of a given code segment (note that this may be impossible with systems that use semiconductor dynamic RAM). The second method is to use a programmable clock derived from a crystal time base, often the crystal used by the CPU for its timing, and to set up the required delays and then watch the clock to determine the elapsed time. The third method is to use interrupt programming. An interrupt is a method of attracting the CPU's attention to a particular peripheral's needs. The CPU responds by terminating the section of code that was undergoing execution and transferring control to a particular service routine for that peripheral. In the service routine, the previous status of the machine must be stored, so that on completion of the service to the peripheral, the CPU can pick up where it left off without loss of data or control. The required code to implement the storage of status and its restoration is spoken of as the "interrupt overhead." An example of the use of interrupts might be to control an ADC with a programmable clock and to have the end of conversion signal from the ADC trigger an interrupt to the CPU. In this way, accurately timed values can be acquired while the CPU is busy processing the previously acquired data. Conflicts can arise when more than one peripheral can interrupt the CPU. In this case, we must establish a method of determining which device caused the interrupt, so that the correct service routine obtains control. (This determination of "who called" is also part of the "overhead.") This may be done relatively automatically by hardware or by software. It is also important that a priority be established about what devices may interrupt the service routines of other peripherals. Typically, in a laboratory computer, either the programmable clock or ADC have the highest priority; they can interrupt any other device, but their routines cannot be interrupted by other devices, while operator I/O, such as a teletype, will have the lowest priority. When more than one device is requesting service at once, the highest priority is serviced first and then the next highest, and so on.

The fourth method is direct memory access (DMA). In DMA operations, the data from or to a peripheral link directly with the memory without the intervention of the CPU. There are two basic forms of DMA: cycle stealing and lockout DMA. In cycle stealing DMA, the DMA peripheral takes over the address and data buss whenever the CPU is not actually using them (typically 25–50% of the time, in most systems), and in a sense, the CPU is viewed as having higher priority than the DMA device. In lock-out DMA, the peripheral is viewed as having highest priority and takes over the busses when it desires. A DMA peripheral must be much more complex (smarter) than a non-DMA device because once the initial information about the number of points involved, their location in memory, and timing are given to the DMA device, it must operate on its own. In our ASV example one might construct a DMA device to handle the entire operation. Such a device would be told the number of data points to acquire, the MPX channel or channels to use, the time between points, and the locations in memory that are to receive the results, and then be left on its own. The DMA device would send a digital signal to the instrument to initiate the stripping operation, load and start an internal clock, and then, when the clock times out, initiate an ADC conversion operation while reloading and restarting the clock to maintain absolute timing synchrony. At the end of the conversion it would insert the correct memory address on the address buss and the data on the data buss and perform a write to memory operation, then release the busses and wait for the next time out. In the example used, the timings are slow enough (i.e., the data rate is low enough) that a cycle-stealing DMA device would be adequate, but there are cases where only a lock-out device would be useful. The absolute limit on the rate that a DMA device can transmit data to memory is the memory cycle time of the computer and a DMA device can acquire data much faster than the CPU can, since the DMA channel is highly specialized and designed for one particular task. A computer with a memory cycle time of 1 μsec might be able to acquire data from a peripheral once every 20 μsec, due to the time required to load and execute the necessary instructions of a read from a peripheral, write to memory, increment the index registers, and test to see if the block is completed. A DMA device could transfer data every microsecond if the device being controlled is fast enough. If one wishes to obtain current transient data at a rate of one million points per second on a machine with a 1 μsec memory cycle time, then the only way to achieve this rate is to use a DMA device.

3. Programming Languages[23-35]

As we have just seen, the CPU operates on instructions that are stored in memory as bit patterns of 1's and 0's. While it is possible to write programs in this form, known as machine language, most humans find the task laborious and time consuming, as well as being highly error prone. The next stage in sophistication is an assembly language program. In assembly language, or assembler as it is often called, short verbal mnemonics such as ADD for add, ADC for add with carry, LDA for load accumulator, etc., are used, and statement labels and name variables are allowed. The program or code is then digested or assembled by another program, either on the machine for which the program is to be written or on another machine, and machine language (binary bit patterns) is generated. If the computer that does the assembly is the same type of machine that will execute the resulting code, the program is called a self-assembler or simply an assembler. If another machine performs the task, the program is known as a cross-assembler, indicating that two different machines are used. Cross-assemblers are often written in higher level languages such as FORTRAN, ALGOL, or BASIC. Assembly language programs give the programmer direct access to the true capabilities of the machine, can produce the most efficient code, and have the best control over timing. However, in many respects they are extremely limited. For example, most laboratory computers do not have a multiply or divide instruction implemented in the CPU, nor can one read or write a message to a teletype with a single command on most CPUs. All of this and many more features that high-level programmers take for granted are unavailable as simple instructions at the assembly language level. High-level languages such as FORTRAN or BASIC allow simple statements (such as READ, WRITE, PRINT, DO, FOR, etc.) that require many assembly language statements to implement. The amount of memory required to implement a programming function will usually be larger if written in a high-level language than in assembler, and will take longer to execute, but the programming time itself will be far less, often by two orders of magnitude. It is important to realize that there are two basic types of high-level languages: interpreter based and compiler based.

An interpreter-based system stores the high-level language in symbolic form. For example, the BASIC language statement LET $A = B + C$ will be stored in memory exactly as written, one eight-bit byte per character, including any blanks. A resident program called the interpreter will decode the instructions, one character at a time, determine what is required, and execute the instruction. At no time does the interpreter generate any machine language code representing the user program. The most common interpreter-based language is BASIC. In contrast, compiler-based languages decode the user's program and generate a machine language version of the original, place this version in memory, and the machine language code representing the symbolic program is what is executed. FORTRAN, PASCAL, and ALGOL are the most common compiler-based languages. Some laboratory computers also support a compiler-based version of BASIC but, in general, BASIC is usually an interpreter-based language.

The distinction between interpreter and compiler languages is important to the user. An interpreter language is much slower in executing a set of instructions than is a compiler language, since the instruction must be decoded each time it is encountered. In fact, the vast majority of the time spent in program execution is in the decoding of the instruction itself, not the execution of the desired function. On the other hand, a compiler language spends a large amount of time setting up a program for execution the first time and requires additional support devices such as magnetic tape or disks to perform the compilation (it is possible to do the job with only a teletype, but it is doubtful that any individual would try this more than once). An interpreter language can function without these support devices. This can be an important consideration since the cost of a modest disk drive and its controller will usually exceed the cost of a 32 k word minicomputer and teletype.

There are three other classes of languages in addition to interpreter and compiler, but there are no generally acceptable labels to describe them. One is what we will refer to as a pre-compiler-interpreter (PCI) and is most commonly encountered as a form of BASIC. In the PCI the BASIC text is scanned (usually on entry), spaces are eliminated, and the key words or instructions (such as FOR, GOTO, =, +, etc.) are turned into single eight-bit tokens, although variable names are generally left alone. This format executes faster than a pure interpreter since some of the work has already been done. However, it still allows the program to be listed back and retains all of the ease of program debugging of the simple interpreter form.

A second form, exemplified by FORTH, is often called a compilation, but this is not strictly true and we will refer to it as a quasi-compiler (QC). In the QC form the program statements are reduced to what is essentially a list of addresses. These are the addresses of the routines that execute an operation and the addresses of the data, if any, to be operated upon. On execution a resident interpreter scans this code and transfers control to the desired routines. The QC is significantly faster than the PCI, yet typically slower than a true compiler and the resulting compiled program cannot be listed back. FORTH can perform its "compilation" in memory and does not actually *require* any form of mass storage as is the case for the true compiler.

The third form, PASCAL being the most prominent example, looks quite similar to the QC on the surface. A compilation step, typically requiring mass storage, is performed on the program. The result of this step is something called P-Code (PC). The PC produced is not directly executable by *any* computer but may best be viewed as instruction for a hypothetical computer. The PC is said to be "machine independent," that is, any computer with the correct software can then process this code. Upon execution the PC is processed by a resident program (indistinguishable from an interpreter to the uninitiated), which translates the PC instructions into operations that can actually be performed by the computer in question. A PC-type language will usually operate at rates comparable to QC types, and, like the QC, the compiled program is nonlistable. In addition, the PC-type languages will require mass storage devices to operate. It should be noted that there are also forms of PASCAL that produce a truly compiled code that is directly executable on the host computer and these are usually denoted by calling them "native-code" compilers. Additionally, there is a computer produced by Western Digital that directly executes PC, i.e., its assembly language is structured for the direct execution of PASCAL P-code.

4. Microprocessors[31,32]

The major barrier to the use of computers in electrochemistry has been the cost of implementing a system. Even a modern minicomputer selling for less than $10 for a 32 k word machine is still too expensive for many laboratories and far too expensive for instrumentation manufacturers to consider incorporating into commercial electrochemical instruments.

There is one commercially available electrochemical instrument that uses a computer, the Princeton Applied Research 374, but it is not based on a minicomputer. This instrument was made possible by a truly revolutionary development in electronics, the development of the single integrated circuit microprocessor. A microprocessor is generally defined as the implementation of all the CPU functions of a computer on a single integrated circuit chip. Typically the circuits are 0.6 in. by 1.5 in. to 2.0 in. and sell for between $15 and $60 in single unit quantities. This development will ultimately have far greater impact on electrochemistry than the advent of the laboratory minicomputer, as the hardware investment required to implement computer control is in the process of being cut by an order of magnitude or more. Microcomputer systems consisting of a CPU, 1 k byte of RAM, 2 k bytes of system ROM, 14 bidirectional I/O lines, teletype interface, and an interface to an audio cassette that can be used for program storage and loading can be purchased for under $250 and only 5 V power supply at 1.5 amps to operate. Production has begun on single-chip systems that contain the CPU, 1 k byte of ROM, 64 bytes of RAM, and 36 I/O lines in a 0.6 in. × 2.0 in. package, and there are reports of the imminent production of systems with far larger internal memory; for example, 2 k bytes of EPROM, 2 k bytes of RAM, and 36 I/O lines along with the CPU in one package. There are also prototype systems that contain DACs and ADCs along with CPU and memory on a single chip.

The first microprocessors were fairly slow (roughly equal to the throughput of the first commercial computers) but in the last five years, their speed has increased over tenfold and now several eight-bit versions are capable of servicing a 12-bit ADC, including testing status, storing the conversion, testing to see if sufficient points have been taken, etc., in under 10 μsec. During the initial revision of this manuscript, a preproduction announcement was made of a microprocessor chip designed to have an instruction execution time of from 20 to 50 nsec and which would be capable of performing the same task in under 1 μsec. It is not yet in production.

Microprocessors come with four-bit, eight-bit, 12-bit, and 16-bit data word sizes. There are microprocessors that execute the instruction set of the DEC PDP-8 (a 12-bit word), as well as the DEC PDP-11, Data General Nova series, and Texas Instruments 990 series (16-bit word) minicomputers. Most of the major development action has been in the eight-bit word size types but the 16-bit machines are now becoming available and some 32-bit machines are in the development stages. In fact, several of the 16-bit machines are actually 32-bit internally.

The first microprocessors were treated by their manufacturers as just another, although more complex, integrated circuit with little if any software support. This situation has now changed. There are FORTRAN or BASIC based cross assemblers for all of the commercial microprocessors; some have BASIC interpreters and even FORTRAN compilers and a number also have specialized mid- or high-level languages peculiar to the individual manufacturers. The chips that execute the instruction sets of the minicomputers can also use essentially all of the software developed for those computers.

A number of microcomputer systems are available with a BASIC interpreter resident in ROM so that the user only has to enter the program and go into execution. A fairly complete BASIC interpreter can usually be written in form 6 to 12 k bytes. In contrast, a practical compiler-based language requires the availability of magnetic tape or disk support facilities in order to function with reasonable ease. For program development one can sometimes use other computers; for example, at least one manufacturer has developed a FORTRAN program that runs on a large computer that will translate a user FORTRAN program into the microprocessor machine language.

The advent of inexpensive mass storage devices in the form of floppy disks has completely changed the picture for microcomputers. The floppy disk (FD) is a magnetic media in the form of a non-rigid disk, typically either 8 or $5\frac{1}{4}$ inches in diameter. The disk is actually homogeneous over its surface and either one or both surfaces will be covered with a magnetic material (these are referred to as either single or double sided disks). The surface is organized into a number of tracks, which are simply constant radius circles around the center with track 0 having the largest radius. Data are written by placing a head *onto* the disk surface and writing magnetic patterns of alternating polarity. There is absolutely no physical indication of the location of a track and a purely arbitrary set of conventions has been established to allow exchange of diskettes between drives produced by different manufacturers. Typically an 8 in. drive will have 77 tracks while a normal $5\frac{1}{4}$ in. drive will have 40 tracks.

Each track is typically subdivided into a number of sectors, which are merely arcs on the track. It is necessary to establish some reference for the beginning of the first sector (sector 1) on a track and this is done by punching a hole in the disk surface near the center of the disk in a region not used for recording data. The location of this hole is sensed optically. The number of sectors on a track is normally arbitrarily established. The state of the art in inscribing the magnetic fields sets a limit to the density of flux changes per radial inch. Early in the development of floppy disks it was decided that there should be the same number of bits on the innermost track as on the outermost track so that the bit density is established by the smallest radius track. Each sector on a track contains essentially five fields, a preliminary synchronization pattern used by the electronics to allow for variations in the true data rate caused by the radial distance of the track. A preamble typically contains the track and sector numbers; the data field; a postamble, which contains a cyclic redundancy check (CRC) pattern; and a pad, which fills out the sector. The CRC pattern is a code generated from all of the data bits, combining them bytewise in a particular manner. The CRC is planted in the write phase and compared to a similar code built in the read back; if the two do not match then an error has definitely occurred.

There are two ways to define the number of sectors on a track. In hard-sectored disks there is a second set of synchronization holes made in the disk surface, which signals the beginning of each sector. In the soft-sectored format there is no physical indication of the beginning or end of any sector except for sector 1, and in principle, a soft-sectored disk may have any number of sectors per track consistent with the maximum number of flux transitions allowed per track. In practice almost all formats in use today assume that the data field will be 128, 256, 512, or 1024 bytes long with the two most common being 128 and 256. For the 128-byte format each sector contains 58 nondata bytes so that only about two-thirds of the actual bit capacity of a track is available as data.

There are basically two common ways of encoding the flux transitions on the disk: FM and MFM. For our purposes the difference is that twice as much data can be stored on an MFM track as on a track encoded in FM. These two schemes are more commonly referred to as single- and double-density recording. In double-density (MFM) formatted systems the 256-byte record is the most common and in this case there are typically 104 nondata bytes per sector so that roughly 5/7 of the available bits on a track are used for data. The number of sectors/track on a disk depends on the diameter of the FD, the density, and sector size. For single density 8 in. drives there are usually 26 sectors of 128 bytes although 15 sectors of 256 bytes is also not uncommon. In double density mode one usually finds 26 sectors of 256 bytes used. For the $5\frac{1}{4}$ in. FD the single density format is usually 12 sectors as 128 bytes although some systems use 16 sectors.

Recently, several manufacturers have begun producing 96-track double-density $5\frac{1}{4}$ in. drives, selling for under $500. The low cost and relatively high storage capacity of the FD has meant that microcomputers could be economically configured with mass storage capabilities. As a direct result of this, it is now feasible to use languages other than BASIC on microcomputers. As of this date one can run one or more of the following languages on most types of microcomputers: BASIC (both interpreter types and true compiler forms), FORTRAN, PASCAL, FORTH, LISP, ALGOL, PILOT, FOCAL, COBOL, and C.

A second result of the availability of inexpensive floppy disk capabilities has been the development of machine independent operating systems for a number of different types of microprocessors. While there is a large number of different types of microprocessors available, we shall arbitrarily limit our discussion to four basic families. The first is the 8080 family, which includes the Z-80 and 8085 as subsets. There are approximately 30 companies manufacturing microcomputers based on these microprocessor chips, each computer having different system structures. An operating system known as CP/M (developed and marketed by Digital Research Corporation) has become essentially a *de facto* standard for this family of processors. It is relatively simple to modify the basement form of CP/M to run on any individual microcomputer (almost all vendors now provide a form of CP/M for their system). Once a microcomputer can run CP/M, then any language or program that runs under CP/M can run on that computer. All of the high-level languages and much more can run under CP/M. The second family is based on the 6800 series of microprocessors including the newer 6809. There are roughly seven companies building computers based on this family, and a machine independent operating system known as FLEX (developed by Technical Systems Consultants, Inc.) has essentially the same capabilities in terms of a "universal operating system" as CP/M, although the total range of available software is not quite as large as for CP/M. The third class of microprocessors is the 6502, which at the time of writing is the most widely produced eight-bit microprocessor chip. There are five companies building computers based on this part, but there is no standard operating system for this family as was the case for the other two. Even with this deficiency, there is still an impressive number of programming languages available for members of this family. The fourth class is a class of one, the Digital Equipment Corporation LSI/11 family of microcomputers. This family is essentially fully supported by DEC and a fairly wide variety of languages and operating systems is available. There are several other excellent microcomputers including, but not limited to, the Western Digital PASCAL Micro-Engine and ADA-Micro-Engine, Texas Instruments 900 Series, and the Hewlett Packard Series 85. These will not be discussed for reasons of brevity, even though several of them are more powerful than some of the systems mentioned above.

5. The Electrochemist and the Computer

The uses of computers in electrochemistry can normally be broken down into five categories, depending on the major use of the computer in the particular application. These are data acquisition, control, analysis, signal averaging, and closed-loop systems. In many applications the primary use of the computer is to provide the experimenter with data that has high accuracy, e.g. $\pm 0.05\%$, is accurately synchronized with respect to time or some other control variable, and that is either impossible or inconvenient to acquire manually. In control applications the computer is used to generate electrochemical wave forms or other control functions that would be either impossible or too difficult to generate otherwise. Signal averaging is a special case of the data acquisition function where one has a system that is reproducible; i.e., the response will be the same each time, but the signal-to-noise ratio is smaller than is desired and the noise is approximately random. Under these special circumstances, which can be approximated in a number of electrochemical techniques if sufficient guile is employed,[5,8] one can obtain a signal-to-noise improvement that is theoretically equal to the square root of the number of samples in the average. The term averaging is somewhat misleading because the data points are summed at identical portions of the response every pass. At the end of the accumulation, the data are divided by the number of accumulation passes. Closed-loop systems are those where the computer alters the experimental environment in response to the behavior of the experimental system.

This classification scheme is not perfect, with many applications cutting across boundaries, but it will serve as a point of departure for a discussion of how computers are used now in the laboratory and how they may be used in the near future.

The hardware aspects of a laboratory computer system only set upper performance limits; it is the programming of software that is usually the limiting feature. In commercial systems the usual role of thumb has been that the software cost for a project will be approximately 10 times the hardware cost, and that a *good* programmer can turn out about forty debugged, working statements a day.

When considering how to implement a particular experiment or technique with a laboratory computer, it is usually advisable to write as much of the program in a high-level language as possible. Essentially all minicomputer BASIC or FORTRAN languages have the capability of either calling assembly language subroutines or accepting assembly language segments imbedded in the higher-level language program. The microprocessor FORTRAN mentioned above has this capability, and most of the microprocessor BASIC languages are structured so that either this approach is possible or so that one can actually reach into the interpreter and define new BASIC commands such as $DAC(I) = A$, where the result of the execution of this statement is that the contents of A are transferred to the DAC specified by the contents of I. It should be noted that modification of a BASIC interpreter is not a task to be undertaken lightly.

To clarify the question of when assembly language should be used and when to use a higher-level language, we will consider several examples.

5.1. Ion Selective Electrodes

The ion selective electrode (ISE) is an extremely convenient sensor for measuring solution activities, but the nonlinearity of the response poses some problems that can be solved by application of computer technology. A computerized argentometric titration of seawater utilizing a real-time Gran extrapolation to the endpoint has been reported by Jagner and Aren.[33] Brand and Rechnitz[34] have described a multiple standard addition method, which uses a nonlinear curve-fitting algorithm that makes no assumptions about the slope of the voltage–log (ion activity) relationship. A very similar computer approach has been applied to potentiometric titrations with an ISE by Isbell et al.[35] and Neff et al.[36] have described an interrupt driven clinical electrode system, and Fleet and Ho[37] have reported on an off-line computerized Gran's plot calculation approach.

Zipper, Fleet, and Perone[38] have reported on a computer-controlled monitoring and data reduction system for multiple ISEs in a flowing system. The instrument is based on an 8 k Hewlett–Packard 2115A minicomputer with a 16-bit DAC and 11-bit ADC. Connection from individual ISEs is through a multiplexer constructed of mercury-wetted relays (normal solid-state multiplexers were not adequate for this particular task because of the high impedance of the ISE) to a very high input impedance varactor bridge operational amplifier. This signal is fed through a second amplifier, which can be offset by means of the 16-bit DAC and then to the ADC. System programming is done in BASIC with assembly language routines to control the peripherals (ADC, DAC, MPX, etc.). The DAC is used to select the ±100 mV electrode potential window, which is converted. The data obtained are signal averaged 10,000 times, providing an S:N improvement of 100:1 and then treated by means of a weighted iterative least-squares algorithm. The program is written so that data can be obtained under operator intervention or under computer control.

It is easy to envisage microprocessor based instruments based on either the Zipper or Brand and Rechnitz[34] procedures. We will examine the Brand and Rechnitz approach since it is a somewhat simpler system.

The Brand and Rechnitz method for extracting concentration information from ion selective electrodes uses a nonlinear curve-fitting algorithm and makes no assumptions about the slope of the voltage–log activity relationship. The procedure involves reading the potential of the unknown solution prior to the addition of standard, manual reading of potentials, and use of a sophisticated programmable calculator that executes a subset of BASIC. A computer-based implementation of this process may be of interest in a number of industrial and medical environments. Ideally, the standard addition should be done by computer control, perhaps by means of a stepping motor attached to a micrometer burrett, with the computer reading the resulting potential either with an ADC or by means of a digital pH meter having digital output that can be read into the computer in BCD form. The generation of the pulses to the stepper motor could be accomplished by an assembly language subroutine that generates the pulses on a digital output line. The inputting of the cell potential could also be handled by an assembly language routine that either services an ADC or reads the BCD coded results from the pH meter. The remainder of the functions including the setup and calling of these assembly language routines would best be handled in a high-level language. Since the times required to ensure stabilization of the electrode potentials and to change samples is fairly long, usually tens of seconds or more, the speed of execution of the analysis code is relatively unimportant and a BASIC interpreter should be perfectly adequate. This function could be implemented with today's micro-processor components using approximately 16 integrated circuits if a digital pH meter is used as the cell monitoring device. These would include one CPU chip; eight PROM or ROM chips (2 k $\times$ 8 bits each); two peripheral interface adapter chips (16 bidirectional digital lines each), one for human I/O through a keyboard and printer and the other for pulse generation and interrogation of the pH meter; two RAMs (1 k $\times$ 4 bits each); and three decoder and buffer chips. This version assumes that the program is written in BASIC and that 12 k byte basic interpreter and 4 k byte basic program is stored in PROM or ROM. In a commercial realization of this fuction the whole operation *might* be written in assembly language, in which case far less ROM would be required. But with the rapidly falling price of ROM and PROM and the increasing price of trained programmers, the cost effectiveness break point is beginning to shift toward the BASIC approach.

If one were to start from scratch on the pH meter, the interface of Zipper *et al.*[38] has a lot of merit. A slight improvement would be to substitute a highly linear VFC in place of the ADC, preferably one with a 100 KHz or better upper limit and to count over one or more multiples of the basic line frequency. Obviously one would need the multiplexer unless one were going to look at multiple electrodes. With the expanded precision of a high-frequency VFC one could also use a cheaper 12-bit DAC for the offsetting control.

5.2. Pulse Polarography[39–41]

The field of pulse polarography provides a second example. In its simplest form, a computer-controlled pulse polarograph consists of a computer, two 12-bit DACs, a 12-bit ADC, a programmable gain amplifier between the cell transducer amplifier and the ADC, and a digital output line controlling a drop detachment device, a potentiostat, some form of human I/O, and an X–Y recorder. The operation aspect of the experiment is the application of a base potential to the cell with one DAC with a dynamic range of ±2.0 V (resolution 1/2 mV at the cell), detachment of the old drop by application of a pulse on the digital line, which is power boosted to drive the actual detachment solenoids, an accurately controlled time delay, application of the potential pulse with the second DAC and a second accurate time delay (nominally 50 msec for most commercial instruments), and acquisition of the cell current information. There are several features that require amplification at this point. The normal pulse height at the cell is usually in the range of from 5 to 100 mV and the use of a second DAC may seem unwarranted. One could generate the pulse with the same DAC that controls the base potential, but an examination of the resolution of the DAC (1/2 mV) indicates that errors in relative pulse amplitude of 1 mV from pulse to pulse could result. This would be catastrophic at the lower pulse amplitudes and unacceptable even at the 100 mV level. One could use a DAC with lower resolution for the pulse generation if a precision of 0.025% in the pulse height is not required; however, the use of a second 12-bit DAC has a secondary advantage. It is desirable to use an X–Y recorder as an output device for the operator. By multiplexing the pulse DAC between the potentiostat and the Y-axis of the recorder and controlling the servo mute of the recorder with a digital line to turn off the Y-axis servo when the pulse is applied, one can both run the recorder and the potentiostat with the same DACs. The X-axis would be connected to the base potential DAC, which is generating the potential staircase. After the experiment the cell could be switched to open circuit by means of digitally controlled relay, thus freeing the two DACs for more complex graphics. The software controls should allow implementation of sampled d.c., normal pulse differential pulse, and the alternate drop forms developed by Osteryoung and coworkers.[40] The critical portions of the software are the timings between the detachment of the earlier drop and initiation of the pulse and the width of that pulse. Variation in either time will produce variations on the output, which will appear as noise (and, in fact, is timing noise) but of the two the most critical is the pulse width. The drop areas grow as roughly $t^{2/3}$ and the current is proportional to the electrodes area; a ±1% variation in drop fall to pulse initiation would result in roughly ±0.7% error or noise component on the signal, ignoring all other sources. For a perfectly reversible process the current decays at $t^{-1/2}$ during the pulse and a 1% error in timing will result in a ±0.5% error.

This formulation of the constraints hides the essential point that timing errors are in fixed time units, not percentages. A 1% drop time error on a typical 1 s drop is 10 msec. The same 10 msec ambiguity during a pulse of nominal duration of 50 msec is a 20% error. The optimum approach for programming this type of system would be to perform all operations during the polarogram in assembly code to minimize the timing errors. Operator input specifying the initial and final potentials for the potential staircase, drop time, and tread height of the staircase or, analogously, drop time scan rate and pulse height can be carried out in a high-level language. The same is true of the mathematical operations required to translate the voltages into the binary values to be transmitted to the DACs and the increment to be added to the DACs for each new potential in the staircase. After the initial setup, one would enter an assembly language subroutine, which will perform the actual experiment as well as controlling the operator's X–Y recorder. After acquisition, one would then perform any other signal processing desired, including correlation of peak potential with material and calculation of concentration from working curves, etc. It is perfectly appropriate and straightforward to carry out simple signal processing during the acquisition phase as long as one is sure that timing problems will not arise. The obvious time for these operations is during the relatively long delay between detachment of the old drop and the application of the new pulse. This delay is rarely less than 0.450 s, which would allow between 100,000 and 1,000,000 useful instructions with typical minicomputers and microprocessors operating on code generated by an assembler. (Remember that an interpreter spends the great majority of its time decoding the instructions to be performed and not actually carrying out the instruction itself.) Operations that could be profitably carried out in this time include subtraction of backgrounds, calculation of approximate derivatives to see if the last ADC value is consistent with preceding readings or may represent an error due to a bad drop, etc. The direct control of the processor will also make possible recapture of a bad point in the polarographic case (note that this is not possible when the system is used in a stripping application, but then there is no such thing as a "bad drop" in stripping either) or the use of signal averaging. Normally, most noise encountered is derived from the line frequency and is thus either 50 or 60 Hz or higher harmonics of these frequencies. The obvious way to eliminate these unwanted components is to use a digitizing algorithm that rejects them. One approach is to use a VFC type ADC that digitizes the data over the final 16.667 or 20.00 msec of the pulse. This has the disadvantage that one must design and construct the interface, but it will give excellent results. A second approach, which is almost as satisfactory, is to use a successive approximation ADC with S&H and to accumulate as many conversions during the 1/60 or 1/50 of a second period as possible and retain the average value of these results. Here one can use a standard ADC-S&H interface rather than a specialized home-built one, although the software required to run the system is now much more complex. The correct way to approach the problem is to determine initially how many conversions are possible; i.e., for a 20 μsec per point system with 60 Hz, one can accumulate $16,667/20$ or 833 points. This is not a perfect emulation of the VFC, since the ideal number of conversions would be 833.3333 points, which is impossible since one cannot carry out a fractional conversion. One should also place a low-pass filter between the current transducer and the programmable gain amplifier to filter out high-frequency spikes. The cut-off frequency for this filter is obtainable by observing that for analytical applications we are only interested in the final 1/50 or 1/60 of a second of the transient and we do not care if the earlier data are distorted. Further, what we really want is an accurate integral of the unfiltered signal over this interval. The use of fast Fourier transformation programs and a 6 db per octave apodization function indicates that a 3 KHz cut-off frequency will give good results. The programmable gain amplifier can be used to keep the signal to be digitized within the range of from full scale to half full scale to maintain maximum resolution. The CPU should examine the previous readings to see if the signal is increasing, essentially stable, or decreasing, and adjust the gain so that the new data will fall within the desired range. Alternately, several conversions can be made just prior to the actual accumulation period to check to see what gain is required, but if this approach is taken, one must make sure that the settling time of the programmable gain amplifier is taken into account in determining the timing for these samples. If this sytem and software are set up to work correctly for the pulse, it will also work at least as well for the prepulse sample required for differential pulse polarography. It may be useful to note here that there is a second way to correct for the current flowing due to the prepulse potential that is a viable alternative to the subtraction usually practiced. This is to subtract the background current at the working electrode[42] rather than after acquisition by the computer. In the system discussed here, this would best be done by means of a DAC and a set of computer-selectable resistors. The computer either monitors the ADC prior to pulse initiation and adjusts the DAC and resistor combination for zero net current, or in the case of the alternate drop approach, employs the valus determined in the previous drop where the pulse was not applied. This method has the advantage that small signals riding on top of large backgrounds can be examined accurately since the transducer amplifier is not saturated.

5.3. Anodic Stripping Voltammetry

The implementation of a computer-based anodic stripping system requires essentially the same hardware as a pulse polarographic instrument; only the software control is different. Some additional digital control lines are required to support the stripping apparatus. For example, if the rotating cell design of Clem[43] is used as commercially implemented by McKee-Pederson, then three lines are required to control the motors and gas solenoid. If a mercury drop is used, no lines are required for human extrusion of the drop. Alternately one line can control a stepping motor for automatic extrusion. Two or more lines would be required to control stirring motors and nitrogen flow control solenoids. The timings in the prestrip operations are not as critical as in the stripping operations; for example, 10 msec ambiguities in the degas, plate and rest, or equilibration times will be trivial so that a high-level language with assembler subroutines for timing and control of the digital lines will be preferable as well as adequate. In stripping voltammetry, one has an additional degree of freedom in dealing with closely spaced redox processes which is not available in the polarographic approach. This problem has been extensively studied by Perone and coworkers,[16,17,44–48] who have advocated the strip and hold approach. Basically what one is doing is to hold the potential fixed for a period of time after the major peak for one species, and to allow the current to decay essentially to baseline before resuming the stripping process for the remaining species. If the peaks are widely separated enough, one can effectively exhaustively deplate the first species before proceeding on to the next; even when the overlap is great, one can achieve some improvement. There are two basic ways to approach the problem: the operator can inform the CPU as to what potentials should be considered rest points (and perhaps even how long to rest), or CPU can determine when the hold should occur on the basis of the current–potential profile being obtained. The latter case is by far the most complex and requires fairly sophisticated and complex programming that involves the determination of local minimas without detecting irregularities due to noise, mechanical vibrations, and background nonlinearities.

An alternate approach, which may eventually be successful, is the deconvolution of the stripping results with both the diffusion process and possibly the charge transfer process shape functions.[50–56] The response for any Faradaic electrochemical process is the result of the natural analog convolution of the diffusion process with the charge-transfer process and the concentration and $E°$ of that process. It is theoretically possible to remove the unwanted components from the response signal, in this case the diffusional information that gives the tailed shape to the stripping signal. This would leave a signal that would appear very much like the differential pulse polarogram of the material, an improvement in terms of removal of overlap, but still nonoptimal. The resulting shape still contains unwanted information about the nature of the charge-transfer process, namely the rate constant, charge-transfer coefficient, and the number of electrons. It is theoretically possible to perform a second deconvolution to remove this information as well. The results of this process would be a series of impulses along the potential axis. The location of the impulse would still contain the identity information and the height of the impulse would contain the concentration information. The implementation of such a system is still in the future, but the hardware for the implementation is available now. One might think that the deconvolution operations would require at least a large minicomputer with a fast Fourier transform processor or a large-scale computer working off-line in a batch mode, but this is no longer true. Recent advances in integrated circuit technology have led to the development of a hardware multiplier chip that is about 3 in. × 0.5 in. × 0.5 in., and can perform a 16×16-bit multiplication in 350 nsec. The chip sells for about \$180 in single units and should allow the implementation of a 2048 word Fourier transform capability for microprocessors for a hardware cost of under $600, which could do the deconvolution in under 1 s. An article discussing the possibility of an essentially single chip Fourier transform capability has already appeared,[57] and it would be rather surprising if such a device were not available at least in prototype form prior to the actual publication of this volume. The real holdup in implementing this type of capability is in the development of the software for such an operation, for finding ways to treat noise (small noise contributions become amplified in deconvolution operations), and finding a way to handle correctly the charge-transfer kinetics shape function problem in multiple species cases where these parameters are not known ahead of time. One must also treat the problem of non-Faradaic shape function contributions, such as absorption and chemical kinetic (complexation) contributions.

5.4. Alternating Current Polarography

Few electroanalytical techniques have benefited as much by the advent of the laboratory minicomputer as alternating current polarography (ACP). The technique has been used for two, heretofore distinctly different purposes. One application is the determination of electrochemical charge-transfer rate constants and/or solution chemical rate constants and mechanisms. The second application has been in the determination of solution species concentrations at low levels. Basically the technique consists of superimposing a small amplitude (typically 1–25 mV) sinusoidal potential excitation on top of a slow potential ramp. The sinusoidal excitation elicits a current response from the cell, which is composed of two components, a double layer response which, in the absence of solution resistance, is 90° out of phase with the excitation but essentially spectrally pure; that is, the response is at the exciting frequency with no significant harmonics. The Faradaic response has its major component at the exciting frequency and also produces higher harmonic components due to the inherent nonlinearity of the Faradaic admittance. The phase angle of the Faradaic response is sensitive to charge transfer and chemical kinetics, being 45° for electrochemically reversible charge-transfer reactions where both members of the couple are present in a stable form with no kinetic links. The peak current of an a-c polarogram is directly proportional to the square root of the exciting frequency and linearly proportional to the concentration and excitation amplitude under these conditions. Under these conditions one can also discriminate against the double-layer charging current by monitoring the current, which is precisely in phase with the exciting potential and frequency coherent with it. When charge-transfer kinetics or chemical kinetics are present the phase angle changes with frequency in a manner that is diagnostic for the mechanism. This is in contrast to the simple case where the phase angle is frequency independent, and the peak current is no longer linear with the square root of the exciting frequency.[58,59] The major interest in computers has been in obtaining the high-quality experimental data needed to fit the theoretical predictions for charge-transfer kinetics and chemical reaction mechanisms, followed by subsequent data reduction, normally on-line to yield the rate and/or equilibrium constants of interest. The original publications utilized an a-c polarograph that required manual control of the frequency and tuning of filters, has a built-in lock-in amplifier with the computer synchronizing drop fall, and was essentially a passive data collection and analysis device.[9,60] Even with a computer the experiments were fairly laborious. The solution to this problem lay in the application of Fourier analysis to ACP.

There are several ways that the practical application of Fourier transforms via the computerized fast Fourier transform (FFT) has appeared in the electrochemical literature. One approach has been to use a sinusoidal oscillator and utilize the FFT to determine the direct-current, first and second harmonic components of the cell response. This has been termed "a-c Polarography in the Harmonic Multiplex Mode."[13] Even this approach is still operator intensive and relatively slow. One elegant solution arose from realizing that by subjecting the cell to an excitation that is a composite of a number of different frequencies and then taking the Fourier transform of the response, one can determine the response of the cell to each frequency component in the excitation signal. This can greatly speed up the data collection procedure and eliminate the manual intervention. Initial investigations centered on voltage impulses,[61–63] which consist of all frequency components and the analyses were done off-line. Random white noise and pseudorandom (computer-generated) white noise signals were used with on-line data reduction by Creason and Smith[65] with data for an individual d-c potential being available within 2 s. In this study it proved necessary to apply signal averaging to minimize the effects of "extraneous" (nonrandom) noise and to minimize "side lobes" due to sampling window errors. These signals are both "constructed" of signals equispaced in frequency space, while the optimal spacing from a theoretical standpoint would be a square root of frequency distribution. A second problem arises from the nonlinearity of the Faradaic impedance, which generates relatively small amplitude responses at multiples of the exciting frequency. Since essentially all possible exciting frequencies are employed, all response frequencies are "contaminated." The solution to this was to tailor an excitation signal so that the data were essentially equidistributed through the square root of frequency space and chosen such that the higher harmonic contributions from one exciting frequency did not fall near another exciting frequency.[15]

This approach has been successfully implemented, and Smith[65] estimates that starting from scratch one can set up a minicomputer system to do the job for $15–20K. The approach allows one to determine the frequency response of the Faradaic process at a large number of d-c potentials through the wave in little more time than is required to accomplish a normal single-frequency ACP experiment, thus vastly increasing the throughput of systems as well as greatly increasing the dynamic limits of the reaction rates (both homogeneous and heterogeneous) that can be successfully studied.

The original system utilized two DACs, one for d-c ramp emulation and the other to construct the excitation signal. This signal is low-pass filtered to remove unwanted higher harmonic contributions. The response of both the working electrode and the reference electrodes is monitored and transformed by the FFT approach. The monitoring of the excitation signal at the reference electrode allows one to eliminate effectively the effects of imperfect cell geometry and potentiostat design.[56] Because of speed limitations, two ADCs are used to increase the throughput.

A recent modification of the system has been described that places the dynamic signal generation DAC and the two ADCs in a separate chassis. The 10-bit DAC has its own 1024-word memory, as do each of the two 10-bit ADCs. The data from each ADC can be added to the current value of the 16-bit word representing past samples, which allows for 256 replicates for time plane signal averaging. The computer's intervention is limited to loading the excitation signal into the DAC buffer, providing the d-c potential to the cell and synchronization pulses, thus freeing the computer for the FFT operation. The data rate of the system is 500 KHz.[66]

The ability of systems of this sort to do highly complex analyses of the cell admittance has opened up new possibilities for electroanalytical applications. One of the problems with analysis of "real" systems is that the media of interest is by definition relatively uncontrolled. A number of otherwise reasonable systems, such as cadmium, exhibit strong charge-transfer rate constant dependences on the concentration of surfactants not inherent to the systems. Since the peak current strongly depends on rate constant, the peak current for a real sample depends not only on the concentration of the material of interest but also on the presence or absence of background materials. This can lead to either the possibility of error or the necessity for fairly complex sample cleanup. The FFT approach described here can be used to remove essentially most of the problems of this sort.[56,67]

One can, in principle at least, analyze a material in almost any matrix, subject the FFT derived results to a comparison of the amplitude and phase angle vs. frequency behavior of the raw data to the predicted response of model systems find the correct model, and then determine what the true concentration is. The implications of this are very far reaching.

In its present form the technique is still relatively expensive and some components are not commercially available,[66] although the final modification is not absolutely necessary for the type of experiment just described. The expense of the approach would indicate that commercial instrumentation will not appear in the immediate future unless some modifications can be made. One strong possibility is that the functions of the control box could be performed by a microcomputer system operating at a 2 MHz clock rate with DMA control of the ADCs and DACs and using hardware signal averaging.[68]

It is also possible for the FFT to be performed under the control of the microcomputer due to the recent availability of some very fast hardware multiplier chips. These devices, some of which have built-in accumulators capable of addition and subtractions, can multiply two 16-bit integers in 350 nsec[57] so that a very fast FFT peripheral for a microcomputer is conceptually possible; since the costs are under $300 and dropping they are economically feasible. We are, however, talking about a very powerful microcomputer system in contrast to our previous examples. With improvements in integrated circuit technology it seems likely that such a unit could be produced for a parts cost under $3K in the very near future (this does not mean, however, that a commercial unit will be available in the same timeframe). In addition, the problem of software development is far more critical in this case than in the previous examples, but the impact on practical electroanalytical chemistry would be hard to overstate. If one were to tackle the problem from scratch, one would obviously have to write the data acquisition and control program segments in assembly language. If a DMA device or an outboard peripheral such as that described by Smith[66] were employed, this software would be fairly minimal. The FFT algorithm would have to be in assembler because of speed requirements, unless a hardware FFT peripheral was available. Even then the setup and control of the peripheral would of necessity be an assembler operation, although it might well be accessed out of a higher-level language. Once the data are acquired and transformed, the problem of determining the mechanism and rate constants could be handled either in assembler or a higher-level language. Unless the postrun processing was very time critical it would be best to perform this data manipulation in a high-level language. For a research laboratory environment the high-level language approach is distinctly preferable because of the volatility of the problems of interest and the relative ease in reprogramming in a higher-level language relative to assembler.

5.5. Conductance Measurement

Other forms of electroanalytical techniques can be implemented by computer, including conductance measurements. Enke and coworkers have shown that the bi-polar pulse technique can lead to a very rapid, accurate determination of solution resistance.[70] Essentially what is done is to apply a short potential pulse to the electrodes, measure the current at the end of the pulse, and then reverse the potential of the pulse for the same amount of time and measure the current at the end of this pulse. The difference in the currents is directly proportional to the solution resistance. With careful cell geometry and attention to electrode design, pulses as short as 20 μsec can be employed. When such a system is implemented by computer, using DACs to generate the pulses and an ADC-S&H and programmable gain amplifier to monitor cell current, the resulting system can follow very rapid changes in cell conductivity, including monitoring chemical kinetic reactions with rate constants on the millisecond time scale.[71] For the highest data rate situations, it is necessary to perform the entire data acquisition operation in assembly language as the inefficiencies of even a compiler-based program will result in code that is too slow in execution to be useful. The setup and analysis portions would not pose any particular timing constraints so that even an interpreter-based language could be used. The possible applications of such an instrument range from following reaction kinetics such as micelle formation and decomposition rates in salt solutions, and rate of complex formation to the monitoring of industrial process streams and effluent. A microprocessor-based instrument has recently been described.[21]

6. The Future

There is an aspect to the use of computers in the laboratory that perhaps does not fit neatly into this chapter, but that is important enough that it should be included. There are two major problem areas in the application of computers to chemical instrumentation: the problem of interfacing devices such as pH meters, voltmeters, counters, computing integrators, etc., to computers, and the problem of linking one computer to another to form hierarchical systems. This latter problem is of particular importance in industrial and governmental laboratories where a number of computer-controlled instruments exist, and it is necessary to transfer records from local laboratory computers to a central system. One solution to both problems lies in a standardized interfacing philosophy first implemented by Hewlett Packard as a protocol for attaching engineering instruments such as counters, voltmeters, programmable power supplies, etc., to their lines of programmable calculators. The protocol involves the use of three sets of bidirectional lines, eight data lines, address lines, and control lines to communicate between peripheral and calculator. The physical implementations of this approach will allow transmission of data at about 500,000 bytes per second over distances of up to 20 m. The buss structure and timing with some minor modifications have been adopted by several international standards agencies including IEEE as RS-488, and interfaces between the buss and a number of different minicomputers/microcomputers are now available. A number of international scientific and engineering instrument manufacturers are producing instruments compatible with the buss and at least one company is manufacturing a general purpose interface that should allow almost any device to be interfaced to the buss. Several manufacturers have developed a software system that allows the communication of high-level language problems directly with the buss.

It seems quite likely that most microprocessor-based instruments developed in the future will possess the ability to communicate over this buss structure to a centrally located minicomputer.

References

1. S. J. Lindenbaum, *Ann. Rev. Nucl. Sci.* **16**, 619 (1966).
2. J. A. Jones, *Trans. Nucl. Sci.* **14**, 576 (1967).
3. B. J. Allen and J. R. Bird, *At. Energy (Aust.)* **9** (1966).
4. G. Lauer, R. Abel, and F. C. Anson, *Anal. Chem.* **39**, 765 (1967).
5. S. C. Creason, R. J. Loyd, and D. E. Smith, *Anal. Chem.* **44**, 1159 (1972).
6. H. W. VandenBorn Whitson and D. H. Evans, *Anal. Chem.* 1298 (1973).
7. S. P. Perone, J. W. Frazer, and A. Kray, *Anal. Chem.* **43**, 1485 (1971).
8. D. E. Smith, In: *Computers in Chemistry and Instrumentation*, J. S. Mattson, H. D. MacDonald, Jr., and H. B. Mark, Jr., Eds., Vol. 2, Marcel Dekker Inc., New York (1972).
9. D. E. Glover and D. E. Smith, *Anal. Chem.* **44**, 1140 (1972).
10. B. J. Huebert and D. E. Smith, *Anal. Chem.* **44**, 1179 (1972).
11. B. J. Huebert and D. E. Smith, *J. Electroanal. Chem.* **31**, 333 (1971).
12. H. Kojima and S. Fujiwara, *Bull. Chem. Soc. Jpn.* **44**, 2158 (1971).
13. D. E. Glover and D. E. Smith, *Anal. Chem.* **45**, 1869 (1973).
14. D. E. Smith, *Anal. Chem.* **47**, 363A (1975).
15. S. C. Creason and D. E. Smith, *J. Electroanal. Chem.* **40**, 1 (1972).
16. S. P. Perone, D. O. Jones, and W. F. Gutknecht, *Anal. Chem.* **41**, 1154 (1969).
17. S. P. Perone, J. E. Harrar, F. B. Stephens, and R. E. Anderson, *Anal. Chem.* **40**, 899 (1968).
18. R. G. Clem, G. Litton, and L. D. Ornelas, *Anal. Chem.* **45**, 1306 (1973).
19. U. Eisner, J. A. Turner, and R. A. Osteryoung, *Anal. Chem.* **48**, 1609 (1976).
20. E. S. Pilkington, C. Meeks, and A. M. Bond, *Anal. Chem.* **48**, 1665 (1976).
21. K. Casserta, Symposium on Instrumental Application of Microprocessors, IEEE-ISA Electronics and Instrumentation Conference, Cincinnati, Ohio, 1977.
22. B. Souceck, *Minicomputers in Data Processing and Simulation*, Wiley-Interscience, New York (1972).
23. C. L. Wilkins, S. P. Perone, C. E. Klopfenstein, R. C. Williams, and D. E. Jones, *Digital Electronics and Laboratory Computer Experiments*, Plenum Press, New York (1975).
24. S. P. Perone and D. O. Jones, *Digital Computers in Scientific Instrumentation*, McGraw-Hill, New York (1973).
25. G. A. Korn, *Minicomputers for Engineers and Scientists*, McGraw-Hill, New York (1973).
26. *The Semiconductor Memory Data Book*, Texas Instruments, Houston, Texas (1975). *Memory Design Handbook*, R. Greene, Intel. Corp., Santa Clara, California (1977). *Semiconductor Memories*, J. Eimbinder, Ed., Wiley-Interscience, *Semiconductor Memories*, D. A. Hodges, Ed., IEEE Press, New York (1972). *Semiconductor Memory Design and Applications*, R. C. Sawyer, Ed., McGraw-Hill, New York (1973).
27. H. Simpson, *Digital Design*, October, p. 28 (1977).
28. C. G. Enke, *Anal. Chem.* **43**, 69A (1971).
29. D. B. Buck, *Data Conversion Handbook*, Hybrid Systems Corp., Burlington, Massachusetts (1974).
30. *Analog–Digital Conversion Notes*, D. H. Sheingold, Ed., Analog Devices, Norwood, Massachusetts (1977).
31. B. Souceck, *Microprocessors and Microcomputers*, Wiley–Interscience, New York (1976).
32. *An Introduction of Microcomputers*, Vol. I and Vol. II, A. Osborne, S. Jacobson, and J. Kane, Adam Osborne and Associates, Berkeley, California (1976).
33. D. Jagner and K. Arens, *Anal. Chim. Acta* **52**, 490 (1970).
34. M. J. D. Brand and G. A. Rechnitz, *Anal. Chem.* **42**, 1172 (1970).
35. A. F. Isbell Jr., R. L. Pecsok, R. H. Davies, and J. H. Purnell, *Anal. Chem.* **45**, 2263 (1973).
36. G. W. Neff, W. A. Radke, C. J. Sambucetti, and G. M. Widdowson, *Clin. Chem.* **16**, 566 (1973).
37. B. Fleet and Y. W. Ho, *Talanta* **20**, 793 (1973).
38. J. J. Zipper, B. Fleet, and S. P. Perone, *Anal. Chem.* **46**, 2111 (1974).

39. H. E. Keller and R. A. Osteryoung, *Anal. Chem.* **43**, 342 (1971).
40. J. H. Christie, L. L. Jackson, and R. A. Osteryoung, *Anal. Chem.* **48**, 242 (1976).
41. S. C. Rifkin and D. H. Evans, *Anal. Chem.* **48**, 2174 (1976).
42. W. S. Woodward, T. H. Ridgway, and C. N. Reilley, *Anal. Chem.* **46**, 1151 (1974).
43. R. G. Clem, F. Jakob, D. H. Anderberg, and L. D. Ornelas, *Anal. Chem.* **43**, 1398 (1971).
44. W. F. Gutknecht and S. P. Perone, *Anal. Chem.* **42**, 906 (1970).
45. L. B. Sybrandt and S. P. Perone, *Anal. Chem.* **44**, 2331 (1972).
46. M. A. Pichler and S. P. Perone, *Anal. Chem.* **46**, 1790 (1974).
47. Q. V. Thomas and S. P. Perone, *Anal. Chem.* **49**, 1369 (1977).
48. Q. V. Thomas, R. A. Depalma, and S. P. Perone, *Anal. Chem.* **49**, 1376 (1977).
49. Q. V. Thomas, L. Kryger, and S. P. Perone, *Anal. Chem.* **48**, 761 (1976).
50. S. C. Rifkin and D. H. Evans, *Anal. Chem.* **48**, 2174 (1976).
51. H. L. Surprenant, T. H. Ridgway, and C. N. Reilley, *J. Electroanal. Chem.* **75**, 125 (1977).
52. J. C. Imbeaux and J. M. Saveant, *J. Electroanal. Chem.* **44**, 169 (1973).
53. F. Ammer and J. M. Saveant, *J. Electroanal. Chem.* **47**, 215 (1973).
54. L. Nafjo, J. M. Saveant, and D. Tressler, *J. Electroanal. Chem.* **52**, 403 (1974).
55. C. N. Reilley, U.S.-Japan Seminar on Computer Assisted Chemical Research Design, Honolulu, Hawaii, July 2–6, 1973.
56. M. Goto and I. Ishii, *J. Electroanal. Chem.* **61**, 361 (1975).
57. D. E. Smith, *Anal. Chem.* **48**, 526A (1976).
58. D. J. Geist, *Electronics*, July 3, p. 113 (1977).
59. D. E. Smith, In: *Electroanalytical Chemistry*, A. J. Bard, Ed., Vol. I, pp. 1–155, Dekker, New York (1966).
60. M. Sluyters-Rehbach and J. H. Sluyters, In: *Electroanalytical Chemistry*, A. J. Bard, Ed., Vol. 4, pp. 1–128, Dekker, New York (1970).
61. J. M. Lawrence and D. Mohilner, *J. Electrochem. Soc.* **118**, 259 (1971).
62. A. A. Pilla, *J. Electrochem. Soc.* **117**, 467 (1970).
63. K. Dobler and A. A. Pilla, *J. Electroanal. Chem.* **39**, 91 (1971).
64. R. L. Birke, *Anal. Chem.* **43**, 1253 (1971).
65. S. C. Creason and D. E. Smith, *J. Electroanal. Chem.* **36**, App. 1 (1972).
66. D. E. Smith, *Anal. Chem.* **48**, 221A (1976).
67. R. J. Schwall, A. M. Bond, R. J. Loyd, J. G. Larsen, and D. E. Smith, *Anal. Chem.* **49**, 1797 (1977).
68. R. J. Schwall, A. M. Bond, and D. E. Smith, *Anal. Chem.* **49**, 1805 (1977).
69. W. S. Woodward, T. H. Ridgway, and C. N. Reilley, *Analyst* **99**, 838 (1974).
70. D. E. Johnson and C. G. Enke, *Anal. Chem.* **42**, 329 (1970).
71. J. K. Caserta, F. J. Holler, S. R. Crouch, and C. G. Enke, *Anal. Chem.* **50**, 1534 (1978).

3

Electrochemistry of Ion-Selective Electrodes

RICHARD P. BUCK

1. Definitions and Principles[1-3]

1.1. Membrane-Format Devices

Ion-Selective Electrodes (ISEs) for measurement of activities of species in aqueous and mixed solvents, or partial pressures of dissolved gases in water, are mainly membrane-based devices, and involve permselective, ion-conducting materials. The electrodes are generally used in the potentiometric (zero-current) mode, and they superficially resemble classical redox electrodes of Types 0 (inert), 1 (Ag/Ag^+), 2 ($Ag/AgCl/Cl^-$), and 3 ($Pb/PbC_2O_4/CaC_2O_4/Ca^{2+}$). The latter, while ion selective, depends on a redox couple (electron exchange), rather than solely on ion exchange as the principal origin of interfacial potential differences (pd's). Ion selective electrodes have the typical shorthand form

$$
\text{Cu} \mid \text{Ag} \mid \text{AgCl} \mid \text{Electrolyte} \mid \overset{\text{I.Ex.}}{\underset{\text{selective to } M^+}{\text{membrane perm-}}} \mid \overset{\text{I.Ex.}}{\text{test sol'n}} \quad \text{(I)}
$$

lead wire   inner filling sol'n

inner reference electrode

RICHARD P. BUCK • Department of Chemistry, University of North Carolina, Chapel Hill, North Carolina 27514.

or

I.Ex.

$$\text{Cu} \mid \text{Ag} \mid \text{membrane permselective to } \text{M}^+ \mid \text{test sol'n} \qquad \text{(II)}$$

where I.Ex. means ion exchange active.

The first is a "membrane" configuration electrode, and the second is the "solid-contact" or "all-solid-state" configuration electrode. In the former, both membrane interfaces are ion exchange-active and the potential response depends on M^+ activities in both the test solution and in the inner filling solution. In the latter, the membrane must possess sufficient electronic conductivity (or other exchange process) to provide a reversible, stable, potential difference at the inner interface, with ion exchange only at the test solution side. An exploded view of these configurations is shown in Figure 1.

1.2. Segmented Potential Model for Membrane Pd's

Calculations of the static membrane potential difference require evaluating the quantity†

$$\Delta \phi_m = \phi_{b.r} - \phi_{b.1.} \qquad (1)$$

where the ϕ are the inner potentials of the electroneutral bathing solutions. The quantity $\Delta \phi_m$ can be calculated in segments in a large number of ways. Two are most important: (I) for membranes with a region of interior electroneutrality and (II) for membranes without electroneutrality

$$\text{(I)} \quad \Delta \phi_m = (\phi_{b.r.} - \bar{\phi}_{m.r.}) + (\bar{\phi}_{m.r.} - \bar{\phi}_{m.l.}) + (\bar{\phi}_{m.l.} - \phi_{b.l.}) \qquad (2)$$

$$\text{(II)} \quad \Delta \phi_m = (\phi_{b.r.} - \bar{\phi}_{s.r.}) + (\bar{\phi}_{s.r.} - \bar{\phi}_{s.l.}) + (\bar{\phi}_{s.l.} - \phi_{b.l.}) \qquad (3)$$

Subdivision of the total membrane potential into a series of differences or segments is a concept of basic importance and was introduced into membrane electrochemistry by Teorell and by Meyer and Sievers, the so-called TMS Theory.[4-9] In the first case (illustrated in Figure 2) potentials just inside the membrane surface beyond the space charge layers are used, since these values can be computed for reversible interfaces using the electrochemical potential concept. On the other hand, if no point in the membrane is electroneutral, or if the precise space-charge distributions arising from fixed and partitioned charges are known, then the second equation can be used. Membranes with preferential solubility for ions of one kind, and membranes with neutral carriers

---

† The following notation is used: *b.r.*, bulk solution on the right side of a membrane; *b.l.*, bulk solution on the left side of a membrane; *m.r.*, inside the right interface of a membrane beyond the space charge region; *m.l.*, inside the left interface of a membrane beyond the space charge region; *s.r.*, inside the right interface of a membrane at the inner surface; *s.l.*, inside the left interface of a membrane at the inner surface.

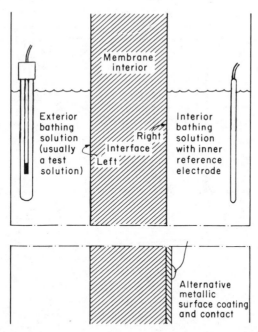

Figure 1. Schematic essentials of membrane cells. Top: Membrane configuration in which the membrane separates two electrolyte bathing solutions. The left side is the exterior or test solution. The right side is the interior or inner reference solution. Also shown is a typical exterior reference electrode (junction type) and interior reference electrode. Bottom: Metal contact or "all-solid-state" configuration in which an electronically reversible contact replaces the interior reference and inner reference electrode.

that enforce preferential solubility of alkali cations have been treated by the second equation.[10,11] TMS theory appears in standard monographs that treat membrane potentials.[12,13]

1.3. Potential Profiles at Single Interfaces[14]

Interfacial transfer of material, single ions, neutral salts, and neutral molecules can be classified in terms of forward and backward rates of transfer. For rapid, reversible behavior, local thermodynamic equilibrium prevails and total interfacial pd (inside and outside a membrane/electrolyte interface) can be calculated to within an additive constant from the equality of the species' electrochemical potential in each phase. For virtually all membranes used in ISEs, including solid and liquid types, rapid ion exchange or extraction makes the activities of transferring ions known or expressible in thermodynamic terms so that the electrochemical potential equality cay be applied to determine net interfacial potential differences between regions of bulk solution and bulk membrane just inside the space-charge region. This calculation is basic for ISEs and is given in detail in 1.4.

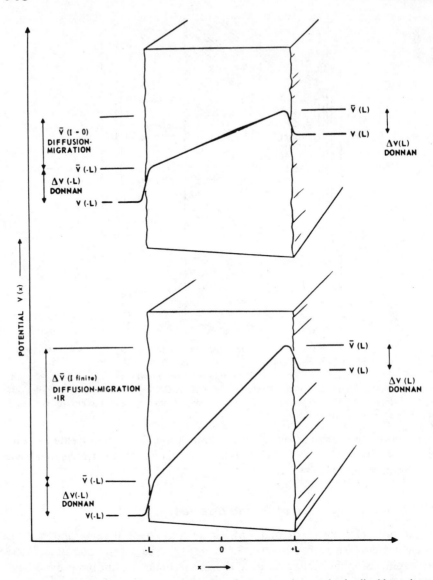

Figure 2. Potential distributions at reversible, homogeneous, cation-conducting liquid membranes under bionic conditions. Upper: condition of zero current but nonzero flux conditions. Lower: condition of negative current flow in the positive direction or positive current flow in the negative direction. Potential increases positive downward. The membrane thickness in this drawing is 2L.

However, a more classical calculation of potential profiles shows some of the problems and limitations of the attempts to calculate single electrode potentials. The electrochemical potential, which relates ion activity and potential in the electroneutral bulk, also applies at all points occupied by charged species from one phase to the other. This assumption underlies the Gouy–Chapman theory and provides an approximation to the distribution of potential from one phase to another. Probable charge density, field, and potential profiles at zero current and constant temperature and pressure are shown schematically in Figure 3. Phases I and II are electroneutral at long distances from the interface and the dielectric constants of the phases are taken as $\varepsilon_I < \varepsilon_{II}$. This very simple model for the interface may be suitable for two structureless fluid phases, such as immiscible liquids containing dissolved electrolytes. Since the energy levels of extractable ions are not likely to be identical in each phase, space charge is presumed to develop spontaneously.

Quantitative descriptions of these profiles, at the simplest level of approximation, make the assumptions that charged species are points subject to central electrical (Coulomb's law) force and that Poisson's equation and Gauss' law are corollaries. The classical Boltzmann distribution follows from Poisson's equation and the Nernst-Planck equations of motion and applies to blocked situations (neither ionic species can pass an interface) and to equilibrium reversible interfaces.

The equations of motion at nonzero current (or zero current but nonzero flux) predict a perturbed Boltzmann distribution that reduces to the usual distribution when current and individual fluxes are zero.[15] The distribution of charged species is considered "diffuse." Charge densities, electric fields, and potentials at zero current and zero flux obey complicated functions of distance from the interface in either direction.[16] However, at small charge and small fields, corresponding to interfacial potential differences less than RT/zF volts, the linearized version of the Gouy–Chapman theory gives the internally consistent simplifications.

$$\rho(x) = \pm \frac{4RT\kappa^2}{zF} \exp(-\kappa x) \tag{4}$$

$$E(x) = \mp \frac{4RT\kappa}{zF} \exp(-\kappa x) \tag{5}$$

$$\phi(x) - \phi_b = \pm \frac{4RT}{zF} \exp(-\kappa x) \tag{6}$$

where x is the distance perpendicular to an interface and is taken to be positive in either direction from an interface, and κ, the reciprocal Debye thickness (the place where field, voltage, and charge decrease to $1/e$ of their surface values in the linearized version), is given by

$$\kappa = \left(\frac{2\sum z^2 F^2 c_b}{\varepsilon RT} \right)^{1/2} \tag{7}$$

In these equations, x cannot be zero and must be greater than the radius of an ion. Equation (7) applies to single, symmetric electrolytes of charge z and

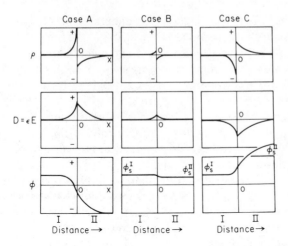

Figure 3. Systemic interfacial electrostatic characteristics for immiscible liquid/liquid electrolyte interfaces. The interface is located at X = 0; electroneutrality holds in the bulk of each phase. Case A: Phase I is positively charged; Case B: nearly 0 charge corresponding to a nearly flat band potential; Case C: Phase I is negatively charged; ρ = space density; D = electric displacement; ϕ = local potential.

concentration c_b in the bulk of either phase. Equations (4)–(7) apply to the diffuse charge in either phase. For electrolyte/metal interfaces, the charge on the metal is localized in a sheet. For electrolyte/semiconductors both phases can have diffuse charge regions.

Unfortunately, the theory cannot be applied to regions of space very near surfaces when solvent dipoles are efficiently aligned, and ions are rejected. The dipole field is accompanied by a linear potential drop. Yet there are no ions in this region whose concentration could adjust appropriately to maintain constancy of the electrochemical potential of the particular ions. In addition to water dipole layers, other dipolar materials in water solvent and oil-soluble dipolar molecules in the membrane will become oriented at the two sides of an interface. Precise direction and extent or orientation is a matter of conjecture. One intuitive approach is to assume that polar ends align with their end of sign opposite to the external phase charge nearest the interface. These orientable molecules compete with ions for space near an interface. Rejection of ions from the first few angstroms defines the distance of closest approach of diffuse-layer ions, the outer Helmholtz plane (OHP).

Some ions in either phase that do compete with oriented solvent or other dipolar molecules at the interface become specifically adsorbed (called "contact" adsorbed). These ions' centers are somewhat closer to the interface than those nonadsorbed ions. The specifically adsorbed plane of ion centers defines the inner Helmholtz plane (IHP) in the Stern model.[14,17] There is one further limitation to the simple Gouy-Chapman theory, which assumes ions have no size. Ions are not point charges and cannot pile up at the surfaces to create space-charge densities greater than that allowed by their finite sizes or by their salt solubilities and ion pairing constants.[18-21]

For insoluble, ionic, solid crystals such as silver halides, the basic model is again the Gouy–Chapman–Stern picture as adapted for ionic solids by

Grimley and Mott.[22,23] Since intrinsic ion-defect crystals cannot be manufactured, the surface properties for a doped, interstitial-rich AgX (Ag$_2$S dopant) exposed to an excess Ag$^+$-containing solution such as AgNO$_3$ is illustrated in Figures 4(a) and (b). The first is a basic case without cation adsorption, corresponding to an inner-crystal potential that is positive with respect to the solution. Figure 4(b) is modified to include specific cation adsorption, while Figure 4(c) is the flatband condition. Figure 4(d) corresponds to a bathing solution with excess anion (KX or NaX, for example) with specific anion adsorption. Figure 4(e) shows superequivalent anion adsorption.

The reasons for the primary importance of the electrostatic features of the membrane interface are fourfold. First, rates of slow ion transfers (analogous to the well-known irreversible electron transfer) may depend on the actual concentration of transferring species at the OHP or the IHP. These values are not the same as bulk values and must be accounted for in surface kinetics. Second, ion-transfer rate constants may be potential-dependent, and

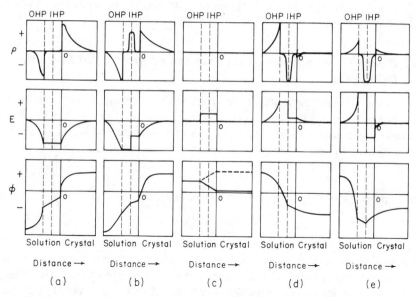

Figure 4. Schematic interfacial electrostatic characteristics for electrolyte/crystal interfaces. The interface is located at $x = 0$. Electroneutrality holds in each bulk phase. Potentials in the compact regions are not drawn "to scale." (a) A hypothetical positively charged crystal in a nonadsorbing electrolyte; (b) a positively charged crystal interface for which there is both specifically adsorbed cations and diffuse charge; (c) a crystal interface at the "null" solution composition corresponding to the potential of zero charge (pzc); aligned solvent dipoles gives rise to small potentials through the compact layer; positive and negative potential differences are illustrated; (d) a crystal interface near the potential of zero charge for strongly adsorbed anions; (e) a crystal interface near the potential of zero charge for superequivalent adsorption of anions, which gives rise to a potential minimum. Surface states, adsorbed vacancies, and interstitial ions on the inner crystal surfaces are not illustrated.

it is necessary to know the local potential at the position of an ion transfer in order to describe the rate accurately. The combination of these factors leads to the so-called Frumkin correction in electrode kinetics. Third, the transport rate of ions is modified in the space-charge region by the high fields. The first and third are considered in electron-transfer kinetics, as double-layer or psi effects.[17] The equivalent theory for ion-transfer kinetics at membrane and immiscible liquid interfaces has been derived recently using the quasithermo-dynamic method.[24,25]

In addition to specific adsorption of charged species, field and potential profiles at interfaces are affected by adsorption of uncharged molecules. This effect is primarily a change in the local dielectric constant of the compact layer. Interpretation of nonionic adsorption as a function of species activities (adsorption isotherm) and interfacial potential or charge can be difficult.[17] The adsorption isotherm is frequently potential-dependent, even though the adsorbed material has no net charge. Finally, the time response of membrane potentials ultimately depends on the buildup or relaxation of space charge in response to activity changes in bathing solutions or upon application of a voltage or current.

1.4. Quasi-Thermodynamic Calculation of Interfacial Potential Differences[26]

At equilibrium, equality of electrochemical potential in two phases, for each charged species in rapid, reversible exchange, defines the net interfacial potential difference within an additive constant. The electrochemical potential is a conventional ionic energy in a region of local potential

$$\tilde{\mu}_i = \mu_i + z_i F \phi \tag{8}$$

for a species with charge z_i. The chemical potential

$$\mu_i = \mu_i^0 + RT \ln a_i \tag{9}$$

is also a conventional quantity referred to a standard state of $a_i = 1$. The activities are so-called surface values determined by mass transport to and from an interface. "Surface" activities at the outer edge of the space-charge region must be known or expressible in terms of flux or current. For systems with permselectivity (see Section 1.5) the zero current condition is also zero flux of each species; the surface activities are bulk values. At zero current with compensating nonzero fluxes,† the activities are those values outside the space charge, but they will differ from bulk values, because of concentration polarization. Even with finite current flow, local equilibrium can persist across interfaces. The so-called surface concentrations are again involved.

---

† Counter-current flow of ions of same sign of charge; or, parallel flow of salt (ions of opposite sign) in site-free systems, or at high activities when Donnan Exclusion Failure occurs.

Reversible interfacial "local" equilibrium means that flux of material through the interface does not perturb local equilibrium activity ratios. However, the local surface activities will not, in general, be bulk values. Surface activities are determined through flux balance by whatever slow transport processes occur in the membrane and in the external solution near the surfaces. For example, Figure 5 shows the membrane part of a cation-permeable, liquid ion exchanger membrane cell, with current flow. The exterior solution will be concentration polarized (depleted on the entering side and raised above bulk values on the existing side of the membrane). Similarly, inside the membrane, species' concentrations profiles under current flow will, in general, be higher than the average on the inside of the entering surface and lowered at the inner side of the exiting surface.[15,27]

Application of equilibrium calculations across an interface requires use of these bulk-transport-determined surface activities. They can be found from flux balance in terms of bulk values in the steady state

$$ J_{\text{left to right}} = D\frac{{}^{c}b, 0 - {}^{c}s, 0}{\delta'} = \bar{D}\frac{\bar{{}^{c}}s, o - \bar{{}^{c}}s, d}{d} $$

$$ = D\frac{c_{s,d} - c_{b,d}}{\delta''} \tag{10} $$

$$ c_{\text{surface left}} = c_{s,0} = c_{b,0} - J\delta'/D \tag{11} $$

$$ c_{\text{surface right}} = c_{s,d} = c_{b,d} + J\delta''/D \tag{12} $$

Super bar quantities apply inside the membrane. These equations apply to neutral species, to charged species moving in the presence of inert electrolyte, and to liquid ion exchanger membranes permeated by a single counter ion, e.g., a calcium electrode bathed by calcium salt solutions. When a permeable cation species is required to obey electroneutrality in the membrane because

Figure 5. Illustration of steady-state diffusion (from left to right) of uncharged species through an inert, uncharged membrane. Although the bathing solutions are stirred, account is taken of exterior surface film diffusion (external concentration polarization) through the Nernst layer. For reversible interfaces, surface concentrations $C_{s,0}$ and $\bar{C}_{s,0}$ and $C_{s,d}$ and $\bar{C}_{s,d}$ are related by partition equilibria. Diffusion coefficients in the external and internal phases are D and $\bar{D}$.

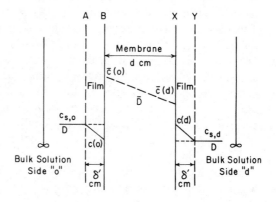

of confined mobile sites, then

$$\bar{c}_{s,0} + \bar{c}_{s,d} = 2\bar{c} \tag{13}$$

and

$$\bar{c}_{s,0} = \bar{c} + Jd/2\bar{D} \tag{14}$$

$$\bar{c}_{s,d} = \bar{c} - Jd/2\bar{D} \tag{15}$$

Steady-state concentration profiles are justified in these cases. However, linearity of all charged species in multi-ion permeable membranes is not generally true.[1,2]

Nernst diffusion layer thicknesses δ' and δ'' depend upon the stirring rate, i.e.

$$\delta \propto \omega^{-1/2} \tag{16}$$

for rotating-disk geometry in the steady state. For other geometries, the exponent of ω will be near $-\frac{1}{2}$.

Prior to the steady state, concentration profiles in the membrane and in the external-concentration-polarized films are not linear. If one increases the external concentration on one side, the surface film concentration must equilibrate to the new bulk solution value. At a distance δ' from the interface (and beyond), a uniform concentration is established at a new high (or low) value when the bulk concentration is varied. However, diffusion/migration brings the surface concentration to the new value. The surface concentration builds up or decreases according to a complicated function dependent on $t^{-1/2}$ at short times less than approximately $\sim \delta^2/2D$. When a current flows and when, at zero current, there is a flux of material in the membrane, the interior surface concentration also readjusts with $t^{-1/2}$ dependence for a comparable time $d^2/2D$. Eventually a new linear profile is established, at least for membranes in which all of the ions have the same absolute charge.

There is an apparent paradox in the use of surface concentration of ionic species. The concentrations derived from fluxes are not literally at the surface, but are extrapolated values that would occur in the absence of space charge. This distinction is nonexistent for neutral species. The surface concentration for a single univalent salt is $(c_+c_-)^{1/2}$ evaluated at $x = 0$. Alternatively, if the profile of salt concentration from bulk to surface is not steep (cases of low current densities so that film diffusion limitation is avoided), then the surface concentration is approximately the value just outside the diffuse double-layer region. Use of the latter surface value is justified because mass transport controls the surface concentration by continuity of flux, and depletion regions occur over a 10^{-4}–10^{-3} cm distance while the space-charge region is only about 10^{-5}–10^{-6} cm thick.

The local equilibrium condition at an interface for neutral species at constant temperature and pressure is simply equality of chemical potential across that interface

$$\mu_i = \bar{\mu}_i \tag{17}$$

with the result that an extraction coefficient K_{ext} is defined as

$$K_{ext} = \bar{a}_s/a_s = \exp[(\mu^0 - \bar{\mu}^0)/RT] \tag{18}$$

For species i of charge z, equality of electrochemical potential is presumed to hold locally near the interface but outside the space-charge region during passage of flux or current, for rapidly exchanging, reversible species

$$\tilde{\mu}_i = \bar{\tilde{\mu}}_i \tag{19}$$

and the single-ion extraction coefficient is defined with the same form, viz.

$$K_{ext,i} = (\bar{a}_s/a_s) \exp[zF(\bar{\phi}_s - \phi_s)/RT] = \exp[(\mu_i^0 - \bar{\mu}_i^0)/RT] \tag{20}$$

This equation is exactly true at equilibrium zero flux conditions where $\bar{a}_s$, a_s, $\bar{\phi}_s$, and ϕ_s become bulk values $\bar{a}_b$, a_b, $\bar{\phi}_b$, and ϕ_b. In the absence of flux, this equation is not limited to activities in the electroneutral region, but also applies in the space-charge region. Generally

$$K_{ext,i} = [\bar{a}(x)/a(x)] \exp\{zF[\bar{\phi}(x) - \phi(x)]/RT\} \tag{21}$$

It is valid to formulate extraction in terms of activities at an interface where the local potential has a single value (in the absence of dipolar layers). Thus

$$K_{ext,i} = \bar{a}(0)/a(0) = \exp[(\mu^0 - \bar{\mu}^0)/RT] \tag{22}$$

This result follows from the general zero flux condition outside a membrane

$$RT \ln[a(x)/a_b] = zF[\phi_b - \phi(x)] \tag{23}$$

and a corresponding expression involving overbar quantities for points within a membrane.

In Figure 6 are shown the interfacial potential differences resulting from an intrinsic difference in the energies of charged species in two phases as reflected by their μ^0 value differences. For metals in contact [Figure 6(a)], a potential difference arises, known as a contact pd because the electrochemical potentials of the electrons in each are not necessarily equal before contacting

$$\phi(I) - \phi(II) = \frac{1}{F}[\mu^0(I) - \mu^0(II)] \tag{24}$$

The activity ratio is nearly unity. The chemical potential difference is a combination of the electronic work function difference and any entropy difference. When a metal ion, such as Ag^+ in solution, contacts Ag metal, the mean Ag^+ electrochemical potentials in each phase are equalized by generation of an interfacial pd and space charge. In Figure 6(b), Ag^+ energies before and after contact are illustrated. The distribution of aqueous energies correspond to the activity dependence of the Ag^+ energy. After defining partial ion

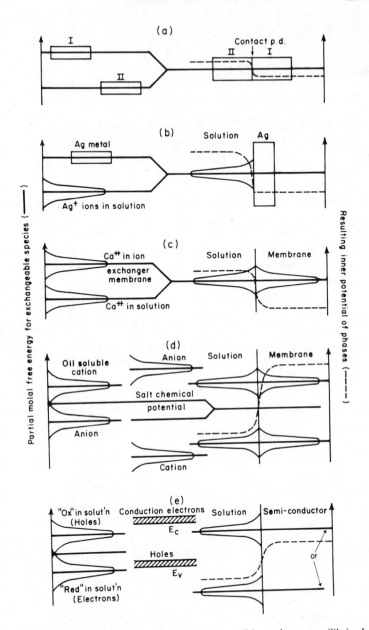

Figure 6. Development of interfacial potentials by reversible exchange equilibria. Left side: energies of exchangeable species prior to contact; right side: energies and developed potentials after contact. (a) Two metals that exchange electrons; (b) Ag metal and Ag^+ containing electrolyte. Gaussian curves indicate activity effect on ionic energy levels in an electrolyte; (c) a cation in an electrolyte (Ca^{2+}) and a liquid or solid ion exchanger containing the same counter ion (Ca^{2+}); (d) an immiscible extraction interface, electrolyte/organic phase across which both cations and anions equilibrate; (e) an intrinsic semiconductor and an electrolyte containing a redox couple.

exchange constants in the same way as single-ion extraction coefficients, the interfacial pd is given by

$$\bar{\phi} - \phi_{soln} = \frac{RT}{F}\left[\frac{\mu^0_{Ag^+,soln} - \bar{\mu}^0_{Ag^+}}{RT} + \ln\left(\frac{a_{Ag^+,soln}}{\bar{a}_{Ag^+}}\right)\right]$$

$$= \frac{RT}{F}\ln\left[\frac{K_{ext,Ag^+}a_{Ag^+,soln}}{\bar{a}_{Ag^+}}\right] \tag{25}$$

Since the solid is in equilibrium with its charged components

$$\bar{\mu}_{Ag} = \bar{\mu}_{Ag^+} + \bar{\mu}_e \tag{26}$$

then

$$\bar{\phi} - \phi_{soln} = \frac{RT}{F}\left[\frac{\mu^0_{Ag^+,soln} + \bar{\mu}_e - \bar{\mu}^0_{Ag}}{RT} + \ln\left(\frac{a_{Ag^+,soln}}{\bar{a}_{Ag}}\right)\right] \tag{27}$$

Equation (27) shows a dependence of pd on metal activity. This phenomenon is general and provides the basis for the analysis of "all-solid-state" electrodes. AgBr backed by Pt and the AgBr contacted by aqueous Ag^+, shows ϕ^0 dependent on the crystal doping. Excess electrons (Ag^0) or excess holes (Br^0) produce widely different ϕ^0 values.[28] The same analysis (Figure 6c) for a liquid ion exchanger such as a calcium phosphonate membrane exposed to a calcium salt solution gives

$$\bar{\phi}_{memb} - \phi_{soln} = \frac{RT}{2F}\left[\frac{\mu^0_{Ca^{2+},soln} - \mu^0_{Ca^{2+},memb}}{RT} + \ln\left(\frac{a_{Ca^{2+},soln}}{a_{Ca^{2+},memb}}\right)\right] \tag{28}$$

For a pH-sensitive glass membrane, the interfacial potential difference can be described.[29] These glasses are chiefly lithium silicates, with lanthanum and barium ions as lattice "tighteners" added to retard silicate hydrolysis and lessen alkali ion, sodium, and potassium mobilities. Lithium ions are the bulk mobile charge carriers under an applied electric field. After the membrane is soaked in water, the surface layer is depleted of Li^+, which is replaced by H^+. Virtually all surface silicate anion sites, "fixed" sites, are neutralized by H^+. Content of H^+ decreases in a complex way with increasing distance into the membrane, while Li^+ content increases in such a way that the sum of positive ions (charge carriers and other cations) balances the presumed uniform fixed-site concentration, which is about 20 M for typical pH-sensitive glasses. This idealized model can be amended to allow for osmotic pressure-driven uptake of water with consequent hydrolysis of the surface silicate chains to yield, perhaps, a lower fixed-site density at the surface than exists in the bulk. These comments apply equally to membranes of sulfonated polystyrene resins or other synthetic membranes containing high concentrations of fixed-site ionic groups homogeneously distributed.

The interfacial potential at one side of a pH-sensitive glass membrane is given by

$$\bar{\phi}_{\text{memb}} - \phi_{\text{soln}} = \frac{RT}{F}\left[\frac{\mu^0_{\text{H}^+,\text{soln}} - \mu^0_{\text{H}^+,\text{memb}}}{RT} + \ln\left(\frac{a_{\text{H}^+,\text{soln}}}{a_{\text{H}^+,\text{memb}}}\right)\right] \tag{29}$$

which is the same form as Eq. (28)

If, in addition to rapid reversible equilibrium of hydrogen ions, the solution and the membrane surface contain other univalent ions, for example, Na^+ at equilibrium, then an expression equivalent to Eq. (29) with subscripts Na^+ applies also. Because the interfacial potential is unique, the single ion partition coefficients are related through the ion-exchange constant for the reaction

$$\text{Na}^+_{\text{soln}} + \text{H}^+_{\text{memb}} \rightleftharpoons \text{Na}^+_{\text{memb}} + \text{H}^+_{\text{soln}} \tag{30}$$

$$K_{\text{H}^+/\text{Na}^+} = \exp\left[\frac{1}{F}(\mu^0_{\text{Na}^+} - \mu^0_{\text{H}^+})_{\text{soln}} - \frac{1}{F}(\mu^0_{\text{Na}^+} - \mu^0_{\text{H}^+})_{\text{memb}}\right] \tag{31}$$

$$K_{\text{H}^+/\text{Na}^+} = K_{\text{ext,Na}^+}/K_{\text{ext,H}^+}$$

By mass balance on concentration,

$$(C_{\text{Na}^+} + C_{\text{H}^+})_{\text{memb}} = \text{fixed sites} = \bar{X} \text{ (conc. units)} \tag{32}$$

The interfacial pd can also be rewritten

$$\bar{\phi}_{\text{memb}} - \phi_{\text{soln}} = \frac{RT}{F}\left[\frac{\mu^0_{\text{H}^+,\text{soln}} - \mu^0_{\text{H}^+,\text{memb}}}{RT}\right.$$

$$\left. + \ln\left(\frac{1}{\bar{\gamma}_{\text{H}^+}\bar{X}}\right)\left(a_{\text{H}^+} + K_{\text{H}^+/\text{Na}^+}\frac{\bar{\gamma}_{\text{H}^+}}{\bar{\gamma}_{\text{Na}^+}}a_{\text{Na}^+}\right)\right] \tag{33}$$

Equation (33) shows that the interfacial pd reflects changes in solution activities for species existing in equilibrium in both phases.

Calculation of the interfacial pd for a simple, single-crystal membrane in a saturated solution follows the same procedure as used for liquid-ion exchanger or metal/metal ion interfaces. For a silver salt AgX with rapid exchange of Ag^+, X^-, or both, equilibrium is described by

$$[\bar{\mu}^0_{\text{Ag}^+} + \bar{\mu}^0_{\text{X}^-}] - [\mu^0_{\text{Ag}^+,\text{soln}} + \mu^0_{\text{X}^-,\text{soln}}] = RT\ln\frac{K_{\text{so,AgX}}}{\bar{a}_{\text{AgX}}} \tag{34}$$

where overbars apply to salt and the defined solid phase activities are related by

$$\bar{a}_{\text{AgX}} = \bar{a}_{\text{Ag}^+}\bar{a}_{\text{X}^-} \tag{35}$$

The interfacial potential is found by equating electrochemical potentials of either Ag^+ or X^- in each phase. Thus

$$\bar{\phi} - \phi_{soln} = \frac{RT}{F} \ln \left[K_{ext,Ag^+} \frac{a_{Ag^+,soln}}{\bar{a}_{Ag^+}} \right]$$

$$= -\frac{RT}{F} \ln \left[K_{ext,X^-} \frac{a_{X^-,soln}}{\bar{a}_{X^-}} \right] \tag{36}$$

and from Eq. (34) and (35), the single ion partition coefficients are related to the thermodynamically significant solubility product

$$K_{ext,Ag^+} K_{ext,X^-} = \bar{a}_{AgX} / K_{so,AgX} \tag{37}$$

There is no evidence to suggest that pressed pellets of single salts or corresponding heterogeneous membranes behave thermodynamically differently from single crystals, provided that no small particles of higher than normal solubility are present. Mixed pellets of two substances may simultaneously equilibrate. If common ions are involved, as in the case of AgX–Ag_2S systems, the interface pd can be expressed in terms of combinations of solubility products and partition coefficients. The more soluble component dominates the equilibrium and the pd obeys the form above.

For electroneutral salt extraction of ions of equal absolute charge into site-free extraction membranes [Figure 6(d)] it follows that

$$K_{ext.} + K_{ext,-} = \frac{\bar{a}_+ \bar{a}_-}{a_+ a_-} = (K')^2 = K_D^2 \left(\frac{\bar{\gamma}_\pm}{\gamma_\pm} \right)^2 \tag{38}$$

$$= \exp \left[(\mu_+^0 + \mu_-^0)/RT - (\bar{\mu}_+^0 + \bar{\mu}_-^0)/RT \right] \tag{39}$$

where K' is the thermodynamic salt extraction coefficient and K_D is the "concentration" coefficient. Because the interfacial pd must be satisfied independently by each equilibrated species according to

$$\bar{\phi}_s = \phi_s = \frac{RT}{zF} \ln \left(\frac{K_{ext,a_s}}{\bar{a}_s} \right) \tag{40}$$

the interfacial pd can be expressed

$$\bar{\phi}_s - \phi_s = \frac{RT}{zF} \ln \left(\frac{K_{ext,+} a_+ + \bar{a}_-}{\bar{a}_+ + K_{ext,-} a_-} \right) \tag{41}$$

regardless of electroneutrality in each phase. When electroneutrality occurs on both sides of the interface, beyond the space-charge regions, substitution of Eq. (38) and (39) into (41) leads to concentration-independent results for a single extracting salt $i^+ x^-$.

$$\bar{\phi}_s - \phi_s = \frac{RT}{2zF} \ln \left(\frac{K_{ext,i} \gamma_i / \bar{\gamma}_i}{K_{ext,x} \gamma_x / \bar{\gamma}_x} \right) \tag{42}$$

When an extraction membrane interface is in equilibrium with two salts and quasielectroneutrality also holds in the bulk of each phase, the interfacial pd is constant for a single salt and is a slowly varying function of concentration for two salts. Writing Eq. (40) for two salts i^+x^- and j^+y^- with the conditions

$$(i^+) + (j^+) = (x^-) + (y^-) \tag{43}$$

$$(\bar{i}^+) + (\bar{j}^+) = (\bar{x}^-) + (\bar{y}^-) \tag{44}$$

then

$$\bar{\phi}_s - \phi_s = \frac{RT}{2F} \ln \left[\frac{K_{\text{ext},j}a_i/\bar{\gamma}_i + K_{\text{ext},j}a_j/\bar{\gamma}_j}{K_{\text{ext},x}a_x/\bar{\gamma}_x + K_{\text{ext},y}a_y/\bar{\gamma}_y} \right] \tag{45}$$

In Figure 6 the majority of the examples involve reversible species of only one sign. However, both crystalline interfaces and extraction interfaces involve equilibration of species of opposite sign. In the latter cases, the salt chemical potentials as well as individual ion electrochemical potentials, are equal in each phase. This point is illustrated in Figure 6(d).

1.5. Potential Profiles and Potential Differences Across the Interior of Membranes[13]

It is well known that a potential difference exists in a homogeneous bulk phase when current, ionic or electronic, passes through the phase. Generally, a voltage is applied to produce current, and one accounts for the applied voltage in terms of interfacial pd's and the so-called *IR* drop in the bulk phase, where R is the resistance. It is important in membrane electrochemistry to identify conditions under which a bulk potential drop occurs, even at zero current.

To develop pd's in electroneutral interior bulk membranes, necessary conditions are:

1. One or more concentration gradients of charged species exists,
2. Fluxes of ions are nonzero,
3. Mobilities of these species are different for uni- or counter-directional flows.

These conditions become apparent from application of the Nernst–Planck flux equations, discussed in this section.

Spontaneous potential differences from point to point in a bulk ionic conductor are not expected for crystalline ionic conductors bathed in solutions of component ions, because there is no charge carrier gradient. Likewise, a permselective liquid or solid ion exchanger bathed in two different concentrations of a single permeable ion salt does not develop an interior pd. A dialysis membrane bathed in two different concentrations of aqueous KCl solutions does not develop a significant pd because K^+ and Cl^- undergoing unidirectional flow have nearly identical mobilities. Diffusion of KCl from high to low concentration does not provide an opportunity for one ion to move appreciably faster than the other. In describing electroneutral bulk phases it is assumed that the pd's will be computed only through the neutral region and that separation of charge and potential curvature at interfaces are not included.

In the following summary it is assumed that membranes have permeability for ions and in some cases for salts and solvents. Quantitative expression of permeability follows later. Three general descriptive terms "semipermeability," "permselectivity," and "permeability" coefficients are frequently used.

Semipermeability is a measure of a membrane's transport or rejection of neutral species, e.g., solvent vs. salt. *Permselectivity* is a measure of the current-carrying ability of ions in a membrane. Ideal cation permselectivity occurs when the cation transference number $t_c = 1$, i.e., all current is carried through the membrane by cations and $t_a = 0$. The converse definition applies to ideal anion permselectivity. This property is closely related to the generation of ideal membrane potentials at zero current in response to bathing activities of permeable ions. Ideal permselectivity for cations, for example, is necessary for the development of full Nernstian potential response to cations. However, full Nernstian response to a specific cation does not require that the transference member of that cation be unity. For example, a pH sensing glass membrane contains surface regions containing protons as the permselective cation. The interior of the glass contains Li^+ as the permselective cation. Current carries protons in, Li^+ and protons, out.

For uncharged materials penetrating a membrane barrier, simple flux equations can be written in terms of net external bathing concentration or activity differences. The quantity relating flux and activity differences is a *permeability coefficient*, which is expressible in terms of a diffusion coefficient, extraction coefficient, and membrane thickness. When the diffusion coefficient is space and concentration independent, the permeability coefficient or simple "permeability" is independent of bathing solution activities. Similarly, for pressure-gradient-driven transport, a concentration-independent permeability depending specifically on membrane properties can usually be observed.

On the contrary, ion permeabilities expressed as a proportionality with external activity differences, in general, are not independent of the bathing solution activities except for certain conditions of bathing solutions surrounding electroneutral membranes with sites, and some cases involving site-free membranes. For electroneutral membranes with sites, the permeability coefficients are generally constant for ions of the same charge and are proportional to the ion mobility and the extraction coefficient.

In an elaborate and thorough analysis, Sandblom and Eisenman[29] demonstrated that permeability ratios for ions can be independent of solution compositions when the temperature is constant, no neutral species, ion pairs, or solvent transport occurs; ions of only one sign are transported, and the standard chemical potentials of ions in each phase are constant. These conclusions apply to fixed-site membranes at all times and to mobile-site membranes in the steady state. This analysis is very basic to the design of permselective membranes and ISEs.

Permeabilities can be true constants (independent of solution activities) for two special cases: 1) when the total ionic concentrations are the same on both sides of site-free membranes (or constrained-liquid junction) and 2) when interfacial potentials are the same by virtue of identical ion extraction coefficients for all cations and for all anions.

The most general driving force for ions in membranes is the gradient of the electrochemical potential. When osmotic pressure is included, Eq. (8) becomes

$$\tilde{\mu}_i = \mu_i^0 + RT \ln a_i + pv_i + z_iF\phi \tag{46}$$

Although this equation is still incomplete, it is suitable for most membrane problems.

The rate at which groups of ions move from regions of high concentration to low is measured in terms of flux J of $\bar{J}$. In dilute solutions and membranes, the flux is a result of a hypothetical force, which is the gradient of the local electrochemical potential. In a membrane out of equilibrium, when concentration gradients exist, the electrochemical potential varies from point to point. In one dimension the flux is given by the usual dilute solution form of the N-P (Nernst–Planck)[12,13,26,30,31] equation

$$\bar{J}_i = -\bar{u}_i\bar{c}_i\frac{\partial\tilde{\bar{u}}_i}{\partial x} \tag{47}$$

This widely applied transport or flux equation is consistent with irreversible thermodynamic flux equations and can be expanded to account for motion of pore liquids (solvent) as well as ions.[32] Coefficients relating forces and fluxes are primitive "friction" coefficients that can be specified in conventional terms using diffusion coefficients, mobilities, or equivalent conductances.[2,33] Irreversible thermodynamic interactions (cross-terms) are omitted by assuming, for membranes containing dissolved neutral and charged species, that the motion of each species occurs under local diffusive, electric field (migration), and pressure gradient forces. Ionic motion is coupled through the local electroneutrality requirement. A recent analysis of these assumptions is by Buck.[33a]

Flux for each species of charge z_i is specified as the product of concentration and velocity, where the latter is proportional to the local force. Thus, at constant temperature

$$\bar{J}_i' = \bar{c}_iv_i = \bar{c}_i\bar{f}_i\sum(\text{forces}) \tag{48}$$

is the flux with respect to the center of mass, while f_i is a friction (proportionality) coefficient, and $\bar{c}_i$ a local concentration. Relations with other transport coefficients are

$$RT\bar{f}_i = \bar{u}_i RT = \bar{D}_i = \bar{u}^* RT/F|z_i| \tag{49}$$

where $\bar{D}_i$ (cm^2/sec) is the diffusion coefficient, $\bar{u}_i^*$ (cm^2 sec^{-1} V^{-1}) is the physical mobility, and $\bar{u}_i$ (cm^2/sec J/mol) is the chemical mobility. The forces are grad $\tilde{\mu}_i$ and $\bar{v}$ grad $\bar{p}$.

In the absence of solvent penetration and with the frequently used approximation that the activity coefficient is not a function of distance

$$\bar{J}_i = -\bar{D}_i \left(\frac{\partial \bar{c}_i}{\partial x} + z \frac{F}{RT} \bar{c} \frac{\partial \bar{\phi}}{\partial x} \right) \tag{50}$$

This is a dilute solution form of the one-dimensional N–P equation. It is nonlinear in distance because both C_i and $\bar{\phi}$ are functions of distance.

Potential profiles (and net potential differences), called *internal diffusion potentials*, are usually computed from the dilute solution form of the N–P equations applied to each mobile, charged species in a membrane, subject to boundary and initial conditions. In using Eq. (47) for the transport of each species, in developing derived equations for total flux, current, and membrane potentials, simplifications are usually made to yield Eq. (50). Activity coefficients are assumed to be either constants or simple functions of concentration, i.e., $\bar{\gamma} = b\bar{c}^{n-1}$, so that integration becomes possible. Transport of solvent is also ignored, as are pressure effects on the transport of other species. For the zero-current condition the simplified N–P equations for each free ion are first summed to express current densities in the absence of interactions between species, viz., ion pairing, to give

$$I = F \sum_i z_i \bar{J}_i \tag{51}$$

This equation applies in the steady state for all values of I. A more complete form is given later to explicitly account for charging current during the transient response. For the potentiometric response, the current density is set equal to zero and the diffusion potential gradient is computed in terms of local concentrations. Finally, the net diffusion potential is computed by integration, in terms of inner surface concentrations excluding space-charge perturbations. Poisson's equation

$$\frac{d\bar{E}}{dx} = \frac{\bar{\rho}}{\varepsilon} = \frac{zF(\bar{c}_+ - \bar{c}_-)}{\varepsilon} \tag{52}$$

which accounts for space-charge buildup at the surface due to transient ionic fluxes is not used to obtain net steady-state pd's. However, exact potential–distance profiles can be calculated for equilibrium and steady-state membrane systems.[15,16]

Steady-state, zero-current, time-independent internal diffusion potentials have similar forms. They are generally logarithmic functions of interior and external surface concentrations of bathing electrolyte. These are concentrations just beyond the space charge for electroneutral membranes, or true interface concentrations for charged membranes.

The logarithmic forms has been called "block," because terms for concentrations on each side of the membrane are separated in blocks. This result means that interfacial pd's can be added logarithmically (i.e., multiplied) to give overall membrane pd's in term of bathing activities. This form is recognized as the Horovitz–Nicolsky–Eisenman equation.[1,2]

For those cases of membranes that can have nonzero diffusion potentials, it is convenient to consider three classes: class I, site-free membranes; class II, fixed-site membranes; and class III, mobile-site membranes.

1.6. Summary of Steady-State Diffusion Potentials and Total Membrane Potentials

At zero current, spontaneously developed bulk potential profiles will occur in charged and uncharged bulk phases when there is asymmetry in the bathing solution (different salts in bathing solutions, or different concentrations of the same salt on two sides) *and* simultaneously "different" terms in the individual ionic flux expressions. These "different" terms may occur as different spatially dependent standard states, different concentration gradients, different concentration profiles and mobilities. In the site-free limit, membranes develop potential profiles in unidirectional flow because of differences of mobilities and possibly differences in local activity coefficients of permeable species. In the limit of high-site density, permselective membranes develop potential profiles when two or more permeable ions of differing mobility are used in bathing solutions *and* they undergo counterdirectional flows.

There is a logical distinction between thick, electroneutral membranes (with or without sites) and thin membranes. The latter, because they are thinner than the Debye thickness, can possess a net charge by preferential solubility of cations or anions.[34–37]

Charged membranes, such as some lipid bilayers, in which a single ion is dissolved, do show interior pd's at zero current because the concentration profiles are nonlinear and there is no bulk region, in the usual sense.[10,11] The problem is one of space charge and is analogous to calculation of potentials at interfaces. There is always a net interior potential difference except in the case of symmetric bathing solutions.

1.6.1. Potentials for Class I, Site-Free membranes†

1.6.1.1. Membrane Potentials for exceedingly Dilute Membranes Containing Ions of One Kind

Condition 1.

$$\text{Debye thickness} = 1/\kappa = \left(\varepsilon RT / z^2 F^2 \sum_i \bar{c}_i \right)^{1/2} > d \tag{53}$$

† Sign convention: x is positive from 0 to d through a membrane of thickness d. A positive current is positive ions moving in a positive x direction.

Condition 2.

$$\bar{\gamma}_i \neq f(x)$$

Condition 3.

$$\bar{\mu}_i^0 \neq f(x)$$

Condition 4.

$$\bar{u} \neq f(x)$$

Condition 5.   All ions of the same charge, z

Condition 6. $I = 0$

When a thin membrane such as a lipid bilayer is exposed to solutions of salts, usually one ion is much more soluble than the other. This solubility is measured in terms of the single ion partition coefficient (or the free energy of transfer). These nonthermodynamic quantities are estimated by Popovych's method[38] or by calculation using the Born equation with image force corrections.[39] In the limit that only one ion dissolves in the membrane (equilibrium extractable ion of one kind), the N–P equation, and Poisson's equation can be solved exactly. The internal pd from interface to interface is

$$\Delta \bar{\phi}_m = \frac{RT}{zF} \ln \left(\frac{\bar{c}_{\text{left}}}{\bar{c}_{\text{right}}} \right) \tag{54}$$

and upon adding in the interfacial potentials Eq. (40)

$$\Delta \phi_m = \frac{RT}{zF} \ln \left(\frac{a_{\text{left}}}{a_{\text{right}}} \right) \tag{55}$$

This is the potential of the right-side bulk-bathing solution less that of the left-side bulk-bathing solution.

When the membrane is permeable to more than one ion of a given charge, the internal diffusion potential is approximately

$$\Delta \phi_m \sim \frac{RT}{zF} \ln \left[\frac{\sum_i \bar{u}_i \bar{c}_i(0)}{\sum_i \bar{u}_i \bar{c}_i(d)} \right] \tag{56}$$

Since this equation is in the block-logarithmic form, interfacial potentials can be added directly without calculating surface concentrations via charge balance; the result (for ions "i" on left, "k" on right) is

$$\Delta \phi_m \sim \frac{RT}{zF} \ln \left[\frac{\sum_i \bar{u}_i K_{\text{ext},i} a_i / \bar{\gamma}_i}{\sum_k \bar{u}_k K_{\text{ext},k} a_k / \bar{\gamma}_k} \right] \tag{57}$$

1.6.1.2. Membrane Potentials for Exceedingly Dilute Membranes That Contain Ions of Both Signs But Are Not Electroneutral and an Approximation for Higher Concentration Electroneutral Membranes (Excluding Space Charge Confined to Surface Regions)

An exact solution for electroneutral membranes with equal total salts of same absolute ion charge:

Conditions 1–6 apply.

Condition 7.   All ions have the same absolute charge

Condition 8.   Nearly constant field given by

$$d\bar{\phi}/dx \sim [\bar{\phi}(d) - \bar{\phi}(0)]/d \tag{58}$$

$$\Delta\phi_m \sim \frac{RT}{zF} \ln \left[\frac{\sum_i \bar{u}_i K_{\text{ext},i} a_i^+ / \bar{\gamma}_i + \sum_k \bar{u}_k K_{\text{ext},k} \bar{a}_k / \bar{\gamma}_k}{\sum_k u_k K_{\text{ext},k} a_k^+ / \gamma_k^+ + \sum_i \bar{u}_i K_{\text{ext},i} \bar{a}_i / \bar{\gamma}_i} \right] \tag{59}$$

This equation, associated with the name Goldman, can be an exact solution for the case in which the total ionic strength of the salt ions on each side of the membrane is the same. It is, in fact, a degenerate case of a Planck constrained junction, when the total inner-ion concentrations $\bar{c}(0)$ and $\bar{c}(d)$ are the same even though they are comprised of different ions.[40] When the interfacial single-ion extraction coefficients happen to be the same for each cation and another constant value for all anions, the interfacial potential contributions are the same and Eq. (59) becomes exact.

1.6.1.3. Electrical Response Functions for High Concentrations of Ions of Both Signs Such That Electroneutrality Exists in the Membrane Bulk

For the membranes of high dielectric constant and those that take up water, such as cellophane, one has an exact analog of the classical constrained junction, treated by Planck,[40] who showed that at zero current

$$\Delta\bar{\phi}_m = (RT/F) \ln \varepsilon \tag{60}$$

where ε obeys the equation

$$\frac{\varepsilon U_d - U_0}{V_d - \varepsilon V_0} = \frac{\ln[\bar{c}(d)/\bar{c}(0)] - \ln \varepsilon}{\ln[\bar{c}(d)/\bar{c}(0)] + \ln \varepsilon} \left[\frac{\varepsilon\bar{c}(d) - \bar{c}(0)}{\bar{c}(d) - \varepsilon\bar{c}(0)} \right] \tag{61}$$

The $\bar{c}$'s are total concentrations, while $U = \sum \bar{c}_i^+ \bar{u}_i$ and $V = \sum \bar{c}_i^- \bar{u}_i$ are evaluated at the surfaces just inside any space charge. The sum $\sum z_i \bar{c}_i = 0$, taken over all ions. This implicit form of Eq. (61) applies to quasielectroneutral membranes in which the surface salt concentrations remain constant, but not necessarily equal on two sides. This situation would prevail when the interfacial exchange current is large and bathing volumes are large. The net interfacial potential from electroneutral solution bulk to electroneutral membrane bulk is virtually constant [see Eq. (45)]. The result means that the two space-charge regions at each interface contain equal and opposite charges and that the net interfacial potentials are about equal on each side. Then

$$\Delta \bar{\phi}_m \simeq \Delta \phi_m \tag{62}$$

A subsequent generalization of the N–P equation integration for ions of arbitrary valence was made by Henderson.[40] He asserted that each ion concentration fraction profile was linear. While this assumption is incorrect, the errors introduced are small for systems with ions of the same absolute charge. Membrane potentials can be evaluated in closed form by this method.

In Figure 7 are illustrated schematically the space charge, fields, and potential profiles for the range of membranes, charged to electroneutral, to which the equations above apply.

1.6.2. Electrical Properties of Class II Fixed-Site Membranes

1.6.2.1. Membrane Potentials for Membranes of High Site Density to Assure Permselectivity (Co-Ion Exclusion)

Analysis of solid ion exchange membrane transport was presented by Helfferich.[12] The majority of published solutions to transport problems ignore solvent transport and use the following assumptions:

Conditions 1–3 above apply.

Condition 4. $\bar{u}_i RT = \bar{D}_i$

Condition 5. $I = 0$

Condition 6.   Inside space-charge regions $\sum z_i \bar{c} + \bar{\omega} \bar{X} = 0$

For the multi-ionic case (i cations of charge z on the left, and k cations of charge z on the right), assuming reversible interfacial ion exchange

$$\Delta \phi_m = \frac{RT}{zF} \ln \left[\frac{\sum_i \bar{u}_i K_{\text{ext},i} a_i / \bar{\gamma}_i}{\sum_k \bar{u}_k K_{\text{ext},k} a_k / \bar{\gamma}_k} \right] \tag{63}$$

The interior membrane potential is an integral of an exact differential so that the potential depends only on interior surface concentrations

$$\Delta \bar{\phi}_m = \frac{RT}{zF} \ln \left[\frac{\sum_i \bar{u}_i \bar{c}_i(0)}{\sum_k \bar{u}_k \bar{c}_k(d)} \right] \tag{64}$$

and is independent of time, i.e., independent of the adjustment of concentration profiles with time. The membrane potential–time response depends only on the time required to charge the surfaces and to establish ion exchange equilibrium, a topic discussed in Section 1.10.

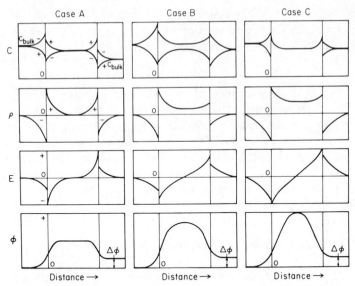

Figure 7. Schematic illustrations of expected electrostatic characteristics for site-free membranes. (Case A) A membrane with preferential solubility of cations over anions, but nonzero values of extraction coefficients, such that electroneutrality prevails in the membrane interior. (Case B) A membrane with preferential solubility of cations over anions and very small anion extraction coefficient so that electroneutrality does not prevail in the membrane interior. (Case C) Membrane with exclusive solubility of cations. Anion extraction coefficient is zero. [From R. P. Buck, *Crit. Rev. Anal. Chem.* **5**, 323 (1975).]

A general expression for $\Delta\phi_m$ at zero current, using $\bar{\gamma}_i = bc_i^{n-1}$, is

$$\delta\phi_m = \frac{nRT}{zF} \ln \left[\frac{a_i^{1/n} + \sum_j (k_{ij}^{\text{pot}} a_j)^{1/n}}{a_k^{1/n} + \sum_l (l_{kl}^{\text{pot}} a_l)^{1/n}} \right] \tag{65}$$

in which

$$k_{ij}^{\text{pot}} = \frac{\bar{u}_j}{u_i} \frac{K_{\text{ext},j}}{K_{\text{ext},i}} \tag{66}$$

and similarly for k_{kl}^{pot} where i and j ions in the left side bathing solution, while k and l are on the right side. This equation has been used to interpret the responses of synthetic ion exchanger resin and glass membranes.[41]

1.6.3. Electrical Properties of Class III, Liquid Ion Exchanger Membranes

1.6.3.1. Membrane Potentials for Permselective Liquid Ion Exchanger Membranes

Condition 1. $\bar{\gamma}_i \neq f(x)$

Condition 2. $\bar{u}_i \neq f(x)$

Condition 3. $\bar{\mu}_0 \neq f(x)$

Condition 4. $\bar{u}_i RT = \bar{D}_i$

Condition 5. $\int_0^d \bar{X}(x)\,dx = \bar{X}_0 d$ (confinement of sites within a membrane)

The most rigorous and comprehensive of the treatments of electroneutral, dissociated liquid ion exchange electrical properties is by Conti and Eisenman.[27] The zero-current, steady-state membrane potential for permeable ions of charge z is exactly the same as that in Eq. (65) ($n = 1$). However, Eq. (64) applies at all times after completion of the surface-ion exchange for fixed-site membranes, but only applies in the steady-state for liquid ion exchanger membranes. The reason for this difference is that, in contrast with fixed-site membranes, the inner-surface concentrations of exchanging ions continue to vary in time as the sites redistribute in space and time. Only in limited cases will Eq. (63) and (65) ($n = 1$) be exact at short times compared with the time required to reach steady state. The trivial case is a permselective membrane bathed by a single salt of a permeable ion and subjected to a step activity change on one side of the membrane. Another case is that involving mixtures of permeable ions, all of which have the same mobility in the membrane. In both examples, the profile of sites is flat at all times after initial double-layer charging.

At zero current, the distribution of mobile sites in space is flat (zero slope) for asymmetric bathing solutions of a single electrolyte, but is linear with finite slope when different permeable ions with different mobilities are present on either side. In that (bi-ionic or multi-ionic) case, the individual ion fluxes are nonzero and can perturb the site profile. Passage of current in the steady state also creates linearly tipped site profiles. Site concentrations are higher than average at the side where permeable ions enter and lower at the exciting side. This feature is a result of the condition that sites, while free to move, are constrained to the interior of the membrane.

Subsequently, Sandblom et al.[42] solved the algebraically difficult problem of membrane complexation for univalent permeable cations and a univalent anion liquid exchanger (subscript 's'). The total steady-state potential can be expressed as the usual log term plus an integral

$$\Delta\phi_m = \frac{RT}{F} \ln \left[\frac{a_n + \sum_m k_{n,m}^{\text{pot}} a_m}{a_k + \sum_i k_{k,l}^{\text{pot}} a_l} \right] + \int_0^d t\,d \ln \left[\frac{\sum_i u_{is} K_{is} c_i}{\sum_i u_i c_i} \right] \tag{67}$$

$$t = \frac{u_s c_s}{(u_s c_s / \sum_i u_{is} c_{is} + 1)(\sum_i u_i c_i) + u_s c_s} \tag{68}$$

In this statement, n and m are ions in the left bathing solution and k and l are on the right. Overbars have been omitted; the u are again mobilities, and other quantities have been defined earlier. When the membrane is fully dissociated $K_{is} = 0$, and only the first term remains.

For strong complexation, three cases must be noted. If only one permselective ion is involved, the response to variation in activity of this ion on one side is normal, viz., the first term in Eq. (67) remains. From potentiometry alone, one cannot tell from the Nernstian response that the permeable ion is strongly complexed or not complexed at all. For a two-permeable-ion system, of which one is complexed but the other is not, the response again follows the logarithmic term. However, the ion forming the ion-paired species is highly favored in the membrane and shows a large potentiometric selectivity coefficient since this quantity, discussed later, is directly proportional to the product of the ion-pairing formation constant and the single-ion extraction coefficient. When both permeable ions are complexed strongly, Sandblom et al.[42] evaluated the integral by further rearrangement and simplification. The responses are not necessarily Nernstian except when the bathing solution activities are widely different for the two ions. Increased selectivity through complexation is only possible when site and complex mobilities are nonzero.[43]

1.7. Time-Dependent Membrane Potential Theory

Time-dependent membrane potentials (profiles and net pd's) are computed from the time-dependent solutions for concentrations and local fields with reversible or irreversible boundary conditions.[44]

The three basic equations to be solved numerically are the Nernst–Planck equation (dilute solution form), the continuity equation, and Poisson's equation

$$J_i(x, t) = -D_i[\partial c_i(x, t)/\partial x - z_i c_i(x, t)(F/RT)E(x, t)] \tag{69}$$

$$\partial c_i(x, t)/\partial t = -\partial J_i(x, t)/\partial x + \text{kinetic terms (optional)} \tag{70}$$

$$\partial E(x, t)/\partial x = (1/\varepsilon)\rho(x, t); \rho(x, t) = \sum_i z_i c_i(x, t) \tag{71}$$

where

$$J_i(x, t) = \text{the flux of the } i\text{-th species}$$
$$c_i(x, t) = \text{the concentration of the } i\text{-th species}$$
$$D_i = \text{the diffusion coefficient of the } i\text{-th species}$$
$$z_i = \text{the valence of the } i\text{-th species}$$
$$E(x, t) = \text{the electric field}$$
$$\varepsilon = \text{the dielectric permittivity (rationalized)}$$
$$\rho(x, t) = \text{the charge density.}$$

For purposes of the simulation it is convenient to replace Poisson's equation [Eq. (71)] by the totally equivalent displacement current equation, as described by Cohen and Cooley[45]

$$I = F \sum_i z_i J_i(x, t) + (\varepsilon)\partial E(x, t)/\partial t \tag{72}$$

Equations (69) and (71) are equivalent to Eq. (50) and (52) but show explicitly time dependences. Because of computer simulation procedures that use fields rather than potentials, these equations are written in terms of field.

The time dependence of the electrical properties of membranes is the least well-understood topic among the fundamental characteristics. The dilute solution, one-dimensional Nernst–Planck equation is nonlinear. Space–time solutions for concentrations, fields, and potential profiles using the Nernst–Planck–Poisson system depend upon the magnitude of the perturbation.

Consequently, there is an important theoretical distinction between linear and nonlinear perturbations. When a membrane and its bathing solutions are initially in a steady state or an equilibrium state, the system can be perturbed by applied voltage, current, or external activity change of a permeable ion (usually on one side). These perturbations are linear, when the applied voltage or resulting voltage change is less than RT/zF V. For reversible interfacial processes, a concentration step increase or decrease of 50% of the initial value (or less) is within a linear range. Total linearization implies that steady-state responses obey ordinary linear partial differential equations. Linearization for calculation of time responses leads to forms in which there is a d-c steady-state solution plus time-dependent terms. It is possible to obtain "mixed" solutions in which the time response is found by linearization, but the solution may itself be the answer to a nonlinear problem. To achieve this form, one assumes a solution for current or voltage in the form

$$V = V_{DC} + \sum_n V_n \exp(-n\omega t) \tag{73}$$

For ordinary sinusoidal excitations, only the $n = 1$ term is retained.

Furthermore, linearized systems obey a quite general form in transform space

$$\Delta\phi_m(s) = Z(s)I(s) \tag{74}$$

where $s = j\omega$ and $\omega = 2\pi f$ (f = frequency in Hz). Given the system response function $Z(s)$ and the perturbation $I(t)$ whose transform is $I(s)$, the membrane potential can be found by inverting the transform. Conversely, a given applied voltage perturbation of any transformable type (step, pulse, sinusoid, etc.) divided by $Z(s)$ allows current vs. time to be computed.

An exact solution for the time-dependent properties of thin, site-free membranes permeable to a single-sign ion has been derived by deLevie et al.[46] under conditions of pure membrane control (exterior film diffusion limitation is ignored). The general problem of the time-dependent behavior of thick, site-free membranes permeable to ions of both signs has been discussed by Michaelis and Chaplain.[47] The most extensive exact solutions for thick and thin membranes containing two ions of opposite charge and arbitary mobilities, valences, and surface rate constants are by Macdonald.[48-50] His general theory is an extension of the uni-univalent cases previously treated.[51-56] The more recent theory does take into account surface adsorption; it allows for generation of charge carriers from the lattice ions or recombination of charge carriers (analogous in liquid membranes to formation of ion pairs). It is limited to the initial state of the system without space charge at either interface (so-called flatband potential or pzc). The results are cast in the form of impedance functions $Z(j\omega)$, which are thoroughly investigated by plots of parallel conductance and capacitance as functions of frequency.

The time-course of internal, interfacial, and total membrane potentials can be acquired by a process known as digital simulation.[57] The membrane and bathing solutions are hypothetically divided into volumes by planes parallel to the membrane surfaces. Each ion's transport is described by dimensionless, finite difference Nernst–Planck equations obtained from Eq. (69) using a set of unique nondimensionalizing factors.

An efficient finite difference simulation procedure for the steady-state and transient solution of the Nernst-Planck and Poisson equation system has been presented by Brumleve and Buck.[44] The procedure is general and extremely efficient. It contains refinements in time and distance scaling. The implicitly formulated nonlinear equations are solved with an iterative Newton–Raphson technique. Steady-state solutions, transient responses, and impedance-frequency responses have been presented for a number of examples from the fields of membrane electrochemistry and solid-state physics.

The efficient numerical solution of systems of coupled nonlinear partial differential equations has been made possible only in the past 10–15 years with the advent of large and fast digital computers. In 1965, Cohen and Cooley[45] developed a simulation procedure for the complete diffusion–migration problem. Although their procedure was explicit (nonpredictive), they introduced the system of reduced units still employed in most electrochemical simulations, and they included space charge via the displacement equation. In 1975 Sandifer and Buck[58] introduced a mixed explicit–implicit method similar to that of Cohen and Cooley. The procedure is implicit in the electric field variable and allows the reduced dielectric constant to assume any value (including zero). In 1974, Joslin and Pletcher[59] demonstrated how distance grid expansion could be used to economize on the total number of grid points and calculations in both explicit and implicit simulations. We have employed this concept in out treatment by tailoring the distance grid system to each specific problem.

In addition to the electrochemically oriented work described above, researchers in the semiconductor field have developed a number of excellent numerical procedures for solving the Nernst–Planck and Poisson equation system.[60-66]

The simulation procedure[44] can determine the impedance–frequency response of virtually any steady-state system (including all of Macdonald's flatband cases).

1.8. Time-Dependent Membrane Potential Parameters

Theories and experiments on polished, homogeneous-membrane electrode systems suggest that at least four widely separated time constants may be observable; however, this statement does not imply that voltage–time responses after a bathing solution activity step or impulse can be resolved into four exponentials. Not all processes necessarily occur, and there may be an overlapping of time constants. Furthermore, the slower processes involving diffusion-migration within a membrane are not precisely represented by a single time constant, but by a distribution, because of the form of response dependent on $t^{-1/2}$ rather than $\exp(-kt)$.

The shortest time constant in the 1–100 MHz region because of dielectric relaxation traceable in the case of liquid electrodes to solvent dipoles and reorganization of dissolved species and to ionic inertia, is not expected to be important for ISEs.

The next longer time constant is caused by charging the external space-charge surface regions coupled to the high-frequency or a-c bulk-electrode resistance. The capacitance is geometric and the high-frequency resistance is that value calculated classically from the concentrations and mobilities of charge-carrying species at their steady-state or equilibrium distributions. This quantity τ_B (or $\tau\infty$) is the usual high-frequency, thickness-independent, space-charge relaxation time in conducting systems. For silver chloride, it is about 0.2 ms at 25°C[67] while for glass 100–200 ms[68] is typical.

Capacitance is geometric and is given by

$$C_B = C_g = \varepsilon A / d \tag{75}$$

ε, the dielectric constant, is $\varepsilon_0 \varepsilon_r$ using ε_0 as the permitivity of free space and ε_r the conventional dielectric constant that is unity in free space.

The high-frequency resistance for an electroneutral, uniform concentration two-ion system of arbitrary charges and mobilities is

$$R_B = R_\infty = \frac{d}{F^2(z_+^2 \bar{u}_+ \bar{c}_+ + z_-^2 \bar{u}_- \bar{c}_-)A} \tag{76}$$

(For additional mobile ions add further terms of the same form into the denominator. Use appropriate charges, mobilities, and concentrations.) The time constant τ_B corresponds to the highest-frequency processes determined by transport of charged species in a medium of constant dielectric and constant resistivity. It is the product $R_\infty C_g$ (or $R_B C_B$) and corresponds to no change in the internal ionic concentration profiles. It is the usual high-frequency, space-charge relaxation time found in conducting systems.

A membrane excited by high-frequency a-c current is well represented by the simple $R_B C_B$ circuit because space charge density *in the membrane* cannot form and disappear in response to the exciting signal. The space charge represented by C_g is outside the membrane in the lower resistance bathing solution. Of course, one could then be more precise and represent the equivalent circuit as in Figure 8. Here the diffuse space-charge capacitances in the membrane are shown, even though they do not store charge or even become apparent at high frequencies.

The value of C_{dl} depends on the conditions of the system. For example, for one blocking electrode, the Gouy–Chapman theory gives

$$C_{dl} = A\varepsilon\kappa \cosh\left(\frac{\phi}{2}\right) = C_R \text{ or } C_F \tag{77}$$

where κ is again the reciprocal Debye length and $\phi = \phi_{dl}(RT/F)$, where ϕ_{dl} is the applied potential difference. ϕ_{dl} is, in general, the actual voltage across the diffuse layer being described. Since there is a C_{dl} for each diffuse layer, then the C_{dl} in Figure 8 is a series combination of an actual portion in the membrane (with corresponding ε_m, κ_m, and ϕ_{dl}) and a portion from the contact bathing electrolyte (with ε_{el}, κ_{el}, and ϕ_{el}). For a two-sided membrane there are four C_{dl} values in series. C_R and C_F are capacitances used in expressions for time constants when a surface resistance or surface film is present. Equation (77) holds only when:

1. The electrode separation is much greater than the Debye length.
2. The frequency of the applied potential is much lower than the frequency of relaxation of the diffuse double layer.
3. Positive and negative species do not recombine.
4. z_n, the charge on the negative species, and z_p the charge on the positive species, are both equal to one.
5. There is no specific adsorption at the electrodes.
6. $\Delta\phi_D$ (the diffusion potential) $= 0$.

The third time constant that may be observed is an intermediate value that arises from coupling of slow surface rates (including parallel processes such as adsorption) and the relaxed, doubly diffuse (or compact or both) capacitance of each interface. When surface rate processes are rapid and reversible, this time constant does not appear. The effective resistance at the

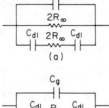

(a)

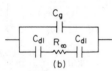

(b)

Figure 8. Some equivalent circuits for membrane cells with physical correspondence. These circuits apply to simple two-ion charge carrier cases; equal mobilities and equal concentration. (a) One ion unblocked—$C_g = C_B$ = geometric capacitance; $R_\infty = R_B$ = high frequency resistance; $R_0(DC) = 2 R_\infty$; $C_{dl} = C_F$ or C_R. (b) Both ions blocked.

equilibrium or steady-state potential, for slow surface rate control is

$$R_R = R_0 = \frac{2RT}{zFi^0} = \frac{2RT}{z^2F^2k_0(\bar{a}_+a_+)^{1/2}} \tag{78}$$

for single-cation transport, and

$$R_R = R_0 = \frac{2RT}{F^2[z_+^2k_0^+(\bar{a}_+a_+)^{1/2} + z_-^2k_0^-(\bar{a}_-a_-)^{1/2}]} \tag{79}$$

when ions of both signs transfer slowly with rate constants k_0^+ and k_0^- at unit activities. The appropriate relaxed capacitance C_{dl} (ignoring the compact value) is related to the doubly diffuse value given by the series combination of Gouy–Chapman capacitances, and proves to be given by

$$C_F \text{ or } C_R = C_{dl} = MC_g = MC_B \tag{80}$$

where

$$M = \frac{d}{2/\kappa} \tag{81}$$

is the number of Debye lengths in half of the membrane. The theory for blocked electrodes contains the capacitance $C_{dl} - C_g$ in series with the membrane resistance. For slow surface rates and thick membranes such that $C_{dl} \gg C_g$, the time constant is $R_\theta C_{dl}$.

It should be noted that in the linearized theory, C_{dl} is independent of interfacial potential. Actually, this cannot be true over a wide bathing solution concentration range. Thus $C_{dl} = C_R$ or C_F is expected to vary (increase) at all potentials away from the pzc according to Eq. (77) and to reach limiting values, given by the capacitance of the external and internal compact layers.

Slow surface rates behave as apparent resistances and can have a significant effect only if the values exceed the bulk resistance. However, it is possible that surface resistances can be identified as long as they are a measurable fraction, say 10% or more, of the bulk resistance. The slow time response of glass electrodes has been attributed to this source because the time constant can be shortened by surface etching. However, it is possible and even likely that the apparent surface resistance in this instance (and possibly OH$^-$ interference at LaF$_3$ electrodes) is associated with mass transport through the surface film of protonated and hydrolized glass. At 25°C, time constants as great as 30 sec for E-2 glass and 7 sec for GP glass were observed.[69] While these measurements were obtained by voltage perturbations across electrodes in homogeneous solutions, the same order of magnitude values were obtained by activity pulses. Markovic and Osburn[69] also cite many other references to long time constant values in the 1–10 sec range for glass electrodes. These time constants for electrodes with glass surface films are longer than expected (and found) for relaxation of concentration polarization at electrode surfaces without surface films.

The longest membrane-controlled time constant arises from concentration polarization of charge carriers within a membrane electrode and is observed when a current passes the membrane. This is the Warburg finite-diffusion process and the appropriate resistance is the d-c value, which is significantly greater than the a-c value because only permeable ions carry current at d.c. The d-c capacitances are large and arise from charge separation due to transport. The Warburg behavior requires at least two charge carriers, both of which contribute to high-frequency current, but one ion is partially or completely blocked at d.c. Thus, Warburg behavior is expected for liquid ion-exchange electrodes and may possibly be observed for solid and glass membranes if transport involves typically vacancy and interstitial motion, but it is not expected for neutral-carrier systems, lipid bilayers, or fixed-site membranes, through which ions of only one sign are permeable and transported by a single mechanism.

In the low-frequency (long time or few second) or finite Warburg region, the membrane-controlled time constant is that required to adjust concentration profiles throughout the membrane. If only cations are permeable, the d-c resistance R_W is

$$R_W = \frac{d}{F^2 z_+^2 \bar{u}_+ \bar{c}_+} = \left[1 + \frac{\bar{u}_-|z_-|}{\bar{u}_+ z_+}\right] R_\infty \tag{82}$$

for ions of charge z_+ and z_-. The d-c capacitance (really a diffusional pseudo-capacitance) is approximately

$$C_W = C_{0,r} = \frac{dF^2}{2RT}(z_+^2 \bar{c}_+ + z_-^2 \bar{c}_-) = M^2 C_g \tag{83}$$

for potentials near the pzc, and the time constant (defined as the reciprocal frequency at maximum imaginary impedance) is given by

$$\tau_W = \frac{M^2}{2.53} \tau_B \text{ for } z_+ = |z_-|, \; \bar{u}_+ = \bar{u}_- \tag{84}$$

The d-c capacitance for unequal ion mobilities

$$C_{0,r} = \frac{dF^2}{2RT} \frac{[z_+^2 \bar{c}_+ + z_-^2 \bar{c}_-]}{[1 + u_-|z_-|/u_+ z_+]} \approx M^2 Cg[1/(1 + u_-|z_-|/u_+ z_+)] \tag{85}$$

and the time constant (defined as the reciprocal frequency at maximum imaginary impedance) is given in terms of $C_{0,r} R_W$ by

$$\tau_W = [1/2.53][z_+ z_-/(z_+ + |z_-|)^2][(u_+ + u_-)^2/u_+ u_-]C_{0,r} R_W \tag{86}$$

This definition coincides with that for any true parallel network, even though Warburg equivalent network is not a simple parallel RC, but a finite transmission line. Time constants τ_∞ and τ_W apply to membranes in which cations and anions move within the membrane at high frequencies, but only the cations or anions pass the membrane surface and carry current at zero frequency. When both cations and anions pass the membrane surface at high and low frequencies, resistances are nearly the same and the finite Warburg region does not exist. Fixed-site membranes permselective to ions of a given charge and lipid bilayers permeable to an ion of a single charge should not show Warburg response. Experimentally, Warburg behavior has not been observed for glass membranes or for lipid bilayers, but the possibility of some mechanism equivalent to transport of two oppositely charged species cannot be ruled out for solid-state systems.

Finally, when concentration polarization (external film diffusion limitation) can occur in the bathing electrolytes, another time constant arises that is of the same form as above: the Nernst thickness δ replaces d, and the mobilities of ions in solution replace those for the membrane in the expressions. Effect from rates of adsorption and from slow generation of mobiles species from lattice positions have not been described here.

The theoretical models that give the derived results are principally found in Macdonald's work. His assumptions, given below, were listed by Archer and Armstrong.[70]

1. All the mobile ions in the electrolyte are point charges.
2. The electrolyte consists of a continuum dielectric and these point charges.
3. The results obtained apply only to samples in the form of a parallelepiped.
4. Acceleration effects are neglected. In doing this, Macdonald assumes that the ions reach their terminal velocity spontaneously when subjected to the applied field. This assumption probably breaks down at high frequency ($>10^8$ Hz).
5. The distribution of counterions and co-ions around the ion under consideration is neglected, i.e., Debye–Hückel effect are neglected.
6. Any compact double-layer effects are neglected. This means that the center of any ion can be located exactly at the electrode surface. This is not the case if ionic size is taken into account, since the distance of closest approach of the ion to the electrode surface is then controlled by the ionic radius. Later work explicitly covers effects of the compact layer.[71]

Armstrong has considered the structure of compact layers, particularly emphasizing how packing of solids near electrode surfaces affects space-charge densities. The review of Archer and Armstrong[70] gives examples of these effects as they apply to molten salt electrolytes and to the double-layer structure in ionic solids.

1.9. Potential–Time Responses after Activity Steps (Responses Outside the Linear Regime)

Time dependences of voltages and currents out of the linear regime have often been measured despite the difficulty in interpreting the results. Using fast-flow techniques to create step activity changes in the bathing solution on one side of a membrane, it is possible to avoid the long-time concentration polarization response. The time constant for readjustment of the membrane potential corresponds well with the high-frequency membrane-controlled charging process. Without special precautions involving fast-flow techniques[72] and rotating-disk methods,[73] observed time constants from "dipping experiments" will correspond to film diffusion values (concentration polarization-determined values) as the stagnant film of initial solution changes by diffusion–migration to the new stepped value shown in Figure 9. Numerous examples of these measurements show time constants that are consistent with either film diffusion on the solution side of the interface or slow surface processes. The latter may actually be the time constant for diffusion-migration through a surface layer, or slower diffusion inside the membrane surface.

For membranes that are permselective to ions of a given sign, time responses to step changes in solution activities depend upon the slowest of the three processes: surface-ion exchange, bulk transport through the stagnant film "Nernst diffusion layer" outside the membrane, or bulk transport through a stagnant film on (or in) the membrane surface region. The latter may be hydrolyzed surface film on glass electrodes, a very slowly encroaching water layer in the surface of liquid ion exchanger membranes, or an adsorbed or occluded layer of material at or near the surface of membranes. Slow adjustment of concentration profiles within the bulk of a fixed-site ion exchanger is not usually considered, because the membrane potential is only dependent on

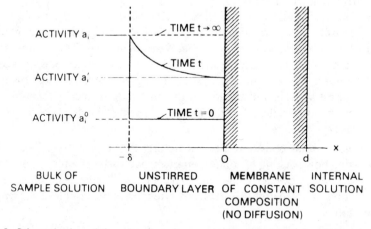

Figure 9. Schematic view of the relaxation of concentration polarization in a Nernst boundary layer film at the surface of a nondiffusive membrane electrode. The initial condition is uniform activity a_i^0. The final condition after a step activity change at $x = -\delta$ is a_i. The activity in the layer moves through a series of profiles of which a_i' is a typical example at time t. In following equations, $a_i' = a_i(t)$; $a_i^0 = a_i(0)$; $a_i = a_i(\infty)$. [From W. E. Morf, E. Lindner, and W. Simon, *Anal. Chem.* **47**, 1596 (1975), with permission.]

surface activities. However, ion exchange membranes with mobile sites, neutral-carrier membranes, and all ion-exchanging membranes exposed to ions of different charges or subject to ion pairing, can show very long time constants because the steady-state concentration profiles through the entire membrane must be achieved before a steady value of membrane potential will be attained.

The simplest case is that in which an external solution film controls the time response.[74] Diffusion of fresh electrolyte into the static Nernst layer determines the time response. This time-dependent problem is analogous to the finite Warburg case at long time where concentration builds up exponentially according to

$$a_i(t) - a_i(0) = [a_i(\infty) - a_i(0)][1 - \exp(-t/\tau)] \tag{87}$$

where

$$\tau \sim \delta^2/2D \tag{88}$$

$a_i(0)$ is the concentration in the stagnant Nernst layer prior to the concentration step, and $a_i(\infty)$ is the new bulk value, which slowly equilibrates by diffusion-migration throughout the layer. D is the diffusion coefficient of the perm-selective, potential-determining ion in the presence of an inert supporting electrolyte. D is replaced by the salt diffusion coefficient when a single, binary electrolyte is used. The measured potential $\phi(t)$ varies monotonically (except possibly at very short times where other processes are effective) to the new equilibrium value according to

$$\phi(t) - \phi(\infty) = \frac{RT}{F} \ln \left\{ 1 - \left[1 - \frac{a_i(0)}{a_i(\infty)} \right] \exp(-t/\tau) \right\} \tag{89}$$

Even though one time constant appears, the response to a step increase is more rapid than a step to a more dilute solution as shown in Figure 10. This curious effect arises from the form the equation takes. As expected, the time response depends on δ and so is sensitive to stirring.

The exponential form of Eq. (89) is a long-time solution to the diffusional filling problem for a confined region of space. At short times, the activity buildup depends on $t^{1/2}$ in a complicated way. This fact has important consequences when two diffusional barriers are adjacent to one another. For example, filling the external electrolyte Nernst layer with ions, some of which pass into a hydrated film on a glass or in a liquid membrane, requires a flux balance calculation at the surface. For high-mobility ions in aqueous solution, with much lower mobilities in the membrane surface film, the outer activities may be in the exponential regime, while the inner film is filling as $t^{1/2}$. Morf et al.[74] have investigated this case as a model for neutral-carrier membrane responses, but it may also apply to liquid ion exchange membranes whose surfaces contain a hydrated layer and relatively low mobile-site densities. In

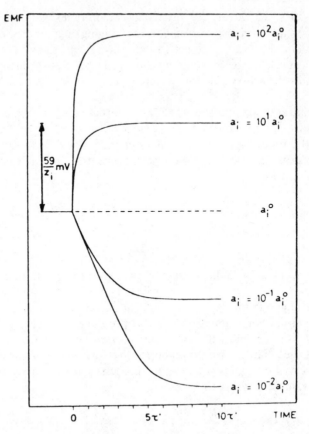

Figure 10. Theoretical EMF response vs. time profiles for ion exchange membrane electrodes, calculated according to Eq. (89). [From W. E. Morf, E. Lindner, and W. Simon, *Anal. Chem.* **47**, 1596 (1975), with permission.]

this case

$$a_i(t) - a_i(0) \sim [a_i(\infty) - a_i(0)] \left[1 - \frac{1}{(t/\tau)^{1/2} + 1} \right] \tag{90}$$

$$\phi(t) - \phi(\infty) = \frac{RT}{F} \ln \left\{ 1 - \left[1 - \frac{a_i(0)}{a_i(\infty)} \right] \frac{1}{(t/\tau)^{1/2} + 1} \right\} \tag{91}$$

This approximation has been tested for neutral-carrier electrodes and found acceptable. It is also observed that time constants are larger for liquid-ion-exchanger membranes subject to step changes of two ions, when one ion at lower concentration, is, in fact, preferred by the membrane ($K^{\text{pot}} > 1$). Step changes of a single ion (or mixture when the second ion is not preferred by the membrane) give smaller time constants.[75,76]

Nonmonotonically changing potential responses to step changes in solution activities have been observed. These are nonequilibrium effects that have not yet been sufficiently well characterized to lead to a general interpretation. In a study by Bagg and Vinen,[77] it was found that nonmonotonic potential changes are sensitive to stirring and are temperature independent on the rising potential (short-time overshoot region), but are stirring insensitive and temperature dependent on the return swing. This result suggests a short-time charging process in which rapid surface ion exchange leads to space charge of the preferred ion during the overshoot. Slow membrane transport of sites to compensate this charge may account for the return to equilibrium and the temperature dependence.

This hypothesis for over- and undershoots using liquid ion exchangers was investigated by Mathis and Buck.[78] When a membrane, e.g., Aliquat nitrate in nitrobenzene, is removed from an NO_3^- bathing solution and placed in another pure electrolyte say KI, or KCl, rapid ion exchange occurs at the interface where NO_3^- is replaced by I^- or Cl^-. There will be an initial, external mass transport-controlled step in interfacial pd. Sign and magnitude depend on ion charge and single ion partition coefficient. Relaxation of site profile creates a diffusion potential that either adds or subtracts (absolutely) from the step. Thus, there can be a maximum (or minimum) in pd or a slow relaxation in the *same* sense as the initial transient. The effect depends on relative mobility of entering ion (I^- or Cl^-) and exiting ion (NO_3^-).

Reports of ion-selective electrode responses frequently include time dependences (measured cell or membrane potentials vs. time) after "step" activity changes are made on one side.[69,75–83] An example is given in Figure 11 using data of Shatkay.[83] If an activity step were, indeed, one with hypothetical infinite slope at the leading edge, then the resulting $\Delta\phi$ vs. t response would be a measure only of electrode properties. However, the "step" is never an ideal, abrupt activity change on the outer test-solution side of a membrane-type electrode. Except for a few studies based on stopped flow and related techniques,[72,84–86] a dip method is used; the measurement is complicated by convolution of a further time constant into the response. This additional time constant reflects establishment of a new bulk concentration at an electrode surface by diffusion within the external static Nernst layer of bathing electrolyte of dissolving membrane species. In our experience, this latter time constant is often longer than the electrical relaxation time or the kinetic relaxation time for slow ion transfers at an electrode. This activity-step time constant can be confused with properties of the electrode itself.

Pure response to ideal, rapid activity steps at one interface can be compared with responses for current steps using digital simulation. Slow attainment of external stepped activity by diffusion relaxation of concentration polarization is not included in the simulation.

It is common experience that an ion-selective electrode exposed suddenly to a fresh solution will change its initial measured potential to a new one through a series of stages.[69,75,83] These stages are not necessarily obvious or resolved steps in time; the transient is often a featureless monotonic with a decreasing slope. However, analysis of the $\Delta\phi$ vs. t transient shows that there are generally several time constants, even though there are no arrests between the segments. The processes and their associated time constants can, for ideal systems, be identified, itemized, and ranked from impedance theory.[1,2,87,88]

The various time constants described above are connected to the physical processes for transport of charged species across interfaces and through the bulk of membranes used as ISEs during current flow. The act of measuring the cell voltage after an activity step of potential altering magnitude, involves also the same transport processes. However, the processes may not affect the outcome (ϕ vs. t) in the same way.

An analysis of this problem was given by Stover, Brumleve, and Buck.[89] The chief result was an observation that fewer time constants are observed in an activity step experiment compared with the current step experiment. The reason is that forced current affects space-charge distributions, electric fields, and ionic motion at both interfaces and throughout the membrane. But the activity step affects fields and space charge at one interface. Thus, slow

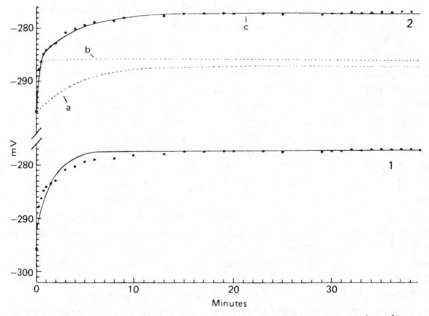

Figure 11. Potential response on transfer of Ag_2S electrode from $AgNO_3$ 10^{-4}–10^{-5} M. Points are experimental: curves are theoretical. $\phi^0_{\text{exp.}} = -296$ mV. Curve 1 best fit from Eq. (92) using $i = 1$, $\phi^0 = -292$ mV, $k_1 = 15$ mV, $\tau_1^{-1} = 0.60$ min^{-1}; best $\phi_{\infty,\text{calc.}} = -277$ mV. Curve 2, using $i = 2$, $\phi^0 = -296$ mV, $k_1 = 10$ mV, $\tau_1^{-1} = 5.0$ min^{-1}, $k_2 = 9$ mV, $\tau_2^{-1} = 0.24$ min^{-1}; best $\phi_{\infty,\text{calc.}} = -277$ mV, $\sigma = 0.3$ mV (s.d.). [From A. Shatkay, *Anal. Chem.* **48**, 1039 (1976), with permission.]

interfacial kinetics limit responses after the activity step. In that case, the bulk semicircle charging processes do not occur on the expected rapid time scale. It was found that those observed processes affecting voltage in the activity step have the same numerical values of time constants found in the impedance experiment.

Resulting characteristic system functions, the impedance function $Z(j\omega)$, contain time constants that appear experimentally in $\Delta\phi$ vs. t plots at constant I or, in a more readily apparent way, in impedance plane plots. Time constants, when widely spaced, give rise to responses to constant I in the form[1,75]

$$\Delta\phi = \sum_i k_i[1 - \exp(-t/\tau_i)] + \Delta\phi^0 \qquad (92)$$

Equation (92) is a monotonic function that can show arrests or general featureless behavior *similar* to experimental results for activity steps on one side of a membrane in an electrochemical cell. The identifiable time constants are sufficiently similar to the time constants expected for equivalent current or voltage perturbations that it seems important to make theoretical comparisons between the origins and magnitudes of the time constants found by the two different experimental methods.

The system under study is a permselective membrane containing totally blocked, mobile, negative sites. Positive counterions are permeable and may be rate-limited at the interfaces. Numerical values of reduced constants are given to achieve the following ends: the membrane parameters are given values such that M, the number of Debye lengths in half the membrane, is 10^5; this results in well-separated time constants for the geometrical and diffusional processes; the reduced geometric capacitance has the value 5×10^{-6}, and the reduced high-frequency resistance R_B has the value 10^{-6}, while R_0 (d-c) is 2×10^{-6} (for reversible counterions). The reduced charging time constant (τ_B) is 5×10^{-12}, and the reduced diffusional time constant ($\tau_W = (M^2/2.53)\tau_B$) is 2×10^{-2}. The initial state is the flatband condition: $I = 0$, $\Delta\phi = 0$ with symmetrical bathing solutions.

Case I. The first case to be considered is the membrane bathed in a single salt whose counterions are completely reversible at the interfaces. The salt activity in the left bathing solution is raised by 1% corresponding to a linear activity step. The final reduced membrane potential $\Delta\phi\infty$ is $\ln(1.01) = 0.01$. For analysis of the time response of the system, a quantity $\ln(\Delta\phi\infty - \Delta\phi(t))$ is calculated; this is plotted in Figure 12, curve (a). It should be noted that the zero time intercept is $\ln(\Delta\phi\infty) = -2 \times 2.303$ or -4.606 reduced units.

For the activity step, the counterion concentration at the membrane surface establishes its equilibrium value instantaneously, and a diffuse space-charge region begins to develop just inside the membrane. At longer times, the electric fields near the interface grow; sites become depleted and a diffuse double layer resembling an equilibrium Gouy–Chapman distribution forms. Eventually, the salt concentration in the bulk of the membrane is raised very

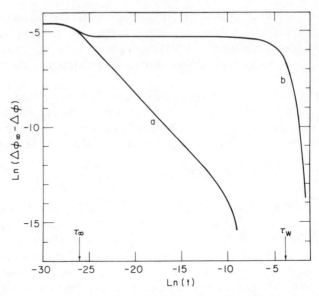

Figure 12. Curve (a): logarithmic potential–time plot for a linear single-salt activity step at a thick permselective liquid ion-exchange membrane with equilibrium surface extraction rates; membrane parameters are $D_+ = D_- = 5.0 \times 10^{12}$, $d = 10^6$, $K = 5.0$, $K_{eq} = 1.0$, $k_b = 10^{15}$, and $I = 0$; membrane salt concentration is 0.1; initial salt activity in the bathing solutions is 0.1; left bathing solution salt activity is stepped to 0.101; final reduced potential is 0.01. Curve (b): logarithmic potential–time plot for linear current step; membrane parameters are the same, except that the left bathing solution salt activity is held at 0.1, and current density is stepped from 0 to 5000, resulting in a final reduced potential of 0.01; low-frequency time constant, $\tau_W = 2.0 \times 10^{-2}$; geometric time constant, $\tau_\infty = 5.0 \times 10^{-12} = \tau_B$.

slightly (in the absence of electric fields), and a small equilibrium diffuse double layer forms at the right interface. The first time constant [when $\Delta\phi(t)$ has approached within $1/e$ of the net change] proves to be 5×10^{-12} reduced time units. This result is expected and demonstrates that the electric space-charge relaxation time has the same value for space charge that occurs across a membrane as for diffuse regions external to the two interfaces, or as for two diffuse regions across a single interface. There is no longer time constant in this case, and the membrane potential continues to approach the steady-state value with a $t^{-1/2}$ dependence.

    Case II. The membrane is the same as in Case I, but a current step is made, which results in the same steady-state potential change as in Case I. For this perturbation, the expected $\tau_\infty = 5 \times 10^{-12}$ is seen. In contrast to the single-salt activity step result, a second time constant also appears in the potential response, as shown in Figure 12, curve (b). Polarization of the bulk concentration gives rise to the long-time diffusional time constant. The current step produces rearrangement in the bulk concentration profiles that the previous activity step cannot produce. The salt concentration profile remains flat at short times, and the electric fields grow uniformly throughout the membrane. After this charging process has been completed and the time scale reaches the diffusion time, the salt concentration profile tips, and equilibrium diffuse double layers containing equal and opposite charge form at the interfaces.

Case III. Although thick membranes exposed to a single-salt linear activity step (as in Case I) do not show a second, diffusional time constant, certain activity steps may produce internal bulk concentration changes and long time-constant responses in two different ways.

First, a membrane exposed to two or more permeable ions will display potential transients with long time-constants when a significant change in the ratio of the external activities of permeable ions occurs, and the mobilities of the permeable ions are different. For membranes exposed to a single permeable ion, variations in external activity result in potential changes through the adjacent space-charge region only. When more than one permeable ion is present, activity variations can lead to substantial changes in the value of the internal electric fields (diffusion potential changes) as well as potential changes through the opposite space-charge region. These latter two processes will occur with a diffusional time constant, τ_W.

An example of the potential-time response is shown in Figure 13, curve (a). The membrane is exposed to a single-salt solution of a second permeable ion. This activity change to a bi-ionic system displays an identical geometric time constant to that seen in Case I, because the mobility of the second species is nearly the same as the first. The magnitude of the geometrical time response accounts for approximately 90% of the total potential change. As can be seen from Figure 13 the final 10% of the potential transient is due to the potential drop through the membrane bulk and the right space-charge region, and is delayed with a time constant τ_W, which agrees with the calculated value.

Membranes exposed to more than one permeable counterion will not always display the long time-constant potential response. The potential–time response to a 1% change in the ratio of permeable counterion activities is shown in Figure 13, curve (b). The geometrical potential transient accounts for the total potential change, and no long time response is seen. In this case, bulk rearrangement of the internal concentration profiles produces insignificant changes in the diffusion potential.

With regard to the second possibility, thin membranes are also capable of displaying diffusional time constants with a single-salt bathing solution. In thick membranes, formation of the diffuse space-charge region at the interface where the activity step occurs causes little change in the bulk ionic concentration profiles. However, formation of diffuse space charge regions in thin membranes can substantially affect bulk concentrations. Activity steps may be large enough to displace significant fractions of the total sites from the interface into the bulk of the membrane. Bulk concentration changes result in variations in the diffuse space-charge region at the opposite interface. The potential drop through the second space-charge region will then account for a substantial

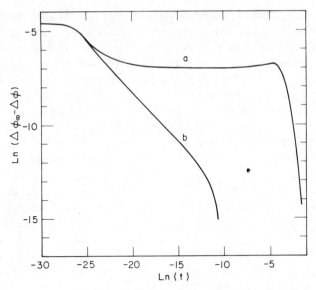

Figure 13. Curve (a): logarithmic potential–time plot for a multi-ion activity step at a thick permselective liquid ion-exchange membrane with equilibrium surface extraction rates; membrane parameters are the same as for Figure 12(a) except for external activities; original permeable ion is at an activity of 0.1 in both bathing solutions; step in left bathing solution is to a pure solution of a second permeable ion at an activity of 0.1005; membrane diffusion coefficient for the second ion is 5.025×10^{12}; final reduced potential is 0.01. Curve (b): logarithmic potential–time plot for a multi-ion activity step displaying no low-frequency time constant; original bathing solution contains two ions with diffusion coefficients of 5.1×10^{12} and 5.0×10^{12} at equal activities of 0.05; step raises the left bathing solution activity of the first ion to 0.051; final reduced potential is 0.01.

portion of the total potential transient, and will change on a time scale equivalent to the membrane diffusion time.

Case IV. Membranes with slow interfacial transfer rates. Single-salt activity steps at permselective membranes result in potential time responses governed by geometrical and, in some cases, diffusional time constants. These membranes have surface rate processes that are rapid and reversible. A different time response is expected when permeable ion transport across the interface is limited by slow surface rates, and the resulting apparent interfacial resistance is larger than the bulk resistance.

An approximate expression for a slow interfacial time response to an activity step may be derived by assuming that the charging process establishes an equilibrium diffuse double layer at each instant of time. Consideration of the linearized Gouy–Chapman concentration distribution yields

$$\bar{c}_0(t) = c_0 K_{\text{ext}} - [c_0 K_{\text{ext}} - \bar{c}_b] \exp\left[-(k_b \kappa/2)t\right] \qquad (93)$$

where $\bar{c}_b$ is the concentration of the ion in the bulk membrane. Since the time-dependent membrane potential is simply $(RT/F) \ln \bar{c}_0(t)/c_b$, and the steady-state potential is $(RT/F) \ln (c_0 K_{\text{ext}}/c_b)$, the potential–time response should be of the form

$$\Delta\phi(t) = (RT/F) \ln \{\exp(F\Delta\phi_\infty/RT)$$
$$-[\exp(F\Delta\phi_\infty/RT) - 1]\exp(-t/\tau_R)\} \qquad (94)$$

This expression is formally equivalent to that given by Morf *et al.*[74] for boundary-layer diffusion-controlled potential transients. In practice surface film and interfacial rate control are distinguishable. The time constant for surface film control is approximately $\delta^2/2D$ (where δ is the Nernst thickness and D is the diffusion coefficient of the bathing solution salt) and should depend on the stirring rate. The time constant used in Eq. (94), $\tau_R = 2/k_b\kappa$, is exactly the value obtained from impedance measurements, $R_\theta \cdot C_{0,r}$ or $(R_R C_R)$.

Rate-limited membrane potential responses from activity-step perturbations are compared with current step responses in Figure 14. The striking feature of the activity step response is that no geometrical charging response is seen. A single surface-rate time constant is displayed, which accounts for the entire potential transient. As in the earlier cases, long-time diffusional response is not observed. The current step time-response displays potential changes that are due to all three membrane processes: geometrical, interfacial, and diffusional. The values of the time constants for the current step responses are in excellent agreement with the theoretical expressions given earlier.

1.10. Potential-Time Responses for Solid Membrane Electrodes

The complete, detailed analysis and derivation of exact equations for the time response of many ISEs is not available. The reason is that the theory of mass transport, concentration, field and potential profiles for single activity steps on one side of a membrane is not completely developed. The nonlinear Nernst–Planck–Poisson system has been partially but not generally solved, even for parallel-faced, homogeneous, conducting phases with charge carriers of constant mobilities. In addition, real systems are not ideal and many factors such as heterogeneities,[90] microcracks,[91,92] high-resistance surface layers (in or on supports),[93-95] time-dependent resistances of supports (e.g., because of hydration),[78,96] slow extraction or dissolving of membranes and crystals into test solution,[97,98] slow adsorption in parallel with diffusion to the electrode surface,[90,99] and other unrecognized processes, conspire to complicate analysis further.[100]

Solid membrane electrodes, including glasses, are not, or do not remain permanently, ideal, atomically smooth ion exchangers. They do not show, exclusively, the theoretical time constants predicted from surface kinetic and mass transport considerations. The reasons for the appearance of additional

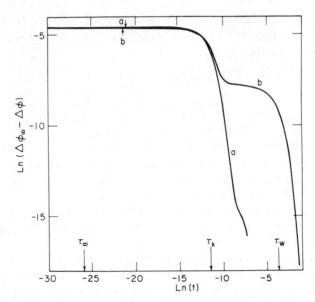

Figure 14. Logarithmic potential–time plots for linear activity and current steps at a membrane with slow surface extraction rates. Curve (a) plots activity step; membrane parameters are $\tau_R = 1.0 \times 10^{-5}$, as calculated for impedances $(R_\theta C_{0,b})$ and the expected time constant $2/k_b\kappa$. Curve (b) plots current step; membrane parameters are the same as for Figure 12 (curve b), except that $k_b = 1.0 \times 10^6$ and current density is stepped to 833. Final reduced potential is 0.01. $\tau_B = \tau_\infty = 5.0 \times 10^{-12}$, $\tau_R = \tau_k = 1.0 \times 10^{-5}$, $\tau_W = 2.0 \times 10^{-2}$.

long time-constant, exponential, potential-time transients seem to include surface chemical inhomogeneity, physical nonuniformity, including roughness and, more important, grain boundaries, crystallite boundaries and crevasses in the surfaces. The resulting slow processes include: slow contact adsorption and slow desorption coupled with double-layer charging to produce a kind of surface rate process;[101,102] surface film heterogeneity that produces barriers (high resistance) regions with surface-rate time constants different from the adjacent lower resistance regions;[103] and slow adjustment of bathing activities in the regions of the surface because of local adsorption, leaching, and dissolution.[104,105] Local uneven leaching may be responsible for the suggestion that there are local cells, e.g., lateral variations of interfacial pd's along an electrode surface.

A series of recent papers using surface analysis[106–108] demonstrate non-uniform surfaces. Time responses after "switched" flow from one concentration jet to another have been brilliantly exploited by Pungor et al.[103–105] to measure potential–time transients.

Presence of extraneous adsorbed material in solid electrode materials has been pointed out by many authors as an explanation of the different behavior of electrodes such as Ag_2S, depending on whether $AgNO_3$ has been added to

Na_2S or conversely, during the precipitation of Ag_2S. Recent evidence that time constants, especially in low-activity bathing solutions, depend upon leaching and adsorption along grain boundaries has been presented by van de Leest and Geven.[102] Time responses of mixed sulfide electrodes have been difficult to explain because of the variety of factors that are not apparent. Certainly redox reagents and complex-forming reagents have been studied in terms of time responses and steady potential development. Reducing agents stabilize responses. Complexing agents sometimes improve response speed and stability. However different complexing agents are not equally effective, for unknown reasons. Directional effects, i.e., negative-going potentials upon addition of cation ligands, are thermodynamically consistent.[109–111]

The effects of adsorption and physical heterogeneity of surfaces are "kinetic." The equilibrium interfacial pd depends on bulk solid activities and bulk solution activities of exchanging ions. Except for surface barriers, chemical and physical, electrode membrane compositions are frequently unit mole fraction, or a significantly large mole fraction, in comparison with impurities. Consequently, equilibrium pds are expected to prevail after slow surface processes have reached equilibrium or steady state. In addition to transport through a resistive film, adsorption and desorption of exchangeable ions and other species in the sample are *parallel* processes with ion exchange. At reversible ion exchanging interfaces, adsorption is not considered to be potential determining, but adsorption and desorption require simultaneous equilibration with ion exchange before the steady pd can be established. Archer and Armstrong[70] point out that a possible long-time process is the rate of adsorption equilibrium coupled to the diffuse double-layer capacitance. Especially in low bathing activity electrolytes, a long time-constant can often be seen that has been identified with slow electrolyte activity equilibration near the electrode surface. Because an electrode only senses activities at its surface, slow adsorption and desorption, especially along deep paths such as grain boundaries, may lead to observed long time constants in the 10–60 sec range.

2. Ion-Selective Electrode and Membrane Electrode Formats and Materials

2.1. Conventional Ion-Selective Electrode ISE Formats—General Comments

ISEs are more often cylindrical cells $6'' \times 1/4''$ or $1/2''$ with a lead wire exiting at the top and a membrane sensor located at the lower end. The membrane with diameter from about 1 cm to about 1/4 cm is held in place by various means including glue, springs, removable "O" rings, and plastic rings that are part of a base or screw-on cap. Membranes of solid material are easily made rigid, but liquid membranes must be supported in inert matrices. Examples of tip construction are shown in Figures 15 and 16. The

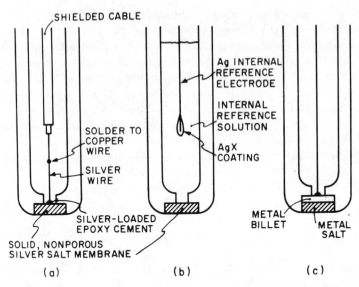

Figure 15. Examples of selective electrodes based on inorganic salts: (a) all solid-state electrode; (b) ion-selective electrode with internal solution and internal reference electrode; (c) electrode of the second kind.

conventional format is intended for "dip" measurements with samples large enough to provide space for an external reference electrode. Single combination electrodes described in Section 3.5 are useful for smaller samples because the external reference electrode is built in, concentrically or off center, but attached to the basic ISE body. Drilled and channel-cap electrodes are intended for use with flowing samples. The solid electrodes can be drilled for flowthrough applications, while the special caps are used to channel flowing samples past supported liquid membranes. Electrodes in a U-shape with the active surface upward are used with microsamples, typically one drop.[3]

   ISEs are useful for monitoring activities in flowing solutions and in stirred, batch solutions because electrodes detect and respond only to activities at their surfaces. Time responses of solid and liquid membrane electrodes to ideal step activity changes of the principal ion (already present in the membrane) can be very rapid: 200 ms for glass electrodes to about 30 μs for AgBr electrodes, provided the time for mixing of fresh solutions with the stagnant electrolyte surface layer is made rapid by stirring.

2.2. Kinds of Cells

   A membrane electrode is a portion of a membrane cell. The entire cell is a two-terminal (or two-electrode) device consisting of the membrane electrode and an external reference electrode. The latter may be physically attached

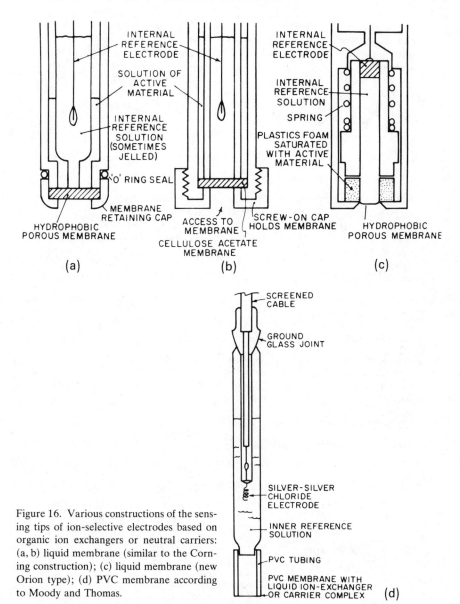

Figure 16. Various constructions of the sensing tips of ion-selective electrodes based on organic ion exchangers or neutral carriers: (a, b) liquid membrane (similar to the Corning construction); (c) liquid membrane (new Orion type); (d) PVC membrane according to Moody and Thomas.

to a single probe shaft, as in the combination electrode format. Complete cells for membrane configuration electrodes (Cell I and II), and all-solid-state configuration (Cell III) are shown in Figure 17. Cell I is cation responsive, while Cell II is anion responsive. Conventional cells (IV and V) are shown for comparison in Figure 17. Also shown in Figures 17 and 18 are sketches of the physical layout of cells. Figure 17 is cell III with a classical salt bridge.

| Cu; | External reference electrode: typically mercury–calomel or silver–silver chloride in fixed-activity Cl^- | Salt bridge | External monitored solution | ; Membrane reversible and perm-selective to cations | ; Inner filling solution containing a fixed activity of perselect-ive cations and fixed Cl^- activity | ; AgCl; Ag; Cu Internal reference electrode |

"Membrane" electrode [Cell I]

| Cu; | External reference electrode typically mercury–calomel or silver–silver chloride in fixed-activity Cl^- | Salt bridge | External monitored solution | ; Membrane reversible and perm-selective to anions | ; Inner filling solution containing a fixed activity of permselect-ive anion and fixed Cl^- activity | ; AgCl; Ag; Cu Internal reference electrode |

"Membrane" electrode [Cell II]

| Cu; | External reference electrode typically mercury–calomel or silver–silver chloride in fixed-activity Cl^- | Salt bridge | External monitored solution | ; Membrane reversible and perm-selective | ; Ag; Cu |

All-solid-state electrode [Cell III]

(a)

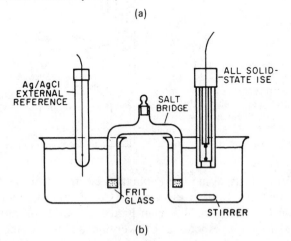

(b)

Figure 17. (a) Cell arrangements for ion-responsive membrane-type electrodes. (b) Sketch of a typical cell embodiment.

Cu; | External reference electrode, typically mercury–calomel or silver–silver chloride in fixed-activity Cl^- | Salt bridge | External monitored solution containing M^+ | ; M; | Cu | [Cell IV]

Cu; | External reference electrode, typically mercury–calomel or silver–silver chloride in fixed activity Cl^- | Salt bridge | External monitored solution containing X^-, saturated with MX | ; MX; M; | Cu | [Cell V]

(a)

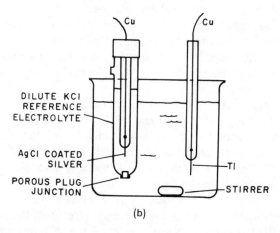

DILUTE KCl REFERENCE ELECTROLYTE

AgCl COATED SILVER

POROUS PLUG JUNCTION

Cu   Cu

Tl

STIRRER

(b)

Figure 18. (a) Cell arrangements for ion-responsive classical electrodes of first (cell IV) and second (cell V) types. (b) Sketch of a typical classical cell embodiment.

Figure 18 shows cell IV with a commercial reference electrode with attached, surrounding electrolyte compartment.

The commonly found organization illustrated here corresponds to classical junction cells. The vertical lines are conventional indications of interfaces involving solid–solid, solid–solution, and immiscible liquid/liquid contact equilibria. The dashed vertical lines are indications of interfaces between miscible liquid-containing electrolytes. In all instances some potential difference may exist. The general form is

Cu | external reference electrode | salt bridge | external monitored solution or "test" solution | membrane electrode | Cu   (Cell VI)

In these junction cells, there are two additional components: the external reference electrode and the salt bridge. The former is a generic name. The most common references electrodes are anion-responsive Type 2 electrodes such as the saturated calomel electrode.

$$\text{Hg} \mid \text{Hg}_2\text{Cl}_2 \mid \text{KCl (saturated in water)} \qquad \text{(III)}$$

or a defined silver/silver chloride electrode

$$\text{Ag} \mid \text{AgCl} \mid \text{NaCl (0.1 M)} \qquad \text{(IV)}$$

These electrodes are also called "half-cells" for obvious reasons. The critical features of these electrodes are their ease of manufacture, their insensitivity to construction, their stability, reproducibility of potential difference, and their reversibility toward small current flow. (See Section 3.5.)

Reference electrodes, reversible to a large number of anions and cations, are now available. Any commercial anion-selective electrode or conveniently fabricated equivalent electrode can function as a reference electrode. Even cation-selective, liquid-membrane, and glass electrodes are possible reference electrodes.

Junction cells are useful and important because test solutions are separated from the reference electrode (metal and metal salt) by an electrolyte solution of constant composition. This electrolyte is dictated by the chemistry of the reference electrode itself. In addition, this electrolyte serves the purpose of stabilizing the reference electrode potential while contacting a variety of test solutions. The electrical contact between this electrolyte and the test solution is the "junction" that may, for example, be a controlled leak through a glass separator. When using a commercial reference electrode, the contact is very small and a small quantity of intermixing occurs. A few, often minor, problems are introduced in addition to the interdiffusion: one is development of a usually small, but unknown, junction potential; another is clogging of the junction pathway. The latter is especially troublesome when using basic, aqueous test solutions and nonaqueous solutions. Many designs for liquid junctions are commercially available or described in the literature.

Complete isolation of the test solution is possible by using a double-junction arrangement in which the test solution is separated from the reference electrode by a compatible inert electrolyte. This type of structure provides two diffusional barriers, which avoid Cl^- leaks when saturated calomel reference electrodes are used in cells for the measurement of K^+ or Cl^- activities. This arrangement introduces the concept of the "salt bridge" that bridges ionically between the reference electrolyte and the test solution. This bridge may contain the reference electrolyte. Some examples are as follows

$$\text{Hg} \mid \text{Hg}_2\text{Cl}_2 \mid \text{KCl (sat'd)} \mid \begin{array}{l} \text{Li}^+\,\text{Cl}_3\text{C}-\text{COO}^- \\ \mid \text{Na}^+\,\text{NO}_3^-,\,\text{or} \\ \mid \text{K}^+\,\text{NO}_3^- \end{array} \mid \text{test sol'n} \qquad \text{(V)}$$

$$\text{Ag} \mid \text{AgCl} \mid \text{NaCl}\,(0.1\,M) \,\vdots\, \text{Na}^+\,\text{ClO}_4^- \,\vdots\, \text{test sol'n} \qquad \text{(VI)}$$

or

other external	other salt	test sol'n	(VII)
reference half cell	bridge		

Junctionless cells contain only a single electrolyte solution, the test solution. Consequently, the test solution must include a constant activity of that ion required for the stable, constant-potential reference electrode. If by plan or by chance, a constant activity of Cl^- is present in a series of test solutions, a Cl^--reversible external reference electrode could be dipped into the test solution. The salt bridge and fixed-activity solution for the external reference electrode could be eliminated. An example of a junctionless cell is Cell VII

Cu	Ag	AgCl	external monitored test solution containing fixed activity of Cl^- (or Ag^+)	membrane electrode selective to cations	Cu

(Cell VII)

Junctionless cells can be composed of a cation- instead of an anion-selective reference electrode. Whenever an anion-responsive membrane cell is used in a junctionless configuration, the reference electrode must be cation reversible. Since junctionless cells are at equilibrium (there is no junction contact and therefore no nonequilibrium diffusion potential), they are required for precise analytical measurements, and for all thermodynamically significant measurements. Determination of stability constants of complex species and dissociation constants of weak acids and bases frequently use electrochemical methods involving junctionless cells with selective electrodes.

2.3. Materials

Membranes for ISEs are as immiscible as conveniently possible with respect to the bathing solutions or solid contacts. Hydrophobic organic liquids and solids and low water-solubility inorganic solids constitute the main materials of membrane construction. Nevertheless, useful membranes are not electrical insulators. They are permeable to an easily measurable extent for ionic species in their immediate environment. Porous membranes are those such as organic liquid and solid, synthetic ion exchangers, which absorb and become saturated by an external solvent, usually water. They also permit water from two bathing solutions with nonidentical ionic strengths (nonidentical osmotic pressures) to pass slowly from one side of the membrane to another. However, many membranes are nonporous and solvent transport is not an important process when considering membrane potential responses. Useful membranes are most often solid or liquid electrolytes, because they are composed of partially or completely ionized acids, bases, or salts, or because they contain potentially ionizable species.

Most widely studied are those membranes of polyelectrolytes ("solid" synthetic ion exchangers), aqueous-immiscible organic liquid electrolytes ("liquid" ion exchangers), and solid, ion-conducting electrolytes including silver halides, silver sulfide, rare-earth fluorides, and alkali silicate and alumino–silicate glasses. All of these materials contain ionic species or ionizable groups whose electrical state depends on the membrane dielectric constant and extent of solvent penetration. A characteristic of these membranes is the presence of charged *sites*. If ionic groups are fixed in space in a membrane as $-SO_3^-$ and $-COO^-$ attached to a cation exchanger resin backbone, the membrane is considered to contain *fixed, charged sites*. Liquid ion exchangers such as salts of phosphoric acids and quaternary ammonium salts possess mobile sites that are free to move, but remain trapped in the membrane because of high oil-solubility. Membranes need not contain sites of only one sign. However, it is frequently necessary to incorporate sites of one sign. Single-crystal Frenkel membranes, silver halides, sulfide, and LaF_3, for example, behave as though they contain fixed, charged sites. At room temperature, impurities determine their mobile ionic species: interstitials or vacancies. A divalent anion impurity in AgX generates mobile cation interstitial silver ions and fixed sites that are the divalent anions.

2.4. Interfaces and Reversible Ion Exchange[12]

The membrane portion of an ISE is separated from the test solution by an interface. Similarly, the interior of the membrane is separated from the inner filling solution or the contacting metal by an interface. Interfaces are geometric barriers between two phases. They may be clearly defined as between two immiscible liquids and between liquids and solids, or they may be diffuse as in the liquid junction between miscible liquids. Some ions in solution permeate the interface of a membrane and others do not. For electrolyte/ion exchanger interfaces, permeable ions are called counterions, and are opposite in sign to those ions that are part of the ion exchanger membrane or crystal structure. The latter are fixed or mobile, trapped sites because they do not leave the membrane or crystal. In contrast to the sites, the permeable ions are able to enter and to transport through a membrane under an applied voltage or, rarely, under a concentration gradient. Occasionally, ions of the same sign as the sites penetrate the membrane. These ions are called co-ions. When ions of only one sign penetrate, the membrane is *permselective*. If solvent, but not salt penetrate, the membrane is semipermeable.

In membrane electrochemistry and in the design of ISEs for their own sake or for chemically sensitive semiconductor devices, the kind, location, and mobility of charged species in membranes and in the exterior phases are of primary importance. It is the distribution of charge that gives rise to the electric field and resulting membrane potentials. The membrane and bathing solutions (or metallic contact) must be overall electrically neutral. Yet, permeable ions are not at the same energy in each phase, and the surface region contains nonelectroneutral (space-charge and adsorbed charge) sections as double layers at the interfaces. The space-charge regions extend out into the bathing or contacting phases, and inward into the membrane. The width of the space-charge region is variable (but 10^{-5} cm is a high side limit) and depends on the activity of charged carriers and their energy (standard ionic chemical potential in each phase). The existence of space charge and potential curvature are synonymous general features of membrane systems. The membrane itself will normally possess a net charge and this charge resides at the inner side of the interfaces. The interior of the membrane will most frequently contain a region of electroneutrality in the bulk. The compensating space charge for the membrane exists in diffuse and adsorbed charges on the bathing solution or metal contact side of each interface.

Ion exchange is a general type of process that describes the reversible and irreversible (activated) transfer of ions from one phase to another. Ion exchange includes transfer of ions across such phase boundaries as an interface between a metal and an electrolyte, two immiscible liquids, a metal and an ionic crystal, an ionic crystal and an electrolyte solution, as well as between liquid and solid ion exchanger resin membranes and bathing solutions. The broad classification of ion exchanger includes phases with ions in common, as well as phases that initially contain different ions. An AgCl wafer is an ion exchanger for Ag^+, as can be demonstrated by exposing the wafer to radioactive Ag^+ and counting the incorporated radiosilver after different lengths of exposure. Similarly, silver metal is an ion exchanger when it is exposed to radiosilver ions. The latter are rapidly incorporated into the metal and an equivalent number of nonradiosilver ions are released to the solution.

Possibly the more characteristic view of ion exchange at zero current is the equilibration of two or more ions of the same charge, or same sign of charge between two phases such as Na^+ exchange with H^+ at a Dowex cation exchange resin. However, ion exchange involving ions of more than one kind is simply a historic case observed with ion exchange resins. The phenomenon is quite general and is a property of all membrane electrode systems and classical electrodes of the first, second, and third types. Ion exchange at zero net flux is characterized by the equal and opposite fluxes of ions across the phase boundary, as shown in Figure 19. The extent of ion exchange is measured by the equilibrium constant for the process. The partitioning of a single ion corresponds to a hypothetical, unmeasurable single-ion extraction equilibrium constant. Two or more ions of equal or same sign of charge will exchange with a measurable ion exchange constant that is the ratio of single-ion partition constants. Ion exchange is closely related to extraction and it is possible to exchange simultaneously ions of opposite charge across an interface. Such a case has a measurable equilibrium constant related by the product of single-ion partition constants.

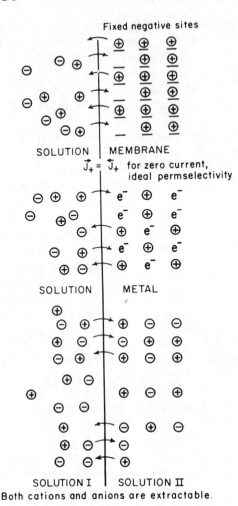

Figure 19. Schematic view of ion exchange at interfaces found in ISE systems. Top: ion exchange of cations between a bathing electrolyte solution and a fixed-site membrane with negatively charged sites. Middle: comparable cation exchange between an electrolyte solution and a metal containing the same cation with electrons, e^-. Bottom: ion exchange of both cations and anions at a constrained junction interface, or at an immiscible organic solvent/aqueous electrolyte extraction interface. In the top and middle diagrams, equilibrium means equal and opposite cation fluxes. In the bottom diagram, equilibrium equal and opposite fluxes of each kind of ion, while steady state means equal cation and anion fluxes.

The quantitative measure of the rate of ion exchange is the exchange current or exchange flux density. It is the number of moles of ions that flow in opposite directions per second per square centimeter. Rapid, reversible ion exchange corresponds to higher than about 0.1 A/cm², while slow or kinetically limited, irreversible ion exchange shows lower than about 10^{-2} A/cm² exchange current density.

Ion exchange is a potential-developing process. A reversible metal such as silver, dipped into a silver-containing solution, provides a commonplace example. The metal consists of silver ions and electrons. The latter have an extremely low solubility in water, while silver ions are readily hydrated and are stable in water. Thus silver ions in the metal attempt to dissolve in water since their concentration in the metal is higher than any saturated aqueous solution. In a very short time, the electrical relaxation time, the excess dissolved silver ions form a space charge in solution very near to the metal surface. The process of space-charge formation is a charge separation that quickly stops itself. The equal and opposite residual charge on the metal is composed of the remaining electrons, which could not follow the ions into solution. This simple picture ignores the fact that solvent and some silver ions are adsorbed at the solution side of the interface. Consequently there may be a variable potential drop through the so-called compact water layer. This component of interfacial pd from this region adds to the curved potential profile through the diffuse space-charge regions. The single interface pd is not measurable. But there is some silver ion activity that corresponds to a zero interfacial pd. Based on the observation that silver ions are exchanging back into the metal in proportion to their solution activities, and that increasing activities of silver ions should allow less space charge to accumulate, pd changes are predictable, in terms of solution activities. Thus pd measured as $\phi_{metal} - \phi_{solution}$ becomes more positive with increasing silver ion activities in solution.

A nearly identical argument is used to show that an interfacial pd develops at other ion exchanger interfaces. For example, Ag^+ in $AgCl$, $AgBr$, Ag_2S or a synthetic cation exchange resin, exchanges Ag^+ with an external electrolyte. Depending on the Ag^+ bathing activity, the membrane or crystal will gain or lose Ag^+. Residual negative charge is not electronic in these cases, but is composed of excess site or crystal anion species. Exchange of both signs of ions is possible for many crystals such as $AgCl$. When the $AgCl$ interface is bathed in excess Ag^+, the interfacial pd $\phi_{crystal} - \phi_{soln}$, becomes more positive for increasing Ag^+-bathing activities. Likewise in excess Cl^-, $\phi_{crystal} - \phi_{soln}$, becomes more negative upon increasing Cl^--bathing activities. This result is consistent with the fact that, at equilibrium

$$a_{Ag^+} \times a_{Cl^-} = \text{constant} \tag{95}$$

At equilibrium in saturated solutions, exchange of both ions occurs at possible widely different rates. However, for conditions away from equilibrium, e.g., during dissolution or during precipitation, the rates of cation and anion exchange are coupled so that additional space charge does not continually develop. This nonequilibrium coupling of fluxes is closely related to the development of junction potentials at miscible liquid interfaces.

As a third example, consider an organic phase containing the calcium salt of an oil-soluble phosphonic acid diester. The anions are, like electrons in silver metal, trapped in the organic phase. Calcium ions are free to exchange with calcium ions in an aqueous bathing solution. In a short time after phase contact, calcium ions are either gained or lost to the aqueous phase to create a space charge and an interfacial pd is developed that increases positively with increasing bathing solution activities of calcium ion. Ion exchange of anions behaves in a similar fashion, but the sign of charge and pd are reversed. For an anion exchange resin, $\phi_{resin} - \phi_{solution}$ becomes more negative with increasing anion activities in a solution in contact with a synthetic anion exchanger, such as quaternary ammonium-polystyrene membranes, or liquid anion exchanger such as supported Aliquat nitrate in nitrobenzene solvent.

As a fourth example, an interfacial pd is produced when a single salt partitions between two immiscible liquid phases such as water/nitrobenzene or between water and polyvinylchloride. The interfacial pd is determined mainly by the single ion partition coefficients. For a given salt, one ion will be more soluble in a given phase than the other. Because of the difference in standard ionic chemical potentials for the cation and the anion in each phase, one ion of the salt attempts to be partitioned across the interface to a different extent relative to the other, i.e., the more oil-soluble ion tries to build up a higher concentration inside the oil-like phase. This process creates a real charge imbalance on the two sides of an interface, with the result that a field and a potential are generated so that the process soon stops itself and the interfacial pd is established.

It is expected for macroextraction processes that the interfacial pd will be independent of bathing activities. However, for systems with one ion highly favored in an insulating membrane, or when one ion is very much more mobile than the other in the membrane, it is possible that potentials related to single ion activities may be measured. A number of examples have been reported by Higuchi's group [112] and by Liteanu's group. [113]

3. Types of Ion-Selective Electrodes

IUPAC has recommended nomenclature for the various kinds of ISEs, based on the physical state of the membrane materials. [114] Each electrode, regardless of complexity of sequential reactions, involves ion exchange at some central point.

3.1. Crystalline, Primary, or Homogeneous Electrodes: Composition, Interferences, and Sensitivities [105-116]

These electrodes are membrane type and are composed of solid salts. The salts may be single crystals, but are more often pressed pellets, from powdered starting materials. Compositions and the ions for which these electrodes are intended (the primary ions) are shown in Table 1. These include single salts and mixtures prepared to give intimate, homogeneous compositions. In addition to these, mention of HPO_4^{2-}- and SO_4^{2-}-sensing electrodes that can be fabricated and may be useful for specific applications are to be found in reviews of Buck [115] and now compiled in Freiser's edited Volume II. [116] Certain liquid ion exchange electrodes can be gelled to give the appearance of solid membrane electrodes. These systems are given later, and are not truly examples of this category.

Table 1
Materials for Electrodes[a]

Primary ion	Orion electrode	Other homogeneous membranes	Heterogeneous membranes
F^-	LaF_3		—
Cl^-	$AgCl/Ag_2S$	$Hg_2Cl_2/HgS, AgCl$	$AgCl$
Br^-	$AgBr/Ag_2S$	$Hg_2Br_2/HgS, AgBr$	$AgBr$
$\left.\begin{array}{l} I^- \\ CN^- \\ Hg^{2+} \end{array}\right\}$	AgI/Ag_2S	AgI	AgI
SCN^-	$AgSCN/Ag_2S$	$Hg_2(SCN)_2/HgS, AgSCN$	$AgSCN$
S^{2-}	Ag_2S	Ag_2S	Ag_2S
Cu^{2+}	Cu_xS/Ag_2S	$CuSe$	$Cu_xS, Cu_xS/Ag_2S$
Pb^{2+}	PbS/Ag_2S		PbS/Ag_2S
Cd^{2+}	CdS/Ag_2S		CdS/Ag_2S

[a] Mainly from R. P. Buck, *Physical Methods of Chemistry*, Part IIA, A. Weissberger and B. W. Rossiter, Eds., Ch. 2, Interscience, New York (1971).

The solid ion exchangers of this section respond to bathing activities of those species that 1) exchange directly and rapidly and 2) influence the activity of those ions that do exchange directly. The latter can operate by complex formation and by solid-state (precipitation) equilibria. For example, the silver salt electrodes are believed to be rapid ion exchangers of both Ag^+ and component anions X^- or S^{2-}. Consequently these materials develop interfacial potentials in relation to log activity of Ag^+, X^-, or S^{2-}, but they are sensitive to interference by species reactive toward the exchanging ions. Although cation interferences are rare, Hg^{2+} and Hg^+ should be recognized as serious interferences at any activity level such that they react with X^- or S^{2-}. Experimentally, all of the sulfide-based electrodes are susceptible to metathesis by Hg^{2+} and the halides are attacked to form soluble halide complexes. These reactions, when controlled, can be arranged to give a method for monitoring Hg^{2+}.

Anion interferences occur with many of the solid electrodes. For example, AgCl, when contracted by Br^-, I^-, OH^-, or SCN^-, will undergo metathesis when the inequality is obeyed

$$(X^-) > \frac{K_{so}(AgX)}{K_{so}(AgCl)}(Cl^-) \tag{96}$$

where the solubility product $K_{so}(AgCl) = 1.5 \times 10^{-10}$ and $K_{so}(AgX)$ values are less than $K_{so}(AgCl)$. Experimentally, Br^- and I^- displace Cl^- from the crystalline electrode surface when the concentrations obey this inequality. This rule is only approximate because activity coefficients were omitted; practical solubility product K_s values used should be those values appropriate for the ionic strength of the electrolyte, rather than K_{so} (infinite dilution thermo-dynamic values). Furthermore this inequality is a simplification of a more complicated expression needed to interpret cases when the solid phase is partially converted from one salt to the other. Sulfide is a particularly important interference ion because its solubility product is so small. Sulfide ion concentrations need to be kept less than

$$(S^{2-}) < \frac{K_{so}(Ag_2S)}{K_{so}^2(AgCl)}(Cl^-)^2 \tag{97}$$

to avoid interferences. Some electrolytes containing two precipitants can do double duty and attack an otherwise stable electrode. For example, Cu^{2+} and Cl^- together etch Ag_2S because of the reaction

$$Ag_2S + Cu^{2+} + 2Cl^- \rightarrow CuS + 2AgCl \tag{98}$$

The successful primary electrodes in Table 1 are composed of one, two, or three species. The salts are those of metals whose aquo-ions are labile. Conspicuously absent are transition metals. The reason is that ion exchange at the salt/electrolyte interface must be rapid. A high degree of lability is required for rapid kinetics. Transition metal ions are frequently inert. The salts are also notably Frenkel-defective, which imparts good ionic conductivity at room temperature. Lanthanum fluoride has a large bandgap and shows no significant electronic conductivity, while the others show easily measured electronic as well as ionic conductivity. The presence of Ag_2S in mixtures is purposeful because it acts as a kind of glue to hold the more soluble salts in a stable matrix and to give both electronic and ionic conductivities throughout the membranes. This effect has been particularly useful in preparing AgI-based electrodes that would normally be difficult to prepare. AgI, when pressed, converts to a high pressure, high density form that disintegrates when the pressure is released. Use of selenides, tellurides, and mercurous salt–mercuric sulfide mixtures are equally permissible. The mercury salt-based electrodes have the advantage of very low detection limits because of the low solubilities, in comparison with corresponding silver salts.

All of these solids have been used in a "membrane" configuration. All except LaF_3 can be made conveniently in the "all-solid-state" configuration, as well. Because of the nobility of silver metal relative to Cu, Cd, and Pb, it is the favored contacting metal. Furthermore, silver salts, subjected to thermal treatments, develop a small nonstoichiometric excess of Ag metal. If that condition does not exist after a fresh precipitation, silver metal contact and exposure to light inevitably converts some silver salt to the metal. This effect stabilizes the metal/salt interfacial potential difference. This conversion also occurs with the mixed heavy metal–silver sulfide systems. The mercury salt-based electrodes are Hg-rich, and should be contacted with mercury metal. LaF_3 is a particularly difficult case, and it has been suggested that metallic contact such as platinum, can be used if the LaF_3 is bonded and the entire system is encapsulated, except for the side of the LaF_3 exposed to the electrolyte under test.

The solid electrodes respond normally, and well, when the ion exchange is rapid. Thus silver salts perform well in solutions containing excess Ag^+-soluble salt, or excess X^-, typically Na^+ or K^+X^-. However, upon dilution, response remains normal until the solubility limit of the membrane material is reached. Numerically this value is about $K_s^{1/2}$ for 1:1 salts, using the appropriate value at the ionic strength of inert species. It is about $K_s^{1/3}$ for Ag_2S or about $K_s^{1/n}$ for salts with n-ionizable species. Activities, by dilution, below the detection limit cannot be measured because the electrode is out-of-equilibrium, and will tend to dissolve slowly into the test solution. Some limits of detection are given in Table 2. Note that LaF_3 is unusual in that the detection limit applies only to F^- salts dilutions. LaF_3 does not respond to solution of excess, soluble La^{3+} because of slow ion exchange.

Determination of detection limits is not so easily done by simple dilution. Dilutions should always be made at constant ionic strength using an inert electrolyte for responsive ion activities down to about $10^{-6} M$. Below that value dilutions are unreliable and either metal ion buffers or precipitates giving

Table 2
Response Ranges for Electrodes at 25°C[a,b]

Electrode	Active membrane	Upper useful limit M	Limit of Nernstian response M	Limit of detection M	pH range
F^-	LaF_3	sat. solns	2×10^{-6}	10^{-7}	5–8
Cl^-	$AgCl$	1	10^{-4}	10^{-5}	2–11
	Hg_2Cl_2	—	2×10^{-6}	5×10^{-7}	0–6
Br^-	$AgBr$	1	10^{-5}	10^{-6}	2–12
	Hg_2Br_2	—	10^{-6}	10^{-7}	1–6
SCN^-	$AgSCN$	1	5×10^{-5}	5×10^{-6}	2–12
	$Hg_2(SCN)_2$	—	5×10^{-6}	5×10^{-7}	1–6
I^-	AgI	0.2	2×10^{-6}	10^{-8}	3–12
CN^-	AgI	10^{-2}	10^{-5}	10^{-6}	11–13
S^{2-}	Ag_2S	sat. solns	5×10^{-6}	10^{-7}	13–14
Ag^+	Ag_2S	sat. solns	2×10^{-6}	10^{-7}	2–9
Pb^{2+}	$PbS–Ag_2S$	10^{-1}	10^{-5}	10^{-7}	3–7
Cd^{2+}	$CdS–Ag_2S$	10^{-1}	10^{-5}	10^{-7}	3–7
Cu^{2+}	$CuS–Ag_2S$	1	10^{-7}	10^{-9}	3–7
Hg^{2+}	AgI	10^{-4}	5×10^{-7}	10^{-8}	4–5

[a] Limit of Nernstian response is the activity of detected ion giving a deviation from extrapolated high-activity response of $17.7/|z|mV$ at 25°C. Limit of detection is the activity of detected ion giving no further change of response with further dilution. Note: These defined limits are based upon calibrations using successive dilution, not metal-ion or other buffers.

[b] From R. P. Buck, *Physical Methods of Chemistry*, Part IIA, A. Weissberger and B. W. Rossiter, Eds., Ch. 2, Interscience, New York (1971); and P. L. Bailey, *Analysis with Ion-Selective Electrodes*, Heyden International Topics in Science, Heyden, London (1976).

saturated solutions with known metal–ion activity values must be used for calibrations. Metal ion buffers are mixtures of metal complexes and free ligands to give labile mixtures containing known, free metal ion activities in equilibrium.

Hydrogen ion and hydroxide ion concentrations (pH) also affect response limits directly and indirectly. The upper pH limits for use of LaF_3 in F^- solutions, or AgCl in Cl^- solutions is determined by the fact that these salts form hydroxides. Higher hydroxide concentrations can be tolerated at higher F^- and Cl^- activities and vice versa. Indirectly H^+ affects limits for those species that are weak acid anions. Dilutions of sulfide or cyanide must be done at constant pH to assure linear dilution. Since the electrode responds only to the component ion F^-, S^{2-}, or CN^-, not the total or formal activity, these species can be detected more effectively in base conditions rather than in acid where protonation converts F^- to HF, S^{2-} to HS^- and H_2S, and CN^- to HCN.

Higher concentration limits for samples are often saturated solutions. However, because of complex formation, equilibria such as AgCl in Cl^- to form soluble species, corrosion of electrodes at very high bathing activities can be expected.

The effects of oxidizing and reducing agents should not be underestimated. Although electrodes of the first and second types are prone to oxidation and reduction interferences by soluble species, the membrane electrodes are not as sensitive. The former types have exposed metal, while the latter present only a salt interface to the test solution. LaF_3 is virtually insensitive, while AgCl is attacked only by the strongest of oxidants and reductants, such as acidic permanganate and basic dithionite. The salts of more readily oxidized anions, Br^-, I^-, S^{2-}, are more sensitive to oxidants. The AgI and Ag_2S electrodes are sensitive to dissolved oxygen and peroxide in acid solutions. Consequently, under conditions of oxidizing agents the lower limit of detection cannot be achieved. These electrodes are much more forgiving of weak reducing agents such as ascorbic acid, but cannot tolerate strong reductants used in photographic development, for instance.

3.2. Crystalline, Primary, Heterogeneous Electrodes

The electrode materials of the previous section, even the mixed salts used in pressed pellet electrodes, are relatively homogeneous on a molecular scale. They are prepared by simultaneous precipitation, which leads to good coprecipitation, and they can be made by the very slow processes of homogeneous precipitation. Surprisingly, electrode function does not seem to be determined exclusively by homogeneity. Consequently, powders suitable for pressing can be mixed with dissolved polyvinylchloride (PVC), unpolymerized silicone rubber, unpolymerized epoxy resins, and collodion, and cast in thin films on glass plates. When these materials are polymerized or cured, they can hold typically 50 wt. percent of the solid powder in film form. After mounting the film, peeled from the glass plate, on the end of a compatible plastic or glass tube, a membrane electrode is formed, as shown in Figure 16(d). When activated by soaking in water, these electrodes behave very much like the pressed pellet electrodes made from the same materials. Powders can be mixed with dry powdered binders such as polyethylene and pressed.

3.3. Noncrystalline, Primary, Heterogeneous, and Homogeneous: Compositions, Interferences, and Sensitivities[117–127]

The electrodes in this category are intrinsically liquid, although they can appear as solids by judicious use of polymeric supports and gelling agents. The liquids in the original group were liquid ion exchangers possessing mobile trapped sites, which could be used in liquid solution form, suspended in inert, porous polymers, or dissolved in microporous polymers. The liquid ion exchangers were commercially available chemicals used in liquid/liquid extraction technology. It became clear in the early exploratory studies that lipophilic ion exchangers (solid and liquid) could also be dissolved in immiscible, nonaqueous solvents to control site concentrations, and that nonionic additives could be included. The immediate advantage was that the dielectric constant and solvating properties of the liquid solvent could be controlled and could provide a means of influencing and moderating ionic selectivity. Furthermore, many insoluble ion exchanger systems could be solubilized. The liquid solvent, or solvent mixture, is called the "mediator." The main liquid ion exchanger electrodes, and the primary active ingredient in the membranes are given in Table 3.

The list in Table 3 is not complete. Virtually any ionic species can be detected and measured by liquid ion exchanger electrodes. The principle for design is clear: to build a membrane electrode response to ion X^-, for example, the salt R^+X^- should be incorporated into a nonvolatile solvent. R^+ must be highly lipophilic, e.g., a high molecular weight cationic dye or a quaternary ammonium ion such as Aliquat, and solvents may be diphenyl ether, p-nitrocymene, or o-dichlorobenzene, for example. Similarly for cations M^+, an oil-soluble salt M^+R^- is used where R^- is lipophilic, such as tetraphenylborate. Generally, the elementary, trial electrodes made from this recipe will function only in the presence of pure salts in aqueous solutions. Selectivity over interferences will be limited to those ions less oil-soluble than the principal ion. The Hofmeister series[128] is frequently obeyed, and serves as a rule for deciding on possible interferences. In decreasing order of lipophilicity, the series is

$$\text{tetraphenylborate} > ClO_4^- > I^- > Br^- > NO_3^- > Cl^- > OH^- > F^-$$

This series means that a typical NO_3^- sensing electrode will not respond well in the presence of those ions to the left (Br^-, I^- etc.), because the latter are actually more oil-soluble and therefore preferred anions in many membranes.

Table 3
Electrodes Based on Liquid Ion Exchangers[a]

Ion	Form of membrane	Active material	Solvent mediator	Comments	Ref.
Ca^{2+}	(a) liquid	calcium di-(n-decyl)-phosphate	di-(n-octylphenyl)-phosphonate	Orion electrode	(117)
	(b) solid (PVC)	calcium di(n-decyl)-phosphate	di-(n-octylphenyl)-phosphonate		(118, 119)
	(c) solid (PVC) "Selectrode"	calcium di-(n-octylphenyl)-phosphate	di-(n-octylphenyl)-phosphonate	Improved sensitivity and pH range	(120)
NO_3^-	(a) liquid	tridodecylhexadecylammonium nitrate	n-octyl-2-nitrophenyl ether	Corning electrode	
	(b) liquid	tris(substituted 1,10-phenanthroline) nickel(II) nitrate	p-nitrocymene	Orion electrode	(121, 122)
	(c) solid (PVC)	tris(substituted 1,10-phenanthroline) nickel(II) nitrate	p-nitrocymene		(123)

Ion	State	Compound	Solvent	Notes	Ref.
ClO_4^-	liquid	tris(substituted 1,10-phenanthroline)iron(II) perchlorate	p-nitrocymene		(121)
BF_4^-	liquid	tris(substituted 1,10-phenanthroline) nickel(II) tetrafluoroborate	p-nitrocymene		(124)
Divalent cations (water hardness)	liquid	calcium di-(n-decyl)-phosphate	decanol		
CO_3^{2-}	liquid	tri(n-octyl)methylammonium chloride	trifluoroacetyl-p-butylbenzene	Not commercially available	(125)
U (VI)	(a) liquid	methylene blue–uranyl tribenzoate	o-dichlorobenzene	Not commercially available	(126)
	(b) solid (PVC)	di(2-ethylhexyl)phosphoric acid	diamylamyl phosphonate	Not commercially available	(127)
Cl^-	liquid	dimethyl-dioctadecylammonium chloride		Quaternary ammonium based electrodes are not highly selective	

---

[a] From R. P. Buck, *Physical Methods of Chemistry*, Part IIA, A. Weissberger and B. W. Rossiter, Eds., Ch. 2, Interscience, New York (1971) and P. L. Bailey, *Analysis with Ion-Selective Electrodes*, Heyden International Topics in Science, Heyden, London (1976).

By consulting the original literature, it is possible that some of these reported membrane compositions will be found useful for specific applications, despite interferences.

Liquid ion exchanger membranes can be made from the free-standing liquid held between aqueous electrolyte-permeable supports: dialysis membranes or cellophane. This construction is, however, fragile and not practical. In Figure 16(a), (b), (c) are illustrated several supported configurations. Originally cellulose acetate "Millipore" or equivalent filters were used as inert, porous media for holding liquid ion exchangers by wick or capillary forces. Other membranes whose surface tensions permit wetting by the liquid phase are also useful. Teflon films are not suitable. Liquid exchangers can be gelled by incorporation of collodion, for example. If the cross section is not too great, typically $\frac{1}{4}''$ or less, the gelled membrane is self-supporting. In the original Orion design, positioning of the membrane to achieve a watertight seal was crucial, and the excess liquid ion exchanger was stored in contiguous chambers. As the liquid exchanger slowly dissolved in samples, the reservoirs were drained, but the membrane retained lipophilic character and maintained a nearly water-free surface exposed to the test solution. The new design replaces the reservoirs with a plastic foam container holding the liquid ion exchanger. Also the membrane is positioned automatically and more reproducibly.

A closely related construction in Figure 16(d) is based on casting liquid mixtures of ion exchanger, mediator, and dissolved polymer (PVC, polyethylene, silicone rubber) to form membrane films.[129] These are generally microporous and the mediator and ion exchanger are more nearly dissolved. In the previously described, nearly ideal physical support membranes, the liquid remains virtually in pure form within open pores. In the microporous cases, the diffusion coefficients or mobilities are less than the parameters for the pure liquids in massive state, or in macropores. The words "homogeneous" or "heterogeneous" presumably could be applied, but they are essentially meaningless because all practical designs call for some kind of support. It is also possible to construct all-solid-state versions of these electrodes using liquid ion exchangers. For example, an Ag/AgCl or Ag/Ag$_2$S half-cell can be coated with a liquid ion exchanger in an inert binder to form a reasonably stable electrode. These electrodes have good response slopes (mV vs. log a_i), but they tend to show drifts in $\Delta\phi°$ (shifts in the entire calibration curve up or down the potential scale) during their lifetimes.

Like solid electrodes, the liquid ion exchanger membranes respond to ion activities at their inner and outer surfaces. Complex formation that reduces sensed-ion activities below the formal activity are interferences. Intrinsic interferences are those species which, by ion exchange, are preferred in the membranes. For anions, the Hofmeister series dominates. For cations (the calcium ion electrode is the main example), interferences are those ions that are preferred by the PO$^-$ group in low dielectric media. Heavy divalent metal ions and other alkaline earths are high on the interference list. Another rule is that ions of lower valence are preferred in the membrane at low dielectric constants, while higher valence ions are preferred at high dielectric constants. This rule primarily results from the greater polarizability and greater cancellation of local fields around ions within membranes when the solvent has a high dielectric constant. Of course, the limit value of about 35 occurs because the membrane must remain water insoluble. Since liquid membranes are not rigid, size criteria do not play a decisive role as they do with synthetic, solid polymer membrane ion exchangers.

The magnitude of interference, measured quantitatively by the selectivity coefficient defined later, is a result of the ion exchange constant for the reaction between the interference ion and the principal ion. This quantity is similar to the ratio of solubility products used above as a measure of interference for solid electrodes. The known selectivity coefficient values are not, unfortunately, known as precisely as the solubility products. In part, the reason for uncertainty in ion exchange constants is the less reliable and reproducible nature of the potential measurements for liquid electrodes relative to solid electrodes. Also the selectivity coefficient values are dependent on the bathing solution concentrations and on the site concentration, especially for ion exchange between ions of different charge.

When membranes containing the primary ions are used to measure activities in the presence of interferences, interfering ions begin immediately to replace principal ions in the membrane. After use, the membrane will contain interfering ions. Eventually, these ions could diffuse through the membrane to the inner filling solution. To restore electrodes that have been once exposed to interferences, the electrodes should be soaked in an interference-free solution of the principal ion salt. By diffusion and ion exchange, the interference will slowly empty out of the membrane phase.

Liquid membrane electrodes show Nernstian responses in the absence of interferences. Typical low limits of a Nernstian response and limits of detection are shown in Table 4. The lower limits are set mainly by the solubility of the compound R^+X^- or M^+R^- in water. One cannot measure activities of X^- or M^+ below the values determined by the thermodynamic extraction equilibria of the exchanger material. It is possible that kinetic factors (slow extraction) may help to provide lower detection limits in a few cases.

pH and pOH have relatively obvious effects on response. In strong acids, cationic exchangers will be protonated and lose their sensitivity to other cation activities. The Ca^{2+} electrode is limited to a range of pH values around neutral. Because OH^- is not an oil-soluble anion, relatively high OH^- concentrations can be tolerated by many X^- sensing electrodes. Once again, indirect effects occur: Ca^{2+} is hydrolyzed to $CaOH^+$ in dilute base, while anions of weak acids are protonated in acidic media.

Another category of liquid membrane electrode, principally for positive ions (although an electrode for HCO_3^- has been reported), is based on lipophilic, neutral carrier complex-formers.[130] These recently discovered

Table 4

Response Ranges of Liquid Ion-Exchange Electrodes at 25°C[a]

Electrode	Upper useful limit M	Limit of Nernstian response M^b	Limit of detection M^b	pH range
Ca^{2+} (a), (b)	1	3×10^{-5}	3×10^{-6}	6–10
(c)	1	3×10^{-6c}	1×10^{-8c}	5–10
NO_3^-	1	1×10^{-4}	5×10^{-6}	3–8
ClO_4^-	10^{-1}	5×10^{-5}	1×10^{-6}	4–10
BF_4^-	10^{-1}	1×10^{-4}	2×10^{-6}	2–12
Divalent cation	1	3×10^{-5}	3×10^{-6}	6–10
CO_3^{2-}	10^{-2}	10^{-7}	10^{-8}	6–9
U (VI) (a)	?	2×10^{-6}	?	4–5
(b)	10^{-1}	1×10^{-4}	1×10^{-5}	2–4
Cl^-	1	3×10^{-4}	1×10^{-5}	3–10

[a] From P. L. Bailey, *Analysis with Ion-Selective Electrodes*, Heyden International Topics in Science, Heyden, London (1976).
[b] Note the definitions of these quantities given in the footnote of Table 2.
[c] In Ca^{2+} buffer.

antibiotic, organic compounds permit extraction of cations into thin, hydrophobic, synthetic liquid membranes. Electrodes can be constructed for the measurement of Na^+, K^+, Li^+, NH_4^+, Ca^{2+}, and Ba^{2+}. Response sensitivity matches or exceeds the liquid ion exchanger and glass equivalents, especially for measurements in F^-, OH^-, and Cl^--containing electrolytes. Responses show reduced slope and maxima in response when oil-soluble anions are present in solution. Responses are radically altered when oil-soluble cations such as quaternary ammonium ions are present. However, alkyl chains need to be about five carbons or greater to yield oil solubility. Tetramethyl ammonium through tetrabutyl ammonium salts can be tolerated, but should be checked if present in unusually high concentrations.

The neutral carriers are multifunctional molecules, but the primary active groups are ether and keto oxygens.[131–136] In each molecule, valinomycin (val) for example, the alternating ether oxygens can conform to produce a cage for the positive ion. Shielding of the ion from solvent is accomplished by the remaining hydrophobic groups, which point away from the central cation. The $Kval^+$ complex is large and the entire species is stabilized in the membrane phase. A typical K^+ salt such as KCl would be insoluble in the membrane phases of low dielectric constant used with valinomycin. There is considerable conjecture on the reason for selectivity of the ether linkage compounds for one unipositive ion relative to others. Size of the bare ion relative to the size of the cage is only one factor. Energy of removal of the hydration layer vs. gain of energy by replacement with ether oxygen linkages appears to be dominant. Calculations for various configurations of the oxygens and X-ray structure analysis show that an octahedral rather than a tetrahedral arrangement is preferred. Synthetic, simpler molecules, cyclic polyethers or crown ethers, possess the important functionality. They are useful for theoretical studies because they behave in an analogous way to the naturally occurring antibiotics. However, these molecules do not have the necessary selectivity for extraction of one unipositive ion relative to others. The $Kval^+$-based electrode is more selective over Na^+ than any of the related crown ether-based electrodes. Composition and ranges of response are given in Tables 5 and 6.

The neutral carriers can be used in liquids, supported in pores (macroporous membranes) and in microporous PVC or equivalent membranes. Selectivities and sensitivities are about the same for all configurations.

This category also includes classical solid ion exchangers made from synthetic and from some natural materials. Probably the best known examples are membranes made from polystyrene with fixed negative sites: sulfonate, carboxylate or phosphonate, or fixed positive sites: quaternary ammonium and quaternary phosphonium. The sulfonate and quaternary ammonium examples have been widely studied as electrode components for cations and anions, respectively. However, these materials are relatively flexible and they take up water. Selectivities are based almost exclusively on charge so that discrimination among ions of the same charge is very slight. These membranes have the advantage that they can be used in adverse environments such as concentrated acid solvents. Although trivalent ions are preferred over divalent ions, both of these high valence species have relatively small mobilities. Consequently, design of membrane electrodes for divalent and trivalent ions using synthetic membranes has not been successful.

3.4. Sensitized Ion-Selective Electrodes

This category makes use of conventional electrodes with an interposed chemical reaction. Thus a gas such as ammonia or carbon dioxide can be measured by using a chemical reaction that converts NH_3 or CO_2 to ions that can be detected directly by ISEs: NH_4^+ and CO_3^{2-}. However, there is a far more subtle interposed chemical reaction that can be used for these species because they are basic and acidic, respectively. Varying quantities, measured as partial pressures, can affect the pH of buffer electrolytes without being entirely converted to detectable ions. For example, out of the buffer region, the pH of a dilute solution of sodium or potassium bicarbonate is very sensitive to even slight changes in CO_2 pressure to form H_2CO_3 with a significant and easily measured increase in acidity. The CO_2 equilibrates with a thin layer of electrolyte, dissolves, and hydrates. The pH decreases when only a small fraction of H_2CO_3 ionizes, and a small fractional increase on HCO_3^- occurs as well. For NH_3-sensing electrodes, NH_4Cl serves as the pH-sensitive electrolyte. Other gas sensors have been designed and tested. If pH change can be used, it provides great sensitivity. However, for measurement of very weakly

Table 5
Neutral Carrier-Based Electrodes[a]

	Form of membrane	Active material	Solvent mediator	Ref.	Comments
K^+	(a) liquid	valinomycin	diphenyl ether	(131)	Philips electrode (similar to several other manufacturers' electrodes)
	(b) solid (PVC)	valinomycin	dioctyladipate	(132)	
	(c) solid (silicone rubber)	valinomycin	—	(133)	No solvent
NH_4^+	liquid	nonactin/monactin	tris(2-ethylhexyl) phosphate	(134)	Philips electrode
Ca^{2+}	solid (PVC)	see Ref. (135)	o-nitrophenyl octyl ether	(135)	Made by Glasblaserei (W. Moller), Zurich, Switzerland
Ba^{2+}	liquid	nonylphenoxy poly (ethylene oxy) ethanol Ba^{2+} (tetraphenylborate)$_2$	p-nitroethylbenzene	(136)	Not commercially available, contact J. D. R. Thomas. UWIST, Cardiff, Wales, UK

[a] Modified from P. L. Bailey, *Analysis with Ion-Selective Electrodes*, Heyden International Topics in Science, Heyden, London (1976).

Table 6
Response Ranges of Neutral Carrier Electrodes at 25°C[a]

Electrode	Upper useful limit M	Limit of Nernstian response M	Limit of detection M	pH range
K$^+$ (a), (c)	1	1×10^{-5}	10^{-6}	3–10
(b)	1	3×10^{-5}	10^{-6}	3–10
NH$_4^+$	10^{-1}	1×10^{-5}	10^{-6}	5–8
Ca^{2+}	1	3×10^{-5}	10^{-6}	4–10
		1×10^{-7} [b]		
Ba^{2+}	10^{-1}	1×10^{-5}	10^{-6}	5–9

[a] Compiled by P. L. Bailey, *Analysis with Ion-Selective Electrodes*, Heyden International Topics in Science, Heyden, London (1976).
[b] In Ca^{2+} buffer.

acidic substances, such as H_2S and HCN, conversion to the anion and direct detection is necessary.

Gas-sensing probes were introduced by Stow, Baer, and Randall,[137] and by Severinghaus and Bradley.[138] The concept was tested by using a normal pH glass electrode covered by a small "sock" that was wetted by the pH-sensitive electrolyte. The electrode in improved forms became well known in clinical chemistry circles. The commercially available electrodes use a flat, or slightly rounded glass pH electrode. By screwing on an electrode tip, the glass sandwiches the sensitive electrolyte between itself and an outer, gas-permeable membrane. The covering membrane equilibrates with the external gas by permitting the gas to dissolve or partition between the external region and the membrane pores at the surface. The membrane is hydrophobic and as porous as is consistent with holding back the aqueous solvent of the electrolyte solution inside. The electrode can be exposed to, and used to measure, gases dissolved in liquids and in ambient atmospheres. The external gas, when equilibrated at the outer membrane surface, diffuses inward, crosses the membrane barrier into the electrolyte, and establishes a new pH value. The gas-permeable membrane can be replaced by an airgap.[139] However, the electrolyte must then be kept slowly flowing over the glass electrode surface and held in place by surface tension. Designs are shown in Figure 20. Obviously, the sensor electrode cannot be dipped into a solution sample. Rather, a gas sample or a solution containing dissolved gas to be measured, must be passed near to the electrode. For solution samples a rapidly stirred cup, separated from the electrode by the airgap, is sufficient to equilibrate the gap with the dissolved gas.

Responses of the gas sensors are linear (mV vs. log partial pressure or log concentration) over a range of typically four powers of ten. The upper limit is determined by the concentration of buffer salt.[139-149] For linearity,

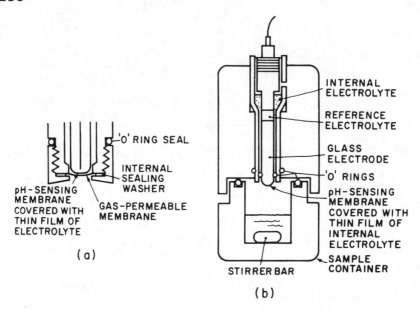

Figure 20. Constructions of gas-sensing probes: (a) cross section of the end of the probe showing the thin film of internal electrolyte (not to scale); (b) a gas-sensing probe without membrane (an "air-gap" electrode).

the dissolved gas concentration in the sensor electrolyte must not cause a large fractional change in the electrolyte concentration. For example, dissolved NH_3 when ionized must produce a small enough concentration of NH_4^+ such that the initial concentration of NH_4Cl is not changed significantly. The lower detection limit is determined by the level of NH_3 or CO_2 in the detecting electrolyte. Also the rate of transfer of gas and the rate of equilibration become slow at low sample concentrations. Lack of equilibrium can be mistaken for leveling off of sensitivity. This limit can sometimes be improved by heating the electrode to remove the last traces of NH_3 or CO_2 from the electrolyte.

Selectivity is primarily determined by the partitioning of gas into the membrane and transport through the membrane. If two gases partition equally and transport at the same rate, they will be found in the membrane electrolyte in proportion to their partial pressures. Fortunately, overlapping of physical properties is rare, and there are many gas-permeable membranes to select and use. However, permeabilities of gases in membrane films are not always available in the literature. Some electrode compositions are given in Table 7. Response ranges and detection limits are in Table 8. It is known from scattered studies of interferences that chemically related molecules can interfere with the monitoring of basic and acidic gases. For the NH_3 electrode, low volatility, high molecular weight basic amines do not affect determinations of NH_3 in water. However, hydrazine, methylamine, ethylamine, morpholine, and cyclo-

Table 7
Examples of Gas-Sensing Probes

	Internal electrolyte	Internal ion-selective electrode	Internal reference electrode	Membrane	Ref.
CO_2	0.01–0.005 M NaHCO$_3$ +0.02–0.1 M NaCl sat'd with AgCl	pH electrode	Ag/AgCl	microporous PTFE	(140), (141)
NH_3	0.1–0.01 M NH$_4$Cl sat'd with AgCl	pH electrode	Ag/AgCl	microporous PTFE	(140), (142)
SO_2	0.1–0.01 M K$_2$S$_2$O$_5$ or NaHSO$_3$ + NaCl	pH electrode	Ag/AgCl	silicone rubber or microporous PTFE	(140), (142)
NO_x	0.1–0.02 N NaNO$_2$ +dil. NaCl or KBr sat'd with AgCl or AgBr	pH electrode	Ag/Ag/Cl or Ag/AgBr	microporous PTFE or polypropylene	(140), (142)
H_2S	Citrate buffer (pH 5)	S^{2-} electrode	no information available		(142)

Table 8
Response Ranges of Gas-Sensing Probes

Probe	Type	Upper useful limit M	Limit of Nernstian response[a]	Limit of detection M^a	pH requirement	Ref.
NH_3	E.I.L./Orion	1	$ca.\ 10^{-6}$	$ca.\ 10^{-7}$	>12	(140), (142–146)
	"air-gap"		10^{-4}	10^{-5}	>12	(139)
CO_2	Radiometer	1	10^{-5}	2×10^{-6}	<3.4	(147)
	"air-gap"	1	2×10^{-3}	10^{-4}	<3.4	(148)
SO_2	E.I.L.	5×10^{-2}	5×10^{-5}	5×10^{-6}	<0.7	(142)
	Orion	10^{-2}	5×10^{-6}	5×10^{-7}	<0.7	(149)
NO_x	Orion	10^{-2}	2×10^{-6}	10^{-7}	<2	(140), (142), (150), (151)
H_2S	Orion	10^{-2}		10^{-8}	<5	(140)

[a] Definitions of these quantities are given in the footnote of Table 2.

hexylamine cause significant errors, typically +30% at equal concentrations of the interference and ammonia. Another subtle interference has been recognized to operate with samples containing very high ionic strengths. Since the membranes do transport water slowly, it is believed that the low activity of water in high ionic strength electrolytes such as seawater, causes slow flow of water from the membrane interior to the outer solution. This transport may account for the slow drift of potential of electrodes in high electrolyte solutions containing ammonia.

There are other examples of sensitized electrodes.[152–159] Enzyme-catalyzed reactions that produce detectable ions or gases can be used. These electrodes are already articles of commerce because they are included in various forms in automated devices. They are not, however, available independently. Simplified sketches of two kinds of enzyme electrodes are shown in Figure 21. In the first case in Figure 21, the immobilized gel is polyacrylamide containing the enzyme urease and NH_4Cl buffer. The electrode is pH sensitive. When it is dipped into a solution containing urea the reaction

$$CO(NH_2)_2 + H_2O \xrightarrow{\text{urease}} CO_3^{2-} + 2NH_4^+ \tag{99}$$

occurs. The ammonium ion produced changes the buffer composition and the pH value. In another version, the NH_4^+ is detected directly using an NH_4^+-sensing neutral carrier or glass electrode. In the second example in Figure 21, the enzyme reaction is accomplished in the same way, but the product NH_3 is detected with a gas-sensing membrane electrode.

Among the tricks in designing and using these electrodes is recognition of the ideal pH ranges for an enzyme reaction in comparison with the optimum

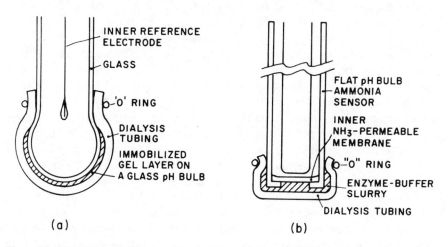

(a)   (b)

Figure 21. Constructions of enzyme-based probes: (a) Immobilized enzyme bonded to surface, immobilized in gel layer or held in place by dialysis tubing; (b) similar construction options on gas-sensor membrane.

range for product detection. For example, the enzyme reaction above is best done at pH 7. However, for detection of NH_3, the reagent should be as basic as possible. For the integrated electrode, this discrepancy poses a problem and some compromise pH value is selected. Alternatively, the reaction and the detection can be done in separate containers.

These electrodes are not stable over the long term. They can be used up to a few weeks if they are stored cold when not in use. Response times tend to be slow because of the long diffusion path between test solution and inner detector surface. Some examples of systems are given in Table 9.

3.5. Reference Electrodes and Combination Electrodes

Reference electrodes, or reference half-cells, are used with ISEs to complete electrochemical cells. These half-cells are considered to be "external." In membrane configuration, another reference half-cell is inside the ISE, and is considered to be "internal." With all-solid-state electrodes only the external reference electrode is used. Pointed out in Section 2.1, many kinds of electrodes can be used for reference purposes, viz. completion of the cell with a source of constant potential difference, pd. Classically, electrodes of type 2 including the calomel (III) and silver chloride (IV) have been widely used. However, almost all ISEs could conceivably be used for reference purposes if the cell were arranged to provide constant activity of the responsive ion.

The primary feature of a good reference electrode is a reproducible potential difference. The pd must also be stable and independent of common variables, such as quantity of dissolved oxygen, and insensitive to small current flow during a potential measurement. Potential differences of type 2 electrodes and ISEs, used as references, depend on the activity of the responsive ion in the bathing solution. In the junction configuration, these solutions are protected from oxidants and reductants by the junction structure. However, in junctionless cells, the presence of mild oxidants and reductants in test solutions must not affect the pd's. It is this reason that reference electrodes based on relatively noble metals are mandatory. Physical properties such as pressure and temperature do affect the pd's, but these effects have been measured and tabulated. Some temperature-dependent reference values are given in Table 10.

Saturated calomel electrodes have temperature hysteresis if cycled from low to high temperatures and back. If this problem arises, an unsaturated electrode, either using mercury or silver salts, can improve this situation.

For precise measurements of ion activities, $\pm 0.4\%$ for univalent ions, reproducibility must be ± 0.1 mV. Consequently, reference electrodes should be stable to at least ± 0.1 mV. By using massive construction in terms of total quantities of oxidizable and reducible species and $0.01\ M$ or more concentrated bathing electrolytes, passage of current (nanoamps) does not affect the reference electrode pd outside of this desired range.

Table 9

Enzyme Electrode and Separated ISE Enzyme Methods

Enzyme	Sensor	Conc'n. range	Reaction pH	Measurement pH	Ref.	Comments
Urea	urease	ammonia-sensing membrane probe	$10^{-4}-10^{-1}$ M	7.4	12	(152)	Continuous flow method
Urea	urease	ammonia "air-gap" electrode	$10^{-4}-10^{-2}$ M	7.0	12	(153)	Manual method
Creatinine	creatininase	ammonia-sensing membrane probe	1–100 mg %	8.5	12	(154)	Procedure semi-automated
Amygdalin	β-glucosidase	cyanide electrode	$10^{-4}-10^{-1}$ M	7	7	(155)	Enzyme electrode method
Cholesterol	(i) cholesterol ester hydrolase (ii) cholesterol oxidase catalyst: Mo^{VI}	iodide electrode	80–420 mg dl^{-1}	6.8	<1	(156)	Three-stage procedure including two enzyme reactions
Penicillin	Penicillinase	pH electrode	3.5–1100 mgl^{-1}	—	—	(157)	Enzyme electrode method (Rel. std. deviation 3% at 100 mgl^{-1})
L-Phenylalanine	L-amino acid oxidase	ammonium electrode (monactin/nonactin)	$5 \times 10^{-5}-10^{-2}$ M	7	7	(158)	Enzyme electrode method. Precision, about 1.7%; accuracy, 2%
Glucose	(i) glucose oxidase (ii) peroxidase	iodide electrode	$10^{-4}-10^{-3}$ M	5 ± 0.5	5 ± 0.5	(159)	Flow system with enzyme electrode, two consecutive reactions. Slow response

Table 10
Potential of Calomel Half-Cell[a]

Temperature (°C)	0.1 N KCl	1 N KCl	Sat'd. KCl
12	+0.3362	—	+0.2528
15	+0.3360	—	+0.2508
20	+0.3358	—	+0.2476
25	+0.3356	+0.2802	+0.2444
30	+0.3354	–	+0.2417
35	+0.3353	—	+0.2391
38	+0.3352	—	+0.2375

[a] From R. P. Buck, *Physical Methods of Chemistry*, A. Weissberger and B. W. Rossiter, Eds., Ch. 2, Interscience, New York (1971).

The most common reference electrodes are based on $Hg/Hg_2Cl_2/Cl^-$ "calomel" systems. The bathing solution is saturated with calomel and may be saturated with KCl, or held at a known concentration value 3.8, 3.5, 1, or 0.1 M. These electrolytes are commercially available, generally in the saturated form. Some examples are shown in Figure 22(a), (b), and (c). The electrodes can be homemade. Place a thin layer of calomel on a mercury surface in a test tube, followed by a layer of saturated KCl solution with a few extra crystals of KCl. Contact is provided by a glass-sealed platinum wire that reaches through to the mercury. A salt bridge is used to make contact with the test solution. Several electrodes of this form, made at the same time, may be shorted together overnight to permit spontaneous internal electrolysis and accelerated aging and potential stabilization. The result is a set of reference half-cells with pd's agreeing to about 0.1 mV. Calomel electrodes are useful to about 100°C provided allowance is made for the spontaneous disproportionation of calomel to mercury and mercuric chloride. This reaction causes drift in measured potentials, but the reference value will stabilize. Also, the electrode

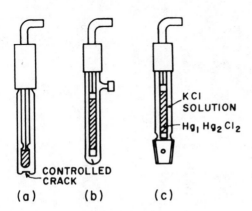

KCl
SOLUTION
$Hg_1 Hg_2 Cl_2$

CONTROLLED
CRACK

(a)   (b)   (c)

Figure 22. Examples of calomel reference electrodes with (a) microcrack as a liquid junction; (b) an asbestos fiber as a liquid junction; and (c) a ground-glass tapered sleeve as a liquid junction.

can be used below 0°C by including "antifreeze" glycerol in the electrolyte mixture.

Commercial calomel electrodes are well protected by the junction. However, they are sensitive to intrusion of Br^-, I^-, S^{2-}, and EDTA since these ions react with the more soluble mercurous salts. Under ideal constant temperature circumstances, the stability of the pd is at least ± 0.1 mV.

The $Ag/AgCl/Cl^-$ reference electrodes are equally as good as the calomel. However, the silver-based electrodes are more frequently used with dilute chloride electrolyte, rather than saturated. Silver chloride is very much more soluble in high concentrations of chloride. Electrodes can be stable to ± 0.02 mV. The standard potential for unit activity is given in Table 11. Electrodes of the form shown in Figure 23 can be purchased, but they are easily made. A silver wire (5–9 s purity) is cleaned in hot nitric acid to remove sulfide. After rinsing in distilled water, it should be dipped in dilute ammonia and rinsed again. With an ordinary 1.5 V battery and a platinum wire connected to the negative pole, the silver wire is anodized in $0.1\,M$ KCl with some trace of HCl to form a layer of AgCl. The voltage should not be so positive that oxygen is evolved. Some workers cathodize to remove the first coat of AgCl and then reanodize. A series of wires should be made simultaneously, shorted together, and stored in $0.1\,M$ HCl or acidified KCl.

Many reference electrodes free from halides are possible. Mercury/HgO, mercury/Hg_2SO_4, mercury/$Hg_2C_2O_4$, and corresponding silver-based electrodes are possible. The electrolytes should be a soluble hydroxide, sulfate, and oxalate, respectively.

ISEs can be used as reference electrodes in junction cells. However, since ISEs do not have surrounding electrolytes, they are more directly applicable to junctionless cells, provided some conditions are met. For example, if it is known that a series of fluoride samples to be measured for fluoride activities

Table 11

Standard Potential of the Silver–Silver Chloride Electrode[a]

Temperature (°C)	$E^0_{AgCl/Ag}$	Temperature (°C)	$E^0_{AgCl/Ag}$
0	+0.23655	40	+0.21208
5	+0.23413	45	+0.20835
10	+0.23142	50	+0.20449
15	+0.22857	55	+0.20056
20	+0.22557	60	+0.19466
25	+0.22234	70	+0.18782
30	+0.21904	80	+0.1787
35	+0.21565	90	+0.1695
		95	+0.1651

[a] From R. P. Buck, *Physical Methods of Chemistry*, A. Weissberger and B. W. Rossiter, Eds., Ch. 2, Interscience, New York (1971).

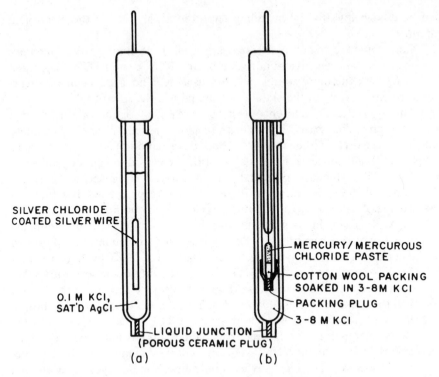

SILVER CHLORIDE
COATED SILVER WIRE

MERCURY/MERCUROUS
CHLORIDE PASTE

COTTON WOOL PACKING
SOAKED IN 3-8M KCl

0.1 M KCl,
SAT'D AgCl

PACKING PLUG

3-8 M KCl

LIQUID JUNCTION
(POROUS CERAMIC PLUG)

(a)   (b)

Figure 23. References Electrodes. (a) Silver/silver chloride and (b) Mercury/Calomel, saturated potassium chloride (3.8 M).

are buffered, contain either potassium ions at constant activity, or contain equal fluoride and potassium ion concentrations, the following junctionless cell is possible

Cu	K^+ selective electrode or pH glass electrode	Test solution	LaF_3 membrane	Inner solution containing F^- and Cl^-	AgCl	Ag	Cu

(Cell VIII)

Finally, when halide-free reference electrodes are needed, but special reference electrodes are not available, or cannot be made, there is an alternative. As mentioned in Section 1.2, a double-junction system can be used to isolate a KCl-containing reference electrode from a KCl-free test solution. Shorthand notation of examples were given in V, VI, and VII. A drawing of a commercial double junction, external reference electrode is shown in Figure 24.

The reference electrode structures shown in Figures 22(a), (b), and (c) and 23(a) and (b) are self-contained, complete "external" half-cells. When constructing a membrane configuration ISE, the "internal" reference electrode

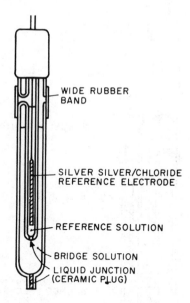

WIDE RUBBER
BAND

SILVER SILVER/CHLORIDE
REFERENCE ELECTRODE

REFERENCE SOLUTION

BRIDGE SOLUTION

LIQUID JUNCTION
(CERAMIC PLUG)

Figure 24. A double junction reference electrode. The inner half-cell is silver/silver chloride in a dilute chloride electrolyte. The outer filling solution contains a compatible electrolyte isolated from the test solution by the ceramic plug.

must also be provided. The most common example is Ag/AgCl/Cl⁻ half-cell shown in Figures 15(b) and 16. The interior or "inner" filling solution contains two principal components: a chloride salt and a salt reversible at the inner membrane surface. For simple electrodes such as an AgCl pellet or a K^+-sensing neutral carrier membrane, the inner filling solution would be typically 0.01–0.1 M KCl. For Br^- and S^{2-}-sensing electrodes, the inner reference half-cell can be Ag/AgBr/0.01–0.1 M KBr, and Ag/Ag$_2$S/Na$_2$S. However, even in these cases, Ag/AgCl/Cl⁻ (saturated with AgCl) is sufficient to poise the inner reference solution with a constant activity of Ag^+. Ordinarily, ISEs are calibrated against standard test solutions. For this reason, the inner references electrode pd value is not measured independently. Thus, even simpler inner reference electrodes of the junction type can be used. For example, when building experimental HPO$_4{}^{2-}$-sensing electrodes, an inner filling solution of Na$_2$HPO$_4$ is required. Rather than adding KCl and fixing in place an Ag/AgCl electrode, a commercial Ag/AgCl/KCl or Hg/Hg$_2$Cl$_2$/KCl reference electrode can be used.

Finally, a popular construction is the so-called "combination" ISE. This type of design was originally suggested and manufactured for pH glass electrodes. The construction combines both ISE sensor and external reference electrode into a single, integrated structure. An example is shown in Figure 25. In this case, an Ag/AgCl/Cl⁻ electrode (inner reference) is combined with a nearly identical "external" reference electrode built on the side. There are separate filling ports and liquid junction of the porous plug type.

Some commercially available combination electrodes stress sealed construction. The advertised advantage is infrequent maintenance. The filling

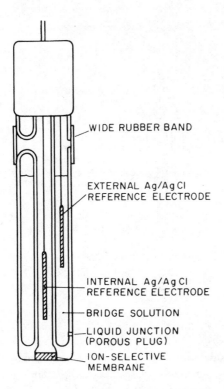

WIDE RUBBER BAND

EXTERNAL Ag/AgCl
REFERENCE ELECTRODE

INTERNAL Ag/AgCl
REFERENCE ELECTRODE

BRIDGE SOLUTION

LIQUID JUNCTION
(POROUS PLUG)

ION-SELECTIVE
MEMBRANE

Figure 25. An example of a self-contained ion-selective electrode and external reference electrode, a so-called "combination" electrode.

solutions are gelled so that the electrodes can be used in virtually any position. The problem, however, is the difficulty in replacing either filling solution. When the electrodes are heated, or they are exposed to a region of low pressure, the external reference solution can be lost through the junction structure. On other occasions, when temperature is lowered or pressure raised, test solutions can be taken into the electrode through the junction port.

4. Response Equations for the Steady State, Response Parameters, and Examples[160–178]

4.1. Noncrystalline, Primary, Heterogeneous, and Homogeneous

The membrane potential across ISEs is one portion of the overall, measured potential difference. For solid electrodes, those whose ion exchange sites do not move, the following formulas describe the initial responses after double-layer charging and equilibration of ion exchange processes at membrane surfaces. But for liquid ion exchanger membrane electrodes, the formulas apply only at steady state with respect to species profiles in the membranes. In every case, one assumes in using these formulas that the electrolyte activities at the surfaces of the membrane electrodes have reached their steady-state values. Membrane configuration electrodes are discussed first.

Membrane potentials $\Delta\phi_m$ for permeable ions of the same charge take a block-logarithmic form where $\Delta\phi_m$ = (potential on inner reference side)—(potential on outer reference side). Assumptions of constancy of mobilities, $\bar{u}_i$ or $\bar{u}_k$, and activity coefficients $\bar{\gamma}_i$ and $\bar{\gamma}_k$ are used to derive this equation. K_{ext} values are single ion partition coefficients. No distinction is made between homogeneous membranes and supported or heterogeneous membranes made from the same ion-exchanging materials.

One ion is frequently specified as "dominant," i.e., its single-ion partition coefficient K_{ext} and/or mobility exceed all others; then one has

$$\Delta\phi_m = \frac{RT}{zF} \ln \left[\frac{a_i + \sum_j k_{ij}^{pot} a_j}{a_k + \sum_l k_{lk}^{pot} a_l} \right] + \frac{RT}{zF} \ln \left[\frac{\bar{u}_i \bar{\gamma}_k K_{ext,i}}{\bar{u}_k \bar{\gamma}_i K_{ext,k}} \right] \quad (100)$$

where

$$k_{ij}^{pot} = \frac{K_{ext,j} \bar{u}_j \bar{\gamma}_i}{K_{ext,i} \bar{u}_i \bar{\gamma}_j} = (K_{exc,ij}) \frac{\bar{u}_j \bar{\gamma}_i}{\bar{u}_i \bar{\gamma}_j} \quad (101)$$

$K_{exc,ij}$ is the measurable ion exchange constant, a ratio of single ion partition coefficients, and the second term in Eq. (101) is frequently zero because the same dominant ion is present on both sides. k_{ij}^{pot} values in Eq. (101) are coefficients of the activities other than the dominant ion activity. Superscript "pot" signifies potentiometric selectivity coefficient. According to IUPAC report No. 43, January 1975, on nomenclature for ion-selective electrodes, the potentiometric selectivity coefficient should be designated k_{AB}^{pot} in the future. However, the present literature uses the nomenclature above almost exclusively. This form for k_{ij}^{pot} in Eq. (101) is one of many depending on the type of membrane. These quantities, given in tabular form later, are not necessarily independent of bathing solution compositions.

Overall responses of membrane electrode cells include several terms in addition to $\Delta\phi_m$

$$\Delta\phi_{meas} = \Delta\phi_m + \phi_{inner\,ref.} - \phi_{ext.\,ref.} + \phi_{jun} \quad (102)$$

where $\phi_{inner\,ref.}$ and $\phi_{ext.\,ref.}$ are generally the interfacial potentials calculated from the Nernst equation for the respective half-cells in their bathing solutions and ϕ_{jun} is the junction potential. The junction potential (or sum of junction potentials) is the least well-known quantity in the equation.

Frequently the junction potential is simply ignored when $\Delta\phi_{meas}$ is calibrated against standard solutions of prescribed activity; or the junction potential can be avoided when $\Delta\phi_{meas}$ is found from cells without liquid junction. The usual measured potential for cells in which a saturated calomel external electrode and an Ag/AgCl internal electrode are used has the form

$$\Delta\phi_{\text{meas}} = \frac{RT}{zF} \ln \left[\frac{a_i + \sum_j k_i^{\text{pot}} a_{j,\text{ext}}}{a_{i,\text{int}}} \right] + \phi^0 \text{AgCl/Cl}^- - \frac{RT}{F} \ln a_{\text{Cl}^-,\text{int}}$$

$$- \phi_{\text{sat'd calomel, ext ref.}} + \phi_{\text{jun}} \tag{103}$$

Gathering the terms together, the equation can be written for electrodes with fixed inside activities and external reference

$$\Delta\phi_{\text{meas}} = \phi^0 + \frac{RT}{zF} \ln \left[a_i + \sum_{ij} k_{ij}^{\text{pot}} a_j \right] \tag{104}$$

Experimental responses illustrating the logarithmic dependence on activities are given in Figures 26 and 27. For mixtures of permeable ions of different valences, but of the same sign, the approximation is sometimes made in the log term

$$\Delta\phi_{\text{meas}} = \phi^0 + \frac{RT}{F} \ln \left(a_i^{1/z_i} + (k_{ij}^{\text{pot}} a_j)^{1/z_j} \right) \tag{105}$$

where a_i has charge z_i and other ions have charge z_j. The modified equation is not exact and k_{ij}^{pot} will depend on concentrations a_i and a_j as well as site concentration. An example of the form of Eq. (104) is given in Figure 28 where $a_i = a_{\text{Ca}^{2+}}$ and $a_j = a_{\text{Ba}^{2+}}$. Barium ion activity is held constant at 5×10^{-2} and 5×10^{-1} M, and $K_{\text{Ca/Ba}}^{\text{pot}} = 0.01$.

4.2. Glass Electrodes

Equation (103) for glass electrodes using inner and external reference electrodes of $\text{Hg/Hg}_2\text{Cl}_2$ or Ag/AgCl is

$$\Delta\phi_{\text{meas}} = \frac{RT}{F} \ln \left[a_{\text{H}^+} + \sum_i k_{\text{H}^+/i}^{\text{pot}} a_i \right] - \frac{RT}{F} \ln a_{\text{H}^+\text{int}} + \frac{RT}{F} \ln \left[\frac{a_{\text{Cl}^-,\text{ext}}}{a_{\text{Cl}^-,\text{int}}} \right] + \phi_{\text{jun}} \tag{106}$$

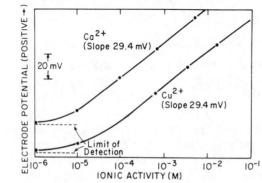

Figure 26. Calibration curves for cupric and calcium ion-selective electrodes.

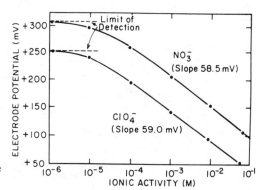

Figure 27. Calibration curves for nitrate and perchlorate ion-selective electrodes.

for mixtures of H^+ and univalent ions at activities a_i. $k^{\text{pot}}_{H^+/i}$ is a combination of parameters: the ion exchange equilibrium constant $k_{H^+/i}$, mobilities $\bar{u}$, and activity coefficients within the membrane phase. The general expression for k^{pot} is

$$k^{\text{pot}}_{H^+/i} = k_{H^+/i} \frac{\bar{u}_i \bar{\gamma}_{H^+}}{\bar{u}_{H^+} \bar{\gamma}_i} \tag{107}$$

and it is also called the selectivity coefficient for ion i relative to hydrogen.

The quantity of $k^{\text{pot}}_{H^+/i}$, defined in this way, will be less than unity. At a given activity level in the test solution, hydrogen ion gives the most positive measured potential of all other monovalent positive ions. Consequently, hydrogen is designated a_1 and the selectivity for other monovalent ions referred to 1 is less than unity. The reciprocal of $k^{\text{pot}}_{1/i}$ is a sensitivity coefficient to describe how much more sensitive the electrode is for ion 1 relative to ion i. For example, a typical general-purpose glass electrode has a selectivity coefficient $k^{\text{pot}}_{H^+/Na^+} \sim 10^{-11}$. The electrode is therefore 10^{11} times more sensitive to H^+ than to Na^+. The potential response of this glass electrode at $a_{H^+} = 10^{-11}$ makes the same contribution to the potential as $a_{Na^+} = 1$. Use of a saturated calomel exterior electrode leads at 25°C to the simple expression

$$\Delta\phi_{\text{meas}} = 0.05914 \log \left(a_{H^+} + \sum_i k^{\text{pot}}_{H^+/i} a_i \right) + \text{const (volts)} \tag{108}$$

because all the terms except ϕ_{jun} are fixed. The latter is a slowly varying function of test solution ionic strength, and its effect must be removed by calibration or by use of constant-ionic-strength solutions. In the absence of interferences

$$\Delta\phi_{\text{meas}} = -0.05914 \, p\text{H} + \text{const (volts)} \tag{109}$$

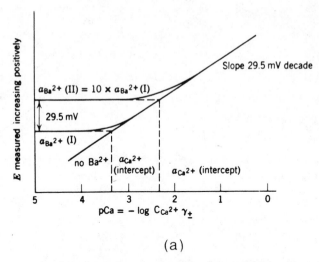

(a)

Figure 28(a). An example of ion interference—the effect of constant barium ion activity levels on a calcium response curve. A theoretical Nernstian response according to Eq. (104) is a straight line with slope 29.5 mV/decade and 25°C. Computation used $a_{Ba^{2+}}(I) = 5 \times 10^{-2}$; $a_{Ba^{2+}}(II) = 5 \times 10^{-1}$. [From R. P. Buck, Potentiometry: pH measurements and ion selective electrodes, in: *Techniques of Chemistry*, A. Weissberger, Ed., Wiley (1971), with permission.]

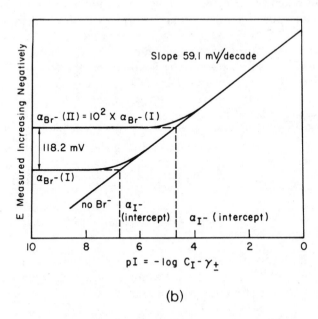

(b)

Figure 28(b). An example of ion interferences—the effect of constant bromide ion activity levels on an iodide response curve.

4.3. Crystalline, Primary, Homogeneous, and Heterogeneous Electrodes (with One Charge Carrier)

Closely related to fixed-site membranes are the solid-state single-crystal pressed-pellet membranes, and heterogeneous or supported membranes in which the solid powders are immobilized in a binder. The block logarithmic form for the membrane potential applies. In this form, this equation is too general because most solid-state membranes have only one charge carrier. Consequently the mobility or mobilities and extraction terms will not usually occur.

For silver salt membranes involving one or more phases, only Ag^+ carries current in the membranes; the second phase, if more insoluble, determines a_{Ag^+} in solution and therefore indirectly determines the membrane potential. Analysis of solid-state membrane phenomena has been accomplished almost exclusively on this basis. Equation (100) reduces to

$$\Delta\phi_m = \frac{RT}{F}\ln\left[\frac{a_{Ag^+}\text{ outer soln}}{a_{Ag^+}\text{ inner soln}}\right] \tag{110}$$

and

$$\Delta\phi_{\text{meas}} = \phi^0 + \frac{RT}{F}\ln\left(a_{Ag^+}\text{ outer soln}\right) \tag{111}$$

For all-solid-state electrodes with Ag inner contact, the same equation applies but ϕ^0 is ϕ^0_{Ag/Ag^+}. This simplified equation applies to all AgX and Ag_2S electrodes bathed in solutions containing excess Ag^+. For more and more dilute solutions, a more complicated expression is required to accommodate solutions so dilute that dissolution of the membrane itself has to be considered. Then in excess anion-containing solutions

$$\Delta\phi_{\text{meas}} = \phi^0 - \frac{RT}{F}\ln\left(a_{X^-}\text{ outer soln}\right) \tag{112}$$

The more general form of response in excess of monovalent anion mixture using iodide as the "dominant" ion

$$\Delta\phi_{\text{meas}} = \phi^0 - \frac{RT}{F}\ln\{a_{I^-} + [K_{so}(AgI)\gamma_{I^-}/K_{so}(AgX)\gamma_{X^-}]a_{X^-}\} \tag{113}$$

is applicable when a_{Ag^+} is known and fixed on one side, while the "test" solution contains I^- and the ion X^-. The apparent selectivity coefficient $k^{pot}_{I^-/X^-}$ does not include the mobility ratio of the anions since mobilities are not involved in the development of a diffusion potential

$$k^{pot}_{ij} = \frac{K_{so}(Agi)\bar{\gamma}_i}{K_{so}(Agj)\bar{\gamma}_j} \tag{114}$$

The Nernstian responses for AgCl are schematically shown in Figure 29 as straight lines. These are "folded" so the same scale of activities for Ag^+

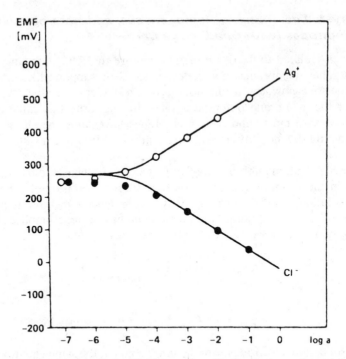

Figure 29. Experimental cell potential (EMF) for a silver chloride membrane electrode in response to activities of Ag^+ and Cl^-, for comparison with theoretical predictions. Calculated: solid lines. Experimental points: Ag^+ ($\bigcirc$); Cl^- ($\bullet$). [From W. E. Morf, G. Kahr, and W. Simon, *Anal. Chem.* **46**, 1538 (1974), with permission.]

and Cl^- can be used. Similarly, the curved lines are folded plots vs. log concentration of added Ag^+ or Cl^-. As added Ag^+ or Cl^- goes to zero, the actual Ag^+ and Cl^- approach $(K_{so})^{1/2}$, so the potential levels off. Experimental data are shown in Figure 29. For the Ag_2S electrode with a saturated calomel reference

$$\Delta\phi_{meas\,at\,25°C} = 0.557 - \frac{0.0591}{2}\log a_{S^{2-}} + \frac{0.0591}{2}\log K_{so}(Ag_2S) + \phi_{jun}$$

(115)

Using $pK_{so} = 49.2$, the log term has the value -1.45 V. Durst[160] and later Vesely *et al.*[161] have tested Eq. (115) over a wide activity range as shown in Figure 30.

For a mixed-sulfide electrode using a reversible sulfide salt, such as CdS, PbS, or CuS

sat'd calomel ref	KNO$_3$	cadmium soln $a_{Cd^{2+}}$	Ag$_2$S–CdS	inner soln	Ag$_2$S	Ag
						(Cell IX)

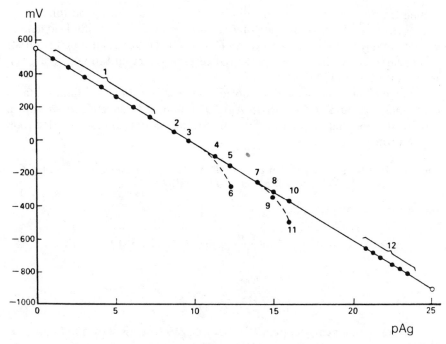

Figure 30. Wide range response of a silver sulfide membrane electrode. (1) Solutions of AgNO$_3$ ($\mu = 0.1$), 10^{-1}–10^{-7} M; (2) 0.1 M NaCl; (3) 1 M NaCl, saturated with AgCl; (4) 0.1 M NaBr; (5) 1 M NaBr, saturated with AgBr; (6) 1 M NaBr, unsaturated; (7) 0.01 M NaI; (8) 0.1 M NaI saturated with AgI; (9) 0.1 M NaI, unsaturated; (10) 1 M NaI saturated with AgI: (11) 1 M NaI, unsaturated; (12) 0.01 M Na$_2$S ($\mu = 0.11$), pH 7.0–12.0. [From J. Vesely, O. J. Jensen, and B. Nicolaisen, *Anal. Chim. Acta* **62**, 1 (1972), with permission.]

$$\Delta\phi_{\text{meas at } 25°} = \frac{0.0591}{2}\log a_{\text{Cd}^{2+}}$$

$$+ \frac{0.0591}{2}\log\frac{K_s\,(\text{Ag}_2\text{S})}{K_s\,(\text{CdS})}$$

$$+ 0.557 + \phi_{\text{jun}}\ (1 \text{ and } 2) \qquad (116)$$

The K_s term has the value -0.68 V using K_{so} (CdS) $= 10^{-26.1}$. These equations contain unknown parameters (ϕ_{jun}) and not very well-known parameters (K_s), and the calculations serve merely as a guide for directional changes in potential. Electrodes must be standardized against solutions of ionic strength comparable with unknowns. All the equations given with these example cells are Nernstian. They depend on the activity of the dominant species in solution. For AgCl membranes in Ag$^+$ solutions at concentrations in excess of $K_s^{1/2}$ (AgCl) the response is logarithmic in a_i (Ag$^+$ in the case of AgX and Ag$_2$S, Ag$_2$Se, Ag$_2$Te membranes) and is determined by (1) C_{Ag^+} added to the solutions and (2) Ag$^+$

arising from dissolving of the membrane. The general expression for a_{Ag^+} at the electrode surface where the potential is generated, is a common-ion effect calculation involving C_{Ag^+} or C_{Cl^-} and K_s (AgCl). This calculation gives rather complicated equations for the potential response. Simplifications were given in Eq. (110) and (111).

The commercially available fluoride electrodes use inner filling solutions containing NaF and NaCl with an Ag/AgCl inner reference electrode. For a typical junction cell configuration with external saturated calomel electrode of the form

$$\text{Cu}\,|\,\text{Pt}\,|\,\text{Hg, Hg}_2\text{Cl}_2\,\Big|\,\begin{array}{c}\text{KCl}\\[2pt]\text{sat'd}\end{array}\,\Big|\,\begin{array}{c}a_{F^-}\\[2pt]\text{test}\end{array}\,\Big|\,\begin{array}{c}\text{LaF}_3\\[2pt]\text{membranes}\end{array}\,\Big|\,\begin{array}{c}0.1\ M\ \text{NaF}\\[2pt]0.1\ M\ \text{NaCl}\end{array}\,\Big|\,\text{AgCl}\,\Big|\,\text{Ag}\,\Big|\,\text{Cu} \qquad \text{(Cell X)}$$

the measured potential obeys

$$\Delta\phi_{\text{meas. }25^\circ\text{C}} = -0.0591\,\log\frac{a_{F^-}}{0.1\,\gamma_{F^-}} + 0.222 - 0.0591\,\log\,(0.1\,\gamma_{Cl^-})$$

$$- 0.242 + \phi_{\text{jun}}$$

$$= \phi^0 - 0.0591\,\log a_{F^-\,\text{test}} \qquad (117)$$

The response of an early commercial fluoride electrode is shown in Figure 31. Different, parallel response curves are mainly ionic strength effects that cause different ϕ^0 values through the ionic strength dependence of γ_{F^-}. Whenever activity coefficients are ignored, ϕ^0 is not constant. It is often indicated to be a "formal" quantity $\phi^{0'}$.

4.4. Junction and Diffusion Potentials

When two electrolytes of different compositions touch, a junction potential arises because ions do not move at the same rates during the spontaneous intermixing. This means that the contact between electrolyte solutions surrounding the external reference electrode, and the test solution will become a source of a potential difference. The ions of highest concentration and mobility carry their charge into the other solution until a restraining field and potential arises in a few microseconds. At this time the junction potential is established, and continues to hold its value as long as the source and sink solutions maintain their concentrations essentially constant. This junction potential adds to the membrane potential and is measured by the potentiometer as a series component. Junction potentials usually run from about -10 to $+10$ mV. Some data are included in Tables 12 and 13 to show typical magnitudes. It is noted that when one solution, but not the other, contains either a strong acid or strong base, the potential is large and can reach 40 mV. Junction potentials exist at both sides of salt bridges when three different electrolytes or different electrolyte concentrations are used.

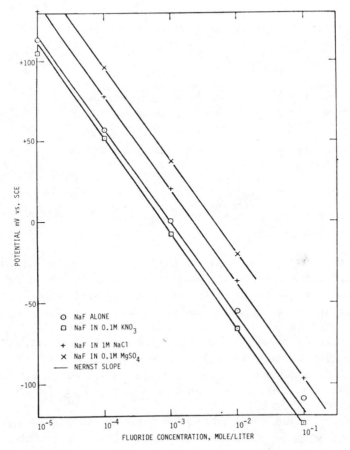

Figure 31. Potential of LaF$_3$ membrane electrode as a function of fluoride concentration in various supporting electrolytes.

Salts with ions of the same mobility (KCl is almost ideal in this regard) and mixtures of salts such that the mobility–concentration product summed over positive ions (+ = cations) equals the same product summed over negative ions (− = anions), are called "equitransferent" salts or "equitransferent" mixtures. More generally the equation

$$\sum_{+} z_{+}c'_{+}u_{+} = \sum_{-} z_{-}c'_{-}u_{-} \quad \text{or} \quad \sum_{+} z_{+}^{2}c_{+}u_{+} = \sum_{-} z_{-}^{2}c_{-}u_{-} \tag{118}$$

must be obeyed for equivalent concentrations c'_i, molar concentrations c_i, and ionic mobilities u_i. When ions have the same absolute charge, several systems have been evaluated. These are noted in Table 14. These mixtures are suitable for use in salt bridges although KCl at a high concentration is typically used alone for this purpose because diffusion of KCl outward into two solutions

Table 12
Liquid Junction Potentials at 25°C[a]

Boundary	E_J (mV)[b]	Boundary	E_J (mV)[b]
LiCl(0.1 F)|KCl(0.1 F)	−8.9	KCl(0.1 F)|KCl(3.5 F)	+0.6
NaCl(0.1 F)|KCl(0.1 F)	−6.4	NaCl(0.1 F)|KCl(3.5 F)	−0.2
NH$_4$Cl(0.1 F)|KCl(0.1 F)	+2.2	NaCl(1 F)|KCl(3.5 F)	−1.9
NaOH(0.1 F)|KCl(0.1 F)	−18.9	NaOH(0.1 F)|KCl(3.5 F)	−2.1
NaOH(1 F)|KCl(0.1 F)	−45	NaOH(1 F)|KCl(3.5 F)	−10.5
KOH(1 F)|KCl(0.1 F)	−34	KOH(1 F)|KCl(3.5 F)	−8.6
HCl(0.1 F)|KCl(0.1 F)	+27	HCl(0.1 F)|KCl(3.5 F)	+3.1
H$_2$SO$_4$(0.05 F)|KCl(0.1 F)	+25	H$_2$SO$_4$(0.05 F)|KCl(3.5 F)	+4

[a] The data are taken from G. Milazzo, *Elektrochemie*, Springer-Verlag, Vienna (1952); of R. G. Bates, *Electrometric pH Determinations*, p. 41, Wiley, New York (1954).
[b] $E_J = \phi_{\text{right}} - \phi_{\text{left}}$.

abutting the salt bridge swamps out diffusion of more dilute salts into the bridge. In this case junction potentials are minimized, but not entirely eliminated. It is also possible to use uniformly high electrolyte solutions throughout an entire cell. If the salt or salts are chosen to provide a constant ionic strength, discussed later, there is little tendency for ions to diffuse, and so junction potentials are minimized. This precaution is not always practical, and would not be useful for natural water analysis *in situ*. However, when these techniques can be used, the junction potential, although not known, remains constant within limits, and so appears as a constant offset voltage, and is contained in the apparent standard potential $\Delta\phi^{0\prime}$.

Junction potentials are part of a larger category of diffusion potentials. In liquid ion exchange membranes, for example, diffusion potentials can arise whenever two counterions are simultaneously present in the membrane. This situation occurs when test solutions contain two permeable ions, and can be avoided by making measurements with high selectivity electrode membranes.

Table 13
Some Calculated Liquid Junction Potentials (mV) between Saturated KCl
and Solutions at 25°C[a]

Concentration/M	HCl	KCl	KOH	KH phthalate	CH$_3$COONa/CH$_3$COOH
10^{-1}				2.8	2.2
10^{-2}	2.9	2.8	1.9	3.5	3.2
10^{-3}	4.0	3.9	3.2	4.1	4.2
10^{-4}	4.8	5.0	4.5	4.9	5.0
10^{-5}	5.7	6.1	5.8	5.8	5.8
10^{-6}	6.7	7.1	6.9	6.7	6.7

[a] From R. G. Picknett, *Trans. Faraday Soc.* **64**, 1050 (1968).

Table 14

Some Electrolytes Reported to be Equitransferent

Source	Composition
1. Grove–Rasmussen	$1.8\ M$ KCl $+ 1.8\ M$ KNO$_3$
2. Kline, Meacham, and Acree	$3\ M$ KCl $+ 1\ M$ KNO$_3$
3. Orion Res. Inc. Type 90-00-01	$1.70\ M$ KNO$_3$ $+ 0.64\ M$ KCl
	$+ 0.06\ M$ NaCl
	$+ 1$ ml/l of 37% HCHO
4. Wilson, Haikala, and Kivalo	$2\ M$ RbCl

1. Ref. K. V. Grove–Ramussen, *Acta Chem. Scand.* **3**, 445 (1949); **5**, 422 (1951).
2. Ref. G. M. Kline, M. R. Meacham, and S. F. Acree, *Bur. Stand. J. Res.* **8**, 101 (1932).
3. Orion Research Corp. Technical Notes.
4. Ref. M. F. Wilson, E. Haikala, and P. Kivalo, *Anal. Chim. Acta* **74**, 396, 411 (1975).

Diffusion potentials within membranes are accounted for in theory and in practice. Because the internal potential has a logarithmic form, it appears as a component of the overall membrane potential difference. Thus, calibration of ISEs automatically include the internal diffusion potential and no corrections are needed or are even appropriate.

A thorough analysis of liquid junction designs and errors in various media has been reported by Brezinski.[179–181]

4.5. Activity, Activity Coefficients, and Ionic Strength

ISEs respond to ions that exchange at their interfaces in a way analogous to redox electrodes whose responses require electron exchange. The potential and charge that develop in both cases are determined by the energies of exchanging species in the two phases. Ion energies are directly proportional to the log of their activities at constant temperature, pressure, and local potential. The activity is numerically equal to concentration at low ionic strengths, or very dilute solutions of a single salt. Ionic activities a_i are always defined to be proportional to a measure of ionic concentration c_i, but the proportionality factor (the activity coefficient) falls from unity to lower values in a complex manner depending on overall ionic strength

$$a_i = c_i \gamma_i \tag{119}$$

From thermodynamic analysis, activity coefficients must depend on the concentration scale used. The latter can be conveniently molarity (c_i moles/liter), molality (m_i moles/1000 g solvent), or mole fraction. Most compiled data on salt activity coefficients are reported on the molality scale. But, for dilute aqueous solutions, c and m are nearly identical. The third scale, mole fractions, is used for electrolytes in mixed solvents, but is closely related to the molality scale for dilute electrolytes in water solvent.

Every cell containing an ISE and reference electrode responds to two or more ion activities. In junctionless cells, these will be cation and anion activities of a component salt. In junction cells, the salt activity measured usually involves separated species, e.g., the cation may be in the test solution, while the anion may be the Cl^- in the reference electrode electrolyte. The junction cell does respond to, and measure pairs of ionic activities, but these should be regarded as nonthermodynamic activities because the results include the nonequilibrium junction potential contribution. All cells respond to products or ratios of single ion activities such as

$$a_+ a_- = c_+ c_- \gamma_+ \gamma_- \qquad (120)$$

There is no thermodynamic way to measure an individual activity or an individual activity coefficient. The few activity standards that exist are based on the Bates–Guggenhein convention, which uses a calculated value for the activity of chloride ions in dilute solutions in which the coefficients are near unity and obey the extended Debye–Hückel equation. Single ion activity coefficients are based on precise salt activity coefficients and γ_{Cl^-}. For example, from cells or isopiestic measurements in solutions of known concentrations of NaCl, the quantity γ_{NaCl}^2 is determined. By identifying $\gamma_{NaCl}^2 = \gamma_+ \gamma_-$ with a reference value for $\gamma_- (\gamma_{Cl^-})$, tables of γ_{Na^+} for different salt concentrations can be compiled using

$$\gamma_{Na^+} = \gamma_{NaCl}^2 / \gamma_{Cl^-} \qquad (121)$$

Then, using solutions of NaBr, γ_{Br^-} values can be compiled. This "daisy chain" method, while tedious, leads to a set of internally consistent activity coefficients for single ions in single salt solutions. Some examples of solutions with precisely characterized activity coefficients, suitable for use as calibrating solutions for ISEs, are given in Tables 15 and 16.

There are many more electrodes than there are standardized electrolytes for their calibration. The choices for calibrating solutions are (1) use of concentration standards in decade (or closer) intervals. The responses on a semilog scale (mV vs. log c) will not be linear; (2) use of known concentrations diluted with a constant ionic strength inert electrolyte. For example, NaBr (0.1 M) can be diluted 1:10 with 0.1 M KNO_3 to yield decade dilutions of Br^-. Since activity coefficients depend mainly on ionic strength, γ_{Br^-} will be nearly constant, although not known, and mV vs. log c_{Br^-} will be nearly linear; (3) use of concentration standards in decade dilutions with computed values of γ_+ and γ_- by assuming that $\gamma_+ = \gamma_- = \gamma_\pm$ from tabulated isopiestic or cell values at each concentration. Although $\gamma_+ \neq \gamma_-$ as a general rule, the corrections from concentration to approximate activity $c_i \gamma_i$ do help to linearize the semilog response plot; (4) when known concentration of mixtures occur, say NaBr in varying, but known concentrations of KNO_3 (or other 1:1 electrolyte), compute the total concentration of salt and use $\gamma_+ = \gamma_- = \gamma_\pm$ from tabulated data for NaBr at the total concentration. This method can be generalized to

Table 15

Solutions of Measured or Calculated Activity[a]

	Salt			
	KF	CaCl$_2$	NH$_4$Cl	KCl
mol kg^{-1}	pF = pK	pCa	pNH$_4$	pCl
0.01	2.044			2.044
0.0333		1.900		
0.05	1.387			
0.1	1.111	1.570	1.112	1.113
0.2	0.837	1.349	0.840	0.840
0.5	0.475	0.991	0.483	0.486
1.0	0.190	0.580	0.208	0.216
2.0	−0.119	−0.198	−0.080	−0.059

$$pX = -\log_{10}a_x.$$

[a] R. G. Bates, B. R. Staples, and R. A. Robinson, *Anal. Chem.* **42**, 867 (1970); R. A. Robinson, W. C. Duer, and R. G. Bates, *Anal. Chem.* **43**, 1862 (1971); J. Bagg and G. A. Rechnitz, *Anal. Chem.* **45**, 271 (1973); Compiled by P. L. Bailey, *Analysis with Ion-Selective Electrodes*, Heyden, London (1976).

Table 16a

pHs of NBS Primary Standards from 0 to 95°C[a]

Temp. (°C)	0.05 M KH$_2$ citrate	0.025 M each NaHCO$_3$, Na$_2$CO$_3$	KH tartrate (sat'd. at 25°C)	0.05 M KH phthalate	0.025 M KH$_2$PO$_4$, 0.025 M Na$_2$HPO$_4$	0.008695 M KH$_2$PO$_4$, 0.03043 M Na$_2$HPO$_4$	0.01 M borax
0	3.863	10.317	—	4.003	6.984	7.534	9.464
5	3.840	10.245	—	3.999	6.951	7.500	9.395
10	3.820	10.179	—	3.998	6.923	7.472	9.332
15	3.802	10.118	—	3.999	6.900	7.448	9.276
20	3.788	10.062	—	4.002	6.881	7.429	9.225
25	3.776	10.012	3.557	4.008	6.865	7.413	9.180
30	3.766	9.966	3.553	4.015	6.853	7.400	9.139
35	3.759	9.925	3.549	4.024	6.844	7.398	9.102
38	—	—	3.548	4.030	6.840	7.384	9.081
40	3.753	9.889	3.547	4.035	6.838	7.380	9.068
45	3.750	9.856	3.547	4.047	6.834	7.373	9.038
50	3.749	9.838	3.549	4.060	6.833	7.367	9.011
55	—	—	3.554	4.075	6.834	—	8.985
60	—	—	3.560	4.091	6.836	—	8.962
70	—	—	3.580	4.126	6.845	—	8.921
80	—	—	3.609	4.164	6.859	—	8.885
90	—	—	3.650	4.205	6.877	—	8.850
95	—	—	3.674	4.227	6.886	—	8.833

[a] These reagents can be purchased in powdered form from the Bureau of Standards. R. P. Buck, Compilation in *Physical Methods of Chemistry*, Part IIA, A. Weissberger and B. W. Rossiter, Eds., Ch. 2, Interscience, New York (1971).

Table 16b
pHs of NBS Secondary Standards

Temperature (°C)	pa_{H^+} values of 0.05 M $KH_3(C_2O_4)_2 \cdot H_2O^a$	pa_{H^+} values of Ca(OH)$_2$ sat'd. at 25°C^a
0	1.666	13.423
5	1.668	13.207
10	1.670	13.003
15	1.672	12.810
20	1.675	12.627
25	1.679	12.454
30	1.683	12.289
35	1.688	12.133
38	1.691	12.043
40	1.694	11.984
45	1.700	11.841
50	1.707	11.705
55	1.715	11.574
60	1.723	11.449
70	1.743	—
80	1.766	—
90	1.792	—
95	1.806	—

Temperature (°C)	pa_{H^+} values of 0.02 M piperazine phosphate^b	pa_{H^+} values of 0.05 M piperazine phosphate^b
0	6.580	6.589
5	6.515	6.525
10	6.453	6.463
15	6.394	6.404
20	6.338	6.348
25	6.284	6.294
30	6.234	6.243
35	6.185	6.195
40	6.140	6.149
45	6.097	6.106
50	6.058	6.066

^{a} R. P. Buck, *Physical Methods of Chemistry*, Ch. 2, Interscience, New York (1971).
^{b} From A. B. Hetzer, R. A. Robinson, and R. G. Bates, *Anal. Chem.* **40**, 634 (1968).

any ionic strength, IS. For 1:1 electrolytes IS = c(salt); for 1:2 and 2:1 salts, IS = $3c$(salt) and for 2:2 salts, IS = $4c$(salt). For single salts and mixtures

$$IS = 1/2 \sum_i z_i^2 c_i \qquad (122)$$

Activity coefficients can be estimated for single ions in mixtures using the mean activity coefficient for the single salt at the ionic strength of the total mixture.

Table 16c
Additional pH Standards with pH Values at 25°C[a]

Solution	pH
0.1 M HCl	1.10
0.01 M HCl, 0.09 M KCL	2.07
0.01 M acetic acid, 0.1 M Na acetate	4.65
0.01 M acetic acid, 0.01 M Na acetate	4.71
0.025 M NaH succinate, 0.025 M Na$_2$ succinate	5.40
0.01 M Na$_3$PO$_4$	11.72
0.05 M NaOH	12.62

[a] R. P. Buck in *Physical Methods of Chemistry*, A. Weissberger and B. W. Rossiter, Eds., Ch. 2, Interscience, New York (1971).

Every junctionless cell configuration will provide equilibrium cell potential measurements involving products and/or ratios of activities and therefore activity coefficients. Junction cells involving ISE's provide potentials dependent often on activity products in different electrolyte regions of the cell! As an elementary example, consider cell XI

$$Pt, H_2|H^+, Cl^-, AgCl|Ag \qquad \text{(Cell XI)}$$

This cell requires only a single electrolyte with both electrodes immersed in the same solution of HCl—there is no salt bridge. The voltage for this cell is obtained from the two half-cell potentials

$$\Delta\phi_{H^+/H} = -\frac{0.059}{2}\log\left[\frac{P_{H_2}}{(H^+)^2}\right] \qquad (123)$$

$$\Delta\phi_{AgCl/Ag} = \Delta\phi^0_{AgCl/Ag} - 0.059\log(Cl^-) \qquad (124)$$

$$\Delta\phi_{cell} = \Delta\phi^0_{AgCl/Ag} + \frac{0.059}{2}\log P_{H_2} - 0.059\log(H^+)(Cl^-) \qquad (125)$$

Here, the product of two activities appears. While we can measure the concentration of each ion (for example, by titration or precipitation) we can measure only the product of their activities and thus only the product of their activity coefficients. If both ions have the same numerical charge, it is customary to assign the same value to each activity coefficient, geometric mean. The mean ionic activity coefficient is defined

$$\gamma_\pm = \sqrt{\gamma_+\gamma_-} \qquad (126)$$

Although the mean ionic activity coefficient is the only one we can measure, the Debye–Hückel limiting law provides a means to estimate the value of an individual ionic activity coefficient

$$-\log\gamma_i = 0.512z_i^2\sqrt{IS} \text{ at } 25°C \qquad (127)$$

where z_i is the charge on the ion and IS is the ionic strength of the solution. Equation (127) as given applies reasonably well to dilute aqueous solutions with concentrations no greater than about 0.05 M. It is a limiting law and has been modified by many workers in an effort to approach the behavior of real systems at higher concentrations. One such equation takes into account the effect of the finite size of the ions

$$-\log \gamma_{\pm} = 0.512 z_i^2 \left[\frac{\sqrt{IS}}{1 + Ba\sqrt{IS}} \right] \text{ at } 25°C \qquad (128)$$

where B is a function of the dielectric constant and the temperature ($B = 0.328$ for water at 25°C) and a is an ion size parameter (Table 17). Equation (128) extends the applicability of the Debye–Hückel limiting law up to concentrations of 0.1 M. Selected values of the individual ionic activity coefficients are given in Table 17.

From the definition of activity coefficients, Eq. (119) can be rewritten

$$\Delta\phi_{cell} = \Delta\phi_{AgCl/Ag}^0 + \frac{0.059}{2} \log P_{H_2} - 0.059 \log C_{HCl}^2 - 0.059 \log \gamma_{\pm}^2 \qquad (129)$$

Clearly the $\gamma_{\pm}^2$ term depends on concentrations through the ionic strength. This means a plot of $\Delta\phi_{cell}$ vs. log C_{HCl} is *nonlinear*. To linearize the plot, one looks up the salt activity coefficient for each ionic strength. This procedure is exact for a single salt solution, but it is approximately correct for mixtures.

Table 17
Single Ion Activity Coefficients[a]

Ion size, a	Ion	Ionic strength			
		0.005	0.01	0.05	0.01
9	H^+	0.933	0.914	0.86	0.83
6	$Li^+, C_6H_5COO^-$	0.929	9.907	0.84	0.80
4	$Na^+, IO_3^-, HSO_3^-, H_2PO_4^-$	0.927	0.901	0.82	0.77
3	$K^+, Cl^-, Br^-, CN^-, NO_3^-$	0.925	0.899	0.81	0.76
2.5	Cs^+, NH_4^+, Ag^+	0.924	0.898	0.80	0.75
8	Mg^{++}, Be^{++}	0.755	0.69	0.52	0.45
6	$Ca^{++}, Cu^{++}, Zn^{++},$ $Mn^{++}, C_6H_4(COO)_2^=$	0.749	0.675	0.49	0.41
4	$Hg_2^{++}, SO_4^=, CrO_4^=, HPO_4^=$	0.740	0.660	0.445	0.355
9	$Al^{+3}, Fe^{+3}, Cr^{+3}$	0.54	0.445	0.245	0.18
4	$PO_4^{-3}, Fe(CN)_6^{-3}$	0.505	0.395	0.16	0.10
5	$Fe(CN)_6^{-4}$	0.31	0.20	0.05	0.02

[a] From Kielland, *J. Am. Chem. Soc.* **59**, 1675 (1937).

Frequently, ISE measurements are done with junction cells and the response depends formally on a single ion activity. In that case, the approximation is made that $\gamma_{\pm}$ for the system ionic strength can be used to convert C_+ or C_- to a_+ or a_-.

4.6. Junction Structure, Cracks, and Leaks

Junction structures have two purposes: (1) to maintain a constant junction potential (or potentials) and (2) to separate the reference electrode electrolyte from the test electrolyte, or at least, to avoid extensive intermixing. The latter is not considered a priority by some workers and so intermixing has been tolerated in designs to keep constant composition and constant junction potential at the region of solution contact. Consequently junction structures either allow, or even emphasize flow, or seek to minimize convective and diffusive flows. An impractical but classical design allows continuous flow of reference and test electrolytes [Figure 32(a)]. A similar design [Figure 32(b)] uses an intermediate compartment with flowing solution to separate reference and test electrolyte, which are also flowing.

Examples of junction structures that restrain flow are shown in Figure 32(c) and (d). These examples use either fritted glass disks (medium or fine) or gas dispersion tubes. In 32(c), the reference compartment might contain an inert electrolyte or the electrolyte of the reference electrode, typically KCl. In 32(d), the salt bridge could similarly contain an inert electrolyte or KCl. Both of these examples restrain convection, but do allow nearly free diffusion of salts and osmotic-pressure driven uptake of water. Two other arrangements for more efficient restraining of intermixing uses ground glass joints (or plastic joints) as shown in Figures 32(e) and (f). The inside and outside may contain the same or different solutions. Ordinarily glass is not preferred when solutions are basic because of corrosion and "freezing" of the joint. Flow rates of intermixing are very low in these designs.

The majority of junction structures, made commercially, are shown in Figure 32(g)–(j). These are typically controlled leaks using porous ceramic or graphite, mismatched glass such that a small crack is formed, asbestos fiber or wood fiber, and palladium or other inert metal wire that is mismatched thermally so that an annular crack is formed. These junctions are usually small (0.5 mm diam), but can be larger. Of the ceramic and asbestos fiber types, leak rates of 0.01–0.1 ml/day per 5 cm head of internal solution can be expected. Stability of junction potentials can be $\pm 40 \ \mu V$ over several hours. However, day-to-day reproducibility is about ± 0.2 mV.

4.7. Selectivity Coefficients and Interferences

Selectivity has both qualitative and quantitative significance. For example, cation-responsive electrodes give potential–log activity plots for several

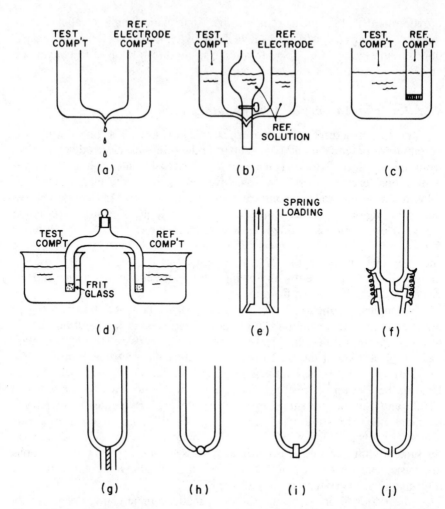

Figure 32. Examples of liquid junctions and salt bridges. (a) and (b) are free-flowing junctions that consume reactants and products in (a), but consume excess reference solution from the middle vessel in (b). (c) and (d) are slow free-flowing junctions involving a glass frit in (c) and two glass frits (with or without a gelling agent in the bridge) (d). (e)–(j) are constricted flow, reagent-conserving junctions. (e) and (f) are typically ground-glass joints wetted with electrolyte; (g) typically a porous ceramic plug; (h) a mismatched thermal expansion glass sphere with natural cracks; (i) may be ceramic, carbon, or asbestos in glass, and (j) is a metal wire, typically platinum in glass.

responsive ions, as shown in Figure 33. If the ions have the same charge, and are reversible, then responses are parallel. One can compare the measured potentials at identical activities. Some ion will give the most positive potential at that activity. The electrode has the greatest selectivity for that ion. Other ions will be less sensitively detected, especially in mixtures. Taking the basic

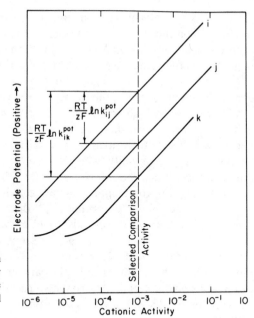

Figure 33. Illustration of the two-solution method for determining selectivity coefficients. Comparison activity must be in the region of known slope and parallel responses.

response equation for mixtures of two ions of the same charge, as might be appropriate for a liquid ion exchanger electrode

$$\Delta\phi_{\text{meas.}} = \phi^0 + \frac{2.303RT}{zF} \log\left(a_i + k_{ij}^{\text{pot}} a_j\right) \qquad (130)$$

The responses of salts of ion i obeys, over some reasonable range

$$\Delta\phi_{\text{meas.}} = \phi^0 + \frac{2.303RT}{zF} \log a_i \qquad (131)$$

and salts of ion j obey

$$\Delta\phi_{\text{meas.}} = \phi^0 + \frac{2.303RT}{zF} \log k_{ij}^{\text{pot}} + \frac{2.303RT}{zF} \log a_j \qquad (132)$$

By noting responses in pure solutions, the spacing between the response curve for i salts and responses of j salts

$$\Delta\phi_{\text{meas.}(i)} - \Delta\phi_{\text{meas.}(j)} = \frac{2.303RT}{zF} \log k_{ij}^{\text{pot}} \qquad (133)$$

Provided the more selectively sensed ion is taken as i, all other salts of ion j will give less positive responses and so

$$k_{ij}^{\text{pot}} < 1 \qquad (134)$$

If ion i were chosen from an ion in the middle of the response sequence, $k_{ij}^{pot} > 1$ for the more selective ions, while $k_{ij}^{pot} < 1$ for the less selectively sensed ions. This is a quantitative view of the membrane or electrode selectivity and k_{ij}^{pot} is the selectivity coefficient of ion j relative to ion i.

When mixtures of ions i and j are studied, typical response curves for variable amounts of i in the presence of a constant amount of j are presented in Figure 28. At high activities of i, Ca^{2+} in the sample, Eq. (131) is obeyed and the normal Nernstian response of Ca^{2+} is seen. Then as a_i is lowered, the activity a_j begins to contribute to total response and Eq. (130) describes the curvature region. Finally, when $a_i \ll k_{ij}^{pot} a_j$, the response is dominated by the interference. Since the interference, Ba^{2+} in this example, was held constant in each solution, the response at low activities of Ca^{2+} becomes constant.

If it is desired to determine k_{ij}^{pot}, the shape of the mixture response can be fit, by mathematical methods, to a best value of k_{ij}^{pot}. However, for a quick check of selectivity, the response difference between single solutions of salts of i and j at equal concentrations, can be used with Eq. (133). This mixture method leads to reliable results, in part because the experiment is equivalent to many single solution experiments. The single solution method will lead to erroneous results whenever (1) the pure solution responses are not parallel; (2) the selectivity coefficients are concentration dependent; and (3) the responses are too slow to reach equilibrium. Figures 34 and 35 illustrate how k_{ij}^{pot} can very by choosing inappropriate single solution concentrations for the comparison determination.

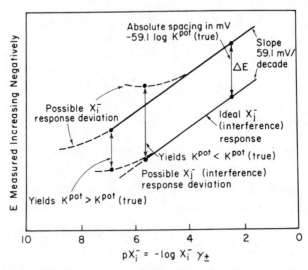

Figure 34. Illustration of k_{ij}^{pot} determination by the two-point method using $\Delta\phi$ measured, 25°C, for equal activities of X_i^- and X_j^- (interference). [Modified from R. P. Buck, *Techniques of Chemistry*, Vol. 1, part 2A, A. Weissberger and B. W. Rossiter, Eds., Interscience, New York (1971), with permission.]

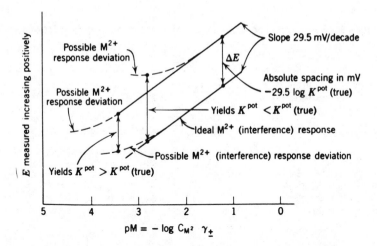

Figure 35. Illustration of k_{ij}^{pot} determination by the two-point method (not recommended) using $\Delta\phi$ measured for equal activities of M^{2+} and M^{2+} (interference). [From R. P. Buck, in Potentiometry: pH measurements and Ion Selective Electrodes in *Techniques of Chemistry*, A. Weissberger, Ed., New York; Wiley (1971), with permission.]

Selectivity coefficients have theoretical bases, in terms of single ion partition coefficient ratios and mobility ratios. The single ion partition coefficient ratio is, for liquid and many solid ion exchanger membranes, an easily measurable thermodynamic quantity—the ion exchange equilibrium constant. Also for solids, it is the well-known solubility product ratio. However, for liquid membranes, mobilities of ions are also involved and are important. These quantities are very difficult to determine unambiguously. Depending on the membrane–solvent dielectric constant, membrane electrolytes may be ion paired so that equilibrium constants for processes within a membrane are also involved. Table 18 shows some theoretical relationships.

For many electrodes, especially liquid membrane systems, selectivity coefficients are not highly reproducible or precise quantities. The first reason is that selectivity coefficients are time dependent. A fresh electrode exposed for the first time to interferences, passes through nonequilibrium states as interferences diffuse toward the electrode surface and replaced ions diffuse away into bulk electrolyte. During this time, selectivity coefficients are much larger than equilibrium or steady-state values.[182] At a later time, the theoretical values are approached when a steady state exists. Especially at low bathing electrolyte concentrations of interferences, the time to reach steady state can be a few minutes, and the unwary may have completed a measurement and gone on to another sample. Using the two solution method, a steady state may not be reached because the membrane surface or membrane interior can require massive reconstitution from ion i to ion j.

Table 18

Selectivities for Different Types of Membrane Electrodes and Restrictions Used[a,b]

Type of membrane	K_{ij}^{Pot}	Restrictions
Liquid membrane, no ligand (negligible complex formation)	$\dfrac{u_j k_j}{u_i k_i}$	$u_x = 0$ or $c_x(x) = 0$
	$\dfrac{(u_j + u_x)k_j}{(u_i + u_x)k_i}$	$c_x(x) = $ const. electroneutrality holds
	$\dfrac{k_j}{k_i}$	Zero diffusion potential $c_i(x) + c_j(x) = $ const.
Liquid membrane, electrically neutral ligand S (complex formation dominant)	$\dfrac{u_{js} K_j}{u_{is} K_i}$	$u_x = 0$ or $c_x(x) = 0$ $c_s(x) = $ const.
	$\dfrac{(u_{js} + u_x)K_j}{(u_{is} + u_x)K_i}$	$c_x(x) = $ const. electroneutrality holds
	$\dfrac{K_j}{K_i}$	Zero diffusion potential $a(x) = $ const.
Liquid membrane, electrically charged ligand S$^-$ (complex formation dominant, flux of anions X$^-$ negligible)	$\dfrac{u_j k_j}{u_i k_i}$	$u_s \ll u_i, u_j$
	$\dfrac{(u_j + u_s)k_j}{(u_i + u_s)k_i}$	$c_s(x) = $ const. electroneutrality holds
	$\dfrac{u_{js} K_j}{u_{is} K_i}$	$u_s \gg u_i, u_j$ total flux of all ligand forms negligible
	$\dfrac{K_j}{K_i}$	Zero diffusion potential $a(x) = $ const.
Solid ion exchangers	$\dfrac{u_j k_j}{u_i k_i}$	Permselective for ions of one sign and magnitude
Solid crystal membranes	$\dfrac{K_{sp,i} \bar\gamma_i}{K_{sp,j} \bar\gamma_j}$	Zero diffusion potential ions of same charge

[a] Reprinted with permission from R. P. Buck, *Crit. Rev. Anal. Chem.* **5**, 323 (1975). Copyright The Chemical Rubber Co., CRC Press, Inc.

[b] Explanations: k_i, k_j = partition coefficients of cations I$^+$ and J$^+$ between outside solutions and membrane; K_i, K_j = partition coefficients of cations I$^+$ and J$^+$ between outside solution and the respective complexes in the membrane; u = mobilities in the membrane; $c(x)$ = concentrations in the membrane; $a(x)$ = degree of dissociation in the membrane, defined as the ratio of free ligand concentration to concentration of all forms of ligand; $k_i = k_{ext}/\bar\gamma_i$; $k_j = k_{ext,j}/\bar\gamma_j$; $k_j/k_i = k_{i,exc} \cdot \bar\gamma_i/\bar\gamma_j$.

Another factor is uptake of water with consequent change in the dielectric constant, which can affect the internal complex or ion pairing equilibria. In addition, electrodes from different manufacturers are not necessarily identical in composition, and therefore may have truly different selectivity coefficient values. As an illustration of the rather high accuracy of steady-state selectivity coefficients for solid-state electrodes, a plot of measured vs. computed values are shown in Figure 36. These are response selectivities for silver salt-based electrodes.

All of the equations for potential responses given in this section are for interferences of the same sign of charge as the principal ion. However, for neutral carrier electrodes, ions of opposite sign cause significant interferences. For example, the potassium selective electrode, based on valinomycin, responds with Nernstian slope only when bathed in solution of KOH, KF, and KCl. As the anion becomes more oil soluble, the salt solubility increases and anions are carried into the membrane electrode. This process is accompanied by a decrease in slope, or by decreased slope and a maximum in response. The theory has not been established, but it was pointed out that response errors correlate with the Hofmeister lipophilicity series. Examples are shown in Figure 37.

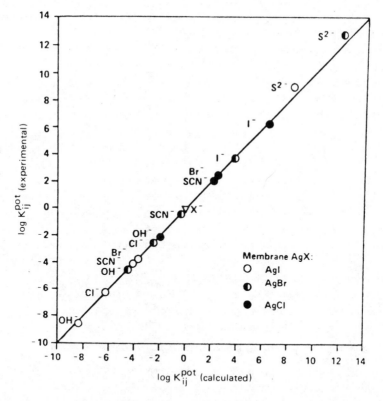

Figure 36. Comparison of the experimental and calculated anion selectivity coefficients of different silver halide membrane electrodes. [From W. E. Morf, G. Kahr, and W. Simon, *Anal. Chem.* **46**, 1538 (1974), with permission.]

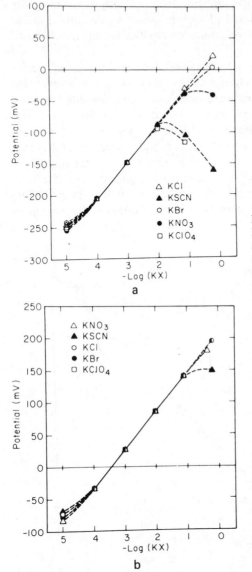

a

Figure 37(a). Responses for the Orion Model 93-19 potassium-selective electrode at 26°C. Deviations from Nernstian response at high activities increase with increasing solubilities in the membrane phase.

b

Figure 37(b). Responses for silicone-coated polycellulose acetate (GA-8) supported potassium-selective electrode, using 1-bromoheptane as solvent at 30°C. Both the Orion and this electrode use valinomycin as solubility and selectivity enforcing neutral carrier. [From S. B. Lewis and R. P. Buck, *Anal. Lett.* **9**, 439 (1976), with permission.]

4.8. Low Activity Limits by Dilution (Detection Limits)

In the absence of identifiable interfering species in "test" solutions, all ISEs deviate from Nernstian response at low activities of sensed species. As a rule, these quoted lower limits are experimental values based on dilution experiments using solvent or solvent plus inert electrolyte at constant ionic strength. For example, silver chloride membrane electrodes appear to give a level response at all activities af Ag^+ below approximately $(K_s)^{1/2}$, when tests are performed by dilution of soluble silver salts. Yet, it is also experimentally feasible to measure Nernstian responses to high Cl^- activities in which Ag^+ activities are clearly much less than the solubility value. Consequently, detection limits are artificial in the sense that they are based on dilution experiments.

Similarly, for liquid ion exchanger membrane electrodes, the lowest limits from dilution experiments are determined by the partitioning of the ion exchanger salt between membrane and bathing electrolyte. For a Ca^{2+} electrode, the partition coefficient is

$$K_D = \frac{[Ca(Phosphon.)_2\ (org.)]}{[Ca(Phosphon.)_2\ (aq.)]} \tag{135}$$

in which the partition species is a calcium salt of a diester of an oil-soluble phosphonic acid. At low bathing activities, extraction of the calcium salt establishes a lower level of Ca^{2+}. However, if bathing solutions are saturated with exchanger, and the Ca^{2+} activity is established by a metal ion buffer, it is possible that the lower limit or detection limit can be somewhat extended, but in an artificial way. By use of metal ion buffers, it has been possible to show that heavy metal-sensing electrodes based on sulfides can also respond to activities in a Nernstian fashion well below the levels established by dilution experiments. Thus detection limits, determined by dilution experiments, represent a special experimental result which is primarily, and ideally, a function of the thermodynamic solubility product or solubility of the active portion of a membrane.

However, there are many experimental factors that are difficult to control and may be determining in specific instances. Naturally occurring substances, coprecipitated impurities, impurities in reagents used in bathing electrolytes, and soluble oxidants can affect the detection limits. Thus for Ag_2S-based electrodes, the measured detection limits tend to be higher than the thermodynamic solubility values by air oxidation and by the presence of leachable impurities in the membrane solids. For liquid ion exchanger membranes such as Aliquat-based membranes for NO_3^- and Cl^-, impurities in the electrolytes, Br^- and I^-, are possible interferences that cause an apparently high detection limit. Even in the absence of impurities, water itself contains H^+ and OH^-, which must play a role in determining ultimate detection limits.

For Further Reading

J. Koryta, *Ion-Selective Electrodes*, Cambridge Monographs in Physical Chemistry, No 2, Cambridge Univ. Press, London (1975).

K. Cammann, *Working with Ion-Selective Electrodes.* 2nd Ed., Springer-Verlag, Berlin (1977).

N. Lakshminarayanaiah, *Membrane Electrodes*, Academic Press, New York (1976).

A. Covington, Ed., *Ion-Selective Electrode Methodology*, Vol. I, CRC Press, W. Palm Beach, Fla. (1979).

P. W. Cheung, D. G. Fleming, W. H. Ko, and M. R. Neuman, *Theory, Design and Biomedical Applications of Solid State Chemical Sensors*, CRC Press, W. Palm Beach, Fla. (1978).

P. L. Bailey, *Analysis with Ion-Selective Electrodes*, 2nd ed., Heyden, London (1983).

W. E. Morf, *The Principles of Ion-Selective Electrodes and of Membrane Transport*, Elsevier Scientific, Amsterdam (1981).

References

1. R. P. Buck, Theory and principles of membrane electrodes, In: *Ion-Selective Electrodes in Analytical Chemistry*, Vol. 1, Plenum Publ. Corp., New York (1978).
2. R. P. Buck, Electroanalytical chemistry of membranes, *Crit. Rev. Anal. Chem.* **5**, 323 (1975), CRC Press, W. Palm Beach, FL. (1975).
3. R. A. Durst, Ed., *Ion Selective Electrodes*, National Bureau of Standards, Spec. Pub. # 314, U.S. Gov't Printing Off., Washington D.C. (1969).
4. T. Teorell, *Proc. Soc. Exp. Biol. Med.* **33**, 282 (1935).
5. T. Teorell, *Proc. Nat'l. Acad. Sci. U.S.A.* **21**, 152 (1935).
6. T. Teorell, *Z. Elektrochem.* **55**, 460 (1951).
7. T. Teorell, *Progr. Biophys. Biophys. Chem.* **3**, 305 (1953).
8. T. Teorell, *Disc. Faraday Soc.* **21**, 9 (1956).
9. K. H. Meyer and J. F. Sievers, *Helv. Chim. Acta* **19**, 649, 665, 987 (1936).
10. R. deLevie and H. Moreira, *J. Membrane Biol.* **9**, 241 (1972).
11. R. deLevie, N. G. Seidah, and H. Moreira, *J. Membrane Biol.* **10**, 171 (1972).
12. F. Helfferich, *Ion Exchange*, McGraw-Hill Book Co., New York (1962).
13. N. K. Lakshminarayanaiah, *Transport Phenomena in Membranes*, Academic Press, New York (1969).
14. R. Parsons, Equilibrium properties of electrified interfaces, In: *Modern Aspects of Electrochemistry*, J. O'M. Bockris and B. E. Conway, Eds., Vol. 1, Chap. 3, Butterworths, London (1954).
15. R. P. Buck, *J. Electroanal. Chem.* **46**, 1 (1973).
16. E. J. W. Verwey and K. F. Niessen, *Philos. Mag.* **28**, 435 (1939).
17. P. Delahay, *Double Layer and Electrode Kinetics*, Interscience, New York (1965).
18. D. Mohilner, In: *Electroanalytical Chemistry—A Series of Advances*, A. Bard, Ed., Vol. 1, p. 241, M. Dekker, New York (1966).
19. R. Payne, In: *Techniques of Electrochemistry*, E. Yeager and A. L. Salkind, Eds., Vol. 1, p. 43, Wiley, New York (1972).
20. R. Payne, In: *Progress in Surface and Membrane Science*, J. F. Danielli, M. D. Rosenberg, and D. A. Cadenhead, Eds., Vol. 6, p. 51, Academic Press, New York (1973).
21. R. M. Reeves, In: *Modern Aspects of Electrochemistry*, J. O'M. Bockris and B. E. Conway, Eds., Vol. 9, p. 239, Plenum Press, New York (1974).
22. T. B. Grimley and N. F. Mott, *Disc. Faraday Soc.* **1**, 3 (1947).
23. T. B. Grimley, Proc. Roy. Soc. **A201**, 40 (1950).
24. R. P. Buck, In: *Proceedings of the Symposium on Ion Exchange—Transport and Interfacial Properties*, R. S. Yeo and R. P. Buck, Eds., Vol. 81-2, p. 16, The Electrochemical Society, Pennington, NJ (1981).
25. O. R. Melroy and R. P. Buck, *J. Electroanal. Chem.* **136**, 19–37 (1982); **143**, 23–36 (1983); **147**, 351–352 (1983); and **151**, 1–9 (1983).
26. J. O'M. Bockris and A. K. N. Reddy, *Modern Electrochemistry*, Plenum Press, New York (1970).
27. F. Conti and G. Eisenman, *Biophys. J.* **6**, 227 (1966).
28. R. P. Buck and V. R. Shepard, Jr., *Anal. Chem.* **46**, 2097 (1974).
29. J. P. Sandblom and G. Eisenman, *Biophys. J.* **7**, 217 (1967).
30. V. G. Levich, *Physicochemical Hydrodynamics*, Prentice-Hall Publ. Co., Englewood Cliffs, NJ (1962).

31. J. Newman, In: *Advances in Electrochemistry and Electrochemical Engineering*, P. Delahay and C. W. Tobias, Eds., Vol. 5, p. 87, Wiley, New York (1967).
32. R. Schlögl, *Stofftransport durch Membranen*, Steinkopff, Darmstadt (1964).
33. R. P. Buck, *Sensors and Actuators* **1**, 137 (1981).
33a. R. P. Buck, *J. Membrane Sci.* **17**, 1–62 (1984).
34. G. Eisenman, Ed., *Membranes*, Vol. 1 1972, Vol. 2 1973, M. Dekker, New York.
35. J. F. Danielli, M. D. Rosenberg, and D. A. Cadenhead, Eds., *Progress in Surface and Membrane Science*, Vol. 6, Academic Press, New York (1973).
36. S. Ciani, G. Eisenman, and G. Szabo, *J. Membrane Biol.* **1**, 1 (1969).
37. G. Szabo, G. Eisenman, S. Ciani, S. McLaughlin, and S. Krasne, *Ann. N.Y. Acad. Sci.* **195**, 273 (1972).
38. O. Popovych, *Crit. Rev. Anal. Chem.* **1**, # 1, 73, Chem. Rubber. Publ. Co., Cleveland, Ohio (1970).
39. R. deLevie, *J. Electroanal. Chem.* **69**, 265 (1976); *Sensors and Actuators* **1**, 97 (1981).
40. See D. A. MacInnes, *The Principles of Electrochemistry*, Dover Publ. Co., New York (1961).
41. G. Eisenman, Ed., *Glass Electrodes for Hydrogen and Other Cations*, Marcel Dekker, New York (1967).
42. J. P. Sandblom, G. Eisenman, and J. L. Walker, Jr., *J. Phys. Chem.* **71**, 3862, 3871 (1967).
43. F. S. Stover and R. P. Buck, *J. Phys. Chem.* **81**, 2105 (1977).
44. T. R. Brumleve and R. P. Buck, *J. Electroanal. Chem.* **90**, 1 (1978).
45. H. Cohen and J. W. Cooley, *Biophys. J.* **5**, 145 (1965).
46. R. deLevie, N. G. Seidah, and H. Moreira, *J. Membrane Biol.* **16**, 17 (1974).
47. B. Michaelis and R. A. Chaplain, *Math. Biosci.* **18**, 285 (1973).
48. J. R. Macdonald, Interpretation of ac impedance measurements in solids, In: *Superionic Conductors*, G. D. Mahan and W. L. Roth, Eds., p. 81, Plenum Publ. Corp., New York (1976).
49. J. R. Macdonald, In: *Electrode Processes in Solid State Ionics*, M. Kleitz and J. Dupuy, Eds., p. 149, Reidel Publ. Co., Dordrecht-Holland (1976).
50. See R. P. Buck, In: *Ion Selective Electrode Reviews*, **4**, 3–74, Pergamon Press (1982).
51. J. R. Macdonald, *Trans. Faraday Soc.* **66**, 943 (1970).
52. J. R. Macdonald, *J. Electroanal. Chem.* **32**, 317 (1971).
53. J. R. Macdonald, *J. Chem. Phys.* **54**, 2026 (1971).
54. J. R. Macdonald, *J. Chem. Phys.* **58**, 4982 (1973).
55. J. R. Macdonald, *J. Chem. Phys.* **60**, 343 (1974).
56. J. R. Macdonald, *J. Electroanal. Chem.* **53**, 1 (1974).
57. S. W. Feldberg, In: A. J. Bard, Ed., *Electroanalytical Chemistry*, Vol. 3, Marcel Dekker, New York (1969).
58. J. R. Sandifer and R. P. Buck, *J. Phys. Chem.* **79**, 384 (1975).
59. T. Joslin and D. Pletcher, *J. Electroanal. Chem.* **49**, 171 (1974).
60. H. K. Gummel, *IEEE Trans. Electron Devices* **ED-11**, 455 (1964).
61. A. de Mari, *Solid-State Electron.* **11**, 1021 (1968).
62. H. L. Stone, *SIAM J. Numer. Anal.* **5**, 530 (1968).
63. G. D. Hachtel, R. C. Joy, and J. W. Cooley, *Proc. IEEE* **60**, 86 (1972).
64. J. W. Slotboom, *IEEE Trans. Electron. Devices* **ED-20**, 669 (1973).
65. C. M. Lee, R. J. Lomax, and G. I. Haddad, *IEEE Trans. Microwave Theory, Tech.* **MTT-22**, 160 (1974).
66. D. L. Scharfetter and H. K. Gummel, *IEEE Trans. Electron Devices* **ED-16**, 64 (1969).
67. R. P. Buck, D. E. Mathis, and R. K. Rhodes, *J. Electroanal. Chem.* **80**, 245 (1977).
68. R. P. Buck and I. Krull, *J. Electroanal. Chem.* **18**, 387 (1968).
69. P. L. Markovic and J. O. Osburn, *A.I. Ch. E. J.* **19**, 503 (1973).
70. W. I. Archer and R. D. Armstrong, *The Application of A.C. Impedance Methods to Solid Electrolytes*, Specialist Periodic Report, p. 157, The Chemical Soc., London (1980).
71. D. R. Franceschetti and J. R. Macdonald, *J. Electroanal. Chem.* **99**, 283 (1979).
72. G. A. Rechnitz, *Acc. Chem. Res.* **3**, 69 (1970).

73. Y. W. Chien, C. L. Olson, and T. D. Sokoloski, *J. Pharm. Sci.* **62**, 435 (1973).
74. W. E. Morf, E. Lindner, and W. Simon, *Anal. Chem.* **47**, 1596 (1975).
75. B. Fleet, T. H. Ryan, and M. J. D. Brand, *Anal. Chem.* **46**, 12 (1974).
76. R. E. Reinsfelder and F. A. Schultz, *Anal. Chim. Acta* **65**, 425 (1973).
77. J. Bagg and R. Vinen, *Anal. Chem.* **44**, 1773 (1972).
78. D. E. Mathis and R. P. Buck, *J. Membrane Sci.* **4**, 379 (1979).
79. K. Toth, L. Gavaller, and E. Pungor, *Anal. Chim. Acta* **57**, 131 (1971).
80. B. Karlberg, *J. Electroanal. Chem.* **42**, 115 (1973).
81. B. Karlberg, *J. Electroanal. Chem.* **49**, 1 (1974).
82. G. J. Moody and J. D. R. Thomas, *Lab. Pract.* **23**, 475 (1974).
83. A. Shatkay, *Anal. Chem.* **48**, 1039 (1976).
84. K. Toth and E. Pungor, *Anal. Chim. Acta* **64**, 417 (1973).
85. E. Lindner, K. Toth, and E. Pungor, *Anal. Chem.* **48**, 1071 (1976).
86. A. Denks and R. Neeb, *Fresenius' Z. Anal. Chem.* **285**, 233 (1977); **297**, 121 (1979).
87. R. P. Buck, *Anal. Chem.* **48**, 23R (1976).
88. R. P. Buck, *Anal. Chem.* **50**, 17R (1978).
89. F. S. Stover, T. R. Brumleve, and R. P. Buck, *Anal. Chim. Acta* **109**, 59 (1979).
90. R. D. Armstrong, T. Dickinson, and P. M. Willis, *J. Electroanal. Chem.* **53**, 389 (1974).
91. R. deLevie, *Adv. Electrochem. Electrochem. Eng.* **6**, 329 (1976).
92. R. D. Armstrong and R. A. Burnham, *J. Electroanal. Chem.* **72**, 257 (1976).
93. W. E. Morf, *Anal. Lett.* **10**, 87 (1977).
94. F. G. A. Baucke, In: E. Pungor, Ed., *Ion Selective Electrodes*, p. 215, Akademiai Kiado, Budapest (1978).
95. Z. Boksay, In: E. Pungor, Ed., *Ion Selective Electrodes*, p. 245, Akademiai Kiado, Budapest (1978).
96. D. E. Mathis, F. S. Stover, and R. P. Buck, *J. Membrane Sci.* **4**, 395 (1979).
97. R. K. Rhodes and R. P. Buck, *Anal. Chim. Acta* **110**, 185 (1979).
98. A. Hulanicki and A. Lewenstam, *Anal. Chem.* **53**, 1401 (1981).
99. R. D. Armstrong, *J. Electroanal. Chem.* **52**, 413 (1974).
100. R. C. Hawkings, L. P. V. Corriveau, S. A. Kushneriuk, and P. Y. Wong, *Anal. Chim. Acta* **102**, 61 (1978).
101. A. Hulanicki, A. Lewenstam, and M. Maj-Zurawska, *Anal. Chim. Acta* **107**, 121 (1979).
102. R. E. van de Leest and A. Geven, *J. Electroanal. Chem.* **90**, 97 (1978).
103. K. Toth, E. Lindner, and E. Pungor, In: *Ion-Selective Electrodes, 3,* E. Pungor and I. Buzas, Eds., Akademiai Kiado, Budapest (1981).
104. E. Lindner, K. Toth, E. Pungor, W. E. Morf, and W. Simon, *Anal. Chem.* **50**, 1627 (1978).
105. E. Lindner, K. Toth, and E. Pungor, *Anal. Chem.* **54**, 72 (1982).
106. H. Malissa, M. Grasserbauer, E. Pungor, K. Toth, M. K. Papay, and L. Polos, *Anal. Chim. Acta* **80**, 223 (1975).
107. E. Pungor, K. Toth, M. K. Papay, L. Polos, H. Malissa, M. Grasserbauer, E. Hoke, M. F. Ebel, and K. Persy, *Anal. Chim. Acta* **109**, 279 (1979).
108. G. J. Moody, N. S. Nassory, J. D. R. Thomas, D. Betteridge, P. Szepesvary, and B. J. Wright, *Analyst* **104**, 348 (1979).
109. W. E. van der Linden and R. Oostervink, *Anal. Chim. Acta* **108**, 169 (1979).
110. G. J. M. Heijne, W. E. van der Linden, and G. den Boef, *Anal. Chim. Acta* **100**, 193 (1978); **98**, 221 (1978).
111. G. J. M. Heijne and W. E. van der Linden, *Anal. Chim. Acta* **93**, 99 (1977); **96**, 13 (1978).
112. T. Higuchi, C. R. Illian, and J. L. Tossounian, *Anal. Chem.* **42**, 1674 (1970).
113. C. Liteanu and E. Hopirtean, *Talanta* **17**, 1067 (1970).
114. IUPAC, *Recommendations for Nomenclature of Ion-Selective Electrodes*, No. 43, IUPAC Secretariat, Oxford (January 1975).
115. R. P. Buck, *Anal. Chem.* **44**, 270R (1972); **46**, 28R (1974); **48**, 23R (1976), and **50**, 17R (1978).

116. H. Freiser, Ed., *Ion-Selective Electrodes in Analytical Chemistry*, Vol. II, p. 175, Plenum Publ. Co., New York (1980).
117. J. W. Ross, *Science N.Y.* **156**, 1378 (1967).
118. G. J. Moody, R. B. Oke, and J. D. R. Thomas, *Analyst* **95**, 910 (1970).
119. G. H. Griffiths, G. J. Moody, and J. D. R. Thomas, *Analyst* **97**, 420 (1972).
120. J. Ruzicka, E. H. Hansen, and J. C. Tjell, *Anal. Chim. Acta* **67**, 155 (1973).
121. J. W. Ross, U.S. Patent No. 3, 483, 112 (9th December 1969).
122. A. Hulanicki, R. Lewandowski, and M. Maj-Zurawska, *Anal. Chim. Acta* **69**, 409 (1974).
123. J. W. Davies, G. J. Moody, and J. D. R. Thomas, *Analyst* **97**, 87 (1972).
124. R. M. Carlson and J. L. Paul, *Anal. Chem.* **40**, 1292 (1968).
125. H. B. Herman and G. A. Rechnitz, *Anal. Chim. Acta* **76**, 155 (1975).
126. J. R. Entwistle and T. J. Hayes, *Proceedings of the I.U.P.A.C. International Symposium on Selective Ion-Sensitive Electrodes*, Paper 17, Cardiff (1973).
127. D. L. Manning, J. R. Stokely, and D. W. Magouyrk, *Anal. Chem.* **46**, 1116 (1974).
128. See S. Glasstone, *Textbook of Physical Chemistry*, Second Ed., p. 1259, D. Van Nostrand Co., Inc., New York (1946).
129. G. J. Moody and J. D. R. Thomas, In: *Ion-Selective Electrodes in Analytical Chemistry*, H. Freiser, Ed., Vol. 1, Plenum Publ. Co., New York (1978).
130. W. E. Morf and W. Simon, In: *Ion-Selective Electrodes in Analytical Chemistry*, H. Freiser, Ed., Vol. 1, Plenum Publ. Co., New York (1978).
131. L. A. R. Pioda, V. Stankova, and W. Simon, *Anal. Lett.* **2**, 665 (1969).
132. U. Fiedler and J. Ruzicka, *Anal. Chim. Acta* **67**, 179 (1973).
133. J. Pick, K. Toth, E. Pungor, M. Vasak, and W. Simon, *Anal. Chim. Acta* **64**, 477 (1973).
134. R. P. Scholer and W. Simon, *Chimia* **24**, 372 (1970).
135. D. Ammann, M. Guggi, E. Pretsch, and W. Simon, *Anal. Lett.* **8**, 709 (1975).
136. R. J. Levins, *Anal. Chem.* **43**, 1045 (1971); **44**, 1544 (1972).
137. R. W. Stow, R. F. Bair, and B. F. Randall, *Arch. Phys. Med. Rehabil.* **38**, 646 (1957).
138. W. Severinghaus and A. F. Bradley, *J. Appl. Physiol.* **13**, 515 (1958).
139. J. Ruzicka and E. H. Hansen, *Anal. Chim. Acta* **69**, 129 (1974).
140. J. W. Ross, J. H. Riseman, and J. A. Krueger, *Pure Appl. Chem.* **36**, 473 (1973).
141. D. Midgley, *Analyst* **100**, 386 (1975).
142. P. L. Bailey and M. Riley, *Analyst* **100**, 145 (1975).
143. D. Midgley and K. Torrance, *Analyst* **97**, 626 (1972).
144. R. F. Thomas and R. L. Booth, *Environ. Sci. Technol.* **7**, 523 (1973).
145. T. R. Gilbert and A. M. Clay, *Anal. Chem.* **45**, 1757 (1973).
146. M. J. Beckett and A. L. Wilson, *Water Res.* **8**, 333 (1974).
147. D. Midgley, *Analyst* **100**, 386 (1975).
148. U. Fiedler, E. H. Hansen, and J. Ruzicka, *Anal. Chim. Acta* **74**, 423 (1975).
149. Orion Research Inc., *Instruction Manual, Sulphur Dioxide Electrode*, 3rd Ed. (1974).
150. M. A. Tabatabai, *Soil Sci. Plant Anal.* **5**, 569 (1974).
151. P. L. Bailey, unpublished work in the E.I.L. laboratory.
152. R. A. Llenado and G. A. Rechnitz, *Anal. Chem.* **46**, 1109 (1974).
153. E. H. Hansen and J. Ruzicka, *Anal. Chim. Acta* **72**, 353 (1974).
154. H. Thompson and G. A. Rechnitz, *Anal. Chem.* **46**, 246 (1974).
155. M. Mascini and A. Liberti, *Anal. Chim. Acta* **68**, 177 (1974).
156. D. S. Papastathopoulos and G. A. Rechnitz, *Anal. Chem.* **47**, 1792 (1975).
157. L. F. Cullen, J. F. Rusling, A. Schleifer, and G. J. Papariello, *Anal. Chem.* **46** (1974).
158. G. G. Guilbault and G. Nagy, *Anal. Lett.* **6**, 301 (1973).
159. G. Nagy, L. H. von Storp, and G. G. Guilbault, *Anal. Chim. Acta* **66**, 443 (1973).
160. R. A. Durst, Ed., *Ion Selective Electrodes*, Ch. 11, p. 375, National Bureau of Standards, Spec. Pub. # 314, U.S. Gov't. Printing Off., Washington, D.C. (1969).
161. J. Vesely, O. J. Jensen, and B. Nicolaisen, *Anal. Chim. Acta* **62**, 1 (1972).
162. J. Pick, K. Toth, and E. Pungor, *Anal. Chim. Acta* **65**, 240 (1973).

163. M. Mascini and A. Liberti, *Anal. Chim. Acta* **64**, 63 (1973).
164. M. J. D. Brand, J. J. Militello, and G. A. Rechnitz, *Anal. Lett.* **2**, 523 (1969).
165. M. Mascini and A. Liberti, *Anal. Chim. Acta* **60**, 405 (1972).
166. Orion Research Inc., *Newsletter* **2**, 41 (1970).
167. W. E. Morf, D. Ammann, and W. Simon, *Chimia* **28**, 65 (1974).
168. M. S. Frant and J. W. Ross, *Science N.Y.* **167**, 987 (1970).
169. J. W. Ross, In: *Ion-Selective Electrodes*, R. A. Durst, Ed., Ch. 2, N.B.S. Spec. Publ. No. 314, Washington, D.C. (1969).
170. K. Srinivasan and G. A. Rechnitz, *Anal. Chem.* **41**, 1203 (1969).
171. G. A. Rechnitz and T. M. Hseu, *Anal. Lett.* **1**, 629 (1968).
172. R. J. Baczuk and R. J. DuBois, *Anal. Chem.* **40**, 685 (1968).
173. R. F. Hirsch and J. D. Portock, *Anal. Lett.* **2**, 295 (1969).
174. G. J. Moody and J. D. R. Thomas, *Lab. Pract.* **23**, 475 (1974).
175. D. Langmuir and R. L. Jacobson, *Env. Sci. Technol.* **10**, 834 (1970).
176. S. S. Potterton and W. D. Shults, *Anal. Lett.* **1**, 11 (1967).
177. E. Pungor, K. Toth, and A. Hrabeczy-Pall, *Pure Appl. Chem.* **51**, 1915 (1979).
178. M. Mascini and F. Pallozzi, *Anal. Chim. Acta* **73**, 375 (1974).
179. D. P. Brezinski, *Analyst* **108**, 425 (1983).
180. D. P. Brezinski, *Anal. Chim. Acta* **134**, 247 (1982).
181. D. P. Brezinski, *Talanta* **30**, 347 (1983).
182. A. Hulanicki and A. Lewenstam, *Anal. Chem.* **53**, 1401 (1981).

Auxiliary Notation

A	area
a_i	activity of an ion in exterior, bathing solution
a_+, a_-	activities of cations and anions
a_b	bulk activity in exterior bathing solution where electroneutrality applies
a_s	"surface" activity as determined by mass transport but outside the space-charge region
$a(0)$	activity of a species at a surface within the space-charge region
$a(x)$	activity of a species as a function of position x
b	parameter in the power-law activity coefficient expression
c_i, c_k, etc.	concentration of a species in a bathing solution
$c_b, c_{b,0}, c_{b,d}$	bulk species concentration where electroneutrality applies, typically side "0" or "d"
$c_s, c_{s,0}, c_{s,d}$	surface species concentration determined by mass transport, but outside the space-charge region
c'	concentration is equivalent units rather than molar, molal
$c(x)$	concentration of a species as a function of position x
c_+, c_-	concentrations of a cation and an anion
$C_{d,1}$ or C_R or C_F	double-layer capacitance at low frequencies for interfaces contacting permeable and nonpermeable ions. Subscripts R and F are used for time constant expressions involving resistive surface layers and films

$C_{0,r}$ or C_W	capacitances, often called "pseudo" for low-frequency time constants involving nonequilibrium and steady-state transport. W is given to the Warburg capacitance
C_g or C_B	capacitance, geometric or bulk
d	membrane thickness
D	diffusion coefficient
E	electric field
f_i	friction coefficient
F	Faraday constant
I	total current density
i^0	exchange current density
J	flux of a solution species
$K_{ext}, K_{ext,i}$	extraction coefficient (partition coefficient)
K'	thermodynamic salt extraction coefficient
K_D	concentration ratio salt extraction coefficient
$K_{i\,exc}$ or $K_{i/j}$	thermodynamic ion exchange constant
K_{so}	solubility product in terms of ion activities
K_s	solubility product in terms of ion concentrations
K_{is}	formation constant of an ion pair from ion i and ligand s
k_{ij}^{pot}	potentiometric selectivity coefficient
k_0, k_0^+ and k_0^-	standard surface rate constants for slow "activation" surface transfer processes
k_b	surface rate constant for a slow transfer process
M	Macdonald's ratio in Eq. (81)
p	local osmotic pressure
R	gas constant
$R_\infty, R_\theta, R_R, R_F,$ and R_W	resistances; high frequency, "activation," surface layer, surface film, and Warburg
R_0	dc resistance
s	Laplace transform variable
s	subscript for neutral carrier or univalent anion exchanger
T	temperature
t	transference number
u_i, u_i^*	mobilities of ions
v_i	partial molar volume of an ion
v	velocity of a species
V, V_c, V_{DC}	cell voltage
x	running distance variable for one-dimensional transport
$\bar{X}$	membrane charged sites concentration
z or z_i	charge of an ion with sign
Z	impedance function
δ	Nernst diffusion layer thickness
δ'	Nernst diffusion layer thickness at left membrane interface

δ''	Nernst diffusion layer thickness at right membrane interface
ε	dielectric permittivity $= \varepsilon_r \varepsilon_0$
ε_0	dielectric permittivity of free space (rationalized)
ε_r	conventional dielectric constant
$\gamma_i, \gamma_\pm, \gamma$	mean activity coefficient
κ	reciprocal Debye thickness
ϕ	local inner potential of a phase
ϕ_0	local inner potential of a phase, specifically bulk where electroneutrality applies
ϕ_s, ϕ_b	local inner potential values near the surface but outside the space-charge region, similarly in the bulk
$\Delta\phi_m, \Delta\phi$	overall membrane potential difference; cell voltage
ρ	space-charge density
ω	frequency in radian/sec
$\bar{\omega}$	charge with sign for sites
μ_i, μ	chemical potential of a species
$\tilde{\mu}_i, \tilde{\mu}$	electrochemical potential of a species
μ_i^0, μ^0	standard chemical potential of a species
τ	time constant
τ_∞ or τ_B	time constant for the high-frequency bulk transport process
τ_0 or τ_W	low frequency Warburg time constant
τ_R or τ_F	surface rate time constants (from "activation," surface resistance, or surface film)

Bars over symbols indicate quantities for the membrane or metal phase

4

Polarography

JAROSLAV KŮTA†

1. Introduction

Polarography, one of the most widespread electrochemical techniques both for the study of electrochemical processes and for analytical applications, was introduced in 1922 by Heyrovský.[1] The name polarography was coined originally from the "electrochemical polarization" of the polarizable dropping mercury electrode (DME) in connection with a reference electrode when increasing successively a d-c voltage applied externally to the given cell. In terms of the later classification of various electrochemical techniques [see, e.g., References (2) and (3)], polarography belongs to the potentiostatic techniques with linearly increasing potential using a dropping mercury electrode as a working electrode.

During the first 20 years attention was directed almost entirely to the interpretation of current–voltage (or current–potential, i.e., $I–E$) curves obtained in electrolysis with the dropping mercury electrode. Later other relationships, e.g., $dI/dE = f(E)$, $I = f(t)$, $E = f(t)$, $dE/dt = f(t)$ or $f(E)$ were studied with the DME or with other mercury electrodes introduced later, such as the streaming mercury electrode, rotated and vibrated mercury electrodes, and hanging mercury drop electrodes (HMDE). Studies with the latter electrode were frequently called "stationary polarography" or sometimes voltammetry with HMDE. According to the recent nomenclature the second name is more appropriate.

When an oscilloscope was used for the readout of the measured quantity, this type of measurement was called oscillographic polarography.

JAROSLAV KŮTA • The J. Heyrovský Institute of Physical Chemistry and Electrochemistry, Czechoslovak Academy of Sciences, 118 40 Prague 1, Czechoslovakia. † Deceased.

Further advances in polarography were achieved through modifying the linearly increasing potential of the DME by superimposing a low-level periodic perturbation (sine wave, square wave, repetitive pulses, etc.).

This rather confusing scope of polarography is due to its historical development over a half a century, which led to an enormous accumulation of electrochemical data for electrode processes and mechanisms of many thousands of inorganic and organic substances. [A rough estimate of the number of polarographic papers in 1978 is about 30,000; according to Reference (4), in 1967 the total number was 21,798.] This was due to an extremely high reproducibility of results on DME and to the rapidity in obtaining these results from the very beginning. Since 1925 an automatic photographic readout of polarographic curves—the polarograph—was available. This was one of the first automated apparatuses in physicochemical instrumentation.

Polarography is sometimes considered merely as a suitable and rapid method of analysis. This is only partly true, however. Polarography, starting from the empirical relations in electrolysis with the DME, continued to build a theoretical basis for observed phenomena and developed, itself, into a scientific branch of electrochemistry having a strong impact on the development of electrochemistry. It is interesting to note, e.g., that the first polarographic results on the influence of cations of alkali metals on hydrogen overvoltage,[5] stimulated Frumkin[6] in 1933 to make the derivation of his well-known correction of charge-transfer kinetics for the double-layer effect and served as the first verification test of his double-layer correlation. A period of polarographic investigation was devoted to the solution of mass transport problems. The established formulations were helpful in solving transport phenomena for other electrochemical techniques introduced later. One very important contribution of conventional polarography was the concept and the mathematical treatment of chemical reactions coupled with charge transfer. This concept is now currently utilized in electrochemical studies employing various techniques.

The first quantitative relations in polarography were derived for redox systems showing so-called Nernstein behavior, where charge-transfer reactions are very fast and the systems appear as reversible under polarographic conditions. A number of important thermodynamic parameters resulted from these studies. In the 1950s polarography was directed toward electrode kinetic studies. The solution of the case of charge transfer combined with mass transfer (e.g., Smuteck[7]) in the early 1950s contributed to the development of measurements of relatively fast electrode reactions. A large variety of systems has been studied polarographically, and now a basis for exploring these systems in more detail by new, more advanced electrochemical techniques and on electrodes other than mercury is possible. In addition, adsorption studies of both reactants and inhibitors by means of polarographic currents established a good fundamental knowledge for these studies, which are now pending in electrode kinetics.

Polarography has been a useful technique in analytical chemistry from the beginning, and it is hoped that the ability and skill of analytical chemists will continue to contribute to the development of polarography. New, simple, elegant, sensitive, and fast polarographic analytical procedures surpassing the other analytical techniques will also be stepping stones in the future for further applications of polarography, especially in advanced techniques.

In this chapter the basic theoretical background of conventional (classical) d-c polarography with a DME polarized by linearly increasing voltage will be presented in some detail, and applications of the method will be described. The techniques derived from polarography will be treated only briefly. Several monographs dealing with the principles of polarography and related techniques are available.[8-15]

2. Principle of Polarographic Measurements

The principle of conventional polarography is very simple. The mean current flowing at the dropping mercury electrode is measured at various potentials; the resulting current–voltage (potential) curve is called a polarogram.[16] The voltage is customarily plotted as the abscissa and the corresponding current as the ordinate. On the polarogram one or several steps or polarographic "waves" are usually observed if small concentrations ($10^{-3}\,M$ or less) of electroactive species are present in the solution of an "inert" supporting electrolyte. The position of the wave on the potential axis is characteristic of the redox reaction of the species in the solution under investigation. The occurrence of several waves can be an indication of the presence of several electroactive substances. However, their existence may be also connected with various oxidation states of a single electroactive compound only, or with some complications in the overall electrode process. The potential corresponding to half of the waveheight, the so-called half-wave potential, is in a simple correlation with the reversible potential or the overvoltage of the charge-transfer reaction. The proportionality between the height of the wave and the concentration of the electroactive species forms the basis for the application of polarography in quantitative analysis. From the magnitude of the current along the wave and on its plateau (limiting current), conclusions can be drawn on the contribution of various kinds of transport control (diffusion, charge transfer, coupled chemical reactions, adsorption). Therefore, from the shape of the polarographic I-E curve and from the magnitude of the current, important information about electrochemical and physicochemical properties of the studied system can be obtained. The theoretical background for such a possibility will be briefly outlined in later sections.

In his original setup, Heyrovský used a cell consisting of a DME (polarizable) and a reference (unpolarizable) electrode. The external circuit was formed by a low-resistance potentiometric wire, both ends of which were

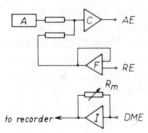

Figure 1. Scheme of a three-electrode polarographic circuitry. A, d-c ramp generator; C, controlling amplifier; F, voltage follower; I, current-to-voltage transducer; R_m, measuring resistance; AE, auxiliary electrode; RE, reference electrode; DME, dropping mercury electrode.

connected to a storage battery. The sliding contact moving along the wire permitted the application of any desired voltage to the cell. A suitably damped galvanometer connecting the sliding contact and the DME served for the measurement of the mean current. In the so-called polarograph constructed by Heyrovský and Shikata[16] in 1925 the relative movement of the sliding contact along the potentiometric wire was achieved by revolving the potentiometric wheel by a motor. The deflection of a light beam, reflected from the mirror of the galvanometer and passed through a narrow slit in the cylindrical housing of a photographic cassette, was recorded on the photographic paper placed inside. A system of gears connected this cassette with the potentiometer so that one full turn of the cassette corresponded to the movement of the sliding contact along the whole length of the potentiometric wire. The scan rate was usually 100 or 200 mV per min.

Polarographs for a two-electrode cell with photographic recordings were used for several decades; in the mid '50s they were slightly modernized by using pen-recording. Presently, as in other electrochemical techniques, a three-electrode cell is preferred in polarography and the mechanically generated d-c ramp is replaced by an integrator. A simple scheme is shown in Figure 1.

For more details concerning stability, bandpass, and additional circuits for IR compensation, see Reference (17). Some disturbances on I-E curves due to the mechanical d-c ramp are minimized and measurements in solutions of lower conductivity can be carried out.

3. Dropping (Streaming and Hanging) Mercury Electrodes

Although the theoretical basis of polarography is well established, and the results on a DME due to renewal of its surface are the most reproducible of all electrochemical measurements, it is possible that the results could be erroneous if the capillary for the DME is not properly chosen.

The dropping mercury is mainly obtained by means of a glass capillary of 3–7 mm outer diameter and 0.05–0.1 mm internal diameter from which mercury flows, forming drops at intervals of about 3 s. The capillary is usually connected by tubing or sometimes by glass tubing to a mercury reservoir.

Two quantities, the so-called capillary constants, are characteristic for a given capillary, namely, the flowrate (outflow velocity) m, i.e., the quantity of mercury leaving the capillary in unit time (g/s) and the droptime t_d(s). The flowrate is directly proportional to the effective pressure P_{eff} acting on the mercury at the orifice of the capillary. The total pressure P_t is equal to the height of the mercury column from the capillary orifice to the level of the mercury in the reservoir h. However, if mercury reaches the solution in single drops, the interfacial tension between the mercury and the solution, γ, prevents the formation of a larger surface area. The pressure perpendicular to the surface of the growing drop increases and opposes the growth of the drop. This is the so-called backpressure, P_b, for which the following relation holds

$$P_b = \frac{2\gamma}{r} \tag{1}$$

and which is in opposition to the total pressure. In the above equation, r denotes the radius of the drop at any instant of its growth. The effective pressure is the difference between the total pressure and the backpressure. For these reasons one must distinguish between the flowrate independent of backpressure, the mean flowrate m and the instantaneous flowrate. For details see Reference (9).

These differences in m values are not very important for most purposes, but have to be taken into account for a precise comparison of magnitudes of theoretically calculated and experimentally measured currents. However, even for fundamental studies in d-c polarography it is sufficient to know the mean flowrate, which is determined from the weight of mercury collected for a certain time in the given solution. Therefore, in further derivations, only the mean outflow velocity of mercury will be considered. The total weight w of the drop is given by the equation

$$w = mt_d g = 2\pi r_c \gamma \tag{2}$$

where g is the gravitational constant and r_c is the radius of the orifice of the capillary. The product mt_d is constant at the given interfacial tension γ, and hence t_d is inversely proportional to the effective pressure or mercury height.

The flowrate is practically independent of the composition of the solution and of the potential (small changes are due to the backpressure), whereas the droptime strongly depends on the composition of the solution and potential.[9,10] The area of the drop at any instant of its growth, for drops having a spherical shape, is given by

$$A = 4\pi r^2 \tag{3}$$

According to film records made by Smith,[18] capillaries with internal radius of 0.017–0.2 mm form drops of practically spherical shape.

From the relation for the volume of the drop

$$V = \frac{mt}{s} = \frac{4}{3} \pi r^3 \tag{4}$$

r can be expressed and substituted in Eq. (3); thus

$$A = 4 \left(\frac{3mt}{4\pi 13.6} \right)^{2/3} = 0.85 m^{2/3} t^{2/3} \tag{5}$$

where the value $s = 13.6$ is the density of mercury and t is any time elapsed from the beginning of drop formation.

For the mean surface of the DME, A can be written

$$\bar{A} = \frac{1}{t_d} \int_0^{t_d} A \, dt = \frac{3}{5} 0.85 m^{2/3} t_d^{2/3} = 0.51 m^{2/3} t_d^{2/3} \tag{6}$$

In this way the area of the electrode can be expressed by means of the easily and relatively precisely measurable experimental parameters m and t or t_d, respectively.

It must be stressed that the magnitude of m is very important for achieving correct results, because at high outflow velocities the so-called maxima of the second kind might occur, which distort polarographic waves (this will be discussed later). This can be avoided simply by using a capillary with a flowrate below 2 mg/s and in the case of an amalgam[19] DME below 1 mg/s.

A regular and reproducible dropping is the prerequisite for obtaining good results with a DME. The regular functioning of the capillary may deteriorate, e.g., by the penetration of the solution inside the capillary followed by the formation of solution "pockets." From this effect, resulting capillary response[20] or capillary noise[21] becomes more important if more sensitive polarographic techniques are used. In addition, impurities from the solution or electrode reaction products may accumulate at the orifice and cause irregularities in drop formation. The penetration between the thread of mercury and the glass walls occurs especially with cylindrical capillaries, which are predominantly used, and the creeping effect increases with increase of the inner radius. Many attempts have been made to minimize the creeping of solution, e.g., by siliconizing the inner walls of the capillary or by using various forms of the end of the glass capillary [e.g., References (20), (22), (23)]. A conically drawnout capillary seems to improve the regularity of drop formation.[24]

For aggressive solutions like hydrofluoric acid, a Teflon capillary[25,26] was applied. Since Teflon is not wetted in aqueous solutions, a Teflon capillary should have less capillary response and capillary noise. There is not enough evidence of a reliable DME made of Teflon working satisfactorily at negative potentials. Some other plastic materials showing no wetting have been proposed.[23] Short conical tips made of plastic can be fixed on a conical end of the glass capillary. The results seem to be very promising.

Most polarographic investigations have been undertaken with capillaries in the vertical position. The outflow velocity and the droptime are mutually dependent; however, for some purposes it is desirable to keep the droptime constant and independent of m. This can be achieved, for example, by a mechanical device tapping on the capillary. With a horizontal or tilted capillary, rather short droptimes for small m values can also be achieved. This capillary practically removes the depletion effect.[27]

For other modifications of the DME, mainly from an analytical point of view, consult the review by Kolthoff and Okinaka in Reference 28.

The streaming mercury electrode (mercury jet electrode) was introduced by Heyrovský and Forejt[29] in 1943. As described by Heyrovský, the mercury streams in an upward direction, reaching the surface of the solution at an angle of about 45°. The continuous stream of mercury breaks up into droplets outside the solution. Hence a constant length of the stream and thus a constant electrode surface for any flowrate of potential is ensured. For further modifications and studies, see references (29)–(33). The flowrate is approximately 100 times higher than with the DME and the time of contact of the mercury element with the solution is very short, 10^{-2}–10^{-3} s. The latter fact is advantageous in some kinetic studies, for it may eliminate some coupled chemical reactions, reduce the influence of adsorption, etc.

Measurements with a hanging mercury drop electrode provide a useful supplement to studies with the DME when the HMDE is applied as the stationary electrode in single-sweep or cyclic voltammetry. This will not be dealt with in this review. This electrode is of practical use in so-called anodic stripping polarography (to be discussed later). For these purposes commercial types are available in which, by means of a micrometric screw, a stainless steel rod squeezes out the mercury from a reservoir into a connected glass capillary until a drop of desired size is formed [see, e.g., References (34), (35)].

The overwhelming majority of experiments in d-c polarography was carried out with the DME. Besides many advantages pertinent to mercury as an electrode material in general (easy purification, high overvoltage), the renewal of the surface is the main advantage of the DME leading to the most reproducible results obtained in electrochemistry until now. During electrolysis with the DME the concentration of the electroactive substance is practically unchanged which, together with the ease of carrying out the polarographic procedure on a miniature scale, predestined polarography for its application in microanalysis. With a thin drawnout capillary, the electrolysis of very small volume (0.01–0.05 ml) can be studied. The accessible potential range in aqueous solution is from +0.4 V (dissolution of mercury) to −2.6 V (NCE) and in nonaqueous even higher (decomposition of solvent).

The use of DME is, however, also restricted. At positive potentials mercury undergoes oxidation and electrodes of other materials must be sought.

From a theoretical point of view, a serious complication in the mathematical formulation is introduced by the necessity of considering the growth of the drop; a number of transport effects and electrode mechanisms have, however, been solved for the DME and will be dealt with in this chapter.

4. Charging Current

As will be discussed in the following sections, important conclusions on the redox properties of systems studied can be obtained from the dependence of the faradaic current on the potential. However, the observed current in polarography, as in other electrochemical techniques, also contains a non-faradaic component, due to double-layer charging. In polarography the faradaic current is generally obtained by subtraction of the double-layer charging (nonfaradaic) current I_{DL} (in absence of electroactive species) from the measured total current of the redox system. For problems related to the coupling of the charging and faradaic currents, especially in the case of adsorption of reactants and products, see References (36) and (37).

For the double-layer charging current the following relation holds

$$I_{DL} = \left(\frac{\partial Q}{\partial t}\right)_E + \left(\frac{\partial Q}{\partial E}\right)_t \frac{\partial E}{\partial t} \tag{7}$$

where Q is the charge on the electrode. For d-c polarography using small potential changes, the expression for the charging current is simple[38]

$$I_{DL} = \frac{dQ}{dt} = E^* \frac{dK'}{dt} = E^* K \frac{dA}{dt} \tag{8}$$

Here the potential E^* relates to the potential of zero charge (pzc) as $E^* = E - pzc$ and K' the capacity of the electrode is given by $K' = (K)(A)$ where K is the integral capacity per cm^2 and A the instantaneous area of the drop. With respect to Eq. (5) one gets for the instantaneous polarographic charging current[38]

$$I_{DL} = \frac{2}{3} 0.85 E^* K m^{2/3} t^{-1/3} \tag{9}$$

and for the mean current

$$\bar{I}_{DL} = \frac{1}{t_d} \int_0^{t_d} I_{DL}\, dt = 0.85 E^* K m^{2/3} t_d^{-1/3} \tag{10}$$

The instantaneous current attains its maximum at the beginning of drop formation and has its maximum value at the end of the droptime $(I_{DL})_{min}$ (Figure 2). The mean current reaches $(3/2)(I_{DL})_{min}$. When expressing the capillary characteristics as the function of mercury pressure, it follows that the mean charging current increases linearly with the height of the mercury column.

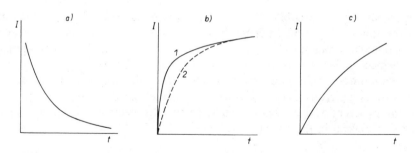

Figure 2. Instantaneous currents in polarography. (a) charging current; (b) diffusion current: 1) on the "first" drop, 2) on the second drop; (c) kinetic current.

The charging current should increase linearly with the potential and should equal zero at E = pzc. At potentials more positive than the potential of zero charge an anodic current flows, and at more negative potentials than the potential of electrocapillary maximum the drop has negative charge and the direction of the current is reversed—a cathodic charging current. Since the double layer capacity for anions is usually approximately twice that for simple cations the anodic current is larger.

The charging current causes limitations in the measurement of faradaic polarographic currents. Already at the concentration of $10^{-5} M$ the diffusion waves are deformed and overlapped by the charging current. A simple circuit proposed by Ilkovič and Semerano[39] for compensation of charging current was included in commercial polarographs. This compensation, based on a simplified assumption of a linear increase of the charging current with potential that is counterbalanced by an opposite current, has to be performed by trial and error and is not effective enough. It had some practical analytical applications, but a compensated wave could hardly be analyzed for theoretical purposes.

New approaches for an effective *in situ* compensation of charging current in d-c polarography proposed quite recently[40,41] seem to be very promising. They are based in the use of a-c superposition for the charging current compensation. The a-c charging current is given by

$$I_{\mathrm{DL}}^{\mathrm{a.c.}} = \left(\frac{\partial Q}{\partial E}\right)_t \frac{d(\delta E)}{dt}$$

where δE is the superposed a-c voltage. Hence $\partial Q/\partial E$ has to be converted to $\partial Q/\partial t$ and subtracted from the total cell response to yield the "d-c response." Various circuits have been suggested and the corresponding mathematical reasoning given. The analysis shows that the rising portion of the curve is practically not distorted by the compensation. Distinct, well-measured, and fully compensated d-c waves for concentration $3 \times 10^{-6} M$ are reported.[41] However, an increase in sensitivity to the limit $5 \times 10^{-8} M$ and even higher was obtained by using superimposed a-c or square-wave signals and by measuring the low-level periodic perturbation response in a short time interval of the droplife in which the charging current is small. This led to the development of advanced polarographic techniques (as discussed later).

Sometimes the charging current is called the residual current.[11] However, this term is generally understood to refer to measured current consisting of charging current and of faradaic currents due to electroactive impurities or traces of oxygen in the base electrolyte.

5. Diffusion Controlled Processes

5.1. Ilkovič Equation

To ensure diffusion control, an excess of the so-called supporting electrolyte (or base electrolyte) is added to the solution of the electroactive substance under investigation to eliminate the contribution due to migration of charged electroactive species.

The migration may lead to an increase or decrease in the observed current. If the direction of migration is the same as the direction of diffusion, the resulting current exceeds the value of the diffusion current. In the opposite case the observed current is smaller than for diffusion control alone. An approximate derivation of the currents with the participation of migration in polarography was given by Heyrovský[42] under the assumption of additivity of both components. The migration current was expressed by the product of transport number t and the total limiting current I_1, i.e., $I_m = tI_1$. A relatively good agreement between calculated and observed currents was found.[9] For more precise derivation, see References (43) and (44). However, the migration component of the current is very sensitive to the value of the transference number, i.e., to the presence of foreign salts. Therefore, in polarography, as in many electrochemical techniques, an excess of the supporting electrolyte is used.

The magnitude of the diffusion-controlled current to a DME was solved theoretically first by Ilkovič.[45,46] McGillavry and Rideal[47] took originally into account the spherical character of diffusion. However, after some approximations the problem simplified to a linear diffusion toward a growing DME and an expression identical with the Ilkovič equation was obtained.

The electroactive particles are diffusing toward the electrode and, because of the drop growth, the electrode is moving toward the incoming particles. Hence the total change in concentration with time can be described by the general equation for convective diffusion

$$\frac{\partial c}{\partial t} = D\nabla^2 c - v \operatorname{grad} c \tag{11}$$

where D is the diffusion coefficient (cm^2/s).

Ilkovič made a simplifying assumption that the DME behaves as a planar electrode with an area equal to that of the surface of the drop, i.e., an area increasing with time. For linear diffusion, the preceding equation becomes

$$\frac{\partial c}{\partial t} = D \frac{\partial^2 c}{\partial x^2} - v_x \frac{\partial c}{\partial x} \tag{12}$$

To express v_x as a function of x and t, Ilkovič considered a shell of incompressible liquid contained between the sphere with inner radius r_1 formed by the drop surface and a second concentric sphere with radius r_2. If the shell thickness $r_2 - r_1 = x$ is very small, the volume of this shell is $V = (A)(x)$ where A is the area of the drop. From the condition of incompressibility

$$\frac{dV}{dt} = 0 = A \frac{dx}{dt} + x \frac{dA}{dt} \tag{13}$$

holds. Thus, according to Eqs. (5) and (13)

$$v = \frac{dx}{dt} = -\frac{x}{A} \frac{dA}{dt} = -\frac{2}{3} \frac{x}{t} \tag{14}$$

which when substituted in Eq. (12) yields

$$\frac{\partial c}{\partial t} = D \frac{\partial^2 c}{\partial x^2} + \frac{2}{3} \frac{x}{t} \frac{\partial c}{\partial x} \tag{15}$$

This equation describes the linear mass transfer toward a growing planar electrode. For its solution the appropriate initial and boundary conditions must be chosen. Ilkovič assumes that at the beginning of electrolysis, i.e., at $t = 0$ and $x = 0$, the concentration at the electrode c equals that in the bulk of the solution c^0. For $t > 0$ and $x > 0$ the concentration at the electrode is constant and equals c (which depends on the potential of the electrode) or equals zero (limiting diffusion current). By solving the preceding equation, one obtains for the concentration gradient at the electrode

$$\left(\frac{\partial c}{\partial x} \right)_{x=0} = (c^0 - c) \left[\frac{3}{7} \pi Dt \right]^{-1/2} \tag{16}$$

where c^0 is the bulk concentration and c the concentration at the electrode. For a planar stationary electrode the well-known relation is valid

$$\left(\frac{\partial c}{\partial x} \right)_{x=0} = (c^0 - c)(\pi Dt)^{-1/2} \tag{16a}$$

The expression for a growing DME differs from the corresponding one for linear diffusion to a stationary planar electrode [Eq. (16a)] by the factor $(3/7)^{-1/2}$, which accounts for the decrease in thickness of the diffusion layer during the growth of the drop. When inserting the above calculated concentration gradient in the equation for current I

$$I = nFAD\left(\frac{\partial c}{\partial x}\right)_{x=0} \tag{17}$$

we obtain

$$I = nFAD(c^0 - c)[\tfrac{3}{7}\pi Dt]^{-1/2} \tag{18}$$

when n is the number of electrons involved in the electrode reaction and F is Faraday's constant (96,500 C).

When substituting $A = 0.85 m^{2/3} t^{2/3}$ and combining all numerical constants, the Ilkovič equation is finally obtained for the instantaneous current as

$$I = 0.732 nFD^{1/2} m^{2/3} t^{1/6}(c^0 - c) \tag{19}$$

and for the instantaneous limiting current as

$$I_d = 0.732 nFD^{1/2} m^{2/3} t^{1/6} c^0 \tag{20}$$

The current for each growing drop increases with time from zero to a maximum value at the end of the drop life (Figure 2b).

In normal d-c polarographic measurements mean currents are generally recorded and, thus, the following relation holds

$$\bar{I} = \frac{1}{t_d} \int_0^{t_d} I\, dt \tag{21}$$

Hence

$$\bar{I} = 0.627 nFD^{1/2} m^{2/3} t_d^{1/6}(c^0 - c) \tag{22}$$

or for the mean limiting diffusion controlled current

$$\bar{I}_d = 0.627 nFD^{1/2} m^{2/3} t_d^{1/6} c^0 \tag{23}$$

The last equation is currently called in the literature the "Ilkovič equation." Ilkovič used the c.g.s. system of units in which the concentration is in g/cm^3 and the current is in amperes. In British and American literature the Ilkovič equation has been written in the form

$$\bar{I}_d = 607 nD^{1/2} m^{2/3} t_d^{1/6} c^0 \tag{23a}$$

where the value of the faraday is included in the numerical constant, the flowrate is expressed in mg/s and the concentration in millimols/liter; the current is then given in μA.

It follows from the Ilkovič equation that for given values of m and t_d the diffusion current is directly proportional to the bulk (analytical) concentration

$$\bar{I} = \mathscr{H}c^0 \tag{24}$$

where $\mathscr{H}$ is called the Ilkovič constant:

$$\mathscr{H} = 0.627 nFD^{1/2} m^{2/3} t_d^{1/6} \tag{25}$$

Equation (24) represents the basis of quantitative determinations in polarographic analysis.

Since m is directly proportional to the height of mercury column h (hydrostatic pressure) and t_d is indirectly proportional to this quantity, and after substitution into the Ilkovič equation, one obtains

$$\bar{I} = kh^{1/2} \tag{26}$$

The linear dependence of $\bar{I}$ on $h^{1/2}$ not only represents a verification of the validity of the Ilkovič equation but also provides an easily accessible test for diffusion-controlled currents.

The influence of temperature on the magnitude of diffusion-controlled currents was also derived by Ilkovič[48] by expressing the temperature coefficients of D, m, and t_d. It follows that the temperature coefficient of diffusion currents amounts to 1.63% per degree for many cases.

The viscosity of the solution (η) has a significant effect on the magnitude of the diffusion current. The largest effect of viscosity is shown on D and t_d. Since in the Ilkovič equation only the 6th root of drop time appears, the contribution of the latter to the viscosity effect is relatively small. The major part is due to the effect of viscosity changes on D following the relation $D \sim 1/\eta$ and hence

$$\bar{I}_d \eta^{1/2} = \text{constant} \tag{27}$$

If no other complications appear (e.g., change in solvation, complexation, etc.), this relation is roughly valid [see, e.g., References (49) and (50)].

5.2. Correction for Spherical Diffusion in the Ilkovič Equation

On the whole, the validity of the Ilkovič equation has been proved satisfactorily. Thus, the linear dependence of the diffusion current $\bar{I}_d$ on concentration c^0 is satisfied within ±1%, the dependence on $m^{2/3}t_d^{1/6}$ within ±3%, and the constancy of the so-called diffusion current constant I^D, defined as

$$I^D = \frac{\bar{I}_d}{c^0 m^{2/3} t_d^{1/6}} = 0.627 nFD^{1/2} \tag{23b}$$

within ±5%. It was this relatively large inaccuracy in the diffusion current constant which led to corrections to the original Ilkovič equation.

Corrections with different degrees of approximation were made using a simplified procedure by Lingane and Loveridge,[51] Strehlow and Stackelberg,[52] and Kambara and Tachi[53] (for a more detailed discussion of the accuracy of these equations see Markowitz and Elving[54] and Koutecký and Stackelberg.[95])

Rigorous derivations of the diffusion current equation for spherical diffusion to the DME have been presented by Koutecký[56] and Matsuda[57] who also considered the shielding effect.

The differential equation for diffusion to the surface of a growing sphere is given with respect to Eq. (11) by

$$\frac{\partial c}{\partial t} = D\frac{\partial^2 c}{\partial r^2} + \frac{2}{r}\frac{\partial c}{\partial r} - v_r\frac{\partial c}{\partial r} \tag{28}$$

The origin of coordinates is placed at the center of the drop. The drop, considered as an ideal sphere, increases in a time unit by $(4/3)a^3$, where a is the radius at $t = 1$. At time t the radius of the drop becomes $r_0 = ((3mt)/13.6)^{1/3} = at^{1/3}$. Let us consider a point at a distance r from the center of the drop. A concentric sphere passing through this point has the volume $V = (4/3)\pi r^3$. If this point of the solution is moving because of the growth of the drop at a velocity $v_r = dr/dt$, we may write

$$\frac{dV}{dt} = 4\pi r^2 v_r = \frac{4}{3}\pi a^3 \tag{29}$$

From this relation it follows for the velocity of convection

$$v_r = \frac{a^3}{3r^2} \tag{30}$$

Substitution of Eq. (30) into Eq. (28) yields

$$\frac{\partial c}{\partial t} = D\frac{\partial^2 c}{\partial r^2} + \frac{2}{r}\frac{\partial c}{\partial r} - \frac{a^3}{3r^2}\frac{\partial c}{\partial r} \tag{31}$$

The appropriate initial conditions express the constancy of the concentrations

$$t = 0, \qquad r \geq 0, \qquad c = c^0, \qquad r \to \infty, \qquad c \to c^0 \tag{31a}$$

The following boundary condition ensures that during electrolysis a certain constant concentration c is present at the electrode surface (which may be zero for the limiting current)

$$t > 0, \qquad r = at^{1/3}, \qquad c = c^0 \tag{31b}$$

On solving this equation Koutecký obtained for the mean limiting current

$$\bar{I}_d = 0.627nFD^{1/2}m^{2/3}t_d^{1/6}c^0\left[1 + 3.4\frac{D^{1/2}t_d^{1/6}}{m^{1/3}} + \left(\frac{D^{1/2}t_d^{1/6}}{m^{1/3}}\right)^2\right] \tag{32}$$

For the instantaneous current, instead of 3.4, the numerical coefficient 3.9 should be applied. The contribution of the quadratic term in brackets is very small.

Matsuda[57] also considered the influence of screening by the tip of the capillary. The reduction in the diffusion volume by the tip of the capillary decreases the numerical constant in the second term to 2.39.

As can be seen from the given formula, the current according to the corrected equation should be larger for a normal DME by 10–13% for usual values of m and t_d, and smaller for an amalgamated DME (the second term has a minus sign). The observed mean limiting currents at the DME were, however, in most cases smaller by about 5–15% than those calculated from the equation concerning the sphericity of the drop and fitted better for the uncorrected Ilkovič equation.

This is due to the depletion effect (transfer of the concentration polarization). The diminution of the concentration in the vicinity of the capillary orifice is caused by electrolysis at the preceding drops and as the solution around the tip of the capillary is not completely renewed with the fall of the drop, the depletion is transferred from one drop to another.

This can be seen very clearly on current–time curves recorded on the so-called first drop, i.e., on the drop unaffected by previous polarization. This can be realized by an electronic device[58] possessing the ability to switch on the potential polarizing the DME at the beginning of its growth, or simply by using a horizontal capillary with which the falling of the drop removes the depleted solution almost completely from the capillary orifice. The difference in the form of I-t curves on the "first" and on successive drops is shown in Figure 2b. The lowering of the current is clearly seen at the beginning of drop formation, which causes mean limiting currents smaller than calculated from the rigorous formula. From a series of studies [for details see References (9) and (59)] it follows that the depletion effect practically counterbalances the influence of spherical diffusion. Limiting diffusion currents and I-t curves on a single drop satisfies the equation corrected for spherical diffusion only if the depletion effect is eliminated. With usually used vertical capillaries a better agreement is found between experimental limiting currents and the uncorrected Ilkovič equation.

In the derivation of both the Ilkovič and Koutecký equation, a concentric growth of the drop was assumed. The effect of eccentric drop growth, which is nearer to the experimental facts, was considered by de Levie.[60] The calculation shows an uneven current distribution at the electrode surface but the effect of the eccentricity calculated for the Ilkovič approach on the total current through the electrode is negligible ($\sim 1.5\%$).

5.3. Diffusion Current at a Streaming Electrode

For the limiting diffusion controlled current the following equation has been derived

$$I_d = 4nF \sqrt{\left(\frac{Dlm}{s_{\text{Hg}}}\right)} c^0 \tag{33}$$

where l is the length of the jet and m the flowrate of mercury.

6. Polarographic Diffusion-Controlled Waves

Let us assume a simple redox couple

$$Ox + ne \rightleftharpoons Red$$

the charge reaction of which proceeds very fast near to the equilibrium conditions—the so-called "reversible"-electrode reaction. In this case the ratio of concentrations at the electrode follows the Nernst equation

$$\frac{c_{Ox}}{c_{Red}} = \exp\left[\frac{(E - E_0)}{RT}\right] nF = \lambda \tag{34}$$

or

$$E = E_0 + \frac{RT}{nF} \ln \frac{c_{Ox}}{c_{Red}} \tag{34a}$$

In the simplest way the equations of the "reversible" polarographic wave, i.e., *I-E* curve, can be obtained by expressing the surface concentrations by means of the Ilkovič equation

$$\bar{I} = \mathscr{H}_{Ox}(c^0_{Ox} - c_{Ox}) = \bar{I}_{d_c} - \mathscr{H}_{Ox}c_{Ox} \tag{35}$$

and from the continuity of diffusion fluxes it is also valid that

$$+\bar{I} = \mathscr{H}_{Red}(c_{Red} - c^0_{Red}) = \mathscr{H}_{Red}c_{Red} + \bar{I}_{d_a} \tag{35a}$$

where $\bar{I}_{d_c}$ is the mean limiting cathodic and $\bar{I}_{d_a}$ the mean limiting anodic current, respectively. After substituting in the Nernst equation and equating $\mathscr{H}_{Ox}$ and $\mathscr{H}_{Red}$, we have[61]

$$E = E_0 + \frac{RT}{nF} \ln \frac{\bar{I}_{d_c} - \bar{I}}{\bar{I} - \bar{I}_{d_a}} \sqrt{\frac{D_{Red}}{D_{Ox}}} \tag{36}$$

This equation can also be written in the form

$$\bar{I} = \frac{\bar{I}_{d_c} + \lambda \bar{I}_{d_a} \sqrt{\dfrac{D_{Ox}}{D_{Red}}}}{1 + \lambda \sqrt{\dfrac{D_{Ox}}{D_{Red}}}} \tag{37}$$

or

$$\bar{I} = \frac{c^0_{Ox} - \lambda c^0_{Red}}{1 + \lambda \sqrt{\dfrac{D_{Ox}}{D_{Red}}}} \mathscr{H}_{Ox} \tag{37a}$$

The potential at which the current reaches one-half of the sum of both limiting currents, $(1/2)(\bar{I}_{d_c} + \bar{I}_{d_a})$, corresponds to the half-wave potential

$(E_{1/2})$ of the wave

$$E_{1/2} = E_0 + \frac{RT}{nF} \ln \sqrt{\frac{D_{Red}}{D_{Ox}}} \tag{38}$$

Hence the potential corresponding to one-half of the wave is a constant that is practically equal to the standard potential of the redox couple.

For a cathodic wave ($c_{Red}^0 = 0$ and $\bar{I}_{d_a} = 0$) one obtains

$$E = E_0 + \frac{RT}{nF} \ln \frac{\bar{I}_{d_c} - \bar{I}}{\bar{I}} \sqrt{\frac{D_{Red}}{D_{Ox}}} \tag{39}$$

This equation can be written

$$\bar{I} = \frac{\bar{I}_{d_c}}{1 + \lambda \sqrt{\dfrac{D_{Ox}}{D_{Red}}}} \tag{39a}$$

or

$$\frac{\bar{I}_{d_c}}{1 + \lambda'} \tag{39b}$$

where

$$\lambda' = \exp \frac{nF}{RT}(E - E_{1/2}) = \lambda \sqrt{\frac{D_{Ox}}{D_{Red}}} \tag{40}$$

Again for $\bar{I} = \bar{I}_d/2$ the potential on the wave corresponds to the half-wave potential $E_{1/2}$.

The equation for the anodic wave can be easily obtained from Eq. (36) by inserting $\bar{I}_{d_c} = 0$ (no oxidized form is present in the bulk).

The equation of the wave corrected for spherical diffusion is somewhat complicated.[56,62] For a reversible cathodic process it is found that

$$\bar{I} = \frac{\bar{I}_{d_c}}{1 + \lambda \sqrt{\dfrac{D_{Ox}}{D_{Red}}}} \left(1 + \frac{3.4 D_{Ox}^{1/2} t_d^{1/6}}{m^{1/3}} \frac{1 \pm \lambda}{1 + \lambda \sqrt{\dfrac{D_{Ox}}{D_{Red}}}} \right) \tag{41}$$

The positive sign in the second term in the brackets applies to diffusion of the reduced form back into the solution, whereas the negative sign refers to the diffusion of the metal inside the mercury drop. This equation can be rearranged into the more commonly known form

$$E = E_{1/2}' + \frac{RT}{nF} \ln \frac{\bar{I}_{d_c} - \bar{I}}{\bar{I}} \tag{42}$$

formally identical with Eq. (39), based on the Ilkovič equation. However, the half-wave potential $E_{1/2}'$ depends on the flowrate of mercury and droptime according to the following relation

$$E'_{1/2} = E_0 - \frac{RT}{nF} 3.4 \frac{t_d^{1/6}}{m^{1/3}} (\sqrt{D_{Ox}} + \sqrt{D_{Red}})$$ (43)

The theoretical changes in the half-wave potential for normal capillaries ($m = 1–2$ mg/s) are in most cases very small and, if necessary, they can be accounted for.

It should be mentioned that the equations of diffusion-controlled current–voltage curves could be obtained in a more rigorous way. The diffusion equation for both oxidized and reduced form must be solved with the boundary condition $c_{Ox}/c_{Red} = \lambda$. After transformation of the corresponding differential equations one gets, e.g., in the case of the Ilkovič approximation, the same Eq. (36).

The described reversible behavior can be complicated by a semiquinone or dimer formation. Depending on the value of the corresponding formation constants, the shape of the essentially reversible wave might be changed and for high values of the formation constant two waves of equal height may appear.[66]

6.1. Differential (Derivative) Polarography

From the differentiation of the equation for a reversible cathodic wave it follows ($D_{Red} = D_{Ox} = D$) that

$$\frac{d\bar{I}}{dE} = -\frac{nF}{RT} \frac{\bar{I}(\bar{I}_{d_c} - \bar{I})}{\bar{I}_{d_c}}$$ (44)

or from Eq. (39b)

$$\frac{d\bar{I}}{dE} = -\frac{nF}{RT} \bar{I}_d \frac{\lambda'}{(1 + \lambda')^2}$$ (44a)

i.e., the slope is zero at $\bar{I} = \bar{I}_{d_c}$ and $\bar{I} = 0$ and reaches a maximum at $\bar{I} = \bar{I}_{d_c}/2$ for which the relation

$$\left(\frac{d\bar{I}}{dE}\right)_{E_{1/2}} = -\frac{nF}{4RT} \bar{I}_{d_c}$$ (45)

holds. Hence the derivative curve exhibits a maximum (Figure 3). A simple circuitry was proposed by Heyrovský[63] in 1946 using two capillaries. A simpler approach with a single capillary using R-C circuitry[64,65] is usually preferred. The main application is in analytical chemistry for the determination of traces of an electroactive substance A in excess of the substance B reduced at a more positive potential; in conventional d-c polarography the wave A would be very small, preceded by the limiting current of B whereas in differential polarography $d\bar{I}/dE$ at the potential of the limiting current of the wave B is zero and the peak for the substance A can be recorded at high sensitivity.

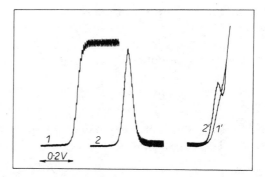

Figure 3. Recorded normal and derivative d-c curve. 1) 5×10^{-4} M CdCl$_2$ in 0.1 M NH$_3$, 0.1 M NH$_4$Cl, normal d-c curve; 2) its derivative; 1') 10^{-3} M KCl in 0.1 M LiOH normal d-c curve; 2') its derivative.

Another advantage of derivative polarography is in better separation of two processes with half-wave potentials near each other so that a single wave is observed on the conventional $\bar{I}$-E curve.

6.2. Diffusion-Controlled Waves of Complexes

By means of polarography, complex formation can be studied and in the case of amalgam formation, stability constants and composition of complexes can be determined.[67-71]

If we assume a series of complexes of the cation Me^{n+} with ligand X being in a mobile equilibrium, the stability constant of jth complex is

$$K_j = k_1, k_2 \ldots k_j = \frac{c_{\mathrm{Me}X_j}}{(c_{\mathrm{Me}^{n+}})c_X^j} \tag{46}$$

where k_j is the consecutive stability constant. Further, it is assumed that the concentration of ligand is in excess so that $c_X = c_X^0$.

For the concentration of free cation $c_{\mathrm{Me}^{n+}}$ at the electrode surface from the preceding mobile equilibrium it holds

$$c_{\mathrm{Me}^{n+}} = \frac{\sum\limits_{j} c_{\mathrm{Me}X_j}}{\sum\limits_{j} K_j(c_X^0)^j} \tag{47}$$

Substitution in Eq. (34a) results in

$$E_c = E_0 + \frac{RT}{nF} \ln \frac{\sum\limits_{j} c_{\mathrm{Me}X_j}}{c_{\mathrm{Me}} \sum\limits_{j} K_j(c_X^0)^j} \tag{48}$$

The following relation holds if the diffusion coefficients for all the complexes are the same

$$\bar{I} = \sum_{j} \bar{I}_j = \bar{I}_d - \mathcal{H}_c \sum_{j} c_{\mathrm{Me}X_j} \tag{49}$$

And for the diffusion of the reduced form in the drop

$$\bar{I} = \mathscr{H}_{\mathrm{Red}} c_{\mathrm{Me}} \tag{49a}$$

Substitution of both diffusion conditions into Eq. (48) leads to

$$E_c = E_0 - \frac{RT}{nF} \ln \frac{\bar{I}_d - \bar{I}}{\bar{I}} \frac{\mathscr{H}_{\mathrm{Red}}}{\mathscr{H}_c} - \frac{RT}{nF} \ln \sum K_j (c_X^0)^j \tag{50}$$

The shift of the half-wave potential of the complex, $E_{1/2} = (E_{1/2})_c - (E_{1/2})_{\mathrm{free}}$, is obtained as

$$E_{1/2} = -\frac{RT}{nF} \ln \sum_j K_j (c_X^0)^j \tag{51}$$

If only one stable complex, e.g., MeX_p is formed, the former equation simplifies to

$$\Delta E_{1/2} = -\frac{RT}{nF} \ln K (c_X^0)^p \tag{51a}$$

Hence from the shift of $E_{1/2}$, the stability constant K is easily obtained and from the dependence on c_X^0 the number p of ligands bound in the complex can be evaluated.

　　If various complexes are formed, differing considerably in stability constants then, as follows from Eq. (51), the $\Delta E_{1/2}$ vs. log c_X^0 plot will consist of linear portions with breaks indicating the number of complexes. From the slopes of the linear portions the number of ligands can be determined; at certain concentrations (c_X^0) the other terms can be neglected and the stability constants of individual complexes determined. However, in most cases the stability constants lie close to one another and the $\Delta E_{1/2}$ vs. log (c_X^0) plot is a smooth curve with no breaks. Then the procedure suggested by De Ford and Hume[68] must be applied. Very precise values of $E_{1/2}$ are necessary for obtaining reliable results.

　　In the case that the reduced form also undergoes complexation and diffuses back into the solution, the complex formation of both oxidized and reduced forms has to be considered.

　　In general, a formula similar to Eq. (51) can be derived

$$E_{1/2} = -\frac{RT}{(n-m)F} \ln \frac{\sum K_j (c_X^0)^j}{\sum K_j' (c_X^0)^j} \tag{51b}$$

where K_j and K_j' denote the stability constants for complexes with the oxidized and reduced forms, respectively, and $(n - m)$ is the number of electrons transferred in the reduction from the higher (n) to the lower oxidation state (m).

　　A great number of stability constants have been determined and many can be found in Reference (72).

6.3. Diffusion-Controlled Waves of Organic Substances

Similar shifts in half-wave potentials occur with a change in the hydrogen ion concentration when hydrogen ions participate in the electrode process. This is an almost general case with organic substances. For the very common case of a two-electron reduction under formation of a dibasic anion, a simple relation for the polarographic half-wave potential can be easily derived, assuming that the oxidized form is not protonated[9]

$$E_{1/2} = (E_{1/2})_0 + \frac{RT}{2F} \ln (c^0_{H_3O^+} + K'_{a_1} c^0_{H_3O^+} + K'_{a_1} K'_{a_2}) \tag{52}$$

where $(E_{1/2})$ is the half-wave potential at $c^0_{H_3O^+} = 1$ and K'_{a_1} and K'_{a_2} are the corresponding dissociation constants of the reduced form. For more complicated cases see Reference (9). In organic polarography studies a well-buffered solution is used. In unbuffered or poorly buffered solutions complications may arise leading to a splitting of the wave or to the occurrence of several waves on the polarogram because of diffusion control of proton donors. The problem was also treated quantitatively.[73] Analogous complications arise in the polarographic studies of complexes if the complexing agent is not in an excess.[74]

7. Polarographic Waves with Diffusion and Charge-Transfer Control

Consider a simple electrode reaction in which both species are diffusing to or away from the electrode

$$Ox + ne \underset{k_{ox}}{\overset{k_{red}}{\rightleftharpoons}} Red$$

The rate constants of the charge-transfer reaction are supposed to fulfill the well-known equations

$$k_{red} = k^0_e \exp \left[\frac{-\alpha nF}{RT} (E - E_0) \right] \tag{53}$$

$$k_{ox} = k^0_e \exp \left[\frac{(1 - \alpha)nF}{RT} (E - E_0) \right] \tag{54}$$

where k^0_e is the rate constant of the charge-transfer reaction at the standard potential E_0 (for a given medium) and α the transfer coefficient.

A rigorous treatment of a polarographic wave with charge-transfer control was first presented by Mejman[75] for a polarographic wave of reduction of hydrogen ions. A more general treatment was given by Koutecký,[76] which is currently used in polarography. The reaction scheme for charge and diffusion control can be described by the following differential equations

$$\frac{\partial c_{Ox}}{\partial t} = D_{Ox} \frac{\partial^2 c_{Ox}}{\partial x^2} + \frac{2x}{3t} \frac{\partial c_{Ox}}{\partial t} \tag{55}$$

$$\frac{\partial c_{Red}}{\partial t} = D_{Red} \frac{\partial^2 c_{Red}}{\partial x^2} + \frac{2x}{3t} \frac{\partial c_{Red}}{\partial t} \tag{56}$$

The initial and boundary conditions are

$$x \geq 0, t = 0 \qquad c_{Ox} = c_{Ox}^0, \qquad c_{Red} = c_{Red}^0 \tag{57}$$

$$x = 0, t > 0 \qquad D_{Ox} \frac{\partial c_{Ox}}{\partial x} \pm D_{Red} \frac{\partial c_{Red}}{\partial x} = 0 \tag{58}$$

$$-D_{Ox} \frac{\partial c_{Ox}}{\partial x} = k_{red} c_{Ox} - k_{ox} c_{Red} \tag{59}$$

$$x \to \infty, t > 0 \qquad c_{Ox} \to c_{Ox}^0, x \to \pm\infty, c_{Red} \to c_{Red}^0 \tag{60}$$

The plus sign applies to the diffusion of the product back into the solution and the minus is used if the product is soluble in mercury.

The most important point is the boundary condition expressing that the concentration gradient which is responsible for the magnitude of the current is given by the difference between the rates of electro-reduction and electro-oxidation.

Note that for large values of rate constants, $k_{red} c_{Ox} - k_{ox} c_{Red} \approx 0$ and $c_{Ox}/c_{Red} = \lambda$ [see Eq. (34)] and hence one obtains the boundary conditions for rigorous solution of the equations for diffusion-controlled waves (see previous section).

On solving this system of equations Koutecký obtained the formula

$$\bar{I} = \bar{I}_{rev} \bar{F}(\mathscr{X}_1) \tag{61}$$

where $\bar{I}$ is the mean observed current, which is controlled both by mass and charge transfer ("irreversible" process) and $\bar{I}_{rev}$ is the current at a given potential, which would result if k_{red} and k_{ox} were increased to infinity, the ratio of rate constants remaining constant. For $\mathscr{X}_1$ the following relationship applies

$$\mathscr{X}_1 = \sqrt{\frac{12}{7}} \left(\frac{k_{red}}{D_{Ox}^{1/2}} + \frac{k_{ox}}{D_{Red}^{1/2}} \right) \sqrt{t_d} \tag{62}$$

For the instantaneous current, an analogous equation is valid having only the arbitrary value of time t instead of t_d. The function for $\bar{F}(\mathscr{X}_1)$ and an analogous one for instantaneous current are tabulated in References (9) and (76).

The mean current predicted by the former rigorous equation may be expressed in a simplified form by the following equation

$$\bar{I} = \bar{I}_{rev} \frac{0.886 \left(\dfrac{k_{red}}{D_{Ox}^{1/2}} + \dfrac{k_{ox}}{D_{Red}^{1/2}} \right) t_d^{1/2}}{1 + 0.886 \left(\dfrac{k_{red}}{D_{Ox}^{1/2}} + \dfrac{k_{ox}}{D_{Red}^{1/2}} \right) t_d^{1/2}} \tag{63}$$

Practically the same formula can be obtained when using an approximate treatment based on the diffusion layer concept. In this case the current is expressed simply by

$$\bar{I} = nF\bar{A}(k_{\text{red}}c_{\text{Ox}} - k_{\text{ox}}c_{\text{Red}}) \tag{64}$$

and the concentrations of both forms at the electrode are evaluated from the Ilkovič equation. In this case a numerical constant 0.81 results instead of 0.886.

For a reversible (diffusion-controlled) anodic–cathodic wave the following relationship applies [with respect to Eqs. (36), (53), and (54)]:

$$\frac{\bar{I}_{d_c} - \bar{I}_{\text{rev}}}{\bar{I}_{\text{rev}} - \bar{I}_{d_a}} \left(\frac{D_{\text{Red}}}{D_{\text{Ox}}}\right)^{1/2} = \exp\left[\frac{(E - E_0)nF}{RT}\right] = \frac{k_{\text{ox}}}{k_{\text{red}}} \tag{65}$$

consequently

$$\bar{I}_{\text{rev}} = \frac{k_{\text{red}}D_{\text{Red}}^{1/2}\bar{I}_{d_c} + k_{\text{ox}}D_{\text{Ox}}^{1/2}\bar{I}_{d_a}}{k_{\text{red}}D_{\text{Red}}^{1/2} + k_{\text{ox}}D_{\text{Ox}}^{1/2}} \tag{66}$$

To obtain an anodic–cathodic wave for the case of both charge and mass transfer control, one substitutes for $\bar{I}_{\text{rev}}$ in Eq. (63) and obtains

$$\left(\frac{D_{\text{Red}}}{D_{\text{Ox}}}\right)^{1/2} \frac{\bar{I}_{d_c} - \bar{I}}{\bar{I} - \bar{I}_{d_a}} - \frac{\bar{I}}{\bar{I} - \bar{I}_{d_a}} \frac{1}{0.886k_{\text{red}}\left(\dfrac{t_d}{D_{\text{Ox}}}\right)^{1/2}} = \frac{k_{\text{ox}}}{k_{\text{red}}} = \lambda \tag{67}$$

Provided the rate constants are large, the second term is dropped and the equation reduces to the diffusion-controlled (reversible) anodic–cathodic wave [see Eq. (36)].

To get some basic features of the charge-controlled wave in a simple way we shall consider a cathodic wave only ($\bar{I}_{d_a} = 0$) (see Figure 4). In this case, from the general Eq. (67), one obtains

$$\frac{\bar{I}_{d_c} - \bar{I}}{\bar{I}}\left(\frac{D_{\text{Red}}}{D_{\text{Ox}}}\right)^{1/2} = \lambda + \frac{1.13}{k_{\text{red}}}\left(\frac{D_{\text{Ox}}}{t_d}\right)^{1/2}$$

$$= \exp\left[\frac{nF(E - E_0)}{RT}\right] + \frac{1.13}{k_e^0} \frac{D^{1/2}}{t_d^{1/2}} \exp\left[\frac{\alpha nF}{RT}(E - E_0)\right] \tag{68}$$

For large rate constants, this equation reduces to that for the cathodic diffusion controlled wave (Figure 4, curve 1). For very small values of the rate of oxidation ($k_{\text{ox}} \ll k_{\text{red}}$), i.e., for a unidirectional cathodic process, $k_{\text{ox}}/k_{\text{red}}$ can be dropped from Eq. (68) and one obtains

$$\frac{\bar{I}}{\bar{I}_{d_c}\bar{I}} = 0.886k_{\text{red}}\frac{t_d^{1/2}}{D_{\text{Ox}}^{1/2}} \tag{69}$$

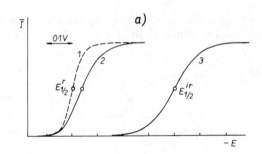

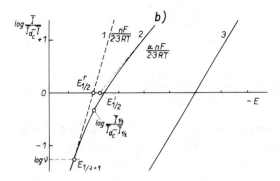

Figure 4. Forms of a cathodic wave (a) with corresponding log-plot analysis; (b) 1) reversible, 2) "quasireversible," 3) irreversible charge transfer reaction ($n = 1$, $\alpha = 1/2$).

After substituting for k_{red} from Eq. (53), the well-known form is obtained

$$E = E_0 - \frac{RT}{\alpha nF} \ln \frac{\bar{I}}{\bar{I}_{d_c}\bar{I}} + \frac{RT}{\alpha nF} \ln 0.886 k_e^0 \frac{t_d^{1/2}}{D_{Ox}^{1/2}} \qquad (70)$$

The preceding equation also follows directly from Eq. (63) setting $k_{ox} = 0$ and $\bar{I}_{rev} = \bar{I}_d$. Such a wave is also called a totally irreversible wave (Figure 4, curve 3). For the half-wave potential, i.e., for $\bar{I} = \bar{I}_d/2$

$$E_{1/2} = E_0 + \frac{2.3RT}{\alpha nF} \log 0.886 k_e^0 \left(\frac{t_d}{D_{Ox}}\right)^{1/2} \qquad (71)$$

Provided α is potential-independent, log-plot analysis of the irreversible wave should be a straight line, the reciprocal slope of which is larger than that for the reversible process. From the slope, the transfer coefficient can be obtained and from the difference of half-wave potential and the standard potential the rate constant k_e^0 can be easily evaluated.

The equation for a totally irreversible wave can be applied to processes with a large polarographic overvoltage defined as $\eta_{1/2} = E_{1/2,ir} - E_0$. For relations between $\eta_{1/2}$ and the overvoltage defined at a given current density see Reference (77). Hence, if $\eta_{1/2} \gg 200$ mV for one electron and >100 mV for a two-electron process, the backward reaction can be neglected and the equation for a totally irreversible cathodic [Eq. (70)] or an analogous one for an anodic wave could be used. In contrast to the reversible wave, the half-wave potential depends on droptime. The exponent of I-t curves changes along the rising portion of the irreversible wave from $2/3$ to approximately $1/6$. On the foot of the wave where practically no concentration polarization occurs (i.e., $\bar{I} \ll \bar{I}_d$) it follows from Eq. (64) that the mean current is proportional to the growth of the area ($\bar{I} \sim m^{2/3}t_d^{2/3}$) and hence it is independent of the height of the mercury column ($\bar{I} \sim h^0$). When approaching the limiting current of the wave, the diffusion comes gradually to force and modifies the previous dependences in an increasing manner.

If both rate constants are comparable, a general equation must be applied [Eq. (67) and (68)]. The form and position of the wave depend on the rate of the charge-transfer reaction k_e^0. It follows from the solution of charge transfer and mass transfer equations for the potentiostatic techniques that the response is insensitive to charge-transfer control if $k_e^0 \geq 5\sqrt{D/t}$. For polarographic conditions, setting $D = 10^{-5}$ cm^2/s and $t_d = 3$ s it follows that rate constants k_e^0 less than $\sim 10^{-2}$ cm/s can be determined. For a polarographic wave departing from Nernstian behavior and not yet fulfilling the criteria of a totally irreversible wave, i.e., for k_e^0 lying approximately in the range $10^{-2} > k_e^0 > 5 \times 10^{-5}$ cm/s, the term "quasireversible" wave was introduced in the literature (Figure 4, curve 2) [References (78) and (79)]. This term, although not well defined, is very often used in d-c polarography and in many other electroanalytical techniques as well.

A very simple graphical method for obtaining charge-transfer parameters follows from the modified equation for the cathodic wave if the rate constants are related not to E_0 but to the reversible half-wave potential $E_{1/2}^r$. In this case Eq. (68) can be written[79]

$$\frac{\bar{I}_{d_c} - \bar{I}}{\bar{I}} = \exp\left[\frac{nF}{RT}(E - E_{1/2}^r)\right] + \frac{1.13}{k_e^* t_d^{1/2}} \exp\left[\frac{\alpha nF}{RT}(E - E_{1/2}^r)\right] \quad (72)$$

where

$$k_e^* = \frac{k_e^0}{(D_{Ox}^{1/2})^{1-\alpha} D_{Red}^{1/2\alpha}} \quad (73)$$

and for $\bar{I}_{1/2}$ at $E = E_{1/2}^r$ it follows for the rate constant[80] that

$$k_e^0 = \frac{\bar{I}_{1/2}}{\bar{I}_{d_c} - 2\bar{I}_{1/2}} \frac{\sqrt{D_{Ox}^{1-\alpha} D_{Red}^{\alpha}}}{0.886\sqrt{t_d}} \quad (74)$$

From the given considerations it follows that, at relatively positive potentials, $k_{ox} \gg k_e^0$ and hence by the extrapolation of the reversible part of the d-c cathodic wave the reversible half-wave potential can be determined (see Figure 4, curve 2) whereas in the upper part of the wave, where $k_{red} \gg k_e^0$, the second term of Eq. (72) prevails.

From the log-plot analysis $E_{1/2}^r$ and $\bar{I}_{1/2}$, corresponding to $E_{1/2}^r$, can be determined and by means of Eq. (74) k_e^0 can be evaluated.

Matsuda and Ayabe[79] suggest for the determination of the slope of the log-plot analysis at sufficiently high potentials,

$$\alpha = 2.303 \frac{RT}{nF} \{\Delta \log [\bar{I}/(\bar{I}_d - \bar{I})]/\Delta(-E)\}_{E=-\infty} \tag{75}$$

For $E'_{1/2}$ Eq. (72) can be rearranged to

$$E'_{1/2} = E_{1/(\nu+1)} - 2.3 \frac{RT}{nF} \left\{ \log \left[\nu - \exp \frac{\alpha nF}{RT} (E_{1/(\nu+1)} - E'_{1/2}) \right] \right\} \tag{76}$$

where ν is some $(\bar{I}_{d_c} - \bar{I})/\bar{I}$ ratio at the foot of the polarographic wave and $E_{1/(\nu+1)}$ is the corresponding potential (Figure 4b). From the available values of $E'_{1/2}$ and $E_{1/(\nu+1)}$, respectively, and from the values of α obtained from log-plot analysis, one can calculate $E'_{1/2}$ and even k^*_e using the relationship

$$\log k^*_e = \frac{\alpha nF}{2.303} (E'_{1/2} - E^r_{1/2}) - \tfrac{1}{2} \log t_d + 0.053 \tag{77}$$

The standard rate constant k^0_e is then calculated from Eq. (73). A comparison of both procedures was made by Ružič et al.[81] who suggested another graphical method.

7.1. Analysis of Polarographic Waves

From equations given in the previous section and here, it follows that the form and position of the wave gives information on the kinetics of the charge-transfer reaction. Hence, the so-called log-plot analysis (Tafel plot) is currently used in polarographic studies. The analysis of the wave is more difficult if the wave under investigation is followed by a second wave having a near half-wave potential, which may lead to overlapping of waves. Various procedures for the case of additivity of the currents were suggested. From recent ones the procedures can be quoted for analysis of two overlapping d-c waves: reversible and irreversible process[82] of a multistep reaction,[83] of three overlapping waves of a reversible and irreversible process, and of two overlapping waves in the case of a quasireversible electrode reaction.[84] Another approach (interferometric-like) for the processing of polarographic waves, although not tested experimentally, was suggested by Barker and Gardner.[85] The graphical procedures are now surpassed by methods using computers for curve fitting.[86–88]

7.2. Irreversible Reduction of Complexes

The equation for totally irreversible reduction of complexes to metal can be derived very simply by an approximation method. Provided the chemical equilibria of different complexes $MeX, MeX_2 \ldots MeX_m$ are mobile and assuming that MeX_k is the reducible form (ligand X being again in excess), the following expression for the current applies[89]

$$\bar{I} = nF\bar{A}k_{red}c_{MeX_k} \tag{78}$$

For the sake of simplicity the ions and ligands have been written without charges. If the conditions for chemical equilibria are satisfied

$$k_m k_{m-1} \ldots k_1 c_X^m \gg k_{m-1} k_{m-2} \ldots k_2 c_X^{m-1} \gg \cdots \gg k_1 c_X \gg 1$$

it follows that

$$c_{MeX_k} = \frac{c_{MeX_m}}{k_m k_{m-1} \ldots k_{k+1}(c_X^0)^{m-k}} \tag{79}$$

where k_k are consecutive constants for complex formation. After substituting in Eq. (78) and expressing the concentration c_{MeX_m} at the electrode by means of the Ilkovič equation the final formula is obtained as

$$\frac{\bar{I}}{I_{d_c} - \bar{I}} = 0.886 k_{red} \left(\frac{t_d}{D}\right)^{1/2} \frac{1}{k_m k_{m-1} \ldots k_{k+1}(c_X^0)^{m-k}} \tag{80}$$

The following expression is useful for the determination of the composition of the reducible complex (at a constant potential)

$$-\frac{\partial}{\partial \ln c_X^0} \ln \frac{I}{I_{d_c} - I} = m - k \tag{81}$$

However, the composition of the prevailing complex in the solution MeX_m must be known. Equation (80), after substituting for k_{red} and for the condition $I = I_{d_c}/2$, can be written in the form

$$E_{1/2,c} = E_0 + \frac{RT}{\alpha nF} \ln 0.886 \, k^0 \left(\frac{t_d}{D}\right)^{1/2} - \frac{RT}{\alpha nF} \ln (k_m k_{m-1} \ldots k_{k+1})$$

$$- \frac{RT}{\alpha nF} \ln (c_X^0)^{m-k} \tag{82}$$

If instead of consecutive stability constants, the usual stability constants are introduced, the following equation is obtained

$$E_{1/2,c} = E_0 + \frac{RT}{\alpha nF} \ln 0.886 k_e^0 \left(\frac{t_d}{D}\right)^{1/2} + \frac{RT}{\alpha nF} \ln \frac{K_k(c_X^0)^k}{\sum\limits_{j=0}^{m} K_j(c_X^0)^j} \tag{83}$$

Equation (83) is very similar to the equation for the reversible wave of metal deposition and differs only by the term $K_k(c_X^0)^k$ and by having αn instead of n. Hence, the stability constant K_j could be determined providing K_k and k are known and α does not change with the concentration of ligand. Equations (81) or (82) are usually applied for the determination of the complex, which is subjected to the charge-transfer reaction providing the composition of the prevailing complex in the solution is known.

The polarographic conditions are more favorable in the case of quasi-reversible waves of complexes.[90,91] It is possible to determine the stability constants and the composition of complexes in the bulk, as in the case of a reversible wave, and in addition the parameters of the charge-transfer kinetics (charge-transfer coefficient and rate constant). By one of the previously outlined procedures for the analysis of quasireversible curves, the reversible half-wave potentials are determined and the corresponding changes in extrapolated reversible half-wave potentials are treated as described for reversible waves of complexes. If, for the determination of charge-transfer parameters, the procedure of Matsuda and Ayabe[79] is applied, the rate constant in Eq. (73) is defined by

$$k_e^* = \frac{(k_e^0)_B}{(D_{MeX}^{1/2})^{(1-\alpha)}(D_{Red}^{1/2})^\alpha} c_X^{0-(1-\alpha)m+k} \tag{84}$$

In the above equation $(k_e^0)_B$ is the rate constant of the overall reaction

$$MeX_m + ne \rightarrow Me + mX$$

The parameters α and k_e^* are determined as described before. The composition of the complex subjected to the transfer reaction can be determined from the plot of $\log k^*$ vs. the concentration of the ligand.

Several examples are given in the literature. Matsuda and Ayabe[79] found for zinc complexes with hydroxyl ions $m = 3.85$ and $k = 2.10$, i.e., in the solution the complex $Zn(OH)_4^{2-}$ prevails having a stability constant $\log K_{Zn(OH)_4^{2-}} = 15.3 \pm 0.1$ at ionic strength 2.0 M; the reducible complex has the composition $Zn(OH)_2$ and its corresponding kinetic parameters were also determined.

As an example of consecutive complex formation, the complexes of Zn^{2+} with tartrate[90] and acetate[91] can be given. In this case the nonlinear plot $E_{1/2}^r$ vs. $\log c_X^0$ points to the presence of a number of complexes and the nonlinear plot of $\log k_e^*$ against $\log c_X^0$ provides evidence for participation of two or more complexes with comparable charge-transfer rate constants in the electrode reaction (for details see original paper). From more recent examples and extension of the treatment see, e.g., References (92)–(95). For older comprehensive reviews on polarography of complexes see References (69)–(71).

7.3. Irreversible Waves of Organic Substances

Hydrogen ions play a similar role as a complexing agent in the electrode processes of most organic substances. For the simplest case, corresponding to protonation of the oxidized form undergoing an irreversible reduction, i.e.,

$Ox + H^+ \xrightarrow{ne} Red$ it can be analogously derived that

$$E_{1/2} = E'_{1/2} + \frac{RT}{\alpha nF} \ln \frac{c^0_{H_3O^+}}{K + c^0_{H_3O^+}} \tag{85}$$

where $E'_{1/2}$ includes the charge-transfer kinetic parameters [see Eq. (71)] and K is the corresponding dissociation constant. This problem, for more complicated cases of protonation both of electroactive and electroinactive groups and for various mechanisms of electrode processes including coupled chemical reactions, was treated very extensively recently.[96] The double-layer effect (see further) on $E_{1/2}$-pH-dependence[97] was also taken into account.

If the pH value is not kept constant by a suitable buffer solution or the solution is not buffered, again complications in the polarograms of an organic substance may yield results similar to those seen in reversible reduction. This effect was treated for irreversible reactions of organic substances by Guidelli et al.[98,99]

7.4. Double-Layer Effects on Polarographic Charge-Transfer Kinetics

As mentioned in the Introduction, the shift of the polarographic reduction potential of the hydrogen ion[5] with the concentration of supporting electrolyte was the first example on which the so-called Frumkin[6] double-layer correction was applied. This correction is based on, as is commonly known, two main assumptions:

(i) The concentration of the ions at the distance of closest approach from the electrode depends on the potential of the double layer and on the charge of the ion according to Boltzmann's law.

(ii) The discharge of the particle is influenced not by the total potential E with respect to the solution, but only by the difference between E and the potential in the double layer at the distance of the closest approach.

Very often the distance of closest approach to the interface is assumed to be the same for all electrolyte reactants and is further considered to correspond to the outer Helmholtz plane. Under this simplifying assumption the potential at the outer Helmholtz plane, φ_2, is calculated from double-layer data usually using Gouy–Chapman theory [for calculations see, e.g., Reference (100)].

The relation between the observed rate constant at the standard potential, k^0_e, (until now denoted in the text simply k^0_l) and the value corrected for double layer, $k_{e,cor}$, is given by the well-known formula[6,100]

$$k^0_e = k^0_{e,cor} \exp{(\alpha n - z)} \frac{F\varphi_2}{RT} \tag{86}$$

where z is the charge of the electroactive substance. If this correction is applied to irreversible reduction, the corresponding equation of the wave could be read as follows

$$E = E_0 + \left(1 - \frac{z}{\alpha n}\right) \varphi_2 - \frac{RT}{nF} \ln \frac{\bar{I}}{\bar{I}_d - \bar{I}} + \frac{RT}{\alpha nF} \ln 0.886 k_e^0 \sqrt{\frac{t_d}{D}} \quad (87)$$

which may also be written in the form

$$\frac{\bar{I}}{1 - \frac{\bar{I}}{\bar{I}_d}} = nF\bar{A}k_e^0 c^0 \exp \left\{\frac{-\alpha nF\left[(E - E_0) - \left(\dfrac{\alpha n - z}{\alpha n}\right)\varphi_2\right]}{RT}\right\} \quad (87a)$$

including an approximate correction for concentration polarization.

Hence, the position and form of an irreversible d-c wave depends on all factors that influence the φ_2-potential, i.e., on the potential of the electrode, on the total concentration of electrolyte, and on the concentration of surface-active substances (see further). Equation (87) explains the shift of irreversible waves with the composition of supporting electrolyte. So, e.g., for cations ($z > 0$) with increasing concentration of supporting electrolytes φ_2 becomes less negative for potentials more negative than the pzc and hence the wave is shifted toward more negative values (diminishing φ_2-effect) whereas for reduction of anions the wave is shifted toward positive potentials (increasing φ_2-effect). The frequently observed dip for polyvalent reducible anions was in some cases explained even quantitatively by means of Eq. (87a). Since the φ_2 changes are largest in the vicinity of the pcz it may happen that the magnitude of the φ_2-term in Eq. (87a) exceeds that of the applied potential, the trend $\bar{I}/\bar{I}_d \to 1$ is reversed and the polarographic current begins to diminish. This effect may be viewed as an electrostatic repulsion of the negative anion by the negatively charged mercury surface beyond the electrocapillary zero. The increase of the current following the dip follows from the modifying part of φ_2 because the changes of φ_2 values with E at negative potentials are small.

In many cases it was found that the simple Frumkin correction using φ_2 potentials calculated from Gouy–Chapman theory explains the experimental results quantitatively. Many examples together with a very suitable graphical procedure can be found in a paper by Gierst[101]; the case of electroreduction of anions was treated by Frumkin and Nikolaeva-Fedorovich[102]; summarizing chapters were written by Heyrovský and Kůta[9] and by Delahay.[100] The differences between experimental results and simple double-layer correction were explained by specific adsorption of ions of supporting electrolytes and/or of reactants and in the case of reduction of anions by ion pairing both in the solution and at the reaction side in the double layer (for discussion and literature survey see Gierst et al.,[103] and for a new approach see Bieman and Fawcett[104]). Recently, several quantitative approaches appeared extending the Frumkin correction for the case of specific adsorption of the ion of supporting electrolytes. Parsons's treatment[105] introduces the interaction between the activated complex of the electrode reaction and the adsorbed ions in the form of an activity coefficient. According to the final formula, the logarithm of rate constant depends approximately linearly on the amount of substance specifically adsorbed.

Another approach was used by Guidelli and Foresti[106] and Guidelli,[107] which formally leads to a very similar expression. When applied to an irreversible d-c reduction, it may be written

$$\frac{RT}{F}\ln \mathscr{X}_1 = \text{const} - \alpha n(E - E_0) - (z_{Ox} - \alpha n)\varphi_2 + Aq_i \qquad (88)$$

where in the $\mathscr{X}_1$ parameter [Eq. (62)] the backreaction is neglected. Here q_i is the charge density due to any specifically adsorbed supporting ions and A is a constant depending on kinetic and double-layer parameters.

7.4.1. Reduction of Cations

For Zn^{2+} reduction Koryta[80] found that the observed standard rate constant decreases with increasing ionic strength in 0.1–1 M $NaNO_3$ due to lowering of the absolute value of the φ_2 potential with increasing ionic strength in fairly good agreement with Eq. (86).

For the reduction of Zn^{2+} in the presence of halides[108,109] (I^-, Br^-, Cl^-) a linear relation between $\log k_{red}$ and q_i was roughly confirmed. In the discussion of the origin of certain discrepancies the possibility of participation of halide complexes was also considered.

The simple Frumkin correction describes the polarographic reduction behavior of Ga(III) in the presence of aluminum perchlorate. Several possible causes of its failure with magnesium and sodium perchlorates were considered, but the final explanation is still lacking. A striking example of double-layer effects on the reduction of cations arises with the redox couple Eu(III)/Eu(II)[110,111] which has served until now as a model system for various double-layer correction studies.

The formal standard oxidation–reduction potential of this system is -0.601 V vs. SCE in 1 M $NaClO_4$ so that the influence of the double layer may be studied on either side of the electrocapillary zero (see Figure 5). The system is polarographically reversible at low concentrations of the supporting electrolyte. The minimum observed on the anodic wave could be explained by considerable changes in the φ_2 values for low concentrations of supporting electrolyte in the vicinity of the pzc [Eq. (87a)]. The minimum is less pronounced or disappears entirely by addition of a fairly high concentration of the anion of the supporting electrolyte exhibiting relatively small specific adsorption (perchlorate) or small amounts of anions with specific adsorption; their effect increases in the sequence $Cl^- < NO_3^- \approx Br^- < I^- < SCN^-$. With

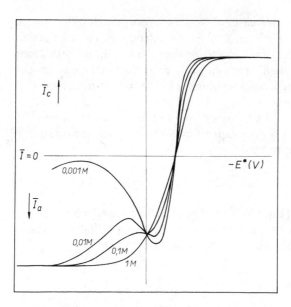

Figure 5. Polarographic curves for the Eu^{3+}/Eu^{2+} couple in different concentrations of supporting electrolyte: $10^{-4}\,M\,Eu^{3+}$, $10^{-4}\,M\,Eu^{2+}$ with variable perchloric acid concentration.

increasing specific adsorption the φ_2 potential at given potentials is becoming less positive or even negative and thus accelerates the oxidation of Eu(II) to Eu(III). The charge-transfer rate constants of Eu^{III}/Eu^{II} in perchlorates, iodides, and chlorides did not differ greatly after the Frumkin double layer correction was made.[112] In thiocyanate solutions complex formation has to be taken into account.[113]

Quite recently the double-layer correction for the systems Eu(III)/Eu(II) and Cr(III)/Cr(II) has been reexamined by Weaver and Anson,[114] for reduction of Cr(III) [see also Reference (115)]. In perchlorate media, the correction of rate constants for the double-layer effect using simple Frumkin approach and φ_2 values from Gouy–Chapmann theory seems to be adequate in relatively dilute lanthanum perchlorate solutions but fails in concentrated perchlorate solutions.

Better but not complete agreement was found when applying a statistical model of the double layer.[116,117] The increase in charge-transfer rate constants in solutions of perchlorate containing iodide and bromide is in relatively good agreement with Frumkin's approach when the experimentally determined rather than when calculated diffuse-layer capacitances were employed in the analysis of results. The effect of ion pairing and ligand bridging of reactants with adsorbing anions seemed rather small. The evidence for the ligand-bridged pathway was established in solutions containing thiocyanate anions.[114,118] The φ_2 potentials found from the influence of dilute solutions of perchlorate, chloride, and bromide on the oxidation rate of Eu^{2+} and V^{2+} differ[119] greatly in magnitude and even in sign from the potentials in the diffuse double layer obtained from electrocapillary data using Guoy–Chapman theory.

7.4.2. Reduction of Anions

The irreversible waves of reducible anions that are reduced at potentials more negative than the pzc are considerably shifted toward more positive potentials with increasing concentration of supporting electrolytes because the φ_2 potential becomes less negative. The shift of the reduction potential (measured at $\bar{I} \ll \bar{I}_d$) of IO_3^- amounts to more than 0.1 V when changing the concentration of potassium chloride from 0.02 to 0.5 M, and the experimental shifts correspond to Frumkin's correction[120] quite well. The less satisfactory validity of this correction in potassium sulfate is ascribed to ion pair formation in the bulk of the solution.[121]

The shift of the half-wave potential of chromate having ionic valency −2 in sodium hydroxide is extremely large,[101] amounting to 0.8 V when the supporting electrolyte concentration varies from 0.003 to 1 M. Since the rate-determining step is pH independent this shift is due to variation of the φ_2 potential. Frumkin's correction holds very well over this wide range of potentials.[101,122]

Generally for the reduction of anions at the same concentration of supporting electrolyte the shifts of the waves toward more positive values increase in the series $Li^+ < Na^+ < K^+ < Rb^+ < Cs^+$ and for divalent cations, $Ca^{2+} < Sr^{2+} < Ba^{2+}$. Hence the very negative reduction of the anion which could not be observed at low concentration of supporting electrolyte can be shifted by an appropriate selection of the concentration of the cation of the base electrolyte to an accessible potential range as was shown, e.g., on reduction of trithionate.[123]

7.4.3. Minima on the Limiting Currents for Anions

For divalent and polyvalent anions yielding diffusion currents at the electrocapillary zero, a decrease of this current may be observed in dilute solutions of supporting electrolytes just beyond the electrocapillary maximum. In general there is a tendency to eliminate the decrease in current with increasing concentration of cation and at a constant salt concentration, with increasing valency and absorbability of the cation. This minimum was first observed by Kryukova[124] in the reduction of persulfate (Figure 6). A similar behavior was found with other anions such as periodate, perbromate, tetrathionate, pentathionate, ferricyanide, and complex anions (particularly $PtCl_6^{2-}$, $PtCl_4^{2-}$, $IrCl_6^{2-}$, $RhCl_6^{2-}$, $PtNO_4^{2-}$, etc.). For details see References (9), (100–103). Whereas for most anions the minimum disappears when the concentration of supporting electrolyte is raised, in some cases the minimum may remain even in 1 M supporting electrolyte (e.g., $PtCl_4^{2-}$). In most cases the double-layer correction according to Frumkin could account for the observed phenomena, even for the electrostatic effect of specifically adsorbed halides; see e.g., the reduction of perbromate[125,126] in the presence of Cl^-, Br^-, and I^-.

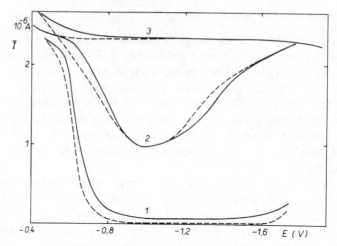

Figure 6. Effect of potassium chloride on the minimum current for persulfate reduction. 5×10^{-4} *M* $K_2S_2O_8$, 1) + 10^{-3} *M* KCl, 2) + 10^{-2} *M* KCl, 3) + 0.1 *M* KCl. Dotted curves calculated from Eq. (87a).

On the other hand, in the case of specifically adsorbed electroinactive ions, according to some authors[127,128] the simple Frumkin correction is not satisfactory and the corrected rate constants of tetrathionate and perbromate still depend on the charge density q_i due to specifically adsorbed halide and pseudohalide ions. This effect is more pronounced[129] at higher concentrations of halides.

8. Kinetic Currents

In more complex electrode processes the electroactive forms of the redox system may take part in some further chemical equilibria. In the course of charge-transfer reactions, the concentration of electroactive forms is changed, which causes perturbation of equilibrium of the chemical reactions. Depending on the rate of the chemical reaction the observed current may be totally or partly controlled by the rate of the corresponding chemical reaction (kinetic currents).

Chemical reactions can be combined with a charge-transfer reaction in three ways:

(i) The chemical reaction precedes the charge-transfer reaction [preceding reaction, chemical–electrochemical mechanism (c.e.)]. In this type, the electroactive form is produced by a chemical reaction from an electroinactive form with which it is in a mobile equilibrium. The measured current is smaller than the diffusion current, which would correspond to reduction of the electroinactive form. Usually, the chemical equilibrium in the solution is shifted in favor of the electro-inactive form, the concentration of which is then practically equal to that in the bulk.

(ii) The chemical reaction runs parallel with the charge transfer (parallel or catalytic reaction). Parallel reactions consist of chemical regeneration of the original electroactive species from the product of the charge transfer reaction. The measured limiting current is always higher than the diffusion-controlled current of the electroactive form.

(iii) The chemical reaction follows the electron-transfer (following or subsequent chemical reaction–e.c. mechanism). In this case the primary product of a reversible charge-transfer reaction is chemically transformed to a less electroactive or inactive form.

Two methods are usually used for computing rate constants of the corresponding chemical reactions from kinetic currents. The approximate method is based on the reaction layer concept[130–136] and a rigorous treatment expresses the changes in concentration at the electrode caused by diffusion and chemical reaction by a system of differential equations in the first instance for a planar[137–139] and later for a growing dropping mercury electrode.[140,141] The principles of both methods will be briefly outlined below. For exhaustive reviews on kinetic currents see References (9), (69), (142–145).

8.1. Preceding Chemical Reaction

The reaction scheme is as follows

$$B \underset{k'}{\overset{k}{\rightleftharpoons}} A \overset{el}{\longrightarrow} P$$

The electroinactive form B is transformed chemically into an electroactive form A (e.g., hydrated form of formaldehyde into nonhydrated reducible formaldehyde), which undergoes an electrochemical transformation to the product P. Here k is the rate constant for the formation of the electroactive form A (it may depend on concentration of other substances participating in excess in the chemical reaction). The equilibrium constant K is defined as the ratio of concentration of the electroinactive to that of the electroactive form $K = [B]/[A] = k'/k$ so that $k' = kK$.

8.1.1. Approximate Treatment of a First-Order Reaction

For the mean limiting current governed by the chemical reaction $\bar{I}_k$ one can write

$$\bar{I}_k = nF\bar{A}D\frac{(c_A)_r}{\mu} \tag{89}$$

where $(c_A)_r$ stands for the concentration at the external boundary of the reaction layer and μ is the thickness of the reaction layer for which it holds[137,146]

$$\mu = \sqrt{D\tau} \tag{90}$$

Here τ is the mean lifetime of the formed electroactive species, which can be expressed as reciprocal of the rate constant of the inactivation reaction k'. Hence

$$\mu = \sqrt{\frac{D}{k'}} \tag{91a}$$

or

$$\mu = \sqrt{\frac{D}{kK}} \tag{91b}$$

The concentration of the electroactive form at a distance equal to or greater than μ can be deduced from the equilibrium constant and hence

$$\bar{I}_k = nF\bar{A}D^{1/2}\frac{k^{1/2}}{K^{-1/2}}(c_B)_r \tag{92}$$

This is the equation of the mean current, which is controlled purely by the rate of the chemical reaction. Since the mean surface area A is independent of the height of the mercury column h, the kinetic current is likewise independent. The time dependence of the instantaneous kinetic current is determined by that of the surface area, i.e., $I_k \sim t^{2/3}$. Both dependences are used as simple criteria for distinguishing currents governed by the rate of chemical reactions.

If the concentration of the electroinactive form at the boundary of the reaction layer is not current it may be expressed approximately by the Ilkovič diffusion equation

$$\bar{I}_k = \bar{I}_{d_B} - \mathscr{H}(c_B)_r \tag{93}$$

and a general approximate formula is obtained by substituting for $\bar{A}$ [Eq. (6)] and $\mathscr{H}$ [Eq. (25)]

$$\frac{I_k}{I_{d_B} - I_k} = 0.81\sqrt{\frac{kt_d}{K}} \tag{94}$$

8.1.2. Rigorous Treatment of a First-Order Reaction

The given reaction scheme is described by the following differential equations[141]

$$\frac{\partial c_A}{\partial t} = D\frac{\partial^2 c_A}{\partial x^2} + \frac{2x}{3t}\frac{\partial c_A}{\partial x} + k(c_B - Kc_A) \tag{95}$$

$$\frac{\partial c_B}{\partial t} = D \frac{\partial^2 c_B}{\partial x^2} + \frac{2x}{3t} \frac{\partial c_B}{\partial x} - k(c_B - Kc_A) \tag{96}$$

$$\text{for } t = 0, x > 0 \qquad c_A = c_A^0 \quad \text{and} \quad c_B = c_B^0 \tag{97}$$

$$\text{for } t > 0, x = 0 \qquad c_A = 0 \qquad \frac{\partial c_B}{\partial x} = 0 \tag{98}$$

$$k \gg 1 \tag{99}$$

Koutecký[141] solved these equations by the method of dimensionless parameters. The ratio of the mean limiting current with the participation of the preceding chemical reaction $\bar{I}_k$ to the limiting diffusion current $\bar{I}_d$ is given by the function

$$\frac{\bar{I}_k}{\bar{I}_d} = \bar{F}(\mathscr{X}_l) \tag{100}$$

where

$$\mathscr{X}_1 = \sqrt{\frac{12}{7} \frac{kt_d}{K}} \tag{101}$$

Similar equations are used for the ratio of instantaneous currents. The tabulated values of these functions[9,141] are equal to those for slow charge transfer in polarography.

Function $\bar{F}(\mathscr{X}_1)$ may be approximated by a simple interpolation formula[141,147]

$$\bar{F}(\mathscr{X}_1) = \frac{\mathscr{X}_1}{1.5 + \mathscr{X}_1} \tag{102}$$

so that for the mean current

$$\frac{\bar{I}_k}{\bar{I}_d - \bar{I}_k} = 0.886 \sqrt{\frac{kt_d}{K}} \tag{103}$$

This formula is almost identical with that resulting from the approximate solution differing only in the numerical coefficient. The determination of the rate constant of the preceding chemical reaction from this expression is straightforward, providing the equilibrium constant K is known. For a more precise value of the rate constant, the solution taking into account the correction for spherical diffusion[148] may be applied.

If the diffusion coefficients of substances A and B differ substantially from each other, the formula for the kinetic current derived by Koutecký[149] is to be used.

Finally it should be mentioned that the previous differential equations are substantially simplified if the chemical reaction proceeds very fast so that a steady state is established between the chemical reaction and diffusion in the immediate vicinity of the electrode. In this case we can write

$$D\frac{d^2c_A}{dx^2} + k(c_B - Kc_A) = 0 \tag{104}$$

$$D\frac{d^2c_B}{dx^2} - k'(c_B - Kc_A) = 0 \tag{105}$$

Provided that the diffusion of substance B can be neglected (a "purely kinetic current") and that the condition $c_A \ll c_B$ holds, Eq. (104) can be easily solved and the following result is obtained

$$c_A = \frac{c_B}{K}\left(1 - \exp\frac{-x}{\sqrt{\dfrac{D}{kK}}}\right) \tag{106}$$

the instantaneous kinetic current is

$$I = nFAD\left(\frac{dc_A}{dx}\right)_{x=0} = nFAc_B\sqrt{\frac{Dk}{K}} \tag{107}$$

This steady-state approach was used in evaluating the double-layer effect on the rate of the chemical reaction step (see further).

8.1.3. Several Examples

From the various types of chemical reactions described in literature some typical examples will be given.

Many organic compounds containing a carbonyl group are present in aqueous solutions predominantly in their hydrated forms, which are usually inactive compared with the free reducible carbonyl group. Thorough kinetic studies were carried out with formaldehyde.[150,154] Further examples studied in detail are, e.g., dehydration of glyoxalic acid and of its anion[155,156] and of pyridine-aldehydes.[157,159] In practically all cases the rate constant is acid-based catalyzed, i.e., it depends on the concentration of all acids and bases present in the solution.

Another very important class of kinetics is represented by the cathodic waves of the anions of many weak reducible organic acids in buffered solutions. Here the current is controlled by the rate of recombination of the anion of the acid with hydroxonium cations forming the undissociated acid, which is reduced at more positive potentials than its anion. The first monobasic acids studied were pyruvic and phenylglyoxalic acids[160] and the first dibasic acids studied were maleic and fumaric acids.[161] For a table of selected measured rate constants, see Reference (9). In the case of boric acid, a cathodic current due to the discharge of the hydrogen ion controlled by the rate of dissociation of the acid was found. This reaction does not fit the reaction scheme of weak reducible acids since the rate of recombinations is bimolecular.[162]

Kinetic currents are frequently encountered in polarographic studies of complexes and are governed by the rate of dissociation of an inactive into an electroactive complex, or of dissociation of a complex into a noncomplexed cation. The pioneering work in this field was carried out by Koryta.[71,142,162–165]

In addition, some monosaccharides[166–168] give rise to kinetic currents where the controlling reaction is the conversion of the nonreducible cyclic hemiacetal form to the free aldehyde form.

Most of the examples given here could be treated as chemical reactions of the first order, i.e., using Eq. (103).

8.1.4. Second-Order Reactions

The approximate method is advantageous in the case of bimolecular reactions, for which a rigorous solution is more difficult. Although the rate constant values may be somewhat in error, the method gives a true picture of the functional dependence of polarographic data on individual parameters.

This will be demonstrated for the case of a bimolecular preceding reaction. The reaction scheme can be written[135]

$$B \underset{k'}{\overset{k}{\rightleftharpoons}} 2A$$

$$A \xrightarrow{el} P$$

For the current the following expression is valid

$$\bar{I}_k = nF\bar{A}\frac{(c_A)_r}{\mu} \tag{108}$$

The thickness of the reaction layer μ for the equilibrium constant $K = (c_B)_r/(c_A)_r^2$ is given by

$$\mu = \sqrt{\frac{D_B}{Kk(c_A)_r}} \tag{109}$$

The current is also given by diffusion of B toward the electrode

$$\bar{I}_k = \bar{I}_d - \mathcal{H}(c_B)_r \tag{110}$$

and the concentration of $(c_A)r$ with regard to the equilibrium relation may be expressed from the preceding equation

$$(c_A)_r = \left(\frac{\bar{I}_d - \bar{I}_k}{\mathcal{H}K}\right)^{1/2} \tag{111}$$

Combination of Eq. (108), (109), and (111) yields the final formula

$$\frac{\bar{I}_k}{\bar{I}_d} = 0.81\alpha \left(1 - \frac{\bar{I}_k}{\bar{I}_d}\right)^{3/4} \tag{112}$$

with

$$\alpha = \left(\frac{kt_d}{(Kc_A)^{1/2}}\right)^{1/2} \tag{113}$$

This reaction scheme was also solved by the rigorous treatment by Koutecký and Hanuš.[169]

As an example of a bimolecular reaction controlled by the rate of decomposition of an electroinactive dimer to an active monomer, the polarographic behavior of dithionite[170–173] and of manganous ions in alkaline tartrate solution[174] can be given.

8.2. Reactions Parallel to the Charge-Transfer Reaction. Catalytic Currents

8.2.1. First-Order Reaction

The reaction scheme is as follows

$$A + ne \rightleftharpoons P$$

$$P + X \xrightarrow{k} A$$

If the product of the fast charge-transfer reaction P reacts chemically sufficiently fast with a compound X in the solution to regenerate the original electroactive form A of the redox-couple A/P, an increase in the limiting current is observed. The redox-couple A/P acts as a catalyst for the reduction or oxidation of X, which is polarographically electroinactive within the given potential range.

In most cases, the formula derived from the rigorous treatment for the evaluation of the rate constant should have the form[140]

$$\frac{\bar{I}_k}{\bar{I}_{d_A}} = \bar{\psi}(\mathscr{X}_1) \tag{114}$$

where

$$\mathscr{X}_1 = kt_d \tag{114a}$$

$\bar{I}_k$ is the total observed current in the presence of X and $\bar{I}_{d_A}$ is the diffusion-controlled current of the substance A alone. The functions are tabulated[9,140] both for the mean as well as for the instantaneous current. For $\mathscr{X}_1 > 10$ the function can be expressed in an asymptotic form

$$\bar{\psi}(\mathscr{X}_1) = 0.81\sqrt{\mathscr{X}_1} \tag{115}$$

Under this assumption, which is equivalent to the condition that the reaction layer thickness is much smaller than the diffusion layer thickness, the rigorous solution is in agreement with that using the approximate method and leads to the equation

$$\frac{\bar{I}_k}{\bar{I}_{d_A}} = 0.81\sqrt{kt_d} \tag{116}$$

In practice this is fulfilled if $\bar{I}_k$ is more than three times larger than $\bar{I}_{d_A}$.

The solution, including the correction for spherical diffusion, is also available.[175]

8.2.2. Second-Order Reaction

If the substance P is not in excess, both reactants are variable and account must be taken of the bimolecular nature of the reaction. The reaction layer concept leads to the formula[176]

$$\frac{\bar{I}_k}{\bar{I}_{d_A}} = 0.81\sqrt{kc_X^0 \left(1 - \frac{\bar{I}_k}{\bar{I}_{d_X}}\right) t_d} \tag{117}$$

The more exact solution is given by Koutecký.[177]

8.2.3. Partial Regeneration of the Electroactive Species

A very frequent reaction scheme involves a disproportionation process of the type

$$2A \underset{}{\overset{el}{\rightleftharpoons}} P \underset{}{\overset{kX}{\rightleftharpoons}} A + B$$

The solution[178,179] allows the determination of the rate constant if the increase of the limiting current does not exceed 13% of the original diffusion controlled current without disproportionation. In this case other techniques such as voltammetry or the RDE are preferred.

8.2.4. Examples

Mechanisms of several mostly cathodic reactions are described in the polarographic literature [for details see References (9), (142–145)]. In most cases, the following redox system was used: Fe^{III}/Fe^{II}, Ti^{IV}/Ti^{III}, Cr^{III}/Cr^{II}, Mo^{VI}/Mo^{V}, W^{VI}/W^{V}, Co^{III}/Co^{II}, Os^{VIII}/Os^{VI} (sometimes aquated or bound in complexes with EDTA, amines, oxalate, etc.). Also used were substances characterized by oxidizing power, mainly hydrogen peroxide, hydroxylamine and chloric, perchloric and nitric acids. As an example of the disproportionation reaction, the chemical reaction of pentavalent uranium[179] formed by the electrochemical reduction of uranium(VI) can be given.

8.3. Subsequent Chemical Reactions (Follow-up Reactions— Postkinetic Waves)

Whereas for preceding and catalyzed volume chemical reactions, the chemical kinetics manifests itself in the magnitude of the limiting currents, in the case of follow-up reactions the limiting current is governed by the diffusion of the electroactive species. The subsequent chemical reaction, in most cases the inactivation reaction, can be revealed polarographically if the electron transfer is fast.

8.3.1. First-Order Reaction

If one considers, e.g., the following reaction scheme

$$A \rightleftharpoons P_r + ne$$

$$P_r \xrightarrow{k} P_i$$

where the product of a reversible oxidation P_r undergoes an irreversible change to the final product P_i, one can write for the mean current

$$\bar{I} = \bar{I}_{d_A} + \mathscr{H}c_A \tag{118}$$

$$-\bar{I} = nF\bar{A}\mu k c_{P_r} \tag{118a}$$

The reaction layer thickness is given by

$$\mu = \sqrt{\frac{D_{Ox}}{k}} \tag{119}$$

If the surface concentrations of A and P_r are substituted in the Nernst equation one gets the equation of the wave as

$$E = E_0 - \frac{RT}{nF} \ln \frac{\bar{I}_{d_A} - \bar{I}}{\bar{I}} - \frac{RT}{nF} \ln 0.81 \sqrt{\frac{D_A}{D_P}} k t_d \tag{120}$$

This equation shows that the wave has a reversible form and that its half-wave potential depends on droptime. The inactivation reaction rate constant can be estimated if the standard potential E_0 of the forms A and P_r is known. This was not the case for the first studied example of oxidation of ascorbic acid.[180–182] However, it was possible in the case of an inactivation reaction during the dissolution of cadmium and amalgam[183] in the presence of EDTA. The more exact solution[184] for a unidirectional inactivation leads to the same expression with the numerical factor 0.886 instead of 0.81.

8.3.2. Second-Order Reaction

For a reduction scheme with an inactivation reaction (dimerization)

$$2A + ne \rightleftharpoons 2P_r \xrightarrow{k_2} P_i$$

and if a reaction layer concept is applied, one gets[135]

$$E = E_0 - \frac{RT}{nF} \ln \frac{\bar{I}^{2/3}}{\bar{I}_d \bar{I}} \bar{I}_d^{1/3} + \frac{RT}{3nF} \ln \frac{c_A^0 k_2 t_d}{1.51} \tag{121}$$

The exact solution[169] can be expressed in the form

$$\frac{\bar{I}_k}{\bar{I}_d} = f(\gamma_1) \tag{122}$$

where $\gamma_1 = \sqrt{k_2 c_A^0 t_d \lambda^3}$ and λ is the potential function given by Eq. (34).

The outlined short survey dividing the kinetic currents into three categories gives only the first information about coupled chemical reactions. The real reaction scheme might be more complicated.

8.4. Double-Layer Effect on Kinetic Currents

This effect was found until now in the case of preceding chemical reactions involving charged particles; it occurs if the thickness of the reaction layer is smaller or comparable with the thickness of the diffuse part of the double layer, i.e., for very fast chemical reactions such as, e.g., protonation of anions of acids.

For solving this problem[101,185,186] Eqs. (104) and (105) are employed assuming a steady state, adding the migration terms for particles B and A [for charged particle A this term is $(z_A F/RT)(c_A)(d^2\varphi_x/dx^2)$; φ_x being the potential in the diffuse part of the double layer at the distance x, commonly replaced by the φ_2 potential]. The solution for the scheme

$$B^{z_B} + \nu X^{z_X} \rightarrow A^{z_A} \xrightarrow{ne} P$$

leads to the final expression

$$\frac{\bar{I}_k}{\bar{I}_d - \bar{I}_k} = \frac{0.886}{G} \sqrt{k \frac{t_d}{K}} \tag{123}$$

The function G for the thickness of the reaction layer larger than the thickness of the double layer approaches 1. In an opposite case it can be simplified to

$$G = \exp[(2p + 1)F|z\varphi_2|/2RT] \tag{124}$$

where $p = (-|z_A| - |z|/2)/|z|$, z_A is the charge of the electroactive species, and z is the charge of the symmetrical supporting electrolyte. The double-layer effect may account for several orders of magnitude in the observed rate constants. Applications of the theoretical equations can be found[187,188] in which the simple correction for the double-layer effect seemed to be quite satisfactory. However, this effect might be more complex depending on the discreteness of charge in the double layer, dissociation field effects, change in the reactivity of polarizable molecules, etc. [for details see Reference (143)].

8.5. Surface Kinetic Currents

So far, in the survey of chemically controlled currents, a direct interaction of the reacting particle with the electrode was not taken into account.

Such types of chemical reactions are called "bulk" or "volume" reactions. However, the majority of organic substances are specifically absorbed at the mercury-solution interface and the relatively high surface concentrations attained at the interface may cause chemical reactions to occur predominantly in the adsorbed state. The adsorption depends mostly on the potential and hence the rate of such a heterogeneous chemical reaction (and correspondingly the limiting kinetic currents) is a function of the potential of the electrode. In order to distinguish these types of reactions from "pure" or "volume" chemical reactions, they are called "surface" chemical reactions and the corresponding currents "surface" kinetic currents.[145,189] These kinds of currents were encountered in the case of preceding reactions, mostly protonations.

A reaction scheme for P proceeding in the adsorbed state can be written as

$$B_{ads} \underset{k_s'}{\overset{k_s}{\rightleftharpoons}} A_{ads} ne \rightarrow P$$

where k_s and k_s' are the rate constants of the surface chemical reaction and $K_s = k_s'/k_s$ is the corresponding equilibrium constant. Supposing that a linear adsorption isotherm applies to the electroinactive form B, one has

$$\Gamma_B = \beta c_B^0 \tag{125}$$

where Γ_B is its surface concentration and β the adsorption coefficient. The instantaneous current under steady-state conditions is expressed by the two equations

$$\frac{I}{nFA} = k_{red}\Gamma_A \tag{126}$$

and

$$\frac{I}{nFA} = k_s\Gamma_B - k_sK_s\Gamma_A \tag{127}$$

By combining these equations one obtains

$$I = nFAk_{red}\frac{k_s\beta c_B^0}{k_{red} + k_sK_s} \tag{128}$$

For the limiting current $k_{red} \rightarrow \infty$

$$I_l = nFAk_s\beta c_B^0 \tag{128a}$$

Since β is potential dependent and k_s may embody the double-layer effect, the limiting current depends on the potential and decreases in the desorption range of the substance B (minima on polarographic waves). For additional details and a more thorough discussion, see Mairanovskii[145] and Guidelli.[143]

8.6. Mixed Volume-Surface Kinetic Currents

In general, the chemical reactions may occur both in the reaction layer ("volume" reactions) and in an adsorbed state ("surface" reactions). The solution for polarographic current under simplified conditions (i.e., if there is no change in their mechanism, in reaction orders, or in the stoichiometry, and assuming that the surface and bulk reactions differ merely by the values of their rate constants) was presented using a diffusion and reaction layer approximation. In this case it is supposed that the kinetic current is the sum of the rates of surface and volume reactions. For details see Guidelli[143] and Mairanovskii.[145]

8.7. Catalytic Currents of Hydrogen

A large group of kinetic currents that was studied by many authors and still has not been solved in full detail are the catalytic currents of hydrogen.[9,145] Some substances are able to lower hydrogen overvoltage on mercury and give rise to enhanced currents, which can be studied under polarographic conditions. These substances can be roughly divided into two groups:

(i)   Electroactive substances (predominantly the salts of platinum metals), which after reduction are deposited on the surface forming some catalytically active centers of atoms:

(ii) Adsorbable organic substances having some groups which are protonated.

Whereas the first case is limited to some metal ions and the interpretation of the catalytic current is rather difficult due to the properties of the deposit in the course of its electrocrystallization,[190,191] the second group consists of many hundreds of organic substances and in several cases quantitative interpretation have been given, e.g., for quinoline alkaloids.[97]

Most results can be explained supposing a reaction scheme based on the treatment of Conway, Bockris, and Lovrecek[97] in which the adsorbed organic base in a protonated form "BH^+" provides an alternative proton source to H_3O^+ in the double-layer, from which preferred proton transfer, leading to H_2 evolution, takes place. Similar ideas were later proposed by Mairanovskii[97a] and given in a review.[145] Schematically, the reaction scheme is

$$B + DH^+ \underset{k_2}{\overset{k_1}{\rightleftharpoons}} BH^+ + D$$

$$BH^+ + e \rightarrow BH$$

$$2BH \overset{k_d}{\longrightarrow} 2B + H_2$$

where B is the basic form of the catalyst (e.g., an alkaloid or pyridine), DH^+ and D represent hydroxonium ions and water, or, in a more general manner, the acidic and basic forms of the solvent, respectively. The protonated form of the catalyst is reduced at a more positive potential than that for the proton donor. The radical BH rapidly dimerizes with regeneration of the catalyst in its original form B and formation of a molecule of hydrogen.

In some cases the reduction of the protonated form of catalyst seems to proceed reversibly (pyridine and its homologues) and the equation of the catalytic hydrogen wave can be derived. Depending on experimental conditions, the rate of the catalytic current may be governed by the rate of protonation or by the rate of dimerization (for very small concentrations of B), which can be in some cases treated as volume reactions.

For strong adsorption and irreversible electrode reaction (e.g., quinine) the corresponding current is controlled by a surface or mixed surface and volume reaction. The limiting catalytic current is then potential dependent.[9,145]

A special subgroup is represented by the catalytic hydrogen currents which arise in the presence of cobalt salts. In this case, a still more pronounced lowering of hydrogen overvoltage and the formation of characteristic limiting currents in buffered solution (mostly ammoniacal) containing cobalt salts and very small traces of organic substances containing nitrogen and sulfur (mostly thiol groups) is observed. In a solution of divalent cobalt a single catalytic wave having the form of a maximum in the presence of relatively simple sulfur containing organic compounds as, e.g., cysteine, cystine, glutathione, thioglycollic acid, etc., is to be found. More complicated substances such as proteins give rise to the formation of a catalytic hydrogen double wave in solutions with trivalent cobalt.[9,145] These catalytic waves (Brdička's protein double wave) were used for the analytical estimation of some substances in biochemistry and medicine.[92]

9. Adsorption in Polarography

9.1. Adsorption of Reactant or Product

9.1.1. Reversible Charge-Transfer Reaction

The phenomenon of reactant or product adsorption has become a problem in electrode kinetics and is relevant to charge-transfer kinetic studies using all electrochemical techniques. A general rigorous treatment of this problem[193,194] appears to be very complex and hardly accessible for calculations, but several approximations can be made. The adsorption of the electroactive form or of the product manifests itself distinctly in d-c polarography.[9,195,196] For a reversible charge-transfer reaction, strong adsorption may cause the appearance[195,196] of a so-called adsorption "postwave" or an adsorption "prewave." The adsorption "postwave" was commonly attributed to the adsorption of the reactant whereas the "prewave" was related to the adsorption of the product in reduction processes (see Figure 7).

If the adsorption is weak, the form of the reversible wave is influenced, and in some cases the slope of the log-plot analysis of the wave might be steeper in certain potential ranges than would correspond to the Nernstian behavior.[197,198]

9.1.2. Reversible Electrode Reaction

The first approximate treatment of reactant or product adsorption in d-c polarography based on the diffusion layer concept was given in the early forties by Brdička.[195,196] His derivation in a somewhat more general form can be outlined as follows.

The electrode reaction $Ox + ne \rightleftharpoons Red$ is reversible with possible adsorption of Ox and Red and the mass transport controlled by diffusion only. Further it is assumed that only Ox is initially present in the solution, whereas the bulk concentration of reduced form is equal to zero, $c_{Red}^0 = 0$. Hence the following equations can be written for the instantaneous current

$$I = nFAD_{Ox}\left(\frac{\partial c_{Ox}}{\partial x}\right)_{x=0} - nF\frac{d(A\Gamma_{Ox})}{dt} \tag{129a}$$

$$I = -nFAD_{Red}\left(\frac{\partial c_{Red}}{\partial x}\right)_{x=0} + nF\frac{d(A\Gamma_{Red})}{dt} \tag{129b}$$

In Brdička's derivation the concentration gradients are expressed from the Ilkovič equation and one obtains

$$I = \mathcal{H}_{Ox}(c_{Ox}^0 - \bar{c}_{Ox}) - nF\Gamma_{Ox}\frac{dA}{dt} \tag{130a}$$

$$I = \mathcal{H}_{Red}\bar{c}_{Red} + nF\Gamma_{Red}\frac{dA}{dt} \tag{130b}$$

Here $\bar{c}_{Ox}$ and $\bar{c}_{Red}$ are supposed to be time independent.

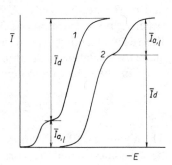

Figure 7. Waveforms for the adsorption of the reduced form (1) and of the oxidized form (2) in cathodic reductions.

The ratio of surface concentration is given by Eq. (34). After solving these equations for c_{Red} one obtains

$$I = \frac{I_{d_c}}{1 + \lambda \sqrt{\dfrac{D_{Ox}}{D_{Red}}}} - nF \frac{\Gamma_{Ox} + \Gamma_{Red}}{1 + \lambda \sqrt{\dfrac{D_{Ox}}{D_{Red}}}} \frac{dA}{dt} \tag{131}$$

The first term represents the reversible cathodic wave in the absence of adsorption whereas the second term must be accounted for in the case of adsorption.

Brdička's original derivation for mean currents distinguishes two extreme cases: (i) adsorption of the oxidized form, (ii) adsorption of the reduced form only. The adsorption was expressed in the form of the Langmuir isotherm, which can be read for the reduced form as follows

$$\Gamma_{Red} = \Gamma_m \frac{\beta_{Red} c_{Red}}{1 + \beta_{Red} c_{Red}} \tag{132}$$

where Γ_m is the maximum surface excess and β_{Red} the adsorption coefficient. Hence for the mean adsorption current it follows (diffusion terms are neglected) that

$$\bar{I}_{ads} = nF\Gamma_m \frac{\beta_{Red} c_{Red}}{1 + \beta_{Red} c_{Red}} \frac{d\bar{A}}{dt} \tag{133}$$

and for the plateau of the prewave at which the drop surface is fully occupied as soon as the drop is formed ($\beta_{Red} c_{Red} \gg 1$) after expressing $d\bar{A}/dt$, one has

$$\bar{I}_{ads,l} = nF\Gamma_m \, 0.85 m^{2/3} t_d^{-1/3} \tag{133a}$$

The instantaneous current at the plateau of the prewave should fulfill the dependence $I \sim t^{-1/3}$. The mean current increases linearly with the height of the mercury column and does not depend on the further increase of the concentration. The derivation leads to a correct formulation of the adsorption limiting current and also Γ_m values and the corresponding areas occupied by the molecules seem to be plausible. Moreover, Brdička also derived the forms of the whole polarographic curve both for the adsorption of the reactant and of the product only. According to his concept, the adsorption prewave occurs if the adsorption coefficient of the product is much larger than that of the reactant and the prewave occurs at a more positive potential than the standard oxidation–reduction potential.

For the half-wave potential of the prewave it can be shown that

$$E_{1/2,ads} = E_o + \frac{RT}{nF} \ln \frac{\beta_{Red} c_{Ox}^0}{2} \tag{134}$$

The higher the β_{Red} the more positive is the adsorption prewave. Its value can be calculated from the corresponding shift of the prewave. This prewave reaches its maximum value at a concentration of 10^{-4} M, and is then followed by the reversible "main" wave with $E_{1/2}$ very nearly equal to the standard potential of the redox system. Typical examples used for verification at that time were methylene blue and riboflavine.

Whereas the occurrence of an adsorption prewave is relatively frequent, adsorption postwaves that are displaced in the case of reduction toward more negative potentials with respect to the "main wave" are rather rare. As typical examples, the reductions of diphenyl disulfide and -diselenide[199] and of salicylidene–iminato azido chelates of iron(III)[200] should be mentioned. Many other examples based on our own results and a critical literature survey of both experimental and theoretical papers are to be found in an exhaustive review by Laviron.[201] An extension of Brdička's theory for electrode reactions that are not completely reversible was presented by Guidelli.[202,203]

It was later shown that typical examples used for the verification of Brdička's treatment (namely methylene blue and riboflavine) should not obey the derived equations because the adsorbabilities of both forms are practically comparable.[204] This is true for most substances giving rise to adsorption prewaves or postwaves, respectively. Under those conditions neither a prewave nor a postwave should appear, since according to Brdička's original concept there is a compensation in the adsorption energies of the two forms.

An interesting approach to explain these discrepancies has been recently presented by Laviron[205] by taking into account the interaction forces between the adsorbed molecules using the Frumkin adsorption isotherm instead of the Langmuir.

It is supposed that the electrode reaction is reversible and both forms are simultaneously adsorbed, obeying Frumkin's adsorption. The adsorption of the oxidized form is described by

$$\beta_{Ox} c_{Ox}(0, t) = \frac{\Theta_{Ox}}{n_{Ox}}(1 - \Theta_{Ox} - \Theta_{Red})^{n_{Ox}} \exp\left(-2a_{Ox}n_{Ox}\Theta'_{Ox} - 2a_{Ox,Red}\Theta_{Red}\right)$$

$$(135)$$

and an analogous expression can be written for the adsorption of the reduced form. β_{Ox} and β_{Red} are the adsorption coefficients; Θ_{Ox} and Θ_{Red} the coverages, $\Theta_{Ox} = \Gamma_{Ox}/\Gamma_{Ox,m}$ and $\Theta_{Red} = \Gamma_{Red}/\Gamma_{Red,m}$. $\Gamma_{Ox,m}$, $\Gamma_{Red,m}$ are the maximum surface excesses, $c_{Ox}(0, t)$ is the concentration of the oxidized form in solution near the surface at time t; the number of molecules of water n_{Ox} displaced by one molecule of Ox; a_{Ox}, a_{Red}, and $a_{Ox,Red}$ are the constants of interaction between molecules of Ox, molecules of Red, and molecules of Ox and Red, respectively.

It is further assumed that the adsorption or the desorption is rapid, so that the equilibrium corresponding to Eq. (135) is established at any time, and the adsorption coefficients and interaction parameters are independent of potential and that the presence of adsorbed molecules has no effect on the diffusion. After some further simplifications the equation can be derived only for the adsorption wave $(\bar{I}_{ads,l} < \bar{I}_d)$. For its halfwave potential the following equation can be written

$$E_{1/2,ads} = E_0 - \frac{RT}{nF} \ln \frac{\beta_{Ox}}{\beta_{Red}} + \frac{RT}{nF}(a_{Red} - a_{Ox})\nu\Theta_T \tag{136}$$

in which ν is the common value of n_{Ox} and n_{Red} ($\nu = n_{Ox} = n_{Red}$). Θ_T is the total coverage of the electrode: $\Theta_T = \Gamma_T/\Gamma_m$, Γ_m being the common value of $\Gamma_{Ox,m}$ and $\Gamma_{Red,m}$. This equation explained the prewaves of methylene blue, riboflavin, and rosinduline GG very satisfactorily. Recent experimental results show that the adsorption of methylene blue is still more complex since two adsorbed layers are formed.[206] This theory thus explains why a prewave is observed, although both Ox and Red are equally adsorbed: the appearance of the prewave is caused by energy effects arising from the interactions between adsorbed molecules, rather than by energy effects caused by the differences in adsorption of the two forms.

An independent approach has been based on the distinction between "surface" and "volume" electron transfer processes, depending on whether the exchange of the electron occurs with or without direct contact of the electroactive species with the electrode.[207,208] It has been suggested that the half-wave potential of a reversible adsorption wave corresponds to a "surface" redox potential and that the difference between it and the "volume" redox potential (i.e., the half-wave potential of the main wave) is determined by interaction of the adsorbable components with the electrode.

The most thorough and quite recent derivation of a reversible adsorption prewave is given by Guidelli and Pergola,[209] based on Brdička's concept, extending the derivation of Laviron,[205] which was applicable for the adsorption wave only, and neglecting product diffusion away from the electrode. However, this derivation fails to account for the so-called "main" wave that accompanies the adsorption wave as soon as the reactant bulk concentration exceeds a certain limiting value.

A derivation based on a generalized approximate procedure worked out previously by Guidelli[210] and using the Eqs. (129a) and (129b) as two boundary conditions and the Nernst equation was carried out supposing Langmuir and Frumkin adsorption behavior. It was possible to derive the form of the whole polarographic curve for Langmuirian adsorption of reactant and product. From the derived equation it follows that for β_{Ox} or β_{Red} equal to zero, Brdička's approximate equation for an adsorption prewave or postwave, respectively, could be obtained. If both reactant and product are strongly adsorbed, an adsorption prewave can only be observed if β_{Red} is at least two orders of magnitude greater than β_{Red}. Analogously, an adsorption postwave is only to be expected if β_{Ox} is at least two orders of magnitude greater than β_{Red}. When β_{Ox} becomes equal to β_{Red} the polarographic wave is practically identical with a reversible wave in the absence of adsorption. The forms of I-t curves can also be calculated. The corresponding current-potential characteristics were also derived using Frumkin's adsorption isotherm for reactants and products.

The shapes of reversible d-c polarographic waves, complicated by reactant and/or product adsorption based on Eqs. (129a) and (129b) and assuming Langmuir and Henry isotherms for various adsorption parameters, were calculated by Sluyters-Rehbach et al.[211] and a comparison with a more general theoretical approach was presented.[212] The simpler diffusion layer approach is sufficient for correct predictions of polarograms; however, for obtaining information about adsorbed quantities the more rigorous approaches are to be preferred.

At the present stage of the development of this problem it seems that Brdička's concept, generalized in recent papers,[206,209,210] which could be mentioned only briefly here, forms a reasonable basis for explaining the unusual pattern of a reversible d-c wave coupled with adsorption of reactants or products. This approach does not involve any assumption as to the mode of electron transfer from the electrode to the reacting particle and vice versa, but simply accounts for the effect of adsorption of reactant and product upon the only rate-determining step, namely the diffusion stage. Regarding the charge-transfer stage and the adsorption–desorption stage, they are both in equilibrium. Hence no further conclusions can be drawn such as, e.g., the position of the reacting particle during the electron transfer from a comparison between theoretical predictions and experimental results.

9.1.3. Irreversible Charge-Transfer Reactions

A still more complex pattern than described previously may occur in the case of irreversible polarographic waves coupled with adsorption of the reactants or products. Only general features can be given here. For an irreversible reduction an adsorption prewave followed by the main wave was observed, which cannot be explained as in the previous case. The prewave is attributed (see further) to inhibition by a film of the reduction product and to reduction on the covered surface. This phenomenon is called autoinhibition.[201] In some cases even several adsorption waves may occur.[213] An almost discontinuous increase in current for some isomeric dipyridylethylenes up to the value of the limiting diffusion-controlled current is accounted for by strong adsorption of the electroactive species.[201,213] The adsorption also influences the steepness of the polarographic wave and the shift of the wave with concentration.

Minima on polarographic curves, which are observed at high coverages, are connected with inhibition by a reactant, assuming that different areas are occupied by the reactant and activated complex, respectively. The derived relations (see Ershler[214] and references therein) are rather complicated and contain many experimentally inaccessible parameters for verification. An exhaustive review dealing with adsorption both in reversible and irreversible processes was published by Laviron[201] who presented (with his collaborators in a series of papers) many examples and derived equations, which are simple if no concentration polarization occurs. The theoretical derivation accounting for the adsorption of reactants and/or products obeying the Frumkin isotherm, with identical parameters for a totally irreversible wave, was given by Guidelli and Pezzatini,[215] expressing the shift of the wave with adsorption parameters. In this paper[215] the theoretical shapes of polarographic irreversible waves were also given, assuming Frumkin adsorption of reactant and the activity coefficients of the oxidized form and that of the activated complex being an exponential function of Γ_{Ox}.

Further developments in the field of adsorption in irreversible processes, which apply to the behavior of the majority of polarographically active substances especially in organic polarography, are to be expected in the near future.

9.2. Adsorption of Electroinactive Substances

Surface-active substances (inhibitors) added into solution even in small quantities might have some of the following effects on the polarographic wave: a successive lowering of the height of the wave or even the elimination of the wave, a shift of a reduction wave toward negative and of an oxidation wave toward positive potentials, often accompanied by a depression of the limiting current or a splitting of a single wave into two waves or a minimum in the limiting currents. A great deal of inhibition data and first attempts for their interpretation were given in a series of papers by Loshkarev and Kryukova.[216] The morphology of the inhibited wave can be treated in some cases by a simple theory, which will be briefly outlined.

First it is considered that the adsorption of surfactants (SAS) is very fast and hence diffusion controlled. The amount of adsorbed substance (M) adsorbed at the surface of the growing drop with area $A = 0.85m^{2/3}t^{2/3}$ is given by the equation[217]

$$M = 0.85m^{2/3}t^{2/3}\Gamma = \int_0^t D_A\left(\frac{\partial c_A}{\partial x}\right)_{x=0} A \, dt \tag{137}$$

where Γ is the surface excess at time t and $D_A(\partial c_A/\partial x)$. A is the diffusion flux under limiting current conditions given by Eq. (19) having the form $0.732c_A D_A^{1/2}m^{2/3}t^{1/6}$. After substitution and integration one obtains

$$\Gamma = 0.736c_A^0 D_A^{1/2}t^{1/2} \tag{138}$$

and for the maximum surface excess

$$\Gamma_m = 0.736c_A^0 D_A^{1/2}\nu^{1/2} \tag{139}$$

where ν is the time corresponding to the full coverage. The coverage is given by

$$\Theta = \frac{\Gamma}{\Gamma_m} = \frac{t^{1/2}}{\nu^{1/2}} \tag{140}$$

Further it is assumed as a first approximation that the charge-transfer rate constant in the presence of uncharged SAS, $k_{e,ef}$, is a linear function of the coverage[218]

$$k_{e,ef} = {}_0k_e(1 - \Theta) + {}_1k_e\Theta \tag{141}$$

where ${}_0k_e$ is the rate constant at the uncovered surface and ${}_1k_e$ on the covered surface ($\Theta = 1$), respectively. For ${}_1k_e < {}_0k_e$, a retardation (inhibition) occurs; in a reverse case, i.e., for ${}_1k_e > {}_0k_e$, an acceleration may occur.

The simplest derivation is for a unidirectional charge-transfer reaction without concentration polarization. For an uninhibited instantaneous current corresponding to an irreversible reduction, the following equation holds [see (64)]

$$I_{ir} = nFA_0k_ec_{Ox}^0 \tag{142}$$

where ${}_0k_e$ is the rate constant of the reduction depending on E according to Eq. (53). For current in the presence of surfactant, $I_{a,s}$ can be written as

$$I_{a,s} = nFAc_{Ox}^0[{}_0k_e(1 - \Theta) + {}_1k_e\Theta] \tag{143}$$

For the ratio of both currents, substituting for Θ from Eq. (140), one gets the expression

$$\frac{I_{a,s}}{I_{ir}} = 1 - \frac{{}_0k_e - {}_1k_e}{{}_0k_e}\left(\frac{t}{\nu}\right)^{1/2} \tag{144}$$

which is valid for $t \leq \nu$. An example corresponding to this equation is given in Figure 8 for ${}_1k_e = 0$. In the case of an acceleration, i.e., ${}_0k_e = 0$ holds for the foot of the wave, Eq. (143) after substituting for A and Θ gives

$$I_a = 0.85nFm^{2/3}{}_1k_ec_{Ox}^0\nu^{-1/2}t^{7/6} \tag{145}$$

The current–time curve with the exponent near 7/6 was found experimentally[219] (Figure 8b, curve 2).

If concentration polarization is to be considered, i.e., on the rising portion of the irreversible d-c wave, the corresponding differential equation must be solved with the boundary condition, including Eq. (141). The solution for the instantaneous current is available in a tabulated form.[220] The shape

Figure 8. Instantaneous currents in the presence of a surfactant for a slow charge-transfer reaction. (a) inhibition: 1) without surfactants, 2) +surfactant, ${}_1k_e = 0, A = 0$; 2') +surfactant, ${}_1k_e = 0, A > 0$; (b) acceleration 1) without surfactant, 2) acceleration ${}_1k_e > 0$, 3) acceleration and inhibition, $A < 0$, Eq. (150).

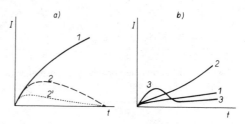

of instantaneous current depends strongly on the constant $_0k_e$ and for high $_0k_e$ values, i.e., at the potential of the limiting current, a sudden decrease of the instantaneous current is found falling either to zero (for $_1k_e = 0$, Figure 9(a)) or to the value of the current corresponding to the rate constant on the covered surface (Figure (9b)). The falling part of the inhibited I-t curves extrapolated to the time axis gives the values of ν. The values of ν for diffusion-controlled adsorption should decrease with the concentration according to Koryta's Eq. (139). The derivation of I-t curves based on an improved model for the adsorbed film was presented by Matsuda.[221] Several examples fulfilling the theoretical expressions were published[222-224] using strongly adsorbable surfactants such as Tritons and polyvinylalcohols as inhibitors. In some cases, by using adsorbable electroactive species such as Tl^+ or halide complexes, the dependence of $_1k_e$ on the concentration of the inhibitor (Figure (9c)) was observed.[224] This was explained recently[225] using a model of a covered surface. The rate constant on this surface was expressed according to transition state theory. The agreement between the derived expression and the experimental results seems to be quite satisfactory.

The mean currents and the shape of the inhibited irreversible wave can also be calculated.[226] The solution is rather simple for the case $_1k_e = 0$. Hence for the shift of the halfwave potential of an inhibited wave

$$\Delta E_{1/2} = \frac{RT}{\alpha nF} \ln 0.155 \sqrt{\frac{\nu}{t_d}} \tag{146}$$

The adsorption influence due to the presence of uncharged surfactants is called blocking or steric or even sieve effect.

Bulky surface active ions may also modify the potential profile in the double layer φ. This electrostatic effect may be expressed by the Frumkin equation [see Eq. (86)]. The value of the φ potential in the double layer depends on the coverage. As in the first approximation it was assumed[226] that the value of φ potential depends linearly on the degree of the coverage of the electrode

$$\varphi = \varphi_0 + \Delta\varphi\Theta \tag{147}$$

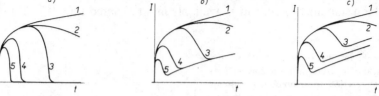

Figure 9. Instantaneous currents in the presence of a surfactant for a fast charge-transfer reaction. (a) $_1k_e = 0$, (b) $_1k_e > 0$, (c) $_1k_e > 0$ and decreasing with the concentration of surfactant. Typical concentrations of surfactant: 1) 0; 2) $5 \times 10^{-5}\,M$; 3) $10^{-4}\,M$; 4) $1.5 \times 10^{-4}\,M$; 5) $2 \times 10^{-4}\,M$, $t_d \approx 5\,s$.

where φ_0 is the value of the φ potential at zero coverage and φ_0 and $\Delta\varphi$ at complete coverage. Hence for $_0k_e$ at the coverage Θ

$$_0k'_e = {_0k_e} \exp(-A\Theta) \tag{148}$$

holds where $_0k_e$ is the rate constant at the uncovered surface, which was supposed to be equal during adsorption of uncharged surfactant. The quantity A is given by[226]

$$A = \left(\frac{z}{\alpha n} - 1\right)\frac{\alpha nF}{RT}\Delta\varphi \tag{149}$$

The instantaneous current on the foot of the wave can be expressed by

$$I'_{a,s} = I_{a,s} \exp(-A\Theta) \tag{150}$$

where $I_{a,s}$ is the current for the adsorption of uncharged particles given by Eq. (143). For positive values of A the inhibition increases; for negative values of A a simultaneous inhibition and acceleration may occur,[227] (Figure 8(b), curve 3). The relations were derived for instantaneous and mean current without concentration polarization.[226,227] The shapes of I-t curves for electrostatic acceleration for concentration polarization were computed,[228] assuming the linear dependence of φ potential on the coverage, but the steric hindrance at high coverages was not considered.

For fast charge transfer electrode processes (reversible d-c wave) the double-layer effects are not to be considered and the instantaneous current has the same shape independent of the potential; i.e., a sharp decrease of the current at the time of the full coverage to zero or to the value corresponding to $_1k_e$ providing that the coverage does not change in the given range with the potential.[229,230] In this case the ratio of the mean current in the presence of the surfactant ($\bar{I}_a$) to that in its absence ($\bar{I}_{rev}$) is potential independent provided that the electrode reaction occurs only on the free electrode surface (i.e., $_1k_e = 0$)

$$\frac{\bar{I}_{a,s}}{\bar{I}_{rev}} = \left(\frac{\nu}{t_d}\right)^{7/6} \tag{151}$$

and that there is neither a shift of the suppressed wave nor a change in the slope.[229,230] If $_1k_e$ is not zero and depends on the potential, the shape of the inhibited wave is complicated.[230] At higher concentration of surfactant and for $_1k_e > 0$, usually a drawn-out inhibited "irreversible" wave can be obtained and the $_1k_e$ can be calculated from the equation of the irreversible wave[225] [see Eq. (71)]

$$\log {_1k_e} = +\frac{\alpha_1 nF}{2.3RT}(E_{1/2,1} - E_0) + \log\frac{1}{0.886}\left(\frac{D}{t_d}\right)^{1/2} \tag{152}$$

where E_0 is the standard potential of the redox couple and $E_{1/2,1}$ the half-wave potential of the inhibited wave. The transfer coefficient at the fully covered surface is determined from the log-plot analysis of the inhibited wave.

The phenomenological theory outlined here cannot explain all the experimental results. Several deviations from a linear dependence of the rate constant upon coverage have been reported.[231,232]

An equation accounting for the blocking effect based on the Frumkin isotherm for the rate constant on the uncovered surface was derived by Parsons.[105] A modified equation by Ayabe[233] for an effective constant (for $\Theta > 0.5$) can be written as follows

$$k_{e,\text{ef}} = \Theta = {}_0k_e(1 - \Theta)^{1+r} \exp(-a\Theta) \tag{153}$$

where r (the number of solvent molecules replaced by the adsorption of one molecule of the surfactant) and a are the constants to be determined.

This equation seems to describe various results in the literature such as, e.g., the adsorption effect of butanol[232] on the reduction of Zn^{2+}, Cu^{2+} with $a = 0$ and $(1 + r)$ equal to 1.74 for zinc,[233] the inhibition of reduction of zinc by polyvinylalcohol with $a = 0$ and $(1 + r)$ equal to 1.4, or the effect of substituted ketones[234] on the reduction of persulfate with $(1 + r)$ equal to one and a equal to 3 or 9 depending on the ketone used.

In this short review no consideration about the structures of the films and the possible interactions of surfactant with electroactive species, and correlation between the adsorptivity and kinetic effects, can be given. The literature in this area is very rich and is not only pertinent to polarography but also to charge-transfer kinetics as a whole. A possible rehomogenization of a film near $t \approx v$ was indicated by Gierst.[235] From extensive studies of inhibition effects in organic polarography, Holleck and Kastening[236,237] concluded that for $\Theta \sim 1$, with the increase of the concentration of the surfactant a displacement of electroactive species by the molecules of inhibitors takes place. The dependence of the rate constant on the concentration of the inhibitor c_A^0 was found experimentally to be

$$k_{e,\text{ef}} \sim \frac{1}{(c_A^0)^r} \tag{154}$$

and was also derived theoretically. The value of r is given by the ratio of the available area for a molecule of the electroactive species and of the inhibitor at the electrode, respectively. For a recent extension of this idea see Reference (238). Holleck and Kastening, in a series of papers [see References (236), (237) and references therein] gave evidence that the film of the inhibitor stabilizes the radicals formed during the charge-transfer reaction. Two typical examples are as follows: the reduction of nitrocompounds[236,237]: RNO_2 in which the radical formed in a fast uptake of the first electron RNO_2^- appears to be stabilized by the film of triphenylphosphine. Some inhibitors (e.g., α-quinoline, triphenylphosphinoxide) lead to the stabilization of the superoxide ion O_2^- formed by the reduction of oxygen molecule after the uptake of the first electron.[239,240]

The modern trend in adsorption studies goes into a deeper understanding of inhibition based on treatments at a molecular level and attempts to establish some correlation and even more predictions for the inhibition effects [see e.g., References (224), (241) for an introduction].

10. Polarographic Maxima

On polarographic curves an increase of current above the plateau of the wave in the form of maxima of various shapes can often be observed. In some cases the origin of the current maximum is connected with the mechanism of the electrode process, especially if an adsorption step is involved. More frequently, however, the polarographic maxima are caused by additional hydrodynamic transport of the solution beyond that due to the diffusion. These so-called streaming maxima are divided into maxima of the first, second, and third kinds, according to their occurrence and properties. They represent a serious drawback in polarography since all theoretical relations forming the basis for the interpretation of polarographic results take into account convective diffusion only in mass transport considerations. Hence some of the most important features of the streaming maxima will be given here and the experimental conditions under which they can be eliminated will be stressed.

10.1. Maxima of the First Kind

Maxima of the first kind, known from the very beginning of polarography, occur in dilute solutions of supporting electrolytes, as shown in Figure 10. A more or less sharp peak formed on the polarographic wave is typical for these maxima. The highest maximum occurs usually at an equal concentration of the electroactive substance and that of the supporting electrolyte and decreases with the concentration of the latter. However, it may be found also at a 100-times excess of the supporting electrolyte; it is quite rare in 1 M solution. Maxima of the first kind can usually be suppressed by traces of surfactants, which are adsorbable in the corresponding potential range. Besides the role

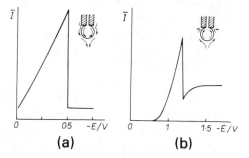

Figure 10. Maxima of the first kind and directions of streaming. (a) positive maxima $5 \times 10^{-4}\, M\, Cu^{2+}$ in $5 \times 10^{-3}\, M\, KClO_4$, (b) negative maxima $5 \times 10^{-4}\, M\, Zn^{2+}$ in $0.1\, N\, KCl$.

(a)   (b)

of conductivity, the polarity is another characteristic feature of these maxima. The maxima occurring at potentials more positive than the potential of the electrocapillary maximum are called positive and those observed on the negative side are called negative maxima. The polar character of the maxima follows also from the suppressing effect of added charged surfactants. The positive maxima are suppressed more easily with the negative charge or the adsorbability of anions or anionic surfactants. The negative maxima are more sensitive toward the valency of the cation of the supporting electrolyte and are more easily suppressed by surfactants of the cationic type. They differ somewhat in their shape. While positive maxima exhibit a relatively broad rising portion, they sometimes also have the form of a rounded peak and fall directly to the limiting current; negative maxima are on the average lower, more acute, and may decrease below the limiting current with a normal vertical capillary. However, most typical for the polarity of the maxima is the direction of the streaming of the solution[243,244] in the vicinity of the drop. In the case of positive maxima of cathodic processes, the solution streams from the neck to the bottom of the mercury drop into the bulk of the solution (Figure 10(a)). With negative maxima the streaming occurs in an opposite direction (Figure 10(b)). The velocity of the streaming usually lies between some tenths of cm or several cm per second.

The quantitative interpretation of all experimental observations is far from being complete. Nevertheless, the qualitative interpretation and some semiquantitative relations for the maxima of the first kind have been given in the theory of Frumkin and Levich.[43,245,246]

The tangential motion of the surface of the drop in dilute solutions is caused by potential differences at the surface arising from an unequal current distribution at the orifice of the capillary (at the neck) and at the bottom of the drop. Three effects may be responsible for this potential difference: the screening of the electrode by the tip of the capillary, unsymmetrical positions of the electrodes in a cell, and possibly an eccentric drop formation. The difference in the potentials on the surface causes a difference in the surface tension leading to a motion of the mercury because the surface contracts to the places of greater surface tension. The contraction sets in motion the surface layer of mercury and the adhering layers of solution. When the DME functions as a cathode, the potential at the bottom of the drop is somewhat more negative than that at the neck because of the greater current density at the bottom. For this reason, in positive maxima, the bottom of the drop possesses a greater surface tension than the neck because the maximum is on the ascending part of the electrocapillary curve. Hence the surface moves from the neck to the bottom and the solution streams in the same direction. The fresh solution transported to the neck shifts its potential towards more positive values while the partially exhausted solution flows to the bottom. Thus the potential difference increases and also the maximum increases. When the mean potential of the electrode reaches the potential of the electrocapillary maximum, the differences in surface tension are nearly equilibrated and the motion stops. When dealing with negative maxima, the potential at the bottom is due to the greatest current density again being more negative, but the drop's surface tension is smaller than at the neck and the streaming has an opposite direction than before. Since in this case, the fresh solution is transported to the bottom, then when the Nernst equation applies, the potential is less negative and the potential differences on the drop surfaces decrease. This explains why, as a rule, negative maxima are lower than the positive ones. When concentration polarization sets in, the concentration of the electroactive species on the whole surface falls to zero, the surface tension becomes uniform over the whole drop, the motion of the surface stops, and the transport drops to the value controlled by diffusion alone. For a somewhat different explanation of negative maxima see References (247), (248).

The quantitative theory of Frumkin and Levich based on the model of a mercury drop falling through a solution can explain certain principal features of the maxima, such as the influence of the conductivity or of the surface tension, in a relatively satisfactory way [see References (9), (245), (246)]. However, the fact that some electroactive substances have a great tendency to form the maxima while other substances at the same potential are reduced or oxidized without a streaming maximum, is still not fully understood.

Another explanation of the origin of the maxima was given by Heyrovský.[9] He explains the streaming of electrolyte in terms of the non-homogeneity of the electric field in the neighborhood of the DME, which is greatest at the orifice.[9] Besides the theoretical treatment of the electric field around the DME screened by the capillary,[249] the quantitative or semiquantitative relations based on this concept are not available.

In order to eliminate the occurrence of maxima it is in many cases sufficient to increase the concentration of the supporting electrolyte up to $1 M$ or to add a small amount of an appropriate surfactant. In the latter case very small concentrations should be used in order to avoid an inhibiting effect on the polarographic wave, as described previously. Because of the high sensitivity to surfactants the gradual decrease of the height of the maximum has been used for the determination of very low concentrations of surfactants.

10.2. Maxima of the Second Kind

These maxima, discovered and in essence explained by Kryukova,[250,251] occur usually in concentrated solutions above $1 M$. They appear on the limiting current and may, depending on the purity of the solution, exceed it approximately five times. They either have the form of a direct increase on the original wave or they form a new additional wave; in extremely pure solutions they have the form of a maximum falling not abruptly to the value of the diffusion limiting current (Figure 11). They are accompanied by one direction of stream-ing of the solution from the bottom to the neck of the drop, and can be easily suppressed by small traces of surfactants. As compared with maxima of the first kind, maxima of the second kind have a purely hydrodynamic origin and are not related to the electrode process. At speeds of flow above 2 cm/s, which correspond to an outflow of mercury $m > 2$ mg/s, the mercury streams directly to the bottom of the drop and so causes a motion of the surface upwards to the top (Figure 11). By this motion the adhering layers of solutions are carried along and the stirring action starts, which transports larger amounts of elec-troactive species to the electrode and this causes an increase of current.

For the velocity of motion v it has been derived[43,252] that

$$v = \frac{1}{3} \frac{(d' - d)gr^2}{2\eta + 3\eta' + g/\mathcal{H}}$$

(155)

where d' and d are the densities of the mercury and of the solution, respectively; g is the gravitational constant, r is the radius of the drop, η is the viscosity of the solution, η' is the viscosity of mercury, q is the charge density, and $\mathcal{H}$ is the specific conductivity of the solution. The maximum streaming occurs at the potential of zero charge ($q = 0$) and decreases with decrease of the conductivity $\mathcal{H}$. The experimental results are in a good agreement with predictions of this equation.[253]

The maxima of the second kind can be easily avoided by choosing a capillary with a small outflow velocity $m < 2$ mg/s (the best being around 1 mg/s).

10.3. Maxima of the Third Kind

Polarographic maxima of the third kind, described[254] in 1959 and studied by Frumkin *et al.* [cf. References (255), (256) and references therein], occur in contrast to the previous ones in the presence of organic substances forming two-dimensional condensed layers on the mercury surface (e.g., cam-phor, borneol, adamantranol, tribenzylamine, etc.). These maxima appear not only at the potential corresponding to the adsorption peak (transition potential)

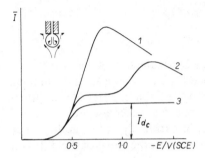

Figure 11. Maximum of the second kind and the direction of streaming. 10^{-3} M CdCl$_2$ in 3 M NaCl, $m = 3.5$ mg/s. 1) purified solution, 2) solution of normal purity, 3) maximum suppressed by addition of the surfactant (0.01% gelatine).

on the curves of differential capacity vs. potential but they may be observed at any potential within the adsorption range of the surfactant for a coverage $\Theta \simeq 0.3$–0.5 (Figure 12). It is established that the velocity of the drop surface is proportional to the quotient $\partial \Gamma / \partial c$.

A very thorough review on all streaming maxima was recently published by Bauer.[257]

11. Techniques Developed from D-C Polarography

The further development of polarography was focused predominantly on increase of sensitivity and on the improvement of resolution for simultaneous determination of several reactants with halfwave potentials close to one another. Another trend was to prove a simple check of reversibility or stability of the electrode reaction product.

One of the early attempts to increase sensitivity was by recording the d-c current only in the final phase of the drop[258] at which the charging current attains its minimum value (tast-polarography). The gain in sensitivity was not very great. The derivative circuitry in d-c polarography improved the resolution factor but lowered the sensitivity.

The first attempt to have a complementary check of reversibility in d-c measurements was made in 1941 by Heyrovský.[259] Basically, it is a galvanostatic method using alternating current at constant amplitude of approximately 0.3 mA. A d-c bias current is added to keep the potential of the DME in an appropriate potential range. Since the corresponding response was read on the screen of an oscilloscope, the term oscillographic polarography at controlled alternating current has been used. In its first version, the dependence of potential vs. time was measured. The relative position of the time delay on cathodic and anodic parts of the curve was taken as a measure of reversibility of the given redox process. Circuitries for other dependencies, e.g., dE/dt vs. t and dE/dt vs. E have also been proposed. The last version found analytical

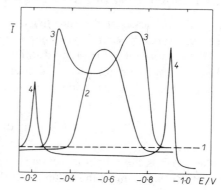

Figure 12. Maxima of the third kind. 1) 10^{-3} M $AgNO_3$ in 1 M Na_2SO_4; conc. of 2-oxo-adamantane. 2) 4×10^{-5} M, 3) 8×10^{-5} M, 4) 3.3×10^{-4} M.

applications predominantly for organic substances. The theoretical background and applications can be found in References (259–261).

For the test of reversibility Kalousek[262] proposed a very simple and effective circuitry known now as Kalousek's commutator (switch). In the simplest of its versions the potential is periodically switched from an auxiliary potential, e.g., from the potential at the limiting current of a cathodic wave, to a more positive potential than $E_{1/2}$, which is scanned in the negative direction. In one of the halfcycles the reduced form is produced, which can be reoxidized in the recording period of the halfcycle. If the charge-transfer reaction is fast and the product is relatively stable a composite reversible anodic-cathodic wave results. The magnitude of the commutated current can be used as a measure of stability of the product or of the intermediate. For irreversible reactions the anodic wave appears at more positive potentials or is absent for very high overvoltage. With modified circuitry the commutated anodic wave may have the form of a mirror image of the cathodic wave. In these procedures a square-wave voltage with high and increasing amplitude at frequencies usually 25 Hz is superimposed on the ramp; however, in contrast with later square-wave techniques the mean direct current is measured. A modification with a constant amplitude was also proposed. The resulting commutated wave then has the form of a peak. A serious problem with Kalousek's commutator is the high charging current which distorts the waves. By its suitable elimination[263] the sensitivity of the method may be increased. For other modifications, outline of the theory, and further details see Reference (9).

A substantial increase in sensitivity was achieved only in later periodic voltage perturbation techniques with lower amplitudes superimposed on the increasing d-c potential in which principally the perturbation of the faradaic current can be treated as first order with respect to the amplitude of voltage perturbation. Three of these techniques, which are also included in commercial instruments, will be treated in some detail below (Figure 13).

11.1. Square-Wave Polarography

The increase of the sensitivity in this technique is due to the practical elimination of the charging current. For the square-wave method, at the instant of the abrupt change in potential at the beginning of each halfcycle the electrode behaves approximately as a condenser of capacity C in series with a resistance R. The charging current in one halfcycle follows the equation

$$I_{DL} = \frac{\Delta E}{R} \exp\left(-\frac{t_c}{RC}\right) \tag{156}$$

where ΔE is the amplitude of the square wave and t_c is the time measured from the beginning of the halfcycle. If the halfcycle of the square-wave voltage

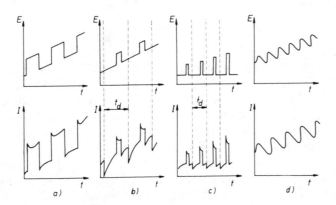

Figure 13. Other polarographic methods: a) square wave, b) pulsed, constant amplitude, c) pulsed, increasing amplitude, d) sine wave.

exceeds several times the value of the product RC the charging current falls to a negligible value. The diffusion-controlled current following the formula $I = kt^{-1/2}$ decreases relatively much less during the same period.

In square-wave polarography[20,264–266] the d-c ramp is square-wave modulated usually at 225 Hz over the entire life of each mercury drop (Figure 13(a)) but the currents are recorded only during a short part of the droplife, e.g., for a 30 ms period of drop formation. Over this short interval the growing mercury drop behaves as an electrode of constant area to a good approximation. The electronic circuitry used in this technique measures only the difference in current between the positive and negative halfcycles of the square wave during a 0.1–0.2 ms period at the end of each halfcycle.

Since this technique is essentially a derivative method, a maximum in the square-wave current occurs at the half-wave potential for reversible waves. According to Barker[20] the cathodic square-wave current-potential dependence is described by the equation

$$I_{sw} = \frac{n^2 F^2}{RT} c^0_{Ox} A \Delta E \left(\frac{D_{Ox}}{\pi t_s}\right)^{1/2} \frac{\lambda'}{(1 + \lambda')^2} L \qquad (157)$$

where L can be considered as a constant for a given apparatus for which it was derived

$$L = \sum_{m=0}^{\infty} (-1)^m \frac{1}{\left(m + \dfrac{t'}{t_s}\right)^{1/2}} \qquad (158)$$

Here $\lambda' = \exp\left(E - E'_{1/2}\right) nF/RT$ [see Eq. (40)], t' is the time elapsed from the beginning of the last halfcycle, t_s is the half period of the square-wave voltage, and ΔE is the peak to peak amplitude of the square-wave voltage.

For $E = E'_{1/2}$ the value of λ' equals 1 and the current reaches its maximum. For small ΔE the width of the peak at half of its maximum height is

$$S = 3.52 \frac{RT}{nF} = \frac{90.5}{n} \text{ mV at } 25°C \tag{159}$$

For reversible processes, square-wave polarography is far more sensitive than classical polarography with a detection limit of $5 \times 10^{-8} M$. Barker[20] and Matsuda[267] have developed equations for participation both of charge-transfer and diffusion control. From these equations it follows that the maximum peak decreases with the diminution of the rate constant k_e^0, most of the decrease being in the range of $k_e^0 = 10^{-1}$–10^{-4} cm/s. For $k_e^0 < 10^{-4}$ cm/s, the peak current is almost independent of k_e^0 and has a value of ~1/20 of the reversible peak height. For an application in determining the charge-transfer rate constant of Zn(II) in various supporting electrolytes, see Reference (268).

11.2. Pulse Polarography

Pulse polarography, which was also introduced by Barker,[266,269] has met with increased interest both in the development of theory and in practical applications in the last decade. The relatively simple circuitry and very high sensitivity attainable, especially in its differential mode enabling the determination of concentrations of the order of $10^{-8} M$, led to a revival of polarography in analytical chemistry.

In this technique a voltage pulse of short duration (e.g., 100 ms), is applied to the DME at a certain time t_1 before the drop falls off. The voltage pulse ΔE_p may be of increasing amplitude with a constant d-c bias potential E_p (usually chosen so that no faradaic reaction occurs), i.e., $E_p + \Delta E_p$ represents the increase of the polarization potential of the DME (Figure 13(c)) or a pulse of constant amplitude ΔE (usually 30 mV) is superimposed on a linearly increasing d-c ramp (Figure 13(b)). In both types of repetitive voltage pulses the current response is measured for a period of several milliseconds (20–40 ms) toward the end of the period of the voltage pulse when the charging current has decayed to a negligible value.

The d-c currents slowly changing with time are filtered off by high-pass filters and only the amplified alternating component associated with the pulse modulation is recorded. When the voltage pulses of increasing amplitude are applied the current-potential response has the shape of conventional d-c polarograms (normal pulse polarography).

The constant amplitude voltage pulse mode yields polarograms that are similar to the derivative of d-c wave (derivative or differential pulse polarography).

The diffusion-controlled current in normal pulse polarography is given in the first approximation by the solution of equations for diffusion toward a planar electrode,[270] viz.

$$I_d = nFAD^{1/2} \frac{c^0}{\sqrt{\pi t_p}} \tag{160}$$

Here t_p is time of voltage application. The area with respect to Eq. (5) is given by

$$A = 0.85 m^{2/3}(t_p + t_1)^{2/3} \tag{161}$$

where t_1 is time of drop growth prior to voltage application. If one denotes $\tau = t_p/t_1$ one can write

$$I_d = 0.48 nFm^{2/3} \frac{t_1^{1/6}}{\tau^{1/2}}(1 + \tau)^{2/3} \tag{162}$$

The influence of sphericity, expansion of the drop, and effects of shielding by the tip of the capillary for the limiting diffusion current were accounted for in a series of papers.[271–274]

The equation of a cathodic current potential curve with diffusion and charge-transfer control in normal pulse polarography, assuming constant area of the drop during the pulse duration, is obtained by application of the well-known equation[2,3] for a planar electrode. If instead of E_0 the value $E'_{1/2}$ is introduced, this equation has the form[275]

$$I = nFAc^0_{\text{Ox}} k^0_e \frac{D_{\text{Ox}}^{1/2}}{D_{\text{Red}}^{1/2}} \exp\left[-\frac{\alpha nF}{RT}(E - E'_{1/2})\right] \exp(Q^2 t)\, \text{erfc}(Qt^{1/2}) \tag{163}$$

where

$$Q = \frac{k^0_e}{D_{\text{Red}}^{\alpha/2} D_{\text{Ox}}^{(1-\alpha)/2}} \exp\left[-\frac{\alpha nF}{RT}(E - E'_{1/2})\right]\left\{1 + \exp\frac{nF}{RT}(E - E'_{1/2})\right\} \tag{164}$$

For high values of k^0_e or long times $\exp(Q^2 t)\, \text{erfc}(Qt^{1/2}) \approx 1/Q(\pi t)^{1/2}$ and the equation expressing the reversible wave is easily obtained. For a finite value of k^0_e Eq. (163) can be transformed[275] into forms from which α and k^0_e can be determined. The fact that the time of measurement is much shorter than in d-c polarography provides the opportunity for measurement of higher charge-transfer rate constants. For a more rigorous approach see Reference (276).

Equations for current coupled with chemical reactions preceding or paralleling the charge-transfer steps were derived[277–279] and various types of chemical reactions studied.[279,280]

In derivative pulse polarography the difference in currents resulting from application of a potential step having an amplitude ΔE is recorded as the function of the potential. The equation of the corresponding polarogram for the reversible process can be obtained very easily. If the equation for a reversible cathodic wave is written in the form[270]

$$I = I_d \left(\frac{1}{1 + \lambda'} \right) \tag{165}$$

where λ' is given by Eq. (40), and is differentiated, then for the change in current ΔI resulting when a potential pulse ΔE is applied (ΔE smaller than RT/nF) one obtains

$$\Delta I_p = \frac{nF}{RT} I_d \Delta E \frac{\lambda'}{(1 + \lambda')^2} \tag{166}$$

The maximum ΔI_{max} is obtained for $\lambda' = 1$. When inserting for I_d from Eq. (160) one obtains

$$\Delta I_{p,max} = \frac{n^2 F^2}{4RT} A c^0 \Delta E \sqrt{\frac{D}{\pi t_p}} \tag{167}$$

Hence a symmetrical peak should be obtained with a maximum near $E_{1/2}^r$.

For analytical purposes pulses of longer amplitude are often applied, leading to an increase of the peak height and hence to higher sensitivity. The term differential pulse polarography is usually used. To obtain the corresponding expression for longer pulses one starts again with Eq. (165). The value of current at the potential E is subtracted from the value of current after the pulse application at the potential E_2, $E_2 = E_1 + \Delta E$. Again the ΔI as the function of the potential can be derived. For its maximum value $\Delta I_{p,max}$ the relation

$$I_{p,max} = nFAc^0 \sqrt{\frac{D}{\pi t_p}} \left(\frac{\lambda^* - 1}{\lambda^* + 1} \right) \tag{167a}$$

can be finally obtained where $\lambda^* = \exp{(E_2 - E_1/2)nF/RT}$.

If $(E_2 - E_1)/2$ is smaller than RT/nF, the exponential can be expanded and the equation simplified to Eq. (167). It can be shown that the peak height increases with amplitude of the pulse.[270] For some further considerations in improvement in sensitivity of differential pulse polarography or other modifications in pulse polarographic technique in general, see e.g., References (281–291). An integrated readout of the current (integrated pulse polarography[292,293]) was also proposed. A rigorous theory of derivative pulse polarography, considering also the charge transfer, was presented by Ružić.[276]

The effect of reactant adsorption leading to an increase in the peak height in the differential mode[294] and modifying normal pulse polarographic waves has also been treated theoretically.[295]

11.3. A-C Polarography

In comparison with the previous two techniques for which the given references represent a large fraction of published papers, the literature for a-c polarography is vast, and has been summarized in a book[296] and two exhaustive reviews.[297,298] This is due to a high level of theoretical development in charge-transfer kinetics, in coupled chemical reactions, and in the treatment of the role of adsorption of reactants and/or products. For the latter case a-c polarography is highly sensitive. The methodology attained a high standard using modern progressive techniques such as automatic recording of several variables or the readout connected to modern data-processing equipment or, e.g., Laplace transform into the frequency domain from the time domain of current response after the voltage application. On-line fast Fourier transform faradaic admittance measurements are in progress [(299) and references therein].

A-c polarography, although historically older, was introduced effectively by Breyer and Gutmann in 1946.[300,301] In a-c polarography the usual procedure is to measure the amplitude of the alternating current produced by a sinusoidal potential of constant amplitude superimposed on linearly increasing d-c voltage (Figure 13(d)). The amplitude of the alternating current is inversely proportional to the absolute magnitude of the complex impedance of the electrode. If phase angle can be measured, the calculation of real and imaginary components of the electrode impedance is possible. When the charging current is small, the a-c amplitude corresponds to the derivative of the d-c wave and for a reversible wave the alternating current exhibits a maximum at the half-wave potential.

The double-layer charging and ohmic losses limit the application of fundamental harmonic a-c polarography in both analysis (the sensitivity is comparable with a-c polarography) and kinetic investigations. A three-electrode system and positive feedback compensation minimize considerably the influence of ohmic resistance. Limitations imposed by double-layer charging are circumvented considerably in phase-selective a-c polarography based on the phase difference between the faradaic and charging currents. The best selectivity would be achieved if ohmic losses did not exist, since charging current would then be exactly 90° out of phase with the applied potential. This leads not only to a considerable increase in sensitivity to the limit of approximately $10^{-6}\,M$ for analytical purposes but also to more precise and convenient kinetic measurements. A very convenient phase-sensitive a-c polarograph was proposed by de Levie and Husovsky.[302] Commercial phase-sensitive instruments are already available.

Since the theoretical treatment for a-c polarography is rather involved, some equations will be derived in a simplified form relying on the more general knowledge of faradaic impedance at equilibrium potential. If the equilibrium potential is perturbed by an a-c signal of small amplitude, a simple relation for diffusion and charge-transfer control can be obtained[2,3,298] as

$$\Delta E = R_s I_m^0 \sin \omega t + \frac{1}{\omega c_s} i_m^0 \cos \omega t \tag{168}$$

where

$$R_s = \frac{RT}{n^2F^2}\left\{\frac{1}{k_e^0 c_{Ox}^{\alpha} c_{Red}^{1-\alpha}} + \frac{1}{2^{1/2}\omega^{1/2}(c_{Ox}^0 D_{Ox}^{1/2} + c_{Red}^0 D_{Red}^{1/2})}\right\}$$

$$= R_{ct} + R_W = \Theta + \frac{\sigma}{\omega^{1/2}} \tag{169}$$

and

$$\frac{1}{\omega C_s} = \frac{RT}{n^2F^2}\frac{1}{2^{1/2}\omega^{1/2}(c_{Ox}^0 D_{Ox}^{1/2} + c_{Red}^0 D_{Red}^{1/2})} = \frac{1}{\omega C_W} = \frac{\sigma}{\omega^{1/2}} \tag{169a}$$

where R_{ct} or Θ are symbols for the frequency independent charge-transfer resistance, R_W and C_W are the resistance and capacitance components of the Warburg impedance per unit area, and i_m^0 is the amplitude of the faradaic component of the current density. The symbols Θ and σ are most frequently used. Since faradaic current is involved the term faradaic impedance is used. The total faradaic impedance in the case of charge transfer is characterized by real Z_f' and imaginary Z_f'' components

$$Z_f' = \Theta + \sigma\omega^{-1/2}, \qquad Z_f'' = \sigma\omega^{-1/2} \tag{170}$$

This behavior of the electrode/solution interface can be represented by the well-known Randles–Ershler equivalent circuit. For the analysis in evaluating kinetic data, see References (2), (3), (298). More complicated equivalent circuits are to be considered for complicated reaction schemes (with coupled chemical reactions or with coupled adsorption).

Based on previous relations the equation of a reversible a-c polarographic wave can be obtained easily. In this case $\Theta = 0$ and the total impedance

$$Z_f = \left(R_s^2 + \frac{1}{\omega^2 C_s^2}\right)^{1/2} \tag{171}$$

for pure diffusion control is

$$|Z_f| = \frac{2^{1/2}\sigma}{\omega^{1/2}} \tag{171a}$$

The amplitude of the a-c current can be expressed by

$$\Delta I_s = A\frac{\Delta E\omega^{1/2}}{2^{1/2}\sigma} \tag{172}$$

However, the concentrations of oxidized and reduced forms are no longer bulk concentrations but concentrations at the electrode at the given d-c potential, i.e., $c_{Ox}(0, t)$ and $c_{Red}(0, t)$. For Nernstian behavior they can be calculated from the equation's for a reversible d-c cathodic wave. From Eqs.

(35) and (39b) it follows[12] that

$$c_{Ox}(0, t) = c_{Ox}^0 \frac{I_d - I}{I_d} = c_{Ox}^0 \frac{\lambda'}{1 + \lambda'} \tag{173}$$

and

$$c_{Red}(0, t) = c_{Ox}^0 \left(\frac{D_{Ox}}{D_{Red}}\right)^{1/2} \frac{I}{I_d} = c_{Ox}^0 \left(\frac{D_{Ox}}{D_{Red}}\right)^{1/2} \frac{1}{1 + \lambda'} \tag{173a}$$

When inserting these concentrations in σ one obtains

$$\Delta I_s = \frac{n^2 F^2 c_{Ox}^0 \Delta E (D_{Ox}\omega)^{1/2} A}{RT} \frac{\lambda'}{(1 + \lambda')^2} \tag{174}$$

It can be easily recognized that the term $\lambda'/(1 + \lambda')^2$ is the absolute value of the derivative dI/dE for a polarographic reversible wave. Hence the a-c response has the form of a peak with its maximum at $E_{1/2}$, i.e., for $\lambda' = 1$

$$I_{s,m} = \frac{n^2 F^2 c_{Ox}^0 E (D_{Ox}\omega)^{1/2} A}{4RT} \tag{174a}$$

For more exact derivation see the reviews in References 297 and 298.

The derivation of the equation of the a-c wave, when charge-transfer control is also operative, is more involved.[267,296,297,304] From the derived equations it follows that the maximum peak height decreases with the rate constant similarly as indicated for square-wave polarography, and that it depends also on the value of α. Therefore the determination of the rate constant from the magnitude of the a-c current is very complicated. For electrode reactions that are not too slow ("quasireversible" wave), the rate constant can be determined from the measurement of the phase angle ϕ from the equation (valid for $E = E_{1/2}$)[296,297]

$$ctg\phi = 1 + \frac{D(\omega)^{1/2}}{k_e^0 2^{1/2}} \tag{175}$$

where $D = D_{Ox}^{1-\alpha} D_{Red}$. The transfer coefficient α can be obtained from the determination of the $ctg\varphi - \omega^{1/2}$ dependence at two different potentials.

For a totally irreversible a-c wave a simple equation for the magnitude of the maximum peak height can be obtained[305,306] as

$$\Delta I_{s,m}^{ir} = \left(\frac{7}{3\pi t}\right)^{1/2} \frac{\alpha n^2 F^2 A c_{Ox}^0 D_{Ox}^{1/2} \Delta E}{RT} \tag{176}$$

For a more precise derivation see Reference (306). The curves also have the form of a peak, but the peak potential $E_{s,m}$ differs from the reversible half-wave potential $E_{1/2}^r$. For a reduction a-c peak[306] the relation is

$$E_{s,m} = E_{1/2}^r + \frac{RT}{\alpha nF} \ln \frac{1.35 k_e^0 t^{1/2}}{D_{Ox}^{1/2}} - \frac{RT}{2\alpha nF} \ln 1.907 \omega^{1/2} t^{1/2} \tag{177}$$

Besides various other approaches[2,3,12,298,303,304] for determining the parameters of the electrode processes, the procedure known in classical impedance measurements as complex plane analysis is very advantageous.[298] Various graphical representations of the electrode admittance (reciprocal value of the impedance), very sensitive for the detection of adsorption coupled reactions, are recommended.[307,308]

A very elegant and simple procedure for the determination of the rate constant, without making any *a priori* assumption as to the dependence of the rate on potential, is suggested in References (307) and (308). The following very simple formula holds

$$\xi = \frac{Y_f'}{Y_f' - Y_f''} = \frac{k_{red}}{(2\omega D_{Red})^{1/2}} + \frac{k_{Ox}}{(2\omega D_{Ox})^{1/2}} \tag{178}$$

where Y_f' is the real (in-phase) and Y_f'' the imaginary (out-of-phase) component of the faradaic admittance contribution. The admittance components can be easily calculated from direct readouts in phase-sensitive a-c polarography (Figure 14). From the dependence of the logarithms of the corresponding rate constants on the applied potential the standard rate constant can be obtained by extrapolation.

The a-c polarography technique in its advanced forms, e.g., phase-sensitive polarography, is very suitable for determination of standard rate constants, surpassing d-c polarography by almost two orders of magnitude; a standard rate constant of approximately 1 cm/s can be determined. In comparison with

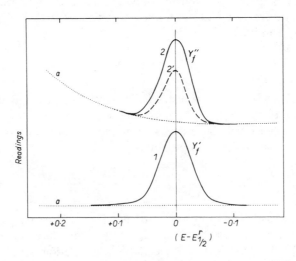

Figure 14. In-phase (Y_f') and out-of-phase (Y_f'') components in phase-selective a-c polarography. (a) Supporting electrolyte 1,2 reversible redox couple at given frequency, (b) 1,2' redox couple with finite charge-transfer rate constant at a given frequency.

other techniques the a-c impedance method is very sensitive for detection of the adsorption of reactants or products.

The problem of a coupled reaction, either heterogeneous or homogeneous, in impedance measurements at equilibrium was treated in 1951 by Gerischer.[309,310] For the a-c technique this problem has been systematically followed by Smith et al. In their review,[297] equations are given for most of the reaction schemes treated previously in d-c polarography. More complicated reaction schemes are being currently solved in some recent papers, usually for both d-c and a-c polarography, using rigorous treatment [see References (311), (312) and references therein]. Although higher rate constants than in d-c polarography can be measured, the analysis of results is more complicated.

In charge-transfer kinetic studies the deviation from the classical Randles equivalent circuit (the anomalous phase angle exceeding the predicted value of 45° for a simple charge transfer) was related from the very beginning to adsorption effects and modified equivalent circuits were proposed.[313–317] The main contribution was made by Delahay[36,37] in 1966, valid for all non-steady-state techniques, who argued that the commonly used a priori separation of the faradaic and charging current is not correct. This concept, often called the coupling effect, modifies the fundamental equations for faradaic current density and fluxes of Ox and Red, which contain $\partial\Gamma/\partial t$ terms for the participants in the charge-transfer reaction. The mathematical solution presented[318] leads to very complicated expressions for the electrode impedances, which at present can be hardly used for analysis of data. The expression for electrode admittance based on Delahay's concept assuming infinite rate of adsorption and very fast charge transfer,[319,320] is relatively simple and very useful for the detection of weak or strong adsorption [for details see Reference (298)].

The theory of faradaic admittance based on Delahay's concept and the coupling between interfacial and transport impedances were presented for the charge-transfer reaction with adsorption of both Ox and Red without making any a priori assumptions regarding the dependences for the rate constant on potential.[321] The expressions are rather too complicated to be verified. For some simpler cases simulation curves based on equations in Reference (321) are available.[322]

The amplitude of the a-c current is proportional to the differential capacity and hence a-c polarography has been used from its very beginning for the determination of electrochemically inactive surfactants. The term "tensammetry" was used at that time for such a type of measurement. The theoretical background with a practically exhaustive survey of obtained results and applications is given in the monograph by Jehring.[323]

A-C alternating potential can be superimposed upon the d-c triangular wave form with a ramp or staircase to a stationary electrode (e.g., hanging mercury drop) and the alternating current admittance is measured as a function of d-c potential. For theoretical background see Reference (324); modern measurements use an on-line fast Fourier transform technique.[325]

11.4. Nonlinear Polarographic Techniques

Since the faradaic current–potential characteristic is nonlinear, when a periodic perturbation signal is applied to the working electrode rectification effects and generation of harmonics may result. When two or more frequency components are present in the signal, intermodulation effects also occur. The double-layer capacitance shows, on the contrary, an approximately linear dependence and therefore the effect of double-layer charging in the total response is substantially lowered, leading to the increase of the sensitivity of these so-called nonlinear or second-order techniques in electrochemical and analytical studies.

The theory and methodology of these techniques usually applied at the equilibrium potential are beyond the scope of this chapter and are thoroughly described in several review articles.[2,3,303] Some of these techniques combined with d-c polarization will be only briefly mentioned here [for reviews see References (326–328)]. The terminology of these techniques seems not to be fully established yet. The method based on the faradaic rectification effect has been called by Barker[20,329,330] radio-frequency polarography. In his version a small amplitude (10 mV) rf signal (100 kHz to 6.4 MHz) is superimposed on the d-c ramp and modulated at low frequency, e.g., 225 Hz. The low-frequency a-c current produced by faradaic rectification at constant mean potential is measured. The current–potential response (faradaic rectification wave) has the form of the second derivative of d-c wave (Figure 15). This technique can be realized by a combination of a square-wave polarograph with further accessories (high-frequency generator and low-pass filter). This approach is also called low-level faradaic rectification polarography.

In high-level rectification polarography (HLFR) high amplitudes (up to 1.5 V) are applied.[327,328,331] The plot of the rectified current vs. mean potential has essentially the form of a d-c wave. Since the measurement of the rectified current is carried out in extremely short times, an HLFR polarogram would correspond to a d-c wave with hypothetical droptimes of 10–1 s. Therefore this technique is very suitable for fast kinetic measurements both in charge-transfer and in coupled chemical reactions.

Although the potentialities of techniques based on the rectification effect are obvious, the results reported thus far are limited[327,328] because of rather involved instrumentation.

In second-harmonic a-c polarography [see References (332–335)] a low-impedance alternating voltage source of extremely low second-harmonic content (0.05%) is required. The second harmonic component is then usually measured with a conventional second harmonic analyzer.[334,335] The second-harmonic polarography readout shows normally two peaks dependent on the frequency, the rate constant, and transfer coefficient α. With phase selective detection the shape qualitatively corresponds to the second derivative of a d-c wave (Figure 16). Second-harmonic polarography has been used for both

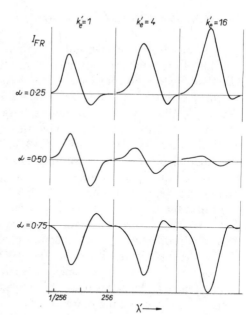

Figure 15. Influence of the frequency ω and of the transfer coefficient α on r.f. polarograms. $k'_e = (\omega D)^{1/2}/k_{1/2}$, $k_{1/2}$ is the rate constant at $E_{1/2}$ of the d-c wave, λ' is given by Eq. (40).

kinetic studies[297,336–339] and analytical applications.[340,341] Concentrations of the order 10^{-6}–10^{-7} M can be determined. The second and third harmonics for reversible behavior were recently reconsidered.[342,343]

Intermodulation polarography,[344–347] using periodic voltages of differing frequencies for polarization of the DME and measuring the a-c component arising from the intermodulation, because of nonlinearity of the electrode reaction, is not yet frequently applied. Intermodulation square-wave polarography has also been developed.[348]

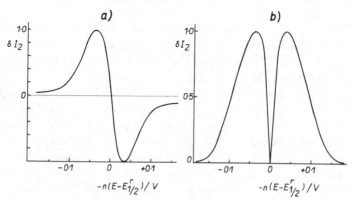

Figure 16. Second harmonic a-c polarography for a reversible reaction. a) phase-sensitive readout, b) nonphase-sensitive readout.

12. Applications of Polarography

As seen from the material presented above, polarography can provide valuable electrochemical data and information. Since the results obtained with a DME are extremely reproducible, the electrochemical investigation of a given system often starts using polarographic techniques and is then extended to solid electrodes. The number of substances, both inorganic and organic, which have been subjected to polarographic investigation is so vast [for basic information see References (8), (10), (13–15), (70), (192), (349–351)], that only tables of fundamental polarographic parameters of these substances (e.g., half-wave potentials, number of electrons, diffusion coefficients, etc.) represent extensive monographs [e.g., References (352–356)].

From the extensive amount of accumulated data, trends were observed that established some correlations between the polarographic behavior and the redox properties and the structure of the compounds investigated.

In organic polarography in the 1930's, Heyrovský[357] deduced that polarographic reduction becomes easier as the number of conjugated bonds in the organic molecule increases. The "electronegativity rule" of Shikata and Tachi,[358] stating that the more negative the substituent the more positive the half-wave potential, has stimulated the later more quantitative correlations between polarographic data and substituent effects in organic polarography. For aromatic compounds substituted in the para- and meta-positions with respect to the reacting group, the Hammett equation applied in the form

$$E_{1/2} - E_{1/2}^0 = \varphi\sigma \tag{179}$$

has been checked in many series of compounds. Superscript o denotes the half-wave potential of the nonsubstituted reference substance, σ is a constant characterizing the substituent, and φ is the reaction coefficient. For aliphatic compounds the analogous Taft equation has been adopted. Exhaustive monographs by Zuman[359,360] dealing with structural and substituent effects in polarography are available.

In the fifties papers started to appear in which relations between the half-wave potentials and the extent of conjugation were treated in terms of quantum mechanics.[361–366] For a very illustrative review see Reference (367). The quantum–mechanical approach in organic polarography or more generally in organic electrochemistry is still being applied even with complicated molecules.[368–370]

At about the same time the pioneering work in establishing basic rules between the redox reactivity and the structure of coordination compounds was carried out by Vlček.[371–375] Only a few can be mentioned here.

Reversible waves are usually obtained with substitution inert complexes having identical atomic configuration in both the oxidized and reduced form, and the electronic configurations differ only in the single electron, which is accepted into the vacant orbital of the oxidized form (e.g., $Fe(CN)_6^{3-}$ +

$e \rightleftharpoons Fe(CN)_6^{4-}$). If the preceding condition for the identity of atomic configurations is not satisfied, the charge-transfer reaction is diffusion controlled (reversible) when the substitution reaction is very fast (e.g., $Fe(Oxal)_3^- + e \rightleftharpoons Fe(Oxal)_2^{2-} + Oxal$). If the substitution reaction is slow the system is irreversible (e.g., $Cr(CNS)_6^{3-} + e \rightarrow Cr(CNS)_5H_2O + CNS^-$).

A slow charge-transfer reaction (or nonreducibility) is to be expected if all the electronic orbitals of the complex are occupied and the lowest vacant orbital is a σ-antibonding one of relatively high energy. Then fission of the bond usually occurs and the vacant orbital is set free since a direct transfer of the electron into an antibonding orbital would require a large amount of energy (oxygen-containing anions or cations of the type MeO_p^{n+}). For further rules and more details refer to Reference (371). For quite recent papers see References (376–378).

In the case of reversible polarographic reactions the determination of molecular orbitals in certain compounds is possible. This determination is based on the fact that substitution of the coordinated ligand causes a shift of the half-wave potential of the complex, similar to substitution of a free ligand, but to a lesser extent. From this lowering of the substitution effects, the nature of the highest occupied orbital and localization of the corresponding electron in the reduced form can be estimated. Some substitution inert complexes react with the electrode as a whole and the orbital from which the electron is removed during the oxidation can be regarded as a molecular orbital ψ formed by linear combination of atomic orbitals of the individual components:

$$\psi = C_1\psi_1 + C_2\psi_2 \tag{180}$$

where ψ_1 is the eigenfunction of the central atom and ψ_2 that of the ligand, and C_1 and C_2 are the mixing coefficients. The electron affinity, E_A, of the oxidized form can be written

$$E_A = -\int \psi^* H\psi dE_A \tag{181}$$

where H is the one-electron Hamiltonian. The electron affinity is related to the standard potential (or reversible half-wave potential) by the relation

$$nFE_0 = nFE'_{1/2} = E_A - \Delta U_S + T\Delta S + K \tag{182}$$

where ΔU_S is the change of the hydration energy and ΔS is the change of entropy. K is a constant converting the potential from the hydrogen scale to the absolute one. If there is a series of structurally close complexes, ΔU_S and ΔS can be considered constant and for the difference in reversible half-wave potentials of two members of the series, the relation

$$nF[(E'_{1/2})_1 - (E'_{1/2})_2] = (E_A)_1 - (E_A)_2 \tag{183}$$

holds.

It follows that the reversible half-wave potential is a function of the molecular orbital and therefore has properties of both atomic orbitals. From the shift of the half-wave potential with substitution, the mixing coefficients can be estimated. In the case of dibenzene chromium it was found that in the highest occupied orbital of the complex a 10–15% electron transfer occurs between the central atom and the aromatic rings, the prevailing portion of the electron being localized at the chromium atom.[372]

Vlček was also the first to recognize some direct correlations between polarographic parameters of complexes and their spectral properties. He found, e.g., correlations between half-wave potentials and the frequency of spectral bonds corresponding to transitions to various states in several series of structurally related complexes. For the theoretical background and for more details, see References (373–375).

From this very incomplete survey of the applications of polarography both in inorganic and organic chemistry, it follows that polarography is a useful tool for gaining deeper understanding of general problems of chemistry. It is extremely advantageous for studying dynamic properties of substances in solutions. Slow chemical changes can be followed from the decrease of the limiting currents or from the increase of a new wave formed in the course of the chemical transformation of the species present in the solution. Very fast chemical reactions giving rise to the so-called kinetic currents as described previously can also be studied. Polarography makes it possible to follow the chemical kinetics either of photoreactions proper[379,380] or reactions subsequent to irradiation[381] (photopolarography[382]).

Systematic polarographic studies of the electrochemical photoeffect have shown that the light-induced transfer of charge across the metal/solution interface can occur either through emission of electrons from the metal to the solution[383] or through a heterogeneous photochemical charge-transfer reaction between the metal and the component of the solution.[384] Research on photoemission of electrons into solutions of electrolytes[385–388] provides valuable information on fundamental electrode kinetics and on properties of short-lived species in solutions. Heterogeneous photochemical charge-transfer reactions, on the other hand, represent a promising field[389,390] for study of elementary electron transfer processes as well as of properties of the metal/solution interface.

Polarographic investigations are not limited to aqueous solutions, and much data have been presented for nonaqueous solvents or for their mixtures.[391] For reversible reactions Gibbs free energies of transfer of various cations from one solvent to another can be determined [see, e.g., Reference (392)]. Polarography, particularly in nonaqueous solvents, led to a great deal of information on the existence of radical and intermediates both in inorganic and organic chemistry.[393–397] Many free radicals formed in radiolytic processes were studied by polarographic techniques with a DME or HMDE dipping into the irradiated solution.[398] Henglein *et al.* developed a method for these studies, which they called pulse radiolysis polarography.[399]

The proportionality between the limiting current and the concentration formed from the very beginning the basis for the application of polarography in analysis.[8,10–15,192,400–404] The lowest concentration that could be estimated at that time by d-c polarography (10^{-5} M) together with the small volume that could be used (even 0.05 ml) predestined polarography to be a powerful microanalytical technique for the determination of traces of substances in excess of the main constituents. The interfering components in the sample are usually eliminated in inorganic polarographic analysis by a suitable selection of complexing agents, which shift the $E_{1/2}$ values of the interfering components. A similar effect can be achieved by a proper choice of pH values. If such an elimination could not be done by a simple chemical operation in the given solution the usual separation procedures of analytical chemistry have to be applied. In the fifties the application of d-c polarography seemed to have reached its maximum, being commonly used for the analysis of ores, steel, drugs, food, biological fluids, etc. Later, other analytical, highly automatized techniques replaced to a large extent the d-c polarographic procedures in analytical practice. However, beginning in 1970 the impact of related polarographic techniques based on new theoretical and instrumental developments can be noticed. These techniques are more than two orders of magnitude more sensitive than classical polarography, enabling analysis on the 10^{-7} M and even on the 10^{-8} M concentration level to be achieved under suitable conditions. In combination with the stripping procedure, concentrations of 10^{-10} M or lower can be estimated.

Stripping polarography or stripping voltammetry is a simple procedure based on an accumulation step of the substance to be analyzed at a microelectrode, followed by electrochemical monitoring of the substance concentrated in this way. In the most frequently used version, anodic stripping voltammetry, the cation is deposited as a metal into a solid microelectrode or forms an amalgam with the mercury of a mercury drop or mercury film electrode and the following step consists of an anodic dissolution or stripping process in which the metal is oxidized. The pre-electrolysis is carried out under reproducible conditions for a definite time (usually 3–10 min) and during the stripping process the corresponding anodic response has a form of waves or peaks depending on the electrochemical technique used. The cathodic variant of stripping voltammetry for the determination of anions that form a strongly adhering insoluble compound or complex is rather rare. A very instructive review[405] and a more exhaustive monograph[406] are available.

The utilization of advanced polarographic techniques and procedures surpassing the sensitivity of other analytical methods is increasing mainly in the field of environmental control and water management.

It is to be expected that further developments in polarographic instrumentation in combination with microprocessor techniques will ensure electrochemical methods a continuing important position in analytical chemistry in the future.

Auxiliary Notation

a	constant of interaction between adsorbed molecules
A	surface area
$\bar{A}$	mean surface area
c	concentration in general or at the electrode
$\bar{c}$	time independent concentration
c^0	concentration in the bulk
C	capacity
C_d	differential capacity
C_s	series capacitance component of faradaic impedance [Eq. (169a)]
C_W	effective capacitance of the Warburg impedance [see Eq. (169a)]
d	density
D	diffusion coefficient
DME	dropping mercury electrode
E	potential
E^*	potential referred to p.z.c.
E_0	standard potential
$E_{1/2}$	half-wave potential
$E_{1/2}^r$	reversible half-wave potential
$E_{1/2}^{ir}$	irreversible half-wave potential
E	pulse or sine voltage amplitude
h	height of the mercury column
H	Hamiltonian
i	current density
i_m	amplitude of a-c current density
I^D	diffusion current constant
I	current in general or instantaneous
I_a	anodic current
$I_{a,s}$	current in presence of surfactant
I_{ads}	adsorption current
$\bar{I}$	mean current
$\bar{I}_{ads,l}$	mean adsorption limiting current
I_c	cathodic current
I_d	limiting diffusion current
$\bar{I}_d$	mean limiting diffusion current
I_{DL}	charging current
I_l	limiting current
I_k	current controlled by the rate of chemical reactions
I_{rev}	current of a reversible charge-transfer reaction
I_s	alternating current
I_{sw}	square-wave current
k	rate constant of forward chemical reaction
$k_{e,ef}$	effective charge-transfer rate constant

k_j consecutive formation constant for the j-th complex

k_{ox} charge-transfer rate constant for oxidation

k_{red} charge-transfer rate constant for reduction

k_s rate constant of the surface chemical reaction

k' rate constant of reverse chemical reaction

k_e^* charge-transfer rate constant given by Eq. (73)

k_e^0 charge-transfer rate constant at $E = E_0$

$k_{e,cor}^0$ charge-transfer rate constant at $E = E_0$ corrected for double-layer effect

$_0k_e$ charge-transfer rate constant on uncovered surface

$_1k_e$ charge-transfer rate constant in covered surface

K integral capacity per unit area

K' integral capacity

K_j formation constant for j-th complex

l length of the mercury jet

m flowrate

n number of electrons involved in an electrode reaction

Ox   oxidized form

pzc   potential of zero charge

P_{eff} effective hydrostatic pressure

P_b back pressure

q charge density

q_i charge density due to specifically adsorbed ions

Q charge at the electrode

r number of solvent molecules replaced by one molecule of a surfactant

r radius of the drop

r_c radius of the capillary

R gas constant

R resistance

R_{ct} charge-transfer resistance

Red   reduced form

R_s resistive component of faradaic impedance [see Eq. (169)]

R_W resistive component of the Warburg impedance [Eq. (169)]

s specific mass

t time

t_d drop time

t_p time of voltage application

t_s half-period of the square-wave voltage

t_1 time of drop growth prior to voltage application

t' time elapsed from the beginning of the last halfcycle

T temperature

v velocity

V volume

x distance from the electrode

Y_f'	real component of faradaic admittance
Y_f''	imaginary component of the faradaic admittance
z	charge of the electroactive species
Z_f	total faradaic impedance
Z_f'	real faradaic impedance component
Z_f''	imaginary faradaic impedance component
α	charge-transfer coefficient of cathodic reaction
β	adsorption coefficient
γ	surface tension
Γ	surface excess
Γ_m	maximum surface excess
η	viscosity
$\eta_{1/2}$	polarographic overvoltage $E_{1/2} - E_0$
ν	time for the complete coverage of the electrode
Θ	fractional coverage
θ	charge-transfer resistance [Eq. (169)]
$\mathcal{H}$	specific conductivity
$\mathcal{H}$	Ilkovič constant [Eq. (25)]
$\bar{\mathcal{H}}$	mean value of Ilkovič constant
λ	potential function, $\lambda = \exp(nF/RT)(E - E_0)$
λ'	potential function, $\lambda' = \exp(nF/RT)(E - E_{1/2}^r)$
μ	thickness of reaction layer
σ	Warburg coefficient
φ	potential in the double layer
φ_2	potential at the outer Helmholtz plane
ϕ	phase angle
$\mathcal{X}_1$	function argument (see Eqs. (62), (101), and (114a))
ψ	eigenfunction
ω	angular frequency

References

1. J. Heyrovský, *Chem. Listy* **16**, 256 (1922); *Philos. Mag.* **45**, 303 (1923).
2. E. Yeager and J. Kůta, In: *Physical Chemistry*, H. Eyring, D. D. Henderson, and W. Jost, eds., vol. 9A, pp. 345–461, Academic Press, New York (1970).
3. J. Kůta and E. Yeager, In: *Techniques of Electrochemistry*, Vol. 1, pp. 141–293, Wiley-Interscience, New York (1972).
4. L. Jellici and L. Griggio, *Bibliographia polarographica*, Suppl. No. 20, Consiglio Nazionale delle Ricerche, Roma (1968).
5. P. Herasymenko and I. Šlendyk, *Z. Phys. Chem.* **A149**, 123 (1930).
6. A. N. Frumkin, *Z. Phys. Chem.* **A164**, 121 (1933).
7. M. Smutek, *Chem. Listy* **45**, 241 (1951); *Collect. Czech. Chem. Commun.* **18**, 171 (1953).
8. J. Heyrovský, *Polarographie*, Springer, Vienna (1941); reprinted Edwards, Ann Arbor, Mich. (1944).
9. J. Heyrovský and J. Kůta, *Principles of Polarography*, Academic Press, New York (1966); in Russian: Izd. MIR, Moscow (1965); in German: Akademie-Verlag, Berlin (1966).

10. I. M. Kolthoff and J. J. Lingane, *Polarography*, Interscience, New York (1941); 2nd ed. (1952).
11. L. Meites, *Polarographic Techniques*, Interscience, New York (1955); 2nd ed. (1965).
12. P. Delahay, *New Instrumental Methods in Electrochemistry*, Interscience, New York (1954).
13. M. v. Stackelberg, *Polarographische Arbeitsmethoden*, W. de Gruyter Co., Berlin (1950); reprinted (1960).
14. G. W. C. Milner, *The Principles and Applications of Polarography and Other Electroanalytical Processes*, Longmans Green Co., London (1957).
15. J. Proszt, V. Cieleszky, and K. Györbiro, *Polarographie* (in German), Akadémiai Kiadó, Budapest (1967).
16. J. Heyrovský and M. Shikata, *Rec. Trav. Chim.* **44**, 496 (1925).
17. R. N. Schroeder, In: *Electrochemistry, Simulation and Instrumentation*, J. S. Mattson, H. B. Mark, Jr., and H. C. MacDonald, Jr., eds., pp. 264–350, Marcel Dekker, New York (1972).
18. G. S. Smith, *Trans. Faraday Soc.* **47**, 63 (1951).
19. J. Kůta and I. Smoler, *Collect. Czech. Chem. Commun.* **28**, 2874 (1963).
20. G. C. Barker, *Anal. Chim. Acta* **18**, 118 (1958).
21. W. D. Cooke, M. T. Kelley, and D. J. Fisher, *Anal. Chem.* **33**, 1209 (1961).
22. R. de Levie, *J. Electroanal. Chem.* **9**, 117 (1965).
23. L. Novotný, J. Kůta, and I. Smoler, *J. Electroanal. Chem.* **88**, 161 (1978).
24. I. Smoler, *J. Electroanal. Chem.* **51**, 452 (1974).
25. H. P. Raaen, *Anal. Chem.* **34**, 1714 (1962); **37**, 677 (1965).
26. A. M. Bond, T. A. O'Donnell, and A. B. Waugh, *J. Electroanal. Chem.* **39**, 137 (1972).
27. I. Smoler, *Collect. Czech. Chem. Commun.* **19**, 238 (1954).
28. I. M. Kolthoff and Y. Okinaka, In: *Progress in Polarography*, P. Zuman and I. M. Kolthoff, eds., Vol. 2, pp. 357–381, Interscience, New York (1962).
29. J. Heyrovský and J. Forejt, *Z. Phys. Chem.* **193**, 77 (1943).
30. A. Rius, J. Llopis, and S. Polo, *Anales Real Soc. Espan. Fis. Quim.* (*Madrid*) **45B**, 1039 (1949).
31. J. Koryta, *Collect. Czech. Chem. Commun.* **19**, 433 (1954).
32. J. R. Weawer and R. W. Parry, *J. Amer. Chem. Soc.* **76**, 6258 (1954).
33. J. R. Weawer and R. W. Parry, *J. Amer. Chem. Soc.* **78**, 5542 (1956).
34. W. Kemula and Z. Kublik, *Anal. Chim. Acta* **18**, 104 (1958).
35. J. Řiha, In: *Progress in Polarography*, P. Zuman and I. M. Kolthoff, eds., Vol. 2, pp. 383–396, Interscience, New York (1962).
36. P. Delahay, *J. Electrochem. Soc.* **113**, 967 (1966).
37. P. Delahay, *J. Phys. Chem.* **70**, 2067, 2373 (1966).
38. D. Ilkovič, *Collect. Czech. Chem. Commun.* **8**, 170 (1936).
39. D. Ilkovič and G. Semerano, *Collect. Czech. Chem. Commun.* **4**, 176 (1932).
40. A. Poojary and S. R. Rajagopalan, *J. Electroanal. Chem.* **62**, 51 (1975).
41. S. R. Rajagopalan, A. Poojary, and S. K. Rangarajan, *J. Electroanal. Chem.* **75**, 135 (1977).
42. J. Heyrovský, *Arh. Hemiju Farmaciju* **8**, 11 (1934).
43. V. G. Levich, *Physico-chemical Hydrodynamics*, Izd. Akad. Nauk SSSR, Moscow (1952): English translation, Prentice Hall (1962).
44. Z. Zembura, A. Fulinski, and M. Bierowski, *Z. Electrochem.* **65**, 887 (1961).
45. D. Ilkovič, *Collect. Czech. Chem. Commun.* **6**, 498 (1934).
46. D. Ilkovič, *J. Chim. Phys.* **35**, 129 (1938).
47. D. MacGillavry and E. K. Rideal, *Rec. Trav. Chim.* **56**, 1013 (1937).
48. D. Ilkovič, *Collect. Czech. Chem. Commun.* **10**, 249 (1938).
49. H. A. McKenzie, *J. Council. Sci. Ind. Res.* **21**, 210 (1948).
50. A. Scholander, *Proc. I. Internat. Polarograph. Congress*, Vol. I, p. 260, Přirod. Vydavatelstvi, Praha (1951).
51. J. J. Lingane and B. A. Loveridge, *J. Amer. Chem. Soc.* **72**, 438 (1950).

52. H. Strehlow and M. v. Stackelberg, *Z. Elektrochem.* **54**, 51 (1950).
53. T. Kambara and I. Tachi, *Bull. Chem. Soc. Jap.* **23**, 226 (1956).
54. J. M. Markowitz and P. J. Elving, *Chem. Rev.* **58**, 1047–1079 (1958).
55. J. Koutecký and M. v. Stackelberg, In: *Progress in Polarography*, P. Zuman and I. M. Kolthoff, eds., Vol. 1, pp. 21–42, Interscience, New York (1962).
56. J. Koutecký, *Czech. J. Phys.* **2**, 50 (1953).
57. H. Matsuda, *Bull. Chem. Soc. Jap.* **36**, 342 (1953).
58. W. Hans, W. Henne, and E. Meurer, *Z. Elektrochem.* **58**, 836 (1954).
59. J. Kůta and I. Smoler, In: *Progress in Polarography*, P. Zuman and I. M. Kolthoff, eds., Vol. 1, pp. 43–63, Interscience, New York (1962).
60. R. de Levie, *J. Electroanal. Chem.* **9**, 311 (1965).
61. J. Heyrovský and D. Ilkovič, *Collect. Czech. Chem. Commun.* **7**, 198 (1935).
62. J. Weber, *Collect. Czech. Chem. Commun.* **24**, 1424 (1959).
63. J. Heyrovský, *Chem. Listy* **45**, 149 (1949).
64. J. Vogel and J. Řiha, *Chem. Listy* **43**, 149 (1949); *J. Chim. Phys.* **47**, 5 (1950).
65. P. Levèque and F. Roth, *J. Chim. Phys.* **46**, 480 (1949); **47**, 623 (1950).
66. R. Brdička, *Z. Elektrochem.* **47**, 314 (1941).
67. M. v. Stackelberg and H. v. Freyhold, *Z. Elektrochem.* **46**, 120 (1940).
68. D. D. De Ford and D. N. Hume, *J. Amer. Chem. Soc.* **73**, 5321 (1951).
69. A. A. Vlček, In: *Progress in Inorganic Chemistry*, F. A. Cotton, ed., Vol. 5, pp. 211–384, Interscience, New York (1963).
70. R. D. Crow, *Polarography of Metal Complexes*, Academic Press, London (1969).
71. J. Kortya, In: *Advances in Electrochemistry and Electrochemical Engineering*, Vol. 6, pp. 289–327, Interscience, New York (1967).
72. L. G. Sillen and A. E. Martell, *Stability Constants of Metal-Ion Complexes*, The Chemical Society, London (1964).
73. P. Rüetschi and G. Trümpler, *Helv. Chim. Acta* **35**, 1021, 1486, 1957 (1952).
74. J. Koryta, *Z. Elektrochem.* **61**, 423 (1957).
75. N. Mejman, *Zh. Fiz. Khim.* **22**, 1454 (1948).
76. J. Koutecký, *Collect. Czech. Chem. Commun.* **18**, 597 (1953).
77. J. Kůta, *Collect. Czech. Chem. Commun.* **23**, 383 (1958).
78. P. Delahay, *J. Amer. Chem. Soc.* **75**, 1430 (1953).
79. H. Matsuda and Y. Ayabe, *Z. Elektrochem.* **63**, 1164 (1959).
80. J. Koryta, *Electrochim. Acta* **6**, 67 (1962).
81. I. Ružić, A. Barić, and M. Branica, *J. Electroanal. Chem.* **29**, 411 (1971).
82. I. Ružić and M. Branica, *J. Electroanal. Chem.* **22**, 242, 422 (1969).
83. I. Ružić, *J. Electroanal. Chem.* **25**, 144 (1970).
84. I. Ružić, *J. Electroanal. Chem.* **36**, 447 (1972).
85. G. C. Barker and A. W. Gardner, *J. Electroanal. Chem.* **46**, 150 (1973).
86. T. Meites and L. Meites, *Talanta* **19**, 1131 (1972).
87. L. Meites and L. Lampugnani, *Anal. Chem.* **45**, 1317 (1973).
88. J. Čipak, I. Ružić, and Lj. Jeftić, *J. Electroanal. Chem.* **75**, 9 (1977).
89. J. Koryta, *Collect. Czech. Chem. Commun.* **24**, 3057 (1957).
90. H. Matsuda and Y. Ayabe, *Z. Elektrochem.* **66**, 469 (1962).
91. H. Matsuda, Y. Ayabe, and K. Adachi, *Z. Elektrochem.* **67**, 593 (1963).
92. C. Nishihira and H. Matsuda, *J. Electroanal. Chem.* **28**, 17 (1970).
93. K. Momoki and H. Ogawa, *Anal. Chem.* **43**, 1664 (1971).
94. M. M. Abou Romia, H. Matsuda, and K. Tokuda, *J. Electroanal. Chem.* **42**, 403 (1973).
95. N. Ohnaka and H. Matsuda, *J. Electroanal. Chem.* **62**, 245 (1975).
96. M. Heyrovský and S. Vavřička, *J. Electroanal. Chem.* **36**, 203 (1972).
97. B. E. Conway, J. O'M. Bockris, and B. Lovrecek, *Proc. 6th Meeting of Intl. Committee of Electrochemical Thermodynamics and Kinetics* [*C.I.T.C.E.*], Poitiers, 1954, pp. 207–230, Butterworths Sci. Publications, London (1955).
97a. S. G. Mairanovskii, *J. Electroanal. Chem.* **4**, 166 (1962).

98. R. Guidelli, G. Pezzatini, and M. L. Foresti, *J. Electroanal. Chem.* **43**, 83 (1973).
99. R. Guidelli, M. L. Foresti, and G. Pezzatini, *J. Electroanal. Chem.* **43**, 95 (1973).
100. P. Delahay, *Double Layer and Electrode Kinetics,* Interscience, New York (1965).
101. L. Gierst, In: *Transactions of the Symposium on Electrode Processes,* E. Yeager, ed., pp. 109–144, Wiley, New York (1961).
102. A. N. Frumkin and N. V. Nikolaeva-Fedorovich, In: *Progress in Polarography,* P. Zuman and I. M. Kolthoff, eds., Vol. 1, pp. 223–241, Interscience, New York (1962).
103. L. Geirst, L. Vandenberghen, E. Nicolas, and A. Fraboni, *J. Electrochem. Soc.* **113**, 1025 (1966).
104. D. J. Bieman and W. R. Fawcett, *J. Electroanal. Chem.* **34**, 27 (1972).
105. R. Parsons, *J. Electroanal. Chem.* **21**, 35 (1969).
106. R. Guidelli and M. L. Foresti, *Electrochim. Acta* **18**, 301 (1973).
107. R. Guidelli, *J. Electroanal. Chem.* **53**, 205 (1974).
108. P. Teppema, M. Sluyters-Rehbach, and J. H. Sluyters, *J. Electroanal. Chem.* **16**, 165 (1968).
109. M. Sluyters-Rehbach, J. S. M. C. Breukel, and J. H. Sluyters, *J. Electroanal. Chem.* **19**, 85 (1968).
110. K. Asada, P. Delahay, and A. K. Sundaram, *J. Am. Chem. Soc.* **83**, 3396 (1961).
111. L. Gierst and P. Cornelissen, *Collect. Czech. Chem. Commun.* **25**, 3004 (1960).
112. C. W. De Kreuk, M. Sluyters-Rehbach, and J. H. Sluyters, *J. Electroanal. Chem.* **28**, 391 (1970).
113. C. W. De Kreuk, M. Sluyters-Rehbach, and J. H. Sluyters, *J. Electroanal. Chem.* **33**, 267 (1971).
114. M. J. Weaver and F. C. Anson, *J. Electroanal. Chem.* **65**, 711, 737, 759 (1975).
115. K. Alias and W. R. Fawcett, *Can. J. Chem.* **52**, 3165 (1974).
116. V. S. Krylov and V. G. Levich, *Zh. Fiz. Khim.* **37**, 106 (1963).
117. V. S. Krylov, *Electrochim. Acta* **9**, 1247 (1964).
118. R. S. Rodgers and F. C. Anson, *J. Electroanal. Chem.* **42**, 381 (1973).
119. F. C. Anson and B. A. Parkinson, *J. Electroanal. Chem.* **85**, 317 (1977).
120. M. Breiter, M. Kleinerman, and P. Delahay, *J. Am. Chem. Soc.* **80**, 5111 (1958).
121. P. Delahay and A. Aramata, *J. Phys. Chem.* **66**, 1194 (1962).
122. J. J. Tondeur, A. Dombret, and L. Gierst, *J. Electroanal. Chem.* **3**, 225 (1962).
123. I. Žežulz, *Chem. Listy* **47**, 1303 (1953).
124. T. A. Kryukova, *Dokl. Akad. Nauk SSSR,* **65**, 517 (1949).
125. A. N. Frumkin, *Trans. Faraday Soc.* **55**, 156 (1959).
126. R. de Levie and M. Nemes, *J. Electroanal. Chem.* **58**, 123 (1975).
127. M. L. Foresti and R. Guidelli, *J. Electroanal. Chem.* **53**, 219 (1974).
128. M. L. Foresti, D. Cozzi, and R. Guidelli, *J. Electroanal. Chem.* **53**, 235 (1974).
129. R. Guidelli and M. L. Foresti, *J. Electroanal. Chem.* **67**, 231 (1976).
130. R. Brdička and K. Wiesner, *Naturwissenschaften* **31**, 247 (1943).
131. R. Brdička and K. Wiesner, *Collect. Czech. Chem. Commun.* **12**, 39, 138 (1947).
132. K. Wiesner, *Chem. Listy* **41**, 6 (1947).
133. R. Brdička, *Chem. Zvesti* **8**, 670 (1954).
134. R. Brdička, *Collect. Czech. Chem. Commun.* **19**, 41 (1954).
135. V. Hanuš, *Chem. Zvesti* **8**, 702 (1954).
136. R. Brdička, *Z. Electrochem.* **64**, 16 (1960).
137. J. Koutecký and R. Brdička, *Collect. Czech. Chem. Commun.* **12**, 337 (1947).
138. J. Koutecký, *Collect. Czech. Chem. Commun.* **18**, 11 (1953).
139. J. Koutecký, *Collect. Czech. Chem. Commun.* **18**, 183 (1953).
140. J. Koutecký, *Collect. Czech. Chem. Commun.* **18**, 311 (1953).
141. J. Koutecký, *Collect. Czech. Chem. Commun.* **18**, 597 (1953).
142. J. Koutecký and J. Koryta, *Electrochim. Acta* **3**, 318 (1961).
143. R. Guidelli, In: *Electroanalytical Chemistry,* A. J. Bard, ed., Vol. 5, pp. 149–374, Marcell Dekker, New York (1971).

144. R. Brdička, V. Hanuš, and J. Koutecký, In: *Progress in Polarography*, P. Zuman and I. M. Kolthoff, eds., Vol. 1, pp. 145–199, Interscience, New York (1962).
145. S. G. Mairanovskii, *Catalytic and Kinetic Waves in Polarography*, Plenum Press, New York (1968).
146. E. Budevski, *Compt. Rend. Acad. Bulg. Sci.* **8**, 25 (1955).
147. J. Weber and J. Koutecký, *Collect. Czech. Chem. Commun.* **20**, 980 (1955).
148. J. Koutecký and J. Čižek, *Collect. Czech. Chem. Commun.* **21**, 836 (1956).
149. J. Koutecký, *Collect. Czech. Chem. Commun.* **19**, 857 (1954).
150. K. Vesely and R. Brdička, *Collect. Czech. Chem. Commun.* **12**, 313 (1947).
151. R. Bieber and G. Trumpler, *Helv. Chim. Acta* **30**, 706, 971, 1109, 1286, 1534, 2000 (1947).
152. R. Brdička, *Collect. Czech. Chem. Commun.* **20**, 387 (1955).
153. R. Brdička, *Z. Electrochem.* **59**, 787 (1955).
154. I. Crisan, A. Calusaru, and J. Kůta, *J. Electroanal. Chem.* **46**, 51 (1973).
155. J. Kůta, *Collect. Czech. Chem. Commun.* **24**, 2532 (1959).
156. J. Kůta and P. Valenta, *Collect. Czech. Chem. Commun.* **28**, 1593 (1963).
157. J. Tirouflet, P. Fournari, and J. P. Chané, *Compt. Rend.* **242**, 1799 (1956).
158. J. Volke, *Collect. Czech. Chem. Commun.* **23**, 1486 (1958).
159. J. Volke and P. Valenta, *Collect. Czech. Chem. Commun.* **25**, 1580 (1960).
160. R. Brdička, *Collect. Czech. Chem. Commun.* **12**, 212 (1947).
161. V. Hanuš and R. Brdička, *Khimyia* **2**, 28 (1951).
162. J. Kůta, *Collect. Czech. Chem. Commun.* **22**, 1411 (1957).
163. J. Koryta and I. Kössler, *Collect. Czech. Chem. Commun.* **15**, 241 (1950).
164. J. Koryta, *Collect. Czech. Chem. Commun.* **24**, 2903, 3057 (1959).
165. J. Koryta, *Electrochim. Acta* **1**, 26 (1959).
166. K. Wiesner, *Collect. Czech. Chem. Commun.* **12**, 64 (1947).
167. J. M. Los and K. Wiesner, *J. Am. Chem. Soc.* **75**, 6346 (1953).
168. J. M. Los, L. B. Simpson, and K. Wiesner, *J. Am. Chem. Soc.* **78**, 1564 (1956).
169. J. Koutecký and V. Hanuš, *Collect. Czech. Chem. Commun.* **20**, 124 (1955).
170. V. Čermák, *Chem. Zvesti* **8**, 714 (1954).
171. V. Čermák, *Collect. Czech. Chem. Commun.* **23**, 1471, 1871 (1958).
172. V. Čermák and M. Smutek, *Collect. Czech. Chem. Commun.* **40**, 3241 (1975).
173. M. Smutek and V. Čermák, *Collect. Czech. Chem. Commun.* **40**, 3265 (1975).
174. Wang-Er-Kong and A. A. Vlček, *Collect. Czech. Chem. Commun.* **25**, 2082 (1960).
175. J. Koutecký and J. Čižek, *Collect. Czech. Chem. Commun.* **21**, 1063 (1956).
176. M. Březina, *Collect. Czech. Chem. Commun.* **22**, 339 (1957).
177. J. Koutecký, *Collect. Czech. Chem. Commun.* **22**, 160 (1957).
178. J. Koutecký and J. Koryta, *Collect. Czech. Chem. Commun.* **19**, 845 (1954).
179. J. Koryta and J. Koutecký, *Collect. Czech. Chem. Commun.* **20**, 423 (1955).
180. Z. Vavřin, *Collect. Czech. Chem. Commun.* **14**, 367 (1949).
181. D. M. H. Kern, *J. Am. Chem. Soc.* **75**, 2473 (1953).
182. D. M. H. Kern, *J. Am. Chem. Soc.* **76**, 1011 (1954).
183. J. Koryta and Z. Zábranský, *Collect. Czech. Chem. Commun.* **25**, 3153 (1960).
184. J. Koutecký, *Collect. Czech. Chem. Commun.* **20**, 116 (1955).
185. H. Matsuda, *J. Phys. Chem.* **64**, 336 (1960).
186. H. Hurwitz, *Z. Elektrochem.* **65**, 178 (1961).
187. R. P. Buck, *Anal. Chem.* **35**, 1853 (1963).
188. L. Pospišil and J. Kůta, *Collect. Czech. Chem. Commun.* **34**, 742 (1969).
189. S. G. Mairanovskii, *J. Electroanal. Chem.* **4**, 166 (1962).
190. M. Fleischmann, J. Koryta, and H. R. Thirsk, *Trans. Faraday Soc.* **63**, 1261 (1967).
191. R. D. Giles, J. A. Harrison, and H. R. Thirsk, *J. Electroanal. Chem.* **20**, 47 (1969).
192. M. Březina and P. Zuman, *Polarography in Medicine, Biochemistry and Pharmacy*, Interscience, New York (1958).
193. W. H. Reinmuth and K. Balasubramanian, *J. Electroanal. Chem.* **38**, 79 (1972).

194. W. H. Reinmuth and K. Balasubramanian, *J. Electroanal. Chem.* **38**, 271 (1972).
195. R. Brdička, *Z. Elektrochem.* **48**, 278 (1942).
196. R. Brdička, *Collect. Czech. Chem. Commun.* **12**, 522 (1947).
197. A. M. Bond and G. Hefter, *J. Electroanal. Chem.* **42**, 1 (1973).
197. A. M. Bond and G. Hefter, *J. Electroanal. Chem.* **42**, 1 (1973).
198. A. M. Bond and G. Hefter, *J. Electroanal. Chem.* **68**, 203 (1976).
199. B. Nygård, *Acta Chem. Scand.* **20**, 1710 (1966).
200. A. Puxeddu and G. Costa, *J. Chem. Soc., Dalton Trans.* **1977**, 2327.
201. E. Laviron, *J. Electroanal. Chem.* **52**, 355 (1974).
202. R. Guidelli, *J. Electroanal. Chem.* **18**, 5 (1968).
203. R. Guidelli, *J. Phys. Chem.* **74**, 95 (1970).
204. G. A. Tedoradze, E. Yu. Khmelnitskaya, and Ya. M. Zolotovitskii, In: *Progress in Electrochemistry of Organic Compounds*, A. N. Frumkin and A. B. Ershler, eds., p. 171, Plenum Press, London (1971).
205. E. Laviron, *J. Electroanal. Chem.* **63**, 245 (1975); see also H. A. Kozlowska, B. E. Conway, and J. Klinger, *J. Electroanal. Chem.* **75**, 45 (1977).
206. G. Piccardi, F. Pergola, M. L. Foresti, and R. Guidelli, *J. Electroanal. Chem.* **84**, 235 (1977).
207. M. Heyrovský, S. Vavřička, and R. Heyrovská, *J. Electroanal. Chem.* **46**, 391 (1973).
208. M. Heyrovský and R. Heyrovská, *J. Electroanal. Chem.* **52**, 141 (1974).
209. R. Guidelli and F. Pergola, *J. Electroanal. Chem.* **84**, 255 (1977).
210. R. Guidelli, *J. Electroanal. Chem.* **33**, 291, 303 (1971).
211. M. Sluyters-Rehbach, C. A. Wijnhorst, and J. H. Sluyters, *J. Electroanal. Chem.* **74**, 3 (1976).
212. M. Sluyters-Rehbach and J. H. Sluyters, *J. Electroanal. Chem.* **75**, 371 (1977).
213. E. Laviron, *Bull. Soc. Chim. Fr.* **1962**, 418.
214. A. B. Ershler, *Elektrokhimiya* **9**, 1595 (1973).
215. R. Guidelli and G. Pezzatini, *J. Electroanal. Chem.* **84**, 211 (1977).
216. M. A. Loshkarev and A. A. Kryukova, *Zh. Fiz. Khim.* **31**, 452 (1957).
217. J. Koryta, *Collect. Czech. Chem. Commun.* **18**, 206 (1953).
218. R. W. Schmid and C. N. Reilley, *J. Am. Chem. Soc.* **80**, 2087 (1958).
219. V. Volková, *Nature* **185**, 743 (1960).
220. J. Weber, J. Koutecký, and J. Koryta, *Z. Electrochem.* **63**, 583 (1959).
221. M. Matsuda, *Rev. Polarogr.* (*Kyoto*) **14**, 87 (1967).
222. J. Kůta and I. Smoler, *Z. Elektrochem.* **64**, 285 (1960).
223. J. Kůta and I. Smoler, *Collect. Czech. Chem. Commun.* **32**, 2691 (1967).
224. J. Kůta and I. Smoler, *Collect. Czech. Chem. Commun.* **33**, 1656 (1968).
225. R. Guidelli and M. L. Foresti, *J. Electroanal. Chem.* **77**, 73 (1977).
226. J. Kůta, J. Weber, and J. Koutecký, *Collect. Czech. Chem. Commun.* **25**, 2376 (1960).
227. J. Kůta and J. Weber, *Electrochim. Acta* **9**, 541 (1964).
228. A. Ja. Gokhstein, *Dokl. Akad. Nauk SSSR*, **137**, 345 (1961).
229. J. Koutecký and J. Weber, *Collect. Czech. Chem. Commun.* **25**, 1423 (1960).
230. J. Kůta and I. Smoler, *Collect. Czech. Chem. Commun.* **27**, 2349 (1962).
231. A. Aramata and P. Delahay, *J. Phys. Chem.* **68**, 880 (1964).
232. S. Sathyanarayana, *J. Electroanal. Chem.* **10**, 119 (1965).
233. Y. Ayabe, *J. Electroanal. Chem.* **81**, 215 (1977).
234. O. Ju. Gusakova, B. B. Damaskin, N. V. Fedorovich, and S. D. Pirozhkov, *Electrokhimiya*, **10**, 1112 (1974).
235. L. Gierst, D. Bermane, and P. Corbusier, *Ric. Sci.* **29**, Suppl. A, Contributi di Polarographia, 75 (1959).
236. B. Kastening and L. Holleck, *Talanta* **12**, 1259 (1965).
237. B. Kastening and L. Holleck, *J. Electroanal. Chem.* **27**, 355 (1970).
238. J. Lipkowski, E. Kosińska, M. Goledzinowski, J. Nieniewska, and Z. Galus, *J. Electroanal. Chem.* **59**, 344 (1975).
239. J. Divisek and B. Kastening, *J. Electroanal. Chem.* **65**, 603 (1975).

240. J. Chevalet, F. Rovelle, L. Gierst, and J. P. Lambert, *J. Electroanal. Chem.* **39**, 201 (1972).
241. S. Sathyanarayana, *J. Electroanal. Chem.* **50**, 195 (1974).
242. J. Heyrovský and R. Šimunek, *Philos. Mag.* **7**, 951 (1929).
243. H. J. Antweiler, *Z. Elektrochem.* **43**, 596 (1937); **44**, 719, 731, 888 (1938).
244. M. v. Stackelberg, H. J. Antweiler, and L. Kieselbach, *Z. Electrochem.* **44**, 663 (1938).
245. A. A. Frumkin and V. G. Levich, *Zh. Fiz. Khim.* **19**, 573 (1945); **21**, 1335 (1947).
246. V. G. Levich, *Zh. Fiz. Khim.* **22**, 721 (1948).
247. M. v. Stackelberg, *Fortschr. Chem. Forsch.* **2**, 229 (1951).
248. M. v. Stackelberg and R. Doppelfeld, In: *Advances in Polarography*, S. Longmuir, ed., p. 68, Pergamon Press, London (1960).
249. Z. Matyáš, *Věstnik Král. Čes. Spol. Nauk, Třida Mat.-Přirodověd.*, p. 1 (1944).
250. T. A. Kryukova, *Zavod. Lab.* **9**, 691, 699 (1940).
251. T. A. Kryukova, *Zh. Fiz. Khim.* **21**, 365 (1947).
252. A. N. Frumkin and V. G. Levich, *Zh. Fiz. Khim.* **21**, 953 (1947).
253. T. A. Kryukova and A. N. Frumkin, *Zh. Fiz. Khim.* **23**, 819 (1949).
254. K. Doss and D. Venkatesan, *Proc. Ind. Acad. Sci.* **49**, 129 (1959).
255. A. N. Frumkin, N. V. Fedorovich, B. B. Damaskin, E. V. Stenina, and V. S. Krylov, *J. Electroanal. Chem.* **50**, 103 (1974).
256. E. V. Stenina, A. N. Frumkin, and N. V. Nikolaeva-Fedorovich, *J. Electroanal. Chem.* **62**, 11 (1975).
257. H. Bauer, In: *Electroanalytical Chemistry*, A. J. Bard, ed., Vol. 8, pp. 169–279, Marcel Dekker, New York (1975).
258. A. Wåhlin and A. Bresle, *Acta Chem. Scand.* **10**, 935 (1956).
259. J. Heyrovský, *Chem. Listy* **35**, 155 (1941).
260. R. Kalvoda, *Techniques of Oscillographic Polarography*, Elsevier, Amsterdam (1965).
261. M. Heyrovský and K. Micka, In: *Electroanalytical Chemistry*, A. J. Bard, ed., Vol. 2, pp. 193–256, Marcel Dekker, New York (1967).
262. M. Kalousek, *Chem. Listy* **40**, 149 (1946); *Collect. Czech. Chem. Commun.* **13**, 105 (1948).
263. R. Kalvoda, J. Macků, and K. Micka, *Z. Phys. Chem. (Leipzig) Sonderheft*, **1958**, 66.
264. G. C. Barker and J. L. Jenkins, *Analyst* **77**, 685 (1952).
265. G. C. Barker, R. L. Faircloth, and A. W. Gardner, *A.E.R.E. Report*, C/R 1786, Harwell (1958).
266. G. C. Barker, In: *Progress in Polarography*, P. Zuman and I. M. Kolthoff, eds., Vol. 2, pp. 411–427, Interscience, New York (1962).
267. H. Matsuda, *Z. Elektrochem.* **62**, 977 (1958).
268. R. Tamamushi and K. Matsuda, *J. Electroanal. Chem.* **80**, 201 (1977).
269. G. C. Barker and A. W. Gardner, *Z. Anal. Chem.* **173**, 79 (1960).
270. E. P. Parry and R. A. Osteryoung, *Anal. Chem.* **37**, 1634 (1965).
271. A. A. A. M. Brinkman and J. M. Los, *J. Electroanal. Chem.* **7**, 171 (1964).
272. A. M. Fonds, A. A. A. M. Brinkman, and J. M. Los, *J. Electroanal. Chem.* **14**, 43 (1967).
273. A. A. A. M. Brinkman, Thesis, Free University, Amsterdam, 1968.
274. J. Galvez and A. Serna, *J. Electroanal. Chem.* **69**, 133 (1976).
275. J. H. Christie, E. P. Parry, and R. A. Osteryoung, *Electrochim. Acta* **11**, 1525 (1966).
276. I. Ružić, *J. Electroanal. Chem.* **75**, 25 (1977).
277. A. A. A. M. Brinkman and J. M. Los, *J. Electroanal. Chem.* **14**, 269, 285 (1967).
278. J. Galvez and A. Serna, *J. Electroanal. Chem.* **69**, 145, 157 (1976).
279. A. W. Fonds, J. L. Molenaar, and J. M. Los, *J. Electroanal. Chem.* **22**, 229 (1969).
280. A. W. Fonds and J. M. Los, *J. Electroanal. Chem.* **36**, 479 (1972).
281. J. H. Christie and R. A. Osteryoung, *J. Electroanal. Chem.* **49**, 301 (1974).
282. J. H. Christie, L. L. Jackson, and R. A. Osteryoung, *Anal. Chem.* **48**, 242 (1976).
283. N. Klein and Ch. Yarnitzky, *J. Electroanal. Chem.* **61**, 1 (1975).
284. W. P. van Bennekom and J. B. Schute, *J. Electroanal. Chem.* **89**, 71 (1977).

285. J. W. Dillard, J. A. Turner, and R. A. Osteryoung, *Anal. Chem.* **49**, 1246 (1977).
286. H. E. Keller and R. A. Osteryoung, *Anal. Chem.* **43**, 342 (1971).
287. J. P. van Dieren, B. G. W. Kaars, J. M. Los, and B. J. C. Wetsema, *J. Electroanal. Chem.* **68**, 129 (1976).
288. M. Križan, *J. Electroanal. Chem.* **80**, 337 (1977).
289. M. Križan, H. Schmidtpott, and H. Strehlow, *J. Electroanal. Chem.* **80**, 345 (1977).
290. S. C. Rifkin and D. H. Evans, *Anal. Chem.* **48**, 1616 (1976).
291. M. Gross and J. Jordan, *J. Electroanal. Chem.* **75**, 163 (1977).
292. F. M. Kimmerle and J. Chevalet, *J. Electroanal. Chem.* **21**, 237 (1969).
293. J. Chevalet and F. M. Kimmerle, *J. Electroanal. Chem.* **25**, 270 (1970).
294. F. C. Anson, J. B. Flanagan, T. Takahashi, and A. Yamada, *J. Electroanal. Chem.* **67**, 253 (1976).
295. J. B. Flanagan, K. Takahashi, and F. C. Anson, *J. Electroanal. Chem.* **85**, 257 (1977).
296. B. Breyer and H. H. Bauer, In: *Chemical Analysis*, P. J. Elving and I. M. Kolthoff, eds., Vol. 13, John Wiley, New York (1963).
297. D. E. Smith, In: *Electroanalytical Chemistry*, A. J. Bard, ed., Vol. 1, pp. 1–155, Marcel Dekker, New York (1966).
298. M. Sluyters-Rehbach and J. H. Sluyters, In: *Electroanalytical Chemistry*, A. J. Bard, ed., Vol. 4, pp. 1–128, Marcel Dekker, New York (1970).
299. R. J. Schwall, A. M. Bond, and D. E. Smith, *J. Electroanal. Chem.* **85**, 217 (1977).
300. B. Breyer and F. Gutman, *Trans. Faraday Soc.* **42**, 645, 650 (1946).
301. B. Breyer and F. Gutman, *Trans. Faraday Soc.* **43**, 785 (1947).
302. R. de Levie and A. A. Husovsky, *J. Electroanal. Chem.* **20**, 181 (1969).
303. P. Delahay, In: *Advances in Electrochemistry and Electrochemical Engineering*, P. Delahay, ed., Vol. 1, pp. 233–318, Interscience, New York (1961).
304. R. Tamamushi and N. Tanaka, *Z. Phys. Chem. (Frankfurt am Main)* **21**, 89 (1959).
305. B. Timmer, M. Sluyters-Rehbach, and J. H. Sluyters, *J. Electroanal. Chem.* **14**, 169, 181 (1967).
306. D. E. Smith and T. G. McCord, *Anal. Chem.* **40**, 474 (1968).
307. R. de Levie and A. A. Husovsky, *J. Electroanal. Chem.* **22**, 29 (1969).
308. R. de Levie and L. Pospíšil, *J. Electroanal. Chem.* **22**, 277 (1969).
309. H. Gerischer, *Z. Phys. Chem.* **198**, 286 (1951).
310. H. Gerischer, *Z. Phys. Chem.* **201**, 55 (1952).
311. I. Ružić, D. E. Smith, and S. W. Feldberg, *J. Electroanal. Chem.* **52**, 157 (1974).
312. I. Ružić and D. E. Smith, *J. Electroanal. Chem.* **58**, 145 (1975).
313. H. A. Laitinen and J. E. B. Randles, *Trans. Faraday Soc.* **51**, 54 (1955).
314. J. Llopis, J. Biarge-Fernandez, and M. Perez-Fernandez, *Electrochim. Acta* **1**, 130 (1959).
315. M. Senda and P. Delahay, *J. Phys. Chem.* **65**, 1580 (1961).
316. W. Lorenz and G. Salié, *Z. Physik. Chem. (Leipzig)* **218**, 259 (1961).
317. W. Lorenz and G. Salié, *Z. Physik. Chem. (Frankfurt am Main)* **29**, 390 (1961).
318. K. Holub, G. Tessari, and P. Delahay, *J. Phys. Chem.* **71**, 2612 (1967).
319. B. Timmer, M. Sluyters-Rehbach, and J. H. Sluyters, *J. Electoanal. Chem.* **18**, 93 (1968).
320. P. Delahay and K. Holub, *J. Electroanal. Chem.* **16**, 131 (1968).
321. H. Moreira and R. de Levie, *J. Electroanal. Chem.* **35**, 103 (1972).
322. L. Pospíšil, *J. Electroanal. Chem.* **74**, 369 (1976).
323. H. Jehring, *Elektrosorptionsanalyse mi der Wechselstrompolarographie*, Akademi-Verlag, Berlin (1974).
324. A. M. Bond, R. J. O'Halloran, I. Ružić, and D. E. Smith, *Anal. Chem.* **48**, 872 (1976).
325. A. M. Bond, R. J. Schwall, and D. E. Smith, *J. Electroanal. Chem.* **85**, 231 (1977).
326. A. H. Schmidt and M. v. Stackelberg, *Modern Polarographic Methods*, Academic Press, New York (1963).
327. H. W. Nürnberg and G. Wolf, *Chem.-Ing.-Tech.* **38**, 160 (1966).

328. H. W. Nürnberg, *Fortschr. Chem. Forsch.* **8**, 241 (1967).

329. G. C. Barker, R. L. Faircloth, and A. W. Gardner, *Nature* **181**, 247 (1958).

330. G. C. Barker, In: *Transaction of Symposium on Electrode Processes*, E. Yeager, ed., pp. 325–365, John Wiley, New York (1961).

331. G. C. Barker and H. W. Nürnberg, *Naturwissenschaften* **51**, 191 (1964).

332. H. H. Bauer, *Aust. J. Chem.* **17**, 591, 715 (1964).

333. H. H. Bauer and P. J. Elving, *Anal. Chem.* **30**, 334 (1958).

334. H. H. Bauer and D. Foo, *Aust. J. Chem.* **19**, 1103 (1966).

335. D. E. Smith, *Anal. Chem.* **35**, 1811 (1963).

336. D. Kooijman and J. H. Sluyters, *Rec. Trav. Chim.* **83**, 587 (1964).

337. T. G. McCord and D. E. Smith, *Anal. Chem.* **40**, 1959 (1968).

338. T. G. McCord and D. E. Smith, *Anal. Chem.* **41**, 116, 131 (1969).

339. T. G. McCord and D. E. Smith, *Anal. Chem.* **42**, 2, 126 (1970).

340. R. Neeb, *Z. Anal. Chem.* **186**, 53 (1962).

341. R. Neeb, *Z. Anal. Chem.* **188**, 401 (1962).

342. A. M. Bond, *J. Electroanal. Chem.* **35**, 343 (1972).

343. A. M. Bond, *J. Electroanal. Chem.* **36**, 235 (1972).

344. R. Neeb, *Naturwissenschaften* **49**, 447 (1962).

345. R. Neeb, *Z. Anal. Chem.* **208**, 168 (1965).

346. R. Neeb, *Z. Anal. Chem.* **216**, 94 (1966).

347. R. Neeb, *Z. Anal. Chem.* **222**, 290 (1966).

348. G. C. Barker, In: *Polarography 1964*, G. J. Hills, ed., pp. 25–47, Macmillan Ltd., London (1965).

349. P. Zuman and Ch. L. Perrin, *Organic Polarography*, Interscience, New York (1969).

350. M. Baizer, Ed., *Organic Electrochemistry*, Marcel Dekker, New York (1973).

351. A. D. Tomilov, S. G. Mairanovskii, M. Ya. Fioshin, and V. A. Smirnov, *Electrochemistry of Organic Compounds* (in Russian), Khimiya, Leningrad (1968).

352. A. A. Vlček, *Tables of Half-Wave Potentials of Inorganic Depolarizers* (in Czech–German), NČSAV, Prague (1956).

353. G. Semerano and L. Griggio, *Selected Values of Polarographic Data*, Supplement, Ric. Sci (1957).

354. K. Schwabe, *Polarographie und chemische Konstitution organischer Vergindungen*, Akademie Verlag, Berlin (1957).

355. L. Meites and P. Zuman, *Electrochemical Data*, Part I, Vol. A, Wiley, New York (1974).

356. L. Meites and P. Zuman, *Handbook Series in Organic Electrochemistry*, Vol. 1 and 2, CRC Press, Cleveland (1977).

357. J. Heyrovský, *Actualités Scientifiques et Industrielles*, No. 90, Hermann, Paris (1934).

358. M. Shikata and I. Tachi, *Collect. Czech. Chem. Commun.* **10**, 368 (1938).

359. P. Zuman, *Substituent Effects in Organic Polarography*, Plenum Press, New York (1967).

360. P. Zuman, *The Elucidation of Organic Electrode Processes*, Academic Press, New York (1969).

361. A. Maccol, *Nature* **163**, 178 (1949).

362. L. E. Lyons, *Nature* **166**, 193 (1950).

363. G. J. Hoijtink and J. Van Schooten, *Rec. Trav. Chim.* **71**, 1089 (1952); **72**, 691, 903 (1953).

364. N. S. Hush, *J. Chem. Phys.* **20**, 1660 (1952).

365. G. J. Hoijtink, *Rec. Trav. Chim.* **74**, 1525 (1955); **77**, 555 (1958).

366. F. A. Matsen, *J. Chem. Phys.* **24**, 602 (1956).

367. A. J. Streitwieser, Jr., *Molecular Orbital Theory of Organic Chemists*, J. Wiley, New York (1961).

368. M. E. Peover, In: *Electroanalytical Chemistry*, A. J. Bard, ed., Vol. 2, pp. 1–51, Marcel Dekker, New York (1967).

369. V. D. Parker, *J. Am. Chem. Soc.* **96**, 5656 (1974); **98**, 98 (1976).

370. T. Kubota, H. Miyazaki, K. Ezumi, and M. Yamaka, *Bull Chem. Soc. Jap.* **47**, 491 (1974).

371. A. A. Vlček, *Collect. Czech. Chem. Commun.* **20**, 894 (1955); **22**, 948, 1736 (1957); **24**, 181, 1748, 3538 (1959); **25**, 3036 (1960).
372. A. A. Vlček, *Z. Anorg. Chem.* **304**, 109 (1960).
373. A. A. Vlček, *Discuss. Faraday Soc.* **48**, 937 (1958).
374. A. A. Vlček, *Electrochim. Acta* **13**, 1063 (1968).
375. A. A. Vlček, *Rev. Chim. Miner,* **5**, 299 (1968).
376. T. Saji and S. Aoyagi, *J. Electroanal. Chem.* **58**, 401 (1975); **60**, 1 (1975); **63**, 31, 405 (1975).
377. T. Saji, T. Yamada, and S. Aoyagi, *J. Electroanal. Chem.* **61**, 147 (1975).
378. H. Yasuda, K. Suga and S. Aoyagi, *J. Electroanal. Chem.* **86**, 259 (1978).
379. H. Berg. and H. Schweiss, *Nature* **191**, 1270 (1961).
380. H. Berg, *Z. Chem.* **2**, 237 (1962).
381. H. Berg. *Naturwissenschaften* **47**, 513 (1960).
382. H. Berg, *Collect. Czech. Chem. Commun.* **27**, 3404 (1960).
383. G. C. Barker, A. W. Gardner, and D. C. Sammon, *J. Electrochem. Soc.* **113**, 1182 (1966).
384. M. Heyrovský, *Proc. Roy. Soc. Ser. A*, **301**, 411 (1967).
385. Ju. V. Pleskov and Z. A. Rotenberg, *J. Electroanal. Chem.* **20**, 1 (1969).
386. G. C. Barker, *Ber. Bunsenges. Phys. Chem.* **75**, 728 (1971).
387. A. M. Brodsky and Ju. V. Pleskov, In: *Progress in Surface Science*, S. G. Davison, ed., Vol. 2, Part 1, Pergamon Press, New York (1972).
388. Z. A. Rotenberg and Yu. Ya. Gurevich, *J. Electroanal. Chem.* **66**, 165 (1975).
389. M. Heyrovský, *Nature* **209**, 708 (1966).
390. M. Heyrovský, *Croat. Chem. Acta* **45**, 247 (1973).
391. C. K. Mann and K. K. Barnes, *Electrochemical Reactions in Nonaqueous Systems*, Marcel Dekker, New York (1970).
392. G. Gritzner, *Inorg. Chim. Acta* **24**, 5 (1977).
393. A. A. Vlček, *Pure Appl. Chem.* **10**, 61 (1965).
394. R. Adams, *J. Electroanal. Chem.* **8**, 151 (1964).
395. B. Kastening, In: *Electroanalytical Chemistry*, H. W. Nurnberg, ed., pp. 421–494, John Wiley, London (1974).
396. B. Kastening, In: *Progress in Polarography*, P. Zuman and L. Meites, eds., Vol. 3, pp. 195–286, Interscience, New York (1972).
397. M. D. Morris, In: *Electroanalytical Chemistry*, A. J. Bard, ed., Vol. 7, pp. 79–160, Marcel Dekker, New York (1974).
398. A. Henglein, In: *Electroanalytical Chemistry*, A. J. Bard, ed., Vol. 9, p. 163, Marcel Dekker, New York (1976).
399. J. Lilie, G. Beck, and A. Henglein, *Ber. Bunsenges, Phys. Chem.* **75**, 458 (1971).
400. T. A. Kryukova, S. I. Sinyakova, and T. W. Arefyeva, *Polarographische Analyse*, in Russian, *Gos. Chim. Izd.*, Moscow (1959); German edition: VEB Deutscher Verlag, Leipzig (1964).
401. P. Zuman, *Organic Polarographic Analysis*, Pergamon Press, Oxford (1964).
402. P. Nangniot, *La polarographie en agronomie et en biologie*, J. Duculot, S.A., Gembloux (1970).
403. Z. Galus, *Theoretical Principles of Electrochemical Analysis*, Polish PWN, Warszawa (1971); Russian Mir, Moscow (1974); English, John Wiley & Sons, Inc., New York (1976).
404. M. Březina and J. Volke, In: *Progress in Medical Chemistry*, G. B. Ellis and G. B. West, eds., Vol. 12, pp. 247–292, North Holland Publ. Company, Amsterdam (1975).
405. E. Barendrecht, In: *Electroanalytical Chemistry*, A. J. Bard, ed., Vol. 2, pp. 53–109, Marcel Dekker, New York (1967).
406. F. Vydra, K. Štulík, and E. Juláková, *Electrochemical Stripping Analysis*, John Wiley, New York (1086).

5

Ellipsometry

ROBERT GREEF

1. Introduction

The term ellipsometry means analysis of the change in the polarization state of a light beam when it is reflected from a surface. It is therefore a kind of reflectance spectroscopy in which complete information about the reflected light is obtained. The term became necessary to distinguish the technique from other more rudimentary kinds of reflectance spectroscopy where the polarization state is ignored or where incomplete information is obtained.

The process of analyzing polarization states may be more familiar to chemists in the polarimeter, where linearly polarized light traverses a solution and is analyzed with a second polarizer. If the solution is transparent and optically active, the process of analysis means turning the analyzing polarizer until a position of extinction is found. However, if the solution is colored and exhibits circular dichroism associated with its optical activity, no clear extinction can be found, because the emerging light is elliptically rather than linearly polarized. A second optical element, a retarder, then has to be brought into use with the second polarizer to obtain all the polarization parameters. The change in the state of the light beam produced by the solution in transmission is in every way similar to the change produced by a surface when linearly polarized light is obliquely incident upon it. The method of analysis might therefore be called polarization spectrometry, and in fact this term was used until Rothen[1] coined the term ellipsometry in 1944, after which the word

ROBERT GREEF • Department of Chemistry, Southampton University, Southampton SO9 5NH, England.

gained popularity. It might be said to have become established with the publication of the Symposium[2] now known as the First International Conference on Ellipsometry in 1963.

While ellipsometry arguably[3] has its origins in the experiments of Rayleigh on the effects of greasy surface films on the polarization of light reflected from water, an overview of the modern development of the subject can be found in the proceedings of the International Conferences on Ellipsometry.[2,4-6] A quick glance at the table of contents of those volumes will show that in the late sixties and seventies there was much attention given to developing automatic instrumentation that would allow scanning in the visible and infrared regions to free the experimenter from the drudgery of manual operation of ellipsometers hitherto available. Most of the instrumentation problems have now been solved, although the number of commercially available machines is still very limited, and the emphasis is shifting toward new applications and to the much more sophisticated mathematical treatment of results that the modern instruments are bringing forth.

A number of excellent reviews and reference works dealing with the theory and practice of ellipsometry have appeared in recent years. For the beginner in the field probably the best introduction is the extended article by Muller,[7] although it is also useful as a general reference on many experimental and theoretical aspects. The standard reference work on ellipsometry, particularly for a comprehensive treatment of the theoretical and mathematical foundation of the subject, is the book by Azzam and Bashara.[8] For a good description of the use of ellipsometry in electrochemistry up to 1973 the article by Kruger[9] is recommended. The most up-to-date summaries of terminology and of instrumental techniques are provided by Hauge and co-authors[10,11] in the proceedings of the 1979 International Conference. Finally, among these general references, two reviews[12,13] providing somewhat different perspectives of recent work are worth mentioning.

This chapter provides a short introduction to modern matrix methods of describing interactions between optical elements and polarized light in Section 2, and a survey of modern instrumentation in Section 3. In Section 4 there is an attempt to indicate the scope of mathematical methods now available with which optical and other data can be used in modeling the structure of interfaces, and Section 5 is devoted to illustrating some of these methods with examples drawn from the recent literature.

2. Theory of Ellipsometry

This section is an introduction to the physical ideas underlying ellipsometry, some understanding of which is necessary for an appreciation of the kind of information about surfaces and interfacial layers it is capable of giving. A comprehensive treatment of ellipsometry is available[8] and a more compact though thorough account[7] provides a basis for detailed study.

2.1. Polarized Light

Light can be regarded as a transverse electrical and magnetic wave, the electric and magnetic vibrations being perpendicular to each other and to the direction of propagation. It is conventional to ignore the magnetic vibration altogether and discuss polarization solely in terms of the electric vibration. A complete description of a monochromatic light beam will therefore include information on its frequency, ω (or wavelength, λ), direction of propagation, amplitude, phase, and a specification of the direction, normal to the direction of propagation, in which the electric vector vibrates. If this last direction is constant in time, the wave is said to be linearly polarized. However, the possibility also exists that at a fixed point in space the electric vector may rotate continuously through 360°. This can be understood by decomposing the electric vector into two perpendicular directions x and y, so that the observed behavior can be represented as the vector sum of two waves of the same frequency but variable relative phase in the two orthogonal planes. Representing the absolute phase of the component wavetrains with respect to a common time origin as δx and δy, the time variation of the electric vectors is then given by

$$E_x = |E_x| \cos(\omega t + \delta_x)$$
$$E_y = |E_y| \cos(\omega t + \delta_y) \tag{1}$$

where E_x, E_y are the instantaneous values and $|E_x|$, $|E_y|$ are the amplitudes in the two directions. The vector sum describes the time variation of a general electric vector

$$\mathbf{E} = |E_x| \cos(\omega t + \delta_x) + |E_y| \cos(\omega t + \delta_y) \tag{2}$$

This is the parametric form of the equation for an ellipse, and if the tip of the electric vector could be plotted on a plane fixed in space, it would trace out this ellipse of polarization. This concept is the basis of the description elliptically polarized light and ellipsometry.

If the two phases δ_x and δ_y approach each other and become equal, the ellipse becomes thinner and eventually degenerates into a line whose inclination to the x-axis is given by arc tan $(|E_y|/|E_x|)$. This, of course, corresponds to linearly polarized light of variable azimuth. The other particular case of interest is circular polarization, where the component amplitudes are equal and the difference between δ_x and δ_y is plus or minus $\pi/2$. These two cases are designated right and left circularly polarized, and a similar direction property is assigned to elliptical states having otherwise identical parameters, as shown in Figure 1.

There are several different choices of parameters available to define elliptical states. One of these is to use a pictorial or geometric pair of parameters, consisting of two angles. These are, firstly, the azimuth of the major axis of the ellipse measured counterclockwise from the positive x-axis having a range 0 to π, and secondly, an angle defining the eccentricity or ellipticity

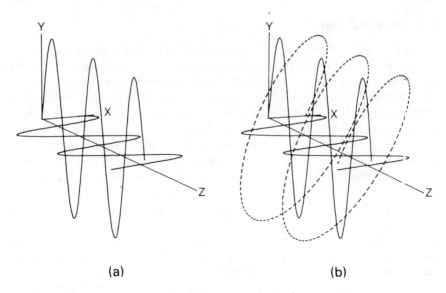

(a)   (b)

Figure 1. (a) Perspective view of the x and y components of an elliptically polarized wave. There is a phase difference of $45°$ between the components, and the amplitude ratio E_y/E_x is 1.5. (b) In this diagram the path of the tip of the electric vector resulting from these two components is plotted as a dotted line.

of the ellipse whose tangent is the length ratio of minor axis to major axis, which can take values from $-\pi/4$ to $+\pi/4$. The sign of the latter also defines the direction or handedness of the ellipse.

These geometric parameters, however, are of limited usefulness, because the physically significant quantities are the amplitudes $|E_x|$ and $|E_y|$ and the phases δ_x and δ_y. What defines the polarization state or ellipticity of a light beam is not the absolute values of these parameters, but the ratio $|E_y|/|E_x|$ and the difference $\delta_x - \delta_y$. Moreover, these two parameters are the ones that usually come directly from the measuring instrument, the ellipsometer, in the form of angles Ψ and Δ defined as follows

$$\tan \Psi = |E_y|/|E_x| \tag{3}$$

$$\Delta = \delta_x - \delta_y \tag{4}$$

2.2. The Jones Calculus

Equations (1) are cumbersome and they can be considerably simplified by converting them to phasor notation. This is a convention in which the time-dependent term ωt is removed, it being assumed that ω is known and fixed, and complex exponential notation replaces the cosine function.

Equations (1) become

$$E_x = |E_x| \, e^{i\delta_x} \qquad\qquad (5)$$
$$E_y = |E_y| \, e^{i\delta_y}$$

with the understanding that the observable quantities E_x, E_y are given by the real part of the complex quantity.

Equation (2) can be further condensed by writing the vector sum of the x and y components as a 2×1 matrix, or vector in the mathematical rather than the physical sense, as follows

$$\mathbf{E} = \begin{pmatrix} |E_x| \, e^{i\delta_x} \\ |E_y| \, e^{i\delta_y} \end{pmatrix} = \begin{pmatrix} E_x \\ E_y \end{pmatrix}$$

This representation is called the Jones vector of a pure polarization state. Insofar as it is just another form of Eq. (1), it can represent any possible pure polarization state.

Not only is this representation compact, but it opens up a powerful methodology in manipulating representations of polarized light: that of matrix algebra. Before describing the connection between physical events such as reflection or transmission through dense media, and the changes in polarization they cause, let us make one more simplification of the Jones matrix, that of normalization. This consists of dividing both elements by whatever quantity gives the neatest result and also causes the sum of the squares of the elements to equal unity. Normalization has to be used with caution when beams are combined in the Jones calculus, but normalized matrices are adequate for the simple examples taken here. The normalized Jones vectors for some simple states are as follows:

linearly polarized in the x direction $\qquad\qquad\qquad \begin{pmatrix} 1 \\ 0 \end{pmatrix}$

linearly polarized in the y direction $\qquad\qquad\qquad \begin{pmatrix} 0 \\ 1 \end{pmatrix}$

linearly polarized at an azimuth α $\qquad\qquad\qquad \begin{pmatrix} \cos \alpha \\ \sin \alpha \end{pmatrix}$

linearly polarized perpendicular to the angle α $\qquad \begin{pmatrix} \sin \alpha \\ -\cos \alpha \end{pmatrix}$

left circularly polarized light $\qquad\qquad\qquad\qquad \dfrac{1}{\sqrt{2}} \begin{pmatrix} 1 \\ -i \end{pmatrix}$

right circularly polarized light $\qquad\qquad\qquad\qquad \dfrac{1}{\sqrt{2}} \begin{pmatrix} 1 \\ i \end{pmatrix}$

Note that if two states are orthogonal in the way that linear states polarized at right angles, or right and left circular states, are orthogonal, the Jones vectors are also orthogonal in the mathematical sense. This is shown by the product of the vector and its Hermitian adjoint equalling zero. The Hermitian adjoint of the vector $\mathbf{E}$ is denoted $\mathbf{E}^{\dagger}$ and defined as the complex conjugate of the transpose. The product is therefore a scalar quantity, and if

$$\mathbf{E} = \begin{pmatrix} E_x \\ E_y \end{pmatrix}$$

then

$$\mathbf{E}^{\dagger}\mathbf{E} = E_x^*E_x + E_y^*E_y$$
$$\mathbf{E}^{\dagger}\mathbf{E} = |E_x|^2 + |E_y|^2$$

It can thus be seen that the product $\mathbf{E}^{\dagger}\mathbf{E}$ is the sum of the squared amplitudes along mutually orthogonal directions, i.e., the intensity of the state.

In any experiment with polarized light, an incident beam will enter the system under investigation, and as long as this is nondepolarizing, another beam in a pure polarization state will leave the system. Because the initial and final states can be represented as Jones vectors, there is a corresponding mathematical operation that transforms the initial vector $\mathbf{E}_{in}$ into the final state $\mathbf{E}_{out}$, and this operation is the matrix multiplication

$$\mathbf{E}_{in} = \mathbf{T}\mathbf{E}_{out}$$

This compact notation is the basis of the Jones matrix formulation. The 2×2 matrix $\mathbf{T}$ is purely characteristic of the optical system with which the light beam interacts. These matrices are calculable for simple devices, and have been tabulated.[8,14] Moreover, all the rules of matrix multiplication apply, so that if an incident beam encounters a train of devices represented by $\mathbf{T}_1$, $\mathbf{T}_2$, $\mathbf{T}_3$, in that order, the final state is given by

$$\mathbf{E}_{out} = \mathbf{T}_3\mathbf{T}_2\mathbf{T}_1\mathbf{E}_{in}$$

The Jones calculus will not be developed any further here, but will be used as a steppingstone to another matrix representation, the Mueller calculus, which is important in that it is more general and, unlike the Jones calculus, can deal with light that is partially polarized. As will be seen later ellipsometry is being increasingly applied to problems where the system may be at least partially depolarizing, so the Mueller calculus and the Stokes vector representation of light upon which it is based are likely to take on increasing importance in the future.

2.3. Partially Polarized Light: Stokes Vectors

In the foregoing discussion of polarization states it has been assumed that pure states were involved, and the state was time invariant. This implies that the light beam was either created polarized (e.g., from a laser with Brewster-angle mirrors) or had passed through a polarizing filter (e.g., a Glan–Thomson prism, which is a highly efficient linear polarizer). Natural light from an incandescent object is unpolarized. This state can be regarded as a random mixture of all possible polarization states, with the composition of the mixture itself changing randomly on a time scale comparable with the frequency of the light.

While it is easy to produce a pure state from natural light, it is not so easy to produce *completely* depolarized light from a pure state. However, a rough surface will introduce a *proportion* of depolarization into an obliquely incident pure state, with a combination of mechanisms including scattering from objects of dimensions of the order of λ, and multiple reflection from surfaces that are inclined to the average plane of the reflecting surface.

Light of any degree or state of polarization can be characterized by its Stokes vector. This is a 4×1 matrix usually given the symbol **S**, where

$$\mathbf{S} = \begin{pmatrix} S_0 \\ S_1 \\ S_2 \\ S_3 \end{pmatrix}$$

The Stokes vector is often written on one line for convenience: $\mathbf{S} = \{S_0, S_1, S_2, S_3\}$, the curly brackets indicating that it is a column rather than a row vector. The elements are scalar quantities, by contrast with the Jones vector, and can be described in terms of quantities already defined.

$$S_0 = \langle E_x^2 + E_y^2 \rangle$$
$$S_1 = \langle E_x^2 - E_y^2 \rangle$$
$$S_2 = \langle 2E_x E_y \cos (\delta_x - \delta_y) \rangle$$
$$S_3 = \langle 2E_x E_y \sin (\delta_x - \delta_y) \rangle$$

where angle brackets indicate time averages.

The parameters are all measurable and can be interpreted in several equivalent ways. S_0 is the *total* intensity of the wave. If, and only if, the state is a pure polarization state, then $S_0^2 = S_1^2 + S_2^2 + S_3^2$. Otherwise, $S_0^2 > S_1^2 + S_2^2 + S_3^2$, and a degree of polarization P can be defined

$$P = (S_1^2 + S_2^2 + S_3^2)^{1/2} / S_0$$

If I_x, I_y, $I_{\pi/4}$, $I_{-\pi/4}$, I_L, I_R are the intensities of light transmitted by polarizers separately placed in the beam and set to transmit x, y, $\frac{1}{4}\pi$, $-\frac{1}{4}\pi$ linear polarizations and left and right circular polarizations, respectively, the

Stokes parameters can be written

$$S_0 = I_x + I_y = I_{\pi/4} + I_{-\pi/4} + I_R$$

$$S_1 = I_x - I_y$$

$$S_2 = I_{\pi/4} - I_{-\pi/4}$$

$$S_3 = I_R - I_L$$

Again for pure polarization states, the relationship between the Jones and Stokes vectors can be expressed in matrix form, using yet another form of representation, the coherency matrix denoted $\mathbf{J}$. This is the other product of a Jones vector and its Hermitian adjoint, this time with the adjoint on the *right*, giving a 2×2 matrix of real elements as the result

$$\mathbf{J} = \langle \mathbf{EE}^\dagger \rangle$$

$$= \begin{pmatrix} \langle E_x E^* \rangle \langle E_x E_y^* \rangle \\ \langle E_y E_x^* \rangle \langle E_y E_y^* \rangle \end{pmatrix}$$

$$= \begin{pmatrix} J_{xx} J_{xy} \\ J_{yx} J_{yy} \end{pmatrix}$$

Since $E_x E_x^* = E_x^2$ and $E_y E_y^* = E_y^2$ the total intensity I is given by the sum of the diagonal elements of $\mathbf{J}$, i.e., the trace of $\mathbf{J}$

$$I = \operatorname{Tr} \mathbf{J}$$

Thus S_0 is defined in terms of $\mathbf{J}$ elements, and the whole Stokes vector comes out as

$$S_0 = J_{xx} + J_{yy}$$

$$S_1 = J_{xx} J_{yy}$$

$$S_2 = J_{xy} - J_{yx}$$

$$S_3 = -i(J_{xy} - J_{yx})$$

or in matrix form

$$\begin{pmatrix} S_0 \\ S_1 \\ S_2 \\ S_3 \end{pmatrix} = \begin{pmatrix} 1 & 0 & 0 & 1 \\ 1 & 0 & 0 & -1 \\ 0 & 1 & 1 & 0 \\ 0 & -i & i & 0 \end{pmatrix} \begin{pmatrix} J_{xx} \\ J_{xy} \\ J_{yx} \\ J_{yy} \end{pmatrix}$$

It might be interesting at this point to mention that the matrix formulations introduced here parallel very closely those used in quantum mechanics, and it would have been possible to use Dirac notation throughout with kets instead of Jones vectors representing states of polarization. The coherency matrix is called the projection operator,[15] and the last relationship can be derived using the Pauli spin matrices. This theme has been developed very thoroughly by Simmons and Guttmann[16] who have used this similarity as a way of introducing the methods of quantum theory.

By analogy with the Jones calculus, the Mueller calculus enables the calculation of the state of the exit beam S_{out} when the input state S_{in} and the Mueller matrix **M** of the optical system are known

$$S_{out} = \mathbf{M}S_{in}$$

The Mueller matrix is a 4×4 matrix of real elements, and **M** has been calculated for all the common optical devices in their principal orientations.[8,14] (**M** depends upon orientation as well as the nature of the device.)

It might be said that the Mueller matrix is wasted on nondepolarizing systems, as many of the elements are redundant and the Jones calculus is more compact. Moreover, there is no specific mention of phase in the Stokes vector, which can be a disadvantage. In reality these two methods are complementary, and they have to compete for popularity with other representations, such as the Poincaré sphere (a kind of three-dimensional Stokes vector), which also have their adherents.

2.4. Reflection of Polarized Light

The material constant controlling reflection and transmission of light is the complex refractive index N. For a substance labelled 1

$$N_1 = n_1 - ik_1$$

where n_1 is the refractive index and k_1 is the extinction coefficient, alternatively the real and complex part of the refractive index. The refractive index n_1 is the quantity that for transparent substances is usually measured with an Abbé refractometer, while k is related to the absorption coefficient α by the relation: $\alpha = 4\pi k/\lambda$.

When light is obliquely incident on a surface, the specular beam is reflected at the same angle of incidence and the direction of the refracted beam is given by Snell's law. The amplitude of these beams is given by Fresnel's laws. The geometry of the situation imposes an essential dissymmetry on the behavior of a pure polarization state, in that vibrations in two orthogonal directions x and y will interact differently with the surface. The two obvious directions to choose as system coordinates are in and perpendicular to the plane of incidence, which is the plane containing the normal to the surface and the incident and reflected beams. The directions called x and y in Eq. (1) are now relabeled p and s (for parallel and perpendicular to the plane of incidence, respectively).

Snell's law for homogeneous media labeled 1 and 2, is

$$n_1 \sin \Phi_1 = n_2 \sin \Phi_2 \tag{6}$$

for transparent media, or for media having nonzero extinction coefficient

$$N_1 \sin \Phi_1 = N_2 \sin \Phi_2 \tag{7}$$

where Φ_1, Φ_2 are the angles of incidence and refraction.

Representing the complex amplitudes of the incident, reflected, and transmitted beams by $\mathbf{E}_i$, $\mathbf{E}_r$, $\mathbf{E}_t$, Fresnel's equations read

$$\frac{\mathbf{E}_{rp}}{\mathbf{E}_{ip}} = \frac{\tan(\Phi_1 - \Phi_2)}{\tan(\Phi_1 + \Phi_2)} = \mathbf{r}_p \tag{8}$$

$$\frac{\mathbf{E}_{rs}}{\mathbf{E}_{is}} = -\frac{\sin(\Phi_1 - \Phi_2)}{\sin(\Phi_1 + \Phi_2)} = \mathbf{r}_s \tag{9}$$

$$\frac{\mathbf{E}_{tp}}{\mathbf{E}_{ip}} = \frac{2\sin\Phi_2\cos\Phi_1}{\sin(\Phi_1 + \Phi_2)\cos(\Phi_1 - \Phi_2)} = \mathbf{t}_p \tag{10}$$

$$\frac{\mathbf{E}_{ts}}{\mathbf{E}_{is}} = \frac{2\sin\Phi_2\cos\Phi_1}{\sin(\Phi_1 + \Phi_2)} = \mathbf{t}_s \tag{11}$$

where $\mathbf{r}_p$, $\mathbf{r}_s$, $\mathbf{t}_p$, $\mathbf{t}_s$ are the complex reflection and transmission coefficients for the parallel and perpendicular components. Why are the coefficients $\mathbf{r}$ and $\mathbf{t}$ complex? This is simply because the vibration undergoes a phase change as well as an amplitude change on reflection. Note that the ofrm of Snell's law[7] and the Fresnel equations implies that the angles Φ_1 and Φ_2 may also be complex—a fact for which there is no simple physical interpretation.

One obvious manifestation of the very different behavior of the s and p components is that of the complete disappearance of the reflected p component on reflection from a transparent substrate at the polarizing or Brewster angle Φ_B. From Eq. (8) it can be seen that this occurs when $\tan\Phi_B = n_1/n_0$.

More generally, let us consider a beam of linearly polarized light impinging at an angle of incidence Φ on an absorbing substrate such as a metal, with the plane of polarization inclined at an angle of 45° to the plane of incidence. This means that the components $|E_{is}|$ and $|E_{ip}|$ are equal and that the absolute phases of the incident components are equal, i.e., $\delta_{is} = \delta_{ip}$. The Fresnel equations show that the reflected beam components $|E_{rs}|$ and $|E_{rp}|$ are no longer equal, nor does $\delta_{rs} = \delta_{rp}$. What does this mean in terms of the polarization state? Simply that the linearly polarization state of the incident beam has been transformed into an elliptically polarized state. This can be described by an extension of our previous notation, using two quantities: the *relative* change in amplitude and the *difference* in the change in phase. The first of these is denoted by the tangent of an angle Ψ, the second by another angle Δ

$$\tan\Psi = \frac{|\mathbf{r}_p|}{|\mathbf{r}_s|}$$

$$\Delta = \delta_{rp} - \delta_{rs}$$

The reason for choosing this pair of parameters is that they come simply from the phasor representation of $\mathbf{r}_p$ and $\mathbf{r}_s$. As before we can write

$$\mathbf{r}_p = |r_p|\, e^{i\delta_{rp}}$$

$$\mathbf{r}_s = |r_s|\, e^{i\delta_{rs}}$$

$$\mathbf{t}_p = |t_p|\, e^{i\delta_{tp}}$$

$$\mathbf{t}_s = |t_s|\, e^{i\delta_{ts}}$$

Defining another complex reflectivity coefficient ρ

$$\rho = \frac{\mathbf{r}_p}{\mathbf{r}_s} = \frac{|r_p|}{|r_s|} e^{i(\delta_{rp}-\delta_{rs})}$$

we see that

$$\rho = \tan\Psi\, e^{i\Delta}$$

This is a fundamental equation of ellipsometry, and it shows that if we can measure the two polarization parameters Δ and Ψ we can find ρ, which in turn can be expressed in terms of the physical parameters of the interface.

3. Analysis of Polarized Light

Production of polarized light from natural light using sheet polarizers is a familiar operation, but is inadequate when very pure states are necessary. One must then turn to a wide range of devices based on crystal optics such as the Glan–Thompson, Glan–Foucault, or Rochon prism.[17,18] Analysis of light known to be in a purely linearly polarized state then reduces to crossing the light with a linear polarizer to obtain minimum transmission and measuring this azimuthal angle as precisely as possible. This is then 90° from the plane of polarization of the light.

With a reasonably strong light source, detection of the position of the null can be carried out visually to a few hundredths of a degree. However, this method is slow and very tiring. An electronic means of finding the null is essential for routine work.

There is an alternative to null detection, however, if the intensity of the light transmitted by the analyzing prism (analyzer) can be reliably measured. This is not as straightforward an operation as it may sound, even with a very stable light source and a good quality photomultiplier (PM) tube, as most PM tubes exhibit a degree of sensitivity to polarization. If a sufficiently large number of intensity measurements is made throughout the 360° range, then Malus's law can be used to find the position of the maximum and minimum (which should be indistinguishable from zero intensity, i.e., a null for a purely linear state). This law states that

$$I = I_{max} \cos^2 (\Theta - \Theta_{max}) \tag{12}$$

where I_{max} is the maximum intensity measured at an angle Θ_{max}.

So far only linear states have been considered, but ellipsometry is all about the analysis of elliptical states with the object of obtaining the two parameters Ψ and Δ. If elliptically polarized light is incident on an analyzer of variable azimuth Θ, the $\cos^2 \Theta$ dependence of the Malus law is retained, but the minimum intensity is no longer zero. As will be seen later, the multiple intensity measurement method can also be used to obtain the required parameters in this case. In order to use the null method another optical component has to be inserted in the beam to introduce an equal degree of ellipticity but of opposite handedness, thus restoring the light to a linearly polarized state before it enters the analyzer. This extra component is called an optical retarder.

3.1. Optical Retarders: The Quarter Wave Plate

When a beam of light passes through an optical retarder, the x and y components of the electric vector are delayed by different amounts, so that the relative phase difference between the components changes progressively within the device. Such a device therefore alters the state of polarization of the beam.

We are interested in a particular type of retarder, the quarter wave plate, where the phase difference introduced into a beam passing through is exactly $\pi/2$. One type of quarter wave plate consists of a thin plate of a birefringent material such as mica. The refractive index of mica is different in two directions at right angles (the optic axes, which are in the cleavage plane of the material), so the velocity of the components of the electric vector in these two directions differs proportionately, leading to the names fast and slow for the optic axes.

If linearly polarized light passes through the plate with direction of propagation perpendicular to the optic axes and with an azimuth of 45° to them so that the amplitude of the x and y components is equal, the light will emerge circularly polarized. If the azimuth of the input beam is any other value between zero and 45° the output beam will be elliptically polarized, while if the plane of polarization is initially aligned along either of the optic axes the state of polarization is unchanged and linearly polarized light emerges, as shown in Figure 2.

Because all refractive indices depend upon wavelength, a birefringent quarter wave retarder is accurately so at only one wavelength. Moreover, this type of device is often optically inhomogeneous and usually exhibits other imperfections of various kinds. An alternative type of retarder exploits the fact that internal reflection causes a differential phase shift between p and s components. Optical retarders based upon this principle can be made virtually

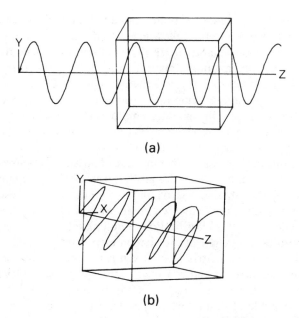

Figure 2. These are two perspective views of a wave, linearly polarized at 45° to the x axis, traveling in air. The wave encounters a slab of transparent birefringent material which has a refractive index of 1.1 in the x direction and 1.15 in the y direction. In view (a) a decrease in wavelength of the wave on entering the material is apparent, while in view (b) it is clear that ellipticity is induced by the birefringence, the degree of ellipticity developing progressively as the wave traverses the slab.

wavelength independent[19,20] and have been incorporated in wavelength-scanning ellipsometers of the null-seeking kind.[21,22]

3.2. Classification of Types of Ellipsometers

An excellent review of instrumentation up to 1979 is available,[11] and the scope and performance of modern ellipsometers can only be roughly sketched here. Hauge[11] summarizes the process of ellipsometry as consisting of five steps: (i) providing an incident beam of known polarization state; (ii) allowing the state to interact with the system under investigation; (iii) measuring the state of the emergent light; (iv) determining the parameters such as Ψ, Δ, or the Mueller matrix of the system; (v) inferring from these parameters the physicochemical changes taking place in the system.

Null-seeking or compensating ellipsometry was the only type of instrument available until around the late 1950s. At its simplest the instrument uses three optical components: a polarizer as incident state generator, a quarter wave plate through which the emergent light passes after reflection and which can be adjusted to restore the light to a linearly polarized state, and a further analyzer. These three components are mounted in divided circles of high accuracy. As will be seen in the following section, this type of instrument can be automated, and many highly developed versions have been described.

Essentially, however, nulling instruments can be applied in theory only to nondepolarizing systems. They give the two parameters Ψ and Δ, which in terms of the Stokes parameters amount to a determination of S_1, S_2, and S_3.[23]

Intensity-measuring ellipsometers employ either oscillating or rotating optical elements in either the incident or reflected beams, or in both beams, with a detector generating intensity data in analog or digital form as a function of azimuthal position. There is thus a multitude of possible designs available.

Hauge[11] has classified these designs in terms of the completeness of the determination of the Mueller matrix of the sytem under investigation. An instrument that permits all sixteen elements to be found is called a Mueller matrix ellipsometer. In this classification, a nulling ellipsometer is incomplete in that it generates only eight of the sixteen elements.

To determine all the matrix elements, it is necessary that the four Stokes parameters for the incident and reflected beam are measured. A polarization generator or a polarization detector [steps (i) and (iii) in the process of ellipsometry] capable of doing this is then described as complete.

While it is obviously desirable to generate as much information as possible on the system under measurement, there is inveitably some trade-off in performance involved in obtaining extra parameters. For the greatest accuracy in Ψ and Δ, and for the greatest sensitivity to changes in these parameters, nulling ellipsometry provides the best solution. The missing elements in this methodology are the Stokes S_0 parameters, i.e., the overall reflectance cannot be obtained. Mueller matrix ellipsometers are bound to involve intensity measurements, which are inherently prone to uncertainty from source and detector noise. Because of this, some kind of signal averaging is usually involved in intensity measurements, and this may involve repeated measurements, or curve fitting, or Fourier analysis of points at finely divided azimuth intervals. The burden of data processing is thus disproportionately greater than in the nulling technique, although the process of collecting the raw data is capable of being made much faster.

3.3. Null-Seeking Ellipsometry

Let us consider a configuration with light from a linear polarizer incident on the surface under test, the reflected beam then passing through a compensator (quarter wave plate) and finally another linear polarizer (i.e., an analyzer). This can be referred to as the PSCA configuration. The process of finding the null involves fixing the aximuth of one component (e.g., the polarizer can be fixed at 45° to the plane of incidence) and then iteratively adjusting the other two components for an absolute minimum in the intensity of light emerging from the analyzer. One way to do this is to find a position on each side of the minimum by turning one of the components, then setting this component to a value midway between these two azimuths, i.e., at the intensity minimum, and continuing by adjusting the other component in the same way until no further reduction in intensity at the minimum can be found.

As has been pointed out, this procedure can be done manually with visual or PM tube detection, but is too slow for production of results at a realistic rate. The process can be automated by substituting stepper motors, driven by a suitable servomechanism, for the experimenter's hands, and a useful ellipsometer can be made in this way.

Because the process is essentially iterative, however, it is rather slow, and a dramatic increase in speed can be obtained by modulating the azimuths of the incident and reflected beams continuously about a mean value, thus permitting the continuous adjustment of both component mountings simultaneously. One way to do this is to use Faraday effect modulators, which consists of a core of glass (or fused silica for better transparency) mounted axially in a coil driven by a sinusoidal alternating current. The plane of polarization of a light beam traversing the core will oscillate about its mean in phase with the current in the core, and will thus generate a corresponding variation in intensity at the output of the analyzer. If the scope of the azimuthal variation includes a minimum, there will be a double-frequency component of the exciting signal in the detected intensity. This can be used in an electronic servomechanism to adjust the mean position of the mounting circle so as to minimize the component at the exciting frequency, at which point the minimum azimuth is read. If the process is carried out simultaneously for both the adjustable components, which can be done by using two nonrelated exciting frequencies, the point at which minima are simultaneously detected corresponds to the required null.

Taking this idea one step further, the process of adjusting the mean azimuth can be carried out by using two more Faraday effect devices and driving them with a steady current to compensate for the error in the mechanical position of the mounting circle, as shown in Figure 3.

The Faraday effect is not the only one that can be applied to modulate the polarization state of a beam. It is possible to use electro-optic effects in crystals in the same way, or to use a retardation plate of electrically or mechanically variable retardation.

Nulling ellipsometers based on these principles are capable of high speed (down to the millisecond level over a restricted range of azimuth variation), high accuracy, and sensitivity to the level of 0.001 degree. The state of the art up to 1975 has been reviewed by Muller.[24]

It should be pointed out that the configuration taken as an example here (PSCA) is not the only one possible, nor is it the best. For reasons connected with the way errors propagate in the approach to the null, there is an advantage in placing the compensator at a fixed azimuth of 45° in the incident beam (i.e., PCSA).

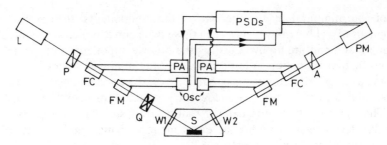

Figure 3. Auto-nulling ellipsometer using magneto-optic (Faraday effect) modulation and compensation. A minimum, manual-nulling ellipsometer consists of a monochromatic source L, a polarizer P, quarter-wave compensator Q, surface S, analyzer A, and photomultiplier detector PM. For auto-nulling, the polarization state of the light is modulated at two fixed frequencies by the power oscillators Osc driving the Faraday modulators FM, and the resulting signals are coherently detected by the phase-sensitive detectors PSD whose reference signals are drawn from the oscillators. The PSD outputs form a feedback connection to the power amplifiers PA which drive current through the Faraday compensators FC in such a sense as to rotate the plane of polarization toward the null setting. For electrochemical work the surface S is the working electrode in a cell which has a counter and reference electrode. Windows W1 and W2 are set perpendicular to the incident and emergent light beams.

3.4. Mueller Matrix Ellipsometry

A very thorough review of this topic is available,[11] and details will not be repeated here. However, to give some idea of the instrumentation involved, the salient features of a Mueller matrix ellipsometer[25] appearing since the time of that review will be described.

The configuration adopted is P M_1 M_2 S M_3 M_4 A, where P and A are linear polarizers with transmission axes perpendicular to the plane of incidence. M_1, M_2, M_3, and M_4 are electro-optic modulators with electrically variable retardance, set at azimuth $45°$, $0°$, $0°$, and $45°$ respectively, with different excitation frequencies in the range 300 Hz to 41 kHz. The modulators are of different types, M_1 and M_2 requiring 750 V and M_3 and M_4 about 3000 V to drive them. The authors show by a Stokes-Mueller analysis of the way the excitation frequencies combine in the output beam that all the Mueller elements can be separately obtained from the amplitude of the a-c component at one of 15 frequencies plus a d-c component, these frequencies being easily generated as sums and differences of the four excitation frequencies.

The electronic instrumentation therefore consists of a set of oscillators and analog multipliers to produce the needed frequencies, with a corresponding set of bandpass pre-amplifiers and phase-sensitive detectors to separate the components from the PM tube output signal, as shown in Figure 4.

This instrument was constructed not for the purposes of reflection ellipsometry but for characterizing scattering systems consisting of particulate matter in suspension, but the principles involved are exactly the same as for

Figure 4. Automated Mueller matrix ellipsometer employing four retardance modulators M1–M4. These are driven at four frequencies from the power oscillators Osc. P and A are linear polarizers in fixed orientation. The resulting signal is very complex, but contains 15 frequencies and a d-c component which are related to the 16 elements of the Mueller matrix representing the surface S. In this implementation (Ref. 25), signals produced from combinations of the four excitation frequencies form the reference signals for the 15 phase-sensitive detectors PSD, the outputs from which are recorded for further processing to yield the matrix elements.

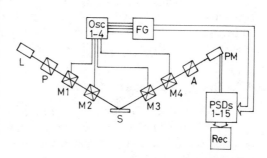

reflection. The qualitative differences from reflection ellipsometry as usually practiced are that there is virtually no pure polarization state in the collected light and that results have to be obtained over a wide range of angles between the incident beam and the collected light (i.e., the scattering angle). In this application complete Mueller matrices were obtained in about 2 mins over scattering angles from 5° to 170°.

This example is a good illustration of the trade-off between speed and accuracy. The accuracy claimed of better than 3% in the matrix elements is very low in comparison with that which is easily achieved in the analysis of specular reflection where the collected light is in a pure or nearly pure polarization state, but is much better than that previously achieved for purely scattering systems, in addition to being much faster.

4. Analysis of Results in Ellipsometry

The process of inferring from measured parameters such as Δ and Ψ, or the Mueller matrix of a surface, the physicochemical changes taking place [step (v) in ellipsometry], involves the construction of a model of the surface and the comparison of the observed parameters with those predicted from changes in optical constants and thickness of overlayers, porosity, roughness, and so on. The simple three-layer model of substrate, overlayer, and immersion medium is susceptible to exact mathematical treatment and therefore is important in this process. This model is described in Section 4.1 and some indication of developments of more complex models is given in Sections 4.2 and 5.

The simplest case of reflection occurs when polarized light traveling in an immersion medium of known refractive index N_1 is incident at an angle Φ_1 on a clean, isotropic, and homogeneous surface of refractive index N_2. The ratio of refractive indices can then be expressed[8]

$$\frac{N_2}{N_1} = \sin \Phi_1 \left[1 + \left(\frac{1 - \rho}{1 + \rho} \right)^2 \tan^2 \Phi_1 \right]^{1/2}$$

Thus it is evident that the complex refractive index of the substrate, involving two unknowns—the real and imaginary part—can be found from a single ellipsometric evaluation of ρ, which also comprises a real and imaginary part, as shown in Figure 5.

4.1. The Three-Layer Model

If the substrate carries a layer of foreign material, obliquely incident light traveling in the immersion medium (medium 1) first strikes and is partially reflected at and partially refracted into the overlayer, medium 2. Light that penetrates this layer then strikes the underlying substrate (medium 3), where it is partially reflected and partially refracted. The reflected part again meets the boundary between media 2 and 1 where the partial refraction and reflection process again takes place. If the media are not strongly absorbing, this multiple

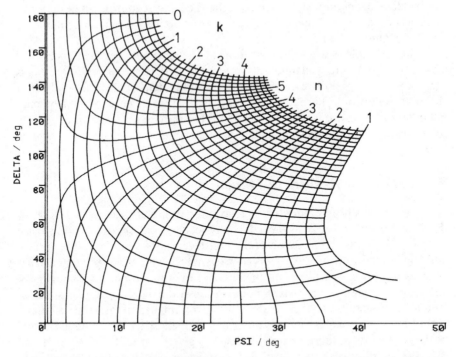

Figure 5. Nomogram showing the one-to-one mapping between the optical constants n and k of a bare substrate and the corresponding delta and psi values. The angle of incidence has been set at 70° and the refractive index of the medium at 1.34 for this calculation.

reflection and refraction process can take place many times; otherwise the amplitude of the light quickly decays effectively to zero after the first few reflections.

Light escaping from the layer–substrate combination will consist of a beam reflected from the immersion medium-to-layer interface plus one or more beams that have traversed the layer, and these beams will combine, according to the laws of vector addition with regard to amplitude and phase, to give a single polarization state in the collection optics.

The Fresnel Eq. (8)–(11) can be applied to each stage of reflection and transmission, and as originally shown by Drude, the overall reflection coefficients for the parallel and perpendicular components result from the summation of an infinite geometric series. Defining a phase delay D due to the travel of light in the film

$$D = \frac{4\pi d N_2 \cos \Phi_2}{\lambda_0} \tag{13}$$

where d is the layer thickness, λ_0 is the vacuum wavelength of the light, and Φ_2 is the complex angle of refraction in the film, the overall reflection coefficients can be written

$$\mathbf{R}_p = \frac{\mathbf{r}_{12p} + \mathbf{r}_{23p}\, e^{-iD}}{1 + \mathbf{r}_{12p}\mathbf{r}_{23p}\, e^{-iD}} \tag{14}$$

$$\mathbf{R}_s = \frac{\mathbf{r}_{12s} + \mathbf{r}_{23s}\, e^{-iD}}{1 + \mathbf{r}_{12s}\mathbf{r}_{23s}\, e^{-iD}} \tag{15}$$

The ellipsometric quantities are defined as before for reflection at a single boundary

$$\rho = \frac{\mathbf{R}_p}{\mathbf{R}_s} = \tan \Psi\, e^{i\Delta}$$

Note that in this case the quantity ρ is dependent on nine real arguments: the layer thickness d, the angle of incidence Φ_1, the wavelength λ, plus the real and imaginary parts of the three refractive indices N_1, N_2, and N_3.

Two of these quantities, Φ_1 and λ, can be fixed by the experimenter, and N_1 can almost always be measured. There will often be circumstances where N_3 can be found from measurements in the absence of the overlayer, and others where the optical constants of the layer are known but not its thickness. There is thus a large range of types of investigation, where the number of parameters to be determined can range from one to five. Since one datum point from an ellipsometric experiment provides only two quantities Δ and Ψ (or three if intensity is also measured), the unambiguous determination of all the unknowns will often involve multiple experiments in which Δ and Ψ are measured with different values for Φ_1, λ, or N_1.

The problem of deciding which type of experiment to carry out is made difficult by the form of the Drude Eqs. (14) and (15), as it is not immediately apparent what the sensitivity of Δ and Ψ will be to changes in Φ_1, λ, or N_1. It is, of course, always possible to compute Eqs. (14) and (15) in a *forward* direction, i.e., to calculate Δ and Ψ from given values of d, N_2, and N_3, and calculations made over ranges of these values will often provide rules of thumb for the behavior of the model in a particular region.

One case where the variation of Δ and Ψ can be predicted explicitly is for changes in d when all the optical constants are fixed. Rewriting Eq. (14) and (15)

$$\mathbf{R}_p = \frac{a + bX}{1 + abX} \tag{16}$$

$$\mathbf{R}_s = \frac{c + dX}{1 + cdX} \tag{17}$$

so that

$$\rho = \frac{(a + bX)(1 + cdX)}{(1 + abX)(c + dX)}$$

or

$$\rho = \frac{A + BX + CX^2}{D + EX + FX^2} \tag{18}$$

where $X = e^{-iD}$; $a = r_{12p}$; $b = r_{23p}$; $c = r_{12s}$; $d = r_{23s}$ and $A = a$; $B = (b + acd)$; $C = bcd$; $D = c$; $E = (d + abc)$; $F = abd$.

Equation (18) can be inverted to give X and hence d in terms of Δ and Ψ and the Fresnel coefficients, which can be calculated knowing N_1, N_2, N_3, and Φ_1. The value of d so obtained is a complex quantity, but if the values of Δ and Ψ obtained were exactly correct the imaginary part would be evaluated as zero. There is also a sign ambiguity in d. Any set of experimental measurements contains errors, so computationally the process of inversion of Eq. (18) consists of finding d and choosing the value that gives d as a positive length with the smallest imaginary part.

If the overlayer is transparent (N_2 is a real number) Eq. (18) predicts that there will be a cyclic variation of Δ and Ψ. This corresponds to the phase delay D moving through multiples of 2π as the thickness d increases, and is shown on a Δ against Ψ plot as a closed curve. If the layer is absorbing, the "tail" of the multiple reflected beams within it is rapidly attenuated and this appears as an inward spiralling of the Δ, Ψ plot. The closed curve is therefore characteristic of a transparent film, and is observed in such systems as the progressive oxidation of silicon and aluminum under certain conditions, as shown in Figure 6.

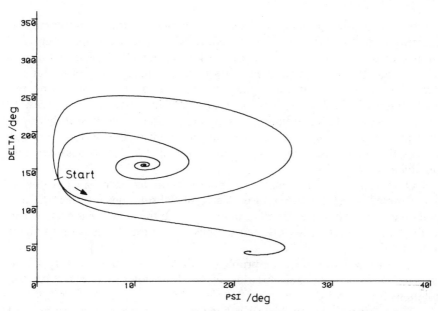

Figure 6. Three-layer model calculations for a film growing on a silicon substrate of refractive index 3.857–0.183i, immersed in a liquid with $n = 1.34$. The closed curve is for a film of silicon dioxide, $n = 1.43$. The curve passes through the starting point at a thickness of 466.8 nm and retraces itself at multiples of this thickness. The inner spiral is for a hypothetical absorbing film of optical constants 5–0.8i. Similarly the outer spiral is for a film of optical constants 2–0.8i. The latter films give a steady state in delta and psi at a thickness above about 400 nm. The angle of incidence is 70° and the wavelength of light is 632.8 nm throughout.

When more than the one unknown d has to be found from Δ and Ψ, one approach is to map regions of Δ and Ψ space over various regions of d, N_2, and N_3 that are thought to be reasonable. Within these regions solutions of Eq. (18)′ can be found using the feasibility criterion for d described above. A computer program developed by McCrackin and Colson[26] capable of carrying out such searches automatically has been in wide use by workers in the field.

Obviously such a procedure is bound to give multiple solutions if the number of parameters sought exceeds the number measured, and the question arises as to which of the other experimental variables should be brought into play to increase the number of independent measurements. Often Φ_1 is the easiest parameter to change, although λ is also an obvious candidate, and if the immersion medium can be changed without affecting the substrate in any way then N_1 could also be varied. There are also circumstances in the study of layers, such as Langmuir–Blodgett films, on metal surfaces where N_3 can be varied with no change in the properties of the layer.

Let us take an example from the effect of varying the angle of incidence in the determination of the thickness and refractive index of a transparent film on a metal surface. Multiple angles of incidence can be used in order to overdetermine the parameters and by an averaging process overcome the effects of experimental error. The procedure is to calculate the more sensitive parameter Δ as a function of thickness d for certain values of n_2 within the likely range of this parameter. Families of these curves are plotted for the different values of Φ_1 at which measurements of Δ are available, and the intersection of the lines with the observed Δ value gives a set of feasible solutions of n_2 and d for each Φ_1. In separate graph of n_2 against d there will then be separate lines for each Φ_1 value, the point of intersection of which defines the n_2, d pair, which is a solution for all the Δ, Φ_1 measurements simultaneously.

Obviously this type of analysis is quite general, as has been shown by So and Vedam,[29] but it will run into difficulty if the solutions intersect not in a point but over a large area. In the example quoted the situation is even worse, as the feasible solution lines do not intersect at all for d values up to about 100 nm.

Another way of stating the problem is that the sensitivity of Eq. (18) to changes in Φ_1 is such that variation of Φ_1 does not generate *independent* solutions for n_2 and d.

The computational problem has been described in detail by Azzam and Bashara[8] who consider the measured parameters Δ and Ψ as functions of n_3, k_3, n_2, k_2, d, and Φ_1, and show that if

$$\frac{(\partial \Delta_i / \partial n_3)}{\partial \Delta_j / \partial n_3} = \frac{(\partial \Delta_i / \partial k_3)}{(\partial \Delta_j / \partial k_3)} = \cdots = \frac{(\partial \Delta_i / \partial d)}{(\partial \Delta_j / \partial d)} \tag{19}$$

where Δ_i and Δ_j are measured Δ values at two angles of incidence, then the measurements *cannot* have a unique solution in parameters n_3, k_3, ... d.

If, however, the problem is well-conditioned, and this can always be checked computationally in the region of parameter space of interest by evaluating equations of the type (19), then a general function–minimization routine can be used to find the optimum solution. This process involves first defining an error function G, for example

$$G = \sum_{i=1}^{M} \{(\Delta_{im} - \Delta_{ic})^2 + (\Psi_{im} - \Psi_{ic})^2\} \tag{20}$$

where Δ_{im}, Ψ_{im} are measured values, Δ_{ic}, Ψ_{ic} are calculated values, and M is the number of sets of measured values. This function is then minimized using a standard least-squares analysis routine (see, for example, the review of algorithms by Pitha and Jones[30]) to give the optimum parameters. Some care has to be taken in choosing the form of G: in the example just quoted the Ψ parameter is extremely insensitive to thickness d, and it would probably be better to minimize the sum of squares of the Δ residuals only.

4.2. Multilayer Models

When the overlayer consists of distinct layers, or of a layer of a single substance but with optical constants that vary perpendicular to the surface, the optical response of the composite structure can be predicted by an extension of the foregoing theory for the three-layer model. For nonabsorbing films, solutions in analytical form have been obtained by Abeles[27] for a film with an exponential dependence in refractive index along its thickness. A more general form of solution has been described by Hayfield and White[28] in the form of a matrix method employing a 2×2 matrix to characterize each film. In this way any form of graded film profile can be approximated by a system of thin sheets of uniform optical properties.

A computer routine to deal with this situation, which is still Fresnel-like in its character in that all the films are uniform and homogeneous, is incorporated in the program of McCrackin and Colson[26] referred to previously. Two other areas where the simple three-layer model is obviously inadequate are in relation to layers that are anisotropic, and to rough surfaces. Theories have been proposed to predict the optical behavior of such systems, and they are being tested and further developed in the current literature. Examples of recent work in these areas are provided in Section 5.

4.3. Experimental Errors

There are many sources of error possible in ellipsometric measurements, and in order to take full advantage of the great sensitivity of the technique it is necessary to be aware of and to minimize as many of these errors as possible. Sources of error can be broadly classed as mechanical, such as azimuth angle errors due to imperfections in mounting circles and bearings, and optical errors such as residual birefringence in all of the components in the light beam. While azimuth errors are obviously important, they can usually be eliminated by zone averaging, i.e., averaging results with the components set at angles that are 180° apart, in the case of polarizers, or 90° apart in the case of quarter wave plates. One mechanical error that cannot be averaged, however, is the angle of incidence Φ. Error here can only be minimized by careful, robust mechanical design. A value correct to 0.01° is a realistic desirable accuracy. Note that this also implies parallelism in the incident beam to at least this tolerance.

Residual strain birefringence in polarizers and retarders is difficult to eliminate, but again the effects of this type of imperfection can be eliminated by four-zone averaging. When strain is present in windows of the measurement apparatus containing the system under test, however, the effects cannot be averaged, and the only circumvention is to measure the effect and eliminate it by calculation.

Azzam and Bashara[8] have developed a theory of the effects of component imperfections that supposes that every real optical component can be regarded as a combination of an ideal component in series with an imperfection plate that accounts for all the deviations from ideality. When represented in Jones or Mueller matrix form, the effects of the imperfection plates can be condensed to give a tractable representation of the imperfections in the instrument as a whole, so that the final values of Δ and Ψ can be corrected. This operation can therefore cut down on what might be the tedious operation of repeating measurements in four zones.

The procedures for calibration of an intensity-measuring Mueller matrix ellipsometer with imperfect compensators have been described by Hauge.[31]

4.4. Why Is Ellipsometry So Sensitive?

It might be worthwhile to make a comment about the extreme sensitivity of ellipsometry toward the presence of, and small changes in, surface films, which allows the detection and measurement of thin films down to, and below, monolayer coverage.[32] Two features of this sensitivity might seem counter-intuitive. Firstly, the technique can evidently show the presence of layers that are small compared with the wavelength of light. Secondly, the polarization change ρ is much more sensitive than the intensity change in either the s or p directions taken separately.

Some understanding of the situation can be obtained by reconsidering the three-layer model and regarding the final beam as the sum of two beams: one is reflected from the front surface of the layer and the other comprises the vector sum of all the multiple reflected beams arising in the film. The way these two beams are affected by growth in the film is quite different. The p and s components of the first beam are changed differently by reflection, but this beam is not changed at all as the film grows. The beam that penetrates the film, however, suffers a phase delay (which is equal for p and s components) proportional to thickness, while there is a relative phase change between p and s components at each reflection. This beam therefore varies in the manner of its interaction with the first beam, thus changing Δ and Ψ.

Seen in this context, the wavelength of light λ is obviously important in determining the size of the effects, but it is not necessary that the layer thickness have any particular relationship to λ in order for the values of Δ and Ψ to change dramatically. This view of ellipsometry has been called the double-beam model by Muller.[7]

5. Examples

Three examples taken from the recent ellipsometry literature are presented here to illustrate how modern instruments coupled with other methods such as computer optimization are being used fully to exploit the sensitivity of the technique.

5.1. Thin Nonabsorbing Anisotropic Layer on a Nonabsorbing Substrate

There have been a number of studies of oriented monolayers on metal and liquid surfaces. For example, Smith[32] made simultaneous ellipsometric, surface tension, and contact potential measurements on the adsorption of long chain fatty acids, α-bromonaphthalene, and pentane on clean mercury confined in a Langmuir trough. It was found that for low coverages, up to a monolayer, the parameter $\Delta - \bar{\Delta}$, where $\bar{\Delta}$ is the clean surface value, was a linear function of Γ, the surface excess of the adsorbate. Moreover, the size of $\Delta - \bar{\Delta}$ observed agreed with the value calculated according to a simple three-layer model using the bulk value of the refractive index and an *average* hypothetical thickness given by the known Γ and surface area of the molecule. At high coverages, molecules such as stearic acid displayed a phase change and accompanying reorientation shown by breaks in the surface pressure, Δ, and contact potential curves. Other assumptions concerning the refractive index, including possible anisotropic effects, were then necessary to explain the $\Delta - \bar{\Delta}$ values.

The theme of the anisotropic, uniaxial nature of large molecules has been developed by a number of authors,[33-37] and mathematical methods appropriate to stratified anisotropic media have been described.[38-42]

den Engelsen and de Koning[43] were able to find the refractive indices of anisotropic Langmuir–Blodgett films by ellipsometric measurements at a single wavelength and angle of incidence, using a Langmuir trough and varying compression of the films. Their calculations relied on data on molecular dimensions and polarizabilities obtained in other ways, and employed the Lorentz–Lorenz formula to calculate the two orthogonal refractive indices.

It has since been shown by Ayoub and Bashara[44] that it is possible to obtain optical constants of the film, independent of a model for its structure, from ellipsometry alone by exploiting the full capabilities of the variable-wavelength method of increasing the number of independent observations. The bulk material of the film (palmitic acid) is transparent at the wavelengths used (632.8, 530.9, 476.2 nm) and gives a nonabsorbing film as shown by the fact that Ψ for the water surface was unchanged after formation of the film on the surface. The wavelength dispersion of the refractive index of water is known with high precision, increasing by about 0.4% from 632 to 476 nm. This is a large effect compared with that for the compressed film. The authors show that this difference is sufficient to make measurements of Δ at different wavelengths *independent*, and demonstrate that their data are consistent with the values of anisotropic optical constants previously reported.[43]

Ayoub and Bashara[44] also emphasize the fact that measurements at varying angles of incidence do not, by contrast, generate independent parameters, a point which has been mentioned in Section 4.1.

5.2. Rough Surfaces

From the point of view of their effect on incident light, rough surfaces can be classified according to the mean height of surface irregularities in comparison with the wavelength of the light. Microscopic roughness is characterized by irregularities that are small in comparison to the wavelength, and gives specular reflection. When the size of the asperities approach or exceed the wavelength, the surface will scatter light, i.e., only a fraction of the intensity will be found at the specular angle, and additionally the reflected beam may be depolarized.

It has long been recognized that ellipsometry can be useful in qualitatively characterizing surfaces that are microscopically rough, and recent work has attempted, by detailed mathematical modeling of such surfaces, to use ellipsometry quantitatively to determine their structure.

A useful summary of earlier work can be found in a paper by Vorburger and Ludema,[45] who have carried out ellipsometry as a function of angle of incidence on metal surfaces with periodic roughness produced by machine scribing. Stylus measurements of surface texture were also carried out to give such parameters as the r.m.s. average of the surface slopes of the grooves, average roughness, and the roughness correlation length.

The authors attempted to use the Kirchhoff theory of scattering as has been done in the past, but found that it was only able to explain qualitatively some of the dramatic diffraction effects, i.e., singularities in Δ and Ψ, observed with periodic surfaces. The behavior for random surfaces was found to be complex, although there were some trends in Δ and Ψ related to surface geometry.

Note that the ellipsometer used was of the nulling type yielding only two parameters, Δ and Ψ. As the authors point out, with such a high degree of scattering the total intensity of the beam is an important parameter, and one might add that the angular dependence of intensity away from the specular angle contains much information related to surface texture.

The effect of microscopic roughness on ellipsometric parameters is much more subtle, and can in many cases be ignored as it often has been in ellipsometric studies in the past. However, recent work, notably by Aspnes and co-workers, has met the challenge of taking proper account of microscopic roughness, and has succeeded in obtaining a wealth of information on surface textures of metal and semiconductor surfaces. These studies do, however, require an impressive array of techniques.

A three-parameter ellipsometer is necessary, and it must be wavelength scanning in the whole of the visible and UV range, a powerful curve-fitting computer program providing proper statistical tests is necessary to allow determination of a meaningful parametric representation of the results.

Aspnes, Theeten, and Hottier[47] have re-examined several theories that describe roughness in terms of a surface layer having dielectric properties that are a simple function of the volume fraction of the constituents.[46] In the case of a surface without contaminants, the two dielectric functions are those of the pure substrate and the immersion medium.

This model, known as the effective medium approximation, can be formulated as follows

$$\frac{\langle \varepsilon \rangle - \varepsilon_h}{\langle \varepsilon \rangle + 2\varepsilon_h} = v_1 \frac{\varepsilon_1 - \varepsilon_h}{\varepsilon_1 + 2\varepsilon_h} + v_2 \frac{\varepsilon_2 - \varepsilon_h}{\varepsilon_2 + 2\varepsilon_h} + \cdots$$

where $\langle \varepsilon \rangle$, ε_h, ε_1, ε_2... are the complex dielectric functions of the effective medium, the host medium, and contaminants or immersion materials of types 1, 2 ... in the host, respectively, and where v_1, and v_2 are the volume fractions of these materials in the total volume of the surface region. This equation summarizes what are in effect three theories, depending on how the host medium is defined.

The first of these, the Lorentz–Lorenz (LL) theory assumes that the material of the surface is embedded in vacuum (or by extension in an immersion medium), and is the theory more usually encountered in the role of predicting dielectric functions from the dimensions of an array of point polarizable entities. The maxwell–Garnett theory (MG) proposes that the host medium can be either the material of the surface or the void itself. That is, in the first Maxwell–Garnett assumption (MG1), the substrate is supposed to be embedded *in vacuo*, which is in effect the same as the LL theory, while in the second (MG2) theory the voids are assumed to be embedded in the bulk material of the substrate. A third theory by Bruggeman and designated EMA by Aspnes, Theeten, and Hottier replaces ε_h with $\langle \varepsilon \rangle$, i.e., the effective medium itself. The differences between the three theories can therefore be seen as reducing to different choices for the host dielectric function ε_h.

If the host and inclusion dielectric functions are substantially different these theories give markedly different predictions for the effective medium dielectric function $\langle \varepsilon \rangle$, and as part of their investigation Aspnes, Theeten, and Hottier set out to discover which, if any, of their theories was sutiable to explain their spectroscopic ellipsometry results on amorphous and polycrystalline silicon films. To do so involved comparing predicted and observed Ψ and Δ values over the whole wavelength range for various models consisting of different thicknesses of void-containing layers, with proper statistical testing of goodness-to-fit as a criterion for acceptance. Silicon dioxide overlayers were also considered and found to give poorer explanations of observed results than those using various volume fractions of silicon alone. In the roughest of the samples used, which under transmission electron microscopy of a shadowed replica showed repeat features at 50–100 nm, a five-parameter fit to a graded three-layer model of the film gave the best representation of the results. This was further shown to be consistent with a model of the topography of the films consisting of hemispherical nuclei rather than triangular ridges, hemicylindrical ridges, or square pyramids.

Of the three theoretical models considered, the LL/MG1 theory was found not to be useful, but EMA and MG2 gave good results with EMA giving best fits most consistently.

A similar battery of techniques has been applied by Aspnes, Kinsbron, and Bacon[48] to the problem of the optical properties of evaporated gold films. They attempted to rationalize previously published studies on gold by relating them to the sample preparation technique. It was found necessary to split the study into the spectral regions below and above the interband transition at around 500 nm (2.5 eV). The EMA model with one free parameter to account for voids or a second parameter to account for contamination layers from the air was found to give good agreement with experiment at incident-light energies >2.5 eV, while experimental results below this threshold could not be explained in this way.

Nevertheless, certain conclusions about the relationship between sample preparation method and shape of the dielectric function were drawn, and it is clear from this work that the methods developed in the previous silicon study[47] are more widely applicable to metals.

5.3. Corrosion of Silver—A Multilayer Problem

Electrochemical corrosion processes tend to be complicated, involving as they often do several different solid phases, the production of diffusion layers, and gaseous products. Analysis of the composition and state of the interface of a corroding material is a challenging problem, and one to which ellipsometry is increasingly being applied. The assertion has sometimes been made,[49,50] and with justification, that while ellipsometry is useful in producing qualitative information on localized corrosion processes, the task of fitting ellipsometric data to a realistic model for such corrosion processes is usually intractable.

A recent paper by Muller and Smith[51] which tackles a complex problem of anodic dissolution using ellipsometry to choose between possible models, may therefore be of particular interest to electrochemists. The authors have proposed a series of eight models as being generally useful in considering anodic film formation, ranging from the simplest single film model (model 1), through more complex multilayer models, some of which have an underlying surface roughness, to the most complex model (model 8) which has six layers. All the models apart from model 1 have a mass transport boundary layer in which the concentration of salts may reach supersaturation level. Model 8 includes a so-called secondary crystal layer formed by a nucleation and growth mechanism from such a supersaturated layer. The models derive in part and are consistent with published work from other laboratories.

Description of a complete cycle from bare metal to the production of surface roughness, growth of one or more layers, creation of supersaturation, nucleation, and growth of secondary crystals, requires many parameters—up to 28 in the case of the most complex model. Not all of these parameters are of equal importance, however, and they are not all independent.

The theory developed takes into account the chemistry of products formed insofar as this is known from other work, the optical constants of these materials separately determined on compressed powders, and the effect on the optical constants of porosity, hydration, and film structure. Equations for mass transport between all the layers is worked out taking into account diffusion, migration, and convection.

In the experiments, either constant current or constant potential measurements were made under controlled mass transport conditions, with simultaneous recording by a nulling ellipsometer of Δ and Ψ. Model 8 was found best to reproduce the observations, and once all of the available chemical and other available information had been fed in, seven of the 28 parameters were left unknown. These parameters included degree of hydration and porosity of the films, time of onset, and rate of crystallization.

The experimental data were fitted to the models by minimizing an error function based on the sum of the squares of the residuals in Δ and Ψ. Two other procedures were applied to the model: error estimates were obtained for all of the parameters, and a correlation analysis was carried out to confirm that none of the parameters were correlated.

The authors point out that despite the large number of fitted parameters in the more complicated models, the quality of fit obtained varied markedly between the models, permitting a clear choice between them. Moreover, it is not claimed that the model selected is incapable of further improvement. The parameters evaluated are, however, physically meaningful and susceptible to checking by further work using other methods such as ion etching and Auger spectroscopy.

6. Outlook

Although the basic theory of ellipsometry is quite old, going back to the turn of the century, the technique has been slow to realize its full potential. This may be due to two main factors: the difficulty of using older manual instruments, and the fact that many high-precision calculations are necessary in even quite simple interpretations of the results. Another point allied to the first of these is that slight imperfections in the optical components, which might be unimportant in other applications, are capable of creating errors in absolute value of the ellipsometric parameters which are far in excess of sensitivity. Working out the effect of these errors was for a long time an involved and rather arcane process.

The last ten years has seen dramatic progress in instrumentation and in the availability of powerful computers, by the application of which these long-standing difficulties in using ellipsometry can be overcome.

The theory of ellipsometry applies not only to light, but to the whole electromagnetic spectrum; not only to reflection, but to transmission and scattering. A current focus of research is the determination of size and shape distribution of particles in aerosols and oceanic hydrosols by light-scattering measurements, which produce the Mueller matrix of the scattered beam over a range of scattering angles. There seems to be every reason to hope that theoretical results from this area will be applicable to scattering by reflection. If so, this will lead to a new role for ellipsometry, that of *in situ* characterization of macroscopically rough surfaces.

Auxiliary Notation

d	thickness of a film
D	phase delay of light
$\mathbf{E}$	Jones vector characterizing a pure polarization state
$\mathbf{E}^\dagger$	Hermitian adjoint of the Jones vector
$\mathbf{E}^*$	complex conjugate of the Jones vector $\mathbf{E}$
E_x, E_y	instantaneous values of the electric field in the direction x and y
$\lvert E_x \rvert, \lvert E_y \rvert$	amplitudes of electric field
$\mathbf{E}_{ip}, \mathbf{E}_{is}$	complex amplitude of a beam incident in the planes parallel and perpendicular to the plane of incidence
$\mathbf{E}_{tp}, \mathbf{E}_{ts}$	complex amplitudes of the transmitted beam
I	intensity of a beam of light
$I_{\max}$	maximum intensity of a light beam
$\mathbf{J}$	coherency matrix derived from the Jones vector
k_1	extinction coefficient i.e., the complex part of the refractive index of substance 1
$\mathbf{M}$	Mueller matrix representing a depolarizing or nondepolarizing system
n_1	real part of the refractive index of substance 1
N_1	complex refractive index of substance 1
P	degree of polarization
$\mathbf{r}_p, \mathbf{r}_s$	complex Fresnel reflection coefficient in directions parallel and perpendicular to the plane of incidence
$\mathbf{S}$	Stokes vector composed of four real elements S_0, S_1, S_2, S_3 representing any state of polarized or unpolarized light
$\mathbf{T}$	Jones matrix representing a non-depolarizing optical system
t	time
$\mathbf{t}_p, \mathbf{t}_s$	complex Fresnel transmission coefficients
v_1	volume fraction of substance 1 in a composite film
α_1	absorption coefficient of substance 1
Γ	surface excess of an adsorbed species

δ_x, δ_y	absolute phase of electric vectors
Δ	absolute phase difference
$\bar{\Delta}$	value of Δ for a clean surface
ε_1	complex dielectric function of substance 1
Θ	azimuthal angle between two linear polarizers
Θ_{max}	azimuthal angle at which the intensity I_{max} is observed
λ	wavelength of light
λ_0	vacuum wavelength
ρ	complex ratio of reflectivity coefficients $\mathbf{r}_p$ to $\mathbf{r}_s$
Φ_1	angle of incidence of a light beam in medium 1
Φ_B	the Brewster angle of incidence
Ψ	relative amplitude ratio of electric fields parallel and perpendicular to the plane of incidence
$\bar{\Psi}$	value of Ψ for a clean surface.

References

1. A. Rothen, The ellipsometer, an apparatus to measure the thickness of thin surface films, *Rev. Sci. Instr.* **16**, 26–30 (1945).
2. *Ellipsometry in the measurement of surfaces and thin films.* U.S. Dept. of Commerce, National Bureau of Standards Miscellaneous Publication 256 (1963).
3. A. C. Hall, A century of ellipsometry, *Surf. Sci.* **16**, 1–13 (1969).
4. Proceedings of the symposium on recent developments in ellipsometry, *Surf. Sci.* **16**, (1969).
5. Proceedings of the Third International Conference on Ellipsometry, *Surf. Sci.* **56**, (1976).
6. Proceedings of the Fourth International Conference on Ellipsometry, *Surf. Sci.* **96** (1980).
7. R. H. Muller, In: *Advances in Electrochemistry and Electrochemical Engineering*, R. H. Muller, ed., Vol. 9, pp. 168–226, Wiley, New York (1973).
8. R. M. A. Azzam and N. M. Bashara, *Ellipsometry and Polarized Light*, North Holland, Amsterdam (1977).
9. J. Kruger, Application of ellipsometry to electrochemistry, In: *Advances in Electrochemistry and Electrochemical Engineering*, R. H. Muller, ed., Vol. 9, pp. 227–280, Wiley, New York (1973).
10. P. S. Hauge, R. H. Muller, and C. G. Smith, Conventions and formulas for using the Mueller–Stokes calculus in ellipsometry, *Surf. Sci.* **96**, 81–106 (1980).
11. P. S. Hauge, Recent developments in instrumentation in ellipsometry, *Surf. Sci.* **96**, 108–140 (1980).
12. W. E. J. Neal, Optical examination and monitoring of surfaces, *Appl. Surf. Sci.* **2**, 445–501 (1979).
13. D. E. Aspnes, Interface ellipsometry; an overview, *Surf. Sci.* **101**, 84–98 (1980).
14. W. A. Shurcliff, *Polarized light: production and use*, O.U.P. London (1962).
15. F. L. Pilar, *Elementary Quantum Chemistry*, McGraw-Hill, New York (1978).
16. J. W. Simmons and M. J. Gutmann, *States waves and photons: a modern introduction to light*, Addison-Wesley, Reading, Mass. (1970).
17. R. J. King and S. P. Talim, Some aspects of polarizer performance, *J. Phys. E: Scientific Instruments* **4**, 93–96 (1971).
18. D. E. Aspnes, Effects of component optical activity in data reduction and calibration of rotating-analyzer ellipsometers, *J. Opt. Soc. Am.* **64**, 812–819 (1974).

19. R. J. King, Quarter-wave retardation systems based on the Fresnel rhomb principle, *J. Sci. Instr.* **43**, 617–622 (1966).

20. J. M. Bennett, A critical evaluation of rhomb-type quarter wave retarders, *Appl. Optics.* **9**, 2123–2129 (1970).

21. D. N. Henty and H. G. Jerrard, A universal ellipsometer, *Surf. Sci.* **56**, 170–181 (1976).

22. A. C. Lowe, Practical limitations to accuracy in a nulling automatic wavelength-scanning ellipsometer, *Surf. Sci.* **56**, 134–147 (1976).

23. W. Budde, Photoelectric analysis of polarized light, *Appl. Optics.* **1**, 201–205 (1962).

24. R. H. Muller, Present status of automatic ellipsometers, *Surf. Sci.* **56**, 19–36 (1976).

25. R. C. Thompson, J. R. Bottiger, and E. S. Fry, Measurement of polarized light interactions via the Mueller matrix, *Appl. Optics* **19**, 1323–1332 (1980).

26. F. L. McCrackin and J. Colson, A Fortran program for analysis of ellipsometer measurements and calculation of reflection coefficients from thin films, *National Bureau of Standards Technical Note*, 242 (1964).

27. F. Abelès, Optical properties of inhomogeneous films, in reference 2, above, pp. 41–58.

28. P. C. S. Hayfield and G. W. T. White, An assessment of the suitability of the Drude–Tronstad polarized light method for the study of film growth on polycrystalline metals, reference 2, pp. 157–199.

29. S. S. So and K. Vedam, Generalized ellipsometric method for the absorbing substrate covered with a transparent film system, *J. Opt. Soc. Am.* **62**, 16–23 (1972).

30. J. Pitha and R. N. Jones, A comparison of optimization methods for fitting curves to infrared envelopes, *Can. J. Chem.* **44**, 3031–3050 (1966).

31. P. S. Hauge, Mueller matrix ellipsometry with imperfect compensators, *J. Opt. Soc. Am.* **68**, 1519–1528 (1978).

32. T. Smith, Ellipsometry for measurements at and below monolayer coverage, *J. Opt. Soc. Am.* **58**, 1069–1079 (1968).

33. R. Steiger, Studies of oriented monolayers on solid surfaces by ellipsometry, *Helv. Chim. Acta.* **54**, 2645–2658 (1971).

34. D. den Engelsen, Ellipsometry of anisotropic films, *J. Opt. Soc. Am.* **61**, 1460–1466 (1971).

35. D. J. de Smet, Ellipsometry of anisotropic surfaces, *J. Opt. Soc. Am.* **63**, 958–964 (1973).

36. D. J. de Smet, Ellipsometry of anisotropic thin films, *J. Opt. Soc. Am.* **64**, 631–638 (1974).

37. D. den Engelsen, Optical anisotropy in ordered systems of lipids, *Surf. Sci.* **56**, 272–280 (1976).

38. R. M. A. Azzam and N. M. Bashara, Application of generalized ellipsometry to anisotropic crystals, *J. Opt. Soc. Am.* **64**, 128–133 (1974).

39. D. W. Berreman, Optics in stratified and anisotropic media: 4×4 matrix formulation, *J. Opt. Soc. Am.* **62** 502–510 (1972).

40. M. Elshazly-Zaghloul, R. M. A. Azzam, and N. M. Bashara, Explicit solution for the optical properties of a uniaxial crystal in generalized ellipsometry, *Surf. Sci.* **56**, 281–292 (1976).

41. D. J. de Smet, Generalized ellipsometry and the 4×4 matrix formalism, *Surf. Sci.* **56**, 293–306 (1976).

42. P. Yeh, Optics of anisotropic layered media: a new 4×4 matrix algebra, *Surf. Sci.* **96**, 41–53 (1980).

43. D. den Engelsen and B. de Koning, Ellipsometric study of organic monolayers *J. Chem. Soc. Faraday I* (1974) 1603–1614.

44. G. T. Ayoub and N. M. Bashara, Characterization of a very thin uniaxial film on a nonabsorbing substrate by multiple wavelength ellipsometry: palmitic acid on water. *J. Opt. Soc. Am.* **68**, 978–983 (1978).

45. T. V. Vorburger and K. C. Ludema, Ellipsometry of rough surfaces, *Appl. Opt.* **19**, 561–573 (1980).

46. C. A. Fenstermaker and F. L. McCrackin, Errors arising from surface roughness in ellipsometric measurement of the refractive index of a surface, *Surf. Sci.* **16**, 85–96 (1969).

47. D. E. Aspnes, J. B. Theeten, and F. Hottier, Investigation of effective-medium models of microscopic surface roughness by spectroscopic ellipsometry, *Phys. Rev. B.* **20**, 3292–3302 (1979).

48. D. E. Aspnes, E. Kinsbron, and D. D. Bacon, Optical properties of Au: sample effects, *Phys. Rev. B* **21** 3290–3299 (1980).

49. J. Kruger and J. R. Ambrose, Qualitative use of ellipsometry to study localized corrosion processes, *Surf. Sci.* **56**, 394–412 (1976).

50. J. A. Petit and F. Dabosi, An ellipsometric approach to localized corrosion processes, *Corros. Sci.* **20**, 745–760 (1980).

51. R. H. Muller and C. G. Smith, Use of film-formation models for the interpretation of ellipsometer observations, *Surf. Sci.* **96**, 375–400 (1980).

6

Raman Spectroscopy

M. FLEISCHMANN and I. R. HILL

1. Introduction

In recent years, research in electrochemistry has turned increasingly to the combination of data obtained by conventional electrochemical techniques with results derived using a range of *ex situ* and *in situ* surface analytical methods[1] (see Table 1). The major reason for this development is that electrochemical techniques inevitably measure the sum of all the processes at the interfaces and, moreover, cannot characterize the molecular species present so that structural information can only be inferred indirectly. *In situ* methods have a special role to play in this search for molecular specificity, as they are able to characterize the nature and structure of both the electrode and solution sides of the interface; by contrast, *ex situ* methods can only give information about strongly adsorbed species. In turn, vibrational spectroscopy has a special role among these *in situ* methods, as vibrational spectra can be used to "finger print" the species present while changes in the spectra of the "telltale" species give information about changes in structure and of the molecular environment of these species.

It is now about eight years since laser Raman vibrational spectroscopy was first used to investigate electrode–solution interfaces. Although infrared spectroscopy is fundamentally more sensitive than Raman spectroscopy, the strong absorption of many polar solvents (and especially of water) throughout most of the infrared spectral region makes measurements in this region difficult.

M. FLEISCHMANN and I. R. HILL • Department of Chemistry, University of Southampton, Southampton SO9 5NH, England.

Table 1
Techniques Used for the Study of the Structure of Electrode-Solution Interfaces and of Adsorption at These Interfaces

In situ methods	*Ex situ* methods
1) Raman vibrational spectroscopy	1) X-ray photoelectron spectroscopy (XPS)
2) Infrared vibrational reflectance spectroscopy	2) Ultraviolet photoelectron spectroscopy (UPS)
3) Ultraviolet-visible reflectance spectroscopy; ellipsometry as a spectroscopic tool	3) Auger spectroscopy
4) Transmission through optically transparent electrodes	4) Electron energy loss spectroscopy
5) Photoelectrochemistry	5) Low energy electron diffraction (LEED)
6) Mössbauer spectroscopy	6) Reflection high energy electron diffraction (RHEED)
7) Electron spin resonance	7) Transmission electron diffraction microscopy
8) Acoustoelectrochemical methods	8) Inelastic electron tunnelling
9) X-ray diffraction	9) Ion scattering spectroscopy
10) EXAFS	10) Secondary ion mass spectroscopy
11) Neutron diffraction	

Promising methods

12. γ-ray induced X-ray fluorescence
13. NMR broadline and high resolution
14. Positron annihilation

The necessary sensitivity for measurements by external modulated reflectance spectroscopy has only recently been achieved[2]; the scope of earlier measurements using the evanescent waves of attenuated total reflectance to probe the electrode–solution interface[3] has remained restricted. The lack of transmittance of cell window materials in the low-frequency region of the vibrational spectrum where the vibrations of metal–adsorbate bonds are expected is another experimental difficulty. For example, KRS-5, one of the best water-insoluble cell window materials from this point of view, becomes opaque near 200 cm^{-1}.† On the other hand, water gives rise to relatively weak Raman vibrational bands (i.e., water has a small Raman scattering cross section) and does not absorb radiation in the visible part of the spectrum so that spectra of species at the electrode–solution interface (or in the solution phase) may

---

† The frequency unit used in vibrational spectroscopy is the wavenumber $\bar{\nu} = \nu/c$ expressed in cm^{-1}, which is equivalent to the reciprocal wavelength (c = velocity of light). This unit is used because of its convenient size, all normal modes of vibration occurring between 4000 and 0 cm^{-1}. Vibrations occurring at higher frequencies than this are overtones (multiples of the normal mode or combinations of the normal modes).

be resolved. Laser beams are now invariably used as high intensity sources for Raman excitation and a narrow beam can be finely focused onto the region of interest. As the Raman scattered radiation is also in the visible region of the spectrum, ordinary hard glass cell windows may be used. Raman spectra can be measured to within about $50 \, \text{cm}^{-1}$ from the exciting line using spectrometers with double monochromators or down to as low as $2 \, \text{cm}^{-1}$ using a triple monochromator; the whole of the vibrational spectroscopic region can therefore be investigated.

The Raman effect is an inelastic light-scattering phenomenon produced typically by one incident photon out of 10^{10}. The total Raman scattered radiation is therefore very small, and this is the major disadvantage of Raman spectroscopy. In a typical experiment on a liquid, the scattered radiation collected by the spectrometer optics is effectively sampling a $\sim 2 \, \text{mm}$ depth of liquid centered on the focused part of the incident beam. The intensity of the spectrum from a monolayer thickness of a sample would therefore be expected to be about 10^6 times weaker; such a spectrum would be very difficult to detect from most Raman scatterers using existing methods. Moreover, for an adsorbed species on an electrode surface, this species would also need to have a Raman spectrum substantially different from that in solution for it to be detectable. Although in the first investigations Raman spectra were obtained from a few monolayers of very strong Raman scatterers on electrode surfaces (such as Hg_2Cl_2 on mercury[4]) the majority of spectra reported to date have involved special enhancement phenomena of the Raman intensity. The first of these uses the resonance Raman effect in which a colored sample absorbs some of the incident laser beam, the electronic transition involved giving rise to very strong Raman scattering (see below). The second enhancement effect is observed on certain metals, specifically silver, copper, and gold. For these metals it has become apparent that a particular degree of surface roughness can lead to the observation of good quality Raman spectra at less than monolayer coverage of adsorbate. The most widely studied system has been pyridine adsorbed at silver electrodes and this will be dealt with fully in a later section. The particular enhancement phenomenon has been termed surface-enhanced Raman scattering (SERS) and, in spite of intensive investigation, the mechanism of enhancement is not yet fully understood. SERS spectra have also been obtained in air and under ultrahigh vacuum conditions, although the original experiments were on electrochemical systems.[5] The SERS spectra of electrochemical species will form a major part of this article and nonelectrochemical systems will only be reviewed where the results are considered to be relevant to this article.

The following section will briefly outline the vibrational Raman spectroscopic technique and the interested reader is referred to one of the many comprehensive texts on vibrational spectroscopy of infrared/Raman spectroscopy, e.g., References (6) and (7), many of which owe much to the classical work by Herzberg.[8]

Succeeding sections cover: a review of the early work; the SERS spectra on silver electrodes; the origins of SERS; and finally, an interim assessment of the results obtained. The field of surface Raman spectroscopy has become so extensive that the scope of this review has necessarily had to be restricted to a small number of illustrative examples; for these examples the main information derived from the experiments is summarized and related, where possible, to models of the scattering species and processes.

2. Raman Vibrational Spectroscopy

2.1. Vibrations and Symmetry

It is well knwon that a molecule composed of N atoms has $3N$-6 fundamental vibrations or normal modes of vibration ($3N$-5 for a linear molecule because there is no rotation about the bond axis).

As an illustration, consider the four normal modes of vibration of carbon dioxide (Table 2). (The arrows refer to motion of atoms in the plane of the page, while + and − are perpendicular to this plane.) The permanent dipole moment of CO_2 is zero but the ν_2, ν_3, and ν_4 vibrational modes involve a periodic fluctuation in the electric dipole, μ, with respect to the normal coordinates expressing the motion of the atoms during the vibration, Q_k. Thus $(\partial\mu/\partial Q_k) \neq 0$ for the ν_2, ν_3, and ν_4 vibrational modes, and with infrared irradiation these oscillating electric dipoles are made to resonate by electromagnetic radiation of the same frequency, leading to absorption of radiation.

Another property that can change during excitation of a mode of vibration is the polarizability of the molecule, α. The polarizability is anisotropic, the electrons forming the bond being more easily displaced by an electric field applied along the bond axis than one across it. The polarizability in different directions can be conveniently represented by a polarizability ellipsoid, this being a three-dimensional surface whose distance from the electrical center of the molecule is proportional to α_A, where α_A is the polarizability along the

Table 2
The Normal Modes of Vibration of Carbon Dioxide

Vibrational mode number	Motion of atoms during the half period	Description of vibration
ν_1	$\bar{O} = C = \bar{O}$	Symmetric stretching mode
ν_2	$\bar{O} = \bar{C} = \bar{O}$	Antisymmetric stretching mode
ν_3	$O = \overset{\uparrow}{C} = O$ $\downarrow \qquad \downarrow$	Degenerate deformation
ν_4	$O = C = O$ $+ \quad - \quad +$	

line joining point A on the ellipsoid to the electrical center. Polarizability ellipsoids for the ν_1 and ν_3 vibrational modes of CO_2 are shown in Figure 1. The polarizability ellipsoid is seen to change in size or shape with these two vibrations and this change can be represented by $\partial\alpha/\partial Q_k$. In order for a vibration to be Raman active the quantity $(\partial\alpha/\partial Q_k)_0$ (the "0" representing the equilibrium position) must be nonzero. This quantity is plotted for small displacements in the modes of vibrations of CO_2 in Figure 2 and it can be seen that $(\partial\alpha/\partial Q_k)_0$ is only nonzero for the ν_1 mode, so only this fundamental vibration is observed in the Raman spectrum of CO_2. Carbon dioxide, being a linear molecule, is a particularly easy example to follow. Water, being a bent planar molecule, is not such a simple case. For example, consider the OH antisymmetric stretching mode of water depicted in Figure 3. The polarizability ellipsoid only appears to change direction during this vibration; it might therefore be supposed that $(\partial\alpha/\partial Q_k)_0$ is equal to zero and hence the vibration is Raman inactive. However, this is not the case because α is a tensor quantity, with components

$$\begin{vmatrix} \alpha_{xx} & \alpha_{xy} & \alpha_{xz} \\ \alpha_{yx} & \alpha_{yy} & \alpha_{yz} \\ \alpha_{zx} & \alpha_{zy} & \alpha_{zz} \end{vmatrix}$$

This polarizability tensor is symmetric with $\alpha_{xz} = \alpha_{zx}$, $\alpha_{xy} = \alpha_{yx}$, and $\alpha_{yz} = \alpha_{zy}$. In order for a vibrational mode to be Raman active, only one component of this tensor, α_{ij}, needs to satisfy the relationship

$$\left(\frac{\partial\alpha_{ij}}{\partial Q_k}\right)_0 \neq 0$$

In the case of the antisymmetric stretching mode of water, this component is α_{xz}. In practice, the activity of a particular vibration of a molecule is readily

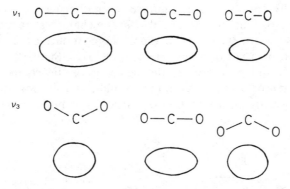

Figure 1. The changes in the polarizability ellipsoid for the symmetric stretching mode (ν_1) and the bending mode (ν_3) of CO_2.

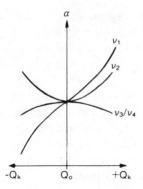

Figure 2. The polarizability α as a function of the normal coordinate Q for the fundamental modes of CO_2.

ascertained by reference to character tables, which can be found in texts on vibrational spectroscopy. A detailed discussion on symmetry and group theory is beyond the scope of this article and can also be found in texts on vibrational spectroscopy. However, in order to familiarize the general reader with some of the nomenclature we consider further the example of the water molecule. This has an axis of rotation whereby a rotation of 180° will lead to superimposition of the molecule into itself. The molecule can also be reflected into itself in two different planes, both parallel to the axis of rotation, which is designated the z axis, and which are the xz and yz planes. The molecule therefore belongs to a group with these three symmetry elements as well as the identity element symbolized by E. C_2 denotes the axis of rotation ($\frac{1}{2}$ rotation), σ_v denotes the reflection plane (where v = vertical): the group that the water molecule belongs to is designated C_{2v}. The character table for the C_{2v} group is shown in Table 3. From the point of view of spectroscopy, the labels A and B denote the symmetry of a particular vibration and the numbers 1 and -1 show whether that vibration is symmetric ($+1$) or antisymmetric (-1) with respect to that particular symmetry element. The labels x, y, and z show the behavior of translations along these axes with respect to these symmetry elements and R_x, R_y, and R_z represent rotations. For infrared excitation a vibration of symmetry A_1, A_2, B_1, or B_2 must belong to a row in the character table containing x, y, or z. This is because the change in dipole moment during the vibration may be represented by a vector along one of these axes. Similarly the labels xy, xz, yz, x^2, y^2, and z^2 represent the behavior of the components of the polarizability tensor with respect to these symmetry elements and a vibration is only Raman active if a vibration of particular symmetry contains one of

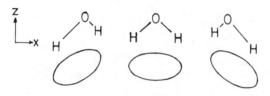

Figure 3. The changes in the polarizability ellipsoid for the antisymmetric stretching mode of water.

Table 3
Character Table of the C_{2v} Group

C_{2v}	E	C_2	$\sigma_v(xz)$	$\sigma_v(yz)$		
A_1	1	1	1	1	z	x^2, y^2, z^2
A_2	1	1	-1	-1	R_z	xy
B_1	1	-1	1	-1	x, R_y	xz
B_2	1	-1	-1	1	y, R_x	yz

these symbols in that row of the character table. For the C_{2v} group it is seen that vibrations of symmetry A_1, B_1, and B_2 will be infrared active and all vibrations will be Raman active. The fact that a vibration is allowed does not necessarily mean that it will be observed in the spectrum, since the Raman or infrared intensity may be very weak. Group theory can be used to show that the $3N$-6 = 3 vibrations of the water molecule have symmetries of A_1, A_1, and B_1, so that all three allowed vibrations are infrared and Raman active.

The top rows of all character tables contain the totally symmetric vibrations, which will variously be symbolized by A, A_1, A', A_g, A_{1g}, and Σ_g^+ depending on the group. B and Σ_u^+ represent antisymmetric vibrations and E and Π degenerate vibrations. For a molecule with an inversion center of symmetry, such as CO_2, modes with a subscript g (=gerade) are only Raman active and those with a subscript u (=ungerade) are only infrared active.

2.2. Classical Interpretation of Raman Scattering

The interaction of an applied electric field, $\mathbf{E}$, with a molecule leads to an induced electric dipole moment, μ, in that molecule such that

$$\mu = \alpha \mathbf{E}$$

where α is the polarizability tensor of that molecule. The components of this equation are

$$\mu_x = \alpha_{xx}E_x + \alpha_{xy}E_y + \alpha_{xz}E_z$$

$$\mu_y = \alpha_{yx}E_x + \alpha_{yy}E_y + \alpha_{yz}E_z$$

$$\mu_z = \alpha_{zx}E_x + \alpha_{zy}E_y + \alpha_{zz}E_z$$

and it is apparent that incident radiation with an electric field in the z direction can induce a dipole moment in the x direction. For electromagnetic radiation of frequency ν_0, the electric field strength is given by $\mathbf{E} = E_0 \cos 2\pi\nu_0 t$ and the induced dipole undergoes oscillations at the same frequency given by

$$\mu = \alpha E_0 \cos 2\pi\nu_0 t$$

The oscillating dipole also emits radiation at this frequency and in directions that can be different from that of the incident radiation; this is called Rayleigh scattering or elastic scattering.

If the molecule possesses normal modes, Q_k, then these can interact with the polarizability of the molecule. Expanding α as a Taylor series with respect to Q_k we have

$$\alpha = \alpha_0 + \sum_k \left(\frac{\partial \alpha}{\partial Q_k}\right)_0 Q_k$$

where α_0 is the polarizability tensor for the equilibrium configuration of the molecule and $(\partial \alpha / \partial Q_k)_0$ is called the derived polarizability for the kth normal mode. We now have

$$\mu = \alpha_0 E_0 \cos 2\pi \nu_0 t + \sum_k \left(\frac{\partial \alpha}{\partial Q_k}\right)_0 Q_k E_0 \cos 2\pi \nu_0 t \qquad (1)$$

A normal mode Q_k will vibrate at a frequency ν_k as $Q_k = Q_k^0 \cos 2\pi \nu_k t$ and substituting this into Eq. (1) we obtain

$$\mu = \alpha_0 E_0 \cos 2\pi \nu_0 t + \sum_k \left(\frac{\partial \alpha}{\partial Q_k}\right)_0 Q_k^0 (\cos 2\pi \nu_k t) \cdot E_0(\cos 2\pi \nu_0 t)$$

or

$$\mu = \alpha_0 E_0 \cos 2\pi \nu_0 t + \tfrac{1}{2} \sum_k \left(\frac{\partial \alpha}{\partial Q_k}\right)_0 Q_k^0 E_0 [\cos 2\pi t (\nu_0 + \nu_k) + \cos 2\pi t (\nu_0 - \nu_k)]$$

The frequencies $(\nu_0 - \nu_k)$ and $(\nu_0 + \nu_k)$ are the Raman scattering frequencies, the former called the Stokes and the latter the anti-Stokes scattering. Raman spectra are always measured relative to the frequency of the exciting line where $\Delta \nu = (\nu_0 - \nu) = \nu_k$ is termed the Raman shift. Because the scattered radiation is emitted at a different frequency from the incident radiation it is also termed inelastic scattering.

The purely classical interpretation gives no indication of the intensities of the lines; in practice the Stokes scattering is more intense than the anti-Stokes scattering and this can be readily understood in terms of the quantum mechanical representation of normal Raman scattering (Figure 4). In simple terms $(\nu_0 - \nu_k)$ represents excitation of the vibration ν_k whereas $(\nu_0 + \nu_k)$ represents de-excitation. At room temperature most molecules are in their lowest vibrational energy state so the probability of Stokes scattering is much higher than that of anti-Stokes scattering. The ratio of population in the upper and lower vibrational states follows a Boltzmann distribution and the Raman intensities on the Stokes side of the exciting line (I_S) are related to those on the anti-Stokes side (I_{AS}) by

$$\frac{I_{AS}}{I_S} = \frac{(\nu_0 + \nu_k)^4}{(\nu_0 - \nu_k)^4} \exp\left(-h\nu_k / kT\right) \qquad (2)$$

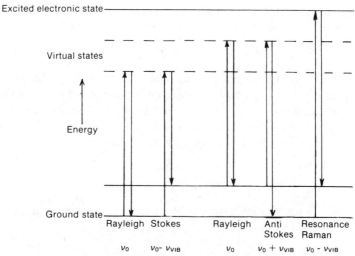

Figure 4. Energy level diagram for the quantum mechanical interpretation of the Raman effect. ν represents any vibrational, rotational, or electronic state.

The factor h/kT is equivalent to $1/200 \text{ cm}^{-1}$ at room temperature, so a vibration such as a C—H stretching mode, at a frequency of about 2800 cm^{-1}, leads to a factor of $\exp(-14)$ or $\sim 10^{-6}$. The exciting radiation has a frequency $\sim 20,000 \text{ cm}^{-1}$ so the first ratio in Eq. (2) does not change appreciably compared with the exponential. Obviously this vibration would be virtually undetectable on the anti-Stokes side of the exciting line, so spectra are usually recorded on the Stokes side.

Figure 4 also shows that if colored samples are used the frequency of the incident radiation can be adjusted to lie within an electronic absorption band of the sample (e.g., using a tunable laser, see below). In this case partial absorption of the laser beam and the consequent population of the relevant excited state can lead to resonant Raman scattering. Band intensities may then be higher by a factor of between 10^4 and 10^6 than those of normal Raman scattering and are dependent on the proximity of the excitation frequency to that of the electronic transition rather than on the usual ν^4 intensity–frequency dependence. However, not all the vibrational bands are enhanced to the same extent, the most intense being those associated with the moiety involved in the electronic transition. Furthermore, resonance Raman spectra are not always obtained as such electronic transitions may lead to fluorescence or even photodecomposition.

2.3. Depolarization Ratios

The Raman scattered radiation produced by the plane-polarized laser beam contains further information about the symmetries of the vibrational modes involved. A depolarization ratio ρ_p is defined as

$$\rho_p = \frac{I_\perp}{I_\parallel}$$

where $I_\perp$ and $I_\parallel$ are the intensities of the Raman scattered radiation with the electric field perpendicular and parallel to the exciting radiation, respectively. $I_\perp$ and $I_\parallel$ are simply determined, for example, by placing a piece of polaroid between the sample and collecting lens and rotating this through the appropriate angles. The theoretical value of ρ_p is given by

$$\rho_p = \frac{3\gamma^2}{5\alpha^2 + 4\gamma^2}$$

where α and γ are the invariants of the polarizability tensor, and which are independent of the axis system

$$\alpha = \tfrac{1}{3}(\alpha_{xx} + \alpha_{yy} + \alpha_{zz})$$
$$\gamma^2 = \tfrac{1}{2}[(\alpha_{xx} - \alpha_{yy})^2 + (\alpha_{xx} - \alpha_{zz})^2 + (\alpha_{yy} - \alpha_{zz})^2]$$
$$+6(\alpha_{xy}^2 + \alpha_{xz}^2 + \alpha_{yz}^2)$$

As the components of the tensor are determined by the symmetry of the molecule, ρ_p gives information about the symmetry. For example, for a spherically symmetric molecule, a symmetric vibrational mode will have $\rho_p = 0$ (the electric vector of the scattered radiation will be parallel to that of the exciting radiation); for symmetric vibrations of anisotropic molecules, ρ_p will lie between 0 and 0.75 while $\rho_p = 0.75$ for antisymmetric vibrations of all molecules. Symmetric modes in resonance Raman spectra usually have $\rho_p \sim 0.3$.

These rules, however, only apply to liquids and gases where molecules are freely rotating. In the crystalline state information can only be obtained by using single crystals with the incident beam parallel to a particular crystal plane. ρ_p is then measured for different orientations of the crystal and interpreted in terms of the symmetry of the vibration and the orientation of the particular moiety in the crystal.

2.4. Experimental Details

The Raman scattered radiation is normally observed at right angles to the incident laser beam and the system employed in electrochemical experiments at Southampton University is depicted in Figure 5. A pre-monochromator is used to eliminate the many background laser lines that are Rayleigh scattered by the sample. Although the intensity of these lines in the incident laser beam is low, Rayleigh scattering is about 10^3 times more intense than Raman scattering, so these lines show up as intense bands in the main

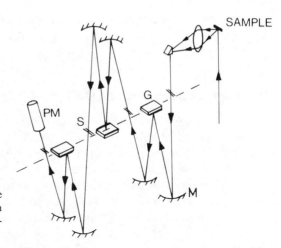

Figure 5. Schematic of a triple monochromator used in Raman spectroscopy. S: slit; PM: photomultiplier; G: grating; M: mirror.

Raman spectrum. The laser beam is normally brought to a focus at the sample and in line with the collecting lens. The scattered radiation passes through four slits and is dispersed by three gratings (additive dispersion) before reaching the photomultiplier. Photomultiplier tubes, sensitive in the red or the blue region, are used. Blue-sensitive tubes can be operated with little background noise without cooling, whereas red-sensitive tubes need cooling to around 0°C. The intensity of Raman scattered radiation is proportional to the concentration of the scattering species, the intensity of illumination at the sample, and the scattering factor $(\nu_0 - \nu_k)^4$. This last factor is important, since it shows that Raman scattering using a blue laser beam is about four times as intense as that using a red laser beam; therefore, from this point of view it is advantageous to use a blue exciting line.

Commonly used continuous wave gas lasers are: argon ion (with laser lines at 457.9, 476.5, 488.0, 496.5, 501.7, and 514.5 nm), krypton ion (530.9, 568.2, and 647.1 nm) and helium–neon (632.8 nm). The outputs from these lasers are highly monochromatic and plane polarized. The range of colors available from these lasers can be increased by adding a tunable dye laser. In these, the output from the gas laser passes through a high velocity jet of a water soluble dye, where it is absorbed and then re-emitted at various wavelengths. This radiation is then made to lase and the lasing frequency can be continuously varied across the emission band of the dye. Obviously this new laser beam is weaker than the original beam (from the "pump" laser).

The ability to change the color of the exciting radiation can be very useful (application to the study of electrode surfaces is discussed later). For white samples, the color of the exciting line should not be important for normal Raman scattering. However, trace impurities of colored species (even vacuum grease) can lead to fluorescence where the radiation is re-emitted in a band typically $\sim500 \ \mathrm{cm}^{-1}$ wide, whereas vibrational bands usually have a width of

$\sim 20\,\mathrm{cm}^{-1}$. By changing the exciting line the broad (and frequently intense) fluorescence background, which is superimposed on the Raman spectrum, can be avoided.

3. The Electrochemical Cell

Figure 6 depicts the design of electrochemical cells usually employed in Raman spectroscopic experiments at Southampton University. The working electrode is mounted in a sheath of Kel-f by heat shrinking. Adhesives such as epoxy–resin should be avoided because they usually fluoresce strongly in the Rayleigh scattered radiation produced by the cell window and the electrolyte. The working electrode is normally placed 1–2 mm behind the cell window, but is sometimes pushed up against the cell window when the Raman spectrum of the electrolyte interferes in the measurements. However, this position is not favored because of the possibility of local heating by the laser beam and also because the spectrum of the cell window is then visible, with a main broad feature near $445\,\mathrm{cm}^{-1}$ and a weaker one near $800\,\mathrm{cm}^{-1}$. The cell is usually mounted at an angle of 40° to the horizontal. In this way the reflected beams (from the cell window and the electrode surface) do not enter the collection optics, which would result in an increase in background stray light (this is particularly serious when using a double monochromator) as well as giving a Raman spectrum of the lens material.

4. Review of Early Work

4.1. Species in Solution

Raman spectroscopy has been used to monitor the changes in concentration of species in diffusion-controlled electrode processes.[9] Measurements

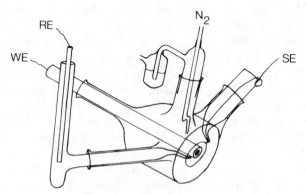

Figure 6. An electrochemical cell for Raman spectroscopy of electrode-solution interfaces. W.E.: working electrode; S.E.: subsidiary electrode; R.E.: reference electrode.

were carried out using a laser beam to sample a thin layer of electrolyte between the cell window and a polished platinum electrode. For the redox system quinol $\rightleftharpoons$ quinone the major Raman bands are at 886 and 1666 cm^{-1}, respectively, and ferrocyanide $\rightleftharpoons$ ferricyanide at 2096 and 2135 cm^{-1}, respectively. Hence, by setting the spectrometer to 1666 cm^{-1} the changes in concentration of quinone can be monitored. Changes in intensity following the application of potential steps lead to calculated diffusion coefficients in good agreement with previously reported data.

Van Duyne and co-workers have used resonance Raman spectroscopy to observe electrochemically generated radical ions,[10-14] both in the bulk electrolyte and the diffusion layer, normal Raman scattering being too weak for measurements in the thin diffusion layer. For example, tetracyanoquinodimethane$^{\mp}$ (TCNQ$^{\mp}$) has a lifetime of several hours when generated in acetonitrile and resonance Raman spectra were reported using 457.9 and 647.1 nm irradiation.[11-13] The spectra showed that the largest shifts following the one-electron transfer reaction TCNQ$^0 \rightarrow$ TCNQ$^{\mp}$ were in the C=N symmetric stretching mode (from 2223 to 2193 cm^{-1}) and in the exocyclic C=C symmetric stretching mode (from 1453 to 1389 cm^{-1}). These shifts reflect the vibrational modes most sensitive to the changes in the electronic structure. In order to observe species present in the diffusion layer the potential was modulated between the oxidized and reduced forms of the species. The oxidation of N,N,N′,N′-tetramethyl-*p*-phenylenediamine (TMPD) to the cation radical, TMPD$^+$, was investigated in this way.[14] The 1628 cm^{-1} benzene ring CC stretching mode of TMPD$^+$ is resonance enhanced using 612 nm excitation and radiation of this frequency was obtained using a Rhodamine 6G dye laser. A repetitive pulse sequence between +0.31 V (SCE) (50 ms) and −0.09 V (950 ms) was applied and the intensity–time variation at 1628 cm^{-1} was accumulated using a multichannel analyzer (Figure 7). Quantitative analysis of the time dependence of the forward and reverse portions of the transient fitted the kinetics of a simple diffusion-controlled electrode reaction.

Ikeshoji *et al.*[15] have also measured the resonance Raman spectrum of the naphthacene dianion generated by electrolytic reduction from the naphthacene anion radical in tetrahydrofuran.

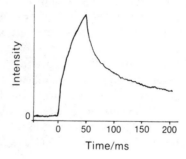

Figure 7. Signal-averaged double-potential step-resonance-Raman intensity vs. time profile for the 1628 cm^{-1} TMPD$^{\pm}$ band during the one-electron oxidation of TMPD. 1000 averages; 3 mM TMPD in 0.1 *M* tetrabutyl ammonium perchlorate in acetonitrile.[14]

4.2. Thin Films on Electrode Surfaces

The first *in situ* Raman spectroelectrochemical measurements were made on thin films of Hg_2Cl_2, Hg_2Br_2, and HgO formed on droplets of mercury electrodeposited onto platinum electrodes.[4] The spectra of as little as two monolayers of Hg_2Cl_2 could be recorded on this high surface area electrode and the spectrum was identical to that of the bulk compound. These mercury compounds have exceptionally high Raman scattering cross-sections but the experiment proved the viability of Raman spectroscopic measurements of species at electrode–solution interfaces.

The electrochemistry of lead in aqueous chloride solutions is a system for which Raman spectroscopy has given information not previously derived using conventional electrochemical techniques. The Pourbaix diagram for the $Pb-H_2O-Cl^-$ system at 0.0 V (NHE) predicts the presence of $PbCl_2$ from $pH = 0-6$, $3PbO \cdot PbCl_2$ from $pH = 6-13$, and PbO from $pH = 13-14$.[16,17] *In situ* Raman spectroscopy has identified species not allowed for in the Pourbaix diagram.[18] Using 0.1 M chloride solutions, the Raman spectrum of the expected species $PbCl_2$ was detected, but at $pH = 6.5$ and +0.6 V the spectrum was that of $[Pb(OH)]_nCl_n$[19] and at $pH = 11$ and −0.11 V the spectrum was that of $[Pb_8(OH)_{12}]_nCl_{4n}$.[19] The latter compound may be alternatively written as $3Pb(OH)_2 \cdot PbCl_2$ and is probably the phase described as $3PbO \cdot PbCl_2$ in the Pourbaix diagram.

Barradas *et al.*[20] have carried out X-ray powder diffraction studies on the surface product for lead in 3% NaCl and have also found Pb(OH)Cl to be the main product in the pH range $\sim 7-11$. The Raman spectra reported in Reference (18) were mainly obtained from dry electrode surfaces, but the spectrum of $[Pb(OH)]_nCl_n$ has since been obtained *in situ* and found to be about ten times more intense using 514.5 nm as compared to 488 nm excitation.[21] Hence $[Pb(OH)]_nCl_n$ is evidently resonance Raman active so that this species is readily observable as a thin layer on the electrode surface.

Similar Raman spectroscopic studies have been carried out by Thibeau *et al.*[17] At $pH = 7$ they reported the formation of orthorhombic PbO between −0.11 V and +0.18 V and tetragonal PbO above +0.18 V. These results may conflict with those of Reference (18), recorded at $pH = 6.5$. At $pH = 10$ tetragonal PbO was detected between −0.07 V and +0.96 V and Pb_3O_4 at potentials above +0.96 V, results which are again in conflict with the Pourbaix diagram. A bicarbonate solution was used as buffer[17] and this may have affected the results; however, the major difference between the work of the two groups is that Reid *et al.*[18] recorded their spectra after a few minutes polarization, whereas Thibeau *et al.*[17] recorded spectra after about 17 hr of current flow. Further work on this system is clearly required.

Thibeau *et al.* have also reported the Raman spectra of the $Pb-H_2O-SO_4^{2-}$ system in relation to the Pourbaix diagram.[22] In this work buffers were avoided, KOH being used where necessary to adjust the pH in the alkaline region. Data obtained by Raman spectroscopy again conflicted with the Pourbaix diagram. At $pH = 1$ after several hours at -0.8 V, the Raman spectrum of tetragonal PbO was detected as well as that of the expected $PbSO_4$. The PbO apparently forms underneath the $PbSO_4$. At $pH = 6.5$, $PbSO_4$ was formed up to $+0.24$ V, then PbO above that potential. At $pH = 11$ the Pourbaix diagram predicts the presence of $3PbO \cdot PbSO_4$ but only the spectrum of tetragonal PbO was detected. Varma et al. have published initial results from the same system.[23]

It is evident, therefore, that in situ Raman spectroscopy can be used to establish the nature of solid species present at working electrode–solution interfaces (for example, on corroding metals) and it is perhaps not surprising that the results obtained differ from those that have been obtained hitherto by means of the ex situ characterization of the species; there must frequently be changes in the nature of these species following the removal of the electrodes from the actual environments in the cells.

4.3. Adsorbed Species at the Electrode/Electrolyte Interface

The first measurements reported[5] were made on pyridine adsorbed at a high surface area silver electrode. Pyridine has a high Raman scattering cross-section, and silver is a metal that would be expected to strongly adsorb highly polarizable molecules (which are good Raman scatterers). The surface of silver can also readily be roughened electrochemically in order to increase the number of adsorbed species. $0.1 M$ KCl/$0.05 M$ pyridine was used as electrolyte and the Raman spectrum was recorded after the application of about 150 linear potential sweeps between $+0.2$ and -0.3 V (SCE). This Raman spectrum proved to be of very high quality (Figure 8) and evidently was due to a surface species in view of its potential dependence. At 0.0 V the ring C—C stretching mode region revealed bands at 1036, 1025, and $1008 \, cm^{-1}$, which all changed markedly in intensity as the potential was changed to -1.0 V. The $1025 \, cm^{-1}$ band was tentatively assigned as pyridine chemisorbed to silver via Lewis acid coordination through nitrogen, whereas the 1036 and $1008 \, cm^{-1}$ bands, being similar in frequency to those in aqueous solution, were assigned as pyridine physisorbed at the electrode/electrolyte interface in association with water. The maximum intensity of these two bands was at -0.6 V, close to the point of zero charge where maximum adsorption is expected for neutral molecules. The two bands also shifted slightly with potential, and this shift was thought to reflect reorientation of both water and pyridine in the electrical double layer.

Further adsorbates on silver were investigated by Fleischmann and co-workers and good quality Raman spectra were obtained for adsorbed formate[24] and thiocyanate ions.[25]

Other metals were also investigated. Raman spectra were obtained from pyridine adsorbed at a copper electrode,[26,27] although these results have since

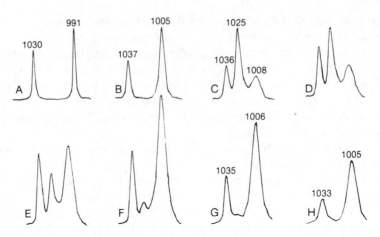

Figure 8. The first measurement of spectra at a silver-electrolyte interface.[5] Electrolyte 0.1 M KCl and 0.05 M pyridine. A: liquid pyridine; B: 0.05 M aqueous pyridine. C—H silver electrode at the following potentials (S.C.E.). C: 0.0 V; D: −0.2 V; E: −0.4 V; F: −0.6 V; G: −0.8 V; H: −1.0 V.

been disputed[28] because the spectra were obtained in the potential region where copper chloride multilayers are formed. Nevertheless, a pyridine band was found at 1015 cm^{-1}, which has no analog in Cu(I) or Cu(II) pyridine chloride complexes and probably corresponds to the 1025 cm^{-1} band of the silver electrode system. In the case of platinum, high surface area electrodes were produced by electrolysis of a lead-free chloroplatinic acid solution. These deposits were grey–bronze in color and Raman spectra were obtained of adsorbed iodine[29] and carbon monoxide.[30] These spectra were very weak and signal averaging was necessary in order to obtain an acceptable signal-to-noise ratio. For iodine, a single band was detected at 174 cm^{-1} and +0.5 V, which was assigned as the molecule in a perturbed crystal-like environment and for carbon monoxide two bands were seen at 2096 and 2081 cm^{-1} at +0.4 V (S.C.E), indicative of end-on coordination. The spectrum of iodine on platinum was later remeasured using a polished electrode[31] and the 174 cm^{-1} band was again detectable with an additional band at 142 cm^{-1}, which was assigned as I_3^-. The absorbed iodine spectra were found to be insensitive to a change in the excitation wavelength from 488 to 514.5 nm, although it has been suggested[32] that the spectra are resonance Raman active.

A different approach to the Raman spectroscopy of electrode surfaces has been made by Fujihira and Osa[33] using optically transparent SnO_2. The SnO_2 electrode was mounted onto the base of a quartz prism such that the incident laser beam was totally internally reflected at the SnO_2/electrolyte interface, the evanescent wave sampling the electrolyte. Methylene blue was used as a resonant Raman scatterer and the effect of changes in potential on the Raman intensity at 1625 cm^{-1} was monitored. Although the penetration depth of the radiation into the electrolyte is expected to be greater than the thickness of a monolayer of methylene blue, a change in Raman intensity was observed and interpreted in terms of the formation of leucomethylene blue.

Following the initial work by Fleischmann *et al.*, Jeanmaire and Van Duyne[34] and Albrecht and Creighton[35] showed that silver electrodes could give even more intense Raman spectra of adsorbed pyridine if the electrodes were subjected to a single roughening cycle; contrary to expectation prolonged roughening leads to a reduction in intensity. Enhancements 10^4–10^6 times compared to the intensities predicted from the scattering cross-sections for bulk pyridine have been claimed, and the phenomenon is now known as surface enhanced Raman scattering or SERS. Many research groups have started work in this field and SERS spectra have been obtained from many different molecules and ions adsorbed to silver, copper, and gold electrodes. SERS spectra have also been obtained in air and under ultrahigh vacuum conditions, a topic that will not be covered in this article; much of this work is covered in the recently published monograph on surface enhanced Raman spectroscopy.[36]

A list of electrochemical systems, investigated using SERS, is given in Table 4. Much work has been done on aqueous Ag/KCl/pyridine and this has been aimed primarily at clarifying the nature of the enhancement process; nevertheless the exact chemical nature of the surface species is still not fully understood. A later section will outline some of the theories put forward to explain SERS but the following sections will summarize the results that have been obtained for adsorbed pyridine, halide, and cyanide ions, the systems that have been most extensively investigated. A number of interpretations of the data is also given.

5. Surface Enhanced Raman Scattering (SERS) of Pyridine on Silver

5.1. Intensity of SERS Spectra

Jeanmaire and Van Duyne[34] have shown that the SERS intensity of pyridine adsorbed on silver depends upon the method of surface preparation, the concentration of pyridine in solution, the nature and concentration of the supporting electrolyte anion, and the electrode potential. The SERS intensity decreases in an order reminiscent of the order of specific anion adsorption, i.e., $I^- > Br^- > Cl^- > NCS^- > HPO_4^- > SO_4^{2-} > ClO_4^-$.[37] Maximum intensity was also obtained using a KCl/pyridine mole ratio of 2:1; this indicates that a surface complex is formed that involves the three ligands but it is not possible to specify how many silver atoms are involved in the complex. No SERS spectra of pyridine adsorbed on silver could be obtained using organic solvents such as methanol, dichloromethane, or acetonitrile, and this

Table 4

SERS Spectra of Species Adsorbed at Silver, Copper, and Gold Electrodes

Silver		Silver	
Adsorbate	Ref.	Adsorbate	Ref.
Acridine	(150)	p-Nitrosodimethylaniline	(43)
Amino Acids	(151)	Oxalmethyline	(34)
Aniline	(34)	Phenol	(102)
Azide	(152)	Piperidine	(34)
Benzylamine	(34)	Pyrazine	(34, 67)
Carbonate	(24)	Pyridine	(5, 24, 27, 34, 35, 38–40, 44,
Crystal Violet	(34)		45, 48–50, 53–55, 57, 59, 62,
Cyanide	(67, 76, 92–95, 97)		65–67, 71, 72, 76, 77, 80, 85,
Cyanopyridines	(34, 129)		87, 100, 102, 156–158)
N,N-Dimethylaniline	(34)	Pyrimidine	(67)
Diphenylthiocarbazone	(153)	Quinoline	(150)
EDTA	(154)	Tetraalkylammonium	(159)
Formate	(24)	(Dissolved in Acetonitrile)	
Halides (Cl, Br, I)	(55, 60, 61)	Tetracyanoplatinate	(42)
n-Heptylviologen	(161)	Tetramethylammonium	(102)
Isoquinoline	(150)	Thiocyanate	(25, 97)
2,6 Lutidine	(155)	2,4,6 Trimethylpyridine	(34)
Methyl Orange	(34)	Water	(60, 61)

Copper		Gold	
Pyridine	(26–28, 41, 49, 102, 160)	Pyridine	(41, 49, 102)

could be due to an absence of adsorption in these solvents. However, later attempts have shown that weak SERS spectra of pyridine derivatives can be obtained in acetonitrile.[38]

The whole of the SERS of pyridine on silver has been recorded at several potentials (Figure 9) and the various bands do not all follow the same intensity trends with potential (Figure 10), a fact that could not be explained.

Experiments have been carried out in which the silver electrode after roughening in the presence of pyridine was maintained at -0.2 V while the pyridine was displaced from the electrolyte. An increase in intensity was then observed on changing the potential to -0.6 V. It follows that the intensity must change with the electric field at the interface, although this may simply reflect a change in orientation of the adsorbed species. Marinyuk et al.[39] have suggested that, because bands at 617, 1005, 1215, and 1590 cm^{-1} all increase in intensity between -0.3 and -0.8 V, the pyridine molecule probably reorients from an end-on coordination to a flat configuration on the surface at -0.8 V, a conclusion somewhat at variance with measurements of adsorption of aromatic systems by direct electrochemical techniques. In the flat configuration

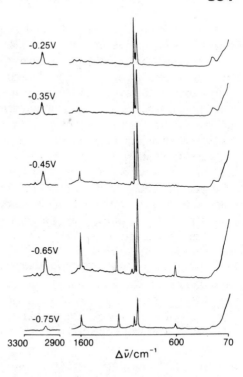

Figure 9. Raman spectra from pyridine adsorbed at a silver electrode surface. Solution 0.1 M KCl and 0.05 M pyridine.

the overlap of wave functions between electrons in the metal and the π electrons of the pyridine ring is at a maximum, leading to higher band intensities.

If F^- ions, which do not specifically absorb, are used over the same potential range, the change in relative intensity is much smaller but is still present. Clearly there must also be a complex interaction between pyridine

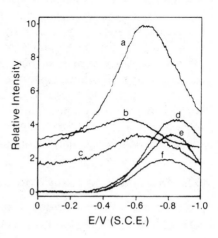

Figure 10. Relative intensity of SERS for pyridine as a function of the applied potential for the following vibrational bands: (a) 1006 cm^{-1}; (b) 1035 cm^{-1}; (c) 3056 cm^{-1}; (d) 1215 cm^{-1}; (e) 1594 cm^{-1}; (f) 623 cm^{-1}. Measurements made with 50 mW laser power at 514.5 nm and 20 cm^{-1} bandpass to minimize errors in intensity measurements due to any changes in vibrational frequencies as a function of potential[34].

and the anions in the double layer. Using piperidine instead of pyridine, no such intensity changes are observed in either F^- or Cl^- containing solutions. It seems, therefore, that π electrons of the pyridine ring must be involved in the interactions giving rise to such changes in band intensities. It has been shown that an observable SERS spectrum of pyridine on silver can be obtained after an electrochemical roughening cycle involving the oxidation/reduction of only a single monolayer of silver, provided an excitation wavelength of ~568 nm or higher is employed[40] (see next section). The SERS intensity increases fairly sharply with the charge passed for about the first ten monolayers and the intensity increase then slows down over the next 100 monolayers. Using copper and gold electrodes with pyridine, the SERS intensity on copper has been seen to increase with the number of monolayers involved in the roughening process; no increase was seen when gold electrodes were used.[41]

Van Duyne and co-workers[38,42] have shown that a weak SERS spectrum (enhanced $\sim 10^4$ rather than 10^6) can be obtained for pyridine adsorbed on a polished silver electrode. Electrodes were polished using alumina powder down to 0.05 μ grain size and then ultrasonically cleaned in de-ionized water and dried in a stream of prepurified nitrogen. With the electrode 0.5 mm away from the cell window, weak Raman bands were detected at 1008 and 1036 cm^{-1} at −0.6 V, which disappeared at 0.0 V showing that they were due to surface species.

Jeanmaire and Van Duyne[34] have reported SERS spectra of several different aromatic and aliphatic amines adsorbed on silver electrodes (see Table 4), showing that pyridine is not a unique adsorbate! Good-quality spectra of adsorbed methyl orange and crystal violet were also obtained using only 1 mW of incident laser power. It follows that the phenomenon of SERS is complementary to the usual resonance Raman scattering. These results are complemented by the work of Hagen et al.[43] on the adsorption of p-nitrosodimethylaniline at both silver and platinum electrodes. This molecule is resonance Raman active, the aqueous solution giving a spectrum using 488 nm excitation, which is about ten times as intense as that obtained using 514.5 nm. When absorbed to a silver electrode the SERS spectra were of similar intensity using 488 and 514.5 nm excitation, whereas on platinum the relative intensities were in the same ratio as those for the solution free species. Although intensities were higher for silver than for platinum, the results show that silver apparently perturbs the electronic energy levels of the adsorbate whereas platinum does not. This is not unreasonable because silver is a soft metal that readily forms covalent bonds with many species, which would lead to a shift and/or broadening of the electronic energy levels of the adsorbate.

5.2. The Effect of Laser Excitation Frequency

In normal Raman scattering the intensity is proportional to the fourth power of the excitation frequency. In resonance Raman scattering the intensity is also dependent on the electronic adsorption band involved, in a way that may follow the shape of this band or be more complex. A departure of the Raman scattered intensity from the ν_{Ex}^4 dependence is normally indicative of resonance Raman scattering. As the sensitivity of the photomultiplier detectors is also frequency dependent it is advantageous to use internal intensity standards that are known to follow the ν_{Ex}^4 law. Creighton *et al.*[44] found that the SERS intensity of pyridine on a silver electrode increased about 15 times, going from green (22,000 cm^{-1}) to red (15,500 cm^{-1}) excitation, relative to internal standards of ClO$_4^-$ (935 cm^{-1}) and water (3450 cm^{-1}). It is evident, therefore, that the ν_{Ex}^4 dependence is not followed. Pettinger *et al.*[45] have found a similar dependence, and in addition, have measured the electro-reflectance spectrum of the electrode before and after roughening. No big changes were seen but a difference spectrum showed an absorption band on the roughened electrode near 750 nm (Figure 11). It appears, therefore, that there is an analogy between SERS and resonance Raman spectroscopy. Furthermore, Van Duyne and co-workers[38] have found a similar dependence for the 1037 and 1008 cm^{-1} bands of pyridine, whereas the 1215 cm^{-1} band has a normal ν_{Ex}^4 dependence. This again suggests that a resonance Raman process is operative where the ring CC modes are selectively enhanced because the electronic absorbance involves these bonds.

A SERS spectrum of pyridine on copper cannot be obtained using excitation wavelengths less than 606 nm because of absorption of the laser beam by the metal.[28] However, using red excitation a SERS spectrum has been obtained from an unanodized surface; anodizing increased this intensity by 40 times.

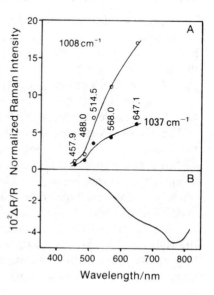

Figure 11. (a) Relative Raman intensities for pyridine at silver (111) surfaces as a function of excitation wavelength. The intensities have been normalized with respect to the ν^4 law using 935 cm^{-1} band of added LiClO$_4$ as a normal Raman scattering internal standard.[45] (b) Relative reflectance change $\Delta R/R$ for silver (111) as a function of wavelength. ΔR = reflectance before anodization − reflectance after anodization. Anodization charge equivalent to eight monolayers of silver. $E = -0.2$ V (S.C.E.).

5.3. SERS from Different Silver Crystal Planes

It has been shown that pyridine adsorbed on polycrystalline and (100) and (111) silver single crystal electrodes, gives different SERS spectra.[45] In order to conserve the single-crystal nature of the surface the roughening cycle was restricted to two monolayers of silver. The results showed that the peak positions of the CC stretching modes changed with the nature of the surface and that these also varied in different ways with the applied potential. The biggest changes were for the 1026 cm^{-1} band on the (100) face, which shifted to 1018 cm^{-1} on the (111) surface, as well as the increase in bandwidth for all three bands in this region on changing from the (100) to the (111) face and to polycrystalline silver.

These results may be interpreted in terms of electrochemical measurements on the Ag/Cl^{-}/pyridine system. Double-layer capacitance studies show that pyridine is most strongly adsorbed on (110) faces, less strongly on (100), and weakest on (111) faces in the presence of chloride ions[46]; the adsorption of chloride ions is in turn stabilized by the adsorption of pyridine. The 1026 cm^{-1} band of chemisorbed pyridine is shifted from 1008 cm^{-1} in solution; hence a band at 1018 cm^{-1} implies a weakened interaction with the (111) face. From our own data we have seen a weak unresolved feature near 1017 cm^{-1} in the SERS spectrum of pyridine on polycrystalline silver. The broadness of the bands for pyridine on the (111) face may also be interpreted as being due to weakened adsorption leading to a wide distribution of adsorption energies and greater molecular motion.

5.4. The Effect of Laser Power

SERS spectra of silver are generally obtained using laser powers ~100 mW, so as to avoid laser damage of the surface. The SERS intensity has been shown to be linear with laser power[38]; in the Ag/pyridine system the surface has been shown to become damaged at potentials positive to −0.2 V (SCE) when laser powers above 100 mW are applied for several hours.

Fleischmann and Hill[47] have seen laser damage even for 100 mW excitation. Using a focused blue 488 nm line at −0.2 V with Ag/KCl/pyridine, the 1025 cm^{-1} band has been shown to decrease in intensity with time, whereas the 1008 and 1037 cm^{-1} bands increased in intensity. Evidently the chemisorbed pyridine species is photolyzed and replaced by the physisorbed species. Using a red exciting line of the same power, there is no, or greatly reduced, photolysis. This result has some bearing on those of Creighton et al.[44] on the excitation frequency dependence of the SERS intensity. These authors found that the 1025 cm^{-1} band of pyridine was enhanced to a greater extent than the 1008 cm^{-1} band using red rather than blue excitation. It has been found[47] that the 1025 cm^{-1} band is enhanced to the same extent as the 1008 cm^{-1} band and the earlier results must have been affected by partial photolysis of the chemisorbed species using blue excitation. Similarly, the conflicting reports in the literature about the formation of chemisorbed pyridine are in part explained by the photolysis of this species.

In order to avoid these effects we have found it advantageous to record SERS spectra without any focusing of the laser beam. In this way the laser power is spread out over a much wider surface area of the sample, thereby minimizing photolysis, whereas only about half of the SERS intensity is lost.

5.5. Depolarization Ratios

The depolarization ratios of the intense bands of aqueous pyridine at 3076, 1037, and 1005 cm^{-1} are all close to zero, showing them to be polarized and hence of A_1 symmetry. In contrast these bands in the SERS spectra on silver, copper, and gold are reported to be depolarized or close to being depolarized.[38,41,44] However, conclusions about symmetry and structure can only be drawn from depolarization ratios measured for rotating molecules in liquids and gases. In the case of adsorbed pyridine, the molecules can only rotate about the coordination bond axis and, in addition, we do not have an ordered two-dimensional lattice as in a single crystal because of surface roughness. Hence, higher depolarization ratios are to be expected.

5.6. Angular Dependence of Incident Radiation

Pettinger *et al.*[48] have shown the SERS intensity to be sensitive to the angle of incidence of the laser beam. With a roughening cycle involving only a monolayer of silver the SERS of pyridine (using 568 nm excitation) was at a maximum at 63°, with a halfwidth of about 5°.[48] With a roughening cycle involving five monolayers of silver, additional peaks were seen and when 20–100 monolayers were involved a broad unstructured curve was produced, extending over the whole range of the angle of incidence (0–90°). Use of s- and p-polarized radiation led only to slight changes in intensity, but not in shape or position of the peaks. The experimental setup did not, however, distinguish whether the surface Raman intensity showed a resonance-like character on the angle of incidence or on the angle of observation, or both. The authors suggested that surface plasmons (group oscillations of electrons) may be involved in the SERS process because surface plasmons can be excited on silver, copper, and gold using visible radiation at specific angles of incidence.

Further work on Cu and Au electrodes using a 647 nm exciting line (copper and gold absorb blue radiation) revealed optimum angles of incidence of 61° and 59°, respectively.[41,49]

Trott and Furtak have observed a similar angular dependence of SERS,[50] but have also employed fiber bundle probes in order to detect the Raman scattered radiation at different angles, although this system leads to rather low-intensity spectra. The SERS was seen to monotonically decrease with the angle of detection, which implies that the detected Raman radiation is coupled out through the surface and not directly from the molecule. The angle-resolved pattern for the SERS spectra of pyridine and cyanide adsorbed to silver was seen to have the same shape as that for the elastic (Rayleigh) scattering and that of the background spectrum. These results add further evidence that there is a direct connection between geometrically defined optical resonances and SERS. Unfortunately, as long as such resonances are required for the detection of Raman scattering associated with adsorbates (owing to experimental limitations), it will be impossible to extract information about the adsorbate orientation from angle resolved data.

5.7. The Halide Ion and Pyridine

In earlier studies on SERS of the Ag/Cl^-/pyridine system, a band was seen near 239 cm^{-1},[34,44] that was not present in liquid or aqueous pyridine. This band was assigned as an $Ag-N$ stretching mode of adsorbed pyridine because of its proximity to bands seen in silver and pyridine complexes.[44] At -0.6 V (SCE), Van Duyne reported a band at 216 cm^{-1}, which was differentiated from that seen at 239 cm^{-1} at potentials more positive than -0.4 V.[38] The former band was assigned as the $Ag-N$ stretching mode of adsorbed pyridine and the latter band as residual AgCl at the surface, because bulk AgCl also has a band at this frequency. Hexter and co-workers[51,52] also predicted that the 240 cm^{-1} band arises from a monolayer of AgCl and Regis and Corset[53] detected a band at 227 cm^{-1} and -0.05 V in the SERS spectrum of Ag/Cl^-/pyridine, which disappeared when the chloride ions were replaced by bromide.

These results are confusing. It is apparent that there is a band arising from a silver and chloride species in this region of the spectrum and possibly a second band arising from an $Ag-N$ stretching mode. More detailed recent work, however, has resolved this problem. Dornhaus and Chang[54] have shown that the band seen at 243 cm^{-1} and -0.05 V for the Ag/Cl^-/pyridine system progressively shifts to 216 cm^{-1} at -0.65 V and to 210 cm^{-1} at -1.0 V. In addition, the 216 cm^{-1} band is not seen in bromide or iodide electrolytes. It would appear, therefore, that the spectra represent an adsorbed chloride species with the $Ag-Cl$ bond being weakened as the potential is taken more negative. Wetzel et al.[55] have seen the band shift from 239 cm^{-1} at -0.2 V to 230 cm^{-1} at -0.5 V for Ag/Cl^- in the absence of pyridine, but the band was too weak to be detected at more negative potentials. It is known from electrochemical data that pyridine increases the specific adsorption of chloride ions[56] and, indeed, the SERS spectra have shown that the chloride ions remain adsorbed at more negative potentials in the presence of pyridine. Atkinson et al.[57] have noted that the intensity ratio of the 1038 and 1011 cm^{-1} bands of pyridine behave in a similar manner to that of the 240 cm^{-1} band on changing the potential, indicating that a surface complex is present at potentials between -0.6 and 0 V that involves both pyridine and chloride. However, the nature of this Ag/Cl^-/pyridine species is still unknown. Finally, Wetzel *et al.*[55] have detected two bands at potentials more negative than -0.7 V, which arise from adsorbed pyridine. At -1.0 V these bands are seen at 168 and 248 cm^{-1} and are both anion and cation independent. We have also observed these bands in our own work using pyridine with fluoride ions.[58] The fact that two bands are seen may point to a surface complex involving one silver atom and two pyridine molecules, such that the two bands represent symmetric and antisymmetric $N-Ag-N$ stretches. Pettinger and Wetzel[59] have accurately measured the low-frequency region (down to 20 cm^{-1}) of the SERS spectrum of the Ag/Cl^-/pyridine system and, at a potential of -0.3 V, report bands at 231, 150, and 121 cm^{-1} with a shoulder at 54 cm^{-1}. This compares with bands reported for the SERS spectrum of Ag/Cl^- at -0.2 V at 240, 150, and 110 cm^{-1},[60] which are also present on the dry electrodes. Hence it appears that at -0.3 V and in the presence of pyridine, adsorbed chloride ions are the main species giving rise to the spectra in this region.

It has also been reported[54] that the band at 216 cm^{-1} in chloride solutions shifts to 182 cm^{-1} in bromide and to 116 cm^{-1} in iodide-containing solutions in accordance with the increase in mass. Dornhaus, however, failed to obtain a SERS spectrum of the halide ion in the absence of pyridine. Such spectra may be obtained by increasing the concentration of the halide ion, e.g., by using 1 M instead of 0.1 M KCl[55,60,61] or by using a red laser beam to obtain a larger degree of enhancement compared with blue excitation.[55] The SERS of the halide species will be the subject of a subsequent section.

5.8. Lewis Acid Adsorbed Pyridine

As has been stated in an earlier section, the SERS spectrum of the Ag/Cl^-/pyridine system originally reported by Fleischmann *et al.*[5] contained a band at 1025 cm^{-1}, which was assigned as pyridine chemisorbed to Lewis acid sites via the nitrogen lone pair. It can be seen in Figure 8 that this band reduces in intensity and disappears from the spectrum between 0.0 and -0.8 V. Furthermore, the band does not reappear when the potential is again increased to 0.0 V. Marinyuk *et al.*[62] employed D_2O in order to determine whether the 1025 cm^{-1} band arises from Lewis acid pyridine rather than the pyridinium ion, which also has a band at this frequency in solution. The SERS spectrum of pyridine in D_2O still shows a band at 1025 cm^{-1}, rather than 1023 cm^{-1}, so that this band does indeed arise from Lewis acid coordinated pyridine. These authors have also pointed out that the adsorbed halide ions could themselves be acting as Lewis acid sites, rather than the metal. The chemisorbed pyridine can, in fact, be separated from physisorbed pyridine by progressively diluting the solution using the base electrolyte alone.[47,63] In this way the bands at 1008 and 1036 cm^{-1} are reduced in intensity while the intensity of the 1025 cm^{-1} band is maintained. The spectrum thus confirms that the pyridinium ion is not present because no band of similar intensity to the $1025\,cm^{-1}$ band is seen near $1010\,cm^{-1}$.

Experiments recently carried out in our own laboratory have shown that the presence of Lewis acid pyridine in the SERS spectrum is markedly dependent on exposure of the silver halide layer to the laser beam.[47] This leads to photolysis of the silver halides, the speed of which is in the order bromide > chloride > iodide. When using these electrolytes with pyridine the Lewis acid band at $1025\,cm^{-1}$ is reduced in intensity by a focused laser beam in the same order. Hence the Lewis acid coordinated pyridine is photolyzed by the laser beam but it is less photosensitive than the silver halides. The Lewis acid pyridine species is obviously dependent on the presence of the halide ion and, in fact, a band at $1025\,cm^{-1}$ has not been detected using fluoride as electrolyte. The $1025\,cm^{-1}$ band is also reduced in intensity when the chloride, bromide, or iodide ions are displaced by fluoride in the electrolyte. It follows that the surface species is more complex than pyridine simply coordinated to the metal.

5.9. pH Dependence of the SERS of Pyridine

The situation is, however, even more complex as it has been found that the pyridinium ion, rather than pyridine, is preferentially adsorbed at the silver electrode surface in the presence of specifically adsorbed chloride ions.[53] The pyridinium ions are strongly bound because, when the pH is increased from 1.2 to 8, the SERS band of the pyridinium species is still observed, which would not be the case if the electrode had been roughened at pH = 8. The pyridinium and chloride ions are adsorbed as an ion pair and are desorbed near $-0.6\,V$. The spectrum of the pyridinium ion does not reappear at more negative potentials, indicating the importance of the nitrogen lone pair in the SERS process, and a spectrum of the pyridinium ion cannot be obtained in fluoride solution.[58]

5.10. Differential Capacitance Measurements

Measurements of the differential capacitance of electropolished (100), (110), and (111) silver single-crystal faces and of polycrystalline silver are strongly dependent on the nature of the surface as well as the composition of the solution. Using aqueous chloride, electrochemical roughening cycles involving 20 or more monolayers of silver result in capacitance/voltage curves that are similar for all four crystal faces. These curves can be interpreted as a superposition of those for the individual single-crystal faces and it follows that the roughened electrode consists of an ensemble of faceted microcrystals.[46]

The differential capacitance curves in the potential region -0.05 to $-1.45\,V$ of a silver microelectrode roughened in the dark in $0.1\,M$ KCl/$0.05\,M$ pyridine has been found to be identical to that measured with a laser beam irradiating the whole the surface. It follows that the laser beam is sampling the species normally present in the double layer.[64]

5.11. Potential Modulation of SERS Spectra

Both Suetaka and co-workers[65] and Van Duyne[38] have shown that difference spectra can readily be obtained for pyridine on silver electrodes. The potential was modulated by a square wave and, using photon counting, the Raman signals recorded over a given period of time at both potentials are subtracted to give directly the difference spectrum. Suetaka and co-workers obtained good quality difference spectra using modulation speeds as high as 20 Hz (Figure 12). This technique enables one to observe potential-dependent species, which are otherwise masked by the background spectrum of the electrolyte.

In our own experiments we have found that, for the Ag/0.1 M KF/pyridine system, the modulated Raman spectrum in the C—C stretching region of pyridine varies with the concentration of pyridine. Hence, increasing surface coverage may lead to changes in the orientation of the pyridine.

5.12. Build-up of SERS during the Roughening Cycle and with Time

Albrecht and Creighton[35,66] originally reported the change in Raman intensity at a fixed frequency while a cyclic potential ramp was applied to a polished silver electrode in 0.1 M KCl/0.01 M pyridine, using a sweep rate of 3.4 mV s^{-1} between the limits of −5 and +145 mV. The Raman intensity rose sharply as the last of the silver chloride was reduced, but after reduction was complete the SERS intensity showed a small fall. The same SERS spectrum of pyridine was obtained when the pyridine was added after the roughening cycle, which implies that the SERS spectrum does not arise from specific complexes containing pyridine, which are formed at positive potentials.

Different results were obtained with KClO$_4$ and KF as supporting electrolyte and applying a potential ramp from −28 to +386 mV at a sweep rate of 0.35 mV s^{-1}. With these electrolytes an intense spectrum of pyridine was obtained at the positive limit where silver goes into solution. This intensity then fell on the cathodic sweep being replaced by a different, but more intense spectrum: the band observed at 1025 cm^{-1} at 386 mV was replaced by bands at 1008 and 1036 cm^{-1} at −28 mV. The authors conceded that the spectrum

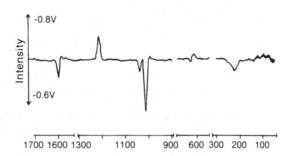

Figure 12. Modulation Raman spectrum of pyridine on a silver electrode. Upward bands correspond to an increase in intensity of a Raman band upon negative shift of the potential, downward bands to an increase upon a positive shift. Electrode potential: −0.7 V ± 0.1 V (S.C.E.); modulation frequency 20 Hz; concentration of pyridine: 0.05 M.[65]

obtained at the positive limit may have been due to species in solution in the vicinity of the electrode rather than species adsorbed to silver. The results did, however, show a maximum SERS intensity as silver was redeposited onto the electrode surface during the cathodic sweep. This implies that very finely divided silver is important in the SERS process (see section on silver sol particles).

A better overall picture of the build-up of the SERS intensity with potential and time can be obtained with the use of an optical multichannel analyzer (OMA). With this system an array of detectors (numbering 1024, for example) or a videcon camera is used to monitor simultaneously a major portion of the Raman spectrum rather than scanning the spectrum past a single detector as in traditional instruments. The OMA system is inserted after the first (or possibly a second) dispersive element of the monochromator and, apart from the advantage of spectrographic rather than spectroscopic operation, these systems can acquire spectra 100 to 1000 times faster than traditional instruments. Another advantage is that changes in laser power or instabilities at the sample are averaged out over the entire spectrum. Dornhaus et al.[67] investigated the SERS of pyridine adsorbed at silver electrodes roughened in 0.1 M KCl. Figure 13 shows a typical presentation of spectra. The Raman bands at 1009, 1037, and 3056 exhibit a pronounced increase in intensity after only 20% of the AgCl layer has been reduced (between t_2 and

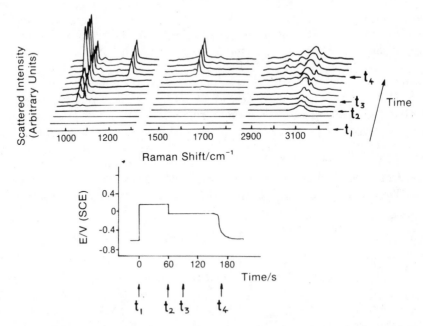

Figure 13. Development of SERS from pyridine adsorbed on silver during the first oxidation–reduction cycle showing the correlation with the voltammogram at various time, t.[67]

t_3), whereas other bands only show a pronounced growth in intensity after completion of the reduction process (t_4) and at more negative potentials. It was reported that by adding pyridine after the roughening cycle the SERS spectrum started to develop after about 5 min and took 20–30 min to reach an intensity similar to that obtained if pyridine is present during the roughening cycle, but the potential at which this study was carried out was not given. Thus the time lag could be due to the displacement of chloride ions by pyridine (at positive potentials near to ~0 V). At more negative potentials (e.g., −0.6 V) where the absorption of chloride ions is decreased, the SERS spectrum of pyridine would be expected to form immediately after addition of pyridine to the electrolyte. We have checked the experiment and have found that a strong SERS spectrum of pyridine does, in fact, appear immediately after the addition of pyridine at −0.6 V but not at 0.0 V. Slow changes in the surface morphology or slow kinetics of formation of particular surface complexes involving pyridine and chloride ions adsorbed to silver species therefore do not appear to be important at negative potentials.

5.13. Observation of SERS Spectra from Metal Sols and the Relationship to Roughened Electrodes

Creighton *et al.*[68] have demonstrated that a SERS spectrum of pyridine can be obtained from silver and gold sols. Silver sols were prepared by reduction of 10^{-3} M silver nitrate with ice-cold 2×10^{-3} M sodium borohydride. The silver sols were yellow with a single absorption band near 400 nm. When pyridine was added a second absorption band appeared to the low-frequency side of this band and shifted with time to ~500–620 nm. This second band is associated with larger sol particles with diameters approaching the wavelength of visible light. A fair correlation was found between the maximum intensity of the SERS obtained from the dispersion using varying excitation frequencies and the absorption profile. The absorption band of such sols is attributed to resonant excitation of surface plasmon oscillations. The SERS spectra of roughened silver electrodes obviously arises from the same phenomenon as those of the sol particles, especially since roughened silver electrodes have been shown to absorb in a similar region of the spectrum.[45] Evans *et al.*[52] have characterized the silver surface before and after electrochemical roughening in aqueous chloride containing pyridine. Electron scanning micrographs showed that the reformed silver surface consisted of nodular deposits, which grew three dimensionally during reduction of the silver chloride. The nodule sizes and spacing were 600–1000 nm, which compares favorably with the diameters of the silver sol particles in the presence of pyridine reported in Reference (68) (500–1000 nm). Secondary ion mass spectroscopy and Auger electron spectroscopy also showed that the reformed silver surface was purer than the original silver surface, the only significant surface contaminant being the adsorbed chloride ions.

Other groups have extended the range of adsorbates on silver sols to include benzene, N-N dimethylaniline, methyl violet, formate, and acetate ions.[69,70] Creighton *et al.* have also used rhodium and platinum sols but these did not absorb in the visible region of the spectrum or give rise to SERS.

5.14. Observation of SERS via Excitation of Surface Plasmons

Pettinger *et al.*[71] have shown that there is a connection between surface plasmons and the SERS effect. Plasmons are group oscillations of electrons and these oscillations are somewhat different near the metal surface, where they are described as surface plasmons as opposed to bulk plasmons. For most metals these surface plasmons are excited (and decay) in the ultraviolet region of the spectrum and the plane of polarization of the incident beam needs to be parallel to the plane of incidence (p-polarized). The metal substrate also needs to be very thin (~ 1000 Å) or roughened in order for excitation to occur. Silver is unusual in that the surface plasmons can be excited using visible radiation. Pettinger *et al.* have deposited a 500 Å-thick layer of silver onto the flat side of a glass hemicylinder and this was used as the electrode in aqueous chloride with pyridine. A p-polarized laser beam was shone through the cylinder to the glass/silver interface and scattered light collected from the silver/electrolyte interface. No SERS spectrum of pyridine was observed without roughening but a strong spectrum was obtained after roughening, which was at a maximum when the angle of incidence of the laser beam with respect to the silver surface was 50°—the angle maximum surface plasmon excitation occurred. In addition no SERS spectrum was obtained at all using s-polarized radiation. Hence surface plasmons must be involved in the SERS phenomenon on silver, but the fact that no SERS spectrum was observed before the roughening cycle shows that there are further factors involved. The roughening cycle may have led to a broadening of the frequency range for surface plasmon decay such that a particular frequency was then in resonance with the charge transfer band of a specific adsorbed species. Similar experiments have been carried out by other groups[72,73] confirming the importance of surface plasmons.

Girlando *et al.*[74] have obtained SERS spectra, in air, of polystyrene that had been coated onto a silver holographic grating. Again, surface plasmons must be involved in generating the enhanced spectra, but it is not known what proportion of the SERS intensity is attributable to surface plasmon interactions.

5.15. Effect of Other Metals in Silver Alloys on SERS

Furtak and Kester[75] have observed the effect of increasing concentrations of palladium in silver–palladium alloys on the SERS spectrum of pyridine. Mechanical abrasion was used to roughen the different alloys; the SERS spectrum of pyridine was already very weak with only 5% Pd present.

6. The Background Spectrum

6.1. The Background Continuum

Several groups have noted that there is, in general, a high baseline in the SERS spectrum of pyridine and of other adsorbates on silver, which has been dubbed the background continuum and extends from the Rayleigh line out to 4000 cm^{-1}.[76-79] This background also increases with the SERS intensity. Chen *et al.*[76] have attributed the background to inelastic light scattering by charge carrier excitations in the metal such as electron–hole pair excitations, whereas Birke *et al.*[77] and Heritage *et al.*[79] favor photoionization of electrons (or holes) from the adsorbate to the metal, followed by radiative recombination. Finally, Timper *et al.*[78] picture electronic Raman scattering caused by atomic scale roughness, because the background continuum and Rayleigh scattering behave differently with potential.

Alternatively, Pettinger and co-workers[59,61] have pointed out that at least some of the background arises from simple fluorescence and also diffuse scattering of the laser beam at the roughened silver surface—a problem that would be more marked using a double rather than a triple monochromator. Some of the background continuum also appears to arise from many weak underlying bands in the SERS of Ag/Cl$^-$/pyridine,[59,80] as can be seen in Figure 14. Some of these weak bands may be attributable to molecular vibrations of impurities adsorbed at the electrode surface, which show large SERS effects. Due to the low concentration of such impurities, their SERS intensity is low and hence only the strongest fundamentals of these species may be detectable, if at all. Other bands may arise from overtone and combination bands of pyridine.

Using a silver grating as substrate in UHV work, Tsang *et al.*[81] have found that the background continuum is angle dependent, which implies that it arises from the excitation of surface plasmons due to the surface roughness.

6.2. The "Carboxy" Background

Another feature in the SERS spectrum of different adsorbates on silver is the so-called carboxy background. Fleischmann *et al.*[27] originally reported seeing this background after roughening a silver electrode in 0.1 M NaF (Figure 15). Two intense and broad bands are seen at 1580 and 1360 cm^{-1}

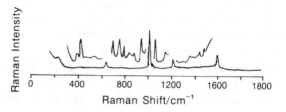

Figure 14. SERS spectra for pyridine on silver from 0 to 4000 cm^{-1}. Several amplified sections are given to show weak SERS lines. Electrolyte: 0.1 M KCl with 0.05 M pyridine; $E = -0.06$ V (S.C.E.).[59]

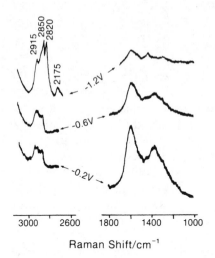

Figure 15. Raman spectra from a silver electrode at various potentials after potential cycling treatment in 0.1 M aqueous sodium fluoride (potentials vs. S.C.E.).[27]

at -0.2 V with a weak feature near 2920 and 2850 cm^{-1}. The former bands are in the region of the carboxylate stretching vibrations of silver formate in solution but displaced to slightly lower frequency, which may be expected for an adsorbed species, and the latter bands correspond to the C—H stretching modes of the formate ion. At potentials more negative than -0.8 V the C—H stretching bands are replaced by features that are characteristic of long chain paraffinic species, and these are accompanied by a new band near 1450 cm^{-1}, characteristic of C—H bending vibrations. All the spectra above were found to increase in intensity by the controlled addition of carbon dioxide, although the voltammetry gave no clear electrochemical evidence for the formation and further reaction of adsorbed intermediates. This carboxy background frequently appears in the SERS spectra of adsorbates on silver. The spectrum also contains a band near 220 cm^{-1}, which is in the region expected for a silver–adsorbate stretching mode.

Mahoney et al.[82] have proposed that the dissolved CO_2 is electroreduced to an unstable silver surface–formate complex, which, in turn, spontaneously decomposes to carbon. Part of their argument lies in the absence of bands near 2100 cm^{-1} in the SERS spectrum when 0.1 M KF in H_2O is replaced by 0.1 M KF in D_2O, indicating that the CO_2 has not been reduced to a C—D containing species. However, Pettinger et al.[61] have detected a band near 2100 cm^{-1} in experiments using D_2O, similar in appearance to the band near 2900 cm^{-1}, so it seems that a C—D containing species is formed after all. Assuming that the bands between 2850 and 2920 cm^{-1} did not arise from a C—H containing species, Mahoney et al. assigned this feature to disordered graphite, which has bands in the region ~2700 cm^{-1} as well as near 1360 and 1595 cm^{-1}.[82–84] This surface graphite is then said to be reduced near -1.2 V to alkyl hydrocarbons, as is implied by the intense bands observed at 2820,

2850, and 2915 cm^{-1}. Howard et al.[85] noted that previous electrochemical studies[86] had indicated that reduction of CO_2 generates formate and that McQuillan and Pope[70] have recently demonstrated that formate coordinated to the surface of precipitated silver particles spontaneously decomposes, over a period of time, to produce surface products that generate a spectrum very similar to that observed on the silver electrode surface.

Howard et al. further consider that the surface graphite acts as a high area adsorbent for pyridine at the silver electrode surface, and that the high intensity of the SERS spectra is partly due to a surface concentration effect.[85] In our laboratory we have found that when a silver electrode is roughened in 0.1 M KF and the SERS spectrum of the carboxy background obtained, addition of pyridine leads to a decrease in intensity of the carboxy background.[58] This would seem to indicate an equilibrium in which pyridine is displacing a proportion of an adsorbate at the surface, which in turn indicates the presence of a mobile carboxy species rather than surface graphite. We suggest that the use of $C^{18}O_2$ dissolved in $H_2^{18}O$ could finally resolve the problem of whether the 1595 and 1360 cm^{-1} bands arise from graphite or a carboxy species.

For the Ag/F^-/pyridine system, the SERS spectrum of pyridine increases in intensity when changing from blue to red excitation but the background carboxy spectrum does not. Hence, either an additional resonance Raman effect is operative with one of these species or else the higher energy blue laser line is perturbing the surface species to produce the carboxy background.

7. Adsorption of Halide Ions with Coadsorbed Water

We have already seen in an earlier section that the SERS spectrum of the Ag/0.1 M KCl/0.05 M pyridine system contains a band near 240 cm^{-1}, which has been assigned as a silver–chloride stretching mode. Attempts to obtain the 240 cm^{-1} in the absence of pyridine were unsuccessful until it was shown that such spectra could be obtained if more concentrated halide solutions (1 M KCl) are used.[60] Wetzel et al.[55] showed that a spectrum could even be obtained using 0.1 M KCl provided red excitation is used with its correspondingly larger surface Raman enhancement factor. SERS bands of adsorbed chloride at -0.2 V have been reported at 240, 150, 110, 18, and 8 cm^{-1}, with other bands present at 3570, 3498, and 1610 cm^{-1} [60] (see Figure 16). The high-frequency modes arise from water, which must be coadsorbed with the chloride ions because the SERS intensities of the two sets of bands follow the same potential dependence, both being very weak at -0.6 V. The bands arisng from adsorbed water are much sharper then their solution counterparts, implying a fairly ordered structure. The frequency of the OH symmetric stretching mode, at 3498 cm^{-1}, is higher than that of bulk water in the solution (3443 cm^{-1}) and is in a frequency region typical of electrostatic interaction as

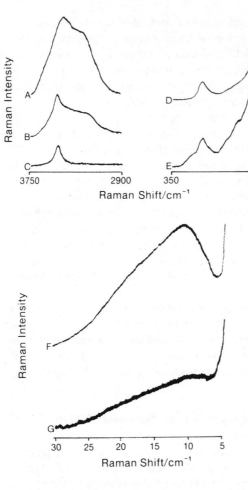

Figure 16. Raman spectra of a silver electrode in 1 M KCl; A–E: 5 cm^{-1} slit width; F and G: 1 cm^{-1} slit width. A: polished electrode (bulk electrolyte only); B and D: roughened electrode at -0.2 V (S.C.E.); C: electrode pushed against cell window; E: dried roughened electrode in air; F: roughened electrode at -0.2 V (S.C.E.); G: roughened electrode at -0.6 V (S.C.E.).[60]

opposed to hydrogen bonding. The adsorbed water rapidly exchanges with added D$_2$O, resulting in bands arising from H$_2$O, D$_2$O, and HDO species.

The band at 8 cm^{-1} was first reported by Genack *et al.*[87] in studies on Ag/0.1 M KCl/pyridine. This band is very intense and was assigned to adsorbed pyridine because the band was not observed in the absence of pyridine. However, as already stated, the SERS spectrum arising from adsorbed chloride using 0.1 M KCl is very weak (unless red excitation is used) so that this band would not be expected to be detected under these conditions. We have measured the SERS spectrum of pyridine adsorbed to Ag from 0.1 M KF/0.1 M pyridine solution and still located a band at 8 cm^{-1}. The band at 18 cm^{-1} [60] is a shoulder on the main 8 cm^{-1} band. Changing to bromide and iodide did not appreciably effect the position of this band. Genack *et al.*[87] considered the band to arise from a rocking or libration of the adsorbed species about the point at which it is bonded to the surface, and calculations have been made to predict the frequency of such a motion for adsorbed cyanide and pyridine.[88] We will return to the origin of this band in the section on adsorbed cyanide ions.

The 240 and 3498 cm^{-1} bands have both been seen to shift to slightly lower frequency as the potential is made more negative.[55,58] A shift of the halide band to lower frequency has already been predicted at lower surface coverage.[89] Similar shifts have also been observed for adsorbed cyanide (see later).

Using bromide and iodide solutions the SERS bands of the adsorbed water have been seen to shift to 3523 and 3493/3553 cm^{-1}, respectively.[60] These shifts may indicate that the water is adsorbed to the halide ion rather than to silver because the charge density is smaller on these larger ions and the OH stretching frequency is closer to that of the free molecule. Pettinger *et al.*[61] have also obtained similar results, but when using 8 *M* NaCl the adsorbed water band was seen at 3520 cm^{-1}, and when using 10 *M* NaBr two bands were seen at 3486 and 3542 cm^{-1}.

In more recent work we have found these adsorbed water bands to be absent when using 1 *M* HCl, but to appear as soon as the metal ion is added to the solution.[90] Hence the adsorbed water appears to be in a bridging position between the chloride and metal ions. Current work shows large changes in the SERS spectrum of water with the nature of the metal ion. For example, using 1 *M* KI two bands arising from coadsorbed water are seen at 3491 and 3549 cm^{-1} (-0.9 V) whose relative intensities invert between -0.9 and -1.1 V. However, changing to NaI results in the observation of only one band at 3594 cm^{-1}; addition of a little KI to the NaI electrolyte again results in the observation of the bands associated with KI.

In addition to the effect of the metal ion on the spectrum of coadsorbed water, it has been established that there is an equilibrium between the co-adsorbed water and pyridine in solution.[90]

(i) Using 1 *M* KCl, electrochemical roughening leads to the observation of the coadsorbed water band at 3498 cm^{-1} and -0.2 V. If pyridine is then added to the electrolyte, at constant potential, the intensity of the 3498 cm^{-1} band is maintained. However, using 0.2 *M* KCl, addition of pyridine at -0.2 V results in a reduction in intensity of the band at 3498 cm^{-1}, while bands due to pyridine appear.

(ii) Using 1 *M* KCl, addition of pyridine and HCl results in the observation of the SERS spectrum of the pyridinium ion, which forms an ion pair with the chloride, but the 3498 cm^{-1} band is not displaced.

We have already suggested that the adsorbed water is forming a bridge between the adsorbed chloride species and the metal ion. At low chloride surface coverage, pyridine is readily adsorbed at the metal surface and apparently forms a complex with the chloride ions, which excludes the water. However, one would expect the pyridinium ion also to displace the adsorbed water molecules at high chloride surface coverage, in order to form an ion pair. The fact that the adsorbed water is still seen may simply be a size effect, where maximum coverage by pyridinium ions still leaves available sites on many of the adsorbed chloride ions.

8. The Adsorption of Cyanide Ions

The adsorption of cyanide to silver electrodes has been studied by several workers (see Table 4), although the original SERS spectrum was obtained in air from a cyanide-corroded silver rod.[91] Furtak,[92] using 0.1 M Na_2SO_4/0.01 KCN, obtained a SERS spectrum of cyanide on silver at −0.8 V with strong bands at 2114 and 226 cm^{-1}, and weaker ones at 315 and 150 cm^{-1}. From a comparison of the frequency of the CN stretching mode in the SERS spectrum with those of di, tri, and tetra-cyano-argentate(I) ions in solution, the surface complex was assigned as tri-cyanoargentate.[93] Dornhaus and co-workers[67,94] used an OMA system to relate the SERS of cyanide on Ag with the voltammogram. A band was seen at 2140 cm^{-1} and −0.3 V, which rapidly shifted to 2110 cm^{-1} by −0.5 V. This shift was interpreted as a change in coordination from 2 to 3 or 4 cyano groups. The non-Lorentzian 2110 cm^{-1} band was fitted with three Lorentzian peaks located at 2095, 2110, and 2140 cm^{-1} (for the three complexes). Furtak et al.[95] have shown that the formation of a Ag(I) compound is important in the SERS activation process and that the potential needs to be changed rapidly between 0.0 and −0.8 V, or the SERS is lost through dissolution of silver as cyanide complexes.

One cannot simply compare the frequencies of the SERS bands with those of the solution species, because the SERS bands shift with potential.[96,97] Thus, the frequency of the cyanide stretch shifts from 2108 cm^{-1} at −1.0 V to 2093 cm^{-1} at −1.5 V (Figure 17). These frequencies may be compared with solution frequencies of 2108 cm^{-1} for $Ag(CN)_3^{2-}$ and 2097 cm^{-1} for $Ag(CN)_4^{3-}$. Hence, if the frequency of the surface species reflects its coordination, then there should be a change in coordination between −1.0 and −1.5 V and a concomitant change in bandshape as this takes place. The band at −1.0 V is, in fact, asymmetric to the low-frequency side, but maintains this same asymmetry through to −1.5 V, and this therefore does not reflect a change in coordination. Moreover, we have detected a strong band arising from coadsorbed water at 3521 cm^{-1} and −1.0 V, which progressively shifts to 3431 cm^{-1} at −1.5 V (Figure 18). Both the 2108 and 3521 cm^{-1} bands shift in a roughly linear fashion over this potential range, so the cyanide stretching frequency would appear to reflect the strength of the metal–cyanide interaction rather than the formation of different complexes.

Substitution of ^{13}CN for ^{12}CN leads to measurable shifts of the bands at 2102.5, 2015, 342, 294, and 221 cm^{-1} (−1.2 V) to bands at 2058.5, 1969, 334, 289, and 213 cm^{-1}, confirming that all arise from cyanide modes. The weak bands at 400 and 154 cm^{-1} in ^{12}CN also appear to shift to lower frequencies

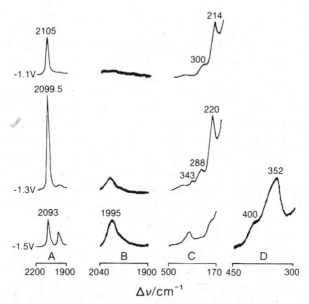

Figure 17. SERS spectra from silver/0.5 M KCN showing the appearance of a new species at negative potentials. Spectra B and D are enlargements of bands in A and C, respectively.[96]

but the shifts are only comparable with the error of the measurements. For complexes involving more than one ligand, isotopic mixtures can be used to determine the coordination number. For example, a 50:50 mixture of ^{12}CN and ^{13}CN in a solution of silver ions of the appropriate concentration will form the di-cyanoargentate ions $(^{12}CN\ Ag\ ^{12}CN)^-$, $(^{12}CN\ Ag\ ^{13}CN)^-$, and $(^{13}CN\ Ag\ ^{13}CN)^-$ in the ratio 1:2:1. Considering vibrations involving Ag—C bonds we would expect bands of 1:2:1 intensity ratios at 221, $\sim(221+213)/2$, and 213 cm^{-1}, respectively. Unfortunately, in practice, this band is too broad to distinguish simple CN$^-$ adsorption from di, tri-, and tetra-cyanoargentate complexes. However, substitution of ^{12}CN to give $(^{12}CN\ Ag\ ^{13}CN)^-$ also leads to a loss of the center of symmetry and does, in fact, lead to a 3.3 cm^{-1}

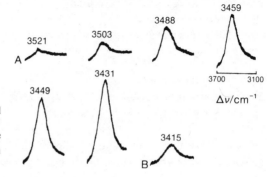

Figure 18. SERS spectra of adsorbed water obtained from silver/ 0.5 M KCN. The potential was made more negative in steps of 0.1 V from -1.0 V (A) to -1.6 V (S.C.E.) (B).[96]

frequency shift of the Raman active cyanide stretching mode.[98] For unresolved bands in our 50:50 mixture the observed frequency shift would only be one half of 3.3 cm^{-1}. The shift we have observed is 1.3 cm^{-1}, in excellent agreement with the predicted 1.6 cm^{-1}. Hence the use of the isotopic mixture proves the presence of a complex. Of course, detailed force constant analysis would need to be carried out to confirm that the complex involves two rather than three or four cyanides. Isotopic mixtures, therefore, will have a useful role to play in the investigation of surface structures.

From a general comparison of the cyanide surface spectrum with the Raman spectra of the di-, tri-, and tetra-cyanoargentate complexes in solution, one sees that the relative intensities of di-cyanoargentate compare favorably with the surface spectrum. Hence it is reasonable to assign the surface complex to di-cyanoargentate(1), with all bands shifted to lower frequency under the influence of the applied potential.[96]

An interesting new finding[96] has been the spectrum of what appears to be the reduced form of the complex at very negative potentials (Figure 17). The bands at 2097, 286, and 218 cm^{-1} at -1.4 V are accompanied by new bands at 2006, 400, and 348 cm^{-1}. The positions of these bands are consistent with the reduction of formally $Ag^I(CN^-)_X \rightarrow Ag^O(CN^-)_X$. The observed bands are all broad with the 2006 cm^{-1} band wider on the low-frequency side and the 348 cm^{-1} band wider on the high-frequency side. This complex is probably only weakly bound to the surface. The importance of this finding is that the previously observed spectra can be seen to be due to essentially Ag^+ species present at the interface even at fairly negative potentials.

The nature of the cyanide complexes raises the issue of whether Cl^- is present as simple adsorbed ions or, for example, in a complex $(AgCl_2)^-$. The low-frequency region of the SERS spectrum of Cl^- on Ag contains several bands,[59,60] which is indicative of a complex adsorbed species. The 240 cm^{-1} band is too broad for mixed isotope analysis to be useful at this stage in identifying coordination, but indirect evidence for a complex comes from the observation of the band at 8 cm^{-1}.[87] A band has also been seen in this frequency region for adsorbed cyanide[87] so, by analogy, if surface cyanide is present as an $(AgX_2)^-$ species then chloride may be as well. This band can possibly be assigned as a rotational oscillation of the two ligands around the silver atom.

SERS spectra of cyanide adsorbed to copper have also been reported, although the spectra were recorded in air.[99]

9. Experiments Relevant to the Physical and Chemical Origins of the SERS Effect

Much of the work reviewed in earlier sections implies that a strong interaction (possibly a covalent bond) between the adsorbate and silver is necessary for the SERS effect to be observed. A number of further experiments relevant to this question is discussed in this section.

Although no SERS spectra of adsorbed pyridine on gold can be obtained using 514.5 nm excitation, deposition at 1.3 monolayers of silver onto gold electrodes does lead to observation of a SERS spectrum.[100] The deposition of silver apparently provides a bridge, through which excited electron wave functions in the gold can communicate with the pyridine states. It is not yet understood why this communication is not effective on pure gold alone, although it would appear that an Ag-pyridine bond is important. Another possibility is that deposition of the silver has led to a large increase in the number of adatoms present at the electrode surface, which Wetzel *et al.*[97] have shown to be important for the observation of SERS. Wetzel *et al.* measured the SERS spectrum of cyanide adsorbed to silver and noted that the spectrum was irreversibly lost at −1.4 V, when the potential was taken through the sequence −0.9 → −1.4 → −0.9 V. This phenomenon has generally been observed for all SERS systems, but with varying cathodic limits. In this case the electrode was held at −1.2 V and 10 ml of 0.01 *M* AgCN added to the electrolyte, the reduction of which resulted in the reappearance of the SERS spectrum. However, when AgCN was added to a polished silver electrode at −1.2 V, no SERS spectrum was obtained. Hence, surface roughness is important but the reduced AgCN is thought to lead to adatoms on the surface, which are of equal importance. Schematically we have the following model (Figure 19).

Our own observations in support of this model relate to adsorbed iodide on silver. The SERS spectrum of iodide is still observable at potentials where hydrogen evolution is occurring.[58] Also, for acidified solutions of KCl, the SERS spectrum of chloride remains strong at potentials where the chloride ion remains adsorbed but hydrogen evolution occurs.

The importance of a strong interaction between the adsorbate and silver has been demonstrated by an experiment in a dry environment. Chen *et al.*[101] adsorbed isonicotinic acid and benzoic acid on glass and then covered this adsorbate with a very thin layer of silver (a metal island film, with aggregate ~50 Å and optical properties similar to silver sols). Figure 20 illustrates the experiment; a SERS spectrum was obtained of the nicotinic acid, but not the

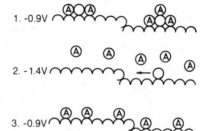

Figure 19. (1) Silver adatoms (O) are stabilized by the adsorbate (A). SERS spectrum is strong. (2) Adatoms migrate to kink sites and become incorporated in the structure. (3) SERS spectrum is now weak or absent.[97]

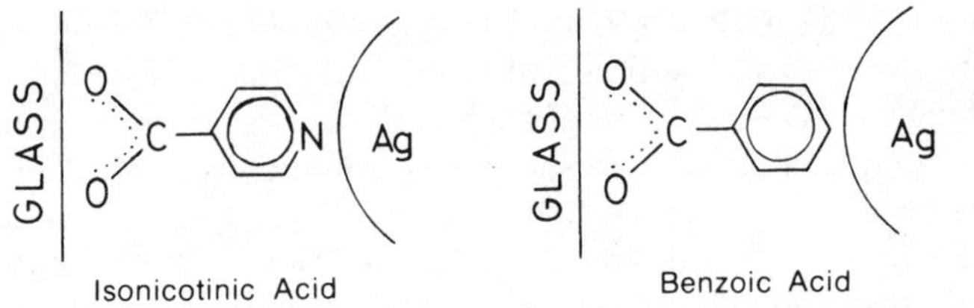

Figure 20. Silver metal island films deposited on isonicotinic acid and benzoic acid adsorbed on glass.[101]

benzoic acid. The result was interpreted as showing that chemisorption of isonicotinic acid to silver via the nitrogen line pair must be important for observation of SERS, whereas close proximity of the benzoic acid to silver was not sufficient for observation of SERS.

In the section on the intensity of the SERS spectrum, it was reported that Van Duyne and co-workers have obtained a spectrum of pyridine adsorbed to a polished silver electrode.[38,42] Assuming a Raman enhancement of $\sim 10^6$ for pyridine adsorbed to an electrochemically roughened silver electrode, then these workers assigned a factor of $\sim 10^4$ enhancement from a roughness independent mechanism and $\sim 10^2$ to a roughness dependent one. Mild electrochemical roughening was assumed to increase the electrode surface area fivefold, thus leaving only a twentyfold increase for the roughness dependent SERS. A review of different theories relating to the SERS phenomenon follows in the succeeding section.

10. The Origins of SERS

In this section we will outline some of the mechanisms that have been proposed to explain the SERS effect. There have been many publications during the past few years that have proposed a number of distinct models. We will not summarize all of these theories but rather relate the essential details to the experimental evidence in order to show their applicability or shortcomings. At the outset we must note that SERS is not only observed at the electrode–solution interfaces but also on dried electrodes and in ultrahigh vacuum experiments. We must also bear in mind that more than one enhancement mechanism may be operative.

A brief summary of experimental findings that must be accounted for in explanations of enhancement mechanisms follows:

1. There is an apparent increase in the Raman scattering cross section of the adsorbate molecules;
2. There are deviations of the Raman intensity from the ν^4 excitation frequency dependence, which is observed in normal Raman scattering (but not in resonance Raman scattering).

3. It has been found that the SERS bands are depolarized (or close to being depolarized) even when the same bands in solution species are completely polarized).
4. There is a strong background continuum.
5. Species not immediately adjacent to the electrode surface appear to be enhanced in intensity, e.g., water coadsorbed with halides.
6. Electrochemical roughening of the silver surface leads to intense SERS spectra, although weak bands have still been observed without this treatment (these weak bands must still be enhanced by a large factor to be observable).

11. The Enhancement Mechanisms

11.1. Conventional Resonance Raman Scattering

Conventional resonance Raman excitation has been suggested as a possible origin of SERS.[53,102] For the Ag/Cl$^-$/pyridine system Regis and Corset[53] have proposed that the observed spectra are of the radicals Cl$_2^-$, PyH$^.$, and Py$^-$, stabilized in an insoluble layer of silver chloride. The Cl$_2^-$ radical, which has been observed in doped crystals, has an optical absorption band at 750 nm and gives a strong resonance Raman effect with the internal stretching vibration between 225 and 264 cm^{-1}, depending on the alkali metal in the M$^+$Cl$_2^-$ ion pair. Such resonance Raman enhancement could therefore account for the increase in SERS intensity going from blue to red excitation. However, using aqueous fluoride, SERS spectra of pyridine are still obtained without the presence of an insoluble halide layer to stabilize the radicals.

Starting from the assumption that the SERS spectra of pyridine adsorbed on Ag, Cu, and Au electrodes arise from a resonance Raman effect, Marinyuk et al.[102] have measured the ratio of the Stokes and anti-Stokes intensities at 1005 cm^{-1} for pyridine in solution and adsorbed to the electrodes. From a comparison of these ratios they have calculated the absorbance frequency for the species on silver to be near 650 nm, in approximate agreement with electroreflectance data.[45] It is thought that this absorption is a charge-transfer band from the adatom adsorbate complex to the metal, giving rise to the resonance Raman spectrum of the adsorbate. Additionally, partial transfer of an s-electron from the adatom to the adsorbate would appear to be necessary in order to observe the resonance Raman spectrum of the adsorbate. Water is thought to play a role in this by increasing the electron affinity of pyridine via hydrogen bonding; hence the SERS spectrum of pyridine would be expected to be much weaker in organic solvents as has been found to be the case.[38]

The interaction between the metal surface and the adsorbate leading to the observation of resonance Raman scattering is considered in the following section.

11.2. Surface-Induced Resonance Raman Scattering (SIRRS)

Metiu and co-workers[103-105] have proposed a mechanism by which the electronic energy levels of an adsorbate could be shifted into the visible region of the spectrum, which would allow resonance Raman excitation to occur. The excitation profile (the intensity of Raman scattering vs. excitation frequency) for pyridine adsorbed to silver rises smoothly from blue to red frequencies, whereas in comparison the excitation profile of a typical resonance Raman scatterer will exhibit one or more "narrow" peaks in the visible region of the spectrum. The theory takes this problem into account in the following manner. The broadening and shifting of the electronic energy levels of the adsorbate is a function of the distance of the adsorbate from the surface. Apart from the fact that there will be a range of adsorption sites on the electrode surface, the vibration of the chemisorbed adsorbate–metal bond will modulate the adsorbate–metal distance. The frequency of this vibration will be in the range 200–400 cm^{-1}, whereas the visible adsorption band will be in the range 15,000–20,000 cm^{-1}—a difference in timescale of about 10^2. Hence the vibration is slow compared with the electronic excitation timescale and a very large broadening of these electronic energy levels will be seen with respect to the vibrational timescale. This in turn will lead to a slowly changing excitation profile for SERS.

The theory further considers that a resonance with the surface plasmons will tend to quench the SIRRS intensity, not to enhance it as has usually been suggested. Experimental SERS excitation profiles could, in fact, be interpreted in this way because the surface plasmon absorption is at the blue end of the visible spectrum where the SERS intensity is least.

Finally we should note that the nearest electronic transition of pyridine to the visible region is at 368 nm (in the U.V.).[106] It seems unlikely, therefore, that adsorption of pyridine could alter the electronic structure of the molecule to such an extent that this absorption band is shifted to the red region of the visible spectrum, while at the same time the vibrational spectrum is virtually unperturbed.

In a somewhat similar manner, Aussenegg and Lippitsh[107] consider the SERS intensity to arise from charge transfer in surface complexes, but with different displaceabilities of the electronic charge transferred between the donor and the acceptor due to vibrations of the adsorbate. During certain vibrations the extent of charge transfer is changed due to alteration of the donor-acceptor distance, leading to an additional change in polarizability during the vibrational motion. An estimation of the order of magnitude of this effect is in good agreement with experimental values.

For pyridine adsorbed to silver, a strong enhancement is expected for vibrations causing the largest changes in the donor-acceptor distance. Hence the 1008 cm^{-1} band (the ring-breathing mode) should be enhanced more than the 1594 cm^{-1} band (in-plane C=C stretching mode). Admittedly, at −0.6 V, the $1008\,\text{cm}^{-1}$ band is more intense than the $1594\,\text{cm}^{-1}$ band, but in comparison with the solution spectrum, the $1594\,\text{cm}^{-1}$ band has been enhanced in intensity more than the $1008\,\text{cm}^{-1}$ band. Hence the validity of this model is unclear.

In contrast, Van Duyne has already shown using dyes such as Crystal Violet adsorbed to silver,[38] that the SERS effect can actually complement the resonance Raman scattering such that a good quality SERS spectrum of a known resonance Raman active species may be obtained using only 1 mW of incident laser power. This does not preclude a contribution to the SERS intensity of pyridine and chloride species from resonance Raman scattering but such a resonance effect is clearly not the sole origin of SERS. A better understanding of the possible contribution of resonance Raman scattering to SERS could be obtained by looking at the SERS spectra of a wide variety of adsorbates on silver using various exciting frequencies. In this way it could be established whether the absorbance at the silver surface is a characteristic of the metal by itself, or a metal–adsorbate interaction. The work done on silver sols would indicate the former.[68] In addition, we have already reported that the relative intensities of the bands at 1036, 1025, and $1008\,\text{cm}^{-1}$ in the SERS spectrum of physisorbed and chemisorbed pyridine remain constant going from blue to red excitation.[47] If resonance Raman excitation of pyridine were involved then it seems unreasonable that the electronic states of physisorbed and chemisorbed pyridine could be so similar as to lead to the same excitation profile. Evidence for a resonance Raman contribution to SERS may lie in the fact that the carboxy background does not significantly change in intensity going from blue to red excitation, although the carboxy background may arise from a photoreaction, which is more efficient using blue irradiation.

11.3. Enhancement via Coupling with Surface Electromagnetic Waves

Philpott[108] predicted that the efficiency of molecular processes such as fluorescence emission or Raman scattering may be increased by interactions between adsorbed molecules and surface electromagnetic waves (such as surface plasmons). This prediction was made before the SERS effect was appreciated and is important because the SERS intensity is known to be dependent on surface roughness, as is the activity of surface plasmons. A mechanism for the formation of a resonance Raman scattering condition is depicted in Figure 21. The energy levels of the collective electron displacements in the metal act as a continuum of virtual states coupled to the adsorbate molecule, thus effectively increasing the probability that the molecule will give rise to Raman scattering.

We have previously mentioned that SERS spectra of pyridine adsorbed to silver have been obtained via excitation of surface plasmons[71–74] and that

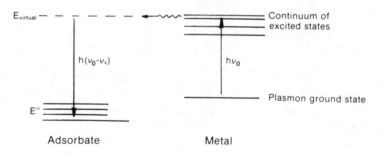

Figure 21. Surface plasmon induced Raman scattering.

silver and gold sols give rise to SERS spectra where it is known that surface plasmons are being excited in these particles.[68]

Hexter and Albrecht[109] have also discussed possible mechanisms by which the normally nonradiative surface plasmons can be coupled to visible radiation. From the electrochemical viewpoint a roughened surface, for which the periodicity of the roughness is of the order of the wavelength of the incident radiation, may act as a diffraction grating providing additional momentum to photons such that their phase velocity is reduced and becomes equal to that of the plasmons. The surface plasmon wave, now radiative, excites an electronic transition of the adsorbed molecule, leading to a resonance Raman effect. Surface plasmon waves are believed to extend some distance beyond the boundaries of the metal so an enhancement mechanism involving these oscillations may give rise to spectra of species not immediately adjacent to the metal surface. Evidence of such spectra may lie in the SERS of coadsorbed water[60–61,96] and condensed multilayers of pyridine in low-temperature UHV studies.[110–112] The UHV studies report that the Raman intensity of the first layer of pyridine is about 100 times more strongly enhanced than the intensity of the subsequent layers.

Aravind and Metiu[113] consider that surface roughness influences the Raman and fluorescence spectra of adsorbed molecules since (i) it allows excitation of surface evanescent fields which increase the induced dipole moment of the molecule, (ii) it allows plasmon radiation modulated by the oscillating nuclei, and (iii) it increases the radiation of the induced dipole. In detail:

(i) The incident field excites a plasmon and the surface field thus created increases the total field incident on the molecule. This will increase the induced molecular dipole moment, hence producing a larger Raman or fluorescence intensity (see also References (114–116).

(ii) The radiation from the rough surface is modulated by the vibrational motion of the nuclei since the effective nuclear charges interact with the surface fields (see also References (114–115). This mechanism is calculated to contribute at the most a factor of 100 to the enhancement.

(iii) The induced molecular dipole moment has a component oscillating at the incident frequency minus the vibrational frequency. This radiates Raman photons and the properties of this radiation are modified by the presence of roughness. This contributes a factor of 10 times and, along with the mirror effect of the metal and the fact that the adsorbed molecules are not tumbling, we get a factor of ~ 300.

(ii) and (iii) together give an enhancement $\sim 10^4$.

Lee and Birman[117-118] have calculated the coupling between discrete dipole-allowed excitations of the molecule and the continuum of excitations of the surface plasmons at a plane metal surface and have concluded that surface plasmon–molecule interactions cannot be a major contribution to SERS on plane surfaces. Surface roughness would change this result, however. Robinson[119] refers to exchange dipole mixing of the electronic states of the adsorbed molecule with those of the surface, where surface roughness allows coupling of the radiation.

Other related enhancement mechanisms have approached the problem from models involving small metal spheres (sols).[120-123] Kerker and co-workers[120-122] predict surface Raman enhancements of adsorbates by 10^6 provided that the metal particles are both considerably smaller than the wavelength of the incident radiation ($a < 0.02\lambda$) and that they consist of media whose optical constants at the excitation or Raman frequency correspond to a particular plasmon frequency. The large enhancements are due to very large local fields near the surface, which occur only for small particles and the condition $m^2 \to -2$, where m is the relative complex refractive index of the particle. The local field induces a dipole at a frequency ω, which interacts with the particle to generate a Raman scattering field that is large for $m^2 \to -2$. The excitation profile predicted for the $1008\ \text{cm}^{-1}$ band of pyridine is sharply peaked at a 382 nm excitation wavelength. For particle sizes comparable to the wavelength the enhancement is $\sim 10^4$ and the excitation profile consists of a broadened peak near 500 nm.

Similar to models postulating coupling with surface plasmons, Moskovits[124] has proposed that the Raman scattering cross section of molecules adsorbed on silver, copper, and gold electrode surfaces may be effectively increased via coupling with optical conduction resonances. Such resonances are the result of cooperative motions of the electrons in the metal surface, which take place on a much larger scale than the motions in a surface plasmon wave. By invoking such large-scale phenomena it has previously been possible to account for certain changes in the electroreflectance profile of gold electrodes in aqueous solution.[125] The electrode surface is modelled as a series of large-scale, colloidal-type, metal "bumps" onto which molecules may be adsorbed; the generation of such surfaces is discussed in terms of the electrochemical treatment necessary to obtain surface spectra. Moskovits has shown that an enhancement mechanism involving optical conduction waves could lead to different excitation profiles for different sized metal clusters.

Hence it would be informative to measure excitation profiles for pyridine adsorbed in silver electrodes that have been roughened to different extents.

11.4. Resonance Raman Scattering via Electron Tunnelling Processes

Gersten, Birke, and Lombardi[126] have proposed that large increases in the effective polarizability of an adsorbate molecule may be produced by photochemically induced electron transfer between metal and adsorbate, or vice versa. A similar mechanism involving electron-hole pair interactions has been put forward by Burstein *et al.*[127]

Figure 22 depicts two possible electron tunnelling mechanisms.

The efficiency of the tunnelling process is clearly dependent on the overlap between the electronic levels of the adsorbate and the Fermi levels of the metal. Electron tunnelling phenomena are believed to take place between artificially roughened metal surfaces due to the breakdown of the conservation of momentum rule with respect to incident photons. Although this mechanism predicts intense Raman scattering, a broad background continuum, and no polarization characteristics of the scattered radiation, it is difficult to envisage such mechanisms as operating for weakly bound adsorbates such as physisorbed pyridine or coadsorbed water.

11.5. Image Field Raman Scattering Mechanisms

In the model of King, Van Duyne, and Schatz,[128] the apparent increase in the Raman scattering cross section of a pyridine molecule adsorbed at a silver electrode surface is thought to be due to an effective increase in the molecular polarizability, via electrostatic imaging of the induced dipole in the metal surface. The total induced dipole for the adsorbed molecule is expressed as

$$\mu_{ind} = \alpha(E + E_{image})$$

where E_{image} is the field at the molecule that is caused by the image of the

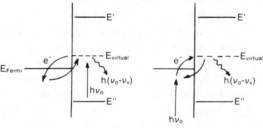

Figure 22. Electron tunnelling mechanisms.

molecule in the metal. As a result of this extra field component, the effective polarizability change accompanying a vibration of the adsorbate must include components produced by the image of the oscillating induced dipole in the metal. By including such components it can be shown that the net polarizability change accompanying a vibration of an adsorbate may be very large indeed. Any molecule that can be brought close enough to the metal surface should give rise to enhanced scattering; this model neglects the apparent need to electrochemically treat the surface prior to the observation of an intense spectrum. In addition, allowing for discontinuities in the electroreflectance profile of the metal, the intensity of the surface spectra should be proportional to ν^4 as in the normal Raman process.

Electrostatic imaging models necessarily predict a severe dependence of the spectral intensity upon the orientation of the molecule. Van Duyne and Allen consider that the results of their experiments using isomeric cyanopyridine species[129] confirm the existence of orientational effects. The results of these experiments are shown in Figure 23 and the intensity of the CN stretching mode has been interpreted as being indicative of a need for an induced adsorbate dipole oriented in a direction perpendicular to the direction of the surface plane. However, the situation is probably more complicated. For example, the intensity of the bands observed in the frequency region 1200–1700 cm^{-1} is significant. In the case of 2-cyanopyridine, several bands in the region 1000–1700 cm^{-1} are remarkably similar to Raman spectra obtained from carboxylate groups adsorbed to silver electrodes.[58] We believe that the cyano group of 2-cyanopyridine, in close proximity to the electrode surface, has been hydrolyzed to the carboxylate, which explains its absence from the spectrum and, in particular, the appearance of new bands similar to adsorbed carboxylates. Hence, these spectra should not be interpreted as supportive of an image field effect.

In a more recent paper, Schatz and Van Duyne[130] have investigated the image field effect in detail and have concluded that it is not a good model, particularly because surface roughness is not dealt with.

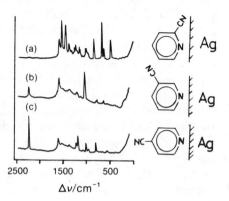

Figure 23. SERS spectrum of (a) 2-cyanopyridine, (b) 3-cyanopyridine, (c) 4-cyanopyridine adsorbed on polycrystalline silver wire electrodes. Excitation: 100 mW at 514.5 nm; slits 2 cm^{-1}; $E = -0.6$ V (S.C.E.).[129]

Eesley and Smith[131] have presented a similar mechanism to that of King *et al.* In this model the possible effects of electrostatic screening of the induced dipole and its image from the incident radiation by adjacent adsorbed species are considered. The inclusion of effects from such screening fields does not appreciably alter the predicted qualitative behavior of the scattered radiation, although this mechanism is coverage dependent.

In addition to the surface induced resonance Raman scattering mechanism,[103-105] Metiu and co-workers have proposed an "image field" model to account for the observation of intense Raman spectra.[132-134] Again, the evaluation of the effective polarizability of a molecule in the presence of a metal surface includes the components arising from the image of the induced dipole in the metal surface. This model further considers the metal surface as a perfect mirror and thus takes into account the effects produced by the reflected field of both the scattered and incident radiation on the adsorbate molecule. Implicit in this model are the operation of orientational and symmetry effects, but no treatment of the role of surface "roughening" is given. Maniv and Metiu[135] have recently suggested a Raman enhancement process, which they refer to as Raman reflection, but they do not say how much this process could contribute to the overall SERS effect. This model predicts that interaction between the vibrating ion-core charges of the adsorbed molecules and the electrons of the metal leads to energy transfer from the electron–hole pair to the oscillating molecule. Since the energy transferred equals the vibrational energy, the photon emitted by subsequent electron-hole recombination is Raman shifted, augmenting the Raman intensity.

11.6. Electric Field Intensity Enhancement Mechanisms

The study of the Raman spectra from gas phase species such as CS_2, or semiconductor materials, under the constraints of a high electric field strength has shown that the presence of such a field may increase the effective Raman scattering cross section of the perturbed species.[136] Van Duyne had proposed that the increase in Raman scattering cross section of molecules adsorbed at an electrode surface could, therefore, result from the presence of the high electric field strength at the electrode/electrolyte interface (10^5–10^6 V/cm),[34] although he later discounted this.[38]

Gersten[137] has produced a modified version of this electric field effect, which takes into account the apparent need for a rough surface. The rough surface is depicted as consisting of metal particles, ellipsoidal in shape, which thus give rise to an inhomogeneous local electric field. Where the particles are needlelike, the effective electric field strength at the tip of the structure will be many times greater than the average field strength. If this much magnified field strength is included in the determination of the effective field strength (surface field + radiation field) seen by the adsorbate molecule at any one time, it can be shown that the net value of the induced molecular dipole (and hence overall scattering intensity) will be many times greater than that for a molecule in the absence of the surface field.

The observation of intense Raman spectra from dry silver surfaces has provoked the following comments on this model: Otto[138] considers that the absence of a high electric field strength at the silver/air interface in the SERS spectrum of cyanide on a dried silver electrode precludes the operation of a scattering mechanism based upon the presence of such a field. However, this does not take into account any electric field induced at the metal surface by the incident radiation. In the presence of surface inhomogeneity, such a field will reach a maximum at the tip of needlelike surface structures and the quantitative effects of the induced field will be less on the dried electrode. Raman spectra from dry silver surfaces are, in fact, considerably weaker than corresponding spectra from wet surfaces. In addition, the SERS spectra on dried surfaces may be weaker because metal clusters present on the electrode surface are stabilized in the electrochemical environment but disappear in the dry state, with the adatom clusters possibly being recincorporated into the lattice. The variation of the intensity of the SERS bands of pyridine with potential for the $Ag/F^-/$pyridine system, however, does not support a significant electric field enhancement.

11.7. Electron-Hole Pair Scattering Mechanisms

Several mechanisms based upon the interaction of electron-hole pairs induced in the metal with the adsorbate molecules have been proposed by Burstein, Chen, Chen, Lundquist, and Tosatti[127] and others.[93,126]

   (i) The first of these mechanisms[127] involves the excitation of an electron producing an electron-hole pair, which may thus give rise to charge transfer from the metal to the adsorbate, or vice versa. This mechanism is basically the same as that of Gersten et al.[126]

  (ii) A second mechanism[127] has been proposed in which the excited electron-hole pairs are inelastically scattered by the vibrating charges of the adsorbate, and then recombine giving rise to scattered radiation containing sidebands at frequencies corresponding to the vibrational frequencies of the adsorbate. A similar phenomenon is believed to occur for species containing mainly infrared active vibrational modes, doped onto $Al/Al_2O_3/Ag$ junctions, in inelastic electron tunnelling spectroscopy (IETS).[139] The difference between the two systems is that for the metal surface/adsorbate system, the tunnelling electron returns to the same metal.

 (iii) The third mechanism involving (e-h) pair interactions[127] treats the effective electric field seen by the adsorbate as a combination of the field of the incident radiation, and the modulating field produced by electron-hole pair excitation. It is possible to mathematically formalize these fields by the introduction of "renormalized" photons or polaritons. As the modulating electric field produced by the (e-h) pair will be directly coupled to the adsorbate molecule via short-range coulombic interactions, the net result is increased adsorbate/radiation interaction and hence increased scattering efficiency. Otto and co-workers[93,140] propose similar mechanisms involving (e-h) pair interactions, but more emphasis is placed on atomic scale roughness and adatoms.

Mechanism (i) has been discussed in a similar form in an earlier section. Mechanism (ii) predicts that metal surface Raman spectra should be similar to the inelastic electron tunnelling (IET) spectra for the same adsorbate. Where such comparative experiments have been carried out there has been no conclusive link between the Raman spectrum and the IET spectrum of the adsorbate[141]; e.g., formic acid gives rise to an intense Raman spectrum corresponding to formate ions adsorbed at a silver electrode.[24] It also gives rise to an IET spectrum, yet no Raman spectrum was observed from formic acid adsorbed onto the IET junction.[142]

Mechanism (iii) relies on coulombic interactions between induced (e-h) pairs and the adsorbate molecule located at some fixed point relative to the surface plane. Theoretical calculations have shown that for an ideal clean surface/vacuum interface, such interactions may extend to some 5 Å beyond the surface plane.[127] In a real situation it is likely that these interactions will be highly attenuated by the presence of a dielectric medium, surface inhomogeneities, etc.

Calculations regarding the operation of (e-h) pair mechanisms in the surface Raman scattering process have not yet been extended to include real experimental observations, except in the case of the mechanism proposed by Otto and co-workers,[93,140] in which the critical dependence on microscopic roughness on the adatom scale is related to the experimentally determined surface structure, for the Ag/CN⁻ system.

11.8. Modulated Reflectance Raman Scattering Mechanisms

Otto has proposed[138] that the effects of the modulation of the electronic polarizability of the metal surface by the vibrating adsorbate give rise to a strong scattered light signal due to the strong reflectance of the bulk metal.

Certain theories of modulated reflectance spectroscopy[143] suggest that changes in the relative reflectivity of the interfacial region $\Delta R/R$, may be correlated with the changes in the surface electron density produced by the presence of oscillating dipoles. Calculations made by Otto and co-workers suggest that the modulation to the local surface electron density produced by a typical infrared active dipole may be so substantial as to produce a dramatic increase in the Raman scattering intensity from the metal. For comparison, a value of $E = 10^5$ V/cm is employed for the surface electric field while the modulation is produced by an infrared active dipole that gives rise to an electric field gradient of $\approx 10^8$ V/cm.

This model therefore represents the electronic Raman scattering from the metal and thus incorporates roughness-induced, inelastic scattering. The pictorial representation employed in these calculations treats the adsorbate-surface charge modulation as arising from weak Van der Waals type forces (Figure 24). The model implicitly predicts severe orientational dependences and also that infrared active modes should be preferentially observed, but there is little experimental data in support of this. Dornhaus *et al.*[67] and Erdheim *et al.*[144] have observed a band arising from an otherwise exclusively IR active mode, in the Raman spectrum of pyrazine adsorbed at a silver electrode. Unlike pyridine, which has C_{2v} symmetry, pyrazine has D_{2h} symmetry by virtue of the center of inversion in the molecule. Bonding to the surface via a ring nitrogen would reduce the symmetry of the species to C_{2v}. However, electrochemical data suggest that pyrazine preferentially adsorbs in a flat configuration[145] and, as a result, Dornhaus *et al.* assume that the symmetry of the adsorbed molecule is the same as the free molecule. This is not the case because by including the formalism of a surface-adsorbate bond, the symmetry of the pyrazine molecule is again reduced to C_{2v}. Thus, for the surface species, the law of mutual exclusion will be relaxed, and infrared active modes may become Raman active also. In the case of pyridine there is no evidence to prove that infrared active modes are enhanced more than Raman ones.

McCall and Platzman[146] have proposed a similar mechanism to that of Otto in which surface charge modulations are produced by short-range dipolar effects arising from a chemisorbed species. Again the Raman scattered radiation is believed to be due to sidebands present on the electroreflectance spectrum from the metal surface, although surface roughness is not an important criterion in this model.

Both models predict orientational dependences and preferential IR mode activity. The McCall–Platzman model, however, predicts that strongly adsorbed ions such as Cl^- at a silver electrode surface should give rise to Raman spectra, by virtue of the chemisorption interactions.

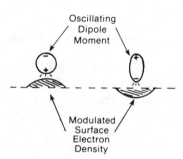

Figure 24. Electroreflectance scattering mechanism.

11.9. Enhancement Models Involving More Than One Mechanism

Several groups believe that a combination of enhancement mechanisms gives rise to the observed SERS effect. Weber and Ford[147] consider that image enhancement of the effective dipole leads to $\sim 10^3$ enhancement of Raman intensity and roughness coupling at the near fields of the dipole and its image leads to another $\sim 10^3$.

McCall and Platzman consider that part of the enhancement is due to modulation of the metal-surface reflectivity by the motion of the atom binding the molecule to the surface, and that a chemisorbed species can have an intrinsic polarizability that is as much as a factor of 10–100 larger than a physisorbed or weakly chemisorbed species. The results we have reported in this article for chemisorbed pyridine being photolyzed to the physisorbed species appear to conflict with the latter effect.

Maniv and Metiu[105] propose a mixture of enhancement effects involving the following phenomena: restricted molecular motion compared with the liquid, mirror effects on molecule illumination and emission, effective polarizability increase, Raman reflectivity, and emission from the metal made possible by roughness. These authors have stated that a more quantitative analysis is needed before the question of the SERS enhancement can be settled.

Gersten and Nitzan[148] believe that the image field effect is important at close range (<4 Å) but that "lightning-rod" and surface plasmon resonance effects also contribute. The lightning-rod effect refers to molecules adsorbed near high-curvature edges of surface irregularities. The polarization of the metal in the external electric field produces a strong local field in the vicinity of such edges, which in turn appears as an enhanced Raman scattering by the molecule experiencing this field.

The "lightning-rod" and surface plasmon resonance effects are thought to contribute to enhancement in two different ways: firstly, by producing strong local fields, as above, and secondly by affecting the polarization of the metal by the molecule dipole. The authors considered that although enhancements of 10^4–10^6 are quoted, only a small number of sites may be favorable for SERS and enhancements up to 10^{11} may be operative.

12. Assessment of Present Position and Prospects

Work on the Raman spectroscopy of electrode–solution interfaces has so far had two essentially separate objectives: firstly, that of characterizing the species at the interface and the nature of the interactions at the molecular level; secondly, that of giving explanations of SERS.

The first approach is now leading to the identification of ionic and molecular adsorbates. Exact identifications of surface complexes, in which several species are coordinated to a metal atom, are being made, and isotopic labelling will clearly be useful in these studies; quantitative information about the nature of interactions between adsorbed species and between these species and the metal is beginning to be defined.

The results obtained are frequently surprising and will complement the information derived by conventional electrochemical measurements. It is now essential that Raman (and, indeed, infrared[149]) spectroscopic data be combined with measurements of surface excesses in the compact double layer. Such combined measurements will show *inter alia* whether the species being detected are the majority species present at the interface. Systematic studies of potential dependent equilibria of surface species could now be carried out.

The development of multiplexing spectrographs and of signal processing techniques applied to the multiplexed outputs (e.g., of cross-correlation methods) as well as improvements in the optical performance of the overall systems (e.g., higher levels of illumination of rotating electrodes) will allow the extension of the range of systems that can be studied by Raman spectroscopy. This will therefore give both a more generally useful surface analytical probe and provide a wider base for the testing of theories of scattering phenomena at interfaces. It is important to develop the technique to the stage where measurements can be made on systems showing no special enhancement. Such measurements would give information on the symmetry of surface complexes and orientation of adsorbates; there is, as yet, little information on these aspects for physisorbed species although some is available for chemisorbed molecules. The development of multiplexing spectrographs will also facilitate the systematic examination of the kinetics of reaction of adsorbed species.

The second approach, that of seeking an explanation of SERS, has also led to much new work even though a definitive explanation of the enhancement mechanism has still not been achieved. It has become apparent that large-scale cooperative phenomena and the detailed nature of the local structure are both important (e.g., SERS and conventional resonance Raman effects are additive). It is not clear at this stage whether several enhancement mechanisms are operating and whether different mechanisms are additive. Comprehensive theories will have to be based on microscopic models of the overall phenomena and this will certainly pose difficult questions. A wide experimental base for testing the theories is a pressing need (see above).

While work in the first field (the spectroscopy of adsorbed species) can proceed to some extent independently of the second (the explanation of SERS) (chemical and structural information can be obtained from spectroscopy independently of the magnitude of an extinction coefficient or scattering cross section), the two fields are naturally closely related. Thus, for example, a comprehensive explanation of SERS must account for the variations of band intensities with potential; equally, these variations cannot be interpreted without an adequate theory of SERS. Both aspects of this field must therefore be developed simultaneously.

Finally, we note that there are further areas of work that remain to be developed, such as the imaging of electrode surfaces using Raman light.

References

1. See, for example, E. Yeager, Recent advances in the science of electrocatalysis, *J. Electrochem. Soc.* **128**, 159C–171C (1981).
2. A. Bewick and S. B. Pons, Chapter in: *Advances in Infrared and Raman Spectroscopy*, R. E. Hexter and R. Clarke, eds., Heyden and Son. To be published in 1984.
3. H. Mark and S. Pons, An *in-situ* spectroscopic method for observing the infrared spectra of species at the electrode surface during electrolysis, *Anal. Chem.* **38**, 119–121 (1966).
4. M. Fleischmann, P. J. Hendra, and A. J. McQuillan, Raman spectra from electrode surfaces, *J. Chem. Soc. Chem. Commun.* 80–81 (1973).
5. M. Fleischmann, P. J. Hendra, and A. J. McQuillan, Raman spectra of pyridine adsorbed at a silver electrode, *Chem. Phys. Lett.* **26**, 163–166 (1974).
6. E. F. Brittain, W. O. George, and C. H. J. Wells, *Introduction to Molecular Spectroscopy—Theory and Experiment*, Academic Press, New York (1970).
7. *Raman Spectroscopy*, Vols. 1 and 2, H. A. Szymanski, Ed., Plenum Press, New York (1967) and (1970), respectively.
8. G. Herzberg, *Infrared and Raman Spectra*, D. Van Nostrand, New York (1945).
9. J. S. Clarke, A. T. Kuhn, and W. J. Orville-Thomas, Laser Raman spectroscopy as a tool for study of diffusion controlled electrochemical processes, *J. Electroanal. Chem.* **54**, 253–262 (1974).
10. D. L. Jeanmaire, M. R. Suchanski, and R. P. Van Duyne, Resonance Raman spectroelectrochemistry. 1. The tetracyanoethylene anion radical, *J. Am. Chem. Soc.* **97**, 1699–1707 (1975).
11. M. R. Suchanski and R. P. Van Duyne, Resonance Raman spectroelectrochemistry. 4. The oxygen decay chemistry of the tetracyanoquinodimethane dianion, *J. Am. Chem. Soc.* **98**, 250–252 (1976).
12. D. L. Jeanmaire and R. P. Van Duyne, Resonance Raman spectroelectrochemistry. 2. Scattering spectroscopy accompanying excitation of the lowest $^2B_{1u}$ excited state of the tetracyanoquinodimethane anion radical, *J. Am. Chem. Soc.* **98**, 4029–4033 (1976).
13. D. L. Jeanmaire and R. P. Van Duyne, Resonance Raman spectroelectrochemistry. 3. Tunable dye laser excitation spectroscopy of the lowest $^2B_{1u}$ excited state of the tetracyanoquinodimethane anion radical, *J. Am. Chem. Soc.* **98**, 4034–4039 (1976).
14. D. L. Jeanmaire and R. P. Van Duyne, Resonance Raman spectroelectrochemistry. 5. Intensity transients on the millisecond time scale following double potential step initiation of a diffusion controlled electrode reaction, *J. Electroanal. Chem.* **66**, 235–247 (1975).
15. T. Ikeshoji, T. Mizuno, and T. Sekine, Resonance Raman spectroscopic detection of electrochemically generated dianion of naphthacene, *Chem. Lett.* **1976**, 1275–1278.
16. K. Appelt, Untersuchung der kristallstruktur und der electrochemischen eigenschaften von akkumulatoren-bleipulvern. Thermodynamische betrachtungen zum system Pb/H_2O bei anwesenheit von Cl^- ionen, *Electrochim. Acta*, **13**, 1521–1532 (1968).
17. R. J. Thibeau, C. W. Brown, A. Z. Goldfarb, and R. H. Heidersbach, Raman and infrared spectroscopy of aqueous corrosion films on lead in 0.1 M chloride solutions, *J. Electrochem. Soc.* **127**, 1702–1706 (1980).
18. E. S. Reid, R. P. Cooney, P. J. Hendra, and M. Fleischmann, A Raman spectroscopic study of corrosion of lead electrodes in aqueous chloride media, *J. Electroanal. Chem.* **80**, 405–408 (1977).
19. P. Tsai and R. P. Cooney, Raman spectra of polynuclear hydroxo compounds of lead(II) chloride, *J. Chem. Soc. Dalton* **1976**, 1631–1634.
20. R. G. Barradas, K. Belinko, and E. Ghibaudi, In: *Chemistry and Physics of Aqueous Gas Solutions*, W. A. Adams, Ed., pp. 357–372, The Electrochemical Society, Princeton, New Jersey (1975).
21. I. R. Hill, Unpublished work.

22. R. J. Thibeau, C. W. Brown, A. Z. Goldfarb, and R. H. Heidersbach, Raman and infrared spectroscopy of aqueous corrosion films on lead in 0.1 M sulphate solutions, *J. Electrochem. Soc.* **127**, 1913–1918 (1980).

23. R. Varma, C. A. Melendres, and N. P. Yao, In situ identification of surface phases on lead electrodes by laser Raman spectroscopy, *J. Electrochem. Soc.* **127**, 1416–1418 (1980).

24. A. J. McQuillan, P. J. Hendra, and M. Fleischmann, Raman spectroscopic investigations of silver electrodes, *J. Electroanal. Chem.* **65**, 933–944 (1975).

25. R. P. Cooney, E. S. Reid, M. Fleischmann, and P. J. Hendra, Thiocyanate adsorption and corrosion at silver electrodes. A Raman spectroscopic study, *J. Chem. Soc. Far. II* **1977**, 1691–1698.

26. R. L. Paul, A. J. McQuillan, P. J. Hendra, and M. Fleischmann, Laser Raman spectroscopy at the surface of a copper electrode, *J. Electroanal. Chem.* **66**, 248–249 (1975).

27. M. Fleischmann, P. J. Hendra, A. J. McQuillan, R. L. Paul, and E. S. Reid, Raman spectroscopy at electrode-electrolyte interfaces, *J. Raman Spectrosc.* **4**, 269–274 (1976).

28. C. S. Allen, G. C. Schatz, and R. P. Van Duyne, Tunable laser excitation profile of surface enhanced Raman scattering from pyridine adsorbed on a copper electrode surface, *Chem. Phys. Lett.* **75**, 201–205 (1980).

29. R. P. Cooney, E. S. Reid, P. J. Hendra, and M. Fleischmann, The Raman spectrum of adsorbed iodine on a platinum electrode surface, *J. Am. Chem. Soc.* **99**, 2002–2003 (1977).

30. R. P. Cooney, M. Fleischmann, and P. J. Hendra, Raman spectrum of carbon monoxide on a platinum electrode surface, *J. Chem. Soc. Chem. Commun.* **1977**, 235–237.

31. R. P. Cooney, P. J. Hendra, and M. Fleischmann, Raman spectra from adsorbed iodine species on an unroughened platinum electrode surface, *J. Raman Spectrosc.* **6**, 264–266 (1977).

32. T. E. Furtak and J. Reyes, A critical analysis of theoretical models for the giant Raman effect from adsorbed molecules, *Surf. Sci.* **93**, 351–382 (1980).

33. M. Fujihira and T. Osa, Internal reflection resonance Raman spectroscopy for studies of adsorbed dye layers at electrode-solution interfaces, *J. Am. Chem. Soc.* **98**, 7850–7851 (1976).

34. D. L. Jeanmaire and R. P. Van Duyne, Surface Raman spectroelectrochemistry Part 1. Heterocyclic aromatic and aliphatic amines adsorbed on the anodized silver electrode, *J. Electroanal. Chem.* **84**, 1–20 (1977).

35. M. G. Albrecht and J. A. Creighton, Anomalously intense Raman spectra of pyridine at a silver electrode, *J. Am. Chem. Soc.* **99**, 5215–5217 (1977).

36. *Surface Enhance Raman Scattering*, R. K. Chang and T. E. Furtak, eds., Plenum Press, New York (1982).

37. R. P. Van Duyne, Applications of Raman spectroscopy in electrochemistry, *J. Phys. (Paris)* **38**, C5-239-C5-252 (1977).

38. R. P. Van Duyne, In: *Chemical and Biochemical Applications of Lasers*, C. Bradley Moore, Ed., Vol. 4, pp. 101–185, Academic Press, New York (1979).

39. V. V. Marinyuk, R. M. Lazorenko-Manevich, and Ya. M. Kolotyrkin, Raman resonance scattering of pyridine adsorbed on silver, *Elektrokhimiya* **14**, 1019–1023 (1978).

40. B. Pettinger and U. Wenning, Raman spectra of pyridine adsorbed on silver (100) and (111) electrode surfaces, *Chem. Phys. Lett.* **56**, 253–257 (1978).

41. U. Wenning, B. Pettinger, and H. Wetzel, Angular resolved Raman spectra of pyridine adsorbed on copper and gold electrodes, *Chem. Phys. Lett.* **70**, 49–54 (1980).

42. S. G. Schultz, M. Janick-Czachor, and R. P. Van Duyne, Surface enhanced Raman spectroscopy: A re-examination of the role of surface roughness and electrochemical anodisation, *Surf. Sci.* **104**, 419–434 (1981).

43. G. Hagen, B. S. Glavaski, and E. Yeager, The Raman spectrum of an adsorbed species on electrode surface, *J. Electroanal. Chem.* **88**, 269–275 (1978).

44. J. A. Creighton, M. G. Albrecht, R. E. Hester, and J. A. D. Matthew, The dependence of

the intensity of Raman bands of pyridine at a silver electrode on the wavelength of excitation, *Chem. Phys. Lett.* **55**, 55–58 (1978).

45. B. Pettinger, U. Wenning, and D. M. Kolb, Raman and reflectance spectroscopy of pyridine adsorbed on single crystalline silver electrodes, *Ber. Bunsenges. Phys. Chem.* **82**, 1326–1331 (1978).

46. M. Fleischmann, J. Robinson, and R. Waser, An electrochemical study of the adsorption of pyridine and chloride ions on smooth and roughened silver surfaces, *J. Electroanal. Chem.* **117**, 257–266 (1981).

47. M. Fleischmann and I. R. Hill, Surface-enhanced Raman scattering from silver electrodes: Formation and photolysis of chemisorbed pyridine species, *J. Electroanal. Chem.* **146**, 353–365 (1983).

48. B. Pettinger, U. Wenning, and H. Wetzel, Angular resolved Raman spectra of pyridine adsorbed on silver electrodes, *Chem. Phys. Lett.* **67**, 192–196 (1979).

49. B. Pettinger, U. Wenning, and H. Wetzel, Surface plasmon enhanced Raman scattering. Frequency and angular resonance of Raman scattered light from pyridine on Au, Ag, and Cu electrodes, *Surf. Sci.* **101**, 409–416 (1980).

50. G. R. Trott and T. E. Furtak, The relationship of angle resolved light scattering from optical resonances to the SERS effect, *Solid State Commun.* **36**, 1011–1015 (1980).

51. R. M. Hexter, Enhanced Raman intensity of molecules adsorbed on metal surfaces. Experiments and theory, *Solid State Commun.* **32**, 55–57 (1979).

52. J. F. Evans, M. G. Albrecht, D. M. Ullevig, and R. M. Hexter, The physical and chemical characterisation of electrochemically reformed silver surfaces, *J. Electroanal. Chem.* **106**, 209–234 (1980).

53. A. Regis and J. Corset, A chemical interpretation of the intense Raman spectra observed at a silver electrode in the presence of chloride ion and pyridine: Formation of radicals, *Chem. Phys. Lett.* **70**, 305–310 (1980).

54. R. Dornhaus and R. K. Chang, Comments on the 210–243 cm^{-1} mode in surface enhanced Raman scattering from the pyridine–Ag system, *Solid State Commun.* **34**, 811–815 (1980).

55. H. Wetzel, H. Gerischer, and B. Pettinger, Surface enhanced Raman scattering from silver–halide and silver–pyridine vibrations and the role of silver adatoms, *Chem. Phys. Lett.* **78**, 392–397 (1981).

56. B. E. Conway, R. G. Barradas, P. G. Hamilton, and J. M. Parry, Electrochemical adsorption of neutral and ionic components in solutions of pyridine and derived ions, *J. Electroanal. Chem.* **10**, 485–502 (1965).

57. G. F. Atkinson, D. A. Guzonas, and D. E. Irish, Raman spectral studies at the silver surface of the Ag/KCl, pyridine electrode, *Chem. Phys. Lett.* **75**, 557–560 (1980).

58. M. Fleischmann and I. R. Hill, unpublished data.

59. B. Pettinger and H. Wetzel, Surface enhanced Raman spectroscopy of pyridine on Ag electrodes. Surface complex formation, *Chem. Phys. Lett.* **78**, 398–403 (1981).

60. M. Fleischmann, P. J. Hendra, I. R. Hill, and M. E. Pemble, Enhanced Raman spectra from species formed by the coadsorption of halide ions and water molecules on silver electrodes, *J. Electroanal. Chem.* **117**, 243–255 (1981).

61. B. Pettinger, M. R. Philpott, and J. G. Gordon, Contribution of specifically adsorbed ions, water and impurities to surface enhanced Raman spectroscopy (SERS) of Ag-electrodes, *J. Chem. Phys.* **74**, 934–940 (1981).

62. V. V. Marinyuk, R. M. Lazorenko-Manevich, and Ya. M. Kolotyrkin, Spectroscopic study of the effect of halide ions on the adsorption properties of silver, *Elektrokhimiya* **14**, 1747–1750 (1978).

63. A. T. Kuhn, private communication.

64. M. Fleischmann, P. R. Graves, I. R. Hill, and J. Robinson, Simultaneous Raman spectroscopic and differential double-layer capacitance measurements of pyridine adsorbed on roughened silver electrodes, *Chem. Phys. Lett.* **98**, 503–506 (1983).

65. W. Suetaka and M. Ohsawa, Potential modulation Raman spectrum of species on metal electrode surface, *Appl. Surf. Sci.* **3**, 118–120 (1979).

66. M. G. Albrecht and J. A. Creighton, Intense Raman spectra at a roughened silver electrode, *Electrochim. Acta* **23**, 1103–1105 (1978).

67. R. Dornhaus, M. B. Long, R. E. Benner, and R. K. Chang, Time development of SERS from pyridine, pyrimidine, pyrazine and cyanide adsorbed on Ag electrodes during an oxidation-reduction cycle, *Surf. Sci.* **93**, 240–262 (1980).

68. J. A. Creighton, C. G. Blatchford, and M. G. Albrecht, Plasma resonance enhancement of Raman scattering by pyridine adsorbed on silver or gold particles of size comparable to the excitation wavelength, *J. Chem. Soc. Far. II* **1979**, 790–798.

69. M. E. Lippitsch, Observation of surface enhanced Raman spectra by adsorption to silver colloids, *Chem. Phys. Lett.* **74**, 125–127 (1980).

70. A. J. McQuillan and C. G. Pope, Raman spectra of formate and acetate ions adsorbed on Ag particles in aqueous solutions, *Chem. Phys. Lett.* **71**, 349–352 (1980).

71. B. Pettinger, A. Tadjeddine, and D. M. Kolb, Enhancement in Raman intensity by use of surface plasmons, *Chem. Phys. Lett.* **66**, 544–548 (1979).

72. R. Dornhaus, R. E. Benner, R. K. Chang, and I. Chabay, Surface plasmon contribution to SERS, *Surf. Sci.* **101**, 367–373 (1980).

73. A. Otto, Investigation of electrode surfaces by surface plasmon polariton spectroscopy, *Surf. Sci.* **101**, 99–108 (1980).

74. A. Girlando, M. R. Philpott, D. Heitmann, J. D. Swalen, and R. Santo, Raman spectra of thin organic films enhanced by plasmon surface polaritons on holographic metal gratings, *J. Chem. Phys.* **72**, 5187–5191 (1980).

75. T. E. Furtak and J. Kester, Do metal alloys work as substrates for surface-enhanced Raman spectroscopy? *Phys. Rev. Lett.* **45**, 1652–1655 (1980).

76. C. Y. Chen, E. Burstein, and S. Lundquist, Giant Raman scattering from pyridine and CN^- adsorbed on silver, *Solid State Commun.* **32**, 63–66 (1979).

77. R. L. Birke, J. R. Lombardi, and J. I. Gersten, Observation of a continuum in enhanced Raman scattering from a metal-solution interface, *Phys. Rev. Lett.* **43**, 71–75 (1979).

78. J. Timper, J. Billmann, A. Otto, and I. Pockrand, Surface enhanced light scattering from Ag electrodes: Background and CN stretch vibrations, *Surf. Sci.* **101**, 348–354 (1980).

79. J. P. Heritage, J. G. Bergmann, A. Pinczuk, and J. M. Worlock, Surface picosecond Raman gain spectroscopy of a cyanide monolayer on silver, *Chem. Phys. Lett.* **67**, 229–232 (1979).

80. B. Pettinger, Surface enhanced Raman spectroscopy of pyridine on Ag electrodes. Evidence for overtones, *Chem. Phys. Lett.* **78**, 404–409 (1981).

81. J. C. Tsang, J. R. Kirtley, and T. N. Theis, Surface plasmon polariton contributions to Stokes emission from molecular monolayers on periodic silver surfaces, *Solid State Commun.* **35**, 667–670 (1980).

82. M. R. Mahoney, M. W. Howard, and R. P. Cooney, Carbon dioxide conversion to hydrocarbons at silver electrode surfaces (1) Raman spectroscopic evidence for surface carbon intermediates, *Chem. Phys. Lett.* **71**, 59–63 (1980).

83. F. Tuinstra and J. L. Koenig, Raman spectrum of graphite, *J. Chem. Phys.* **53**, 1126–1130 (1970).

84. R. Vidano and D. B. Fischbach, New lines in the Raman spectra of carbons and graphite, *J. Am. Ceram. Soc.* **61**, 13–17 (1978).

85. M. W. Howard, R. P. Cooney, and A. J. McQuillan, The origin of intense Raman spectra from pyridine at silver electrode surfaces: The role of surface carbon, *J. Raman Spectrosc.* **9**, 273–278 (1980).

86. W. Paik, T. N. Andersen, and H. Eyring, Kinetic studies of the electrolytic reduction of carbon dioxide on the mercury electrode, *Electrochim. Acta*, **14**, 1217–1232 (1969).

87. A. Z. Genack, D. A. Weitz, and T. J. Gramila, Observations of new low frequency modes with surface enhanced Raman scattering, *Surf. Sci.* **101**, 381–386 (1980).

88. H. Morawitz and T. R. Koehler, A model for Raman active librational modes on a metal surface. Pyridine and CN^- on silver, *Chem. Phys. Lett.* **71**, 64–67 (1980).

89. H. Nichols and R. M. Hexter, Vibrational frequencies of halogen atoms adsorbed on silver metal surfaces, *J. Chem. Phys.* **74**, 2059–2063 (1981).

90. M. Fleischmann and I. R. Hill, The observation of solvated metal ions in the double-layer region at silver electrodes using surface enhanced Raman scattering, *J. Electroanal. Chem.* **146**, 367–376 (1983).

91. A. Otto, Raman spectra of $(CN)^-$ adsorbed at a silver surface, *Surf. Sci.* **75**, L392–L396 (1978).

92. T. E. Furtak, Anomalously intense Raman scattering at the solid-electrolyte interface, *Solid State Commun.* **28**, 903–906 (1978).

93. J. Billmann, G. Kovacs, and A. Otto, Enhanced Raman effect from cyanide adsorbed on a silver electrode, *Surf. Sci.* **92**, 153–173 (1980).

94. R. E. Benner, R. Dornhaus, R. K. Chang, and B. L. Laube, Correlations in the Raman spectra of cyanide complexes adsorbed on Ag electrodes, with voltammograms, *Surf. Sci.* **101**, 341–347 (1980).

95. T. E. Furtak, G. Trott, and B. H. Loo, Enhanced light scattering from the metal/solution interface: Chemical Origins, *Surf. Sci.* **101**, 374–380 (1980).

96. M. Fleischmann, I. R. Hill, and M. E. Pemble, Surface enhanced Raman spectroscopy of $^{12}CN^-$ and $^{13}CN^-$ adsorbed at silver electrodes, *J. Electroanal. Chem.* **136**, 361–370 (1981).

97. H. Wetzel, H. Gerischer, and B. Pettinger, Surface-enhanced Raman scattering from silver–cyanide and silver–thiocyanate vibrations and the importance of adatoms, *Chem. Phys. Lett.* **80**, 159–162 (1981).

98. L. H. Jones, Vibrational spectrum and structure of metal–cyanide complexes in the solid state. I. K $Ag(CN)_2$, *J. Chem. Phys.* **26**, 1578–1584 (1957).

99. G. Laufer, T. F. Schaaf, and J. T. Huneke, Surface enhanced Raman scattering from cyanide adsorbed on copper, *J. Chem. Phys.* **73**, 2973–2976 (1980).

100. B. H. Loo and T. E. Furtak, The giant Raman effect from pyridine on a chemically modified gold substrate, *Chem. Phys. Lett.* **71**, 68–71 (1980).

101. C. Y. Chen, I. Davoli, G. Ritchie, and E. Burstein, Giant Raman scattering and luminescence by molecules adsorbed on Ag and Au metal island films, *Surf. Sci.* **101**, 363–366 (1980).

102. V. V. Marinyuk, R. M. Lazorenko-Manevich, and Ya. M. Kolotyrkin, Nature of the interaction of adsorbate molecules with metal adatoms, *J. Electroanal. Chem.* **110**, 111–118 (1980).

103. S. Efrima and H. Metiu, Surface induced resonant Raman scattering, *Surf. Sci.* **92**, 417–432 (1980).

104. S. Efrima and H. Metiu, Resonant Raman scattering by adsorbed molecules, *J. Chem. Phys.* **70**, 1939–1947 (1979).

105. T. Maniv and H. Metiu, Some comments concerning the microscopic theory of Raman scattering by adsorbed molecules, *Surf. Sci.* **101**, 399–408 (1980).

106. H. H. Jaffé and M. Orchin, In: *Theory and Applications of Ultra-Violet Spectroscopy*, Wiley and Sons, New York (1962).

107. F. R. Aussenegg and M. E. Lippitsch, On Raman scattering in molecular complexes involving charge transfer, *Chem. Phys. Lett.* **59**, 214–216 (1978).

108. M. R. Philpott, Effect of surface plasmons on transitions in molecules, *J. Chem. Phys.* **62**, 1812–1817 (1975).

109. R. M. Hexter and M. G. Albrecht, Metal surface Raman spectroscopy: Theory, *Spectrochim. Acta* **35A**, 233–251 (1979).

110. D. A. Zwemer, C. V. Shank, and J. E. Rowe, Surface enhanced Raman scattering as a function of molecule-surface separation, *Chem. Phys. Lett.* **73**, 201–204 (1980).

111. J. E. Rowe, C. V. Shank, D. A. Zwemer, and C. A. Murray, Ultrahigh vacuum studies of enhanced Raman scattering from pyridine on Ag surfaces, *Phys. Rev. Lett.* **44**, 1770–1773 (1980).

112. P. N. Sanda, J. M. Warlaumont, J. E. Demuth, J. C. Tsang, K. Christmann, and J. A. Bradley, Surface enhanced Raman scattering from pyridine on Ag(111), *Phys. Rev. Lett.* **45**, 1519–1523 (1980).

113. P. K. Aravind and H. Metiu, The enhancement of Raman and fluorescent intensity by small surface roughness, *Chem. Phys. Lett.* **74**, 301–305 (1980).

114. S. S. Jha, J. R. Kirtley, and J. C. Tsang, Intensity of Raman scattering from molecules adsorbed on a metallic grating, *Phys. Rev. B Cond. Matt.* **22**, 3973–3982 (1980).

115. J. R. Kirtley, S. S. Jha, and J. C. Tsang, Surface plasmon model of surface enhanced Raman scattering, *Solid State Commun.* **35**, 509–512 (1980).

116. F. J. Adrian, Surface enhanced Raman scattering by surface plasmon enhancement of electromagnetic fields near spheroidal particles on a roughened metal surface, *Chem. Phys. Lett.* **78**, 45–49 (1981).

117. T. K. Lee and J. L. Birman, Molecule adsorbed on plane metal surface: Coupled system eigenstates, *Phys. Rev. B. Cond. Matt.* **22**, 5953–5960 (1980).

118. T. K. Lee and J. L. Birman, Quantum theory of enhanced Raman scattering by molecules on metals: Surface-plasmon mechanism for plane metal surface, *Phys. Rev. B. Cond. Matt.* **22**, 5961–5966 (1980).

119. G. W. Robinson, Surface-enhanced Raman effect, *Chem. Phys. Lett.* **76**, 191–195 (1980).

120. M. Kerker, D. S. Wang, and H. Chew, Surface enhanced Raman scattering by molecules adsorbed at spherical particles, *Appl. Opt.* **19**, 3373–3388 (1980).

121. M. Kerker, D. S. Wang, and H. Chew, Surface enhanced Raman scattering (SERS) by molecules adsorbed at spherical particles: Errata, *Appl. Opt.* **19**, 4159–4174 (1980).

122. D. S. Wang, H. Chew, and M. Kerker, Enhanced Raman scattering at the surface (SERS) of a spherical particle, *Appl. Opt.* **19**, 2256–2257 (1980).

123. S. L. McCall, P. M. Platzman, and P. A. Wolff, Surface enhanced Raman scattering, *Phys. Lett.* **77A**, 381–383 (1980).

124. M. Moskovits, Surface roughness and the enhanced intensity of Raman scattering by molecules adsorbed on metals, *J. Chem. Phys.* **69**, 4159–4161 (1978).

125. M. J. Dignam and M. Moskovits, Influence of surface roughness on the transmission and reflectance spectra of adsorbed species, *J. Chem. Soc. Far. II* **1973**, 65–78.

126. J. I. Gersten, R. L. Birke, and J. R. Lombardi, Theory of enhanced light scattering from molecules adsorbed at the metal-solution interface, *Phys. Rev. Lett.* **43**, 147–150 (1979).

127. E. Burstein, Y. J. Chen, C. Y. Chen, S. Lundquist, and E. Tosatti, Giant Raman scattering by adsorbed molecules on metal surfaces, *Solid State Commun.* **29**, 567–570 (1979).

128. F. W. King, R. P. Van Duyne, and G. C. Schatz, Theory of Raman scattering by molecules adsorbed on electrode surfaces, *J. Chem. Phys.* **69**, 4472–4481 (1978).

129. C. S. Allen and R. P. Van Duyne, Orientational specificity of Raman scattering from molecules adsorbed on silver electrodes, *Chem. Phys. Lett.* **63**, 455–459 (1979).

130. G. C. Schatz and R. P. Van Duyne, Image field theory of enhanced Raman scattering by molecules adsorbed on metal surfaces: Detailed comparison with experimental results, *Surf. Sci.* **101**, 425–438 (1980).

131. G. L. Eesley and J. R. Smith, Enhanced Raman scattering on metal surfaces, *Solid State Commun.* **31**, 815–819 (1979).

132. S. Efrima and H. Metiu, Classical theory of light scattering near a solid surface, *Chem. Phys. Lett.* **60**, 59–64 (1978).

133. S. Efrima and H. Metiu, Classical theory of light scattering by an adsorbed molecule, *J. Chem. Phys.* **70**, 1602–1613 (1979).

134. S. Efrima and H. Metiu, Classical theory of light scattering by a molecule located near a solid surface, *J. Chem. Phys.* **70**, 2297–2309 (1979).

135. T. Maniv and H. Metiu, Raman reflection, a possible mechanism for enhancement of Raman scattering by an adsorbed molecule, *Chem. Phys. Lett.* **79**, 79–85 (1981).

136. F. Aussenegg, M. Lippitsch, R. Moller, and J. Wagner, Measurement of Raman scattering in high field electric strength, *Phys. Lett.* **50A**, 233–234 (1974).

137. J. I. Gersten, The effect of surface roughness on surface enhanced Raman scattering, *J. Chem. Phys.* **72**, 5779–5780 (1980).

138. A. Otto, Raman scattering from adsorbates on silver, *Surf. Sci.* **92**, 145–152 (1980).

139. J. Kirtley, D. J. Scalapino, and P. K. Hansma, Theory of vibrational mode intensities in inelastic electron tunnelling spectroscopy, *Phys. Rev.* **B14**, 3177–3184 (1976).

140. A. Otto, J. Timper, J. Billmann, and I. Pockrand, Enhanced inelastic light scattering from metal electrodes caused by adatoms, *Phys. Rev. Lett.* **45**, 46–49 (1980).

141. J. C. Tsang, J. R. Kirtley, and J. A. Bradley, Surface-enhanced Raman spectroscopy and surface plasmons, *Phys. Rev. Lett.* **43**, 772–775 (1979).

142. J. C. Tsang and J. Kirtley, Anomalous surface enhanced molecular Raman scattering from inelastic tunnelling spectroscopy junctions, *Solid State Commun.* **30**, 617–620 (1979).

143. J. D. E. McIntyre, In: *Advances in Electrochemistry and Electrochemical Engineering*, P. Delahay and C. W. Tobias, eds., Vol. 9, pp. 61–166, Wiley and Sons, New York (1973).

144. G. R. Erdheim, R. L. Birke, and J. R. Lombardi, Surface enhanced Raman spectrum of pyrazine. Observation of forbidden lines at the electrode surface, *Chem. Phys. Lett.* **69**, 495–498 (1980).

145. B. E. Conway, J. G. Mathieson, and H. P. Dhar, Orientation behaviour of adsorbed pyridine and pyrazine at the mercury-water interface in relation to solution thermodynamic properties, *J. Phys. Chem.* **78**, 1226–1234 (1974).

146. S. L. McCall and P. M. Platzman, Raman scattering from chemisorbed molecules at surfaces, *Phys. Rev. B. Cond. Matt.* **22**, 1660–1662 (1980).

147. W. H. Weber and G. W. Ford, Enhanced Raman scattering by adsorbates including the nonlocal response of the metal and the excitation of nonradiative modes, *Phys. Rev. Lett.* **44**, 1774–1777 (1980).

148. J. Gersten and A. Nitzan, Electromagnetic theory of enhanced Raman scattering by molecules adsorbed on rough surfaces, *J. Chem. Phys.* **73**, 3023–3037 (1980).

149. A. Bewick, K. Kunimatsu, and B. S. Pons, Infrared spectroscopy of the electrode-electrolyte interface, *Electrochim. Acta* **25**, 465–468 (1980).

150. A. Girlando, J. G. Gordon II, D. Heitmann, M. R. Philpott, H. Seki, and J. D. Swalen, Raman spectra of molecules on metal surfaces, *Surf. Sci.* **101**, 417–424 (1980).

151. S. Venkatesan, G. Erdheim, J. R. Lombardi, and R. L. Birke, Voltage dependence of the surface-molecule line in the enhanced Raman spectra of several nitrogen containing compounds, *Surf. Sci.* **101**, 387–398 (1980).

152. R. E. Kunz, J. G. Gordon II, M. R. Philpott, and A. Girlando, Surface enhanced Raman spectra from silver electrodes in azide solution, *J. Electroanal. Chem.* **112**, 391–395 (1980).

153. J. E. Pemberton and R. P. Buck, Dithizone adsorption at metal electrodes. 2. Raman spectroelectrochemical investigation of effect of applied potential at a silver electrode. *J. Phys. Chem.* **85**, 248–262 (1981).

154. H. Wetzel, B. Pettinger, and U. Wenning, Surface enhanced Raman scattering from ethylenediaminetetraacetic-disodium salt and nitrate ions on silver electrodes, *Chem. Phys. Lett.* **75**, 173–178 (1980).

155. K. A. Bunding, J. R. Lombardi, and R. L. Birke, Surface-enhanced Raman spectra of methylpyridines, *Chem. Phys.* **54**, 115 (1980).

156. M. G. Albrecht, J. F. Evans, and J. A. Creighton, The nature of an electrochemically roughened silver surface and its role in promoting anomalous Raman scattering intensity, *Surf. Sci.* **75**, L777–L780 (1978).

157. R. L. Paul and P. J. Hendra, Laser Raman spectroscopy in surface chemical problems, *Min. Sci. Eng.* **8**, 171–186 (1976).

158. V. V. Marinyuk and R. M. Lazorenko-Manevich, Raman scattering cross section of pyridine adsorbed on silver, *Elektrokhimiya* **14**, 452–455 (1978).

159. V. V. Marinyuk, R. M. Lazorenko-Manevich, and Ya. M. Kolotyrkin, Resonance Raman effect of organic cations adsorbed on silver, *Dokl. Acad. Sci. U.S.S.R.* **242**, 1382–1385 (1978).

160. M. L. A. Temperini, H. C. Chagas, and O. Sala, Raman spectra of pyridine adsorbed on a copper electrode, *Chem. Phys. Lett.* **79**, 75–78 (1981).

161. M. Ohsawa, K. Nishijima, and W. Suetaka, Potential modulation Raman spectroscopy for in situ observation of electrode/electrolyte interface, *Surf. Sci.* **104**, 270–281 (1981).

7

Electron Spin Resonance

BERTEL KASTENING

1. Introduction

Electron spin resonance (ESR) is a spectroscopic technique that is specific for paramagnetic material. Various books and articles deal with the theory and experimental techniques of the method; a selection of three monographs may be cited here.[1-3] ESR furnishes valuable and detailed information about free radicals and radical ions. These species play an important role in various electrode processes, because electrons are in general transferred singly, particularly in organic electrode processes. (Under suitable conditions, inorganic substrates exhibit similar behavior, e.g., in the reduction processes $O_2 + e^- \rightarrow O_2^-$ or $SO_2 + e^- \rightarrow SO_2^-$.) Since 1957 excessive use has been made of ESR spectroscopy in electrochemical investigations as well as of electrochemical radical generation for ESR investigations of such species. Some early papers[4-7] and some reviews on the application of ESR to electrochemistry[8-12] may be mentioned.

While most investigations concern electrogenerated radicals in the dissolved state, radicals adsorbed at the electrode have only sporadically been observed; in general the intensity is too small at ambient temperatures. Some studies have been made of the paramagnetic properties of the electrode material proper (with and without electrogenerated adsorbed species, like hydrogen atoms, being involved); this concerns particularly carbon electrodes but also substoichiometric WO_3 (see, e.g., Alquié-Redon *et al.*, 1975).

DR. B. KASTENING • Institut für Physikalische Chemie der Universität Hamburg, Laufgraben 24, D-2000 Hamburg 13, West Germany.

2. Principles of ESR Spectroscopy

Due to the uncompensated spin of the odd electron, radicals are paramagnetic and exhibit, therefore, two quantum states (corresponding to quantum numbers +1/2 and −1/2) which differ, in an external magnetic field, by a small amount of energy. If an electromagnetic wave of corresponding energy is employed to the species, then the distribution to the two quantum states is disturbed and energy is absorbed and transmitted, via a relaxation mechanism, into the environment. The resonance condition is

$$h \cdot \nu = g \cdot \mu_B \cdot B_0 \tag{1}$$

where h is Planck's constant, ν is the frequency of the electromagnetic wave, μ_B is the Bohr magneton ($\mu_B = 9.2732 \times 10^{-24}\ m^2\ A$), B_0 is the external magnetic flux density, and g is the gyromagnetic ratio. The latter is $g = 2.0023$ for the free electron and exhibits, due to spin–orbit coupling, slight deviations (of the order of 10^{-3}) for organic radicals. Most measurements are made at (fixed) X band microwave frequencies (9.3 GHz), the corresponding resonant value B_0 of the (variable) external magnetic flux density B being about $0.33\ T$ (=3,300 Gauss).

Instead of one single absorption line at the resonant field value B_0, most radicals in solution show a spectrum of lines around B_0. This spectrum (hyperfine splitting, HFS) arises from the interaction of the electronic spin with the various nuclear spins present in the molecule, particularly those of protons and of nitrogen nuclei, while ^{12}C and ^{16}O nuclei do not exhibit magnetic momenta. At constant microwave frequency, the presence of a single proton results in two lines at $B = B_0 - \frac{1}{2}a_H$ and $B_0 + \frac{1}{2}a_H$, respectively, where a_H is the coupling constant depending on the bonding situation of the H atom (i.e., the "spin density" at the site of the proton). A nitrogen nucleus creates three lines at $B_0 - a_N$, B_0, and $B_0 + a_N$, respectively. Equal nuclei at equivalent bonding situations have the same coupling constant, thus resulting in lines of different statistical weight, i.e., different intensities. An illustration of a multiline spectrum is shown in Figure 1; it results from the nitrobenzene anion radical in aqueous solution. As usual, the record represents the first derivative of the energy absorption spectrum; this is obtained by superimposing a periodic field modulation to the straight field scanning and measuring the transmitted modulation.

Apart from the special effect of "alternating linewidths" (see below), all the lines of a spectrum exhibit approximately the same width, which is determined by the strength of the interaction with the environment: the stronger the interaction and the shorter the relaxation time, the broader the lines. Spectra of good resolution often show linewidths of about 10^{-6}–$10^{-5}\ T$ (10–100 mG). The information that can be derived from the number, position, and intensity of the lines, from the linewidths as well as from the position of the center of the spectrum (corresponding to the individual g value), will be discussed below.

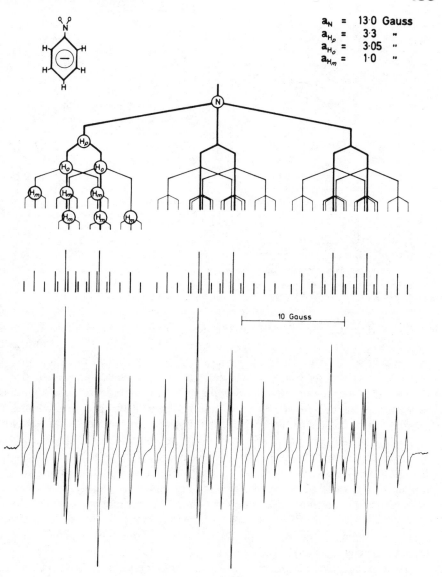

a_N = 13·0 Gauss
a_{H_p} = 3·3 "
a_{H_o} = 3·05 "
a_{H_m} = 1·0 "

10 Gauss

Figure 1. ESR spectrum of electrogenerated nitrobenzene anion radicals in aqueous solution. Above: Schematic representation of the generation of the various lines by the combination of nuclear spin states. From: Kastening (1974), courtesy of John Wiley & Sons, New York.

A schematic representation of an ESR spectrometer is shown in Figure 2 where auxiliary equipment (such as a means for microwave frequency measurements matched to the waveguide system or a means for measuring the magnetic field strength, e.g., by inserting an NMR probe in the field, as well as computers for the analysis or simulation of ESR spectra) are omitted.

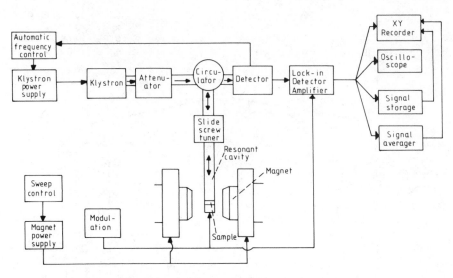

Figure 2. Block diagram of a typical ESR spectrometer.

The frequently applied rectangular resonant cavity shows a tubular opening of about 1 cm diameter for inserting the sample. In order to avoid drastic decrease of sensitivity, the sample, including the sample cell or tube, must be restricted to a flat section, perpendicular to the waveguide axis and close to the center plane of the opening. Sample cells are made from quartz glass or pyrex glass or even from PTFE. Either flat cells with an active volume of about $0.04 \times 1 \times 2$ cm or capillaries with 0.08 cm inner diameter are frequently applied. This geometry causes severe difficulties if the generating electrode is to be inserted in the cell (for "*in situ*" generation, see below). The active volume is small (10^{-5}–10^{-4} liter) and, therefore, in spite of the comparatively fair sensitivity of ESR measurements, the concentration must not be too low, particularly if a multiline spectrum is present. The best range of concentrations is about 10^{-5}–10^{-3} mol/liter; larger concentrations may result in disturbances due to spin–spin interactions while lower concentrations may require enhancing techniques such as signal averaging.

Instead of the rectangular microwave cavity mentioned above, cylindrical cavities and even a helix (simultaneously working as the generating electrode) instead of a cavity have been applied in a few cases. Further development of such techniques as well as of sample cells seems promising for achieving enhanced sensitivity and for the study of short-lived radicals.

Spectra are recorded on an X-Y recorder during slow scanning of the magnetic field. If only short measuring times are available (e.g., of the order of one second), then rapid scanning and observation on an oscilloscope or via signal storage can be applied; high resolution may require multiple scanning and signal averaging. For kinetic measurements, one may observe the decay of intensity by either repeatedly recording one particular absorption line or by matching the magnetic field to the peak of a line. Other experimental methods with the application of flow techniques will be mentioned below.

3. Information Drawn from ESR Data

The appearance of an ESR spectrum upon electrolysis is unambiguous proof that radicals are involved in the electrode process. Beyond this fundamental statement, the details of the ESR spectrum supply further information about the type of radical that appears as an intermediate or product of the process. Thus, an analysis of the number, relative position, and relative intensity of the lines of Figure 1 shows that the spectrum consists of three almost equal groups of lines, each group being composed from two equal subgroups penetrating each other and consisting of nine lines each. These nine lines represent a triplet of triplets with relative intensities $1:2:1$, $2:4:2$, and $1:2:1$. This behavior strongly suggests the presence of one ^{14}N nucleus (generating three equal groups), one single proton (generating two subgroups), and two sets each of two equivalent protons (generating triplets).

The reduction of certain halonitrobenzenes may serve as an example for the elucidation of electrode mechanisms by applying ESR. According to the solvent and potential, one will observe the ESR spectra of either the corresponding anion radical $X-C_6H_4-NO_2^{\cdot-}$ or, upon halide ion elimination, of the nitrobenzene anion radical $H-C_6H_4-NO_2^{\cdot-}$. These can easily be distinguished: while the latter radical exhibits a 54-line spectrum as shown in Figure 1, the p-halonitrobenzene anion radical either gives rise to only 27 lines, because the para-proton splitting is missing, or exhibits additional splitting due to the halogen, if the conditions are suitable. With Cl or Br, this results in the splitting of each line into a quartet, giving a total of 108 lines. In various cases, the observed spectrum is clearly composed from contributions of two types of radical species, which can separately be recognized.

The evaluation of absolute concentrations of radicals from ESR spectra is difficult and will in general require a comparison with the absorption due to a species that is present at a known concentration. For this purpose, it has been suggested to generate electrolytically a known amount of a stable paramagnetic ion such as Cu^{2+}. Relative concentrations of two simultaneously present species or, in kinetic studies of formation or of decay, of one species at different times can more readily be obtained by making use of the relation

$$c_{rad} = \text{const} \cdot \int_{B=0}^{\infty} I \cdot dB \tag{2}$$

where c_{rad} means the concentration of radicals assumed to be homogeneously distributed throughout the sample, and I is the absorption intensity. The constant depends on the sensitivity (including effects of the sample and cell), the sample volume, and the temperature. Peak heights of the derivative curves, which are in general recorded, can directly be compared only if the lineshape is the same (e.g., exact Gaussian or Lorentzian type lines). Then, the relation

$$c_{\text{rad}} = \text{const}' \cdot \left(\frac{dI}{dB}\right)_{\text{max}} \cdot (\Delta B)^2 \tag{3}$$

holds, where $(dI/dB)_{\text{max}}$ is the peak height and ΔB the linewidth (distance of the two peaks of a single line). The constant depends on the conditions mentioned above for "const" and additionally on the particular lineshape. A possible variation of the linewidth upon changing the concentration of the radical or the precursor molecule has to be taken into account.

Fast kinetics of simple homogeneous electron transfer processes such as

$$M + M^{\overline{\cdot}} \rightleftharpoons M^{\overline{\cdot}} + M \tag{4}$$

where M is the precursor molecule and $M^{\overline{\cdot}}$ the corresponding anion radical (or, correspondingly, with cation radicals), can be evaluated from an investigation of linewidth variations upon changing the precursor concentration. Large concentrations result in fast electron exchange, i.e., short lifetimes of spin states. As a consequence, the HFS lines become broader and resolution becomes poorer until the HFS collapses, leaving only a single line that becomes narrower on further increase of concentration due to progressive averaging of the nuclear spin interaction.

In general, the various lines of a given spectrum have approximately equal widths and, therefore, the peak height ratios are approximately integers reflecting the statistical weight of the corresponding coupling nucleus. (Slight deviations, as exhibited in Figure 1 by the high-field group of lines as compared with the center and low-field groups, are associated with anisotropic g tensors and molecular tumbling within the surrounding solvent molecules.) In special cases, however, substantial differences of linewidths occur within a given spectrum due to intramolecular motions or specific interactions with solvent molecules or other solutes. This "alternating linewidth effect" (because, in simple cases, small broad and large narrow lines alternate) is observed, if there are two, in principle, equivalent (ESR-active) atoms or groups, these groups being in two different though rapidly exchanging situations exhibiting different coupling constants. For example, one of the two nitro groups of an aromatic dinitro compound is locked to the aromatic ring plane while the other is twisted out of that plane. If the exchange of the two situations proceeds sufficiently fast, then those lines become broad that correspond to the two nuclei having different spin states during the exchange process.

Because HFS constants (a_{H}, a_{N}, etc.) are directly correlated to "spin densities" at the corresponding atom, an evaluation of these constants supplies valuable information with respect to the electron distribution throughout the radical. The mutual assistance of ESR investigations on the one hand and calculations from molecular orbital theories (Hückel MO theory and corresponding modifications) on the other, results, therefore, in more detailed knowledge of the electronic structure of electrogenerated radicals. Several approaches have been made of correlating electrochemical data (redox or half-wave potentials) directly with ESR data such as coupling constants or g values. Thus, a linear relation exists between g values and half-wave potentials for a group of similar compounds such as aromatic hydrocarbons. These correlations arise from the fact that both ESR and electrochemical data are related to MO energy parameters.

Detailed information may be drawn from ESR data with respect to the conformation of radicals. Thus, HFS data analysis frequently allows one to distinguish whether a particular group of the molecule is freely rotating or whether it is locked in a certain position. In the first case, even the rotation frequency may be evaluated under certain conditions. In the locked case, the ESR spectrum will allow the derivation of the locking position. The spin distribution throughout an aromatic ring system will allow one to conclude whether two rings are coplanar or not and, if not, the twist angle between the two rings. Two simultaneously present conformational isomers have frequently been observed by the presence of two superimposed spectra having different coupling constants; equilibrium constants for the transformation between both forms as well as rate constants of the conformational jump have been evaluated in such cases.

Solvation as well as ion-pairing substantially affect the electron distribution within a radical. Hence, the change of coupling constants in the presence of different solvents or electrolytes furnishes information about the interaction between the radical and its environment. Moreover, ion-pairing may cause additional hyperfine structure due to the nuclear magnetic momentum of the corresponding counter-ion; e.g., alkali metal ions as companions of anion radicals, or halide ions in the case of cation radicals.

Additional information may be drawn from the application of isotopes. Thus, the spin quantum number of deuterons is one (instead of $1/2$ for protons), giving rise to three equidistant lines, instead of two from protons, with corresponding changes in the case of two or more equivalent deuterons. The nuclear magnetic momentum, however, is smaller so that $a_D/a_H = 0.1535$ and HFS from deuterons is more likely to be unresolved. The change of the spectrum upon substituting D for H is a valuable aid in the assignment of coupling constants. No magnetic momenta are shown by ^{12}C and ^{16}O nuclei so that the direct evaluation of spin densities at carbon or oxygen atoms requires the application of isotopes, viz. ^{13}C (giving rise to a doublet) and ^{17}O (creating a sextet of lines), respectively. The natural abundance of ^{13}C (1.1%) is, under favorable conditions, sufficient to observe the corresponding small satellite lines. In general, however, the observation of such effects requires the application of artificially enriched samples. Such samples are, of course, inevitably necessary if a particular position is to be marked. Isotopes from several other elements are applicable as well.

4. Techniques of External Radical Generation

Depending on whether the radical generating electrode is placed somewhere in an electrolysis cell outside the ESR cavity or whether it is placed inside the ESR sample cell, the techniques of electrochemical ESR investigations are referred to as external or internal ("*in situ*", "*intra muros*").

The advantages of the external generation will become more evident from a discussion of the drawbacks of the internal technique (see below). These advantages result from the fact that external generation of radicals can be carried out under optimal conditions in a cell suitable for, e.g., potential-controlled or, if this is advantageous, current-controlled preparative electrolysis with proper removal of interfering impurities, etc. Moreover, radicals will be homogeneously distributed throughout the ESR sample cell.

There is one major disadvantage of the external technique that arises from the necessity of transferring the radical solution from the electrolysis cell to the ESR sample cell. The time that is spent during the transfer of the radical from the electrode to the bulk and during the transport of the solution to the sample cell constitutes the limits for observing ESR spectra of short-lived radicals. The transport from the electrolysis cell to the ESR cavity would require considerable effort if it should proceed faster than within some tenths of a second. At the same time, a lifetime τ_{rad} of at least 0.1 s is necessary in order to obtain radical concentrations c_{rad} of, e.g., 3×10^{-6} mol/liter or more when applying a precursor concentration of $c_0 \simeq 10^{-3}$ mol/liter and an electrolysis cell with a time constant $\tau_{cell} \simeq 30$ s, according to the steady-state condition $c_{rad}/c_0 \simeq \tau_{rad}/\tau_{cell}$. To construct cells exhibiting markedly shorter time constants as well as to obtain highly resolved ESR spectra at lower concentrations would meet with considerable difficulty, while larger precursor concentrations would result in perturbations of the ESR spectrum. From both considerations, 0.1 s appears to be a lower limit of the lifetime of radicals observed with the external generation technique.

In early experiments, samples for ESR investigations have been withdrawn from the electrolysis cell manually or by evacuating the sample cell connected to the electrolysis cell within a closed system; occasionally, samples have been chilled to low temperatures. Later, different kinds of flow systems were used (cf. Figure 3a and c). The solution from the electrolysis cell may either pass the ESR sample cell and flow off or be cycled, as shown in Figure 3a; circulation may be effected either by a pump or by bubbling inert gas through the sidearm. Such flow systems have been applied to kinetic investigations, e.g., by making use of a stopped-flow technique. Another experimental technique is shown in Figure 3c, where the radical solution and a separate reactant solution are mixed before entering a capillary serving as an ESR sample cell; fast reactions between the radical and the reactant can be followed by measuring the variation of the ESR line intensity across the length of the capillary during continuous flow. (In this case, the time constant of the investigated reaction may be as low as 0.001 s.)

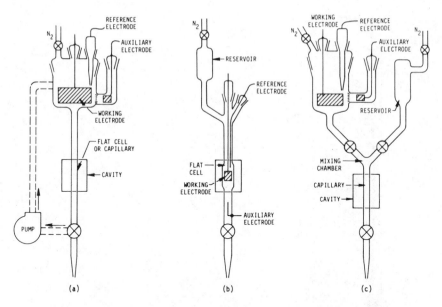

Figure 3. Combinations of electrochemical and ESR flow cells: (a) external electrolysis cell, which may be used with recirculation; (b) *in situ* electrolysis cell; (c) external electrolysis cell with provision for mixing reactant before the sample enters the cavity. From: Goldberg and Bard (1975), courtesy of Marcel Dekker, New York.

5. Internal Generation of Radicals

The main advantage of the internal technique, in which radicals are generated by an electrode placed inside the ESR sample cell, results from the fact that ESR spectra from short-lived radicals are more likely to be observed than with external techniques, because no transportation is necessary. The lower limit of lifetimes is given by the necessity of having a sufficient amount of radicals under steady-state conditions, set up by the balance of production by current and decay. The production by current is limited by the precursor concentration, which should not exceed about 10^{-3} mol/liter for obtaining good resolution, as well as by the size of the effective electrode surface, which is in general not more than about $0.5 \, \text{cm}^2$. This limitation arises from the unfavorable geometry of the cell (a typical design is shown in Figure 4). Since the counter electrode is in almost all instances placed outside the sample cell, the current in the flat portion generates a considerable potential drop across the electrode surface. The length of the electrode must therefore be limited in the direction of the sample axis. (Some attempts have been made to overcome this difficulty, but these create other drawbacks.) From practical experience, one may derive that it is readily possible to observe radicals with lifetimes as low as 0.01 s and that this limit may be extended, with some effort, to about

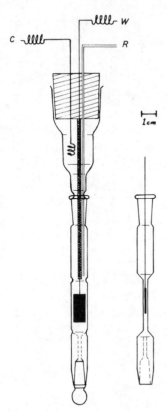

Figure 4. Typical flat ESR sample cell with equipment for *in situ* electrolysis. C: counter, W: working, R: reference electrode connections. The working electrode lead is mantled for insulation.

10^{-3} s. Shorter lifetimes expected from theoretical calculations seem to have been overestimated; it appears that ESR spectra due to radicals of lower stability have not been observed in practice. Another advantage of the internal generation is the small amount of the sample necessary for the experiment.

If radicals are too short-lived for direct observation, they may nevertheless be traced by spin trapping. In this technique, a suitable reactant (e.g., *t*-butyl nitroxide or a nitrone) present during the electrode process reacts readily with the intermediate radical to form a more stable radical adduct. The ESR spectra of the latter species, however, contain only poor information about the radical in question, since the odd electron is essentially located at the reactant part of the adduct. Improvement of the trapping technique would be of great use for mechanistic studies.

The internal technique is preferable for short-lived radicals as well as for the qualitative detection of radicals. However, it is not as advantageous for investigations requiring the determination of, at least relative, quantities, e.g., for kinetic studies. This drawback is mainly due to the inhomogeneous distribution of radicals as well as to the nonuniform sensitivity across the sample volume. The inhomogeneous and time-dependent distribution may be predicted to some extent from theoretical calculations. Such calculations make use of digital simulation techniques and take into account the local currents, diffusion, and reaction processes, making reasonable assumptions with respect to the prevailing mechanism. Much effort, however, is necessary in order to maintain sufficient agreement between experimental and theoretical conditions.

Better defined experimental conditions are present under steady-state conditions with the solution flowing through the sample cell. Figure 3b shows schematically an example of a flow cell with internal generation. In this case, however, very pure solutions are necessary in order to avoid initial trapping of radicals by impurities. This difficulty does not exist if either internal generation with resting solutions or external generation is applied, because some pre-electrolysis would then remove the impurities, e.g., by their reaction with the first portion of electrogenerated radicals.

References

1. P. B. Ayscough, *Electron Spin Resonance in Chemistry*, Methuen and Co., London (1967).
2. R. S. Alger, *Electron Paramagnetic Resonance*, Interscience Publishers, New York (1968).
3. J. E. Wertz and J. R. Bolton, *Electron Spin Resonance—Elementary Theory and Practical Applications*, McGraw-Hill, New York (1972).
4. A. A. Galkin, Ya. L. Shamfarov, and A. V. Stefanishina, *J. Exptl. Theor. Phys.* (*USSR*) **32**, 1581 (1957).
5. D. E. G. Austen, P. H. Given, D. I. E. Ingram, and M. E. Peover, *Nature* **182**, 1784 (1958).
6. A. H. Maki and D. H. Geske, *J. Chem. Phys.* **30**, 1356 (1959).
7. J. P. Billon, J. Cauquis, J. Combrisson, and A. M. Li, *Bull. Soc. Chim. France*, 2062 (1960).
8. R. N. Adams, *J. Electroanal. Chem.* **8**, 151 (1964).
9. B. Kastening, *Chem. Ing. Tech.* **42**, 190 (1970).
10. B. Kastening, In: *Electroanalytical Chemistry*, H. W. Nürnberg, ed., (Advances in Analytical Chemistry and Instrumentation, Vol. 10), pp. 421–494, John Wiley and Sons, New York (1974).
11. I. B. Goldberg and A. J. Bard, In: *Magnetic Resonance in Chemistry and Biology*, J. N. Herak and K. J. Adamic, eds., pp. 255–314, Marcel Dekker, New York (1975).
12. T. M. McKinney, in: *Electroanalytical Chemistry*, A. J. Bard, ed., Vol. 10, pp. 97–278, Marcel Dekker, New York, (1977).
13. A. M. Alquié-Redon, A. Aldaz, and C. Lamy, *Surf. Sci.* **49**, 627 (1975).

8

Electron Spectroscopy for Chemical Analysis (ESCA) and Electrode Surface Chemistry

J. S. HAMMOND AND N. WINOGRAD

1. Introduction

The experimental objective of electrodic studies is to define electrode surface reactions through the measurable parameters of current, electrode potential, and reactant concentration at the electrode–electrolyte interface. However, the chemical characterization of electrosorbed and electrodeposited products on the electrode surface has been hampered by the lack of molecular specificity in the electrochemical measurements. In addition, the electro-catalytic behavior of electrode surfaces has remained a difficult parameter to quantify with conventional electrochemical measurements. The application of spectroscopic techniques such as ESCA, which can yield molecular information relevant to the electrode surface chemistry, is providing new insight into these problems.

2. Spectral Interpretation

The development of ESCA (electron spectroscopy for chemical analysis and also referred to as X-ray photoelectron spectroscopy) can be traced back

J. S. HAMMOND • Physical Electronic Div., Perkin-Elmer Corp., 6509 Flying Cloud Dr., Eden Prairie, Minn. 55344. **N. WINOGRAD** • Department of Chemistry, The Pennsylvania State University, University Park, Pennsylvania 16802.

to the pioneering work of Siegbahn and co-workers.[1,2] When a photon of appropriate energy $h\nu$ impinges on an atom, a photoelectron will be ejected. This electron will have kinetic energy

$$E_{kin} = h\nu - E_B \tag{1}$$

where E_B is the electron binding energy. The photons are usually Al K_α or Mg K_α X-rays with energies of 1487 or 1254 eV, respectively. The E_{kin} or measured by an electron spectrometer and a spectrum is obtained of intensity vs. E_B of the sample photoelectrons.

The E_B of a photoelectron is first of all determined by the electron orbital from which the photoelectron originates. Although the ionization cross-section differs with the element and orbital, measurement of E_B permits qualitative analysis of the sample surface for all elements in the periodic chart except hydrogen. In addition, small shifts of the core electron energy levels are induced by changes in chemical environment of the observed atom. These energy shifts influence E_B through the following relation

$$E_B = -(\varepsilon_i^0 - \varepsilon_f^+) \tag{2}$$

where ε_i^0 is the initial state total energy of the atom and ε_f^+ is the total energy of the final photoionized state. Many approximate calculational methods are available for predicting E_B. The simplistic ground state potential model for estimating ESCA chemical shifts (the shift in binding energy with changes in chemical environment) assumes that, to a first approximation, relaxation occurring in the final state ε_f^+ will not influence chemical shifts.[3] This binding energy can be predicted from the following relationship

$$E_B = K_m q_i + \sum_{i \neq j} q_j/r_{ij} + l \tag{3}$$

where q_i is the initial state charge on atom i, K_m is the binding energy shift per unit charge for the core level m, q_j is the initial state charge on atom j, r_{ij} is the interatomic distance between i and j, and l is a constant related to the reference energy level. The values of K and l are usually determined from empirical fits using a series of compounds of similar structure and charges calculated from semi-empirical energy schemes like CNDO/2 and MINDO/3.

Several studies[4] have shown that significant atomic and extra-atomic relaxation of surrounding electrons toward the core photohole does occur. For similar compounds, the relaxation energy is reasonably constant and, therefore, is included in the empirically determined K and l terms of Eq. (3). The magnitude of the relaxation effect can generally be estimated by an SCF calculation on both the ground state atom and on the excited ion as indicated in Eq. (2). Usually, the energy for the excited ion is obtained using the equivalent cores approximation.[4a] Many of the errors inherent in an SCF calculation for large molecules then tend to cancel when the energy difference is computed.

In the absence of reliable theoretical predictions, assignments of binding energy peaks to specific compounds remain an inexact science. Probably the best way to employ ESCA as a tool for identifying compounds on electrodes is still the use of known reference compounds for comparison with the unknown samples—when the known reference compounds can be prepared with confidence.

Semiquantitative elemental analysis (optimum of ±10% down to <1% elemental composition) can be performed readily using measured photoelectron intensity ratios. For an infinitely thick sample the intensity of a photoelectron j, is

$$I_j(\infty) = F\alpha_j N_j \lambda_j K_j \tag{4}$$

where F is the X-ray photon flux, α_j is the atomic photoelectron cross-section, N_j is the number of such atoms per cubic volume element, λ_j is the escape depth of photoelectrons from the solid, and K_j is the instrumental transmission function for the j photoelectrons. The α's can be obtained both theoretically[5] and experimentally.[6] The λ values are a function of the photoelectron kinetic energy[6,7] with values predictable to ±15% for several inorganic compounds.[6] The limiting accuracy of ESCA measurements for determining quantitative elemental composition is controlled primarily by uncertainty in background subtraction of inelastically scattered electrons[8] and by the presence of various energy loss processes in the solid, which can bleed intensity from the main core level photoemission peak.

The surface specificity of ESCA is apparent since typical values for λ_j range from 10 to 50 Å. Inferences about the thickness of overlayer films on bulk substrates can then be estimated from model equations using the relative intensity of photoelectron peaks arising from the substrate and the overlayer. For a homogeneous overlayer of thickness, d, the photoelectron intensity from the substrate, I_s, is given as

$$I_s = I_s(\infty) \exp[-d(\lambda_j(o) \sin \theta)] \tag{5}$$

where $\lambda_j(o)$ is the attenuation depth of the j photoelectrons through the overlayer and $\sin \theta$ is a geometrical factor related to the departure angle θ of the measured photoelectron with the sample surface. If the element of interest is in a thin overlayer of thickness, d, then the intensity is given as

$$I_0 = I_0(\infty)\{1 - \exp[-d/\lambda_j(o) \sin \theta)]\} \tag{6}$$

If confidence is placed on the values of $I_j(\infty)$ for the overlayer and underlayer materials and in the λ values, the ratio of Eq. (6) to Eq. (5) can be written as

$$\frac{I_0}{I_0} = \frac{I_0(\infty)1 - \exp[-d/\lambda(o) \sin \theta)]}{I_s(\infty) \exp[-d/\lambda(o) \sin \theta)]} \tag{7}$$

and the thickness, d, of the overlayer can be obtained.

The instrumentation necessary for ESCA measurements has been reviewed by several authors.[9] For questions regarding vacuum requirements, X-ray sources and monochromators, electron energy analyzers, resolution, sensitivity, and transmission functions, the reader is advised to consult these references or many other excellent reviews of the area.

3. Coupling of Electrochemical and ESCA Studies

From the above discussion, several key pieces of information seem feasible to be extracted from the combination of electrochemical and ESCA studies. These include (a) semiquantitative elemental analyses, (b) identification of oxidation state and molecular speciation, and (c) estimates of film thicknesses of deposited overlayers. Since ESCA is a spectroscopic technique that requires high vacuum,[9] the problem of sample transfer from the electrochemical environment is of primary concern. Different approaches have been suggested in the literature but all require that the electrode be removed from solution under potentiostatic control. The prepared electrode cannot be left in contact with the electrolyte for even a short time as decomposition and dissolution of the surface can occur as the electrode seeks its rest potential. After rapid removal of solution from the electrode, transfer to the spectrometer either under an inert atmosphere or in vacuum should be completed as quickly as possible. In each of these steps, the role of contamination and/or decomposition of the surface layer should be carefully monitored such that the measured ESCA spectrum most nearly reflects the nature of the electrode surface prepared under potentiostatic control. These precautions should involve careful coulometric measurements on the electrochemically prepared surface both before and after ESCA analysis.

4. Anodic Film Formation on Metal Electrodes

A major application of ESCA in electrochemistry is aimed toward characterizing anodic oxides formed on metal surfaces. An important example of this kind involved the platinum electrode in various acidic media. Results from the first study on this system showed specific evidence for Pt^{2+} oxide and a Pt^{4+} oxide from analysis of the Pt 4f chemical shift.[10] The intensity ratios of these platinum species were determined at ± 0.7, $+1.2$, and $+2.2$ V vs. SCE, which showed that the more highly oxidized species grew in intensity with the more positive applied potential.

Later publications concerning the anodic oxidation of platinum electrodes presented differing binding energy assignments for the Pt^{2+} oxide.[11-13] However, the correlation of *in situ* coulometric measurements with ESCA oxide thickness measurements provides strong evidence for the assignment of the Pt^{2+} oxide.[13] The ESCA Pt 4f spectra for the gradual growth of this Pt^{2+} phase (oxide I) as a function of increasing anodic potential is presented in Figure 1. ESCA measurements have also been able to elucidate the important effect of the acid concentration on the anodic oxidation mechanism of Pt electrodes. In low concentrations of sulfuric and perchloric acid, the platinum electrode can be oxidized to form a thick, highly hydrated $PtO_2 \cdot xH_2O$ phase. In high concentrations of sulfuric and perchloric acid, a limiting growth oxide of less than 10 Å thickness is observed. A significant incorporation of the acid anion into this limiting growth oxide layer is then detected from the ESCA spectra.[13]

The electrochemical growth of WO_3 on the surface of W metal electrodes has been used to develop uniform oxide layers to test the validity of Eq. (4)–(6).[14] The thickness of WO_3 as a function of the anodic voltage can be determined from radiochemical analysis. The ratios of W 4f oxide to W 4f substrate photoelectron intensity can then be measured and the escape depth

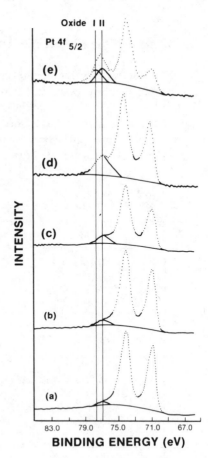

Figure 1. Pt $4f_{7/2,5/2}$ spectra for potentiostatic oxidation of Pt electrodes in 0.2 M $HClO_4$. (a) +1.6 V for 2 min; (b) +1.8 V for 3 min; (c) +2.0 V for 3 min; (d) +2.2 V for 3 min; (e) +2.2 V for $3\frac{1}{2}$ min.

of the 1450 eV photoelectrons calculated for oxide thicknesses of 5–105 Å. The experimental intensity ratios as a function of thickness follow Eq. (6) to within 5%, demonstrating ESCA's applicability for thin film thickness measurements.

The anodic oxidation products of Au, Pd, Ir, Ni, and Al have all been studied with ESCA. The oxide species identified on each electrode surface, as well as the electrolyte used, are summarized in Table 1. It is noteworthy that most anodized oxide layers are found to be highly hydrated structures. The incorporation of the acid anion in an oxide lattice is also commonly observed and often found to be an important key in the rate of oxide growth.

ESCA has also been applied to the study of electrodes exposed to molten electrolytes as well as room temperature liquid solutions. In a KF-2HF system at 100°C, carbon and graphite electrodes were held at anodic potentials to study the formation of surface films blocking the evolution of F_2 gas. These results demonstrate the formation of $(CF)_n$ films on the electrode surface, which is responsible for the passivation of the electrode surface.[19]

5. Anodic Film Formation on Alloys

The ESCA analysis of the anodic oxidation of single component electrodes has provided many powerful examples of how the technique can provide molecular structure and thickness information about the electrode surface layer. Similar analyses of the anodic oxidation of alloys test the technique's ability to elucidate more complex surface structure. A study of the protective oxide film on copper–nickel alloys reveals that at potentials more negative than

Table 1
ESCA Study of Anodized Metal Electrodes

Electrode	Identified oxides	Electrolytes	Ref.
Pt	PtO_{ads}, Pto, PtO_2	$1\ M\ HClO_4$	10
Pt	PtO, PtO_2	$0.5\ M\ H_2SO_4$	11
Pt	Pt oxide I, PtO_2	$0.5\ M\ H_2SO_4$	12
Pt	Pt oxide I, PtO_2, PtO_x (acid anion)y, $Pt(OH)_2$, and/or PtO	$0.1\ M\ H_2SO_4$, $0.2\ M\ HClO_4$, $1.0\ M\ H_2SO_4$, $0.2\ M\ H_2SO_4$	13
W	WO_3	$0.4\ M\ KNO_3 - 0.04\ M\ HNO_3$	14
Au	$Aù(OH)_3$	$0.5\ M\ H_2SO_4$	15
Au	$Au_2O_3 \cdot xH_2O$	$0.5\ M\ H_2SO_4$	12
Pd	$PdO \cdot xH_2O$, $PdO_2 \cdot xH_2O$	$0.5\ M\ H_2SO_4$	16
Ir	$IrO_2 \cdot xH_2O$	$0.1\ M\ KOH$	15
Ni	$Ni(OH)_2$ and/or Ni_2O_3	$0.1\ M\ KOH$	15
Al	AlO_x (acid anion)$_y$	H_2SO_4, Na_2SO_4, $K\ Al(SO_4)_2$, $NaHSO_4$, Na_2SO_3	17, 18

the passivation potential, the surface layer has the same Cu/Ni intensity ratio as the bulk $70:30$ copper/nickel alloy. However, at the passivation potential of -0.35 V vs. SCE, in 1.24 M NaCl a nickel-enriched surface layer is observed with the same Ni $2p_{3/2}$ spectrum as pure passivated Ni foil.[20] Further studies have been reported for various Ni concentrations of CuNi alloys in 0.5 M $HClO_4$[21] and $H_2SO_4-H_3PO_4-H_2O$ electrolytes.[22] A linear increase in the Flade potential of passivation is associated with the decrease in Ni concentration of the alloy. The surface Cu/Ni intensity ratios in the passive layer are found to be the same as the bulk alloy.[21] "Shake-up" satellites in both the Ni $2p_{3/2}$ and Cu $2p_{3/2}$ spectra indicate that NiO and CuO type oxides are present in the passivated layer. The linear variation in Flade potential with the changes in alloy composition results from a mixed potential defined by NiO, Cu_2O, and/or CuO oxide formation in the passive layer with the oxide ratio reflecting the bulk alloy composition.[21] The formation of the transpassive region at potentials more positive than the passivation potential involves a slight enrichment in the Cu oxide content of the oxide layer. The spectra also suggest a possible transformation of the Ni oxide to a thinner Ni_2O_3 with the increase in anodic potential in the transpassive region.[22]

Additional studies of the oxidation products of several alloys have been completed using ESCA.[23-30] The primary interest of this research has been the identification of oxidation products that produce a passive surface layer. The ESCA results confirm the assignment of the oxidation state and many times the structure of these passive layers. The results of these studies are summarized on Table 2.

Table 2
ESCA Study of Anodized Alloy Electrodes

Electrode	Identified oxides	Electrolyte	Ref.
CuNi	Ni oxide, Cu oxide	1.35 M NaCl	20
CuNi	NiO, Cu_2O, and/or CuO	0.5 M $HClO_4$	21
CuNi	NiO, Ni_2O_3, Cu_2O, and/or CuO	$H_2SO_4-H_3PO_4-H_2O$	22
NiSi	NiO + Si^{+x} + SO_4^{-2}	1.0 M H_2SO_4	23
18-8 Stainless Steel	Ni oxide, Fe oxide Cr oxide$\cdot xH_2O$	0.5 M H_2SO_4	24, 25
AlMg	Al_2O_3 + Mg^{+x}	H_2SO_4	26
AlSi	Al_2O_3	H_2SO_4	26
Aluminum Brass	$Mg_6Al_2(OH)_{16}CO_34H_2O^a$	Seawater	27, 28
Aluminum Brass	Cu^{+1} oxide, Fe^{3+} oxide^b	Seawater	27, 28
FeCd	Fe_3O_4, CdO, $Cd(OH)_2$	KOH	29
FeCr	Fe_2O_3, Fe_3O_4, Cr oxide	1 M H_2SO_4	30

^{a} non-protective coating
^{b} protective coating

6. Electrochemical Deposition

Microgram or submicrogram quantities of soluble metal ions that are concentrated by electrochemical deposition on electrode surfaces[31-33] have been characterized with ESCA. Initial experiments demonstrate the capability of depositing microgram quantities of Pb, Cd, and Bi on Hg-coated platinum electrodes. The intensity ratios are not quantitative and significant air oxidation of the metal deposits is observed.[31] Further experiments have been conducted using a glassy carbon electrode and a rotating disk electrode configuration in a sub-ppm solution of Pb^{2+}. A pentanol solution overlaying the aqueous layer allows the working electrode to be removed under potentiostatic control and the water to be removed from the electrode surface without loss of deposited metal. Anodic stripping coulometry has been used to verify that 70–90% of the deposited Pb remains on the electrode surface during removal and reinsertion through the alcohol overlayer. A linear relationship between coulombs of deposited lead and ESCA Pb 4f intensity has been obtained for 0.1–0.5 micrograms of deposited Pb.[32] Further studies with the simultaneous deposition of 0.5 ppm concentrations of Mn, Co, Ni, Zn, Cd, Tl, Pb and Bi ions reveal the capability of ESCA to resolve the complex metal surface matrix on the carbon electrode. Anodic stripping voltammetry could not resolve this matrix due to formation of intermetallic compounds such as Ni-Zn, Cu-Ni, and Co-Ni.[33] The electrochemical-ESCA approach to quantitative analysis does suffer from the simultaneous deposition at cathodic potentials of organic contaminants.[32] In addition, significant oxidation of the deposited metal films during electrode transfer will change the ESCA sensitivity factors [see Eq. (4)] and add additional errors to any quantitative measurement.[33]

The study of the electronic structure of deposited metal atoms on metal electrodes serves as an important application of ESCA due to the sensitivity of the measured E_B on the atomic charge density. For example, many metal ions will deposit on foreign metal substrates at potentials more positive than that predicted by bulk Nernstian thermodynamics. Coulometric measurements of this underpotential deposition (UPD) has revealed film thicknesses of one to two atomic layers. This observation indicates a substantial stabilization in the metal adatom to metal electrode bonding compared to the bulk deposit's intermetallic cohesive energy. The underpotential deposition of Pb^{2+} on an Au electrode has been investigated with ESCA.[34] The Pb 4f binding energy value for these UPD films was comparable to bulk Pb oxide, indicating that the electrode surface was oxidized during electrode transfer. The UPD of Ag and Cu on Pt electrodes has also been studied with ESCA.[35,36] These UPD films have been quantitatively transferred from the electrochemical cell to the ESCA spectrometer in an inert atmosphere box to prevent atmospheric oxidation. The Ag and Cu deposits are not metal oxides or residual metal ion salts from solution. The binding energy shifts for core photoelectrons and Auger electrons indicate enhanced adatom–electrode substrate bonding.[36] Results for the Cu UPD system are shown in Figure 2.

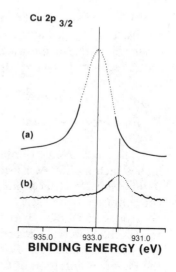

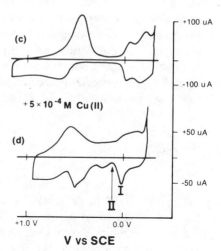

Figure 2. ESCA and electrochemical spectra for the Cu UPD system. (a)XPS Cu $2p_{3/2}$ spectrum of bulk Cu metals, Ar^+ cleaned *in situ*; (b) XPS Cu $2p_{3/2}$ spectrum of Cu UPD on Pt, aspiration of electrolyte $(5 \times 10^{-4}\ M\ CuSO_4 + 0.1\ M\ H_2SO_4)$ at a potential +0.08 V relative to bulk Cu anodic stripping peak; (c) cyclic voltammogram Pt electrode in 0.1 M H_2SO_4, 100 mV/sec scan rate; (d) cyclic voltammogram Pt electrode in 0.1 M $H_2SO_4 + 5 \times 10^{-4}\ M\ CuSO_4$, 20 mV/sec scan rate, aspiration potential is indicated by the arrow.

The electrochemical deposition of other insoluble products on foreign metal substrates has also been an active field for ESCA research. The oxidation of $Pb(NO_3)_2$ at +1.5 V vs. SCE on a Pt electrode produces an insoluble product, which is identified by ESCA as a hydrated layer of PbO_2.[37] The electrochemical deposition of chromium species from chromic acid on Au electrodes has been studied with ESCA. The ESCA intensity ratios of Cr^{+6}, $Cr^{+5\ and/or\ +4}$, Cr^{+3}, and Cr^0 species vary as a function of applied potential. This ESCA analysis of the rather complex surface matrix has clarified the mechanisms of the reduction process.[38] ESCA has also demonstrated that H_2S can react with Pt electrodes in acetonitrile to form a surface layer of PtS. From these results, the kinetics and mechanisms of oxidative addition of H_2S to 9,10 diphenylanthracene must include the reaction of the PtS surface layer as well as the reaction of soluble H_2S.[39]

7. Electrocatalysis

The study of electrocatalysis has important implications for surface techniques such as ESCA. Electron transfer rates are often found to be related to the electrode surface state, a state that is usually independent of the solution concentration of any reactant. For example, two types of iron phthaloxyanine electrodes can be prepared by vacuum deposition on gold substrates. A remarkable difference in activity toward O_2 reduction is observed for the two types of electrodes. An ESCA study has confirmed a significant increase in the Fe^{3+} concentration of the more active electrode.[40] Chemically modified electrodes are also being used to prepare more predictable electrocatalytic surfaces.[41,42] Organosilane chemistry is used to prepare electrode surface states on SnO_2[41] and carbon electrodes.[42] ESCA can then verify the covalent bonding of these derivatives on the electrode surfaces and elucidate the surface properties, such as degree of protonation after exposure to different pH solutions. The covalently bound ligands, such as ethylenediamine, are found to function as ligand complexation centers. The presence of Cu^{2+} on the surface of carbon/ethylenediamine electrodes can be detected, indicating the formation of metal complexes on the modified electrode surface.[42]

8. Conclusions

The summary of published literature has shown that ESCA is being extensively used for elucidation of oxide and passive films on metals and alloys.[10–30] In these papers, ESCA has been shown to be an effective tool for both qualitative and semiquantitative elemental analysis. The binding energy shift information has provided a powerful insight into the oxidation state of many of these observed elements. However, in most of this work, emphasis is placed on structural information as a function of applied potential; the charge transfer associated with the film formation is infrequently mentioned.

There are several studies that have sought to correlate ESCA intensity and thickness measurements with integrated charge transfer.[12–15,32,35,36,38] The results are encouraging but the problems of X-ray damage and contamination during electrode transfer remain as obstacles to effective quantitative ESCA-electrochemistry.[12,13,25,36,38,41]

In projecting the future of ESCA-electrochemistry, two foreseeable problems should be addressed. The first is the need for more satisfactory integration of electrochemical sample preparation with the techniques for electrode transfer. Several results have shown that atmospheric exposure of active electrode surfaces during transfer can result in significant oxidation or degradation of the desired electrode surface.[32–34] There are probably several technological paths for eliminating this problem. The second problem is ultimately the narrow scope of ESCA analysis. For instance, ESCA is insensitive to H in any state, many elemental oxidation states have insignificant binding energy chemical shifts, and the substrate electrode photoelectron binding energies are insensitive to surface chemisorption. The integration of several techniques such as ESCA, LEED (low energy electron diffraction), AES (auger electron spectroscopy), SIMS (secondary ion mass spectrometry), and thermal desorption mass spectrometry into a combined approach for electrode analysis could greatly enhance the chemical information available for interpretation.

References

1. K. Siegbahn, C. Nordling, A. Falham, R. Nordberg, K. Hamrin, J. Hedman, G. Johansson, T. Bergmark, S. Karlsson, I. Lindgreen, and B. Lindberg, *ESCA-Atomic, Molecular and Solid State Structure Studied by Means of Electron Spectroscopy*, Almqvist and Wiksell, Uppsala, Sweden (1967).
2. K. Siegbahn, C. Nordling, G. Johansson, J. Hedman, P. F. Heden, K. Hamrin, U. Gelius, T. Bergmark, L. O. Wermer, R. Manne, and Y. Baer, *Esca Applied to Free Molecules*, American Elsevier, New York (1969).
3. D. A. Shirley, Ed., *Electron Spectroscopy*, p. 47, North Holland Publishing, Amsterdam (1972).
4. (a) D. W. Davis, M. S. Banna, and D. A. Shirley, *J. Chem. Phys.* **60**, 237 (1974). (b) P. H. Citrin and D. R. Hamann, *Phys. Rev. B* **10**, 4948 (1974). (c) F. O. Ellison and M. G. White, *J. Chem. Educ.* **53**, 430 (1976).
5. J. H. Scofield, *J. Electron Spectrosc.* **8**, 129 (1976).
6. R. C. G. Lecky, *Phys. Rev. A* **13**, 1043 (1976).
7. C. J. Powell, *Surf. Sci.* **44**, 29 (1974).
8. R. S. Swingle II, *Anal. Chem.* **47**, 21 (1975).
9. (a) D. M. Hercules, *Anal. Chem.* **42**, 20A (1970). (b) C. A. Luchesi and J. E. Lester, *J. Chem. Educ.* **A205**, **A269** (1973). (c) R. S. Swingle II and W. M. Riggs, *CRC Crit. Rev. Anal. Chem.* 267 (1975).
10. K. S. Kim, N. Winograd, and R. E. Davis, *J. Am. Chem. Soc.* **93**, 6296 (1971).
11. G. C. Allen, P. M. Tucker, A. Capon, and R. Parsons, *J. Electroanal. Chem.* **50**, 335 (1974).
12. T. Dickinson, A. F. Povey, and P. M. A. Sherwood, *J. Chem. Soc. Faraday I* **71**, 298 (1975).
13. J. S. Hammond and N. Winograd, *J. Electroanal. Chem.*, in press.
14. T. A. Carlson and G. E. McGuire, *J. Electron Spectrosc.* **1**, 161 (1972/73).
15. K. S. Kim, C. D. Sell, and N. Winograd, *Proc. Symp. Electrocatal.* 242 (1974).
16. K. S. Kim, A. F. Gossmann, and N. Winograd, *Anal. Chem.* **46**, 197 (1974).
17. S. Eguchi, T. Hamaguchi, S. Sawada, A. Aoki, and Y. Sato, *Kinzoku Hyomen Gijutsu* **25**, 428 (1974), *C. A.* **82**, 130892 (1975).
18. S. Eguchi, T. Hamaguchi, S. Sawada, T. Minzuno, and Y. Sato, *Kinzoku Hyomen Gijutsu* **25**, 437 (1974), *C. A.* **82**, 130893 (1975).

19. H. Imoto, T. Nakajima, and N. Watanabe, *Bull. Chem. Soc. Jpn.* **48**, 1633 (1975).
20. L. D. Hulett, A. L. Bacarella, L. LiDonnici, and J. C. Griess, *J. Electron Spectrosc,* **1**, 169 (1972/73).
21. Y. Takasu, S. Maru, Y. Matsuda, and H. Shimizu, *Bull. Chem. Soc. Jpn.* **48**, 219 (1975)
22. Y. Takasu, H. Shimizu, S. Maru, M. Tomori, and Y. Matsuda, *Corrosion Sci.* **16**, 159 (1976).
23. G. Blondeau, M. Froelicher, M. Froment, A. H. Goff, and C. Vignaud, *C. R. Acad. Sci. Ser. C* **282**, 407 (1976).
24. G. Okamoto, *Corrosion Sci.* **13**, 471 (1973).
25. G. Okamoto, K. Tachibana, T. Shibato, and K. Hoshino, *Nippon Kinzoku Gakkaishi* **38**, 117 (1974), *C.A.* **81**, 67065 (1974).
26. S. Eguchi, T. Hamaguchi, S. Sawada, T. Mizuno, and Y. Sato, *Kinzoku Hyomen Gijutsu* **25**, 492 (1974), *C.A.* **83**, 68716 (1975).
27. J. E. Castle and D. C. Epler, *Surf. Sci.* **53**, 286 (1975).
28. J. E. Castle, D. C. Epler, and D. B. Peplow, *Corrosion Sci.* **16**, 145 (1976).
29. I. Olefjord, *J. Appl. Electrochem.* **5**, 145 (1975).
30. K. Asami, K. Hashimoto, and S. Shimodaira, *Corrosion Sci.* **16**, 387 (1976).
31. J. S. Brinen and J. E. McClure, *Anal. Lett.* **5**, 737 (1972).
32. J. S. Brinen and J. E. McClure, *J. Electron Spectrosc.* **4**, 243 (1974).
33. J. S. Brinen, *J. Electron Spectrosc.* **5**, 377 (1974).
34. R. Adzic, E. Yeager, and B. D. Cahan, *J. Electrochem. Soc.* **121**, 474 (1974).
35. J. S. Hammond and N. Winograd, *J. Electroanal. Chem.*, submitted for publication.
36. J. S. Hammond and N. Winograd, unpublished results.
37. K. S. Kim, T. J. O'Leary, and N. Winograd, *Anal. Chem.* **45**, 2214 (1973).
38. T. Dickinson, A. F. Povey, and P. M. A. Sherwood, *J. Chem. Soc. Faraday Trans. I* **72**, 686 (1976).
39. J. F. Evans, N. N. Blount, and C. R. Ginnard, *J. Electroanal. Chem.*, **59**, 169 (1975).
40. M. Savy, C. Bernard, and G. Magner, *Electrochim. Acta* **20**, 383 (1975).
41. P. R. Moses, L. Wier, and R. W. Murray, *Anal. Chem.* **47**, 1881 (1975).
42. C. M. Elliot and R. W. Murray, *Anal. Chem.* **48**, 1247 (1975).

9

Field Ion Microscopy

LEONARD NANIS

1. Introduction

The field ionization microscope (FIM) is, at present, the only available experimental device that routinely permits direct observation of atomic detail on surfaces. Accordingly, the FIM affords an opportunity to examine materials of electrochemical interest on a scale in which the elementary processes of electrodic reactions take place. The invention and steady improvement of the FIM has taken place over the past few decades, principally under the direction of Müller.[1] In his laboratory and, more recently, elsewhere, effort has been devoted to perfection of equipment and to studies of the basic processes of field ionization and field evaporation of the substrate. The invention by Müller in 1968 of the atom probe—a combination of FIM and time of flight mass spectrometry—has permitted fresh insights regarding the fundamental mechanisms of field ionization and field evaporation.[2] At present, the FIM is a well-developed tool for surface studies.

There is ample experimental evidence linking FIM observations to the behavior expected of surface atom arrays. By and large, the FIM has been applied to problems of metallurgical interest.[3] Some of the notable accomplishments are the demonstration of surface diffusion by individual ad-atoms,[4–6] epitaxial growth of vapor-deposited Pt on Ir,[7,8] lattice

LEONARD NANIS • Consultant, Electrochemical Engineering, 627 Georgia Avenue, Palo Alto, California 94306.

imperfections (vacancies, dislocations),[9,10] and the lattice rearrangements that accompany ordering in alloys,[11] Applications of interest to the science of catalysis are steadily increasing, typically in direct measurement of surface enrichment by alloy components[12] and the resolution of adsorbed compounds by the use of field ion mass spectrometry (FIMS).[13]

There have been very few applications of the FIM technique directly related to the electrochemical interface. Some features of electrodeposits have been observed by Rendulic and Müller,[14] although electrochemical parameters were incompletely defined or controlled. Nanis and Javet[15] have reported atom rearrangements produced at the Ir metal–electrolyte interface in the absence of current and have correlated these with exchange current density. Inal and Torma have used FIM to examine electrodeposits of Ni on stainless steel and Cr on Ni[16] and Cu on W, produced by cementation.[17]

Schubert, Page, and Ralph have examined films formed on Ir and Pt at anodic potentials in $0.1 \, N \, H_2SO_4$,[18] whilst Morikawa and Yashiro made FIM observations on W tips corroded in dilute nitric acid.[19] Further details of these early links between FIM and electrochemistry will be discussed in following sections.

These first studies and the advances made in applications related to metallurgy and catalysis may be considered as providing a firm basis for the as-yet untapped possibilities for the use of the FIM for electrochemistry. In the last few years, the FIM technique has advanced rapidly with the introduction of the channel-plate image amplifier, and the use of Xe, Ne, and Ar as image gases and for gas mixtures to promote ionization at fields lower than for He, thus enabling the study of metals of intermediate strength such as Au and Cu.

It is clearly recognized that the FIM technique cannot be used to examine surfaces during electrolysis. Instead, the FIM may be used to prepare and define a surface, which may then be removed from the FIM and used as an electrode for controlled electrochemical experiments. By subsequently returning the treated surface to the FIM, re-imaging will provide direct observation of the effects of electrode processes on a truly atomic scale. The ability of the FIM to permit atom-by-atom removal of the surface structure by the process of field evaporation allows profiles of electrochemically produced effects to be followed in depth. Due attention must be given to control experiments to distinguish possible artifacts produced by interaction of the tip with air and due to stress relaxation during field interruption and warming.

It is hoped that this section will serve to promote further interest among electrochemists by identifying some promising pathways for combined FIM–electrochemical studies and, ultimately, FIMS for element identification of individual atoms on electrochemically treated tips.

2. Operation of the FIM

2.1. Field Ionization

The basic features of FIM operation include a very fine tip of metal to be investigated (typical radius: 500 Å) brought to a high positive potential (5–20 kV), in a chamber filled by an imaging gas at low pressure (typically He at 10^{-3} Torr). Cryogenic cooling promotes adsorption of image gas molecules, which collide with the tip, attracted in part by the interaction of field-induced polarization. In the high field region very close to the tip, the image gas molecules can lose electrons, generally believed to occur by tunneling ionization to the Fermi level of the tip metal atoms. Ionization is more likely to happen in the locally enhanced field above a protruding atom, most recently believed to be associated with a polarized gas molecule on the apex of an atom with field-polarized electron distribution. The gas ion, e.g., He^+, formed in the local field is accelerated in a radial direction from the tip, which is at a large positive potential. The gas ions striking a fluorescent screen give the FIM image of atomic structural details of the tip surface. The FIM chamber pressure is sufficiently low as to allow the positively charged image gas ions to travel without collision and also without lateral interaction. The texts by Tsong and Müller[20] and Bowkett and Smith[21] discuss in detail the theory of the field ionization process; modifications to account for adsorption effects are discussed by Müller[2,22] and Rendulic.[23,24]

An important and very useful feature of the FIM is the ability to evaporate atoms from the positive metal surface by increased field-induced evaporation. This process is ideally self-regulating and allows atomically smooth surfaces to be obtained during tip preparation (endform) and permits subsequent precise control of metal removal, equivalent to anodic dissolution into vacuum. The atoms that image first (at lower potential) are also those that evaporate first in higher fields. Stable FIM images are obtained when the gas ionization field is *less* than the field required to evaporate atoms from the metal tip.

2.2. Field Evaporation

As mentioned above, the surface atoms of the tip can be made to evaporate by ionization. Field evaporation from atomically smooth net planes may be continued by selectively raising the field voltage in stages. The rings in Figure 1 are produced by enhanced field ionization of the image gas (He, 0.5 mTorr) at a tip of Ir maintained at liquid nitrogen temperature (77 K). When the field is increased by raising the potential above the best image potential of 17.3 kV (4.4 V/Å for He on a tip of 800 Å radius), the rings shrink inward since evaporation occurs first at the step edge of net planes, marked by the rings of extra brightness. The last few atoms remaining in the center of a net plane

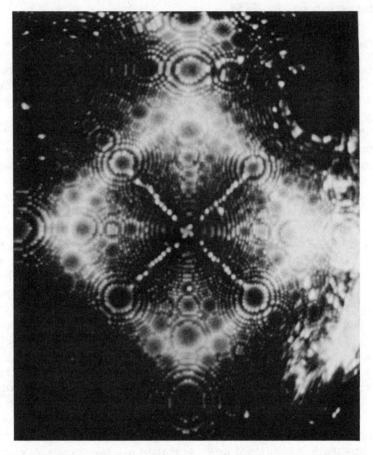

Figure 1. Smooth, field-evaporated Ir tip before electrochemical treatment, (100) plane in center. Bright zone lines are characteristic for Ir and terminate near (113) planes. He image, 17 kV.

tend to field evaporate simultaneously. The removal of atoms by field evaporation may be stopped by decreasing the field to be intermediate between that needed for gas imaging and tip atom evaporation.

Early applications of FIM were limited to high strength and/or refractory metals (W, Mo, Ir) that could be used with He, which images at 4.4 V/Å. Lower field image gases are now being used since image intensification by channel-plate amplifiers avoids the damage to phosphor screens caused by ions of Ne, Xe, and Ar. This convenience requires ultrahigh vacuum techniques since the lower fields may not ionize and repel contaminating molecules of O_2, N_2, and H_2O. By the use of low field image gas and gas mixtures, intermediate strength metals have been successfully examined in the FIM and offer new opportunities for electrochemical studies.

In order to calculate the evaporation field for metals, an energy barrier model has been developed to include terms for image forces and polarization terms.[2,23] A simplified scheme conveys the basic concept underlying the computations shown in Table 1. By neglecting image potential and adsorption terms, the activation barrier, Q, which must be overcome, is

$$Q = \Lambda + I_n - n\Phi - neF_{ev}X_0 \tag{1}$$

where Λ is sublimation energy for vaporization of metal, I_n is ionization energy required to remove n electrons, and Φ is the average surface work function. Energy in eV units is convenient for use in Eq. (1). The product $eF_{ev}X_0$ is the energy added to electron of charge e by raising it through a potential difference produced by the field F_{ev} over the distance X_0 measured from the average tip surface. The cycle of removal of the metal as a gas atom (Λ), adding ionization energy I_n, and recovering the energy $n\Phi$ by returning the n electrons to the metal is closely related to calculation (Born–Haber) of the theoretical potential difference at an electrochemical interface. For Ir, values are $\Lambda = 6.5$ eV, $I_1 = 9$ eV, $I_2 = 27.6$ eV, $\Phi = 5$ eV (average). For the known value of 5.03 V/Å for F_{ev} for Ir, a distance X_0 of 2.4 Å brings the activation energy in Eq. (1) to $Q = 0$. This distance is reasonable, being slightly less than the single-bond diameter 2.5 Å or the atom diameter 2.6 Å for Ir. The lowest evaporation field for several metals, including Ir, is calculated for doubly charged ions. Recent results obtained with FIMS[2] indicate that Mo and W evaporate as +3 and +4 ions and also complexes with He, i.e., WHe^{+3}.

Despite the many assumptions underlying the evaporation field calculation (omission of dipole terms, use of average work function, assumption of constant field), the order of metals is found to be in agreement with experiment. Actually determined values are included in Table 1, together with lowest evaporation fields computed by Müller.[20a] Although the prediction of evaporation field is of theoretical interest, direct observation of evaporating atoms indicates that there is no single value that can be applied since not all surface atoms evaporate simultaneously. In part, this is because the field itself varies considerably due to local differences in radius of curvature of the tip endform. For prediction of suitable materials for study, Table 1 is followed by a listing (Table 2) of image fields for inert gases used in the FIM.[25] For example, Zr cannot be imaged with He since F_{ev} is 3.5 V/Å, less than the He ionization field of 4.4 V/Å. However, Zr has been successfully imaged with Ne.[26] Images of Au have been obtained in Ne but are reported to be unstable, with slow field evaporation of the Au,[27] as might be expected from the similar fields, 3.5 V/Å. Even though relatively low-melting Al (m.p. 660°C) has the smallest calculated F_{ev} in Table 1, it has been successfully imaged[25] in argon. Mechanical features of the tip contribute to successful imaging; a cone angle of less than 10° was found to be unstable. Blunt shanks for the tip endform and heat treatment of starting wire to minimize grain boundaries or twins at the tip were recommended by Boyes and Southon[25] for Al.

Table 1
Evaporation Fields (V/Å)

Metal	Lowest field (calculated, 0 K)	Observed
Be	3.84	
Al	1.61	
Ti	2.33	2.50
V	2.50	
Cr	2.64	
Fe	3.18	3.60
Co	3.49	3.70
Ni	3.30	3.60
Cu	3.08	3.00
Zn	2.87	
Zr	2.84	3.50
Nd	3.48	4.00
Mo	4.52	4.50
Ru	3.66	4.50
Rh	4.07	
Pd	3.63	
Ag	2.31	
Sn	2.24	
W	5.50	5.70
Re	4.34	4.80
Os	4.82	
Ir	5.03	5.00
Pt	4.42	4.75
Au	4.02	3.50

Table 2
Gas Image Fields (V/Å)

Gas	Field	Optimum temperature (K)	Resolution (Å)
He	4.4	20	<3
Ne	3.5	20	<3
Ar	2.2	55	<5
Kr	1.5	~80	6–8
Xe	1.2	~80	6–8
H_2	2.2		
N_2	1.7		
O_2	1.5		

Special considerations for image formation from alloys where F_{ev} is different for the component atoms are treated by Southworth and Ralph[28] and are discussed by Tsong and Müller.[20b]

2.3. Field-Induced Stresses

The tensile stress applied to the tip by the field can, if sufficiently great, produce fracture. During field evaporation to the endform, the tensile stress induces dislocation movement and the transport of imperfections from the tip volume, thus strengthening the material in the tip region. A rough estimate indicates that stresses comparable to the theoretical yield strength of metals are produced. From the theory of electrostatic energy,[42] the mechanical force per unit area (stress) on the surface of a conductor in an electric field is

$$S = \frac{\varepsilon_0}{2}(F_n)^2 \tag{2}$$

where ε_0 is permittivity, $8.85 \times 10^{-12}\,C^2\,N^{-1}\,m^{-2}$ (for free space) and F_n is the field normal to the surface. Using appropriate conversion factors, Eq. (2) can be stated in hybrid (but useful) units as

$$S,\text{psi} = 6.4 \times 10^4 \left(\frac{V}{\text{Å}}\right)^2 \tag{2a}$$

For the He image field, $4.4\,V/\text{Å}$, the tensile stress applied to a tip, according to Eq. (2a), is 1.2×10^6 psi. Technical yield stress of metals is generally much less so that tips that survive the field evaporation process to achieve the end form are either initially of a volume sufficiently small to avoid strength-reducing imperfections or have had the imperfections removed by the asymmetrical stresses applied to the tip because of local variations in the field. The very high strength of dislocation-free whiskers is analogous.

The possibility of introducing artifacts because of the high field-induced stress should be considered in electrochemical experiments. As pointed out by Müller,[20c] dislocations can develop due to strain release during the warming to room temperature of previously field evaporated tips. Müller recommended that, for comparison where external procedures are used, a suitable control is to keep a tip in vacuum for the same time and temperature experienced by the externally treated tip. For electrochemical studies, the control should include immersion in electrolyte for times comparable to those used for electrodic studies on similar tips or, preferably, the same tip. Control Ir tips used by Nanis and Javet[15] were immersed in supporting electrolyte without Ir ion to obtain a measure of the (minimal) background changes that disturb the perfection of the previously imaged FIM tip.

The use of image gases with low ionization fields (Table 2) reduces the stress level significantly because of the squared dependence on field, shown in Eq. (2). However, the lower field requires the use of high vacuum techniques to remove possible contaminant gases, which may react with the tip since they may not become field ionized and repelled.

2.4. Similarity of FIM and Electrochemical Interface

Useful comparison may be made between the tip–gas interface of the FIM and present models of the electrode–electrolyte interface. The similarity between field ionization and electrodic oxidation has been briefly noted by Conway.[29a] Some order of magnitude estimates are instructive.

In models of the double layer at the electrochemical interface, the greatest variation of potential with distance is considered to occur within the inner Helmholtz layer, which extends into the electrolyte by a distance that is roughly equal to the size of a water molecule, i.e., about 2 Å. An overpotential of 1 V magnitude is already quite substantial but corresponds to a gradient of 0.5 V/Å, clearly less than the fields that produce ionization and evaporation in the FIM. However, overpotential is superimposed on the field that already exists at the reversible potential. The magnitude of this field may be estimated as follows for a total picture (although approximate) of the field when electrodic currents are made to occur.

For a symmetrical 1:1 electrolyte, the Gouy–Helmholtz[29b] approach predicts a field just at the outer Helmholtz plane (for 25°C) of

$$F_{OHP}, (V/Å) = 0.0169 C^{1/2} \sinh 19.46 Z(\Psi_0 - \Psi_s) \qquad (3)$$

where C is the bulk concentration, in moles/liter of the 1:1 symmetrical electrolyte with ionic charge Z, equivalents per mole, and $(\Psi_0 - \Psi_s)$ is the potential difference of the OHP relative to the potential in the bulk of the solution. The magnitude of $(\Psi_0 - \Psi_s)$ is at least less than or equal to the unmeasurable absolute potential difference at the electrode–electrolyte interface. Accordingly, reasonable values of $(\Psi_0 - \Psi_s)$ have been assumed in order to compute numerical values from Eq. (3) for representative electrolyte concentrations (Table 3).

As shown in Table 3, fields comparable to FIM ionization fields are attained at the OHP for relatively modest values of C and $(\Psi_0 - \Psi_s)$. In the derivation of Eq. (3), the Gouy–Chapman theory requires distributed space

Table 3
Field (V/Å) at OHP [Eq. (3)]

C mol/liter	$(\Psi_0 - \Psi_s)$, V		
	0.2	0.4	0.6
10^{-4}	0.004	0.2	10
10^{-2}	0.04	2	100
1.0	0.4	20	1000

charge assumptions and constant effective dielectric constant; thus, Eq. (3) (and Table 3) are not intended for application in the region between the OHP and metal surface. It is the field right at the surface of the electrode that is of interest. However, present-day treatments of the double layer assume a constant potential gradient across the Stern layer or, in other models, from the OHP to the metal surface. As a consequence of the models of the double layer close to the metal, the field should be no less than those shown in Table 3 (for 1:1 electrolyte, 25°C).

In summary, for large C, $(\Psi_0 - \Psi_s)$, no current, the field at the electrode–electrolyte interface is comparable with FIM ionization fields. Electrode overpotential adds marginally to the already existing field. Given that the fields are comparable, surface features revealed in FIM images may be considered to be present at surfaces of electrochemical interest. An improved picture of electrodic events on surfaces is possible through comparison with concepts and results of charge transfer at the electrified metal–adsorbed gas–vacuum interface of the FIM.

3. Experimental Features of FIM

3.1. FIM System Requirements

The FIM may be used with a relatively poor vacuum (0.5 mTorr) when He is the image gas. Generally, however, high vacuum (10^{-8}–10^{-10} Torr) is needed for the cleanup of residual gases. Titanium gettering is an ideal means to remove residual O_2, N_2, and H_2O for systems where the vacuum pump may be closed off from the FIM tube. Several FIM designs are reviewed by Müller and Tsong.[20a] A simplified design of FIM tube due to Reed and Graham[30] may be adapted for electrochemical studies, as shown in Figure 2. The demountable cold finger permits ready replacement of tips and is a necessary feature for external electrochemical studies. Cryogenically cooled gas provides heat exchange to the inside of the cold finger, with the tip being cooled by thermal conduction through sealed-in-glass tungsten pins (0.125 cm diam). The tungsten pins provide mechanical support for a hairpin to which the tip material has been spot-welded. Some typical temperatures are indicated in Table 2, together with associated achievable features.

The field is applied through a high voltage lead (not shown in Figure 2) located within the cold finger. A stable high voltage power supply capable of delivering up to 20 kV with smooth variation control is required. As shown in Figure 2, the W pins provide two lead wires for the high voltage supply, whereas only one lead is necessary. The pair of W conductors permits current passage for resistive heating of the wire hairpin supporting the tip, if desired. Ground return contact is made internally to the phosphor through a conductive coating of SnO_2. Since the actual voltages to the tip are generally in the range

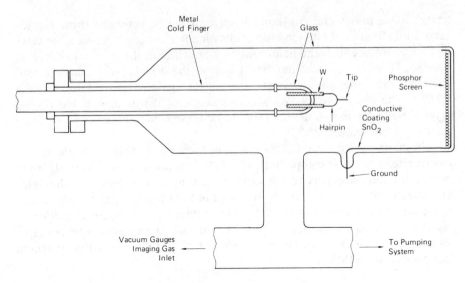

Figure 2. Schematic drawing of field ion microscope.

of 15–20 kV, due caution should be exercised to avoid accidental grounding and electrical shock hazard.

3.2. Tip Preparation

Care is needed in tip preparation because of the very small volumes involved, but with practice, tips can be repeatedly made that produce excellent images. Several electrolytes have been developed for the preparation of tips by anodic dissolution of wire, typically 0.003 in. diam. The solutions are essentially those used for electropolishing. Aqueous and molten salt electrolytes for tip shaping of specific metals are tabulated by Müller and Tsong[20b] and by Bowkett and Smith.[21a] Slow withdrawal of the wire from the electrolyte assures a more prolonged anodic dissolution at the tip. No elaborate fixtures are needed to hold the wire; with practice, a steady hand will suffice. The ideal tip shape tapers to a point that is somewhat blunt. Undercutting of the shank behind the tip region is undesirable and promotes tip breakage by field-induced stresses. Care must be taken to avoid local heating or excessive mechanical agitation during the last stages of dissolution because of the possibility of introducing stress effects in the small volume of material at the tip. For convenience in handling and in mounting in the FIM, tip materials are spot welded as starting wires onto a hairpin-shaped crosswire, typically of W. The radius region of a well-prepared tip is well below the size for optical resolution. However, it is helpful to examine the shank taper in a microscope as polishing proceeds in order to assure uniformity of shaping into a pencil form. The starting wire may also be shielded by capillary glass tubing so that current density decreases along the wire from the tip region.[20c] A scanning or transmission electron microscope is very helpful for determining gross effects of tip shape, which influence image quality. Final tip shaping is accomplished by gradual field evaporation in the FIM. A tip radius of about 500–700 Å is readily achieved, resulting in He ionization with applied potentials of 11–16 kV.

It may be safely stated that no systematic studies have been reported for the electrode kinetics of shaping and polishing of tips for the FIM. Studies of the anodic process (potential control, use of reference electrode, etc.) may prove useful in correlating surface details on the atomic scale with corrosion of metals and with practical anodic processes such as electromachining.

3.3. Image Amplification and Photography

The fluorescent phosphor type of FIM screen (Figure 2) is used with an optically flat glass view surface. Renewal of the screen is required occasionally. A too thinly coated phosphor will "burn off" with frequent use whereas a thick screen will scatter the image intensity. Manganese-activated zinc orthosilicate phosphor gives good performance. Photography may be readily accomplished with Polaroid 3000 ASA film or with 35 mm film with lens systems adjusted for close-up focusing. For convenience in identifying image sequences, a built-in imprint mechanism in the camera is useful. Long exposure times may be used to compensate for poor image brightness. It should be noted that a tip image obtained well below the evaporation field will not change for several hours when He image gas is used. However, image intensification procedures are favored and the relatively new technique of image amplification with microchannel plates has been widely adopted. The microchannel plate (MCP) image amplifier is basically an array of miniature photomultiplier tubes constructed by microcircuitry techniques on a geometry that resembles a porous electrode with well-defined closely packed cylindrical pores. Typically, each cylindrical hole may have a diameter of 14 μm and length of 0.5 mm. The thin (e.g., 0.5 mm, 2.5 cm diam) wafer-shaped MCP and its associated phosphor screen form a combined package that may be readily sealed into the FIM tube with external leads for its separate circuitry. In typical operation, an MCP with applied voltage of 1000 V will produce image intensification by a factor of 10 000.

3.4. Image Interpretation

The interpretation of photographed images forms a major task of FIM work. The featureless region in the center of a net plane image results from there being no protruding field due to electronic orbital overlap for these surface atoms. At the boundary of the net plane, the edge ring atoms form an image because of the sudden variation in local field. An atom missing from an otherwise perfect net plane, such as a vacancy, also perturbs the local field and the site becomes visible as a dark spot against the featureless background. Where the bright edge ring is interrupted by a vacancy, the contrast is more pronounced, as may be seen in Figure 3.

For alloys, the presence of dark spots may be due either to missing solute atoms that have been field evaporated or to the lack of gas ionization by the solute atom. Special attention has been given to this effect because of its importance to phenomena of metallurgical interest (ordering and clustering of solute atoms). Early work is reviewed by Southworth and Ralph[28] for solid–solution alloys in which the components may have significantly differing imaging characteristics. The issue of whether the solute atom is missing or

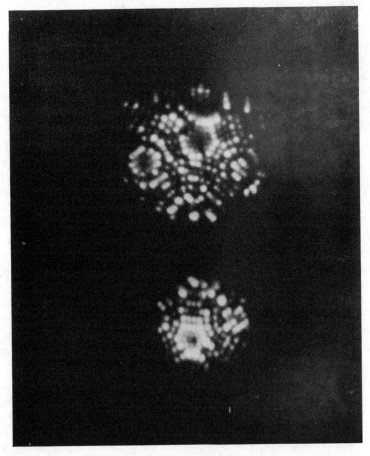

Figure 3. Ir tip (initially as in Figure 1) after 10 min in 2 N H_2SO_4 plus 10^{-4} M Ir^{+4} ion, rinsed in water, alcohol, returned to FIM. Growth region centered on [310] directions. He image, 11 kV.

retained but inactive is still topical. There may be related features for alloy electrodes in electrocatalysis and for corrosion of alloys.

Dislocations can generally be located by the slightly blurred image in such regions. Grain boundaries may be located by the mismatch of rings bounding net planes. Defect location by inspection of image photographs is a skill acquired and improved by practice.

The crystallography of FIM images has been thoroughly covered.[20d,21b] The local radius varies considerably over an endform tip because the surface is actually composed of flat planes and steps. A simple geometrical construction may be used to determine radius by counting the number of net rings between two known crystallographic poles. The angular separation between selected poles is known from X-ray crystallography, as is the lattice spacing perpendicular to a given pole for a given metal. Details and convenient tabulations of geometrical factors are given by Müller and Tsong[20d] and by Bowkett and Smith.[21c] The geometrical properties of FIM image projections have been treated by Fortes.[32]

3.5. Electrochemical Operations on FIM Tips

The combination of FIM–electrochemistry is novel; thus, there are few techniques that have been developed specifically for such studies. The following procedures were useful in exchange current density studies of Ir/Ir^{+4} and should prove to be of general applicability.[15]

Artifact effects are always to be avoided in FIM studies. Accordingly, care should be taken to avoid spurious charge effects when introducing a prepared tip to the electrolyte for controlled electrochemical study. The buildup of static electricity on the experimenter depends on ambient humidity, insulation from ground, and even on type of clothing worn. The discharge of accumulated charge through the tip during immersion can produce uncontrolled electrochemical effects. The problem may be avoided simply by grounding the tip by first immersing a wire of the same material, connected to the tip hairpin support. In this manner, any unwanted coulombic effects are produced on the sacrificial ground wire.

Controlled electrochemical current densities on the small tip area may be achieved by placing the prepared tip in parallel with a large electrode (1 cm^2) of the same material. Since the tip has a negligible surface area compared with a 1 cm^2 sheet electrode, the current may be adjusted on the basis of the immersed large area to produce a required current density on the tip.

Rendulic and Müller[14] reported difficulty in obtaining controllable current densities below 10 A/cm^2 because of the small currents needed with FIM tips serving as the only substrate for cathodic deposition on Ir.

The need for control experiments has been mentioned (Section 2.3). Weakly bound adsorbates produced by external electrochemical treatment may not survive to be imaged but will probably be desorbed either during re-evacuation of the FIM or during raising of the field to the image condition. The extent of surface damage produced by the venting, immersion, washing, and return to the FIM has been determined to be minimal, extending for no more than two or three layers in a few orientations only, for Ir tips immersed in $2 N H_2SO_4$ for 10 min at 25°C. Iridium is thus a very convenient substrate for electrochemical studies, particularly since Ir is easy to image with He. Control immersion studies should be performed for other supporting electrolytes, however, to verify the absence of damage due to transfer between FIM and electrochemical cell.

4. Applications and Suggested Experiments

At present, there have been but a handful of applications of FIM to electrochemistry. Although limited in number, the success of these few point the way for more detailed studies and for new approaches to the understanding of electrode kinetic behavior. Suggestions for additional research have been incorporated in the discussion of the existing studies.

4.1. Electrodeposition: Previous Work

Electrodeposition has received the most attention. Rendulic[14] deposited Pt on Ir and W from a solution of $H_2Pt(NO_2)_2SO_4$ (5 g Pt/liter) at 50°C. Alternating currents were used for convenience and actual current densities were not well defined. However, it was determined that Pt grows epitaxially on Ir cathode substrate. Nucleation began on specific Ir planes, (012) and (135), with nuclei of about 40 Pt atoms. Higher currents produced [111] oriented Pt crystallites ranging in size from 50 to 500 Å. At current densities ($8 A/cm^2$) where H_2 may have been co-deposited, heavily deformed small (100 Å) crystals of Pt were deposited. These early experiments produced important results and bear repetition with attention to control of electrochemical parameters.

Nanis and Javet[15] determined that a perfect tip surface of Ir is severely disturbed by immersion in $10^{-4} M Ir^{+4}$ ion in $2 N H_2SO_4$ supporting electrolyte in the absence of external current, as expected for the existence of exchange current density. By removing the disturbed layer quantitatively by field evaporation of Ir, it was determined that a general roughening of the entire surface occurs during the first 5 min of immersion, followed by selective deposition centered on [310] directions. Deposit regions, 50 to 100 Å diameter (Figure 3), contained many vacancies. On the assumption that the deposits are the result of the cathodic component of the Ir/Ir^{+4} exchange current density, a value of $j_0 = 10^{-4} A/cm^2$ was determined from a direct count of Ir atoms removed. Electrodic studies of Ir are needed to provide j_0 for comparison.

Future studies of electrodeposition with other metals should establish the baseline of natural surface rearrangement caused by j_0 before electrodeposition (or anodic dissolution) FIM results can be properly interpreted.

Inal and Torma[16] have studied Ni plated on stainless steel tips and black chrome plated on Ni tips. They found vacancies and vacancy clusters in the growth regions, which ranged in size from 50 to 100 Å. Inal and Torma[17] have also examined copper deposited on W as a result of a cementation replacement reaction produced by immersion of W endform tips in $CuSO_4$ electrolyte. Combined SEM and FIM studies indicated W dissolution as well as epitaxial growth of about five atom layers of Cu on W. Image gas was 90% He–10% Ne.

4.2. Suggested Electrodeposition Studies

The early stages of electrodeposition can be studied by using coulombic control of deposits on the scale of monolayers. It will be instructive to determine which features on the prepared substrate serve as initial growth centers. Vacancy production may prove to be a natural consequence of exchange current so that nucleation concepts of electrodeposition should be re-examined. The origin of epitaxial overgrowths and preferred deposit orientations may be examined. Matching studies of electrode kinetics are desirable.

Several metals for which aqueous plating baths are well known may be FIM studied, including Ni, Pt, Au, Fe, and Cu. Electroless Ni plating solutions are also suitable. Recent development of a sulfamate electrolyte for Ir deposition[33] offers excellent opportunities since Ir is one of the most thoroughly studied FIM tip materials. Good deposits have been reported for Ir deposition from a bath prepared by the heating of $IrCl_3$ with sulfamic acid at 100°C. Ammonium sulfamate is also added, giving a bath with Ir content in the range 3–20 g per liter for use at 50°–75°C at current densities up to 80 mA/cm^2.

Special substrates for electrodeposition may be obtained by vacuum annealing Ir at 990°C to produce tips with large flat facets in (100, 111, 113) crystallographic regions.[34]

The effect of organic addition agents in promoting desirable electrodeposit properties (stress reducers, brighteners) is generally attributed to surface adsorption and interaction with "active" sites. Certain addition agents or their decomposition products are known to be incorporated in the deposit structure, e.g., p-toluene sulfonamide and saccharin in nickel plating baths.[35] The field evaporation of a deposit may reveal the location of incorporated additives and their localized effects on deposit morphology. Atom-probe FIMS might be used to identify molecular weight of incorporated material. Platinum deposits were reported[20e] to be noticeably more free of defects and to have a preferred [111] orientation when either coumarin or p-toluene sulfonamide was added to the Pt plating bath used for FIM studies by Rendulic and Müller.[14]

4.3. Electrocatalysis

Stable oxide, IrO_2, has been FIM imaged after having been grown by an external anodic electrochemical treatment on a prepared Ir FIM tip.[18] Amorphous IrO_2 was produced by potentiostatically controlled oxidation in 0.1 N H_2SO_4 over a range of potentials from +0.5 to +2.2 V (E_h) at 25°C. By way of contrast, thermally grown IrO_2 was crystallographically regular and also stable in the FIM, readily imaged in an He–Ne gas mixture. When Pt was oxidized for long times (80 min at 2.12 V, E_h, 35°C) the oxide film was estimated to be about 25 Å thick with an amorphous type of structure. However, oxidation in the potential range 0.7–1.0 V, E_h, did not produce evidence of incipient formation of oxygen-containing films determined by other methods, e.g., ellipsometry.[36] This lack of agreement should be resolved. The Pt–0.1 N H_2SO_4 interface is one of the most widely studied electrochemical systems and several techniques (potentiodynamic sweep, cathodic reduction) can be combined with FIM studies to add new insight to the many studies of surface oxides on Pt and related metals. The atomic scale result of various electrode pretreatments used for improving activity of platinum, such as alternating anode–cathode pulsing, may be examined in the FIM.

Ruthenium is easily imaged in the FIM with He at liquid N_2 temperature. Thermal oxidation of Ru has been studied by Cranstoun and Pyke using FIM[37] and FIMS.[38] Crystalline RuO_2 is formed with a region with oxygen deficiency close to the Ru. Well-characterized RuO_2 surfaces could be used as Cl_2 electrodes in order to relate structure and stoichiometry to the behavior of the dimensionally stable anode used in the chlor-alkali industry.

As shown in Table 1, titanium has a particularly low evaporation field. Little FIM work appears to have been done on Ti, its alloys and oxide layers, although Ne imaging of Ti is mentioned in a study of Zr by Carroll and Melmed.[26] A coating of RuO_2, prepared by methods recommended for DSA formation,[39] might be imaged if the substrate Ti–TiO_2 is sufficiently strong to withstand field stress. Atomic features could then be related to the behavior of the surface as a Cl_2 anode.

Several alloy catalysts can be FIM imaged and structures may be correlated with electrochemical behavior determined on the actual tip. Pretreatment by anode–cathode pulsing has been reported[40] to give activity increase for oxygen reduction by two orders of magnitude (10^{-6}–10^{-4} A/cm^2 at 0.8 V, E_h) for 80 Au–20 Pt (at. pct.) alloy. The significant difference of catalytic activity and Tafel slope compared with thermally and chemically activated surfaces could be correlated with the distribution of the alloy components at the surface. Polanschütz and Krautz[41] have demonstrated ordering in FIM studies of 75 Au–25 Pt alloys using Ne image gas. FIM—electrochemical study of Pt–Au alloy activity could be a useful combination.

Loss of catalyst activity of the oxygen electrode occurs in the phosphoric acid H_2–O_2 fuel cell. "Sintering" at temperatures above 160°C is a term used to describe the coalescence of Pt crystallites. Prepared Pt FIM tips could be subjected to fuel cell conditions in order to evaluate sintering mechanisms. For example, Pt loss by formation of ionic intermediates should be distinguishable from redistribution of Pt by diffusion on surface planes. The FIM tip affords the opportunity for study of catalyst on a scale approaching the sizes actually used in practical systems.

4.4. Corrosion

Morikawa and Yashiro[19] have examined the effects of attack by dilute nitric acid on W tips previously cleaned by field evaporation in the FIM. From control experiments, they determined that exposure of the tip to air for more than a day greatly decreased the rate of attack.

A large number of tip materials of technical interest may be subjected to corrosion conditions and examined in the FIM to determine the early stages of corrosion behavior. The atom probe FIM offers the possibility of detecting individual corrosion reaction products or intermediates, if they are sufficiently stable. Thin passivation films may be studied, particularly at the early stages of formation. Recent[12] application of atom-probe FIMS to Cu–Ni alloys showed that, for a 5 at.pct. Cu alloy annealed at 500°C, the surface composition was actually 50 act.pct. Cu. Depth profiling by field evaporation was used to determine that enrichment occurred only in the very first atom layer. Tentative theoretical models based on thermodynamics of solutions are being considered to account for the unexpectecly steep Cu profile. The enrichment of Cu in the surface as a result of heat-treatment may be of importance in the corrosion behavior of Cu–Ni alloys by affecting electrode kinetic behavior and nature of corrosion product. Similar considerations apply to initial corrosion and also electrocatalytic behavior of alloys.

Acknowledgment

Grateful recognition is made of the efforts of colleagues at the University of Pennsylvania who aided in the early combination of FIM and electrochemistry, principally Dr. Philippe Javet (ETH, Lausanne) and Dr. Joseph Varimbi (Bryn Mawr). The late Professor Erwin Müller (Pennsylvania State Univ.) is remembered with gratitude for helpful discussions and encouragement.

References

1. E. W. Müller, *Science* **149**, 591–601 (1965).
2. E. W. Müller, Proceedings Second International Conference on Solid Surfaces 1974, *Jpn. J. Appl. Phys.*, Supplement 2, Part 2, 1–10 (1974).

3. J. J. Southon, E. D. Boyes, P. J. Turner, and A. R. Waugh, *Surf. Sci.* **53**, 554–580.
4. W. R. Graham and G. Ehrlich, *Phys. Rev. Lett.* **31**, 1407–1408 (1973).
5. D. W. Bassett and P. R. Webber, *Surf. Sci.* **70**, 520–531 (1978).
6. G. L. Kellogg, T. T. Tsong and P. Cowan, *Surf. Sci.* **70**, 485–519 (1978).
7. W. R. Graham, F. Hutchinson, J. J. Nadakavukaren, D. A. Reed, and S. W. Schwenterly, *J. Appl. Phys.* **40**, 3931 (1969).
8. W. R. Graham, D. A. Reed, and F. Hutchinson, *J. Appl. Phys.* **43**, 2951–2956 (1972).
9. U. T. Son and J. J. Hren, *Philos. Mag.* **22**, 675–687 (1970).
10. K. Stolt and J. Washburn, *Philos. Mag.* **34**, 1169–1184 (1976).
11. T. T. Tsong and E. W. Müller, *J. Appl. Phys.* **38**, 545–549 (1967).
12. Y. S. Ng., T. T. Tsong, and S. B. McLane, *Phys. Rev. Lett.* **42**, 588–591 (1979).
13. D. L. Cocke and J. H. Block, *Surf. Sci.* **70**, 363–391 (1978).
14. K. D. Rendulic and E. W. Müller, *J. Appl. Phys.* **38**, 550–553 (1967).
15. L. Nanis and P. Javet, *J. Electrochem. Soc.* **115**, 509–511 (1968).
16. O. T. Inal and A. E. Torma, *Thin Solid Films* **54**, 161–169 (1978).
17. O. T. Inal and A. E. Torma, *J. Metals*, **30**, 16–19 (1978).
18. C. C. Schubert, C. L. Page, and B. Ralph, *Electrochim. Acta* **18**, 33–38 (1972).
19. H. Morikawa and Y. Yashiro, Proceedings Second International Conference on Solid Surfaces, *Jpn. J. Appl. Phys.*, Supplement 2, 67 (1974).
20. E. W. Müller and T. T. Tsong, *Field Ion Microscopy, Principles and Applications*, Elsevier, New York (1969): (a) Ch. IV, pp. 99–111; (b) pp. 119–120; (c) p. 123; (d) Ch. VI; (e) p. 289.
21. K. M. Bowkett and D. A. Smith, *Field Ion Microscopy*, Elsevier, New York, 1970: (a) Appendix 3; (b) Appendix 5; (c) p. 30.
22. E. W. Müller, *Surf. Sci.* **8**, 462–473 (1967).
23. K. D. Rendulic and Z. Knor, *Surf. Sci.* **7**, 205–214 (1967).
24. K. D. Rendulic, *Surf. Sci.* **28**, 285–298 (1971).
25. E. D. Boyes and M. J. Southon, *Vacuum* **22**, 447–451 (1972).
26. J. J. Carroll and A. J. Melmed, *Surf. Sci.* **58**, 601–604 (1976).
27. D. G. Ast and D. N. Seidman, *Surf. Sci.* **28**, 19–31 (1971).
28. H. N. Southworth and B. Ralph, *J. Microsc.* **90**, 167–197 (1969).
29. B. E. Conway, *Theory and Principles of Electrode Processes*, Ronald Press, New York (1975): (a) pp. 277–282; (b) p. 30.
30. D. A. Reed and W. R. Graham, *Rev. Sci. Instrum.* **43**, 1365–1367 (1972).
31. Varian Corp Micro-channel Plate Model VUW-8911K.
32. M. A. Fortes, *Surf. Sci.* **28**, 117–131 (1971).
33. R. M. Skomoroski, U.S. Patent 3,639,219 (1972).
34. S. S. Brenner, *Surf. Sci.* **2**, 496–508 (1964).
35. A. K. Graham, *Electroplating Engineering Handbook*, p. 524, Van Nostrand-Reinhold, New York (1971).
36. A. Damjanovic and H. Brusic, *Electrochim. Acta* **12**, 615–628 (1967).
37. G. K. L. Cranstoun and D. R. Pyke, *Appl. Surf. Sci.* **2**, 359–374 (1979).
38. G. K. L. Cranstoun and D. R. Pyke, *Appl. Surf. Sci.* **2**, 375–381 (1979).
39. K. O'Leary, U.S. Patent 3,776,834 (1973).
40. A. Damjanovic and V. Brusic, *Electrochim. Acta* **12**, 1171 (1967).
41. W. Polanschütz and E. Krautz, *Surf. Sci.* **46**, 602–610 (1974).
42. G. P. Harnwell, *Principles of Electricity and Electromagnetism*, p. 51, 2nd ed., McGraw-Hill, New York (1949).

10

Application of Electron Microscopy to Electrochemical Analysis

D. J. KAMPE

1. Introduction

It has often been said that "a picture is worth a thousand words," or, in a similar vein, "seeing is believing." The electron microscope is a tool used not merely for seeing small objects, but also for obtaining pictures of them that tell the observer something worthwhile about the physical world. Such a picture is informative because, like any other image, it exists in relation to an observer; it presents objects in a particular context and invites evaluation, which becomes the basis of some belief. The striking images obtained by electron microscopy that adorn the walls of many laboratories and corporate offices of high-technology companies clearly have an esthetic appeal that qualifies them as works of art as well as artifacts of science. The artistic element in electron microscopy has been greatly increased in recent years by the development of a capability for spectral and three-dimensional analysis, which has transformed an imagery once based on shadows and silhouettes into one as vibrant and chromatic as any work of Ansel Adams or Stanley Kubrick. Because of the artistic appeal, however, some words of warning must be issued. Depending on the operator's experience and skill, as well as the sample and

D. J. KAMPE • Union Carbide Corporation, Carbon Products Division, Parma Technical Center, Cleveland, Ohio 44101.

methods of preparation, either scientifically valuable micrographs or merely "pretty pictures" may be produced. The second possibility does not satisfy the scientist, or, by itself, the artist. Fundamental questions, which the researcher must continue to bear in mind, are: 1) What are you seeing?, and 2) How representative and realistic are these features?

These questions are particularly important ones for electrochemists to consider because the circumstances in which electrochemists are likely to turn to electron microscopy and related techniques may involve abstractions from practical situations occurring outside the laboratory. Electrochemical systems are inherently complex, and operation under power makes them more so. The cathode matrix of a partially discharged alkaline battery, for example, consists of a disorderly mixture of protonated, poorly crystalline manganese dioxide reduction products, an electrically conductive carbon black, precipitated zinc–alkali hydroxides, and an aqueous phase. In order to obtain a representative description of the sample, a sufficient magnification must be used to image all the phases, yet simultaneously be capable of distinguishing among them and resolving components in each phase. The researcher must then decide how complete or realistic the analysis can be, and appreciate the consequences of this decision.

Microscopic analysis of systems far simpler than batteries may require unrealistic idealization. Consider the case of catalyst particles located on a carbon support, immersed in a liquid medium for use in a fuel cell. The resulting micrograph, Figure 1, depicting silver particles dispersed on a carbon black, is well imaged; however, nothing in the micrograph suggests the differences in the wettabilities of the carbon and silver, probes the structural integrity of the wetted carbon–silver interface, which becomes the site of intensive electrochemical activity in the presence of oxygen, or discloses the tendency of carbon to swell and distort in shape when wet and to resume its previous shape when dried. To a judicious investigator, the interpretive limits to such an appealing photograph may be as important as the information itself.

The nature of the catalyst particles and support matrix may dictate further compromises to improve sensitivity and resolution or to permit easier examination of structural features. Light elements, such as carbon, are best imaged by use of biological electron microscopy techniques, whereas silver and other heavy metals require techniques appropriate for metallurgical specimens. Each choice may allow a more straightforward interpretation of the micrographs produced if it is not idealized too far from "real world" operation. Choices of this type are the essence of applying electron optics to electrochemical problems.

The purpose of this chapter is to describe how electron microscopy can be used to study materials and devices of interest to electrochemists. Difficulties in applying standard sample preparation techniques to battery components are noted and certain practical limitations of these methods discussed. This article touches only the surface of the more general electron microscopy and

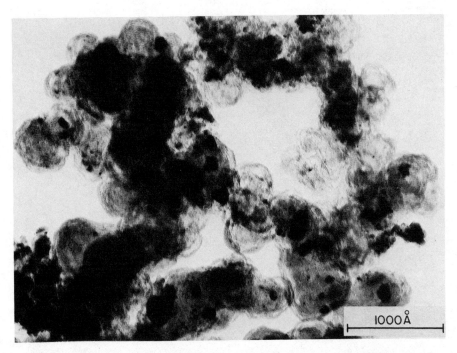

Figure 1. Transmission electron micrograph of 10 weight percent silver crystallites on a carbon black substrate (field of view 0.4×0.5 μm).

spectroscopy techniques. Detailed descriptions of equipment, procedures, and results are available in many excellent texts, a representative number of which are listed in the reference section.

2. Survey of Methods

Close similarities exist between transmission electron microscopy (TEM) and optical microscopy (OM) (see Figure 2). Both use incident radiation to magnify and, thus, resolve small structural features. This wavelike radiation, which is ultimately focused to create the image, may be that transmitted through the specimen or that reflected from it. The focused radiation may consist of a broad range of frequencies or may be a narrow range, perhaps monochromatic; further information may be obtained from changes in the phase of the incident radiation, as in holography, from total reconstruction of the diffraction pattern. Differences in the types of incident radiation, light rays vs. electrons, require environments, focusing units and conditions, and sample types and preparations specific to each.

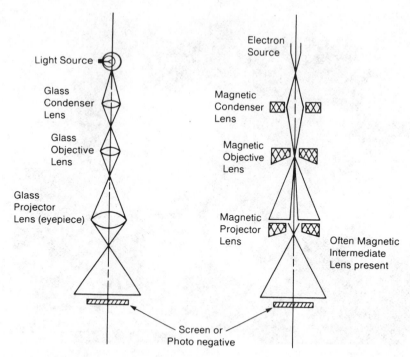

Figure 2. Similarities in the lens arrangements and ray paths for an optical microscope and a conventional transmission electron microscope.

In conventional electron microscopy, a thin beam of thermally excited electrons is collimated and magnetically focussed before striking the sample. These incident electrons can: a) penetrate through a thin film (100 μm or less), b) be scattered by the sample, c) excite X-rays and d) eject additional lower energy electrons. The disparate fates and results of these incident electrons describe, respectively, a) transmission electron microscopy (TEM), b) the back-scattered mode of scanning electron microscopy (BS-SEM), c) energy or wavelength dispersive X-ray analysis, d) and the secondary-electron emission mode of scanning electron microscopy (SE-SEM).

Historically, TEM was the first technique to be developed. Conventional electron microscopes have been available for the past four decades. Although carbon blacks were among the first samples examined, only in the last 10–15 years have techniques perfected for metals been applied routinely to analyze electrochemical components. Each of the other electron optical techniques evolved as the need developed to study high-relief surfaces of bulk articles on a microscale, to analyze surface compositions, and to gain information about crystal structure in selected areas of the surface. The truly revolutionary change, which occurred in the mid-1960's, was the incorporation of all standard electron optical techniques in a single instrument, thus permitting simultaneous examination of surface relief bulk morphology and the quantitative distribution of elements in a sample. This innovation is particularly useful for studies of electrochemical systems that frequently involve a complex interplay between techniques designed to study light element systems, such as polymers, and those that work best with heavy metals.

The electron microscope possesses several advantages when compared with traditional optical microscopy: thousandfold better resolution, greater depth of field at equal magnifications, and the ability to determine elemental composition using self-generated X-rays. However, there are certain limitations associated with use of these potential advantages. One is the need to eliminate solvents in order to protect electron microscopy vacuum systems. Another is that the ultrathin sections required for TEM limit the three-dimensional view of samples and result in a loss of contrast, which is very useful in optical microscopy work.

Similar instrumentally imposed problems of sample size and conditioning are found when the microprobe or scanning electron microscope is used. Since the use of any one electron optical method involved experimental techniques that place some restrictions on the amount of information that can be obtained, it is especially desirable to use instruments that permit the employment of several methods in rapid succession—in fact, nearly simultaneously as far as the results are concerned. The examples in Section 4 are all obtained by use of such instrumentation.

3. Sample Preparations

A fundamental problem in electron microscopy is the interplay of sample selection, preparation, and analysis. The more complex the electrochemical system is and the more elaborately it has been prepared, the more likely it is that sample preparation artifacts will influence any subsequent analysis. A choice must be made between how closely the sample represents the practical situation and how difficult it will be to analyze each component.

Suppose, for example, that one wants to examine the gross structure and the distribution of phases in a simple two-component system consisting of carbon blacks in a polytetrafluorethylene (PTFE) binder. This situation arises frequently in studies of fuel cell electrodes. How might one conduct the examination by TEM and by SEM? In order to observe the sample in a TEM mode of operation, its thickness must be reduced to less than 1000 Å, typically 200–300 Å without loss of integrity of the structure. Grinding is not adequate because PTFE smears when abraded, and the sample probably has to be cleaved in an ultramicrotome. It is helpful to prepare the ultramicrotome specimen at low temperatures; in most other respects, however, microtome technique is similar to that which would be employed in the analysis of biological specimens. Cleavage by an ultramicrotome can introduce substantial mechanical damage in samples of electrochemical interest, since these generally possess higher modulus values and smaller values of breaking strain than do biological samples. They are therefore more likely to retain the history of their deformation in the form of cleavage cracking, dislocation, and structural deformation. On the other hand, the electron microscope beam heats the specimen when it passes through, and beam-induced thermal degradation of specimens such as carbon–PTFE, which have low-heat deformation temperatures, or of highly dispersed catalyst particles, which are subject to sintering, may produce significant changes in the specimen. Thus, great care must be exercised in extrapolating properties obtained from micrographs obtained from thinned microregions of a 3 mm grid size sample to the bulk article.

One can minimize mechanical damage by starting with a relatively thick microtomed section and thinning it by cathodic etching or ion bombardment. Time may be involved in obtaining such specimens, but it is ultimately possible to prepare thin sections that represent bulk structure well. Morphology of these thin sections can be examined at the highest resolution of any technique discussed in this article, and the crystal structure and chemical constitution can also be examined by auxillary techniques such as X-ray diffraction from selected areas or electron beam microprobe analysis.

Other samples of electrochemical interest, such as metals, can be prepared in thin sections by techniques not appropriate for use with carbon–PTFE specimens. These include chemical and electrolyte polishing, chemical reaction, gaseous ion thinning, the previously mentioned cleavage, and microtomy. If appropriate, one can vacuum–evaporate or deposit onto a suitable substrate (typically NaCl) a thin film of the material, dissolve the salt, and use the freed thin film. The morphology of a carbon–PTFE layer can also be studied by forming a film on the layer, stripping off the film, and observing on the film the replicated morphology of the surface if the original surface is not too porous. The advantage of making a replica is that it may not be necessary to destroy the original bulk sample. Formvar or collodion replicas, having resolutions of ~700 Å, can be prepared simply by stripping, and strippable carbon films capable of resolving features less than 100 Å can be floated from the specimen by depositing a layer of soluble salt or soap, then depositing the replica layer, and soaking the replicated article to release the film. If the substrate can be sacrificed by dissolving it, one may easily deposit a carbon film directly on the surface. Replicas of the surface, which are capable of 10 Å resolution, provide information about texture and homogeneity, but replication involves a number of preparation steps. It is not applicable to all types of bulk samples; the replica is difficult to separate, for example, from porous surfaces such as those of carbon–PTFE mixtures. Replication does not permit analyses of crystal structure and chemical composition to be performed.

The scanning electron microscopy (SEM) mode of operation, in which electrons are collected from the sample surface, requires much less sample preparation, offers greater resolution than optical microscopy, and can be combined with elemental analysis to give compositional information. It is frequently the method of choice because magnification can be varied over several orders of magnitude; one can obtain sufficient resolution to image individual particles yet still see the ways these same particles agglomerate with each other and with the PTFE binder. SEM also has the large depth of field required to obtain images of very rough surfaces, such as the carbon black–PTFE tensile fracture surface of Figure 3. These thin PTFE fibrils, resulting from the plastic deformation of the PTFE-bonded electrode, extend for considerable length and can serve as reinforcing ties to hold the carbon-rich matrix together, as well as offer channels or straws of hydrophobic character.

Sample preparation for the SEM, although easier than that for TEM, is not without its pitfalls. Because of the high vacuum required in the electron microscope column, artifacts resulting from electrolyte drying must be avoided. Freeze drying or critical point drying to remove the aqueous media without changing the surface morphology should be considered if the influence of liquid environment on electrode structure and properties is important. This

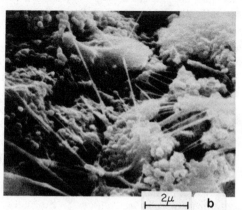

Figure 3. Scanning electron micrographs (secondary electron mode) of a carbon black/PTFE layer. (a) Field of view 40 × 50 μm; (b) field of view 10 × 12 μm.

environmental aspect is often overlooked, resulting in pretty pictures, but not scientifically valuable micrographs.

For SEM back-scattered mode analysis, a thin layer of a metal (Ag, Al, Pt, Pt/Pd) is often evaporated onto the SEM sample. This step is taken to improve the number of electrons excited per unit area and to reduce possible charging effects on poorly conducting portions of a sample. This layer is generally applied either via a vacuum hot filament operation, or by ion sputtering of the coating material. For a heat-sensitive material, such as carbon–PTFE, sufficient distance between sample and a filament must be used to prevent degradation of the structure by thermal bombardment. Although this layer usually uniformly coats the sample surface, this is not always the case. These thin films may have their own structure, which can cause misinterpretation of the structure of the underlying substrate. Before the sample is coated, one should consider whether elemental analysis should be done first, for fluorescence or diffraction from the underlying substrate may be rendered undetectable. If elemental analysis is to be performed by the electron microprobe, sample preparation is further complicated by the necessity of having flat (i.e., polished) surfaces. This condition is required because the elemental sensors operate by diffraction and thus use the angle of diffraction as the discriminator between radiations from different elements.

4. Sample Analysis

A number of examples have been chosen to illustrate how these different microscopy techniques can be used to show structural aspects of plastic-bonded carbon black electrodes found in batteries and fuel cells. Although these samples are specific for one particular application, they serve to indicate the diversity of information available from these tools.

The TEM offers a number of different modes of operation that can be used to characterize suitably thin samples. The most conventional are the so-called bright field, dark field, and phase-contrast modes. In the bright field operation, those electrons that are scattered little after passing through the sample are imaged. This mode most closely corresponds to optical microscopy viewing of a transparent sample in which opaque particles are embedded. Particles that are sufficiently thin will permit partial electron penetration, and thus scatter and cast an electron shadow. These areas will show varying gray levels depending on the degree of penetration of the incident beam. Attenuation of the beam increases as the atomic number of the substrate increases, so that a 100 kV beam, which can be imaged through several thousand angstroms of carbon, may be almost completely attenuated by 500 Å of a heavy metal.

Carbon blacks are especially amenable to TEM analysis. Particle and agglomerate size and shapes can be easily obtained by dusting of these carbon blacks onto an ultrathin carbon formvar support. Figures 4 and 5 show typical results. Particle size distributions, agglomeration, and surface modifications caused by processing changes are also easily obtained by examination of these micrographs. Selected examples of various types of carbon blacks are given in Figures 4a–c, and Figures 5a–d show progressively increasing magnification of an acetylene black.

TEM can also be used to analyze size, shape, and distribution of catalyst particles deposited on various substrates. These may include precious metals

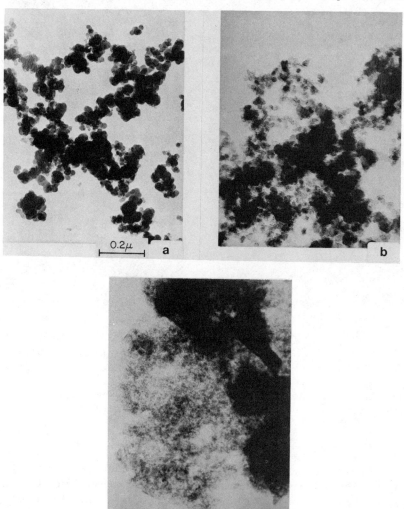

Figure 4. Similar magnification transmission electron micrographs of three carbon blacks. (a) Cabot Vulcan XC72R (all field of view $0.7 \times 1.0 \ \mu m$); (b) Calgon RB; (c) Norit 211.

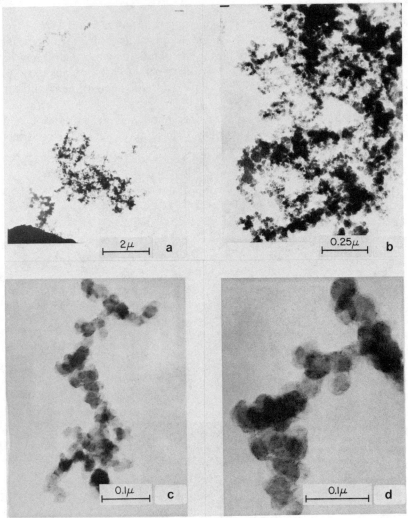

Figure 5. A series of transmission electron micrographs of as-received Shawinigan acetylene black particles and agglomerates. (a) Field of view 8×10 μm; (b) field of view 2×2.5; (c) field of view 0.4×0.5; (d) field of view 0.25×0.3.

and/or manganese dioxide on carbon blacks. Oxide phases, catalyst and poisons, and crystal structure can be identified by using selected area diffraction. Silver and platinum crystallites on carbon blacks are shown in Figures 1 and 6, respectively. Differences in average particle sizes are evident, as well as the degree of homogeneity of each on the carbon substrates.

Dark field illumination, which produces a "negative" imaging of the area, is accomplished by inserting an aperture to physically block the transmitted

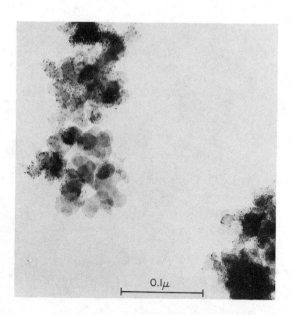

Figure 6. Transmission electron micrograph of 5 weight percent platinum crystallites on a carbon black substrate (field of view 0.4 × 0.5 μm).

beam, thus collecting only the strongly scattered and diffracted electrons. Often, this technique results in improved contrast for surface features not apparent in the bright field case. If the material is crystalline, the scattered transmitted beam can be refocused by the objective aperture so that only those regions so oriented to diffract the transmitted beam into the aperture are highlighted in the dark field mode of operation. An example of this technique is shown in Figure 7. In this photomicrograph, 60 × 1000 Å crystalline structure cells not apparent in bright field illumination are made visible by focusing the (002) diffraction beam of a film of RuO_2. The film is of electrochemical interest because it is the electrocatalyst for chlorine discharge on metal anodes used in chlor–alkali cells. The film was prepared by thermal decomposition of $RuCl_3$ on a smooth substrate, which was dissolved, leaving a film suitable for TEM.

TEM can be coupled to a related mode of operation, selected area diffraction (SAD). This extremely powerful technique allows the observer to determine diffraction pattern in imaged areas as small as 200 Å diameter. In this mode of operation, the focal position is magnetically moved from the objective plane, and an electron diffraction pattern is obtained instead of an image. These diffraction patterns can be interpreted in much the same fashion as X-ray diffraction patterns, except that the electron wavelength in a 100 kV beam is 0.04 Å, much smaller than typical X-ray wavelengths of 1–2 Å, and the relative intensities of diffraction peaks differ from those observed in X-ray diffraction because the scattering factors for electrons differ from those of X-rays. Electron diffraction cameras are best calibrated by comparison with

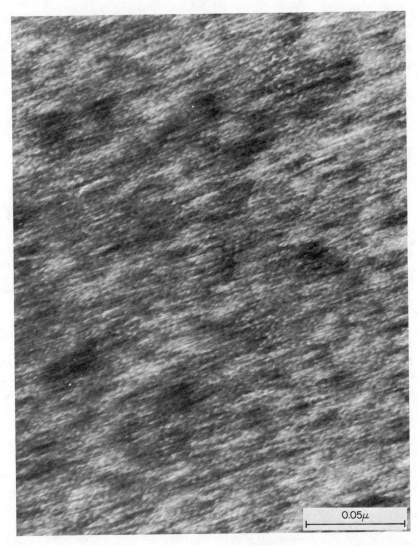

Figure 7. Oblique illumination transmission electron micrograph of a thinned RuO_2 film (field of view 0.2×0.3 μm).

the pattern of a standard material, either an Au or Al vapor-deposited polycrystalline foil. Examples of selected area diffraction patterns illustrating differences among various polycrystalline materials are shown in Figure 8. One should note the improvements seen in the ring patterns of b and c resulting from the high-temperature graphitization treatment given the latter.

In many respects, SEM bears a closer resemblance to optical microscopy than either TEM or microprobe analysis. This relationship occurs because the

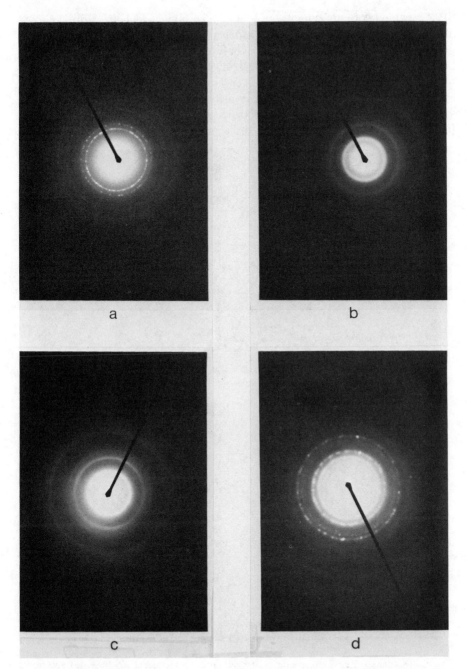

Figure 8. Selected area diffraction micrographs of: (a) well-oriented polycrystalline Al; (b) calcined (1000°C) Shawinigan acetylene black; (c) graphitized (2850°C) Shawinigan acetylene black; (d) amorphous carbon black containing 2 weight percent platinum crystallites.

SEM is capable of being run at lower magnifications, which are more comparable with OM; three-dimensional aspects are present in each, and sample preparation is easier than for TEM analyses. Generally, one uses either the back-scattered (BS) or secondary-electron (SE) mode of operation for conventional analyses.

In BS-SEM mode, a 10–30 kV primary beam is rastered across the samples, and those high energy electrons that have been back-scattered are focused, amplified, and displayed on a cathode ray tube. Since different

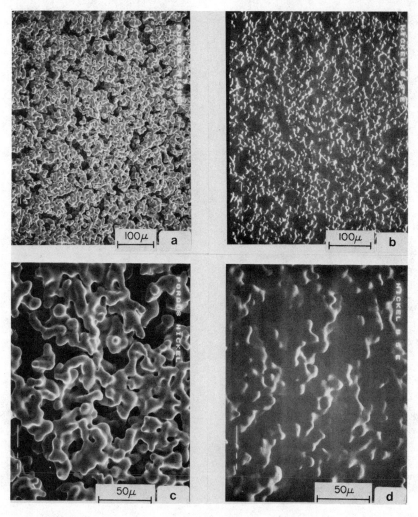

Figure 9. Scanning Electron micrographs of a sintered porous nickel substrate. (a) Secondary electron mode (field of view $500 \times 600 \ \mu$m); (b) back-scattered mode (field of view $500 \times 600 \ \mu$m); (c) secondary electron mode (field of view $160 \times 200 \ \mu$m); (d) back-scattered mode (field of view $160 \times 200 \ \mu$m).

elements have different primary electron-scattering coefficients, the back-scattered mode is sensitive to elemental differences in the sample, and elemental differences appear as differences in contrast or intensity. The secondary-electron mode of operation relies on the ability of the target material to produce a large number of lower energy electrons when struck by the high-energy primary beam. Since a greater number of electrons can be collected in the secondary-electron mode than in the back-scattered mode, better signal-to-noise ratios and better resolution can be obtained. Information from both modes of operation complement each other.

The depth of field at high magnifications offered by electron microscopy often permits the surface topography of electrochemical components to be determined. A different perspective of the degree and type of porosity and tortuosity present in a porous nickel plaque may be obtained after viewing the SEM micrographs shown in Figure 9. Differences in the appearance of the porous nickel plaque due to either the secondary-electron or back-scattered mode of collection should also be noted. Individual PTFE particles and

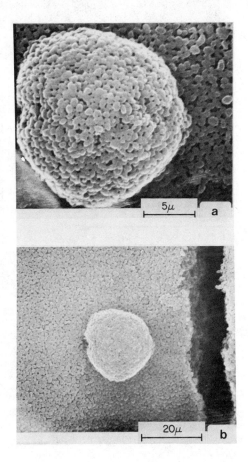

Figure 10. Scanning electron micrograph (secondary electron mode) of a PTFE T30B film coating. (a) (Field of view) 16 × 20 μm; (b) (field of view) 50 × 75 μm.

agglomerates can be identified in the higher magnification SEMs shown in Figure 10. This sample was gold coated to prevent electron charging and illustrates the three-dimensionality and excellent resolution easily obtained in SEM.

The use of energy dispersive attachments on TEM/SEM instruments has become common. This equipment uses the energy of the X-rays generated by the bombardment of the incident beam to identify the elements with atomic number $Z \geq 11$ present in the sample. The so-called "Analytical TEM" incorporates these dual features of microscopy and spectroscopy to image and identify differences within multicomponent samples. In Figure 11, a second type of porous nickel plaque is viewed in the secondary-electron mode of operation. The accompanying energy dispersive profile shows that very little contamination, either sample or equipment related, can be detected in the nickel plaque. However, if concentrations of minor elements present can be

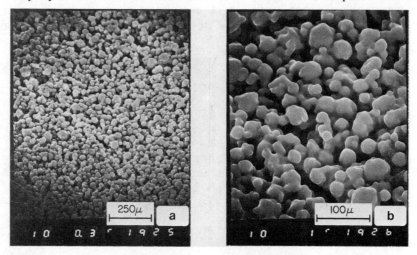

Figure 11. A typical energy dispersive X-ray spectrum of a porous nickel substrate and accompanying secondary electron scanning electron micrographs. (a) Field of view 1000×1250 μm; (b) field of view 300×400 μm; (c) X-ray spectrum.

measured, corrections for the effects of X-ray generation, absorption, and fluorescence can be made and accurate weight fraction values assigned.

A related X-ray mode uses wavelength selective spectrometers to analyze the X-rays generated when the incident beam strikes the sample. These

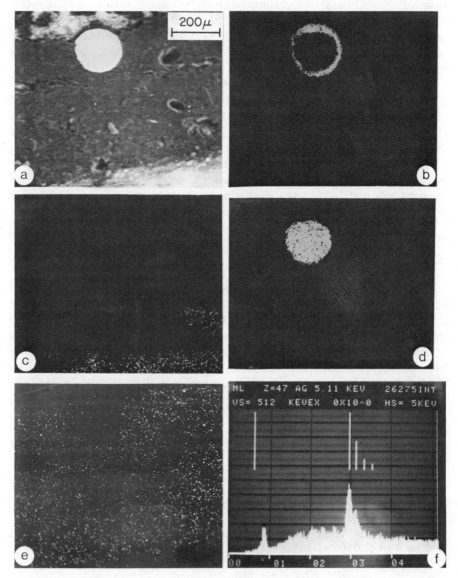

Figure 12. Electron microprobe micrographs of a Pt-catalyzed carbon in a PTFE matrix embedded in a silver-plated nickel screen with a second pure PTFE backing (all fields of view 1000 × 1250 μm). (a) Sample current micrograph; (b) silver map; (c) fluorine map (for PTFE); (d) nickel map; (e) platinum map; (f) X-ray energy dispersive spectrum.

microprobes offer a lower elemental cutoff, can analyze for $Z \geq 4$, possess better signal-to-noise ratios than energy dispersive systems, and can detect lower concentrations of the element being sought. A typical usage is shown in Figure 12, in which identification and location of platinum and PTFE in a thin wetproofed catalyzed carbon layer embedded in a silver-plated nickel screen is shown. The location of the PTFE-coated area, both around the screen and in a separate homogeneous layer, were found by probing for fluorine while the carbon, which cannot itself be detected, could be located by using either the platinum or sulfur peaks.

Generally, four aspects of electron microprobe analysis limit its utility. First, magnifications seldom exceed 1000×. Second, the beam current needed to produce these maps may cause degradation of the sample. Third, wavelength dispersive systems, in contrast to energy dispersive systems, detect only one element at a time; thus, one needs to reorient the crystal each time in order to map all elements of interest. Finally, flat samples are needed. This limitation is necessitated by the need to set accurately the angle of diffraction for the element selected, or if used in a scanning mode, to determine radiation wavelength and, thus, the element response for the X-ray radiation.

5. Summary and Conclusions

Electron microscopy and electron spectroscopy are powerful, simple tools for the study of materials used in electrochemical devices. They can provide compositional and structural analysis of microsamples. Bulk materials and devices are amenable to these techniques. Electron microscopy now comprises a multitude of techniques that provide detailed knowledge of the structure and chemical processes that occur even in very small regions of materials. The application of these techniques to electrochemical analyses is fruitful indeed.

Standard Reference Texts

Transmission Electron Microscopy Related:

1. J. Hren, D. Joy, and J. Goldstein, eds., *Introduction to Analytical Electron Microscopy*, Plenum Press, New York (1979).
2. G. Thomas and M. J. Gounge, *Transmission Electron Microscopy of Materials*, John Wiley, New York (1979).
3. J. W. Edington, *Practical Electron Microscopy in Materials Science*, Van Nostrand Reinhold, New York (1976).
4. G. W. Bailey, ed., *Proceedings of the Meetings of the Electron Microscopy Society of America EMSA*, Claitor, Baton Rouge, Louisiana.

Scanning Electron Microscopy Related:

5. J. Goldstein and H. Yakowitz, eds., *Practical Scanning Electron Microscopy*, Plenum Press, New York (1975).

6. O. C. Wells, *Scanning Electron Microscopy*, McGraw Hill, New York (1974).
7. O. Johari, ed., *Proceeding of SEM* through 1977 ITT Research Institute, Chicago, IL 1978-present SEM, Inc. AMF O'Hare, Ill.

Energy Dispersive X-ray Fluorescence Related:

8. E. Bertin, *Principles and Practice of X-ray Spectrometric Analysis*, Plenum Press, New York (1975).
9. R. Woldseth, *All You Ever Wanted to Know About XES*, Kevex Corp. Burlingame, California (1973).

Electron Microprobe Related:

10. L. S. Birks, *Electron Probe Microanalysis*, Interscience Publishers, New York (1963).
11. S. J. B. Reed, *Electron Microprobe Analysis*, Cambridge University Press, Cambridge, Great Britain (1975).
12. D. Wittry, *Microbeam Analysis*, San Francisco Press, San Francisco, California (1980).

11

Classified Bibliography of Electroanalytical Applications

V. K. VENKATESAN

1. Polarography

Classical polarographic technique, invented by the late Prof. J. Heyrovsky in 1922, is one of the few important electrochemical techniques that finds extensive applications in analytical chemistry.

In the last six decades several new polarographic techniques have been proposed and developed by various researchers and these attempts have the following main objectives: (a) increase in sensitivity; (b) increased resolubility; and (c) ease and speed of operation.

The important references pertaining to theory, instrumentation and applications for the different polarographic techniques, viz, classical d-c polarography, differential and derivative polarography, oscillographic polarography, a-c polarography, square-wave polarography, radio frequency polarography, and pulse and differential polarography, are given in the following. The various works published up to December 1979 have been considered and the relevant ones are included.

1.1. D-C Polarography

The polarographic theory for (1) reversible electrode processes involving (a) simple metal ions and metals soluble in mercury, (b) salts of mercury and

V. K. VENKATESAN • Central Electrochemical Research Institute, Karaikudi-623006, Tamil-nadu, India.

complexation with mercury, and (c) complex ions; (2) irreversible electrode processes; and (3) organic electrode processes have been developed by many electrochemists. Heyrovsky *et al.*,[1-3,20,31] Tomes,[4,16] Lingane,[5] Stackelberg *et al.*,[6,8,22] Kolthoff and co-workers,[17,19] Hume *et al.*,[24,18] Delahay *et al.*,[10,13,35,37] Koutecky,[36,38] Koryta,[26-29,42] Meites,[14-43] Matsuda and Ayabe,[41] and Vlcek,[12,40] have made notable contributions to the theory of current–potential curves for reversible and irreversible electrode processes. The theory of organic electrode processes has been presented and discussed by Kolthoff and Lingane,[46,47] Elving *et al.*,[48,49,51,53,54,56,62] and Zuman.[55,57,61,63,64] The theory of stationary electrode polarography for organic compounds has been presented by Nicholson *et al.*[67,72-74,76-78] The mechanism of organic electrode processes has been reviewed by Perrin.[79] The theory for polarographic kinetic currents has been developed by Weisner and later extended by Koutecky.[60] The theory of polarographic catalytic currents has been presented first by Brdicka, and Mairanovsky has studied this aspect in greater detail.[81] The theories developed in *d-c polarography* have been employed in understanding the kinetics of those electrode processes that are not fast reactions.

The theories of limiting currents, which include contributions from several different processes such as charging or residual current, diffusion current, kinetic current, or catalytic current, are discussed in References (86–139). The need for eliminating or minimizing the residual or charging current has been discussed from the point of increasing the sensitivity of polarographic technique, that is, in the range 10^{-5}–10^{-3} M for d-c polarography. The various procedures employed in the d-c polarographic analysis of inorganic ions and organic substances, as well as the use in medicine and biochemistry are discussed in References (140–179).

The polarographic instrumentation, including those employed to control the drop times of dropping mercury electrodes, have been very well developed and many instruments based on these are available commercially. The polarographic recorders developed in the early period are based on two-electrode systems. With the development of potentiostats based on operational amplifiers, I.C. chips, and X–Y recorders, polarographs employing three electrodes have been developed in many commercial multipurpose instruments available at present. One of the important modes of operation is d-c polarography and tast polarography.

1.1.1. Theory of Current-Potential Curves

1.1.1.1. Reversible Processes Involving Simple Metal Ions and Metals Soluble in Mercury

1. Electrolysis with drops of mercury as the electrode, J. Heyrovsky, *Chem. Listy* **16**, 256–264 (1922).
2. Electrolysis with a dropping mercury cathode. I. Deposition of alkali and alkaline-earth metals, J. Heyrovsky, *Phil. Mag.* **45**, 303–315 (1923).

3. Polarographic studies with the dropping mercury electrode. Part II. The absolute determination of reduction and depolarization potentials, J. Heyrovsky and D. Ilkovic, *Collect. Czech. Chem. Commun.* **7**, 198–214 (1935).
4. Polarographic studies with the dropping mercury cathode. Part LXIII. Verification of the equation of the polarographic wave in the reversible electro-deposition of free cations, J. Tomes, *Collect. Czech. Chem. Commun.* **9**, 12–21 (1937).
5. Thermodynamic significance of polarographic half wave potentials of simple metal ions at the dropping mercury electrode, J. J. Lingane, *J. Am. Chem. Soc.* **61**, 2099–2103 (1939).
6. Die wissenschaftlichen grundlagen der polarographie, M. V. Stackelberg, *Z. Elektrochem.* **45**, 466–491 (1939).
7. Polarographic limiting currents, J. K. Taylor and S. W. Smith, *J. Res. Nat. Bur. Stand.* **42**, 387–395 (1949).
8. Zur theorie der polarographischen kurve, H. Strehlow and M. V. Stackelberg, *Z. Elektrochem.* **54**, 51–62 (1950).
9. Accurate potentials with the dropping mercury cathode, H. J. Gardner, *Nature (London)* **167**, 158–159 (1951).
10. Unified theory of polarographic waves, P. Delahay, *J. Am. Chem. Soc.* **75**, 1430–1435 (1953).
11. Verification of a theory of irreversible polarographic waves, P. Kivalo, K. B. Oldham, and H. A. Laitinen, *J. Am. Chem. Soc.* **75**, 4148–4152 (1953).
12. Polarographic half wave potentials. Method of measurements; half wave potentials of thallium, A. A. Vlcek, *Collect. Czech. Chem. Commun.* **19**, 862–867 (1954).
13. *New Instrumental Methods in Electrochemistry*, P. Delahay, Chapt. 3 and 4, pp. 46–48, Interscience Publishers, Inc., New York (1954).
14. Theory of the current potential curve, L. Meites, *Polarographic Technique*, Chapt. 4, pp. 203–266, Interscience Publishers, New York (1965).

1.1.1.2. Processes Involving Salts of Mercury and Complexation with Mercury

15. Polarographic studies with the dropping mercury electrode. Part I. Anodic polarisation and the influence of anions, J. Revenda, *Collect. Czech. Chem. Commun.* **6**, 453–467 (1934).
16. Polarographic studies with the dropping mercury cathode. Part LXIV. Equations of current voltage curves in the reversible electroreduction of a weak electrolyte, $Hg(CN)_2$, J. Tomes, *Collect. Czech. Chem. Commun.* **9**, 81–103 (1937).
17. Anodic waves involving electrooxidation of mercury at the dropping mercury electrode, I. M. Kolthoff and C. S. Miller, *J. Am. Chem. Soc.* **63**, 1405–1411 (1941).
18. A polarographic study of mercuric cyanide and the stability of cyanomercuriate ions, L. Newman, J. Deo Cabral, and D. N. Hume, *J. Am. Chem. Soc.* **80**, 1814–1819 (1958).
19. Effects on polarographic waves of the formation of insoluble films on dropping mercury electrode, I. M. Kolthoff and Y. Okinaka, *J. Am. Chem. Soc.* **83**, 47–53 (1961).
20. Deposition of mercury ions, In: *Principles of Polarography*, J. Heyrovsky and J. Kuta, Chapter X, 167–179, Academic Press, New York (1966).

1.1.1.3. Reversible Processes of Complex Ions

21. Polarographic studies with the dropping mercury electrode. Part II. The absolute determination of reduction and depolarization potentials, J. Heyrovsky and D. Ilkovic, *Collect. Czech. Chem. Commun.* **7**, 198–214 (1935).
22. Polarographische untersuchungen an komplexen in waszriger losung, M. V. Stackelberg and H. V. Freyhold, *Z. Elektrochem.* **46**, 120–129 (1940).
23. Complex ions. XV. New derivatives of iminodiacetic acid and their alkaline earth complexes. Connection between acidity and complex formation, G. Schwarzenbach, G. Anderegg, W. Schneider, and H. Senn, *Helv. Chim. Acta*, **32**, 1175–1186 (1948).
24. The determination of consecutive formation constants of complex ions from polarographic data, D. D. Defrod and D. N. Hume, *J. Am. Chem. Soc.* **73**, 5321–5322 (1951).

25. The polarographic determination of relative formation constants of metal complexes of ethylenediaminetetra acetic acid, K. Bril and P. Krumholz, *J. Phys. Chem.* **57**, 874–879 (1953).

26. Polarography of complex compounds, J. Koryta, *Chem. Tech (Berlin)*, **7**, 464–470 (1955).

27. Kinetik der elektordenvorgange bei komplexen in der polarographic. I. uber Gewisse polarographische methoden fur die ermittlung des mechanisms der metallabscheidung aus komplexen, J. Koryta, *Collect. Czech. Chem. Commun.* **23**, 1408–1411 (1958).

28. II. Bestimmung der komplexibildungs konstanten aus den halbstufen potentialen kinetischer strome, J. Koryta, *Collect. Czech. Chem. Commun.* **24**, 2903–2917 (1959).

29. III. Durchtritts und dissoziationsreaktion des komplexes, J. Koryta, *Collect. Czech. Chem. Commun.* **24**, 3057–3074 (1959).

30. New methods of estimating the stability constants of complexes, P. K. Kamalkar, *Z. Phys. Chem.* **218**, 189–196 (1961).

31. *Principles of Polarography*, J. Heyrovsky and J. Kuta, Chapter VIII, 147–160, Academic Press, New York and London (1966).

1.1.1.4. Irreversible Electrode Processes: Reduction of Complexes

32. Theory of concentration polarization of a dropping mercury electrode. I. N. Meiman, *Zh. Fiz. Khim.* **22**, 1454–1465 (1948).

33. Theory of polarographic currents controlled by rate of reaction and by diffusion, P. Delahay, *J. Am. Chem. Soc.* **73**, 4944–4949 (1951).

34. Slow electrode reactions, M. Smutek, *Proc. Ist Internatl. Polarog. Congress*, Prague, Vol. III, 677–683 (1951).

35. Unified theory of polarographic waves, P. Delahay, *J. Am. Chem. Soc.* **75**, 1430–1435 (1953).

36. Theorie langsamer elektroden reaktionen in der polarographie und polarographischs verhalten eines systems, bei-welchem der depolarisator durch eine schnelle chemische reaktion aus einem elektroinaktivenstoff entsteht, J. Koutecky, *Collect. Czech. Chem. Commun.* **18**, 597–610 (1953).

37. Kinetic analysis of the discharge mechanism of complex ions, H. Gerisher, *Z. Physik. Chem. (Leipzig)* **202**, 292–301 (1953).

38. Uber die kinetik der elektrodenvorgange XV. Tabellen der funktion fur den polarographischen strom bei einem depolarisations vorgang mit vorgeschalteten oder nachfolgenden sehr schnelle monomolekularen chemischen reactionen, J. Weber and J. Koutecky, *Collect. Czech. Chem. Commun.* **20**, 980–982 (1955).

39. The "exchange current" on an amalgam drop electrode and the composition of the discharged complexes, A. G. Stromberg and M. K. Ivantsova, *Dokl. Akad. Nauk SSSR* **100**, 303–306 (1955).

40. Relation between the electronic structure of inorganic depolarisers and their polarographic behaviour. I. Basic rules, A. A. Vleck, *Collect. Czech. Chem. Commun.* **20**, 894–901 (1955).

41. Theoretical analysis of polarographic waves. II. Reduction of complex metal ions, H. Matsuda and Y. Ayabe, *Bull. Chem. Soc. Jap.* **29**, 134–140 (1956).

42. Kinetik der elektroden vorgange von komplexen in der polarographic. III. Durchtritts und dissozaiations reaktion der komplexes, J. Koryta, *Collect. Czech. Chem. Commun.* **24**, 3057–3074 (1959).

43. The calculation of electrochemical kinetic parameters from polarographic current potential curves, L. Meites and Y. Israel, *J. Am. Chem. Soc.* **83**, 4903–4906 (1961).

44. Electrolysis with constant potential: Irreversible reactions at a hanging mercury drop electrode, I. Shain, K. J. Martin, and J. W. Ross, *J. Phys. Chem.* **65**, 259–261 (1961).

45. *Principles of Polarography*, J. Heyrovsky and J. Kuta, Chapter XIV, 205–266, Academic Press, New York (1966).

1.1.1.5. Organic Electrode Processes

46. Polarographic waves of organic substances, I. M. Kolthoff and J. J. Lingane, In: *Polarography*, Vol. I, Chap. XIV, pp. 246–267. Interscience Publishers, New York (1952).
47. Oxidation and reduction of organic compounds—General characteristics of current-voltage curves of organic compounds, I. M. Kolthoff, J. J. Lingane, and S. Wawzonek, In: *Polarography*, Vol. II, Chap. XXXVI, pp. 623–634, Interscience Publishers, New York (1952).
48. Effect of structure on the stereo chemistry of electrode reaction. Unsaturated C_4 dibasic acids and esters. Stereospecific reduction of the double bond, I. Rosenthal, J. R. Hayes, A. J. Martin, and P. J. Elving, *J. Am. Chem. Soc.* **80**, 3050–3055 (1958).
49. Correlation of polarographic data with structure. Use of the Hammett–Taft relation, P. J. Elving and J. M. Makowitz, *J. Org. Chem.* **25**, 18–20 (1960).
50. *Oxidation-reduction potentials of organic compounds*, W. M. Clark, Williams & Wilkins, Baltimore (1960).
51. Structural effects on the electrochemical reduction mechanism of organic compounds, P. J. Elving, *Ricerca Sci.* **30**, Suppl. No. 5, 205–215 (1960).
52. Controlled potential electrolysis, L. Meites, In: *Physical Methods of Organic Chemistry*, Part IV, A. Weissberger, ed, Interscience Publishers Inc. New York 3281–3333 (1960).
53. Mechanism of organic electrode reactions, P. J. Elving and B. Pullman, In: *Advances in Chem. Phys.*, Vol. III, Prigogine, ed., pp. 1–31, Interscience Publishers, New York (1961).
54. Effect of structure on the stereochemistry of electrode reactions. Monobromo C_4 dibasic acids and esters, P. J. Elving, I. Rosenthal, J. R. Hayes, and A. J. Martin, *Anal. Chem.* **33**, 330–334 (1961).
55. Current trends in the study of the influence of structure on the polarographic behaviour of organic substances, P. Zuman, In: *Progress in Polarography* P. Zuman and I. M. Kolthoff, eds., Vol. I, pp. 319–332, Interscience Publishers, New York (1962).
56. Polarography in organic analysis, P. J. Elving, In: *Progress in Polarography*, Vol. III, P. Zuman and I. M. Kolthoff, eds., pp. 625–648, Interscience Publishers, New York (1962).
57. Quantitative treatments of substituent effects in polarography. II. Free energy relationship in monocyclic heterocyclic series, P. Zuman, *Collect. Czech. Chem. Commun.* **27**, 630–647 (1962).
58. Effect of the double layer structure and of the adsorption of electrode reaction participants upon polarographic waves in the reduction of organic substances, S. G. Mairanovsky, *J. Electronal. Chem.* **4**, 166–181 (1962).
59. Influence of adsorption on the polarographic behaviour of reducible organic substances. An example of auto inhibition of the discharge reaction, E. Laviron, *Bull. Soc. Chim. France* 418–422 (1962).
60. General theoretical treatment of the polarographic kinetic currents, R. Brdicka, V. Hanus, and J. Koutecky, In: *Progress in Polarography*, Vol. I, P. Zuman and I. M. Kolthoff, eds., 145–199, Interscience Publishers, New York (1962).
61. Applications of polarography to study the constitution of organic compounds and intermediate products in their reactions, P. Zuman, *Z. Chem.* 3 (**5**) 161–171 (1963).
62. Variation of the half wave potential of organic compounds with pH, P. J. Elving, *Pure Appl. Chem.* **7**, (2–3) 423–454 (1963).
63. *Organic Polarographic Analysis*, P. Zuman, Pergamon Press, London (1964).
64. *Recent trends in organic polarography*, P. Zuman In: *Polarography 1964*, G. J. Hill, ed., pp. 687–710, Macmillan (London).
65. Electrolytic reductive coupling I. Acrylonitrile, M. M. Baizer, *J. Electrochem Soc.* **111**, (2), 215–222 (1964).
66. The effect of the composition of aqueous organic solvents on the polarographic behaviour of organic compounds, S. G. Mairanovski, In: *Polarography 1964*, G. J. Hill, ed., pp. 719–730, Macmillan, London (1966).

67. Theory of stationary electrode polarography single scan and cyclic methods applied to reversible irreversible and kinetic systems, R. S. Nicholson and I. Shain, *Anal. Chem.* **36**, (4), 706–723 (1964).

68. Adsorption effect of the participants in the electrode and pre-electrode reactions upon the kinetics of electrochemical processes, S. G. Marianovskii, *Electrochim. Acta* **9**, 803–815 (1964).

69. Electrochemical relaxation techniques, W. H. Reinmuth, *Anal. Chem.* **36**, 211R–219R (1964).

70. Polarography of aliphatic compounds, H. Lund, *Talanta* **12**, 1065–1079 (1965).

71. Polarography of aromatic compounds. J. Tirouflet and E. Laviron, *Talanta* **12**, 1105–1126 (1965).

72. Theory and application of cyclic voltammetry for measurement of electrode reaction kinetics, R. S. Nicholson, *Anal. Chem.* **37**, 1351–1355 (1965).

73. Theory of stationary electrode polarography for a chemical reaction coupled between two charge transfers, R. S. Nicholson and I. Shain, *Anal. Chem.* **37** (2), 178–190 (1965).

74. Experimental verification on an ECE mechanism for the reduction of p-nitrosophenol using stationary electrode polarography, R. S. Nicholson and I. Shain, *Anal. Chem.* **37(2)**, 190–195 (1965).

75. Polarography of organic compounds in aprotic solvents, S. Wawzonek, *Talanta* **12**, 1229–1235 (1965).

76. Semiempirical procedure for measuring with stationary electrode polarography rates of chemical reactions involving the product of electron transfer, R. S. Nicholson, *Anal. Chem.* **38**, 1406 (1966).

77. Polarographic theory for an ECE mechanism. Application to reduction of p-nitrophenol, R. S. Nicholson, J. M. Wilson, and M. L. Olmstead, *Anal. Chem.* **38**, (4), 542–545 (1966).

78. Experimental evaluation of cyclic stationary electrode polarography for reversible electron transfer, M. L. Olmstead and R. S. Nicholson, *Anal. Chem.* **38**, 150 (1966).

79. Mechanisms of organic polarography, C. L. Perrin, In: *Progress in Physical Organic Chemistry*, Vol. III, S. G. Cohen, A. Streitwieser, Jr., and R. W. Taft, eds., pp. 165–316, Interscience Publishers, New York (1965).

80. Electrochemical relaxation techniques, W. H. Reinmuth, *Anal. Chem.* **38**, 270R–277R (1966).

81. Chapter II: Currents limited by the rate of chemical reaction, pp. 7–55; Chapter III: Adsorption on the electrode and the electrode processes, pp. 57–114; Chapter IV: Electrode processes with antecedent protonation reaction, pp. 115–153; Chapter V: The effect of the double layer structure on electrode processes, pp. 155–186, In: *Catalytic and kinetic waves in polarography*, S. G. Mairanovskii, Plenum Press, New York (1968).

82. *Topics in organic polarography*, P. Zuman, Plenum Press (1970).

83. Oxidation and reduction of aromatic hydrocarbon molecules at electrodes, M. E. Peover, In: *Reactions of Molecules at Electrodes*, N. S. Hush, ed., pp. 259–281, Wiley-Interscience, London (1971).

84. Reduction potentials and orbital energies of azaheteromolecules, J. Tabner and J. R. Yandle, In: *Reactions of Molecules at Electrodes*, N. S. Hush, ed., pp. 283–303, Wiley-Interscience, London (1971).

85. The electrode reactions of organic molecules, M. Fleischmann and D. Pletcher, In: *Reactions of Molecules at Electrodes*, N. S. Hush, Ed., pp. 347–402, Wiley-Interscience, London (1971).

1.1.1.6. Residual or Charging Current, Migration Current, Diffusion Current, and Instantaneous Current

86. Polarographic studies with the dropping mercury cathode. Part II. Influence of temperature, V. Nejedly, *Collect. Czech. Chem. Commun.* **1**, 319–333 (1929).

87. Polarographic studies with the dropping mercury cathode. Part XXV. Increased sensitivity of micro-analytical estimations by a compensation of current, D. Ilkovic and G. Semerano, *Collect. Czech. Chem. Commun.* **4**, 176–180 (1932).

88. Limiting currents in electrolysis with the dropping mercury cathode, J. Heyrovsky, *Arhiv Hem. Farm* **8**, 11–16 (1934).

89. Polarographic studies with the dropping mercury cathode. Part XLIV. The dependence of limiting currents on the diffusion constant, on the rate of dropping, and on the size of drops, D. Ilkovic, *Collect. Czech. Chem. Commun.* **6**, 498–513 (1934).

90. Polarographic studies with the dropping mercury electrode. Part IV. The measurement of the polarization capacity, D. Ilkovic, *Collect. Czech. Chem. Commun.* **8**, 170–177 (1936).

91. Theory of limiting (diffusion) currents. I. Polarographic limiting current, D. MacGillavry and E. K. Rideal, *Rec. Trav. Chim.* **56**, 1013–1021 (1937).

92. Theory of limiting currents. II. Limiting currents of cells without and with an indifferent electrolyte, D. MacGillavry, *Rec. Trav. Chim.* **56**, 1039–1046 (1937).

93. Theory of limiting currents. III. General solutions with excess of one indifferent electrolyte, D. MacGillavry, *Rec. Trav. Chim.* **57**, 33–40 (1938).

94. The value of diffusion currents observed in electrolysis by the dropping mercury electrode. Polarographic study, D. Ilkovic, *J. Chim. Phys.* **35**, 129–135 (1938).

95. Fundamental studies with the dropping mercury electrode. II. The migration current, J. J. Lingane and I. M. Kolthoff, *J. Am. Chem. Soc.* **61**, 1045–1051 (1939).

96. A study of diffusion processes by electrolysis with micro-electrodes, H. A. Laithinen and I. M. Kolthoff, *J. Am. Chem. Soc.* **61**, 3344–3349 (1939).

97. Capacity phenomena displayed at mercury capillary electrodes, J. Heyrosky, F. Sorm, and J. Forejt, *Collect. Czech. Chem. Commun.* **12**, 11–38 (1947).

98. Polarographic discussion panel on "The validity of the Ilkovic equation in polarographic analysis", F. L. Steghart, Chemistry and Industry, p. 157 (1948).

99. Polarographic current time curves, H. A. McKenzie, *J. Am. Chem. Soc.* **70**, 3147–3148 (1948).

100. Polarographic limiting currents, J. K. Taylor, R. E. Smith, and I. L. Cooter, *J. Res. Natl. Bur. Standards*, **42**, 387–395 (1949).

101. Polarography with slowly forming mercury drops, G. S. Smith, *Nature* **163**, 290–291 (1949).

102. A new polarographic diffusion current equation, J. J. Lingane and B. A. Loveridge, *J. Am. Chem. Soc.* **72**, 438–441 (1950).

103. Zur theorie der polarographischem kurve, H. Strehlow and M. V. Stackelberg, *Z. Elektrochem.* **54**, 51–62 (1950).

104. Studies in the theory of the polarographic diffusion current, I. The effects of gelatin on the diffusion current constants of cadmium and bismuth, L. Meites and T. Meites, *J. Am. Chem. Soc.* **72**, 3686–3691 (1950).

105. Studies in the theory of the polarographic diffusion current. II. The instantaneous diffusion current and the Strehlow–Von stackelberg equation, L. Meites and T. Meites, *J. Am. Chem. Soc.* **72**, 4843–4844 (1950).

106. Mercury drop control: Application to derivative and differential polarography, L. Airey and A. A. Smales, *Analyst.* **75**, 287–304 (1950).

107. Studies in the theory of the polarographic diffusion current, IV. Diffusion current constants of some ions in the absence of gelatin, L. Meites, *J. Am. Chem. Soc.* **73**, 1581–1583 (1951).

108. Studies on the theory of the polarographic diffusion current. VII. The effect of drop weight on the relationship between the diffusion current constant of lead and the drop time, L. Meites, *J. Am. Chem. Soc.* **73**, 3724–3727 (1951).

109. Zur theorie der polarographischen kurve II. Bestimmung der diffusion koeffizienten von ionen in elektrolytlosungen, H. Strehlow, O. Madrich, and M. V. Stackelberg, *Z. Electrochem.* **55**, 244–250 (1951).

110. Studium der an der stromenden quecksilber elektrode auftretenden stromidskontinuitat, P. Valenta, *Collect. Czech. Chem. Commun.* **16**, 239–251 (1951).

111. Zur theorie der polarographischem kurve. IV. Untersuchungen uber die gultigkeit der korrigierten Ilkovic–Gleichung, W. Hans and W. Jensch, *Z. Elektrochem* **56**, 648–662 (1952).

112. Developments of constants in polarography: A correction factor for the Ilkovic equation, O. H. Muller, Natl. Bur. Standards Circ. No. 524, 289–303 (1953).

113. Zur theorie der polarographischen kurve VII. Diffusionsbedingte polarographische strom-starke, W. Hans, W. Henne, and E. Meurer, *Z. Elektrochem.* **58**, 836–849 (1954).

114. Eine neue form der quecksilbertropfelektrode, I. Smoler, *Collect. Czech. Chem. Commun.* **19**, 238–240 (1954).

115. Instantaneous polarographic current. III. Accurate measurements of the residual currents, A. Bresle, *Acta Chem. Scand.* **10**, 947–950 (1956).

116. Current time curves on single drops and the polarographic diffusion current. I. Smoler and J. Kuta, *Z. Physik. Chem. (Leipzig) Sonderheft*, 58–65 (1958).

117. Principles of polarography, M. V. Stackelberg, *Z. Anal. Chem.* **173**, 89–90 (1960).

118. The effect of the "back pressure" on the diffusion current, G. C. Barker and A. W. Gardner, In: *Advances in Polarography*, Vol. I, I. S. Longmuir, ed., pp. 330–339, Pergamon Press, London (1960).

119. The diffusion equation in d.c. polarography. Part I. Current-time curves without depletion effect, pp. 408–424 (1960); Part II. The mass-time relationship of dropping mercury electrodes, pp. 425–436 (1960), J. M. Los and D. W. Murray, In: *Advances in Polarography*, Vol. 2, I. S. Longmuir, ed., Pergamon Press.

120. Grenzstrom ander Hg-trop felektrode bei niedrigen konzentrationen des fremdelektrolyts, Z. Zambora, A. Fulinski, and M. Bierowski, *Z. Elektrochem.* **65**, 887–891 (1961).

121. Die gleichung fur polarographische diffusionsstrome und die grenzen ihrer gultigkeit, J. Koutecky and M. V. Stackelberg, In: *Progress in Polarography*, Vol. I, P. Zuman and I. M. Kolthoff, eds., pp. 21–42, Interscience Publishers, (1962).

122. The instantaneous currents (i-t curves) on single drops, J. Kuta and I. Smoler, In: *Progress in Polarography*, Vol. I, P. Zuman and I. M. Kolthoff, eds., pp. 43–63, Interscience Pulbishers (1962).

123. Effect of the position of the capillary on the transfer of concentration polarization in polarography, I. Smoler, *J. Electroanal Chem.* **6**, 465–479 (1963).

124. Resistance compensation in polarography. Application to high resistance nonaqueous systems and to high current density aqueous systems, W. B. Schaap and P. S. McKinney, *Anal. Chem.* **36**, 1251–1258 (1964).

125. The limiting current, L. Meites, In: *Polarographic Techniques* Chap. 3, pp. 150–176, Interscience Publishers, New York (1965).

1.1.1.7. Kinetic Currents and Catalytic Currents

126. Uber durch wasserstoffatome katalysierte depolarisations vorgange ander tropfenden queck-silbrelektrode, K. Wiesner, *Z. Elektrochem.* **49**, 164–166 (1943).

127. The polarography of uranium I. Reduction in moderately acid solutions. Polarographic determination of uranium, W. E. Harris and I. M. Kolthoff, *J. Am. Chem. Soc.* **67**, 1484–1490 (1945).

128. Polarographic determination of the rate of the reaction between ferrohem and hydrogen peroxide, R. Brdicka and K. Wiesner, *Collect. Czech. Chem. Commun.* **12**, 39–63 (1947).

129. Rate of recombination of ions derived from polarographic limiting currents due to the reduction of acids, R. Brdicka and K. Wiesner, *Collect. Czech. Chem. Commun.* **12**, 138–149 (1947).

130. Kinetic (catalytic) (polarographic) currents for systems containing hydrogen peroxide, I. M. Kolthoff and E. P. Parry, *Proc. 1st Internat. Polarography Cong. Prague*, Vol. I, 145–154 (1951).

131. Catalysis of the polarographic reduction of hydrogen peroxide by compounds of iron in dilute sulphuric acid solutions, Z. Pospisil, *Collect. Czech. Chem. Commun.* **18**, 337–349 (1953).

132. Uber dei kinetik der elektroden vorgange VII a. Theorie einiger katalytischer strome in der polarographie, J. Koutecky, *Collect. Czech. Chem. Commun.* **18**, 311–325 (1953).

133. Evaluation of the rate constant of the decomposition of hydrogen peroxide by catalase from the polarographic limiting current of oxygen, J. Koutecky, R. Brdicka, and V. Hanus, *Collect. Czech. Chem. Commun.* **18**, 611–628 (1953).

134. Anwendungsmoglichkeiten und begrenzungen der polarographischen methods zur verfolgung schneller chemischer reaktionen in losunger, J. Koryta, *Z. Elektrochem.* **64**, 23–29 (1960).

135. General theoretical treatment of the polarographic kinetic currents, R. Brdicka, V. Hanus, and J. Koutecky, In: *Progress in Polarography*, Vol. I, P. Zuman and I. M. Kolthoff, eds., pp. 145–199, Interscience Publishers (1962).

136. The theory of catalytic hydrogen waves in organic polarography, S. G. Mairanovskii, *J. Electroanal Chem.* **6**, 77–118 (1963).

137. Kinetic currents, J. Heyrovsky and J. Kuta, In: *Principles of Polarography*, pp. 380–393, Academic Press, New York (1966).

138. Currents limited by the rate of chemical reaction Chapter II, pp. 7–55, "Electrode processes with antecedent protonation, Chapter IV, pp. 115–153, S. G. Mairanovkii, In: *Catalytic and Kinetic Currents in Polarography*, Plenum Press, New York (1968).

139. Catalytic Processes, Chapter II, Sec. D.1, pp. 17–25, Catalytic hydrogen evolution at the dropping mercury electrode caused by organic catalysts, Chapter IX, pp. 241–285 S. G. Mairanovskii, In: *Catalytic and Kinetic Currents in Polarography*, Plenum Press, New York (1968).

1.1.1.8. Polarographic Maxima

140. Maxima of the polarization curves of mercury cathodes, A. N. Frumkin and B. Bruns, *Acta Phys. Chim. URSS* **1**, 232–246 (1934).

141. Maxima on current-voltage curves, B. Bruns, A. N. Frumkin, S. Zofa, L. Vanyukova, and S. Zolotarevskaya, *Acta Phys. Chim. URSS*, **9**, 359–372 (1938).

142. Polarographic maxima, In: *Polarography*, I. M. Kolthoff and J. J. Lingane, Vol. I, Chapter X, pp. 156–188, Interscience Publishers, New York (1952).

143. Polarographic maxima, In: *Polarographic Analysis* (Russ), T. A. Kruykova, S. I. Sinyakova, and J. V. Arefeva, pp. 94–102, pp. 573–582, State Chemical Technical Publication, Moscow (1959).

144. Polarographic maxima, V. G. Levich, In: *Physico Chemical Hydrodynamica*, Prentice Hall Inc. Englwood Cliffs, 561–589 (1962).

145. Streaming maxima in polarography, H. H. Bauer, In: *Electroanalytical Chemistry*, A. J. Bard, ed., Vol. 8, pp. 169–279, Marcel Dekker Inc., New York (1975).

146. Polarographic Maxima of the Third Kind I., A. N. Frumkin, N. Fedorovich, B. B. Damsakin, and E. V. Stenina, *Sov. Electrochem.* **6**, 1–5 (1970).

147. Polarographic maxima of the third kind, A. N. Frumkin, N. V. Fedorovich, B. B. Damaskin, E. V. Stenina, and V. S. Krylov, *J. Electroanal, Chem. International Electrochem.* **50**, 103–111 (1974).

1.1.1.9. Analysis

148. *Chemische Analysen mit dem Polarographen*, H. Horn, J. Springel, Berlin (1937).

149. Simplification of the polarographic method by introduction of "step quotients", H. E. Forche, *Mikrochemie* **25**, 217–224 (1938).

150. The determination of dissolved oxygen by means of the dropping mercury electrode, with applications in biology, H. G. Petering and F. Daniels, *J. Am. Chem. Soc.* **60**, 2796–2802 (1938).

151. Polarographic determination of nickel and cobalt. Simultaneous determination in presence of iron, copper, chromium and manganese and determination of small amounts of nickel in cobalt compounds, J. J. Lingane and H. Kerlinger, *Ind. Eng. Chem. (Anal. Ed)* **13**, 77–80 (1941).

152. The polarographic method of analysis. V. Applications, O. H. Muller, *J. Chem. Ed.* **18**, 320–329 (1941).

153. Systematic polarographic metal analysis. Characteristics of arsenic, antimony, bismuth, tin, lead, cadmium, zinc and copper in various electrolytes, J. J. Lingane, *Ind. Eng. Chem. (Anal. Ed.)* **15**, 583–590 (1943).

154. Examination of absolute and comparative methods of polarographic analysis, K. Taylor, (Natl. Bur. Standards, Washington, D.C.) *Anal. Chem.* **19**, 368–372 (1947).

155. Device for estimating the height of polarographic waves, J. K. Taylor, (Natl. Bur. Std., Washington D.C) *Anal. Chem.* **19**, 478–481 (1947).

156. Precise measurements of polarographic half wave potential, L. Meites, *J. Am. Chem. Soc.* **72**, 2293–2294 (1950).

157. A new polarographic diffusion current equation, J. J. Lingane and B. A. Loveridge, *J. Am. Chem. Soc.* **72**, 438–441 (1950).

158. Theory of the polarographic curve, H. Strehlow and M. V. Stackelberg, *Z. Elektrochem.* **54**, 51–62 (1950).

159. Method of standardization for polarographic determination of lead in zinc and zinc-bearing materials, P. Rutherford and L. A. Cha, *Anal. Chem.* **23**, 1714–1715 (1951).

160. Common operations in polarographic analysis, I. M. Kolthoff and J. J. Lingane, In: *Polarography* pp. 372–398, Vol. I, Interscience Publishers, New York (1952).

161. High sensitivity recording polarography, M. T. Kelley and H. H. Miller, *Anal. Chem.* **24**, 1895–1899 (1952).

162. Minimum error polarographic analysis of binary mixtures, A. Frisque, V. W. Meloche, and I. Shain, *Anal. Chem.* **26**, 471–473 (1954).

163. Determination of traces of nickel and zinc in copper and its salts, L. Meites, *Anal. Chem.* **27**, 977–979 (1955).

164. Analysis of solutions containing two reducible substances by polarography and coulometry at controlled potential, L. Meites, *Anal. Chem.* **27**, 1114–1116 (1955).

165. "Standard addition" method of polarographic analysis—alternate interpretation, W. H. Reinmuth, *Anal. Chem.* **28**, 1356–1357 (1956).

166. A polarographic method for the study of glycol fission by periodic acid, P. Zuman and J. Krupicka, *Collect. Czech. Chem. Commun.* **23**, 598–607 (1958).

167. Polarographische bestimmung des pyridoxals und pyridoxal 5-phosphats, O. Manousek and P. Zuman, *J. Electroanal. Chem.* **1**, 324–330 (1960).

168. Error analysis for polarographic analytical methods, V. Pliska, *Z. Anal. Chem.* **191**, 241–248 (1962).

169. Direct polarographic determination of pyridoxal in the presence of pyridoxal-5-phosphate, O. Manousek and P. Zuman, *Collect. Czech. Chem. Commun.* **27**, 486–487 (1962).

170. A new method for the study of intermetallic compound formation in mixed amalgams, H. K. Ficker and L. Meites, *Anal. Chim. Acta* **26**, 172–179 (1962).

171. Chromato polarography, W. Kemula, In: *Progress in Polarography*, Vol. II, pp. 397–409, P. Zuman and I. M. Kolthoff, eds., Interscience, New York (1962).

172. Important factors in classical polarography, P. Zuman, In: *Progress in Polarography* P. Zuman and I. M. Kolthoff, eds., Chap. XXIX, pp. 583–600, Vol. II, Interscience Publishers, New York (1962).

173. Treatment of the sample in inorganic polarography G. W. C. Milner, In: *Progress in*

Polarography, P. Zuman and I. M. Kolthoff, eds., Chap. XXX, pp. 601–616, Vol. II, Interscience Publishers, New York (1962).

174. Organic reagents in inorganic polarographic analysis, M. Shinagawa, In: *Progress in Polarography*, P. Zuman and I. M. Kolthoff, eds., Chap. XXXI, pp. 617–24, Vol. II, Interscience Publishers, New York (1962).

175. Polarography in organic analysis, P. J. Elving, In: *Progress in Polarography*, P. Zuman and I. M. Kolthoff, eds., Chap. XXXII, pp. 625–648, Vol. II, Interscience Publishers, New York (1962).

176. Polarography in metallurgy, M. Spalenka, In: *Progress in Polarography*, P. Zuman and I. M. Kolthoff, eds., Chap. XXXIV, pp. 661–665, Vol. II, Interscience Publishers, New York (1962).

177. Polarography in medicine and biochemistry, M. Brezina, In: *Progress in Polarography*, P. Zuman and I. M. Kolthoff, eds., Chap. XXV, pp. 667–686, Vol. II, Interscience Publishers, New York (1962).

178. Polarographic analysis in pharmacy, P. Zuman and M. Brezina, In: *Progress in Polarography*, P. Zuman and I. M. Kolthoff, eds., Chap. XXXVI, pp. 687–701, Vol. II, Interscience Publishers, New York (1962).

179. Quantitative polarographic analysis, L. Meites, *Polarographic Techniques*, pp. 334–410 Interscience Publishers, New York (1965).

1.1.1.10. Polarographic Instrumentation

Manual D.C. Polarographic Instruments

180. Polarographic theory, instrumentation and methodology, J. J. Lingane, *Anal. Chem.* **21**, 45–60 (1949).

181. Polarographic instrumentation, I. M. Kolthoff and J. J. Lingane, *Polarography*, pp. 297–349, Vol. I, Interscience Publishers, New York (1952).

182. Polarographs and other apparatus, L. Meites, *Polarographic Techniques*, pp. 7–94, Interscience Publishers (1965).

Recording Instruments

183. Polarographic analysis with the dropping mercury cathode. XXV. Increased sensitivity of microanalytical estimation by a compensation of current, D. Ilkovic and G. Semerano, *Collect. Czech. Chem. Commun.* **4**, 176–180 (1932).

184. Die polarographische analyse des messings die grondlagen serienma biger schnellanalysen von legier ungen gusge fuhrt mittels quecksilbertropi kathoden, H. Hohn, *Z. Electrochem.* **43**, 127–139 (1937).

185. Use of the polarograph in steel plant laboratory. II. Determination of vanadium, chromium, and molybdenum in steel after removal of iron with alkali hydroxide, G. Thanheiser and J. Willems, *Mitt. Kaiser Wilhelm Inst. Eisenforsch. Dusseldorf.* **21**, 65–78 (1939).

186. Use of a condenser to reduce galvanometer oscillations in polarographic measurements with particular application to compensation method of measuring small diffusion current, J. J. Lingane and H. Kerlinger, *Ind. Eng. Chem.* (Analytical Edition) **12**, 750–753 (1940).

187. Fundamental studies with the dropping mercury electrode. III. Influence of capillary characteristics on the diffusion current and residual current, J. J. Lingane and B. A. Loveridge, *J. Am. Chem. Soc.* **66**, 1425–1431 (1944).

188. Apparatus for testing liquids with a dropping Hg electrode, E. D. Coleman, U.S. Patent 2,343,885 (1944).

189. Dropping Hg electrode apparatus for analyzing liquids, O. Kanner and E. D. Coleman, U.S. Patent 2,361,295 (1944).

190. Application of the Ilkovic equation to quantitative polarography, F. Buckley and J. K. Taylor, *J. Res. Natl. Bur. Stand.* **34**, 97–114 (1945).
191. Application of multicapillary tube mercury electrode in polarography, S. Stankoviansky, *Chem. Zvesti* **2**, 133–142 (1948).
192. Dropping mercury electrode for polarography with enforced removal of the drop, V. A. Tsimmergakl, *Zavod. Lab.* **15**, 1370 (1949).
193. Mercury drop control: application to derivative and differential polarography, L. Airey and A. S. Smales, *Analyst* **75**, 287–304 (1950).
194. Polarographic instrumentation, I. M. Kolthoff and J. J. Lingane, *Polarography*, pp. 297–349, Vol. I, Interscience Publishers, New York (1952).
195. Controlled-potential and derivative polarograph, M. T. Kelley, H. C. Jones, and D. J. Fisher, *Anal. Chem.* **31**, 1475–1485 (1959).
196. Electronic controlled-potential coulometric titrator, M. T. Kelley, H. C. Jones, and D. J. R. Fisher, *Anal. Chem.* **31**, 488–494 (1959).
197. Controlled-potential polarographic polarizing unit with electronic scan and linear residual current compensation, M. T. Kelley, D. J. Fisher, and H. C. Jones, *Anal. Chem.* **32**, 1262–1265 (1960).
198. The diffusion equation in d.c. polarography, In: J. M. Los and D. W. Murray, *Advances in Polarography* Vol. 2, I. S. Longmuir, ed., pp. 408–424, Pergamon Press, Oxford (1960).
199. A device for the synchronization of two dropping mercury electrodes, H. H. Lockwood, H. M. Stationery Office, London, AERER, 3521, 1960.
200. Controlled potential and derivative polarography, In: M. T. Kelley, D. J. Fisher, W. D. Cooke, and H. C. Jones, *Advances in Polarography*, Vol. I, pp. 158–182, Pergamon Press (1960).
201. Incremental approach to derivative polarography, C. Auerbach, H. L. Finston, G. Kissel, and J. Glickstein, *Anal. Chem.* **33**, 1480–1484 (1961).
202. Automatic polarograph for use with solutions of high resistance, P. Arthur, P. A. Lewis, N. A. Lloyd, and R. K. Vanderkam, *Anal. Chem.* **33**, 488–491 (1961).
203. An IR compensator for nonaqueous polarography and amperometric titrations, P. Arthur and R. H. Vanderkam, *Anal. Chem.* **33**, 765–767 (1961).
204. Electroanalytical Controlled—Potential Instrumentation, G. L. Booman and W. B. Holbrook, *Anal. Chem.* **35**, 1793–1809 (1963).
205. Polarographische untersuchungen bei regulierung der tropfzeit durch abschlagen des tropfens. I. Die tropfzeitabhangigkeit polarographischer strome, D. Wolf, *J. Electroanal Chem.* **5**, 186–194 (1963).
206. A multipurpose operational amplifier instrument for electroanalytical studies, W. L. Underkofler and I. Shain, *Anal. Chem.* **35**, 1778–1783 (1963).
207. A simple electronic polarographic voltage compensator, R. Annino and K. J. Hagler, *Anal. Chem.* **35**, 1555–1556 (1963).
208. A simple electronic-scan controlled potential polarograph, R. A. Durst, J. W. Ross, and D. N. Hume, *J. Electroanal Chem.* **7**, 245–248 (1964).
209. Polarographs and other apparatus, L. Meites, In: *Polarographic Techniques*, pp. 7–94, Interscience Publishers (1965).
210. Recent developments in d.c. polarography, D. J. Fisher, W. L. Belew, and M. T. Kelley, *Polarography* 1964 Vol. I, G. J. Mills, ed., pp. 89–134 Macmillan (1966).
211. *Principles of Polarography*, J. Heyrovsky and J. Kuta, Chap. VI, pp. 99–108, Academic Press, New York (1966).

1.2. Differential and Derivative Polarography

These two polarographic techniques have been proposed primarily to improve the sensitivity and resolution of d-c polarography. Derivative polarography has an edge over differential polarography since it is sufficient to employ one DME only in the case of the former while two DMEs may be needed for the latter. Further, with the advancement in the field of electronics, the electronic circuitry employed for obtaining derivative polarograms has improved considerably[22-26] (compared to the resistor capacitor differentiating circuits employed by earlier researchers). It has been reported that the resolution of the first derivative polarographic technique is comparable to that of sine wave a-c and square-wave polarography. In recent years the resolution power has been improved further[30-33] by the use of computer programmed fast sweep interrupted derivative polarographic techniques. The second derivative polarogram has been found to improve the sensitivity of the technique for the analysis of mixtures. Perone *et al.* have shown that by employing "real-time computer interaction" methods it is possible to improve the electroanalytical measurement techniques.[30,35]

1.2.1. Theory of Differential and Derivative Polarography

1. Polarographic and differential polarimetry, G. Semerano and L. Riccoboni, *Gazz. Chim. Ital.* **72**, 297–304 (1942).
2. Differential method of polarographic analysis, E. A. Kanevskii, *J. Appl. Chem. (U.S.S.R)* **17**, 514–519 (1944).
3. The fundamental laws of polarography, J. Heyrovsky, *Analyst* **72**, 229–234 (1947).
4. Modern trends of polarographic analysis, J. Heyrovsky, *Anal. Chim. Acta* **2**, 533–541 (1948).
5. The significance of derivative curves in polarography, J. Heyrovsky, *Chem. Listy* **43**, 149–154 (1949).
6. Differential polarography with unique dropping electrode, P. Leveque and F. Roth, *J. Chim. Phys.* **46**, 480–484 (1949).
7. Polarographic determination of halogen derivatives, M. B. Neimen, A. V. Ryabov, and E. M. Sheyanova, *Dokl. Akad. Nauk. SSSR* **68**, 1065–1068 (1949).
8. Dropping-mercury electrode with enforced separation of the drop, E. M. Skobets and N. S. Kavetski, *Zavod Lab.* **15**, 1299–1305 (1949).
9. Note on differential polarography with a single dropping electrode, J. Vogel and J. Ritha, *J. Chim. Phys.* **47**, 5–7 (1950).
10. Vibrating dropping mercury electrode for polarographic analysis of agitated solutions, D. A. Berman, P. R. Saunders, and R. J. Winzier, *Anal. Chem.* **23**, 1040–1041 (1951).
11. Derivative polarography. I. Characteristics of the Leveque-Roth circuit, J. J. Lingane and R. Williams, *J. Am. Chem. Soc.* **74**, 790–796 (1952).
12. Unique polarographic damping circuit for selective elimination of current fluctuations due to the dropping mercury electrode, M. T. Kelley and D. J. Fisher, *Anal. Chem.* **28**, 1130–1132 (1956).
13. Instrumental methods of derivative polarography, M. T. Kelley and D. J. Fisher, *Anal. Chem.* **30**, 929–932 (1958).
14. Controlled potential and derivative polarograph, M. T. Kelley, H. C. Jones, and D. J. Fisher, *Anal. Chem.* **31**, 1475–1485 (1959).
15. Deformations in polarographic derivative curves obtained with R-C wiring, T. Jackel, *Z. Anal. Chem.* **173**, 59–65 (1960).
16. Controlled potential and derivative polarography, M. T. Kelley, D. J. Fisher, W. D. Cooke, and H. C. Jones, In: *Advances in Polarography* Vol. I, I. S. Longmuir, ed., pp. 158–182, Pergamon Press, Oxford (1960).

17. Rapid polarography, S. Wolf, *Agnew Chem.* **72**, 449–454 (1960).
18. A differential cathode ray polarograph, H. M. Davis and J. E. Seaborn, *Advances in Polarography*, Vol. I, I. S. Longmuir, ed., pp. 239–250 Pergamon Press (1960).
19. The performance of the differential cathode ray polarograph, In: H. M. Davis and H. I. Shalgosky, *Advances in Polarography*, Vol. 2, I. S. Longmuir, ed., Pergamon Press (1960).
20. Recent developments in d.c. polarography, D. J. Fisher, W. L. Belew, and M. T. Kelley, In: *Polarography 1964*, Vol. I, G. J. Hills, ed., pp. 89–134, Macmillan (1966).
21. Studies in subtractive and continuous polarography, In: K. G. Powell and G. F. Reynolds, *Polarography 1964*, Vol. I., G. J. Hills, ed., pp. 249–260, Macmillan (1966).
22. Application of derivative techniques to stationary electrode polarography, S. P. Perone and T. R. Mueller, *Anal. Chem.* **37**, 2–9 (1965).
23. Application of derivative techniques to anodic stripping voltammetry, S. P. Perone and J. R. Birk, *Anal. Chem.* **37**, 9–12 (1965).
24. Theory of derivative voltammetry with irreversible systems, S. P. Perone and C. V. Evins, *Anal. Chem.* **37**, 1061–1063 (1965).
25. Derivative voltammetry with irreversible systems, Application to spherical electrodes, S. P. Perone and C. V. Evins, *Anal. Chem.* **37**, 1643–1646 (1965).
26. Application of derivative read out techniques to stationary electrode polarography with kinetic systems, C. V. Evins and S. P. Perone, *Anal. Chem.* **39(3)**, 309–315 (1967).
27. A solid-state controlled potential dc polarograph with cyclic scanning and calibrated first and second derivative scales, H. C. Jones, W. L. Belew, R. W. Stelzner, T. R. Mueller, and D. J. Fisher, *Anal. Chem.* **41**, 772–779 (1969).
28. Apparatus for precision control of drop time of dropping mercury electrode in polarography, W. L. Belew, D. J. Fisher, H. C. Jones, and M. T. Kelley, *Anal. Chem.* **41**, 779–786 (1969).
29. Differential and derivative polarography for continuous analysis in a flow system: application to alloy dissolution, B. Cahan and R. Haynes, *J. Electroanal Chem.* **22**, 339–345 (1969).
30. Real-time computer optimization of stationary electrode polarographic measurements, S. P. Perone, D. O. Jones and W. F. Gufknecht, *Anal. Chem.* **41**, 1154–1162 (1969).
31. Interactive electronic analytical instrumentation based on computerized experimental design, D. O. Jones and S. P. Perone, *Anal. Chem.* **42**, 1151–1157 (1970).
32. Numerical deconvolution of overlapping stationary electrode polarographic curves with an on-line digital computer, W. F. Gutknecht and S. P. Perone, *Anal. Chem.* **42**, 906–917 (1970).
33. Computerized learning machine applied to qualitative analysis of mixtures by stationary electrode polarography, L. B. Sybrandt and S. P. Perone, *Anal. Chem.* **43**, 382–388 (1971).
34. On-line interactive data processing. II. Processing voltammetric electrochemical data, S. P. Perone, J. W. Farzer, and A. Kray, *Anal. Chem.* **43**, 1485–1490 (1971).
35. Enhancement of electroanalytical measurement techniques by real-time computer interaction, S. P. Perone, In: *Electrochemistry*, J. S. Mattson, H. B. Mark, and H. C. MacDonald, eds., Chap. 13, pp. 423–447, Marcel Dekker, New York (1972).

1.3. Oscillographic Polarography

Oscillographic polarography was developed in 1943 by Heyrovsky. The theory for oscillographic polarography was first developed by Randles[2,17] and Sevick.[19] Delahay *et al.*[4,8] extended the theory for irreversible waves, and presented improved methods of measurements and causes of errors. The Randles–Sevick theory has been further improved by Matsuda and Ayabe,[10] Saveant and Vianello,[26] and Gokshtein *et al.*[12,13] The oscillographic polarographs have been described by Reynolds and Davis,[23] Snowden and Page,[21] Airey,[18] Loveland and Elving,[22] Favero and Vianello,[25] Powell and

Reynolds,[44] and Kalvoda.[42] Two types of oscillographic polarographs have been described in literature. One employs alternating current sweeps described by Heyrovsky *et al.*,[3,4,42] and the second uses single sweep (or triangular or saw tooth voltage pulse) or multisweep (of triangular voltage pulses).[16,17,19,22,25]

1.3.1. Theory of Oscillographic Polarography

1. The cathode ray oscillograph applied to the dropping mercury electrode, L. A. Matheson and N. Nichols, *Trans. Electrochem. Soc.* **73**, 193–206 (1938).
2. A cathode ray polarography. Part II—The current voltage curves, J. E. B. Randles, *Trans. Faraday Soc.* **44**, 327–338 (1938).
3. Oszillographische polarographie, J. Heyrovsky and J. Forejt, *Z. Phys. Chem.* **193**, 77–96 (1943).
4. Reversibility and irreversibility of electrode reactions in oscillographic polarography, P. Delahay, *J. Phys. Colloid Chem.* **54**, 630–639 (1950).
5. An oscillographic polarograph for high rates of potential variation, P. Delahay, *J. Phys. Colloid Chem.* **54**, 402–411 (1950).
6. Oscillographic polarography improved method of measurement and causes of errors, P. Delahay and G. L. Stiehl, *J. Phys. and Colloid Chem.* **55**, 570–585 (1951).
7. Oscillographic polarography. Phenomena occurring during the quiescent period of the voltage wave, P. Delahay and G. Perkins, *J. Phys. and Colloid Chem.* **55**, 586–591 (1951).
8. Theory of irreversible waves in oscillographic polarography, P. Delahay, *J. Am. Chem. Soc.* **75**, 1190–1196 (1953).
9. Voltammetry and polarography with continuously changing potential, P. Delahay, In: *New Instrumental Methods in Electrochemistry*, Interscience Publishers, New York (1954).
10. Zur theorie der Randles–Sevikschen kathodenstrahl-Polarographie, M. Matsuda and Y. Ayabe, *Z. Elektrochem.* **59**, 494–503 (1955).
11. Theorie der polarisation der quecksilberelektrode durch wechselstrom, K. Micka, *Z. Phys. Chem. (Leipzig)* **206**, 345–368 (1956).
12. Multistage electrochemical reactions in oscillographic polarography, Y. P. Gokhshtein and A. Y. Gokhshtein, In: *Advances in Polarography*, I. S. Longmuir, ed., Vol. 2, pp. 465–481 Pergamon Press (1960).
13. Oscillographic polarography—Equation for the descending branch of the polarographic wave and its approximation, Ya. P. Gokhshtein and A. Ya. Gokhshtein, *Zh. Fiz. Khim.* **34**, 1654–1657 (1960).
14. Oscillographic polarography, J. Heyrovsky, In: *Advances in Polarography*, I. S. Longmuir, ed., Vol. I, pp. 1–25, Pergamon Press, Oxford (1960).
15. *Oscillographische Polarographie mit Wechselstrom*, J. Heyrovsky and R. Kalvoda, Akademie Verlag, Berlin (1960).
16. Single sweep method, J. Vogel, In: *Progress in Polarography*, P. Zuman and I. M. Kolthoff, eds., Vol. II, pp. 429–448, Interscience Publishers, New York (1962).

1.3.2. Instrumentation

17. The application of the cathode ray oscillograph to polarography: underlying principles, J. E. B. Randles, *Analyst* **72**, 301–304 (1947).
18. The application of the cathode ray oscillograph to polarography, A. Airey, *Analyst* **72**, 304–307 (1947).
19. Oscillographic polarography with periodical triangular voltage, A. Sevick, *Collect. Czech. Chem. Commun.* **13**, 349–377 (1948).

20. An experimental study of the characteristic features of oscillographic polarography, P. Delahay, *J. Phys. Colloid Chem.* **53**, 1279–1301 (1949).

21. A cathode ray polarograph, F. C. Snowden and H. T. Page, *Anal. Chem.* **22**, 969–981 (1950).

22. Application of the cathode ray oscilloscope to polarographic Phenomena I. Differential capacity of electrical double layer, I. W. Loveland and P. J. Elving, *J. Phys. Chem.* **56**, 250–255 (1952).

23. An improved Randles-type cathode-ray polarograph, G. F. Reynolds and H. M. Davis, *Analyst* **78**, 314–319 (1953).

24. Quantitative cathode-ray polarography, K. Cruse and W. Herberles, *J. Phys. Chem.* **57**, 579–590 (1953).

25. A new single sweep oscillographic polarograph, P. Favero and E. Vianello, *Ricerca Sci.* **25**, 1415 (1955).

26. Recharches sur les courants catalytiques en polarographie-Oscillographique a balayage lineaire de tension etude theorique, J. M. Saveant and E. Vianello, In: *Advances in Polarography*, I. S. Longmuir, ed., pp. 367–374, Pergamon Press, London (1960).

27. Multisweep cathode ray polarograph with two streaming mercury electrodes in a differential circuit, E. Gorlich, J. Srzednicki, and Z. Kowalski, *Zh. Fiz. Khim.* **36**, 449–454 (1962).

28. Polarographie in Salzschmelzen-II, oscillographische wechselstrompolarographie in kalium-chlorid-lithium chlorid eutektikum, E. Schmidt, *Electrochim. Acta* **8**, 23–35 (1963).

1.3.3. Application

29. Polarographic reduction of aliphatic aldehydes II. Oscillographic investigations of formaldehyde, R. Bieber and S. Trümpler, *Helv. Chim. Acta* **30**, 971–990 (1947).

30. The polarographic behaviour of some elements in concentrated calcium chloride solution. I. General introduction—Certain problems arising from the use of 5M calcium chloride, G. F. Reynolds, H. I. Shalagosky, and T. J. Webber, *Anal. Chim. Acta* **8**, 558–563 (1953).

31. Anwendung der oszillographischem polarographie in de quantitativen analyse. IV. Ein gerat zur messung der tiefe de einschnitte auf den kurven, R. Kalvoda, *Collect. Czech. Chem. Commun.* **20**, 1503–1507 (1955).

32. Stabiliserte oszillogramme mit der tropfelektrode, R. Kalvoda and J. Macku, *Collect. Czech. Chem. Commun.* **20**, 257–260 (1955).

33. A comparative study of three recently developed polarographs, D. J. Ferret, G. W. C. Milner, H. I. Shalagosky, and L. J. Slee, *Analyst* **81**, 506–512 (1956).

34. Gerat fur schnelle oszillographische quantitative analyse, P. Valenta and J. Vogel, *Collect. Czech. Chem. Commun.* **21**, 502–508 (1956).

35. Die anwending der oszillographischen polarographie in dei quantitativen analyse. VII. Mikroanalytische bestimmung einiger metalle, R. Kalvoda, *Collect. Czech. Chem. Commun.* **22**, 1390–1399 (1957).

36. Oscillographische mikroanalyse, R. Kalvoda, *Anal. Chim. Acta* **18**, 132–139 (1958).

37. Polarographic determination of dibutyl phthalate in propellant compositions containing nitroglycerin, A. F. Williams and D. Kenyon, *Talanta* **2**, 79–87 (1959).

38. The determination of trace amounts of lead and bismuth in cast iron, R. C. Rooney, *Analyst* **83**, 83–88 (1958); Aluminium in cast iron, *Analyst* **83**, 546–554 (1958).

39. The application of cathode-ray polarography to the analysis of semiconductors, F. A. Pohl, In: *Advances in Polarography*, I. S. Longmuir, ed., Vol. II, pp. 517–523, Pergamon Press, Oxford (1960).

40. The application of the cathode-ray polarograph to the analysis of blasting explosives, A. F. Williams and D. Kenyon, In: *Advances in Polarography*, I. S. Longmuir, ed., Vol. II, pp. 565–574, Pergamon Press, Oxford (1960).

41. Application of the cathode ray polarograph to the analysis of explosives, J. S. Hetman, In: *Advances in Polarography*, I. S. Longmuir, ed., Vol. II, pp. 640–656 Pergamon Press, Oxford (1960).

42. Oscillographic polarography with alternating current, In: R. Kalvoda, *Progress in Polarography*, I. M. Kolthoff and P. Zuman, eds., Vol. II, Chap. XXI, pp. 449–486 Interscience Publishers (1962).

43. Oscillographic polarography of the interrupted mercury electrode and the accumulation effect, Ya. P. Gokhshtein, In: *Polarography 1964*, G. J. Hills, ed., Vol. I, pp. 215–221 Macmillan, London (1966).

44. Studies in subtractive and continuous polarography. Part I. Preliminary studies. The design and construction of a subtractive circuit and associated drop rate controllers for a Tinsley Mark 19 polarograph, K. G. Powell and G. F. Reynolds, In: *Polarography 1964*, G. J. Hills, ed., Vol. 1, 249, Macmillan, London (1966).

1.4. A-C Polarography

1.4.1. Fundamental Harmonic A-C Polarography

Sine wave methods have been proposed and developed mainly to study fast electrode reactions. The theory of faradaic impedance has been worked out in a detailed manner by Randles,[4] Gerischer,[6] and Grahame.[9] The a-c polarographic theory has been formulated by Breyer and his co-workers[3–5,13–15] and later modified and extended by Delahay,[11,12] Koutecky,[20] and Smith.[28,31] The theory of complex plane analysis has been presented by Sluyters and co-workers.[22,32] The application of conventional a-c polarography in analytical chemistry has been exhaustively studied by Breyer and his school.[68–71] Although conventional a-c polarography has the advantage of better resolution of peaks over conventional d-c polarography, its range of applicability in trace analysis is, however, limited by the capacitative current (associated with the charging of the electrical double layer). The use of phase-sensitive detection in a-c polarography, which minimizes the non-faradaic effects, has been found to extend the limit of detection of metallic ions. The sensitivity of phase selective a-c polarography has been reported as $10^{-6} M$ at the dme. The theory of phase selective a-c polarography and instrumentation have been well presented by Reinmuth[29] and Smith.[26,31,49] The instrumentation for phase selective A.C.P. has made rapid advances over the last two decades with the development of highly selective amplifiers. Interfacing of polarography with digital data-acquisition devices other than recorders has also been successfully employed by many researchers.[55,56] The application of advanced a-c polarographic techniques in chemical analysis has been extensively studied by Bond and co-workers.[34–39]

A-C polarography of that class of organic compounds that do not undergo any electroreduction but influence the electrical double layer capacity through adsorption at the interface, has been studied in detail by Breyer and co-workers,[95,96,100] Doss and co-workers,[97,98] and Sharma *et al.*[104–107] This method has been termed as Tensammetry by Breyer. Jehring has made an exhaustive study of the behaviour of organic surface active compounds at the mercury/solution interface by a-c polarography.[101,102,108–110,112–118]

1.4.1.1. Theory

1. Properties of the electrical double layer at mercury surface. I. Methods of measurement and interpretation of results, D. C. Grahame, *J. Am. Chem. Soc.* **63**, 1207–1215 (1941).

2. The capacity of a mercury cathode in the presence of multivalent cations, M. Vorsina and A. N. Frumkin, *Acta Phys. Chim. USSR* **18**, 249 (1943).

3. (a) Reversible electrode reactions in alternating fields. I. Theory of the reversible depolarising process in an alternating field, B. Breyer and F. Gutman, *Trans. Faraday Soc.* **42**, 645–650 (1946). (b) II. Experimental verification of the capacity and dynamic resistance terms in the equation governing the reversible depolarising process in an alternating field, B. Breyer and F. Gutman, *Trans. Faraday Soc.* **42**, 650–654 (1946).

4. Kinetics of rapid electrode reactions, J. E. B. Randles, *Disc. Faraday Soc.* **1**, 11–19 (1947).

5. (a) The behaviour of reversible electrodes in alternating fields, B. Breyer and F. Gutmann, *Disc. Farad. Soc.* **1**, 19–26 (1947). (b) Electrode reactions in alternating fields. III. The dynamic resistance, B. Breyer and F. Gutmann, *Trans. Faraday Soc.* **43**, 785–791 (1947).

6. (a) Wechselstrompolarisation von elektrodes mit eine in potentialbeshmmenden schritt beim gleichgewichts potential-I, H. Gerischer, *Z. Phys. Chem.* **198**, 286–313 (1951). (b) Alternating current polarisation on electrodes with one potential determining step at the equilibrium potential. II, The influence of heterogeneous reaction on the kinetic polarisation resistance, H. Gerischer, *Z. Phys. Chem.* **201**, 55–67 (1952). (c) Bestimmung der austausch geschuindig keit beim gleichgewichts potential durch polarisations messungen mit gleich—und wechselstrom, Wechselstrommez, H. Gerischer, *Z. Elektrochem.* **55**, 98–104 (1951). (d) Methoden in der elektrochemie, H. Gerischer, *Z. Elektrochem.* **58**, 9–24 (1954).

7. Half-wave potentials of polarographic curves by means of alternating current calculation of the transfer coefficient, J. Van Cakenbergeh, *Bull. Soc. Chem. Belges* **60**, 3–10 (1951).

8. Gleich- und wechselstrom widerstand der diffusions-polarisation bei ortlich variables austansch stromdichte an der elektrode, K. J. Vetter, *Z. Phys. Chem.* **199**, 300 (1952).

9. Mathematical theory of faradaic admittance (pseudo capacity and polarisation resistance), D. C. Grahame, *J. Electrochem. Soc.* **99**, 370c–385c (1952).

10. Theory of alternating current polarography. I. Equation of the reversible a. c. polarographic wave, B. Breyer and S. Hacobian, *Aus. J. Chem.* **7**, 225–238 (1954).

11. Theory of alternating polarographic currents. Case of reversible waves, P. Delahay and T. J. Adams, *J. Am. Chem. Soc.* **74**, 5740–5744 (1952).

12. Voltammetry and polarography with periodically changing potential, P. Delahay, In: New Instrumental Methods in Electrochemistry, Chap. 7, pp. 147–198, Interscience Publishers Inc., New York (1954).

13. Alternating current polarography of organic compounds. I. General introduction and theory, B. Breyer, H. H. Bauer, and S. Hacobian, *Aus. J. Chem.* **7**, 305–311 (1954).

14. Ein gerat zum registriereu von impedanzen bei elektrochemischen untersuchungen, J. Schon, W. Mehl, and H. Gerischer, *Z. Elektrochem* **59**, 144–146 (1955).

15. (a) Theory of alternating current polarography. II. Significance of the Heyrovsky–Ilkovic equation and its relation to the production of a.c. polarographic waves, B. Breyer, H. H. Bauer, and S. Hacobian, *Aus. J. Chem.* **8**, 312–321 (1955). (b) Theory of alternating current polarography. III. Frequency of alternating field and reaction rate, B. Breyer, H. H. Bauer, and S. Hacobian, *Aus. J. Chem.* **8**, 322–328 (1955).

16. Studies on a.c. polarography. I. Theoretical treatment, I. Tachi and T. Kambara, *Bull. Chem. Soc. Jpn.* **28**, 25–31 (1955).

17. Studies on a.c. polarography. V. Theory of reversible wave, I. Tachi and M. Senda, *Bull. Chem. Soc. Jpn.* **28**, 632–636 (1955).

18. Studies on a.c. polarography. III. Reversible wave, M. Okerda and I. Tachi, *Bull. Chem. Soc. Jpn.* **28**, 37–41 (1955).

19. Verallgemeinerung der Randles–Gerischerschen theorie der wechgelspannungs polarisation und ihre polarographische anwendung, T. Kambara, *Z. Phys. Chem. N.F.* **5**, 52–65 (1955).

20. Theorie polarographischer strome bei periodisch wechselnder spannung, J. Koutecky, *Collect. Czech. Chem. Commun.* **21**, 433–446 (1956).

21. Zur theorie der wechselspannungs-polarographic, H. Matshuda, *Z. Elektrochem.* **62**, 977–989 (1958).

22. 'On the Impedance of Galvanic Cells,' *Electrode Kinetics Complex plane Polarography and Double layer Phenomena*, M. Sluyters Rehbach, Proefschrift (1963).

23. A generalised equation for the a.c. polarographic wave, H. H. Bauer, *J. Electroanal. Chem.* **1**, 2–7 (1959/60).

24. General theory of alternating current polarography for electrode reactions preceded by slow chemical reactions, S. Satyanarayana, A. K. N. Reddy, and K. S. G. Doss, *Aus. J. Chem.* **13**, 177–179 (1960).

25. (a) The faradaic admittance of electrochemical processes. I. Apparatus suitable for phase angle measurement, H. H. Bauer and P. J. Elving, *J. Am. Chem. Soc.* **82**, 2091–2094 (1960). (b) The faradaic admittance of electrochemical processes. II. Experimental test of the theoretical equations, H. H. Bauer, P. J. Elving, and D. L. Smith, *J. Am. Chem. Soc.* **82**, 2094–2098 (1960). (c) The faradaic admittance of electrochemical processes. III. The frequency dependence, H. H. Bauer and P. J. Elving, *J. Electroanal. Chem.* **2**, 53–59 (1961).

26. D. E. Smith, "*Polarography with periodically varying potentials,*" Ph.D. Thesis, Columbia University (1961), pp. 5–36 (Univ. Microfilm Inc., Ann. Arbor, Michigan).

27. A.C. Polarographic wave-modified equations, S. K. Rangarajan and K. S. G. Doss, *J. Electroanal. Chem.* **3**, 217–218 (1962).

28. (a) Alternating current polarography of electrode processes with coupled homogeneous chemical reactions. I. Theory for systems with first order preceding, following and catalytic chemical reactions, D. E. Smith, *Anal. Chem.* **35**, 602–609 (1963). (b) Alternating current polarography of electrode processes with coupled homogeneous chemical reactions II. Experimental results with catalytic reductions, D. E. Smith, *Anal. Chem.* **35**, 610–614 (1963).

29. Electrochemical relaxation techniques, K. Reinmuth, *Anal. Chem.* **36**, 211R–219R (1964).

30. Complex plane analysis of impedances, M. Slyters-Rehbach, D. J. Koajman, and J. H. Sluyters, In: *Polarography 1964*, G. J. Hills, ed., pp. 135–147, MacMillan (1966).

31. A.C. polarography and related techniques. D. E. Smith, Theory and Practice: Theory, In: *Electroanalytical Chemistry*, A. J. Bard, ed., Vol. 1, pp. 13–92, Marcel Dekker, Inc. (1966).

32. Sine wave methods in the study of electrode processes principles of alternating current electrodynamics. The cell impedance in the case of a simple electrode reactions. The complex plane analysis of cell impedances, M. Sluyters-Rehbach and J. H. Sluyters, In: *Electroanalytical Chemistry*, A. J. Bard, ed., Vol. IV, pp. 5–49, Marcel Dekker, Inc. (1966).

33. Reasons for multiplexing theory (of frequency multiplexing a.c. polarography) B. J. Heubert, In: *A Study of a.c. Polarography in the Frequency Multiplex Mode*' Ph.D. Thesis, pp. 16–27, Northwestern University, Illinois (1971).

34. Some data for the electroanalytical use of fundamental, second and third harmonic alternating current polarography, A. M. Bond, *J. Electroanal. Chem. Interfacial Electrochem.* **35**, 343–361 (1972).

35. Some experimental and theoretical correlations for the use of fundamental harmonic a.c. polarography, A. M. Bond, *Anal. Chem.* **44**, 315–335 (1972).

36. A study of the validity of the Ilkovic and other standard direct and alternating current polarographic equations at short drop time, A. M. Bond and R. J. O'Halloran, *J. Phys. Chem.* **77**, 915–922 (1973).

37. Fundamental and second harmonic alternating current cyclic voltammetric theory and experimental results for simple electrode reactions involving soluble-soluble redox couples, A. M. Bond, R. J. O'Halloran, I. Ruzic, and D. E. Smith, *Anal. Chem.* **48**, 872 (1976).

38. On-line FFT faradaic admittance measurements application to a.c. cyclic voltammetry, A. M. Bond, R. J. Schwall, and D. E. Smith, *J. Electroanal. Chem. Interfacial Electrochem.* 231–247 (1977).

39. A.C. cyclic voltammetry: A digital simulation study of the slow scan limit condition for a reversible electrode process, A. M. Bond, R. J. O'Halloran, I. Ruzic, and D. E. Smith, *J. Electroanal. Chem. Interfacial Electrochem.*, **90**, 381–388 (1978).

40. Influence of adsorption of electroactive species on the interfacial admittance measured in a.c. polarography, C. I. Mooring, M. Sluyters Rehbach, and J. H. Sluyters, *J. Electroanal. Chem. Interfacial Electrochem.* **87**, 1–16 (1978).

41. The admittance of an electrochemical cell with adsorption of electroactive species at the DME, C. A. Wignhorst, M. S. Rehbach, and J. H. Sluyters, *J. Electroanal Chem. Interfacial Electrochem.* **87**, 17–29 (1978).

1.4.1.2. Instrumentation

42. Instrumentation, B. Breyer and H. H. Bauer, *Alternating Current Polarography and Tensammetry*, Chap. 3, pp. 94–112, Interscience Publ., New York (1963).

43. Alternating current polarography and equivalent circuit, E. Niki, *Rev. Polarography* **3**, 41–49 (1955).

44. Studies on a.c. polarography. II. Fundamental circuit and some experimental results, M. Senda, M. Okuda, and I. Tachi, *Bull. Chem. Soc. Jpn.* **28**, 31–36 (1955).

45. A comparative study of three recently developed polarographs, D. J. Ferret, G. W. C. Milner, H. I. Shlagosky, and L. J. Slee, *Analyst*, **81**, 506–512 (1956).

46. The alternating current polarographic method, Y. Yasumori, *Polarography* **2**, 72–78 (1954).

47. An improved alternating current polarograph, T. Takahashi and E. Nicki, *Talenta* **1**, 245–248 (1958).

48. Controlled potential and derivative polarograph, M. T. Kelley, H. C. Jones, and D. J. Fischer, *Anal. Chem.* **31**, 1475–1485 (1959).

49. Phase selective alternating current polarography, D. E. Smith, In: *Polarography with Periodically Varying Potentials*, Ph.D. Thesis, pp. 56–75, Columbia Univ. (1961).

50. Interpretation of the results obtained with the Cambridge univector a.c. polarographic unit, J. W. Hayes and H. H. Bauer, *J. Electroanal. Chem. Interfacial Electrochem.* **3**, 336–347 (1962).

51. The influence of pen response time on recorded a.c. polarograms, E. M. C. and N. G. Lordi, *J. Electroanal. Chem. Interfacial Electrochem.* **4**, 251–255 (1962).

52. A.C. polarography employing operational amplifier, instrumentation. Evaluation of instrument performance and application to some new a.c. polarographic techniques, D. E. Smith, *Anal. Chem.* **35**, 18A–20 (1963).

53. A.C. polarography and related techniques, theory and practice, instrumentation, D. E. Smith, In: *Electroanalytical Chemistry*, A. J. Bard, ed., Vol. I, pp. 102–132, Marcel Dekker, New York (1966).

54. Some investigations on instrumental compensation of nonfaradaic effects in voltammetric techniques, E. R. Brown, T. G. McGord, D. E. Smith, and D. D. Deford, *Anal. Chem.* **38**, 1119–1130 (1966).

55. Instrumentation for digital data acquisition in voltammetric techniques, d.c. and a.c. polarography, E. R. Brown, D. E. Smith, and D. D. Deford, *Anal. Chem.* **38**, 1130–1136 (1966).

56. D. E. Smith, Application of on-line digital computers in a.c. polarography and related techniques, In: *Electrochemistry, Calculations, Simulations, and Instrumentation*, J. S. Mattson, H. B. Mark, and H. C. MacDonald, Jr. eds., Chap. 12, pp. 369–422, Marcel Dekker, New York (1972).

57. (a) B. J. Huebert, *A study of a.c. Polarography in the Frequency Multiplex Mode*, Ph.D. Thesis, pp. 28, 153, Instrumentation, Northwestern University (1971). (b) Alternating current polarography in the noncoherent wave frequency multiplex mode, B. J. Huebert and D. E. Smith, *Anal. Chem.* **44**, 1179 (1972).

58. Principles and applications of a.c. and d.c. rapid polarography with short controlled drop times, A. M. Bond, *J. Electrochem. Soc.* **118**, 1588–1595 (1971).

59. Fourier analysis of alternating current polarography, amplitude and phase of fundamental and second harmonic a.c. polarographic waves, H. Kojima and S. Fujuvara, *Bull. Chem. Soc. Jpn.* **44**, 2158–2162 (1971).

60. A.C. polarography in harmonic multiplex mode, D. E. Glover and D. E. Smith, *Anal. Chem.* **114**, 1140–1145 (1972).

61. Computer-assisted in phase a.c. polarography determination of Cd(II) Ions, S. Fujiuena, M. Hiroba, K. Sawantari, H. Kojima, and Y. Umezawa, *Bull. Chem. Soc. Jpn.* **47**, 499–50 (1974).

62. Use of pulsed direct current potential to minimize charging current in alternating current polarography, A. M. Bond and R. J. O'Halloran, *Anal. Chem.* **47**, 1906–1909 (1975).

63. Alternating current polarography in low frequency (Tieremennotokibov malagabaritnvi polarograph huzkin. yacprtbi), Yu. A. Ivanov, A. I. Plotnikov, and E. I. Chubaker, *Zavod. Lab.* **44**, 401–402 (1978).

64. Low-frequency alternating current small-scale polarograph, Yu. A. Ivanov, A. I. Plotnikov, and E. I. Chubakova, *Zavod. Lab.* **44**, 401–402 (1978).

65. Simultaneous measurement of the inphase and quadratic components of the signals in a.c. polarography using multiplex circuitry, H. Blustain, A. M. Bond, and A. Norris, *J. Electroanal. Chem. Interfacial Electrochem.* **89**, 75–81 (1978).

66. A.C. polarography using a non-linear, potential time ramp to generate the d.c. potential, A. M. Bond and B. S. Grabaric, *J. Electroanal. Chem. Interfacial Electrochem.* **87**, 251–260 (1978).

67. A.C. polarography using an applied d.c. pulse programme of the normal pulse polarographic type: availability of the theoretical rate laws, D. E. Smith, A. M. Bond, and B. S. Grabaric, *J. Electroanal. Chem. Interfacial Electrochem.* **95**, 237–240 (1979).

1.4.1.3. Analysis

68. Analytical applications—methodology, B. Breyer and H. H. Bauer, In: *A.C. Polarography and Tensammetry*, pp. 128–135, Chap. 4, Interscience Publishers, New York (1963).

69. Analytical applications: A.C. polarography of inorganic depolarisers, pp. 135–195, Chap. 4, B. Breyer and H. H. Bauer, In: *Alternating Current Polarography and Tensammetry*, Interscience Publishers, New York (1963).

70. Analytical applications: A.C. polarography of organic compounds, pp. 195–221, Chap. 4, B. Breyer and H. H. Bauer, In: *Alternating Current Polarography and Tensammetry*, Interscience Publishers, New York (1963).

71. Increasing the sensitivity of alternating current polarographic analytical procedures, R. Neeb, *Z. Anal. Chem.* **186**, 53–63 (1962).

72. Electrochemical masking of indium in a.c. polarographic determination of cadmium, E. Jacobsen and G. Tandberg, *Anal. Chim. Acta* **47**, 285–290 (1969).

73. Supporting electrolyte effects in tensammetry, H. H. Bauer, H. S. Campbell, and A. K. Shallal, *J. Electroanal. Chem. Interfacial Electrochem.* **21**, 45–48 (1969).

74. Alternating current polarographic determination of unsaturation, B. Fleet and R. D. Jee, *Talanta* **16**, 1561–1569 (1969).

75. Polarographic studies in aqueous hydrofluoric acid using a.c. and d.c. rapid techniques, A. M. Bond, J. A. O'Donnell, and R. J. Taylor, *Anal. Chem.* **41**, 1804–1806 (1909).

76. Determination of some methyl carbamate insecticides by a.c. polarography and cyclic voltammetry, M. D. Booth and B. Fleet, *Talanta* **17**, 401 (1970).

77. Direct current and alternating current polarographic response of some pharmaceuticals in an aprotic organic solvent system, A. L. Woodson and D. E. Smith, *Anal. Chem.* **42**, 242–248 (1970).

78. Utilisation of surfactants in a.c. polarographic analysis, N. Gunderson and E. Jacobson, *Anal. Chim. Acta* **45**, 346 (1971).

79. Alternating current polarographic method of analysis in the presence of oxygen and other irreversibly reduced species, A. M. Bond and J. H. Canterford, *Anal. Chem.* **43**, 228–234 (1971).

80. Simultaneous determination of two electroactive species by alternating current polarography, A. M. Bond and J. H. Canterford, *Anal. Chem.* **43**, 392–397 (1971).

81. Alternating current polarographic determination of electroactive species more negatively reduced than the major component, A. M. Bond and J. H. Canterford, *Anal. Chem.* **43**7, 1658 (1971).

82. An evaluation of integration procedures for improving the precision of a.c. polarography, D. Fleet and R. D. Jee, *J. Appl. Electrochem.* **1**, 269–274 (1971).

83. A theoretical comparison of the resolution of fundamental second and third harmonic a.c. polarography, A. M. Bond, *J. Electroanal. Chem. Interfacial Electrochem.* **36**, 235–242 (1972).

84. Alternating current and direct current voltammetry with a mercury pool electrode in concentrated hydrofluoric acid, A. M. Bond, J. A. O'Donnel, and R. J. Taylor, *Anal. Chem.* **44**, 464–67 (1972).

85. Alternating current and direct current polarography in concentrated hydrofluoric acid solution with a teflon dropping mercury electrode, A. M. Bond and J. A. O'Donnel, *Anal. Chem.* **44**, 590–592 (1972).

86. Comparative study of a wide variety of polarographic techniques with multifunctional instrumentation (PAR Model 170 Electrochem System), A. M. Bond and D. R. Canterford, *Anal. Chem.* **44**, 721–731 (1972).

87. Theoretical and experimental evaluation of multielement analysis by fundamental harmonic alternating current polarography, A. M. Bond and J. H. Canterford, *Anal. Chem.* **44**, 732–736 (1972).

88. Stability constant determination in precipitating systems by rapid alternating current polarography, A. M. Bond and G. Helter, *J. Electroanal. Chem. Interfacial Electrochem.* **34**, 227 (1972).

89. Analytical applications of high frequency, phase-sensitive short controlled drop time alternating current polarography, A. M. Bond, *Anal. Chem.* **45**, 2026–2031 (1973).

90. Instrumentation for electrochemical trace analysis, A. Poozari, S. R. Rajagopalan, and S. K. Rangarajan, *Trans. Soc. Adv. Electrochem. Sci. Tech.* **8**, 147–153 (1973).

91. Simultaneous determination of cadmium, copper, lead and zinc concentrates by a.c. polarographic method, comparison with atomic absorption spectrometry, M. E. Beyer and A. M. Bond, *Anal. Chim. Acta* **75**, 409 (1975).

92. Characteristics of a.c. polarograms at high sweep rates, C. I. Mooring and H. L. Kies, *Anal. Chim. Acta* **94**, 135–147 (1977).

93. Use of alternating current polarography in studying soils: Determination of lead in soils, L. A. Verobava and R. F. Davletehina, *Biol. Nauki (Moscov)* **8**, 130–142 (1978) (*CA* **90**, 37956a, 1979).

94. Fundamental and second harmonic alternating current cyclic voltammetric theory and experimental results for simple electrode reactions involving amalgam formation, A. M. Bond, R. J. O'Halloran, I. Ruzic, and D. E. Smith, *Anal. Chem.* **50**, 216–223 (1978).

1.4.1.4. Tensammetry: Theory and Analysis

95. Tensammetry: A method investigating surface phenomena by a.c. current measurements, B. Breyer and S. Hacobian, *Aus. J. Sci. Res. Ser. A* **5**, 500–520 (1952).

96. A.C. Polarographic-Tensammetric transition waves, B. Breyer and S. Hacobian, *Aus. J. Chem.* **6**, 186–188 (1953).

97. Effect of surface active substances on the capacity of the electrical double layer, K. S. G. Doss and A. Kalyanasundaram, *Proc. Ind. Acad. Sci.* **35A**, 27–33 (1952).

98. Behaviour of surface active substances at the dropping mercury electrode, K. S. G. Doss and S. L. Gupta, *Proc. Ind. Acad. Sci.* **A36**, 493–500 (1952).

99. The influence of the supporting electrolyte on tensammetric waves, B. Breyer and S. Hacobian, *Aus. J. Chem.* **9**, 7–13 (1956).

100. Theory IV tensammetry, B. Breyer and H. H. Bauer, In: *A.C. Polarography and Tensammetry*, pp. 85–93, Chap. 2, Interscience Publishers, New York (1963).
101. The adsorption in oscillographic polarography and alternating current polarography, H. Jehring, *Chem. Zovesh.* **18**, 313–323 (1964).
102. Investigations of surface phenomena by alternating current polarography, H. Jehring, *Abhanell Deut. Akad. Wiss. Berlin, Kl. Chem. Geol. Biol.* 472–480 (1964).
103. Investigations of alternating current polarography (polarography with superimposed alternating voltage). I. Influence of concentration and rate of outflow of Hg on the adsorption process, H. Jehring, *Z. Phys. Chem. (Leipzig)* **225**, 116–124 (1964).
104. A.C. polarographic studies on the nature of the capacity peaks observed with organic compounds at the dropping mercury electrode. Part I: Effect of pH, nature of the buffer and buffer capacity, S. L. Gupta and S. K. Sharma, *J. Ind. Chem. Soc.* **41**, 384–388 (1964).
105. A.C. polarographic studies on the nature of the capacity peaks observed with organic compounds at the dropping mercury electrode. Part II. Effect of change of the indifferent electrolyte, S. L. Gupta and S. K. Sharma, *J. Ind. Chem. Soc.* **41**, 663–672 (1964).
106. A.C. polarographic studies on the nature of the capacity peaks observed with organic compounds at the dropping mercury electrode. Part III. Effect of medium, S. L. Gupta and S. K. Sharma, *J. Ind. Chem. Soc.* **41**, 668–672 (1964).
107. Organic a.c. polarography and tensammetry in nonaqueous media, S. L. Gupta, M. K. Chatterjee, and S. K. Sharma, *J. Electroanal. Chem.* **7**, 81–84 (1964).
108. Investigations by alternating current polarography (polarography with alternating current voltage) II. Adsorptions of inhibitors, H. Jehring, *Z. Phys. Chem. (Leipzig)* **226**, 59–70 (1964).
109. Alternating current polarography (polarography with superimposed alternating current voltage) III. Time dependence of the decrease of capacity current caused by adsorption, H. Jehring, *Z. Phys. Chem. (Leipzig)* **229**, 39–48 (1965).
110. Structure of the adsorption layer on the surface of the DME, H. Jehring and G. Palyi, *Magy. Kem. Folyvirat* **71**, 427–432 (1965).
111. A.C. polarographic studies of the influence of tensammetric waves on one another, S. L. Gupta and S. K. Sharma, *Electrochim. Acta* **10**, 151–158 (1965).
112. Modern electrochemical methods for analysis of surface-active substances, H. Jehring, *Abhand. Deut. Akad. Wiss. Berlin, Kl. Chem. Geol. Biol.* 197–207 (1966).
113. Electrochemical studies on the adsorption of surface active substances at the mercury/electrolyte phase boundary, H. Jehring, *Akad. Wiss. Berlin. Kl. Chem. Geol. Biol.* **6**, 652–658 (1966) (Ger).
114. Increase of sensitivity in tensammetry I. Simple phase selective, difference, square wave and rectangular wave a.c. polarography with normal quasistationary dropping mercury electrode, H. Jehring, E. Horn, A. Reklat, and W. Stolle, *Collect. Czech. Chem. Commun.* **33**, 1038–1048 (1968).
115. Increase of sensitivity in tensammetry II. Simple and super wave a.c. polarography with a stationary mercury drop, H. Jehring and W. Stolle, *Collect. Czech. Chem. Commun.* **33**, 1670–1677 (1968).
116. Double layer capacity measurements with Breyer alternating current polarography, H. Jehring, *J. Electroanal. Chem. Interfacial Electrochem.* **21**, 77–98 (1969).
117. A.C. polarography IV. Lowering the instantaneous and average current capacity by diffusion controlled and uncontrolled adsorption, H. Jehring, *J. Electroanal. Chem. Interfacial Electrochem.* **20**, 33–46 (1969).
118. Alternating current polarography. VII. Effect of potential and time on the double layer capacity during mixed adsorption, H. Jehring, *Z. Phys. Chem. (Leipzig)* **246**, 1–24 (1971) (Ger).
119. Tensammetric investigations of non-reducible substances, B. Breyer and H. H. Bauer, eds. In: *A.C. Polarography and Tensammetry*, Chap. 4, pp. 221–252.

1.4.2. Second Harmonic A-C Polarography and Intermodulation Polarography

The measurement of second harmonic component of a-c was first attempted by Van Cakenberghe.[1] Later detailed studies were carried out by Bauer and Elving,[5,6] Reinmuth,[13] Smith,[9,15-19] Paynter,[14] Devay *et al.*,[21,22] and Bond.[30,32] The theory of second harmonic a-c polarography has been developed by Gerischer,[2] Matsuda,[4] Tachi and Senda,[3] Smith,[9] Reinmuth and Smith,[10] and Paynter.[14] The instrumentation has been developed to a high degree of accuracy.[33-35] Second harmonic a-c polarography has been compared in a detailed manner, with the fundamental harmonic a-c polarography, both from the point of view of resolution and sensitivity, by Bond and co-workers.[49,50,54-56] Intermodulation polarography, which employs either frequency modulation or amplitude modulation, has also been studied in detail by Paynter,[14] Neeb *et al.*,[12] and Zheleztsov *et al.*[27] Intermodulation polarography has been found to have distinct advantages[14] over the second harmonic a-c polarography for kinetic studies, since the applicability of maximum higher frequency for the former is higher than that for the latter. It has also been reported that the electronic circuitry required for intermodulation polarography is comparatively less complex in nature. The sensitivity of the two techniques in trace analysis is nearly the same, i.e., up to 10^{-6} M for reversible charge transfers.

1.4.2.1. Theory

1. Half-wave potentials of polarographic curves by means of alternating current calculation of the transfer coefficient, J. Van Cakenberghe, *Bull. Soc. Chim. Belg.* **60**, 3–10 (1951).
2. Wechselstrom polarisation von elektroden mit einem potentialbestimmenden schritt beim gleich gewichts potential I, H. Gerischer, *Z. Phys. Chem.* **198**, 286–313 (1951).
3. Studies on a.c. polarography. V. Theory of reversible wave, I. Tachi and M. Senda, *Bull. Chem. Soc. Jpn.* **28**, 632–636 (1955).
4. Beitrage zur theorie der polarographischen stromstarke allgemeine formel der diffusions bedingten stromstarke und ihre anwendung, H. Matsuda, *Z. Electrochem.* **61**, 489–506 (1957).
5. Alternating current polarography determination of transfer coefficient of electrochemical processes, H. H. Bauer and P. J. Elving, *Anal. Chem.* **30**, 341–346 (1958).
6. The faradaic admittance of electrochemical processes. Part I. Apparatus suitable for phase angle measurement, H. H. Bauer and P. J. Elving, *J. Am. Chem. Soc.* **82**, 2091–2094 (1960).
7. The faradaic admittance of electrochemical processes. Part II. Experimental test of the theoretical equations, H. H. Bauer, P. J. Elving, and D. L. Smith, *J. Am. Chem. Soc.* **82**, 2094–2098 (1960).
8. Theory of faradaic distortion, K. B. Oldham, *J. Electrochem Soc.* **107**, 766–772 (1960).
9. *Polarography with periodically varying potentials*, D. E. Smith, Ph.D. Thesis, Columbia University (1961).
10. Second harmonic alternating current polarography with a reversible electrode process, D. E. Smith and W. H. Reinmuth, *Anal. Chem.* **33**, 482–485 (1962).
11. Harmonic alternating current polarography, R. Neeb, *Z. Anal. Chem.* **188**, 401–416 (1962).
12. Intermodulation polarography, R. Neeb, *Naturwissenschaften* **49**, 447 (1962).
13. Electrochemical relaxation techniques, W. H. Reinmuth, *Anal. Chem.* **36**, 211R–219R (1964).
14. *Polarographic techniques based on the faradaic non-linearity*, J. Paynter, Ph.D. Thesis, Columbia Univ. (1964) II. Theory: pp. 27.

15. Alternating current polarography: An extension of the general theory for systems with coupled first order homogeneous chemical reactions, T. G. McCord and D. E. Smith, *Anal. Chem.* **40**, 1959–1966 (1968).
16. Second harmonic alternating current polarography: A general theory for systems with coupled first order homogeneous chemical reactions, T. G. McCord and D. E. Smith, *Anal. Chem.* **40**, 1967–1970 (1968).
17. Alternating current polarography: theoretical prediction for systems with first order chemical reactions preceding the charge transfer step, T. G. McCord and D. E. Smith, *Anal. Chem.* **40**, 116–130 (1969).
18. Second harmonic alternating current polarography: Some experimental observations with quasi-reversible processes, T. G. McCord and D. E. Smith, *Anal. Chem.* **40**, 131–136 (1969).
19. Second harmonic alternating polarography. Experimental observation with a system involving a very rapid chemical reaction following charge transfer, T. G. McCord and D. E. Smith, *Anal. Chem.* **42**, 2–6 (1970).
20. Method for the study of asymmetry in fast electrode processes, J. E. B. Randles and D. R. Whitehouse, *Trans. Faraday Soc.* **64**, 1376–1387 (1968).
21. Second harmonic alternating current polarography, J. Devay, T. Garai, L. Meszaros, and E. Pungor, *Hung. Scient. Instrument.* **12**, 1–9 (1968).
22. Second harmonic polarography of multicomponent systems, J. Devay, T. Garai, L. Meszaros, and Palagyi-Fenyes, *Hung. Scient. Instrument.* **15**, 1–7 (1969).
23. Calculation of the influence of the ohmic resistance of the cell in a.c. polarography in the case of reversible electrode reaction, J. Devay, L. Meszaros, and T. Garai, *Acta. Chim. (Budapest)* **60**, 67–85 (1969).
24. Calculation of the influence of the ohmic resistance of the cell on the third harmonic a.c. polarographic current in the case of a reversible electrode reaction, J. Devay, L. Meszaros, and T. Garai, *Acta. Chim. (Budapest)* **61**, 279–287 (1969).
25. Alternating current polarography: Theoretical predictions for systems with first-order chemical reactions following the charge transfer step, T. J. McCord, H. L. Hung, and D. E. Smith, *J. Electroanal. Chem. Interfacial Electrochem.* **38**, 883 (1969).
26. Second and third harmonic a.c. polarography, J. Devay, T. Garai, L. Meszaros, and B. Palagyi-Fenyes, *Magy. Kem. Foly.* **75**, 460–475 (1969) (Hung.).
27. Alternating current polarograph with amplitude modulated potential, A. V. Zheleztsov, *Pril. Sist. Upr.* **11**, 47–48 (1970).
28. Resolving power of an a.c. polarograph with sinusoidal voltage, A. V. Zheleztsov, *Zh. Anal. Khim.* **26**, 869–874 (1971).
29. Principles and applications of a.c. and d.c. rapid polarography with short controlled drop times, A. M. Bond, *J. Electrochem. Soc.* **118**, 1588–1595 (1971).
30. Fundamental and second harmonic alternating current cyclic voltammetric theory and experimental results for simple electrode reactions involving solution-soluble redox couples, A. M. bond, R. J. O' Halloran, I. Ruzic, and D. E. Smith, *Anal. Chem.* **40**, 872 (1976).
31. Intermodulation a.c. polarography. H. Blutstein and A. M. Bond, *Anal. Chem.* **48**, 1975–1979 (1976).
32. Fundamental and second harmonic alternating current cyclic voltammetric theory and experimental results for simple electrode involving amalgam formation, A. M. Bond and R. J. O'Halloran, *Anal. Chem.* **50**, 216–223 (1978).

1.4.2.2. Instrumentation and Analysis

33. *Polarography with periodically varying potentials*, D. E. Smith, Ph.D. Thesis, pp. 41–55, Columbia Univ. (1961).
34. Tuned alternating current polarography, N. G. Lordi, *Anal. Chem.* **34**, 1832–1833 (1962).
35. A.C. polarography employing operational amplifier instrumentation, D. E. Smith, *Anal. Chem.* **35**, 1811–1820 (1963).

36. Recent polarographic and voltammetric procedure for trace analysis, R. Neeb, *Fortschr. Chem. Forsch.* **4(2)** 333–458 (1963).

37. Double-tone polarography. I. Experimental fundamentals, R. Neeb, *Z. Anal. Chem.* **208**, 168–187 (1965).

38. A.C. polarography and related techniques. Theory and practice: Theory, D. E. Smith, In: *Electroanalytical Chemistry*, A. J. Bard, ed., Vol. 1, pp. 13–92, Marcel Dekker, Inc. (1966).

39. Double-tone polarography II. Effect of cationic surface active agents and of the temperature dependence of peak currents on the alternating current polarography of antimony, R. Neeb, *Z. Anal. Chem.* **216**, 94–96 (1966).

40. Double tone polarography III. Alternating current polarographic behaviour of tin in hydrochloric acid solutions, R. Neeb, *Z. Anal. Chem.* **222**, 290–310 (1966).

41. Second harmonic a.c. polarography of irreversible systems, J. Devay, T. Garai, L. Meszaros, and J. P. Nityanandan, *Hung. Sci. Instrum.* **17**, 1–10 (1970).

42. Resolving power of some polarographic methods, V. V. Senkevich, *Zh. Anal. Khim.* **26**, 461 (1971).

43. Resolving power of an a.c. polarograph with sinusoidal voltage, A. V. Zheleztsov, *Zh. Anal. Khim.* **26**, 869–874 (1971).

44. A.C. polarography based on demodulation of a sinusoidal signal by a reversible electrode process, W. A. Brooke, *J. Electroanal. Chem. Interfacial Electrochem.* **30**, 237–257 (1971).

45. Demodulation polarography with triangularly modulated polarising voltage, W. A. Brooke, *J. Electroanal. Chem. Interfacial Electrochem.* **33**, App. 1–3 (1971).

46. Use of a commercial lock-in amplifier in phase-selective second-harmonic a.c. polarography, H. H. Bauer and D. Britz, *Chem. Instrument.* **2**, 361 (1970).

47. Direct method for phase-angle recording in a.c. polarography with a commercial lock-in amplifier. Analytical utility of phase angle measurements, D. M. McAllister and G. Dryhurst, *Anal. Chim. Acta* **58**, 373–382 (1972).

48. A.C. polarography in harmonic multiplex mode, D. E. Glover and D. E. Smith, *Anal. Chem.* **44**, 1140–1145 (1972).

49. Some data for the electro analytical use of fundamental, second and third harmonic alternating current polarography, A. M. Bond, *J. Electroanal. Chem. Interfacial Electrochem.* **35**, 343–361 (1972).

50. A theoretical comparison of the resolution of fundamental, second and third harmonic a.c. polarography, A. M. Bond, *J. Electroanal. Chem. Interfacial Electrochem.* **36**, 235–242 (1972).

51. On the theory of Kalovsek commutator, square wave and related techniques I. Equation for current-potential curves, I. Ruzic, *J. Electroanal. Chem. Interfacial Electrochem.* **39**, 111–121 (1972).

53. Fast sweep a.c. polarography, R. D. Jee, *Fresenius Z. Anal. Chem.* **264**, 143–146 (1973).

54. New instrumental approach in phase-selective second harmonic alternating current polarography, H. Blustein, A. M. Bond, and A. Norris, *Anal. Chem.* **46**, 1754–1758 (1974).

55. Short communication cyclic fundamental and second harmonic a.c. voltammetry with phase selective detection, A. M. Bond, *J. Electroanal. Chem. Interfacial Electrochem.* **50**, 285–291 (1974).

56. Measurement of higher harmonics with a lock-in amplifier: Phase-selective and other forms of sinusoidal, saw tooth, square wave, triangular wave and white noise alternating current polarography, A. M. Bond and V. S. Flego, *Anal. Chem.* **47**, 2321–2334 (1975).

57. Polarographic measurement of high substance concentration in solutions, A. V. Zheleztsov, *Zh. Anal. Khim.* **32**, 1083–1087 (1977).

58. Optimization of double tone polarography, D. Saur and R. Neeb, *J. Electroanal. Chem. Interfacial Electrochem.* **75**, 171–180 (1977).

59. Polarographic and voltammetric methods for the determination of elements, R. Neeb, *Mikrochim. Acta* (*Wien*) **1**, 305–318 (1978).

60. Methods for the control of phase sensitive detectors in polarographic measurement techniques, S. Dietrich, *Fresnius Z. Anal. Chem.* **290**, 217–219 (1978).

61. Simultaneous measurements of the in-phase and quadrature components of the signal in a.c. polarography using multiplier circuitry, H. Blutstein, A. M. Bond, and A. Norris, *J. Electroanal. Chem. Interfacial Electrochem.* **89**, 75–81 (1978).
62. Fundamentals of non-linear a.c. polarographic analytical method, D. Saur and R. Neeb, *Fresenius Z. Anal. Chem.* **290**, 374–381 (1978).
63. Instrumentation and detection sensitivity in higher harmonics polarographic analysis methods, D. Saur and R. Neeb, *Fresenius Z. Anal. Chem.* **290**, 220–229 (1978).

1.5. Square-Wave Polarography

Square-wave polarography was proposed first by Barker *et al.*,[1] in an effort to eliminate the double-layer charging current and thereby improve the lower limit of detection (sensitivity) of the polarographic technique. The theory of square-wave polarography has been presented by Barker,[3,5] Matsuda,[6] and Kambara.[2] It has been pointed out by Barker[8] that although normal square-wave polarography offers no specific advantage compared to the faradaic impedance method for the study of moderately rapid electrode processes (i.e., those having rate constants between 10^{-3} cm s^{-1} and 10^{-1} cm s^{-1}), the square-wave technique is very suitable for studying the kinetics of totally irreversible processes. The instrumentation for square-wave polarography is well developed and a few polarographs are commercially available.[29–31] The suitability of square-wave polarography in trace analysis has been examined by Ferret and Milner,[10,13] Sturm *et al.*,[15–18,22] Okamoto,[23] and Mizuike *et al.*[25]

1.5.1. Theory

1. Square wave polarography, G. C. Barker and I. L. Jenkins, *Analyst* **77**, 685–695 (1952).
2. (a) Polarographic diffusion current observed with square wave voltage I. Effect produced by the sudden change of electrode potential, T. Kambara, *Bull. Chem. Soc. Jpn.* **27**, 523–526 (1954). (b) Polarographic diffusion current observed with square wave voltage II. Basic theory for a reversible electrode, T. Kambara, *Bull. Chem. Soc. Jpn.* **27**, 527–529 (1954). (c) Polarographic diffusion current observed with square wave voltage III. Applications of the theory, T. Kambara, *Bull. Chem. Soc. Jpn.* **27**, 529–534 (1954).
3. Square wave polarography. Part IV. An introduction to the theoretical aspects of square wave polarography, G. C. Barker, R. L. Faircloth, and A. W. Gardner, Atomic Energy Res. Establish. Report C/R 1786. H. M. Stationery Office, London (1954).
4. Theorie polarographischer strome bei periodisch wechselnder spannung, J. Koutecky, *Collect. Czech. Chem. Commun.* **21**, 433–446 (1956).
5. Square wave polarography and some related techniques, G. C. Barker, *Anal. Chim. Acta* **18**, 118–131 (1958).
6. Zur. theorie der wechelspannungs polarography, H. Matsuda, *Z. Elektrochem.* **62**, 977–989 (1958).
7. Faradaic rectification, G. C. Barker, In: *Transactions of the Symposium on Electrode Processes*, E. Yeager, ed., p. 325, Philadelphia (1959).
8. Square wave and pulse polarography, G. C. Barker, In: *Progress in Polarography*, I. M. Kolthoff and P. Zuman, eds., Vol. II, Chap. XIX, pp. 411–423, Interscience Publishers, New York (1962).

1.5.2. Instrumentation and Analysis

9. Square wave polarography, G. C. Barker and I. L. Jenkins, *Analyst* **77**, 685–695 (1952).

10. The determination of lead in $CoCo_4$ with a square-wave polarography, D. J. Ferret, G. W. C. Milner, and A. A. Smales, *Analyst* **79**, 731–734 (1954).

11. (a) Analytical applications of the Barker square wave polarograph, D. J. Ferret and G. W. C. Milner, *Analyst* **80**, 132–140 (1955). (b) Analytical applications of the Barker square-wave polarograph. Part II. The analysis of copper-base alloys and steels, D. J. Ferret and G. W. C. Milner, *Analyst* **81**, 193–203 (1956).

12. A comparative study of three recently developed polarographs, D. J. Ferret, G. W. C. Milner, H. I. Shalgosky, and L. S. Slee, *Analyst* **81**, 506–512 (1956).

13. Square wave polarography and some related techniques, G. C. Barker, *Anal. Chim. Acta* **18**, 118–131 (1958).

14. Square wave polarography, R. E. Haurm, *Anal. Chem.* **30**, 350–354 (1958).

15. Reproducibility of polarographic measurements, F. Von Sturm, *Z. Anal. Chem.* **166**, 100–114 (1959).

16. Limits of usability of polarographic methods in inorganic analysis, F. Von Sturm, *Z. Anal. Chem.* **173**, 11–17 (1960).

17. Moglichkeiten der polarographischen method in der anorganischen spurenanalyse, In: M. Kankeintsch and F. Von Sturm, *Advances in Polarography*, I. S. Longmuir, ed., Vol. II, pp. 551–564, Pergamon Press, Oxford (1960).

18. Influence of foreign electrolyte concentration in square wave polarography, F. Von Sturm and M. Ressel, *J. Microchem.* **5**, 53 (1961).

19. Studies in the field of quadratic-wave polarography. General principles and the method, B. Ya. Kalpan and I. I. Sorokovskaya, *Zavod. Lab.* **28**, 1053 (1962).

20. Polarographic a ondes carrees, G. Geerinek, H. Hilderson, C. Vanttalle, and F. Verbeck, *J. Electroanal. Chem.* **5**, 48–56 (1963).

21. The Mervyn Merck IV square wave polarographic analyses, Application Bulletin, Matheson Scientific Inc. (1964).

22. Square wave polarography in trace analysis, F. Von Sturm and M. Ressel, In: *Proc. Aust. First Conf. on Electrochemistry*, F. Gutmann, ed., pp. 310–322, Pergamon Press, Oxford (1965).

23. Oscillographic square wave polarographic behaviour of some inorganic and organic compound, K. Okamoto, In: *Modern Aspects of Polarography*, T. Kambara, ed., pp. 225–232, Plenum Press, New York (1966).

24. Theory of square wave polarography, L. Rainaley and M. S. Krause, *Anal. Chem.* **41**, 1362–1365 (1969).

25. Polarographic determination of traces of zinc in bismuth after preconcentration by solid liquid extraction, A. Mizuike and T. Kono, *Mikrochim. Acta* 665–669 (1970).

26. Automatic continuous analyzer for lead in the atmosphere, N. Yamate, Y. Matsumura, and M. Tonomura, *Eisei Shikenjo Hokku* **87**, 28–31 (1969).

27. Square wave polarographic determination of lead as pollutant in river water, E. B. Buchaman, T. D. Schroeder, and B. Novosel, *Anal. Chem.* **42**, 370–373 (1970).

28. Polarographic studies on uranium VI compounds. I. Square-wave polarographic determination of micro amounts of uranium, O. Guertler and Chu-Xuan-Arah, *Mikrochim. Acta* 941–949 (1970).

29. Simple square wave polarograph, R. Kalvoda and I. Holub, *Chem. Listy* **67**, 302–307 (1973).

30. A multimode polarograph, G. C. Barker and A. W. Gardner. *J. Electroanal. Chem. Interfacial Electrochem.* **42**, App. 21–26 (1973).

31. Measurement of higher harmonics with a lock-in-amplifier: Phase-selective and other forms of sinusoidal, sawtooth, square wave, triangular wave and white noise alternating current polarography, A. M. Bond and V. S. Flego, *Anal. Chem.* **47**, 2321–2324 (1975).

32. Modern polarographic techniques, G. C. Barker, *Proc. Anal. Div. Chem. Soc.* **12**, 179–181 (1975).

33. Square wave polarography and related techniques, P. E. Sturrock and R. J. Carter, *Crit. Rev. Anal. Chem.* **5**, 201–223 (1975).
34. A flow detector based on square wave polarography on the dropping mercury electrode, J. Wang and E. Quziel, *Anal. Chim. Acta* **102**, 99–112 (1978).
35. The direct evaluation of overlapping signals in square wave polarography, B. Kuhrig, *Z. Chem.* **18**, 415–417 (1978).
36. Quantitative resolution of overlapped peaks in programmed potential step voltammetry, P. A. Boudreau and S. P. Perone, *Anal. Chem.* **51**, 811–817 (1979).

1.6. Radio Frequency Polarography

Radio frequency polarography (r.f. polarography) was developed by Barker[5-7] based on the faradaic rectification effect, first discovered by Doss and Agarwal,[1,2] for the study of rapid electrode processes. The theory for the faradaic rectification effect, which arises due to the nonlinearity of the electrode processes, was developed by Doss and Agarwal,[1,2] Barker,[7,9] Vdovin,[8] and Oldham.[4,12] The theory has been further improved by Delahay et al.[10,12-17,20] and Rangarajan.[11] The theory developed has been used by Barker et al.,[22-24] Delahay et al.,[17-20] and Nurnberg[25] in the study of fast electrode processes. The applicability of r.f. polarography in trace analysis has been shown by Barker.[30,31] The application of r.f. polarography has been extended by Furatani,[32] Roughton et al.,[36] Bruck and Sternberg,[37] and others in trace analysis of metallic ions. The sensitivity level of r.f. polarography has been reported to be $2 \times 10^{-8}\ M$ for reversibly reduced metallic ions and $10^{-7}\ M$ for many irreversibly reduced metallic ions. It has been reported by Barker that the important advantage of r.f. polarography is its reproducibility.

The instrument needed for r.f. polarography is quite complex in nature. Radio frequency polarographs are commercially available.[41]

1.6.1. Theory

1. A new polarization effect: Redoxokinetic potential, K. S. G. Doss and H. P. Agarwal, *J. Sci. Ind. Res. (India)* **9B**, 280 (1950).
2. (a) The redoxokinetics effect—A general phenomenon, K. S. G. Doss and H. P. Agarwal, *Proc. Ind. Acad. Sci.* **34**, 229–235 (1951). (b) The theory of redoxokinetic effect and a general method for the determination of "a" of absolute reaction rates, K. S. G. Doss and H. P. Agarwal, *Proc. Ind. Acad. Sci.* **34A**, 263–271 (1951). (c) The theory of redoxo-kinetic effect—A correction, *Proc. Ind. Acad. Sci.* **35A**, 45 (1952).
3. The mechanism of the reation at a Cu/Cu^{2+} electrode, P. J. Hillson, *Trans. Faraday Soc.* **50**, 385–393 (1954).
4. Faradaic rectification: Theory and application to the Hg$_2^{2+}$/Hg electrode, K. B. Oldham, *Trans. Faraday Soc.* **53**, 80–90 (1957).
5. G. C. Barker, *Proc. Congr. on Modern Anal. Chem. in Ind.* St. Andrews, p. 199 (1957).
6. Use of faradaic rectification for the study of rapid electrode processes, G. C. Barker, R. L. Faircloth, and A. W. Gardner, *Nature* **181**, 247–248 (1958).
7. Square wave polarography and some related techniques, G. C. Barker, *Anal. Chim. Acta* **18**, 118–131 (1958).

8. The theory of faradaic rectification, Iu. A. Vdovin, *Dok. Akad. Nauk, SSSR* **120**, 554 (1958).

9. Faradaic rectification, G. C. Barker, In: *Trans. of the Symp. on Electrode Processes*, Philadelphia, E. Yeager, ed., p. 325, John Wiley & Sons, New York (1961).

10. Faradaic rectification with control of alternating potential variations. Application to electrode kinetics for fast processes, H. Matsuda and P. Delahay, *J. Amer. Chem. Soc.* **82**, 1547–1550 (1960).

11. Derivation of the general equation for redoxokinetic effect, S. K. Rangarajan, *J. Electroanal. Chem.* **1**, 396–402 (1960).

12. Theory of faradaic distortion, K. B. Oldham, *J. Electrochem. Soc.* **107**, 766–772 (1960).

13. Faradaic rectification and electrode processes, P. Delahay, M. Senda, and C. H. Weis, *J. Phys. Chem.* **64**, 960 (1970).

14. Electrode processes with specific or non-specific adsorption: Faradaic impedance and rectification, M. Senda and P. Delahay, *J. Phys. Chem.* **65**, 1580–1588 (1961).

15. Rectification by the electrical double layer and adsorption kinetics, M. Senda and P. Delahay, *J. Am. Chem. Soc.* **83**, 3763–3766 (1961).

16. Faradaic rectification and electrode processes, P. Delahay, M. Senda, and C. H. Weis, *J. Amer. Chem. Soc.* **83**, 312–322 (1961).

17. Faradaic rectification and electrode processes II, M. Senda, H. Imai, and P. Delahay, *J. Phys. Chem.* **65**, 1253–1256 (1961).

18. VI. Faradaic rectification method: Quantitative discussion, P. Delahay, In: *Advances in Electrochemistry and Electrochemical Engineering*, P. Delahay and C. W. Tobias, eds., Vol. 1, pp. 279–300 (1961).

19. Faradaic rectification and electrode processes IV, H. Imai, *J. Phys. Chem.* **66**, 1744–1746 (1962).

20. (a) Faradaic rectification and electrode processes. III. Experimental methods for high frequencies and application to the discharge of mercurous ion, H. Imai and P. Delahay, *J. Phys. Chem.* **66**, 1108–1113 (1962). (b) Kinetics of discharge of the alkali metals on their amalgams as studied by faradaic rectification, H. Imai and P. Delahay, *J. Phys. Chem.* **66**, 1683–1686 (1962).

21. Faradaic rectification and electrode processes—Experimental review, H. Imai, *Rev. Polarography (Japan)* **10**, 209 (1962).

22. Irreversible electrode processes by modern instrumental methods, G. C. Barker, H. W. Nürnberg, and J. A. Bolzan, *Ber. K.F.A. Julich* No. 137 (1963).

23. Determination of the rate constants for the dissociation and recombination of weak acids by high level faradaic rectification, G. C. Barker and H. W. Nürnberg, *Naturwiss.* **51**, 191 (1964).

24. Non-linear relaxation methods for the study of very fast electrode processes, G. C. Barker, In: *Polarography 1964*, G. J. Hills, ed., Vol. I, pp. 25–47, Macmillan, London (1966).

25. (a) The determination of the rate constants of dissociation and recombination for carboxylic acids by high level Faradaic rectification, H. W. Nürnberg, In: *Polarography 1964*, G. J. Hills, ed., Vol. I, pp. 149–155. Macmillan, London (1966). (b) The method of high level faradaic rectification, H. W. Nürnberg, In: *Polarography 1964*, G. J. Hills, ed., Vol. I, pp. 155–156, Macmillan, London (1966).

26. Faradaic rectification: An amended treatment, R. De Leeuwe, M. Sluyters-Rehbach, and J. H. Sluyters, *Electrochim. Acta* **12**, 1593–1599 (1967).

27. Effect of the external resistance on the high frequency polarographic wave height, T. Kambara, S. Tanaka, and K. Harebe, *J. Electroanal. Chem. Interfacial Electrochem.* **21**, 49 (1969).

28. Faradaic rectification, E. Yeager and J. Kuta, In: *Techniques in Electrochemistry*, E. Yeager and A. Salkind, eds., Vol. I, pp. 245–256, Wiley Interscience, New York (1972).

29. Faradaic rectification method and its applications in the study of electrode processes, H. P. Agarwal, In: *Electroanalytical Chemistry*, A. J. Bard, ed., Vol. 7, pp. 161–199, Marcel Dekker, Inc. New York (1974).

1.6.2. Instrumentation and Analysis

30. Use of faradaic rectification for the study of rapid electrode processes, G. C. Barker, R. L. Faircloth, and A. W. Gardner, *Nature* **181**, 247–248 (1958).
31. Square wave polarography and some related techniques, G. C. Barker, *Anal. Chim. Acta* **18**, 118–131 (1958).
32. Trace elements in food. I. Simultaneous determination of zinc and cobalt as nickel by high frequency polarography, S. Furutani, *Japan Analyst* **16**, 103 (1967).
33. Effect of the external resistance on the high frequency polarographic wave height, T. Kambara, S. Tanaka, and K. Hasebe, *J. Electroanal. Chem. Interfacial Electrochem.* **21**, 49 (1969).
34. High frequency polarography, L. N. Vasileva and N. V. Lukashenkova, *Zh. Anal. Khim.* **25**, 412 (1970).
35. Effect of the rate of voltage supply on the shape of the high frequency polarogram for a stationary mercury electrode, L. N. Vasileva and N. B. Kogan, *Zh. Anal. Khim.* **26**, 1932 (1971).
36. A transistorised radio frequency polarograph: Its use for the determination of tin and lead, C. L. Roughton, M. Hanison, and B. Surfleet, *Analyst* **95**, 894–901 (1970).
37. High frequency polarograph and its use, S. S. Bruk and B. M. Sternberg, *Zavod. Lab.* **36**, 365 (1970).
38. Detector polarography and its use, R. M. Salikhdzhanova and I. E. Bryksin, *Zavod. Lab.* **37**, 765 (1971).
39. Trace elements in food. IV. High frequency polarographic determination of p-toluene sulfonic acid, Y. Osajima, K. Matsumoto, M. Nakashima, F. Hashinage, and S. Furutani, *Benseki Kagaku* **20**, 1292–1297 (1971).
40. Analytical possibilities of high frequency polarography, L. P. Chernega, V. I. Bodyn, and Yu. S. Syalikov, *Zh. Anal. Khim.* **26**, 1686–1690 (1971).
41. A multimode polarograph, G. C. Barker and A. W. Gardner, *J. Electroanal. Chem. Interfacial Electrochem.* **42**, 21–26 (1973).
42. Equivalent circuits for a cell affected by space charge in the solution and simple reactions at the electrodes, G. C. Barker, *J. Electroanal. Chem. Interfacial Electrochem.* **44**, 473–479 (1973).

1.7. Pulse, Differential Pulse Polarography

The pulse polarograph was first developed by Barker[1,2] to overcome instability arising due to capillary response in the dropping mercury electrode, which leads to difficulty in achieving maximum sensitivity with square wave polarograph. The theory of pulse polarography was presented by Barker,[1,2] Osteryoung et al.,[4,7,15,24,43] Oldham,[8] Galvez et al.,[19,22] Kaplan et al.,[26,33,41] Van Bennekem et al.,[34] Anson et al.,[37,39] and Ruzic.[35,44] The instrumentation for pulse polarography has been developed by Osteryoung et al.[49–56,75] and Kalvoda et al.[64,78–80] Osteryoung et al. have also presented studies on digital simulation of differential pulse polarography[70] and a new digital-to-analog pulse polarograph has been developed by Van diern et al.[62] A pulse polarograph with microprocessor control has been employed by Bond et al.,[61,66] while computerized instrumentation has been employed for normal pulse polarography by Osteryoung et al.[50] and Bond et al.[73,74,77,81] It has been reported that the pulse polarograph is more sensitive than the square-wave polarograph for reversibly reduced metallic ions and can employ lower concentrations of supporting electrolyte in trace analysis. The capillary response is allowed to die away more completely in the pulse polarograph than in the square-wave polarograph. The pulse polarograph has also been reported to be more sensitive for irreversible reactions.

Pulse and differential pulse polarographs have found wide application in the analysis of toxic heavy metals,[82,95,100,108,110,114,120,122] pharmaceutical chemicals,[84–87,103–107] and in many other areas. Differential pulse polarography has been extensively used in anodic stripping analysis (see Section 3).

1.7.1. Theory

1. Pulse polarography, G. C. Barker and A. W. Gardner, Atomic Energy Res. Establ. AERE Harwell C/R 2297 (1958).
2. Pulse polarography, G. C. Barker and A. W. Gardner, *Z. Anal. Chem.* **173**, 79–83 (1960).
3. Square wave and pulse polarography, G. C. Barker, In: *Progress in Polarography*, I. M. Kolthoff and P. Zuman, eds., Vol. II, pp. 411–427, Interscience Publ., New York (1962).
4. Evaluation of analytical pulse polarography, E. P. Parry and R. A. Osteryoung, *Anal. Chem.* **37**, 1634–1637 (1965).
5. Stand der polarographischen methoden und ihren instrumentation, H. W. Nürnberg and G. Wolf, *Chem. Ing. Tech.* **37**, 977 (1965).
6. Stand der polarographischen methoden und ihren instrumentation Teil II. Wechsel spannungsverfahren, H. W. Nürnberg and G. Wolf, *Chem. Ing. Tech.* **38**, 160–180 (1966).
7. The use of normal pulse polarography in the study of electrode kinetics, J. H. Christie, E. P. Parry, and R. A. Osteryoung, *Electrochim. Acta* **11**, 1525 (1966).
8. Characterisation of electrode reversibility by pulse polarography, K. Oldham and E. P. Parry, *Anal. Chem.* **42**, 229–233 (1970).
9. Pulse polarography, D. E. Burge, *J. Chem. Educ.* **47**, A81–88, A84, A86, A88, A90, A93–A94 (1970).
10. Anodic dissolution in pulse polarography, G. Donadey, R. Rosset, and G. Charlot, *Chem. Anal. (Warsaw)* **17**, 575–602 (1972).
11. A multimode polarograph, G. C. Barker, A. W. Gardner, and M. J. Williams, *J. Electroanal. Chem. Interfacial Electrochem.* **42**, App. 21–26 (1973).
12. Effect of pulse polarity on the shape of pulse polarograms, B. Ya. Kaplan and T. N. Sevastyanova, *Zh. Anal. Khim.* **28**, 28–32 (1973).
13. Fundamental study of differential pulse polarography, A. Saito and S. Himeno, *Nippon Kagakukaishi* **10**, 1909–1914 (1973).
14. *Innovations in pulse polarography*, J. H. Christie, Thesis, Colorado State University (1974), Ann. Arbor. Mich. Order No. 74-27940.
15. Theoretical treatment of pulsed voltammetric stripping at the thin film mercury electrode, R. A. Osteryoung and J. H. Christie, *Anal. Chem.* **46**, 351–355 (1974).
16. Pulse voltammetry at rotated electrodes, D. J. Myers, R. A. Osteryoung, and J. Osteryoung, *Anal. Chem.* **46**, 2089–2092 (1974).
17. Direct current effects in pulse polarography at the dropping mercury electrode, J. H. Christie and R. A. Osteryoung, *J. Electroanal. Chem. Interfacial Electrochem.* **49**, 301–311 (1974).
18. Pulse polarography. II. Anamolous waves observed on a normal pulse polarograph, A. Saito and S. Himeno, Nippon Kagaku Kaishi, 2340–2345 (1974).
19. Pulse polarography. I. Revised equation for diffusion current, J. Galvez and A. Sena, *J. Electroanal. Chem. Interfacial Electrochem.* **69**, 133–143 (1976).
20. Pulse polarography. II. Kinetic currents of an electrode reaction coupled to a preceding

first order reaction, J. Galvez and A. Serma, *J. Electroanal. Chem. Interfacial Electrochem.* **69**, 145–156 (1976).

21. Pulse polarography. III. Catalytic currents, J. Galvez and A. Serna, *J. Electroanal. Chem. Interfacial Electrochem.* **69**, 157–164 (1976).
22. Pulse polarography. Part IV. Contribution to the theory of catalytic current, J. Galvez, A. Serna, and T. Fuente, *J. Electroanal. Chem. Interfacial Electrochem.* **96**, 1–6 (1979).
23. Pulse polarography—theory and applications, J. Osteryoung and K. Hasebe, *Rev. Polarogr.* **22**, (1, 2), 1–25 (1976).
24. Alternate drop pulse polarography, J. H. Christie, L. I. Jackson, and R. A. Osteryoung, *Anal. Chem.* **48**, 242–247 (1976).
25. Digital simulation of differential pulse polarography, J. W. Dillard and K. W. Hauell, *Anal. Chem.* **48**, 218–222 (1976).
26. Effects of electrical transformations in normal pulse polarography on stationary electrodes, T. J. Varavko and B. Ya. Kaplan, *Zh. Anal. Khim.* **31**, 429–432 (1976).
27. Difference in half-wave potential in normal pulse polarograms in solutions of oxidised and reduced forms of depolariser in case of a completely irreversible electrochemical reaction, B. Ya. Kaplan, *Zh. Anal. Khim.* **31**, 1220 (1976).
28. Normal pulse polarography—application to the analysis of electrochemical reactions, M. Gross, *Bull. Soc. Chim. Fr.* **11–12**, Pt. 1, 1803–1811 (1976).
29. Fast sweep differential pulse voltammetry at a dropping mercury electrode, H. Blutstein and A. M. Bond, *Anal. Chem.* **48**, 248–252 (1976).
30. Some virtues of differential pulse polarography in examining adsorbed reactants, F. C. Anson, J. B. Flangen, K. Takahashi, and A. Yamada, *J. Electroanal. Chem. Interfacial Electrochem.* **67**, 253–259 (1976).
31. Pseudo derivative and pulse polarographic methods at short drop time, A. M. Bond and R. J. O'Halloran, *J. Electroanal. Chem. Interfacial Electrochem.* **68**, 257–272 (1976).
32. General equations for voltammetry with step functional potential changes to differential pulse voltammetry, S. C. Rifkins and D. H. Evans, *Anal. Chem.* **48**, 1616–1618 (1976).
33. Choice of pulse size in differential pulse polarography, B. Ya. Kaplan and T. N. Varavko, *Zh. Anal. Khim.* **32**, 639 (1977).
34. High performance pulse and differential pulse polarography. Part I. Theoretical considerations, W. P. Van Bennekom and J. B. Schulte, *Anal. Chim. Acta* **89**, 71–82 (1977).
35. Theory of pulse polarography and related chronoamperometric and chronocoulometric techniques. I. Influence of mass transport regime and heterogeneous kinetics on current potential curves, I. Ruzic, *J. Electroanal. Chem. Interfacial Electrochem.* **75**, 25–44 (1977).
36. The polarographic current "constant" in pulse polarography, J. A. Bolzan, *J. Electroanal. Chem. Interfacial Electrochem.* **75**, 157–162 (1977).
37. Effect of reactants and products adsorption in pulse polarography, J. B. Flanagan, K. Takahaski, and F. C. Anson, *J. Electroanal. Chem. Interfacial Electrochem.* **85**, 257–266 (1977).
38. Normal pulse voltammetry in electrochemically poised solutions, J. L. Morris, Jr. and L. R. Faulkner, *Anal. Chem.* **49**, 489–494 (1977).
39. Cyclic and differential pulse voltammetric behavior of reactants confined in the electrode surface, A. P. Brown and F. C. Anson, *Anal. Chem.* **49**, 1589–1595 (1977).
40. The current potential relationship in differential pulse polarography—A revision, G. J. Heigne and W. E. Van der Linden, *Anal. Chim. Acta* **99**, 183–187 (1978).
41. *Pulsed Polarography*, B. Ya Kaplan, Khimiya, Moscow USSR, 240 pp (1978) (Russ.).
42. Differential pulse polarography (review), R. F. Salikhdzhanova, N. A. Romanov, N. A. Sobina, and L. Ya. Kheifets, *Zavod. Lab.* **44**, 1171–1173 (1978).
43. Constant potential pulse polarography, J. H. Christie, L. L. Jackson, and R. A. Osteryoung, *Anal. Chem.* **48**, 561–564 (1978).
44. The current potential relationship for differential pulse polarography, I. Ruzic and M. Sluyters Rehbach, *Anal. Chim. Acta* **99**, 177–182 (1978).

45. Current potential–time relationship in differential pulse polarography: Theory of reversible, quasireversible and irreversible electrode processes, R. L. Birke, *Anal. Chem.* **50**, 1489–1496 (1978).

46. Pulse radio polarography—A new method for study of electrode kinetics from traces level to mM concentration, A. J. Andriamantena, R. Garlier, and M. Plissonier, *J. Electroanal. Chem. Interfacial Electrochem.* **97**, 77–83 (1979).

47. Theoretical aspects of new polarographic methods, Z. Galus, In: *Fundamentals of Electrochemical Analysis*, pp. 489–498, Ellis Horwood Ltd. (1976).

48. Pulse polarography. Part XI. Some problems in solving kinetic equations, L. F. Roeleveld, B. J. C. Wetsema, and J. M. Los, *J. Electroanal. Chem. Interfacial Electrochem.* **75**, 839–844 (1977).

1.7.2. Instrumentation

49. Preliminary evaluation of an electronic polarographic instrument, E. P. Parry and R. A. Osteryoung, *Anal. Chem.* **36**, 1366–1367 (1964).

50. Application of a computerized electrochemical system to pulse polarography at a hanging mercury drop electrode, H. E. Keller and R. A. Osteryoung, *Anal. Chem.* **43** (3), 342–348 (1971).

51. Instrumental artifacts in differential pulse polarography, J. H. Christie, J. Osteryoung, and R. A. Osteryoung, *Anal. Chem.* **45**, 210–215 (1973).

52. Study of dropping mercury electrode in normal pulse polarography, J. H. Christie, J. Osteryoung, and R. A. Osteryoung, *Anal. Chem.* **45**, 210–215 (1973).

53. Low noise pulse polarograph suitable for automation, B. H. Vassos and R. A. Osteryoung, *Chem. Instrum.* **5**, 257–270 (1974).

54. Inexpensive solid state modification of the Heath kit polarograph, R. W. Andrew, *Chem. Instrum.* **6**, 163–172 (1975).

55. Improved differential pulse polarography, N. Klein and Ch. Yarnitzky, *J. Electroanal. Chem. Interfacial Electrochem.* **61**, 1–9 (1975).

56. The non-faradaic background in pulse polarography, R. A. Osteryoung and J. H. Christie, *Natl. Bur. Stand (US) Specl. Publ.* **422**, 871–879 (1976).

57. Digital simulation of differential pulse polarography, J. W. Dillard and K. W. Hauck, *Anal. Chem.* **48**, 218–222 (1976).

58. R. H. Abel, J. H. Christie, L. L. Jackson, J. G. Osteryoung, and R. A. Osteryoung, *Chem. Inst.* **7**, 123–135 (1976).

59. Fast sweep differential pulse voltammetry at a dropping mercury electrode, H. Blustein and A. M. Bond, *Anal. Chem.* **48(2)**, 248–252 (1976).

60. Modification of pulse polarograph for rapid scanning and its use with stationary electrodes, K. C. Burrows, M. P. Brindle, and M. C. Hughes, *Anal. Chem.* **49**, 1459–1461 (1977).

61. Simple approach to the problem of overlapping waves using a microprocessor controlled polarograph, A. M. Bond and B. S. Grabaric, *Anal. Chem.* **48**, 1624–1628 (1976).

62. Pulse polarography. VIII. A new digital to analogue pulse polarograph, J. P. Van Dieren, B. G. W. Kaars, J. M. Los, and B. J. C. Wetsema, *J. Electroanal. Chem. Interfacial Electrochem.* **68(2)**, 129–137 (1976).

63. Dynamic compensation of the overall and uncompensated cell resistance in a two or three electrode system transient techniques, Ch. Yarmitzky and N. Klein, *Anal. Chem.* **47**, 850–854 (1976).

64. Improved low noise differential pulse polarography, R. Kalvoda and A. Trojanek, *J. Electroanal. Chem. Interfacial Electrochem.* **75**, 151–155 (1977).

65. Differential pulse anodic stripping voltammetry in a thin layer electrochemical cell, T. P. DeAngelis and W. R. Heineman, *Anal. Chem.* **48**, 2262–2263 (1976).

66. Differential pulse polarography and voltammetry with a microprocessor-controlled polarograph and a pressurised mercury electrode, A. M. Bond and B. S. Grabaric, *Anal. Chim. Acta* **88**, 227–236 (1977).
67. Thin layer differential pulse voltammetry, T. P. De Angelis, R. E. Bond, E. E. Brooks, and W. R. Heineman, *Anal. Chem.* **49**, 1792–1797 (1977).
68. Methods of relationship of short rise times of cell voltage in pulse polarography, M. Krizan, *J. Electroanal. Chem. Interfacial Electrochem.* **80**, 337–344 (1977).
69. A pulse polarograph for the measurement of fast chemical reactions, M. Krizan, H. Schmidt, and H. Strehlow, *J. Electroanal. Chem. Interfacial Electrochem.* **80**, 345–356 (1977).
70. Digital simulation of differential pulse polarography with incremental time change, J. W. Dillard, J. A. Turner, and R. A. Osteryoung, *Anal. Chem.* **49**, 1246–1250 (1977).
71. Pulse differential polarography, P. Blanquet, Fr.235/412, Cl. GOI N 27/48 Dec. 1977.
72. Differential pulse voltammetry with slow dropping mercury electrodes: A new electroanalytical technique, L. Grifore, *Cron. Chim.* **54**, 13–17 (1977).
73. Cyclic differential pulse voltammetry: A versatile instrumental approach using a computerized system, K. F. Drake, R. P. Van Duyne, and A. M. Bond, *J. Electroanal. Chem. Interfacial Electrochem.* **89**, 231–246 (1978).
74. Fast sweep differential pulse voltammetry at a dme with computerized instruments, A. M. Bond and B. S. Grabaric, *Anal. Chem.* **51**, 126–128 (1979).
75. Rapid scan alternate drop pulse polarography, J. A. Turner and R. Osteryoung, *Anal. Chem.* **50**, 1496–1500 (1978).
76. High performance pulse and differential pulse polarography, W. P. Van Bennelcom, *Anal. Chim. Acta* **101**, 283–307 (1978).
77. Use of computerised instrumentation for pseudo derivative direct current and normal pulse polarography with correction for charging current, A. M. Bond and B. S. Grabaric, *Anal. Chim. Acta* **101**, 309–318 (1978).
78. Differential pulse polarography, A. Trojanek and R. Kalvoda, Czech. 172 742 (U.GOI N 27/40) 15 May 1978.
79. Differential pulse polarograph, R. Kalvoda and A. Trojanek, Czech. 175913 (U. GOI N 27/48).
80. Differential pulse polarography, A. Trojanek and R. Kalvoda, Czech. 17914 (U GOI N 27/48).
81. Correction for background current in differential pulse alternating current and related polarographic techniques in the determination of low concentration with computerised instrumentation, A. M. Bond and B. S. Grabaric, *Anal. Chem.* **51**, 337–341 (1979).

1.7.3. Analysis

82. Pulse polarographic analysis of toxic heavy metals, J. G. Osteryoung and R. A. Osteryoung, *Am. Lab.* **4**, 8–12; 14–16 (1972).
83. Pulse stripping analysis. Determination of some metals in aerosols and other limited size samples, E. P. Parry and D. H. Hern, Joint Conf. Sensing Environment Pollutants Collec. Tech. Papers, 1, 6 pp. (1971).
84. Determination of 2,4,diamino-5-(3,4,5-timethoxy benzyl) pyrimidine (Trimethoprim) in blood and urine by differential pulse polarography, M. A. Brooks, J. A. F. De Silva, L. M. D'Arconte, *Anal. Chem.* **45**, 263–266 (1973).
85. Determination of phenobarbital and diphenylhydanation in blood by differential pulse polarography, M. A. Brooks, J. A. F. DeSilva, and M. R. Hackman, *Anal. Chim. Acta* **64**, 165–175 (1973).
86. Determination of diazepam in serum by differential pulse polarography, E. Jacobsen, T. V. Jacobsen, and T. Rojahn, *Anal. Chim. Acta* **64**, 473–476 (1973).

87. Differential pulse polarographic determination of drugs in biological fluids, M. A. Brooks, J. A. F. De Silva and M. R. Hackman, *Amer. Lab.* **5(9)**, 23–26, 28–30, 32, 34–38 (1973).

88. Application of differential pulse polarography to anodic electrode processes involving mercury compound formation, D. R. Canterford and A. S. Buchaman, *J. Electroanal. Chem. Interfacial Electrochem.* **44**, 291–298 (1973).

89. Determination of arsenic (III) at the parts per billion level by differential pulse polarography, D. J. Myers and J. Osteryoung, *Anal. Chem.* **45**, 267–271 (1973).

90. Pulse polarography in process analysis, determination of ferri, ferrous and cuprous ions, E. P. Parry and D. P. Anderson, *Anal. Chem.* **45**, 458–463 (1973).

91. Differential pulse polarographic and second harmonic a.c. polarography are used, G. E. Batley and T. M. Florence, *J. Electroanal. Chem. Interfacial Electrochem.* **55**, 23–43 (1974).

92. H. W. Nürnberg and B. Kastening, Polarographic and voltammetric techniques, In: *Methodium Chimicum*, F. Korte, ed., Vol. 1/A, pp. 584–607, Academic Press, New York (1974).

93. Determination of submicromolar concentrations of sulfonamides by differential pulse polarography after diazotisation and coupling with 1-naphthol, A. G. Fogg and Y. Z. Ahmed, *Anal. Chim. Acta* **70**, 241–244 (1974).

94. Normal pulse polarography of double helical DNA: Dependence of the wave height on starting potential, E. Palecek, *Collect. Czech. Chem. Commun.* **39**, 3449–3455 (1974).

95. Microanalysis of heavy metals by differential pulse polarography, S. Hayano and N. Shinozuka, *Seisan-Kenkyu*, **26**, 78–81 (1974).

96. Effect of film formation on the normal pulse polarographic behaviour of sulphide, D. R. Cauterford, *J. Electroanal. Chem. Interfacial Electrochem.* **52**, 144–147 (1974).

97. Electroreduction and pulse polarographic determination of nicotinamide in multivitamin tablets, E. Jacobson and K. B. Thorgersen, *Anal. Chim. Acta* **71**, 175–184 (1974).

98. Determination of chlorodiazepoxide hydrochloride (librium) and its major metabolites in plasma by differential pulse polarography, M. R. Hackman, M. A. Brooks, and J. A. F. DeSilva, *Anal. Chem.* **46**, 1075–1082 (1974).

99. Amperometric titration employing differential pulse polarography, D. J. Myer and J. Osteryoung, *Anal. Chem.* **46**, 356–359 (1974).

100. *Differential pulse polarography in trace analysis*, Bagheri Ababar Ph.D. Thesis, Howard Univ., Washington, D.C. (1975).

101. Application of differential pulse polarography to the assay of vitamins, J. Lindquist and S. M. Farroha, *Analyst* **100**, 377–385 (1975).

102. Differential pulse polarographic determination of some carcinogenic nitrosamines, K. Hasebe and J. Osteryoung, *Anal. Chem.* **47**, 2412–2418 (1975).

103. Trace level determination of 1,4-benzodiazepines in blood by differential pulse polarography, M. A. Brooks and M. R. Hackerman, *Anal. Chem.* **47**, 2059–2062 (1975).

104. Determination of 1,4 benzodiazepines in biological fluids by differential pulse polarography, M. A. Brooks and J. A. F. DeSilva, *Talanta* **22**, 849–860 (1975).

105. Determination of thiols by pulse polarography, F. Peter and R. Rosset, *Anal. Chim. Acta* **79**, 47–58 (1975).

106. Electrochemistry of thiols and disulphides II. The d.c., a.c., and differential pulse polarography of glutathione, G. A. Mairesse-Ducarmois, G. J. Patriarche, and J. L. Vandenbalck, *Anal. Chim. Acta* **76**, 299–308 (1975).

107. Organic functional group analysis at the micro level. I. Determination of the N-oxide function by differential pulse polarography, T. S. Ma, M. R. Hackman and M. A. Brooks, *Mikrochim. Acta* **2**, 617–625 (1975).

108. Accuracy and reliability of trace metal determination in environmental samples by advanced polarographic and spectroscopic techniques, H. W. Nurnberg, M. Stoeppler, and P. Valenta, *Thalassia Jugose* **11**, 85–100 (1975) (Eng).

109. Comparison of differential pulse and d.c. sampled polarography for the determination of ferrous and manganous ions in lake water, W. Davison, *J. Electroanal. Chem. Interfacial Electrochem.* **72**, 229–237 (1976).

110. Determination of trace elements in zinc plant electrolyte by differential pulse polarography and anodic stripping voltammetry, E. S. Pilkington and C. Weeks, *Anal. Chem.* **48**, 1665–1669 (1976).

111. Determination of uranium in plutonium-238 metal and oxide by differential pulse polarography, N. C. Fawcet, *Anal. Chem.* **48**, 215–218 (1976).

112. Analytical evaluation of differential pulse voltammetry at stationary electrodes using computer based instrumentation, S. C. Rifkin and D. H. Evans, *Anal. Chem.* **48**, 2174–2180 (1976).

113. Determination trace mercury (II) in 0.1 M perchloric acid by differential pulse stripping voltammetry at a rotating gold disc electrode, R. W. Andrews, J. H. Larochdle, and D. C. Johnson, *Anal. Chem.* **48**, 212–214 (1976).

114. Applications of polarography and voltammetry to marine and aquatic chemistry. Part II. The polarographic approach to the determination and speciation of metals in the marine environment, Fresenius, *Z. Anal. Chem.* **282**, 357–367 (1976).

115. Effects of surfactants in differential pulse polarography, E. Jacobson and H. Lindseth, *Anal. Chim. Acta*, **86**, 123–127 (1976).

116. Differential pulse polarographic determination of acrolein in water samples, L. H. Howe, *Anal. Chem.* **48**, 2167–2169 (1976).

117. Determination of acrylamide monomer by differential pulse polarography, S. R. Betso and J. D. McLean, *Anal. Chem.* **48**, 766–770 (1976).

118. Amperometric and differential pulse voltammetric detection in high performance liquid chromatography, D. G. Swartzfager, *Anal. Chem.* **48**, 2189–2192 (1976).

119. Applications of modern polarographic methods to analytical and physical chemistry, H. W. Nürnberg, *Kem. Kozl.* **45**, 13–46 (1976) (Hung.).

120. Potentialities and applications of advanced polarographic and voltammetric methods in environmental research and surveillance of toxic metals, H. W. Nürnberg, *Electrochim. Acta* **22**, 935–949 (1977).

121. Applications of polarography and voltammetry to marine aquatic chemistry. IV. A new voltammetric method for the study of mercury traces in seawater and in land water, L. Sipos, P. Valenta, H. W. Nürnberg, and M. Branica, *J. Electroanal. Chem. Interfacial Electrochem.* **77**, 263–267 (1977).

122. New potentialities in ultra trace analysis with differential pulse anodic stripping voltammetry, P. Valenta, L. Mart, and H. Rutzel, *J. Electroanal. Chem. Interfacial Electrochem.* **82**, 327–343 (1977).

123. Pulse polarographic analysis of nucleic acids and proteins, E. Paleck, V. Brabec, F. Jelen, and Z. Pechan, *J. Electroanal. Chem. Interfacial Electrochem.* **75**, 471–485 (1977).

124. Differential pulse polarographic determination of nitrate and nitrite, S. W. Boese and V. S. Archer, *Anal. Chem.* **49**, 479–484 (1977).

125. Determination of nitrite ion using differential pulse polarography, S. K. Chang, R. Kozeniauskas, and G. W. Harrington, *Anal. Chem.* **49**, 2272–2275 (1977).

126. Polarographic study of certain progestogens and their determination in oral contraceptive tablets by differential pulse polarography, L. G. Chatten, R. N. Yadev, S. Binnigton, R. E. Moskalyk, *Analyst* **102**, 323–327 (1977).

127. Cathodic pulse stripping analysis of iodine at the parts per billion level, R. C. Propst, *Anal. Chem.* **49**, 1199–1205 (1977).

128. Direct and titrimetric determination of hydroxide using normal pulse polarography at mercury electrodes, E. Kirowa-Eisner and J. Osteryoung, *Anal. Chem.* **50**, 1062–1066 (1978).

129. Differential pulse polarographic determination of molybdenum at parts per billion levels, P. Bosserman, D. T. Sawyer, and A. L. Page, *Anal. Chem.* **50**, 1300–1303 (1978).

130. Differential pulse polarography of phenylarsine oxide, J. H. Lowry, R. B. Smart, and K. H. Mancy, *Anal. Chem.* **50**, 1303–1309 (1978).

2. Linear Potential Sweep and Cyclic Voltammetry

Linear sweep voltammetry was first used by Matheson and Nichols.[1] Many researchers have contributed to the theory of this method. Randles[2,3] and Sevick have considered the theory for the single scan method for a reversible reaction at a plane electrode. Delahay[4-6] extended the theory for totally irreversible reactions and also developed a unified theory. Matsuda and Ayabe[9,11] reexamined the Randles–Sevick theory as well as Delahay's theory, and then extended it to the quasireversible case. The theory of stationary electrode polarography has been further extended by Reinmuth[13,21,37] for spherical electrodes, and Nicholson[12] for cylindrical electrodes. The theory of stationary electrode polarography has been reexamined by Nicholson and co-workers.[27-31] Shain *et al.*[36,37,42,43,47,48] have presented the theory for stationary electrode polarography for multistep charge transfer electrode processes with a preceding chemical reaction and involving adsorption of electroactive species. Saveant *et al.*[20,35-41,45,54,55] have developed the theory for linear sweep voltammetry for electrochemical reactions involving dimerization and disproportionation and also extended the convolution and finite difference approach to cyclic voltammetry. Thin layer potential sweep voltammetry has been studied in detail by Laviron.[50,57] Kemula and his co-workers[14,15] have employed cyclic voltammetry with hanging mercury drop for studying the mechanism of reduction of some organic compounds and also the formation of intermetallic compounds. Linear potential sweep and cyclic voltammetry have been reviewed by Vogel,[24] Yeager and Kuta,[49] and MacDonald.[53]

The instrumentation for potentiostat, linear, and triangular pulse generator integrators have been well developed and presented.[63-65,69,70,78,80] The instrumentation for digital data acquisition has been developed by many researchers.[66,67,71,79] The application of computers in these two voltammetric studies have also been described.[75,77,81-84]

Linear potential sweep and cyclic voltammetric techniques find extensive application not only in the mechanistic studies on electrode processes[91,97,98,101,103-105] but also in chemical analysis.[85-87,90,94,102,107-122] The method is more sensitive than polarography when the linear potential sweep voltammetry is combined with stripping analysis, employing hanging mercury drop, glassy carbon, or other suitable electrodes. The method can be extended to trace analysis determinations.[88,89,92,100]

2.1. General Principles and Theory

1. The cathode ray oscillograph applied to the dropping mercury electrode, L. A. Matheson and N. Nichols, *Trans. Electrochem.* 193–210 (1938).
2. Cathode ray polarograph. Verification of the theoretical equation, J. E. Randles, *Trans. Faraday Soc.* **44**, 334 (1948).
3. A cathode polarograph. Part II. Current-voltage curves, J. E. B. Randles, *Trans. Faraday Soc.* **44**, 327–338 (1948).

4. Theory of irreversible waves in oscillographic polarography, P. Delahay, *J. Am. Chem. Soc.* **75**, 1190–1196 (1953).
5. Unified theory of polarographic waves, P. Delahay, *J. Am. Chem. Soc.* **75**, 1430–1435 (1953).
6. Oscillographic polarographic waves for the reversible deposition of metals on solid electrodes, T. Berzins and P. Delahay, *J. Am. Chem. Soc.* **75**, 555–559 (1953).
7. Diffusion currents at cylindrical electrodes. A study of organic sulfides, M. M. Nicholson, *J. Am. Chem. Soc.* **76**, 2539–2545 (1954).
8. Voltammetry and polarography with continuously changing potential, P. Delahay, In: *New Instrumental Methods in Electrochemistry*, Chap. 6, pp. 115–139, Interscience Pub., New York (1954).
9. Zur theorie der Randles-Sevick kathodenstrahl-polarographie, H. Matsuda and Y. Ayabe, *Z. Elektrochem.* **59**, 494–503 (1958).
10. Diffusion currents at spherical electrodes, R. P. Frankenthal and I. Shain, *J. Am. Chem. Soc.* **78**, 2969–2979 (1956).
11. The theory of polarographic currents. A general formula for the diffusion-caused current and its application, H. Matsuda, *Z. Elektrochem.* **61**, 489–506 (1957).
12. Polarography of metallic monolayers, M. M. Nicholson, *J. Am. Chem. Soc.* **79**, 7–12 (1957).
13. Nernst controlled currents in hanging drop polarography, W. H. Reinmuth, *J. Am. Chem. Soc.* **79**, 6358–6360 (1957).
14. Cyclic voltammetry with application of the hanging mercury drop electrode. II. Investigation of the mechanism of the reduction of p-nitroaniline, W. Kemula and Z. Kublic, *Bull. Acad. Polon. Sci. Ser. Sci. Chim.* **6**, 653–659 (1958).
15. A new voltammetric method of investigation of the formation of intermetallic compounds using the hanging mercury electrode, W. Kemula, Z. Galus, and Z. Kertlik, *Bull Acad. Polon. Ser. Sci. Chim. Geol.* **6**, 661–668 (1958).
16. Voltammetry with linearly varying potential case of irreversible waves at spherical electrodes, R. D. DeMars and I. Shain, *J. Am. Chem. Soc.* **81**, 2654–2659 (1959).
17. Voltammetry with the hanging mercury drop electrode, W. Kemula, In: *Advances in Polarography*, I. S. Longmuir, ed., Vol. 1, pp. 105–143, Pergamon Press, Oxford (1960).
18. Kinetic equation for irreversible reactions in oscillographic polarography, A. Ya. Gokhshtein and Ya. P. Gokhshtein, *Dokl. Akad. Nauk SSSR* **131**, 601–604 (1960).
19. Voltammetry with the hanging mercury drop electrode, Y. P. Gokstein and A. Y. Gokshtein, In: *Advances in Polarography*, I. S. Longmuir, ed., pp. 105–143, Pergamon Press, New York (1960).
20. Recherches sur les courants catalytiques en polarographie-oscillographique a'balayage lineaire de tension, J. M. Saveant and E. Vianello, In: *Advances in Polarography*, I. S. Longmuir, ed. Vol. I, pp. 367–374, Pergamon Press, New York (1960).
21. Theory of stationary electrode polarography, W. H. Reinmuth, *Anal. Chem.* **33**, 1793–1794 (1961).
22. Theory of stripping voltammetry with spherical electrodes, W. H. Reinmuth, *Anal. Chem.* **33**, 185–187 (1961).
23. Theory of diffusion limited charge transfer processes in electroanalytical techniques, W. H. Reinmuth, *Anal. Chem.* **34**, 1446–1454 (1962).
24. Single sweep method, J. Vogel, In: *Progress in Polarography*, I. M. Kolthoff and P. Zuman, eds., Vol. II, pp. 429–448, Interscience Publishers, New York (1962).
26. Triangular wave cyclic voltammetry, Z. Galus, H. Y. Lee, and R. N. Adams, *J. Electroanal. Chem.* **5**, 17–22 (1963).
27. Theory of stationary electrode polarography: Single scan and cyclic methods applied to reversible, irreversible, and kinetics system, R. S. Nicholson and I. Shain, *Anal. Chem.* **36**, 706–723 (1964).
28. Theory of stationary electrode polarography for a chemical reaction coupled between two charge transfer, R. S. Nicholson and I. Shain, *Anal. Chem.* **37**, 178–190 (1965).
29. Experimental verification of an ECE mechanism for the reduction of p-nitrosophenol, using

stationary electrode polarography, R. S. Nicholson and I. Shain, *Anal. Chem.* **37**, 190–195 (1965).

30. Some examples of the numerical solutions of non-linear integral equations, R. S. Nicholson, *Anal. Chem.* **37**, 667–671 (1965).

31. Theory and application of cyclic voltammetry for measurement electrode reaction kinetics, R. S. Nicholson, *Anal. Chem.* **37**, 1351–1355 (1965).

32. Significance of non-steady state a.c. and d.c. measurements in electrochemical adsorption kinetics. Application to galvanostatic and voltage sweep methods, B. E. Conway, E. Gileadi, and H. A. Kozlowska, *J. Electrochem. Soc.* **112**, 341 (1965).

33. A potential step-linear scan method for investigating chemical reactions initiated by a charge transfer, W. H. Schwarz and I. Shain, *J. Phys. Chem.* **70**, 845–852 (1966).

34. Evaluation of stationary electrode polarography and cyclic voltammetry for the study of electrode processes, S. P. Perone, *Anal. Chem.* **38**, 1958–1963 (1966).

35. The potential sweep method: A theoretical analysis, S. Srinivasan and E. Gileadi, *Electrochim. Acta* **11**, 34–35 (1966).

36. Multistep charge transfers in stationary electrode polarography, D. S. Polcyn and I. S. Shain, *Anal. Chem.* **38**, 370–375 (1968).

37. Theory of stationary electrode polarography for a multistep charge transfer with catalytic (cyclic) regeneration of the reactant, D. S. Polcyn and I. Shain, *Anal. Chem.* **38**, 376–382 (1966).

38. Potential sweep voltammetry: Theoretical analysis of monomerization and dimerization mechanisms, J. M. Saveant and E. Vianello, *Electrochim. Acta* **12**, 1545–1561 (1967).

39. Potential sweep voltammetry: General theory of chemical polarization, J. M. Saveant and E. Vianello, *Electrochim. Acta* **12**, 629–646 (1967).

40. ECE mechanism as studied by polarography and linear sweep voltammetry, J. M. Saveant, *Electrochim. Acta* **12**, 753–766 (1967).

41. Cyclic voltammetry with asymmetrical potential scan: A simple approach to mechanisms involving moderately fast chemical reactions, J. M. Saveant, *Electrochim. Acta* **12**, 999–1030 (1967).

42. Effects of adsorption of electroactive species in stationary electrode polarography, R. H. Wopschal and I. Shain, *Anal. Chem.* **39**, 1514–1527 (1967).

43. Adsorption effects in stationary electrode polarography with a chemical reaction following charge transfer, R. H. Wopschal and I. Shain, *Anal. Chem.* **39**, 1535–1542 (1967).

44. The interpretation of adsorption pseudo capacitance curves as measured by the potential sweep method, I. J. M. Hale and R. Greef, *Electrochim. Acta* **12**, 1409–1420 (1967).

45. Disproportionation and ECE-mechanisms. I. Theoretical analysis. Relationships for linear sweep voltammetry, M. Mastsagostino, L. Nadjo, and J. M. Saveant, *Electrochim. Acta* **13**, 721–749 (1968).

46. Nonunity electrode reaction orders and stationary electrode polarography, M. S. Shuman, *Anal. Chem.* **41**, 142–146 (1969).

48. Rate controlled adsorption of products in stationary electrode polarography, M. H. Hulbert and I. Shain, *Anal. Chem.* **42**, 162–171 (1970).

49. Single and cyclic linear sweep voltammetry, E. Yeager and J. Kuta, In: *Techniques of Electrochemistry*, E. Yeager and A. J. Salkind, eds., vol. 1, pp. 198–202, Wiley–Interscience, New York (1972).

50. Theoretical study of a reversible reaction followed by a chemical reaction in thin layer linear potential sweep voltammetry, E. Laviron, *J. Electroanal. Chem., Interfacial Electrochem.* **39**, 1–23 (1972).

51. Theory of potential scan voltammetry with finite diffusion kinetics and other complications, H. E. Keller and W. H. Reinmuth, *Anal. Chem.* **44**, 1167–1178 (1972).

52. Low temperature studies of electrochemical kinetics. I. Cyclic voltammetry of diethyl fumarate, R. D. Gryfa and J. T. Maloy, *J. Electrochem. Soc.* **122**, 377–383 (1975).

53. Linear potential sweep and cyclic voltammetry, D. D. MacDonald, In: *Transient Techniques in Electrochemistry*, Chap. 6, pp. 185–228, Plenum Press, New York (1977).

54. ECE and disproportionation. Part V. Stationary state general solution application to LSV, C. Amatore and J. M. Saveant, *J. Electroanal. Chem. Interfacial Electrochem.* **85**, 27–46 (1977).

55. Convolution and finite difference approach: Application to cyclic voltammetry and spectroelectrochemistry, C. Amatore, L. Nadjo, and J. M. Saveant, *J. Electroanal. Chem. Interfacial Electrochem.* **90**, 321–331 (1978).

56. Limiting diffusion currents in hydrodynamic voltammetry. Part IV. Stationary disc and ring electrodes in a rotational flow produced by a rotating disc, S. Hamada, M. Itoh, M. Matsuda, and J. Yamada, *J. Electroanal. Chem. Interfacial Electrochem.* **91**, 107–114 (1978).

57. Theory of regeneration mechanisms in thin layer potential sweep voltammetry, E. Laviron, *J. Electroanal. Chem. Interfacial Electrochem.* **87**, 31–37 (1978).

58. The effect of slow two electron transfers and disproportionation on cyclic voltammograms, M. D. Ray, *J. Electroanal. Chem. Interfacial Electrochem.* **125**, 547–555 (1978).

59. Quantitative resolution of overlapped peaks in programmed potential step voltammetry, P. A. Boudreau and S. P. Perone, *Anal. Chem.* **51**, 811–817 (1979).

2.2. Instrumentation

60. Gerat fur schnelle oscillographische quantitative analyse, J. Vogel and P. Valenta, *Collect. Czech. Chem. Commun.* **21**, 502–508 (1956).

61. Arbeits methoden und anevendung der gleichspannung polarographie. I. Apparatives, methoden und elektroden und II. Theorie de polarographischen kurve, H. W. Nürnberg and M. Stackelberg, *J. Electroanal. Chem.* **2**, 181–229 (1961).

62. Modern methods of current voltage polarography, H. W. Nürnberg, *Z. Anal. Chem.* **186**, 1–53 (1962).

63. A digital readout device for analog integrators, E. C. Toren Jr. and C. P. Driscoll, *Anal. Chem.* **35**, 1809–1810 (1963).

64. Operational amplifier circuits for controlled potential cyclic voltammetry. II. J. R. Alden, J. Q. Chambers, and R. N. Adams, *J. Electroanal. Chem.* **5**, 152–157 (1963).

65. Electroanalytical controlled potential instrumentation, G. L. Booman and W. B. Holbrook, *Anal. Chem.* **35**, 1793–1809 (1963).

66. A digital data collection, wave form generation, and timing instrument for electrochemical measurements, G. L. Boosman, *Anal. Chem.* **38**, 1141–1148 (1966).

67. Instrumentation for digital data acquisition systems in voltammetric techniques, E. R. Brown, D. E. Smith, and D. Deford, *Anal. Chem.* **38**, 1130–1136 (1966).

68. Some investigations on instrumental compensation of non-faradaic effects in voltammetric techniques, E. R. Brown, D. E. Smith, T. G. McCord, and D. Deford, *Anal. Chem.* **38**, 1119–1130 (1966).

69. Continuous ohmic polarisation compensator for a voltammetric apparatus utilising operational amplifiers, D. Pouli, J. R. Huff, and J. C. Pearson, *Anal. Chem.* **38**, 382–384 (1966).

70. Chemical instrumentation, G. W. Ewing, *J. Chem. Educ.* **46**, A717 (1967).

71. Electrochemical data acquisition and analysis system based on a digital computer, G. Lauer, R. Abel, and F. C. Anson, *Anal. Chem.* **40**, 765–769 (1967).

72. Chemical instrumentation, J. B. Flato, Princeton Applied Res. Tech. Note, T-193, Amer. Lab. Feb. 1969, p. 10.

73. Solid-state polarographic instrumentation—A unique approach to versatility, R. Bezman and P. S. Mckinney, *Anal. Chem.* **41**, 1560–1567 (1969).

74. Solid state signal generators for electroanalytical experiments, R. L. Myers and I. Shain, *Chem. Instrum.* **2**, 203–211 (1969).

75. Real time computer optimisation of stationary electrode polarographic measurements, S. P. Perone, D. O. Jones, and W. F. Gutknechr, *Anal. Chem.* **41**, 1154–1162 (1969).

76. Modularized digitizing time-synchronizing current-sampling system for electroanalytical studies, R. G. Glem and W. W. Goldsworthy, *Anal. Chem.* **43**, 918–928 (1971).

77. Computerised learning machine applied to qualitative analysis of mixtures by stationary electrode polarography, S. P. Perone and L. B. Sybrandt, *Anal. Chem.* **43**, 382 (1971).

78. Linear sweep voltammetry—Compensation of cell resistance and stability. Determination of the residual uncompensated resistance, D. Garrean and J. M. Saveant, *J. Electroanal. Chem. Interfacial Electrochem.* **35**, 309–331 (1972).

79. Evaluation of a computerised sampling technique for digital data acquisition of high speed transient wave forms. Application to cyclic voltammetry, D. E. Smith, G. Greason, and R. J. Lloyd, *Anal. Chem.* **44**, 1159–1166 (1972).

80. Operational amplifier instruments for electrochemistry, R. R. Schroeder, In: *Electrochemistry*, J. S. Mattson, H. B. Mark, Jr., and H. C. MacDonald, Jr. eds., Chap. 10, pp. 293–301, 314–350, Marcel Dekker Inc., New York (1972).

81. Interactive laboratory computer applications to electroanalytical research, S. P. Perone, *Ing. Chem. Comput. Assisted Chem. Res. Des.* (Pop. Discuss Jt. jPR, U.S. Seminar) (1973) (Pat. 1975), pp. 51–75.

82. Fast fourier transform based interpolation of sampled electrochem. data, R. J. O'Halloram and D. E. Smith, *Anal. Chem.* **50**, 1391–1394 (1978).

83. Computerised pattern recognition for classification of organic compounds from voltammetric data, D. R. Burgard and S. P. Perone, *Anal. Chem.* **50**, 1366–1371 (1978).

84. Quantitative resolution of overlapped peaks in programmed potential step voltammetry, P. A. Bourdeau and S. P. Perone, *Anal. Chem.* **51**, 811–817 (1979).

2.3. Applications

85. Voltammetry and polarography with continuously changing potential: Applications to analytical, P. Delahay, In: *New Instrumental Methods in Electrochemistry*, Chap. 6, pp. 139–143, Interscience Publishers, New York (1954).

86. Analytical applications of the hanging mercury drop electrode, J. W. Ross, R. D. DeMars, and I. Shain, *Anal. Chem.* **28**, 1768–1771 (1956).

88. Anodic stripping voltammetry using the hanging mercury drop electrode, R. D. DeMars and I. Shain, *Anal. Chem.* **29**, 1825–1827 (1957).

90. Application de la goute pendaute de mercure a la determination d.c. minimes quantites de different ions, W. Kemula and Z. Kublic, *Anal. Chem. Acta* **18**, 104–111 (1958).

91. Cyclic voltammetry with applications of the hanging mercury drop electrode. II. Investigation of the mechanism of the reduction of p-nitroaniline, W. Kemula and Z. Kublic, *Bull. Acad. Polon. Sci. Sev. Sci. Chim.* **6**, 653–659 (1958).

92. A new voltammetric method of investigation of the formation of intermetallic compounds using the hanging mercury electrode, W. Kemula, Z. Galus, and Z. Kublic, *Bull. Acad. Polon. Ser. Sci. Chim. Geol.* **6**, 661–668 (1958).

93. The influence of gold in mercury electrodes on certain electrode processes, W. Kemula, Z. Galus, and Z. Kublic, *Bull. Acad. Sci. Ser. Sci. Chim.* **7**, 613–618 (1959).

94. Voltammetry at inert electrodes. I. Analytical applications of boron carbide electrodes, T. R. Mueller and R. N. Adams, *Anal. Chim. Acta* **23**, 467–479 (1960).

95. Voltammetry with the hanging mercury drop electrode, W. Kemula, In: *Advances in Polarography*, I. S. Longmuir, ed., Vol. 1, p. 105, Pergamon Press (1960).

96. Unter suching von adsorption serschein angen an rhodium iridium palladium und gold mit der potentiostatischen dreieck methode, F. G. Will and C. A. Knorr, *Z. Elektrochem.* **64**, 270–275 (1960).

97. Voltammetry at inert electrodes. II. Correlation of experimental results with theory for voltage and controlled potential scanning, controlled potential electrolysis and chronopotentiometric techniques. Oxidation of ferrocyanide and o-dianisdine at boron carbide electrodes, T. R. Mueller and R. Adams, *Anal. Chim. Acta* **25**, 482–497 (1961).

98. Electrolytic generation of radical anions in aqueous solutions, L. H. Piette, P. Luding, and R. N. Adams, *J. Am. Chem. Soc.* **83**, 3909–3910 (1961).

99. The hanging mercury drop in polarography, J. Riha, In: *Progress in Polarography*, I. M. Kolthoff and P. Zuman, eds., Vol. II, pp. 383–396, Interscience Publishers, New York (1962).

100. Stripping analysis, I. Shain, In: *Treatise on Analytical Chemistry*, I. M. Kolthoff and P. J. Elving, eds., Part I, Sec. D-2, Chap. 50, pp. 2534–2564, Interscience, New York (1963).

101. Integration of single sweep oscillopolarograms. I. Determination of reactant adsorption and oxide film reduction at platinum electrodes, R. A. Osteryoung, G. Lauer, and F. C. Anson, *J. Electrochem. Soc.* **110**, 926 (1963).

102. Determination of mixtures by single sweep oscillography, R. A. Osteryoung and E. P. Parry, *J. Electroanal. Chem.* **9**, 299–304 (1965).

103. Effect of surface state of platinum on electrochemical adsorption of oxygen in acid solutions, V. Lukyanycheva, V. Tikhomirova, and V. S. Bagotskii, *Elektrokhimiya* **1**, 262–266 (1975).

104. Application of controlled potential techniques to study of rapid succeeding chemical reaction coupled to electro-oxidation of ascorbic acid, S. P. Perone and J. W. Kretlow, *Anal. Chem.* **38**, 1760–1763 (1966).

105. Potential sweep voltammetry—Adsorption of chloranalic acid on a mercury electrode, S. Roffia and E. Vianello, *J. Electroanal. Chem. Interfacial Electrochem.* **15**, 405–413 (1967).

106. A comparison of radiotracer and electrochemical methods for the measurement of the electrosorption of organic molecules, E. Gileadi, L. Duic, and J. O'M Bockris, *Electrochim. Acta* **13**, 1915–1935 (1968).

107. Voltammetric determination of neptunium at the glassy carbon electrode, C. E. Plock, *J. Electroanal. Chem. Interfacial Electrochem.* **18**, 289–293 (1968).

108. Ferrate (VI) analysis by cyclic voltammetry, A. S. Venkatadri, W. F. Wagner, and H. H. Bauer, *Anal. Chem.* **43**, 1115–1119 (1971).

109. Voltammetry with disk electrodes and the analytical application, M. Kopanica and F. Vydra, *J. Electroanal. Chem. Interfacial Electrochem.* **31**, 175 (1971).

110. Voltammetry with disk electrodes and its analytical applications, F. Vydra and M. Stulikova, *Collect. Czech. Chem. Commun.* **37**, 123 (1972).

111. Voltammetry with disk electrodes and its analytical applications. VIII. The cyclic voltammetry of copper (II) at the glassy carbon rotating disk electrode, M. Stulikova and F. Vydra, *J. Electroanal. Chem. Interfacial Electrochem.* **44**, 117–127 (1973).

112. Determination of nanogram quantities of carbonyl compound using twin cell potential sweep voltammetry, B. K. Afghan, A. V. Kulkarni, and J. F. Ryan, *Anal. Chem.* **47**, 488–494 (1975).

113. A comparison of the analytical utility of three different potential ramp techniques in voltammetry using a carbon paste electrode, P. Söderhjelm, *J. Electroanal. Chem. Interfacial Electrochem.* **71**, 109–115 (1976).

114. Use of coelectrocrystallisation for expanding the possibilities of voltammetric analysis, T. A. Krapivkima, E. M. Roizenblat, V. V. Nosacheva, G. A. Kalambet, and G. N. Veretina, *Metody Polych. Anal. Mater Electron Tekh.*, 95–101 (1976).

115. Determination of catecholamines by thin-layer linear sweep voltammetry, R. R. Fike and D. J. Curran, *Anal. Chem.* **49**, 1205–1210 (1977).

116. Turbulent hydrodynamic voltammetry. III. Analytical studies using a turbulent voltammetric cell, M. Varadi and E. Pungor, *Magy. Kem. Foly.* **84**, 58–61 (1978).

117. Data treatment in cyclic voltammetry, G. Gunzburg, *Anal. Chem.* **50**, 375–376 (1978).

118. Micromolar voltammetric analysis by ring electrode studding at a rotating ring-disc electrode, S. Bruckenstein and P. R. Gifford, *Anal. Chem.* **51**, 250–255 (1979).

119. A comparison of radioatracer and electrochemical methods for the measurement of the

electrosorption of organic molecules, E. Gileadi, L. Duic, and J. O'M. Bockris, *Electrochim. Acta* **13**, 1915–1935 (1968).

120. Cyclic and stripping voltammetry of mercury at impregnated graphite electrode in chloride thiocyanate media, R. Bilewicz, Z. Stojek, and Z. Kublile, *J. Electroanal. Chem. Interfacial Electrochem.* **96**, 29–44 (1979).
121. Determination of nitrogen oxides and sulphur dioxide absorbed into iron (II) EDTA solution by voltammetry, ACS/CSJ-Chem. Congr. No. 2412 CTE No. 7 (1979).
122. Voltammetric behaviour of chlorine/chloride system and detection of chloride ions in molten nitrates, E. Desimoni and F. Palmisano, *Anal. Chem.* **51**, 122 (1979).

3. Stripping Voltammetry

Stripping voltammetry (sv), first applied by Zbinden as early as 1931,[1] developed as an electroanalytical technique only after Rogers,[5,8,9] Barker and Jenkins,[63] DeMars and Shain,[80,82] and Kemula and Kublik,[16,173] have showed its remarkable sensitivity.

A number of reviews and books on sv have appeared and some of them are by Kemula,[16] Barendrecht,[54] Neeb,[36] Brainina,[43] Nürnberg,[204] and Vydra *et al.*[45] The theory of sv has been discussed by a number of researchers.[19,29,33,34,47–51,57–64,69–75,78,208,209,214] The use of the hanging mercury drop electrode for sv has been studied by Kemula *et al.*, DeMars *et al.*, and others.[80,82,84,87–90,97,98] The employment of electrodes such as thin mercury film electrodes,[29,91,111–113,116,120,128,130,132,140,145,147,156,160,169] graphite, and other solid electrodes[99,111,112,114,115,117,121–123,163,172] have also been described.

The application of different polarographic and voltammetric techniques for stripping analysis has been studied in great detail. While the application of linear potential sweep and cyclic voltammetry for sv is discussed in References (16), (17), (33), (36), (43), (45), (54), (93), (132), (215), (221), and (245), the use of pulse polarography (PP), differential pulse polarography (dpp), a-c polarography (acp) and square-wave polarography (sqp) are discussed in References (pp) (43), (45), (109), (147), (198), (204), (214), (227), (229), and (240); (dpp) (43), (45), (105), (124), (131), (134), (137), (154), (199), (204), (208), (214), (219), (222), (225), (227), (239); (a.c.p.) (36), (43), (45), (65), (103), (114), (119), (161); and (sq.p) (36), (43), (45), and (94), respectively. The use of computerized/programmable instrumentation for stripping voltammetry has been discussed in References (125), (137), (138), (149), and (168).

The applications of cathodic stripping voltammetry are discussed in References (268)–(285).

3.1. General Principles and Theory

1. New method of microdetermination of the copper ion, C. Zbinden, *Bull. Soc. Chim. Biol.* **13**, 35–40 (1931).

2. The theory of concentration polarization, V. G. Levich, *Acta Phys. Chim. USSR* **17**, 257–307 (1942).
3. Theory of concentration polarization. II. Steady-state regime, V. G. Levich, *Acta Phys. Chim. USSR* **19**, 117–132 (1944).
4. Theory of concentration polarization. III. The transition regime, V. G. Levich, *Acta Phys. Chim. USSR* **19**, 133–138 (1944).
5. Coulometric determination of submicrogram amounts of silver, S. S. Lord, Jr., R. C. O'Neill, and L. B. Rogers, *Anal. Chem.* **24**, 209–213 (1952).
6. Square wave polarography, G. C. Barker and I. L. Jenkin, *Analyst* **77**, 685–696 (1952).
7. Polarographic waves of simple metal ions. Significance of the half-wave potential, I. M. Kolthoff and J. J. Lingane, In: *Polarography*, Vol. 1, pp. 190–203, Wiley–Interscience, New York (1952).
8. Coulometric determinations of submicrogram amounts of cadmium and zinc with stationary-mercury plated platinum electrodes, K. W. Gardner and L. B. Rogers, *Anal. Chem.* **25**, 1393–1397 (1953).
9. Coulometric and polarographic determination of trace amounts of lead, T. L. Marple and L. B. Rogers, *Anal. Chim. Acta* **11**, 574–585 (1954).
10. Currents controlled by the rate of semi-infinite linear diffusion, P. Delahay, In: *New Instrumental Methods in Electrochemistry*, pp. 47–57, Wiley Interscience, New York (1954).
11. Inverse polarography with stationary amalgam anodes. I. Basic principles and technique, A. Hickling, J. Maxwell, and J. V. Shennan, *Anal. Chim. Acta* **14**, 287–295 (1956).
12. Polarography of metallic monolayers, M. M. Nicholson, *J. Am. Chem. Soc.* **79**, 7–12 (1957).
13. Anodic amalgam voltammetry. I. Principles of the method, R. Neeb, *Z. Anal. Chem.* **171**, 321 (1959).
14. Classification and nomenclature of electroanalytical methods, P. Delahay, G. Charlot, and H. A. Laitinen, *Anal. Chem.* **32**, 103A (1960).
15. Polarographie im durchfliessenden elektrolyt I. Ein fuhrungsmitteilung, F. Strafelda, *Collect. Czech. Chem. Commun.* **25**, 862–870 (1960).
16. Voltammetry with the hanging mercury drop electrode, In: W. Kermula, *Advances in Polarography*, Vol. I, I. S. Longmuir, ed., p. 105, Pergamon, Oxford (1960).
17. Electrolysis with constant potential: Reversible processes at a hanging mercury drop electrode, I. Shain and K. J. Martin, *J. Phys. Chem.* **65**, 254–258 (1961).
18. Distribution of a metal concentration inside the mercury drop during the electrolytic separation of such a metal on a stationary mercury electrode, L. N. Vasileva and E. N. Vinogradova, *Zavodsk. Lab.* **27**, 1079–1086 (1961).
19. Theory of stripping of voltammetry with spherical electrodes, W. H. Reinmuth, *Anal. Chem.* **33**, 185–187 (1961).
20. Inverse polarography and voltammetry, new methods in trace analysis, R. Neeb, *Angew. Chem.* **74**, 203–213 (1962).
21. Dependence of metal concentration in a mercury drop on the duration of enrichment, L. N. Vasileva and E. N. Vinogradova, *Zavod. Lab.* **28**, 1427–1428 (1962).
22. Influence of surface active substances on the electrodeposition and electrooxidation of metals on the hanging mercury drop electrode, W. Kemula and S. Glodowski, *Roczniki. Chem.* **36**, 1203–1216 (1962).
23. Sensitivity enhancement of amalgam polarography with concentration in the stationary mercury drop, A. G. Stromberg, *Zavod. Lab.* **29**, 387–390 (1963).
24. Polarographie im durchfliessenden elektrolyt II. Stationare spharische elektrode, F. Strafelda and A. Kimla, *Collect. Czech. Chem. Commun.* **28**, 1516–1523 (1963).
25. Polarographie im durchfliessenden elektrolyt III. Diffusion zur unflossenen spharischen elektrode, F. Strafelda and A. Kimla, *Collect. Czech. Chem. Commun.* **28**, 2696–2705 (1963).
26. Anodic current constants and the calculation of amalgam-polarography, M. S. Zakharov and A. G. Stromberg, *J. Phys. Chem. (Russian) Zh. Fiz. Khim.* **38** (1), 130–135 (1964).
27. Effects of electrolysis potentials on depth of anodic indentations in amalgam polarography with stationary mercury electrodes, M. S. Zakharov, *Zavod. Lab.* **30** (1), 14–17 (1964).

28. Influence of solution on the depth of the anodic minimum in amalgam polarography, A. G. Stromberg and N. A. Kaplin, *Zavod. Lab.* **30**, 525–527 (1964).

29. Theory of anodic stripping voltammetry with a plane, thin mercury-film electrode, W. T. Devries and E. Van Dalen, *J. Electroanal. Chem.* **8**, 366–377 (1964).

30. Electrode kinetics with adsorbed foreign neutral substance, A. Aramata and P. Delahay, *J. Phys. Chem.* **68**, 880–883 (1964).

31. Adsorption of dependent stripping analysis at the graphite electrode, S. P. Perone and T. J. Oyster, *Anal. Chem.* **36**, 235–236 (1964).

32. Stripping voltammetry, E. Barendrecht, *Chem. Weekblad.* **60**, 345–358 (1964).

33. Stripping analysis: I. Introduction. II. Theory and application of stripping analysis, I. Shain, *Treatise Anal. Chem.* **4**, 2534–2558 (1964).

34. Exact treatment of anodic stripping voltammetry with a plane mercury film electrode, W. T. de Vries, *J. Electroanal. Chem.* **9**, 448–456 (1965).

35. Ohmic drop distortion of anodic stripping curves from a thin mercury film electrode, W. T. de Vries and E. Van Dalen, *J. Electroanal. Chem.* **12**, 9–14 (1966).

36. *Inverse Polarographie und Voltammetrie*, R. Neeb, Verlag Chemie, Weinheim (1969).

37. Problem of increasing the sensitivity of anodic stripping voltammetry, L. Huderova and K. Stulik, *Talanta* **19**, 1285 (1972).

38. Intermetallic compounds in amalgams, M. T. Kozlovskii and A. I. Zebreva, *Progr. Polarogr.* **3**, 157 (1972).

39. Study of the mutual effect of different metals on a mercury-graphite electrode by inverse voltammetry, E. Ya. Neiman and Kh. Z. Brainina, *Zh. Anal. Khim.* **28**, 886 (1973).

40. Studies of exchange reactions involving electrogenerated mercurous halid films, G. Colovs, G. S. Wilson, and J. L. Moyers, *Anal. Chem.* **46**, 1045 (1974).

41. Evaluation and comparison of some techniques of anodic stripping voltammetry, G. E. Batley and T. M. Florence, *J. Electroanal. Chem. Interfacial Electrochem.* **55**, 23 (1974).

42. Elimination of copper-zinc intermetallic interferences in anodic stripping voltammetry, T. R. Copeland, R. A. Osteryoung, and R. K. Skogerboe, *Anal. Chem.* **46**, 2093 (1974).

43. *Stripping Voltammetry in Chemical Analysis*, Kh. Z. Brainina, Wiley, New York (1974).

44. Diffusion coefficients of metals in mercury, A. Baranski, S. Fitak, and Z. Galus, *J. Electroanal. Chem. Interfacial Electrochem.* **60**, 175 (1975).

45. *Electrochemical Stripping Analysis*, F. Vydra, K. Stulik, and E. Julakova, Wiley, New York (1976).

46. Trace analysis by atomic absorption spectroscopy and anodic stripping voltammetry, F. P. J. Cahill and G. W. Van Loon, *Am. Lab.* **8** (8), 11 (1976).

47. Theoretical treatment of staircase voltammetric stripping from the thin film mercury electrode, J. H. Christie and R. A. Osteryoung, *Anal. Chem.* **48**, 869 (1976).

48. Kalman filter applied to anodic stripping voltammetry: Theory, P. F. Seeling and H. N. Blount, *Anal. Chem.* **48**, 252 (1976).

49. Intermetallic compound formation between copper and zinc in mercury and its effects on anodic stripping voltammetry, M. S. Shuman and G. P. Woodward, Jr., *Anal. Chem.* **48**, 1979 (1976).

50. Interfering influence of silver on the voltammetric behaviour of Cd^{2+} and Zn^{2+} ions on HMDE in supporting electrolytes containing nitrates, P. Ostapczuk and Z. Kublik, *J. Electroanal. Chem. Interfacial Electrochem.* **68**, 193 (1976).

51. The effect of dissolved organics on the stripping voltammetry of sea water, G. E. Batley and T. M. Florence, *J. Electroanal. Chem. Interfacial Electrochem.* **72**, 121 (1976).

52. Trace metal analysis by anodic stripping voltammetry. Effect of sorption by natural and model organic compounds, P. L. Brezonik, P. A. Brauner, and W. Stumm, *Water Res.* **10**, 605 (1976).

53. Determination of the chemical forms of dissolved cadmium lead and copper in sea water, G. E. Batley and T. M. Florence, *Mar. Chem.* **4**, 347 (1976).

54. Stripping voltammetry, E. Barendrecht, In: *Electroanalytical Chemistry*, A. J. Bard, ed., Vol. 2, pp. 53–109, Marcel Dekker, New York (1967).
55. Voltammetric and potentiometric comparison of tendencies of cadmium and zinc to intermetallic compound formation with silver, copper, and gold in mercury, P. Ostapczuk and Z. Kublik, *J. Electroanal. Chem. Interfacial Electrochem.* **83**, 1 (1977).
56. *Electrochemical Stripping Analyses*, F. Vydra, K. Stulik, and E. Julakova, Halsted Press, New York (1977).
57. Law of distribution of anodic peak values in amalgam and film anodic stripping voltammetry, Z. A. Kaplin and B. A. Kubrak, *Zh. Anal. Khim.* **32**, 1838–1841 (1977).
58. Effect of activity coefficient of metal in amalgam on anodic peak characteristics in a.c. stripping analysis, L. V. V. Bashkatova and V. Z. Bashkatov, *Zh. Anal. Khim.* **32**, 1495–1502 (1977).
59. Potentials of current maxima in inverse voltammetry of Pb and Cd on rotating disc carbon-glass and glass carbon electrodes, O. L. Kabonova, Yu. A. Gorcharov, and S. K. Beuiaminok, *Elektrokhimiya*, **13**, 1448–1453 (1977).
60. Study of a residual current of a mercury film electrode used in stripping analysis, B. F. Nazarov, A. D. Rakhmonbadyov, and A. G. Stromberg, *Elektrokhimiya* **13**, 1575–1577 (1977).
61. Application of semidifferential electroanalysis to anodic stripping voltammetry, M. Goto, K. Ikenoya, M. Kajihara, and D. Ishi, *Anal. Chim. Acta* **101**, 131–138 (1978).
62. Intermetallic compounds formed in mixed (complex) amalgams. I. The systems: Copper-mercury, zinc-mercury and copper-zinc-mercury, A. H. I. Ben-Bassat and A. Azrad, *Electrochim. Acta* **23**, 63–69 (1978).
63. Distribution law for anodic current in stripping analysis, B. A. Kubrak, A. A. Kaplin, A. N. Pokrovskaya, and V. N. Polyakova, *Izvest. Tomsk. Politekh, Inst. ta* **61**, 3 (1976). [*CA* **88**, 68491 (1978).]
64. Statistical criteria of sensitivity and resolving power, A. A. Kaplin, A. A. Zheltonozhko, A. A. Pokrovskaya, and B. A. Kubrole, *Izvest. Tomsk. Politekh. Inst. ta*, 56–60 (1976). [*CA* **88**, 68489 (1978).]
65. Dependence of metal concentration on an electrode surface on a.c. voltammetry in a.c. stripping polarography, V. Z. Bashkatov, V. A. Bramin, and L. V. Bashkatova, *Zh. Anal. Khim.* **33**, 230–235 (1978).
66. Minimum determinable anodic peak height at a steep slope of residual current in stripping polarography. Model of the triangular peak, A. G. Stromberg and N. Pibula, *Zh. Anal. Khim.* **33** (3), 436–441 (1978).
67. Stripping polarography, A. A. Kaplin, V. E. Katyulhim, A. E. Geineman, and V. E. Saraeva, USSR 615406, July 1978.
68. Improving the resolution of inverse voltammetry by anodic passivation of amalgams, V. A. Zgadova, B. F. Nazarov, and E. A. Zakharova, *Zh. Anal. Khim.* **33**, 1631–1633 (1978).
69. Ways of increasing the sensitivity of stripping voltammetry methods, A. G. Stromberg and A. A. Kaplin, *Izvest. Sib. Osd. Akad. Nauk SSSR, Ser. Khim. Nauk* **4**, 31–38 (1978). [*CA* **89**, 190261a (1978).]
70. Application of semidifferential electroanalysis to anodic stripping voltammetry, M. Goto, K. Ibenoya, M. Kajuhara, and D. Ishii, *Anal. Chim. Acta* **101**, 131–138 (1978).
71. Study of a quasi reversible electrode process with a consequent first order chemical reaction in stripping analysis on a mercury film electrode, Yu. A. Karbainov, N. N. Chernysheva, and A. G. Stromberg, *Zh. Fiz. Khim.* **52**, 2701–2702 (1978).
72. Study of quasi reversible processes on electrodes of different geometric form in inverse voltammetry, A. E. Geineman, A. A. Kaplin, and A. G. Stromberg, *Zh. Anal. Khim.* **33**, 1510–1516 (1978).
73. Pseudopolarograms: Applied potential-anodic stripping peak current relationships, M. S. Shuman and J. L. Cromer, *Anal. Chem.* **51**, 1546–1550 (1979).

74. Anodic stripping semidifferential electroanalysis with thin mercury film electrode formed in situ, M. Goto, K. Ikenoya, and D. Ishii, *Anal. Chem.* **51**, 110–115 (1979).

75. Elimination of intermetallic compound interface in twin electrode thin layer anodic stripping voltammetry, D. A. Roston, E. E. Brooks, and W. R. Heineman, *Anal. Chem.* **51**, 1728–1732 (1979).

76. The reversible electrodeposition of trace metal ions from multi-ligand systems Part I Theory, Part II. Calculations on the electrochemical availability of lead at trace levels in sea water, D. R. Turner and M. Whitfield, *J. Electroanal. Chem. Interfacial Electrochem.* **103**, 43–60; 61–79 (1979).

77. Flame atomic absorption analysis for selenium after electrochemical preconcentration, W. Lund and R. Bye, *Anal. Chim. Acta* **110**, 279–284 (1979).

78. Faradaic and non-faradaic currents due to reactant stripping at metal surfaces or from thin film electrodes, G. C. Barker and A. W. Gardner, *J. Electroanal. Chem. Interfacial Electrochem.* **100**, 641–656 (1979).

3.2. Instrumentation, Electrodes, and Cells

79. Instrumental approach to potentiometric stripping analysis of some heavy metals, D. Jagner, *Anal. Chem.* **50**, 1924–1929 (1978).

80. Analytical applications of hanging mercury drop electrode, J. W. Ross, R. D. Demars, and I. Shain, *Anal. Chem.* **28**, 1768–1771 (1956).

81. Anodic stripping voltammetry with mercury electrodes—potential-step and current-step methods, G. Mamantov, P. Papoff, and P. Delahay, *J. Am. Chem. Soc.* **79**, 4034–4040 (1957).

82. Anodic stripping voltammetry using the hanging mercury drop electrode, R. DeMars and I. Shain, *Anal. Chem.* **29**, 1825–1827 (1957).

83. A rotating hanging mercury-drop electrode, E. Barendrecht, *Nature* **181**, 764–765 (1958).

84. Application of hanging mercury drop electrode to an investigation of intermetallic compounds in mercury, W. Kemula, Z. Galus, and Z. Kublik, *Nature* **182**, 1228–1229 (1958).

85. Mercury chloride film anode. I. Study of the characteristics of a mercury chloride film anode by chronopotentiometric methods, T. Kuwana and R. N. Adams, *Anal. Chim. Acta* **20**, 51–60 (1959).

86. Mercury chloride film anode. II. Investigation of the characteristics of the mercury chloride film anode by voltage scan method, T. Kuwana and R. N. Adams, *Anal. Chim. Acta* **20**, 60–67 (1959).

87. Analytical application of the hanging mercury drop electrode. Analysis of traces of impurities in uranium salts, W. Kemula, E. Rakowska, and Z. Kublik, *J. Electroanal. Chem.* **1**, 205–217 (1959/1960).

88. Trace analysis with a rotating hanging mercury drop, M. Van Swaay and R. S. Deedler, *Nature* **191**, 241–242 (1961).

89. Use of the rotating hanging mercury drop in analysis. Anodic stripping using a rotating mercury drop, J. J. Engelsman and A. M. J. M. Claassems, *Nature* **191**, 240–241 (1961).

90. Stripping analysis with spherical mercury electrodes, I. Shain and J. Lewinson, *Anal. Chem.* **33**, 187–189 (1961).

91. The rotated, mercury-coated platinum electrode. Preparation and behaviour of continuously deposited mercury coatings and applications to stripping analysis, S. Bruckenstein and T. Nagai, *Anal. Chem.* **33**, 1201–1209 (1961).

92. Anodic amalgam voltammetry. III. Comparative investigations on the use of amalgamated platinum electrodes and stationary mercury electrodes, R. Neeb, *Z. Anal. Chem.* **180**, 161–168 (1961).

93. Microcell for voltammetry with hanging mercury drop electrode, W. L. Underkofler and I. Shain, *Anal. Chem.* **33**, 1966–1967 (1961).

94. Square wave polarography after prior electrolysis at a stationary electrode, F. Von Sturm and M. Ressel, *Fresnius Z. Anal. Chem.* **186**, 63–79 (1962).

95. Anodic amalgam voltammetry. IV. Chronopotentiometric determination of Tl, R. Neeb, *Z. Anal. Chem.* **190**, 98–111 (1962).

96. Coulostatic anodic stripping with a mercury electrode, P. Delahay, *Anal. Chem.* **34**, 1662–1663 (1962).

97. The hanging mercury drop in polarography, J. Riha, In: *Progress in Polarography*, Vol. II, I. M. Kolthoff and P. Zuman, eds., pp. 383–396 (1962).

98. Application of hanging mercury drop electrodes in analytical chemistry, W. Kemula and Z. Kublik, *Adv. Anal. Chem. Instrum.* **2**, 123–177 (1963).

99. Carbon paste electrode's application to cathodic reduction and anodic stripping voltammetry, C. Olson and R. N. Adams, *Anal. Chim. Acta* **29**, 358–363 (1963).

100. New construction of an electrolytic cell and electrodes, V. I. Kuleshov and A. G. Stromberg, *Metody Analize Khim.* **56**, 37–41 (1963).

101. Vector polarography on a stationary drop, S. B. Isfasman and R. M. Salikhdzhanova, *Zavod. Lab.* **30**, 133–140 (1964).

102. Application of derivative techniques to anodic stripping voltammetry, S. P. Perone and J. R. Birk, *Anal. Chem.* **37**, 9–12 (1965).

103. Investigation of a.c. polarography at stationary electrodes, with application to stripping analysis, W. Underkofler and I. Shain, *Anal. Chem.* **37**, 218–222 (1965).

104. Determination of mixtures by single sweep oscillopolarography, R. A. Osteryoung and E. P. Parry, *J. Electroanal. Chem.* **9**, 299–304 (1965).

105. Differential pulse anodic stripping of trace metals, H. Siegerman and G. O'Dom, *Amer. Lab.* **4(6)** (1972).

106. Use of carbon-graphite electrodes in inverse voltammetry (Review), A. A. Antsiferov and S. I. Sinyakova, *Elektrokhim. Metody. Anal. Mater.* **115** (1972).

107. Apparatus for automatic analysis by stripping voltammetry, G. Reusmann and J. Westphalen, *Fresenjus' Z. Anal. Chem.* **259**, 127 (1972).

108. New cell for rapid anodic stripping analysis user-interactive computer program for analysis of anodic stripping data, R. G. Clem, G. Litton, and L. D. Ornelas, *Anal. Chem.* **45**, 1306 (1973).

109. Analytical applications of pulsed voltammetric stripping at thin film mercury electrodes, T. R. Copeland, J. H. Christie, R. A. Osteryoung, and R. K. Skogerboe, *Anal. Chem.* **45**, 2171 (1973).

110. Stripping voltammetry with collection at a rotating ring-disc electrode, D. C. Johnson and R. E. Allen, *Talanta* **20**, 305 (1973).

111. Use of graphite and mercury-plated graphite electrodes in alternating current voltammetry, E. M. Roizenblat, L. F. Levchenko, and G. N. Veretina, *Zh. Anal. Khim.* **28**, 33 (1973).

112. Anodic stripping voltammetry at a tubular mercury-covered graphite electrode, W. R. Seitz, R. Jones, L. N. Klatt, and W. D. Mason, *Anal. Chem.* **45**, 840 (1973).

113. Microcell for stripping analysis with glassy carbon and mercury-plated glassy carbon electrodes, K. Stulik and M. Stulikova, *Anal. Lett.* **6**, 441 (1973).

114. Voltammetry at disc electrodes and its analytical application. VII. Use of the second harmonic alternating current technique for stripping analysis at rotating disc electrodes, M. Stulikova and I. Vydra, *J. Electroanal. Chem. Interfacial Electrochem.* **42**, 127 (1973).

115. Carbon glass-ceramic as a new material for solid electrodes in inverse voltammetry, B. A. Anokhim and V. I. Ignatov, *Zh. Anal. Chem.* **29**, 1221 (1974).

116. Anodic stripping with collection, using thin mercury films, D. Laser and M. Ariel, *J. Electroanal. Chem. Interfacial Electrochem.* **49**, 123 (1974).

117. Determination of trace metals in aqueous environments by anodic stripping voltammetry with a vitreous carbon rotating electrode, A. H. Miguel and C. M. Tankowski, *Anal. Chem.* **46**, 1832 (1974).

118. Electrode for anodic stripping voltammetry, W. R. Matson, *Interface Newsletter*, January 10, 1974.

119. Use of linear sweep, superimposed alternating current and pulse voltammetry in the study of anodic stripping at a rotating mercury-copper electrode, J. P. Roux, O. Vittori, and M. Porthault, *Analysis* **3**, 411 (1975).

120. Silver based mercury film electrode, I. General characteristics and stability of the electrode, Z. Stojek and Z. Kublik, *J. Electroanal. Chem. Interfacial Electrochem.* **60**, 349 (1975).

121. Styrene impregnated, cobalt-60 irradiated graphite electrode for anodic stripping analysis, R. G. Clem and A. F. Sciamanna, *Anal. Chem.* **47**, 276 (1975).

122. Cause of loss of hydrogen-overvoltage on graphite electrodes used for anodic stripping voltammetry, R. G. Clem, *Anal. Chem.* **47**, 1778 (1975).

123. Determination of some metals and their mixtures by stripping chronopotentiometry on a glassy carbon rotating disc electrode, L. Luong and F. Vydra, *Collect. Czech. Chem. Commun.* **40**, 2961 (1975).

124. Differential pulse anodic stripping voltammetry in a thin layer electrochemical cell, T. P. De Angelis and W. R. Heineman, *Anal. Chem.* **48**, 2262 (1976).

125. Computer assisted optimization of anodic stripping voltammetry, Q. V. Thomas, L. Kryger, and S. P. Perone, *Anal. Chem.* **48**, 761 (1976).

126. A novel scheme for the classification of heavy metal species in natural waters, G. E. Batley and T. M. Florence, *Anal. Lett.* **9**, 379 (1976).

127. Experimental factors affecting the phase-selective reversible anodic stripping determination of gallium from 1.0 molar ammonium thiocyanate electrolytes at 60°C and collation of results with sodium thiocyanate/sodium perchlorate-based room temperature measurements, E. D. Moorhead and G. A. Forsberg, *Anal. Chem.* **48**, 751 (1976).

128. Graphite-epoxy mercury thin film working electrode for anodic stripping voltammetry, J. E. Anderson and D. E. Tallman, *Anal. Chem.* **48**, 209 (1976).

129. Cyclic and stripping voltammetry with graphite based thin mercury film electrodes prepared "in situ," Z. Stojek, B. Stepmik, and Z. Kublik, *J. Electroanal. Chem. Interfacial Electrochem.* **74**, 227 (1976).

130. Silver based mercury film electrode. Part II. The influence of aging of electrode on the electrode processes of various metal ions, Z. Stojek, P. Ostapczuk, and Z. Kublik, *J. Electroanal. Chem. Interfacial Electrochem.* **67**, 301 (1976).

131. A multiple cell system for differential pulse anodic stripping voltammetry at a hanging mercury drop electrode, J. H. Lowry and R. B. Smart, *Am. Lab.* **8(12)**, 47 (1976).

132. Staircase voltammetric stripping analysis at thin film mercury electrodes, U. Eisner, J. A. Turner, and R. A. Osteryoung, *Anal. Chem.* **48**, 1608 (1976).

133. Digital microcoulometric measurements of cadmium anodic stripping at the micrometer hanging mercury drop electrode, E. D. Moorhead and W. H. Doub, Jr., *Anal. Chem.* **49**, 199 (1977).

134. Thin layer differential pulse voltammetry, T. P. DeAngelis, R. E. Bond, E. E. Brooks, and W. R. Heinemann, *Anal. Chem.* **49**, 1792 (1977).

135. Programmable electrochemistry, W. R. Matson, E. Zink, and R. Vitukevitch, *Am. Lab.* **9(7)**, 59 (1977).

136. Stripping determination of traces of lead using second harmonic alternating voltammetry and rotating disc electrode, M. Kopanica and V. Stara, *J. Electroanal. Chem. Interfacial Electrochem.* **77**, 57 (1977).

137. Investigation by automated differential pulse anodic stripping voltammetry of the problems of storage of dilute solutions, A. M. Bond and B. W. Kelly, *Talanta* **24**, 453–457 (1977).

138. A programmable device for electrochemical stripping analysis, F. Opebar and M. Herout, *Chem. Listy* **71**, 867–870 (1977).

139. Radiation-cured polymer-impregnated graphite electrodes for anodic stripping voltammetry, K. G. McLaren and G. E. Batley, *J. Electroanal. Chem. Interfacial Electrochem.* **79**, 169 (1977).

140. Silver based mercury film electrode. Part III. Comparison of theoretical and experimental anodic stripping results obtained for lead and copper, Z. Stojek and Z. Kublik, *J. Electroanal. Chem. Interfacial Electrochem.* **77**, 205 (1977).

141. Electrodeposition and stripping at graphite cloth electrodes, D. Yaniv and M. Ariel, *J. Electroanal. Chem. Interfacial Electrochem.* **79**, 159 (1977).

142. Anodic stripping coulometry at a thin-film mercury electrode, R. Eggli, *Anal. Chim. Acta* **91**, 129 (1977).

143. A new electrode system with efficient mixing of electrolyte, T. Magjer and M. Branica, *Croat. Chem. Acta* **49(1)**, 11 (1977).

144. Application of a rotating disc electrode and a rotating cell with stationary electrode in stripping voltammetry for the determination of lead and zinc, F. Vydra and T. V. Nghi, *Anal. Chim. Acta* **91**, 335 (1977).

145. Anodic stripping voltammetry in a flow-through cell with a fixed mercury film glassy carbon disc electrode. Part I. J. Wang and M. Ariel, *J. Electroanal. Chem. Interfacial Electrochem.* **83**, 217 (1977).

146. Study of anodic stripping voltammetry with collection at tubular electrodes, G. W. Schieffer and W. J. Blaedel, *Anal. Chem.* **49**, 49 (1977).

147. Pulsed voltammetric stripping at the thin-film mercury electrode, J. A. Turner, U. Eisner, and R. A. Osteryong, *Anal. Chim. Acta* **90**, 25–34 (1977).

148. Derivative potentiometric stripping analysis with a thin film of mercury on a glassy carbon electrode, D. Jagner and K. Aren, *Anal. Chim. Acta* **100**, 375–388 (1978).

149. A micro-computer system for potentiometric stripping analysis, T. Anfalt and M. Stramberg, *Anal. Chim. Acta* **103**, 379–388 (1978).

150. Subtractive anodic stripping voltammetry at twin mercury film electrode, E. Steeman, E. Temmerman, and R. Verbinnen, *Anal. Chim. Acta* **96**, 177–181 (1978).

151. A survey of instrumentation used for monitoring metals in water, M. S. Quinby-Hunt, *Am. Lab.* **17**, 37 (1978).

152. Applications of staircase voltammetry to trace metal analysis, J. S. Jasinski, EDRO-SARAP, *Res. Tech. Rep.* **1**, 21 pp (1976). [*CA* **88**, 83121f (1978).]

153. Possibility of using factorial design for lowering the detection limit of elements by stripping voltammetry, O. N. Markymenko, A. A. Kaplin, and V. M. Pichugina, Deposited Doc. VINITI 3970–3976, 8 pp (1976). [*CA* **89**, 99021 (1978).]

154. Differential pulse anodic stripping voltammetry as a rapid screening technique for heavy metal intoxicants, J. P. France and R. A. Ze Zeeuw, *Arch. Toxicol.* **37**, 47–55 (1976). [*CA* **89**, 141316 (1978).]

155. Choice analytical signal in inverse voltammetry, Kh. Z. Brainina, G. T. Shishin, L. I. Roitman, and L. N. Kalinishevskaya, *Zh. Anal. Khim.* **33**, 1708–1719 (1978).

156. Substrate anodic stripping voltammetry with rotating mercury coated glassy carbon electrode, L. Sipos, S. Kozar, I. Kontusic, and M. Branica, *J. Electroanal. Chem. Interfacial Electrochem.* **87**, 347–352 (1978).

157. Anodic stripping voltammetry with collection at tubular electrodes for the analysis of tap water, G. W. Schieffer and W. Blaedel, *Anal. Chem.* **50**, 99–102 (1978).

158. The rotating disc electrode in flowing systems. Part 1. An anodic stripping monitoring system for trace metals in natural waters, J. Wang and M. Ariel, *Anal. Chim. Acta* **99**, 89–98 (1978).

159. The rotating disc electrode in flowing systems. Part 2. A flow system for automated anodic stripping voltammetry of discrete samples, J. Wang and M. Ariel, *Anal. Chim. Acta* **101**, 1–8 (1978).

160. Silver based mercury film electrode. Part IV. Comparison of theoretical and and experimental voltammetric results obtained for cadmium, P. Ostapczuk and Z. Kublik, *J. Electroanal. Chem. Interfacial Electrochem.* **93**, 195–212 (1978).

161. Electrode for a.c. inverse voltammetry, B. S. Bruk and M. N. Bogoslovskaya, *Zavod. Lab.* **44**, 395–398 (1978).

163. Application of chemically modified electrodes to analysis of metal ions, G. T. Cheek and R. F. Nelson, *Anal. Lett. A* **11(5)**, 383–402 (1978).

164. Multiple-scanning potentiometric stripping analysis, J. Mortensen, E. Ouziel, H. J. Skov, and L. Kryger, *Anal. Chim. Acta* **112**, 297–312 (1979).

165. Anodic stripping voltammetry with a symmetric double-step wave form, J. J. Kankare and K. E. Haapakka, *Anal. Chim. Acta* **111**, 79–87 (1979).

166. Experimental evaluation of recursive estimation applied to linear sweep anodic stripping voltammetry for real time analysis, P. F. Seelig and H. N. Blunt, *Anal. Chem.* **51**, 327–337 (1979).

167. Application of recursive estimation to the real time analysis of trace metal analyses by linear sweep, pulse and differential pulse anodic stripping voltammetry, P. F. Seelig and H. N. Blunt, *Anal. Chem.* **51**, 1129–1134 (1979).

168. Minicomputer-controlled, background subtracted anodic stripping voltammetry: Evaluation of parameters and performance, S. D. Brown and B. R. Kowalski, *Anal. Chim. Acta* **107**, 13–27 (1979).

169. Anodic stripping voltammetry at a reticulated mercury vitreous carbon electrode, W. J. Blaedel and J. Wang, *Anal. Chem.* **51**, 1724–1728 (1979).

170. Modification of a commercial micrometer hanging mercury drop electrode, J. E. Bonelli, H. E. Taylor, and R. K. Skogerboe, *Anal. Chem.* **51**, 2412–2413 (1979).

171. Origin and elimination of interferences from siliconization procedures in anodic stripping voltammetry, M. Ochme, *Anal. Chim. Acta* **107**, 67–73 (1979).

172. Cyclic and stripping voltammetry of mercury at impregnated graphite electrode in chloride and thiocyanate media, R. Oilewicz, Z. Stojek, and Z. Kublik, *J. Electroanal. Chem. Interfacial Electrochem.* **96**, 29–44 (1979).

3.3. Applications—Anodic

173. Application de la goute pendante de mercure a la determination de minimes quantites de differents ions, W. Kemula and Z. Kublik, *Anal. Chim. Acta* **18**, 104–111 (1958).

174. Determination of small amounts of cadmium, lead, bismuth and thallium, R. Neeb, *Z. Anal. Chem.* **171**, 330 (1959).

175. Anodic stripping voltammetry of nickel at solid electrodes, M. M. Nicholson, *Anal. Chem.* **32**, 1058–1062 (1960).

176. Application of stripping analysis to the determination of iodide with silver microelectrodes, I. Shain and S. P. Perone, *Anal. Chem.* **33**, 325–329 (1961).

177. Application of stripping analysis to the trace determination of tin, S. L. Philips and I. Shain, *Anal. Chem.* **34**, 262–265 (1962).

178. Chronopotentiometric deposition and stripping of silver, lead and copper at platinum electrode, A. R. Nisbet and A. J. Bard, *J. Electroanal. Chem.* **6**, 332–343 (1963).

179. The application of stripping analysis to the determination of silver (1) using graphite electrodes, S. P. Perone, *Anal. Chem.* **35**, 2091–2094 (1963).

180. Anodic stripping voltammetry of gold and silver with carbon paste electrodes, E. C. Jacobs, *Anal. Chem.* **35**, 2112–2115 (1963).

181. Determination of silver in alkali and alkaline earth metals, Ku. Z. Brainina, T. A. Rygailo, and V. B. Belyavskaya, *Metody Analiza Khim.* **5–6**, 124–128 (1963).

182. Trace analysis by anodic stripping and voltammetry. I. Trace metals in dead sea brine. 1. Zinc and cadmium, M. Ariel and U. Eisner, *J. Electroanal. Chem.* **5**, 362–374 (1963).

183. Polarography with accumulation at a stationary electrode, A. G. Stromberg and E. A. Zakharova, *Zavods. Lab.* **30**, 261–267 (1964).

184. Polarographic determination of ultramicro traces by concentration and stripping on mercury

and solid electrode, S. I. Sinyakova and Yu. I. Vainshtein, *Metody. Analiza Khim.* 5–6, 5–15 (1963). [*CA* **61**, 1233g (1964).]

185. Polarography with accumulation on a stationary electrode, A. G. Stromberg and E. A. Zakharova, *Zavod. Lab.* **30**, 261–267 (1964).

186. Stripping analysis. III. Experimental procedures, I. Shain, *Treatise Anal. Chem.* **4**, 2560–2564 (1964).

187. Filtration as a source of error in anodic stripping voltammetry, H. Specker and G. Schieve, *Z. Anal. Chem.* **204**, 1–3 (1964).

188. Trace analysis by anodic stripping voltammetry. II. The method of medium exchange, M. Ariel, U. Eisner, and S. Gottesfeld, *J. Electroanal. Chem.* **7**, 307 (1964).

189. Voltammetry of nickel in molten lithium fluoride, *J. Electroanal. Chem.* **7**, 302–306 (1964).

190. The determination of ionic zinc in sea-water by anodic stripping voltammetry using ordinary capillary electrode, G. Macchi, *J. Electroanal. Chem.* **9**, 290–298 (1965).

191. Anodic stripping voltammetry of mercury(II) at the graphite electrode, S. P. Perone and W. J. Kretlow, *Anal. Chem.* **37**, 968–970 (1965).

192. Chemical stripping analysis. Determination of cerium(IV), permanganate and iron(III) in the micromolar concentration range, S. Bruckenstein and J. W. Bixler, *Anal. Chem.* **37**, 786–790 (1965).

193. Electroanalytical method for amalgam forming metals, Ch. Yarnitsky and M. Ariel, *J. Electroanal. Chem.* **10**, 110–118 (1965).

194. Redissolution anodique par polarographie a impulsions, G. Donadey, R. Rosset, and G. Charlot, *Chem. Anal.* **17**, 575 (1972).

195. Determination of lead, cadmium, copper, thallium, bismuth and zinc in serum by anodic stripping polarography, I. Sinko and S. Gomicsek, *Mikrochim. Acta* **2**, 163 (1972).

196. Determination of mercury(II) in acidic media by stripping voltammetry with collection, R. E. Allen and D. C. Johnson, *Talanta* **20**, 799 (1973).

197. Determination of trace amounts of zinc, cadmium, lead and copper in airborne particulate matter by anodic stripping voltammetry, G. Colovos, G. S. Wilson, and J. Moyers, *Anal. Chim. Acta* **64**, 457 (1973).

198. Effect of supporting electrolyte concentration in pulsed stripping voltammetry at the thin film mercury electrode, T. R. Copeland, J. H. Christie, R. K. Skogerboe, and R. A. Osteryoung, *Anal. Chem.* **45**, 995 (1973).

199. Microdetermination of lead in blood: A derivative pulse stripping voltammetric method, L. Duic, S. Szechter, and S. Srinivasan, *J. Electroanal. Chem. Interfacial Electrochem.* **41**, 89 (1973).

200. Determination of lead in evaporated milk by atomic absorption spectrophotometry and anodic stripping voltammetry collaborative study, J. A. Fiorino, R. A. Moffitt, A. L. Woodson, R. J. Gajan, G. E. Huskey, and R. G. Scholz, *J. Assoc. Offic. Anal. Chem.* **56**, 1246 (1973).

201. Determination of traces of thallium in urine by anodic stripping a.c. voltammetry, D. L. Levit, *Anal. Chem.* **45**, 1291 (1973).

202. Determination of gold in drugs and serum by use of anodic stripping voltammetry, G. M. Schmid and G. W. Bolger, *Clin. Chem.* **19**, 1002 (1973).

203. Determination of lead in blood and urine by anodic stripping voltammetry, B. Searle, W. Chan, and B. Davidow, *Clin. Chem.* **19**, 76 (1973).

204. Polarographie and Voltammetric Techniques, H. W. Nürnberg and B. Kastening, In: *Methodicum Chimicum*, F. Korte, ed., vol. 1/A, pp. 584–607, Academic Press, New York (1974).

205. Electroanalytical determination and characterisation of some heavy metals in sea water, M. Branica, L. Sipos, S. Bubie, and S. Kozar, *Proc. 7th Materials Research Symp.*, Gaithersburg, Md. (1974).

206. Voltammetric deposition and stripping of selenium(IV) at a rotating gold-disc electrode in 0.1 *M* perchloric acid, R. W. Andrews and D. C. Johnson, *Anal. Chem.* **47**, 294 (1975).

207. Determination of zinc in human eye tissues by anodic stripping voltammetry, T. R. Williams, D. R. Foy, and C. Benson, *Anal. Chim. Acta* **75**, 250 (1975).

208. Polarography and Voltammetry in Marine Chemistry, H. W. Nürnberg and P. Valenta, In: *The Nature of Sea Water*, E. D. Goldberg, ed., pp. 87–136, Dahlem Konferenzen, Berlin (1975).

209. The Electroanalytical Chemistry of Sea Water, M. Whitefield, In: *Chemical Oceanography*, J. P. Riley and G. Skirrow, ed., 2nd Ed., Vol. 4, Chap. 20, Academic Press, London (1975).

210. Zero-current inverse chronopotentiometry of lead at a carbon–glass–ceramic electrode, Yu. A. Goncharov and A. N. Doronin, *Zh. Anal. Khim.* **31**, 897 (1976).

211. Inverse voltammetry and coulometry of lead at a carbon–glass–ceramic electrode, O. L. Kabanova and Yu. A. Gancharov, *Zh. Anal. Khim.* **31**, 902 (1976).

212. Selective stripping voltammetric determinations employing cells for electrolysis with simultaneous ion-exchange or solvent extraction, K. Stulik and P. Bedros, *Talanta* **23**, 563 (1976).

213. The electrochemical stripping determination of bismuth in aqueous and nonaqueous media after its extraction, T. V. Nghi and F. Vydra, *J. Electroanal. Chem. Interfacial Electrochem.* **71**, 325 (1976).

214. Applications of polarography and voltammetry to marine and aquatic chemistry. II. The polarographic approach to the determination and speciation of toxic trace metals in the marine environment, H. W. Nurnberg, P. Valenta, L. Mart, B. Raspor, and L. Sipos, *Fresenius Z. Anal. Chem.* **282**, 357–367 (1976).

215. Capabilities of voltammetric techniques for water quality control problems, J. Buffle, F. L. Greter, G. Nembrini, J. Paul, and W. Haerdi, *Fresenius Z. Anal. Chem.* **282**, 339 (1976).

216. A novel scheme for the classification of heavy metal species in natural waters, G. E. Batley and T. M. Florence, *Anal. Lett.* **9**, 379 (1976).

217. Characterization of trace metal-organic interactions by anodic stripping voltammetry, T. A. O'Shea and K. H. Mancy, *Anal. Chem.* **48**, 1603 (1976).

218. Determination of the stability constants of some hydroxo and carbonato complexes of Pb(II), Cu(II), Cd(II) and Zn(II) in dilute solutions by anodic stripping voltammetry and differential pulse polarography, H. Bilinski, R. Huston, and W. Stumm, *Anal. Chim. Acta* **84**, 157 (1976).

219. The determination of zinc, cadmium, lead and copper in a single sea-water sample by differential pulse anodic stripping voltammetry, M. I. Abdullah, B. R. Berg, and R. Klimek, *Anal. Chim. Acta* **84**, 307 (1976).

220. Electroanalytical determination and characterization of some heavy metals in seawater, M. Branica, L. Sipos, S. Bubic, and S. Kozar, *Natl. Bur. Stand (U.S.) Spec. Publ.* 442, 917 (1976).

221. Direct determination of mercury in sea-water by anodic stripping voltammetry with a graphite electrode, R. Fukai and L. Huynh-Ngoc, *Anal. Chim. Acta* **83**, 375 (1976).

222. Determination of trace mercury(II) in 0.1 *M* perchloric acid by differential pulse stripping voltammetry at a rotating gold disk electrode, R. W. Andrews, J. H. Larochello, and D. C. Johnson, *Anal. Chem.* **48**, 212 (1976).

223. Electroanalysis of toxic metals in blood and urine, A. A. Cernik, *Proc. Anal. Div. Chem. Soc.* **13**, 227 (1976).

224. Determination of traces of cadmium in alloy steels by anodic stripping voltammetry, K. Stulik and K. Marik, *Talanta* **23**, 131 (1976).

225. Determination of trace elements in zinc plant electrolyte by differential pulse polarography and anodic stripping voltammetry, E. S. Pilkington, C. Weeks, and A. M. Bond, *Anal. Chem.* **48**, 1665 (1976).

226. Cathodic stripping voltammetry of nanogram amounts of selenium in biological material, M. W. Bles, J. A. Dalziel, and C. M. Elson, *J. Assoc. Off. Anal. Chem.* **59**, 1234 (1976).

227. Application of polarography and voltammetry to marine aquatic chemistry. Part II. The polarographic approach to the determination and speciation of metals in the marine environ-

ment, H. W. Nurnberg, P. Valenta, L. Mart, B. Raspov, and L. Sipos, *Fresenius Z. Anal. Chem.* **282**, 357–367 (1976).

228. Use of stripping analysis (inverse voltammetry) in the analysis of macro- and microsubstances (Review), A. A. Kaplin, A. G. Stromberg, and N. P. Pikula, *Zavod. Lab.* **43**, 385 (1977).

229. Comparative study of the analytical characteristics of thin-film flowing and pulsed system, V. A. Igolinskii and G. I. Zheleznyak, *Zavod. Lab.* **43**, 140 (1977).

230. New potentialities in ultra trace analysis with differential pulse anodic stripping voltammetry, P. Valenta, L. Mart, and H. Rutzel, *J. Electroanal. Chem. Interfacial Electrochem.* **82**, 327 (1977).

231. The stripping voltammetric determination of metals in non-aqueous media on the mercury film electrode prepared in situ. Determination of lead after its extraction with dithizone using a substitution reaction with Hg-salt, F. Vydra and T. V. Nghi, *J. Electroanal. Chem. Interfacial Electrochem.* **78**, 167 (1977).

232. Potentialities and applications of advanced polarographic and voltammetric methods in environmental research and surveillance of toxic metals, H. W. Nurnberg, *Electrochim. Acta* **22**, 935 (1977).

233. An experimental study on the speciation of dissolved zinc, cadmium, lead and copper in river Rhine and north sea water, by differential pulsed anodic stripping volummetry, J. C. Duinker and C. J. M. Kramer, *Mar. Chem.* **5**, 207 (1977).

234. Application of anodic stripping voltammetry to determination of the state of complexation of traces of metal ions at low concentration levels, M. Branica, D. M. Novak, and S. Bubic, *Croat. Chem. Acta* **49**, 539 (1977).

235. Applications of polarography and voltammetry to marine and aquatic chemistry. IV. A new voltammetric method for the study of mercury traces in sea water and island waters, L. Sipos, P. Valenta, H. W. Nurnberg, and M. Branica, *J. Electroanal. Chem. Interfacial Electrochem.* **77**, 263 (1977).

236. Trace chemistry of toxic metals in biomatrixes. II. Voltammetric determination of the trace content of cadmium and other toxic metals in human whole blood, P. Valenta, H. Ruetzel, H. W. Nuernberg, and M. Stoeppler, *Fresenius Z. Anal. Chem.* **285**, 25 (1977).

237. Inverse voltammetric determination of gold in a sulphuric acid solution on a solid electrode, V. A. Zarinskii, L. S. Chulkina, and N. N. Baranova, *Zh. Anal. Khim.* **32**, 530 (1977).

238. Determination of ruthenium as ruthenate by stripping voltammetry, A. Trojnek, *J. Electroanal. Chem. Interfacial Electrochem.* **81**, 189 (1977).

239. Investigation by automated differential pulse anodic stripping voltammetry of the problem of storage of dilute solutions, A. M. Bond and B. W. Kelly, *Talanta*, **24**, 453 (1977).

240. Determination of some thiourea-containing pesticides by pulse voltammetric methods of analysis, M. R. Smyth and J. G. Osteryoung, *Anal. Chem.* **49**, 2310 (1977).

241. Flow injection analysis, Part X. Theory, techniques and trends, J. Ruzicka and E. H. Hansen, *Anal. Chim. Acta* **99**, 37–76 (1978).

242. Defining the electroanalytically measured species in a natural water sample, W. Davison, *J. Electroanal. Chem. Interfacial Electrochem.* **87**, 395–404 (1978).

243. Anodic stripping voltammetry at a mercury film electrode: Baseline concentrations of cadmium, lead and copper in selected natural waters, J. E. Poldoski and G. E. Glass, *Anal. Chim. Acta* **101**, 79–88 (1978).

244. The determination of copper, lead, cadmium and zinc in human teeth by anodic stripping voltammetry, W. Lund, M. Oehme, and J. Jonsen, *Anal. Chim. Acta* **100**, 389–398 (1978).

245. Determination of total arsenic at the nanogram level by high-sweep anodic stripping voltammetry, P. H. Davis, G. R. Dulude, R. M. Griffin, W. R. Matson, and E. W. Zink, *Anal. Chem.* **50**, 137–143 (1978).

246. Anodic stripping voltammetry of tellurium(IV) in aqueous solution and mixed solvents, M. Kopanica and V. Stara, *J. Electroanal. Chem. Interfacial Electrochem.* **91**, 351–357 (1978).

247. Electrochemical stripping determination of traces of copper, lead, cadmium and zinc in zirconium metal and zirconium dioxide, K. Stulik, P. Baran, J. Dolezal, and F. Opekar, *Talanta* **25**, 363–369 (1978).

248. A new voltammetric stripping method applied to the determination of the brightener concentration in copper pyrophosphate plating baths, D. Tench and C. Ogden, *J. Electrochem. Soc.* **125**, 194–198 (1978).

249. Determination of extremely small amounts of substances with the microprobe after electrolytical enrichment on small surfaces, R. Bock, E. Zimmer, and G. Weichbrodt, *Fresenius Z. Anal. Chem.* **293**, 377–387 (1978).

250. Subtractive anodic stripping voltammetry at twin mercury film electrodes, E. Steeman, E. Timmerman, and R. Valinnen, *Anal. Chim. Acta* **96**, 177–181 (1978).

251. Evaluation and optimization of the standard addition method for absorption spectrometry and anodic stripping voltammetry, J. P. Franke, R. A. DeZeeus, and R. Hakkert, *Anal. Chem.* **50**, 1374–1380 (1978).

252. Determination of the detection limit for twenty elements of the periodic system by stripping polarography, A. G. Stromberg, A. A. Kaplin, N. P. Pikula, and B. A. Kubrak, *Fiz.-Khim. Metody. Anal.* **2**, 6–10 (1977).

253. Determination of trace impurities in high purity reagents by mercury-thin film anodic stripping voltammetry, Y. Israel, T. Ofir, and J. Rejek, *Mikrochim. Acta* **1**, 151–163 (1978).

254. Polarographic and voltammetric methods for the determination of elements, R. Neeb, *Mikrochim. Acta* **1** (3–4), 305–318 (1978).

255. Potentiometric stripping analysis for lead in urine, D. Jagner, L. G. Danielsson, and K. Aren, *Anal. Chim. Acta* **106**, 15–21 (1979).

256. Potentiometric stripping analysis for zinc, cadmium lead and copper in sea water, D. Jagner and K. Aren, *Anal. Chim. Acta* **107**, 29–35 (1979).

257. Potentiometric stripping analysis in non-deaerated samples, D. Jagner, *Anal. Chem.* **51**, 342–345 (1979).

258. Pseudo polarographic determination of metal complex stability constants in dilute solution by rapid scan anodic stripping voltammetry, S. D. Brown and B. R. Kowalski, *Anal. Chem.* **51**, 2133–2139 (1979).

259. Interference in anodic stripping voltammetry from the teflon vessels used in pressurized digestion, M. Ochme, *Talanta* **26**, 913–916 (1979).

260. Use of chelex resin for determination of labile trace metal fractions in aqueous ligand media and comparison of the method with anodic stripping voltammetry, P. Figura and B. McDuffie, *Anal. Chem.* **51**, 120–125 (1979).

261. The determination of cadmium, lead and copper in urine by differential pulse anodic stripping voltammetry, W. Lund and R. Eriksen, *Anal. Chim. Acta* **107**, 37–46 (1979).

262. A rapid high performance analytical procedure with simultaneous voltammetric determination of toxic trace metals in urine, J. Golimowski, P. Valenta, M. Stoeppler, and H. W. Nürnberg, *Talanta* **26**, 649–656 (1979).

263. Determination of cobalt by anodic stripping voltammetry at a mercury film electrode, H. Bloom, B. N. Noller, and D. E. Richardson, *Anal. Chim. Acta* **109**, 157–160 (1979).

264. Cyclic and stripping voltammetry of tin in the presence of lead in pyrogallol medium at hanging and film mercury electrodes, S. Glodowski and W. Kublic, *Anal. Chim. Acta* **104**, 55–65 (1979).

265. Determination of traces of arsenic in zinc sulphate industrial solutions by voltammetry and cathodic redissolution, T. Monama and G. Duyckaerts, *Anal. Lett.* **12**, 219–229 (1979).

266. Target preparation for x-ray emission analysis by anodic electrodeposition of cyano metalates from 2-propanol-water mixtures, K. Wundt, H. Duschner, and K. Starke, *Anal. Chem.* **51**, 1487–1492 (1979).

267. Comparison of digestion procedures for the determination of heavy metals (cadmium, copper, lead) in blood by anodic stripping voltammetry, M. Oehme and W. Lund, *Fresenius Z. Anal. Chem.* **298**, 260–268 (1979).

3.4. Applications—Cathodic

268. Determination of chloride by cathodic stripping polarography, R. G. Ball, D. L. Manning, and O. Menis, *Anal. Chem.* **32**, 621–623 (1960).
269. Determination of micro amounts of chloride ions at a stationary mercury electrode, Kh. Brainina and E. M. Roizenblat, *Zavod. Lab.* **28**, 21–23 (1962).
270. Determination of chloride ion in dilute solutions by cathodic stripping voltammetry, W. L. Maddox, M. T. Kelley, and J. A. Dean, *J. Electroanal. Chem.* **4**, 96–104 (1962).
271. Inverse polarographic determination of chloride in ppm range in concentrated mineral acids, H. Specker and G. Schieve, *Z. Anal. Chem.* **196**, 1–5 (1963).
272. Simultaneous determination of bromide and chloride by cathodic stripping voltammetry, G. Colovos, G. S. Wilson, and J. L. Moyers, *Anal. Chem.* **46**, 1051 (1974).
273. Determination of phosphate by cathodic stripping voltammetry at a glassy carbon electrode, J. A. Cox and K. H. Cheng, *Anal. Lett.* **7**, 659 (1974).
274. Cathodic stripping coulometry of lead, H. A. Laitinen and N. H. Watkins, *Anal. Chem.* **47**, 1352 (1975).
275. Determination of the sodium salt of 2-mercaptopyridine-N-oxide by differential pulse cathodic stripping voltammetry, D. A. Csejka, S. T. Nakos, and E. W. DuBord, *Anal. Chem.* **47**, 322 (1975).
276. Mechanism of anodic deposition and cathodic stripping of PbO_2 on conductive tin oxide, H. A. Laitinen and N. H. Witkins, *J. Electrochem. Soc.* **123**, 804 (1976).
277. The determination of bromide, chloride and lead in airborne particulate matter by stripping voltammetry, B. L. Dennis, G. S. Wilson, and J. L. Moyers, *Anal. Chim. Acta* **86**, 27 (1976).
278. Direct determination of thioamide drugs in biological fluids by cathodic stripping voltammetry, I. E. Davidson and W. F. Smyth, *Anal. Chem.* **49**, 1195 (1977).
279. Phase selective cathodic stripping voltammetry for determination of water-soluble mercaptans, W. M. Moore and V. F. Gaylor, *Anal. Chem.* **49**, 1386 (1977).
280. Determination of halide ions by cathodic stripping analysis, K. Manandhar and D. Pletcher, *Talanta* **24**, 387 (1977).
281. Cathodic stripping voltammetric study of the release of inorganic sulphide from proteins in alkaline media, T. M. Florence, *Anal. Lett. B* **11**, 913–924 (1978).
282. Cathodic stripping voltammetry. Part I. Determination of organic sulphur compounds, flavins and porphyrins at the sub-micromolar level, T. M. Florence, *J. Electroanal. Chem. Interfacial Electrochem.* **97**, 219–236 (1979).
283. Cathodic stripping voltammetry. Part II. Study of the release of inorganic sulphide from proteins during denaturation in alkaline media, T. M. Florence, *J. Electroanal. Chem. Interfacial Electrochem.* **97**, 237–255 (1979).
284. Cathodic stripping voltammetric determination of organic halides in drug dissolution studies, I. E. Davidson and W. F. Smyth, *Anal. Chem.* **51**, 2127–2133 (1979).
285. Determination of selenium in soils and plants by differential pulse cathodic-stripping voltammetry, S. Forbes, G. P. Bounds, and T. S. West, *Talanta* **26**, 473–477 (1979).

4. Ion-Selective Electrodes (ISE)

The glass electrode was the first ion selective electrode to be discovered (in the beginning of this century). Teorell[A3] and Meyer and Sievers[A1–2] were the first to develop the theory of membrane potentials of porous membranes. Eisenman and co-workers have developed special glass compositions, selective to Na^+, K^+, NH_4^+, Ag^+, Li^+, Cs^+, and Tl^+ ions.[A11] The theory of membrane potentials, originally developed by Eisenman and co-workers for glass electrodes,[A6] has been improved and generalized to cover other membranes as well.[A10,11] Eisenman *et al.*[A10,A34] have shown that the selectivity coefficient of ISE depends on the ratio of ionic mobility, ion exchange equilibrium constant, ionic strength, and the ion concentration ratio. The theory of membrane potentials has been reviewed and discussed by Buck.[A19,A30,A37]

The solid membrane electrodes of homogeneous and heterogeneous type have been introduced by Frant and Ross[B1,A12] and Pungor,[A7,9,13] respectively. The liquid ion exchange membrane electrodes were first conceived by Sollner and Shean,[C1-2] while Ross developed a liquid ion exchange membrane, selective for cations.[C3,A12] The enzyme electrode was discovered first by Moore and Pressman.[D1] Many enzyme electrodes were later developed and these are discussed in detail by Vadgama.[A55]

The applications for ISE of all types of purposes have been reviewed by a number of researchers.[A74-110] The instrumentation for the use of ISEs including those for continuous analysis in industries as well as the methods of preparation of ISEs are discussed in References (A56)–(73).

While references pertaining to the general principles, theory, instrumentation, and applications of all types of electrodes are given in Section "A," the theory, instrumentation, and application concerned with the three types of ISEs, viz. solid membrane electrodes, liquid ion exchange membrane electrodes, and enzyme electrodes are discussed in the references given in Sections B, C, and D, respectively.

The ISEs based on neutral carriers that are specific to cations have been discussed by Morf and Simon.[C16] The applications of ISE ranging from the determination of concentration of ions, individual ionic activity measurements, "in vitro" studies in biological systems, to continuous monitoring of various ions in industries, have been discussed by a large number of researchers and are given in (A74–110), (B69–122), (C37–64), and (D13–23), and (77) and (78) in Section 8.3.

The development of ISEs in the past two decades has been very rapid and the number of publications on ISE has increased. For example, from the biannual reviews of *Anal. Chem.* published in the years 1974 and 1980, it can be seen that the number has increased by two or threefold. The papers published during the period 1979–1980 are over 719, while during 1975–1978 it is about 800 only. This rapid growth of ISE can be compared with that of polarography during the fifties and sixties.

4.A.1. Books and Reviews

1. Permeability of membranes. I. Theory of ionic permeability, K. H. Meyer and J. F. Sievers, *Helv. Chim. Acta* **19**, 649–664 (1936).
2. Permeability of membranes. II. Studies with artificial selective membranes, K. H. Meyer and J. F. Sievers, *Helv. Chim. Acta* **19**, 665–677 (1936).

3. General discussions on papers pertaining to biomembranes and bioelectrochemistry, T. Teorell, *Trans. Faraday Soc.* **33**, 1053–1055 (1937).

4. Permeability of membranes. II. Effect of boric anhydride and aluminium oxide on the electrode properties of glass, B. P. Nikolski and T. A. Tolmacheva, *Zh. Fiz. Khim.* **10**, 504–512 (1937).

5. Electric potentials of crystal surfaces and at silver halid surfaces in particular, I. M. Kolthoff and H. L. Sanders, *J. Am. Chem. Soc.* **59**, 416–420 (1937).

6. Glass electrode for measuring sodium ion, G. Eisenman, D. O. Rudin, and J. U. Casty, *Science* **126**, 831–834 (1957).

7. Membranes of heterogeneous structure for the determination of the activity anions, E. Pungor, J. Havas, and K. Toth, *Acta Chim. Acad. Sci. Hung.* **41** (1–2), 239–255 (1964) (Eng).

8. Theory and application of anion selective membrane electrodes, E. Pungor, *Anal. Chem.* **39**, (No. 53), 28A–45A (1967).

9. Recent developments in the theory and application of some ion selective membrane electrodes, E. Pungor and K. Toth, *Hung. Sci. Instrum.* **14**, 15–20 (1968).

10. Similarities between liquid and solid ion exchangers and their usefulness as ion specific electrodes, G. Eisenman, *Anal. Chem.* **40**, 310–320 (1968).

11. Theory of membrane electrode potentials: An examination of the parameters determining the selectivity of solid and liquid ion exchanges and of neutral ion-sequestering molecules, G. Eisenman, In: *Ion Selective Electrodes*, R. A. Durst, ed., Chap. 1, pp. 1–56, National Bureau of Standards Pub. No. 314 (1969).

12. Solid-state and liquid membrane ion selective electrodes, J. W. Ross, Jr., In: *Ion Selective Electrodes*, R. A. Durst, ed., National Bur. of Stand. Publ. No. 314, Chap. 2, pp. 57–88 (1969).

13. Heterogeneous membrane electrodes, A. K. Covington, In: *Ion Selective Electrodes*, R. A. Durst, ed., National Bur. of Stand. Pub. No. 314, Chap. 3, pp. 89–106 (1969).

14. Thermodynamic studies, J. N. Butler, In: *Ion Selective Electrodes*, R. A. Durst, ed., Natl. Bur. of Stand. Pub. No. 314, Chap. 5, pp. 143–189 (1969).

15. New directions for ion selective electrodes, G. A. Rechnitz, *Anal. Chem.* **41** (No. 12), 109A–113A (1969).

16. Ion selective membrane electrodes. Part I. Glass electrodes, G. J. Moody, R. B. Oke, and J. D. R. Thomas, *Lab. Prac.* **18**, 941–945 (1969).

17. Ion Selective Membrane Electrodes, E. Pungor and K. Toth, *Analyst* **95**, 625–648 (1970).

18. *Selective Ion Sensitive Electrodes*, G. J. Moody and J. D. R. Thomas, Merrow, Watford (England) (1971).

19. R. P. Buck, *Physical Methods of Chemistry*, A. Weissberger and B. Rossiter, eds., Vol. I, Part II, Chap. II, pp. 61–162, Wiley, New York (1971).

20. Ion selective electrodes in science, medicine and technology, R. A. Durst, *Am. Sci.* **59** (3), 353–361 (1971).

21. Ion selective sensors, W. Simon, *Pure Appl. Chem.* **25**, 811–823 (1971).

22. Function potential and selectivity rating of selective ion-sensitive membrane electrodes, G. J. Moody and J. D. R. Thomas, *Lab. Pract.* **20**, 307–311 (1971).

23. Theory and applications of ion-selective electrodes, J. Koryta, *Anal. Chim. Acta* **61** (3), 329–411 (Eng) (1972).

24. Development and publication of work with selective ion sensitive electrodes, G. J. Moody and J. D. R. Thomas, *Talanta* **19**, 623–639 (1972) (Eng).

25. Ion selective electrodes, potentiometry and potentiometric titrations, R. P. Buck, *Anal. Chem.* **44**, 270R–295R (1972).

26. Some problems concerning the behaviour of ion selective membrane electrodes under current, C. Liteanu, I. C. Popescu, and E. Hopirtean, In: *Ion Selective Electrodes*, E. Pungor, ed., pp. 51–96, Akademiai Kiado, Budapest (1973).

27. Selectivity and sensitivity of ion selective electrodes, G. J. Moody and J. D. R. Thomas, In: *Ion Selective Electrodes*, E. Pungor, ed., pp. 97–113, Akademiai Kiado, Budapest (1973).

28. An examination of the temperature coefficients of ion selective electrodes in non-isothermal galvanic cells, E. Lindner, K. Toth, and E. Pungor, In: *Ion Selective Electrodes*, E. Pungor, ed., pp. 205–217, Akademiai Kiado, Budapest (1973).

29. Selectivity of coated wire and liquid ion sensitive electrodes, T. Stworzewicz, J. Czapkiewicz, and M. Leszko, In: *Ion Selective Electrodes*, E. Pungor, ed., pp. 259–267, Akademiai Kiado, Budapest (1973).

30. Steady-state space charge effects in symmetric cells with concentration polarized electrodes, R. P. Buck, *J. Electroanal. Chem. Interfacial Electrochem.* **46**, 1–23 (1973).

31. K. Cammaun, *Das Arbeiten mit Ion Selektiven Elektroden*, Springer Verlag, Berlin (1973).

32. Ion selective electrodes, potentiometry and potentiometric titrations, R. P. Buck, *Anal. Chem.* **46**, 28R (1974).

33. Ion selective electrodes, R. P. Buck, *Anal. Chem.* **48**, 23R (1976).

34. *Applications of ion-selective membrane electrodes in organic analysis; Part I. Theoretical consideration*, G. E. Bainlescu and V. V. Cosofret (Eng. Trans), Chap. 1–5, pp. 9–49, Ellis Horwood Ltd. Pub., Chichester (1977).

35. *Analysis with Ion Selective Electrodes*, J. Vesely, D. Weiss, and K. Stulik, ed., Wiley, New York, 220 pp. (1977).

36. Determination of selectivity coefficients of ion selective Electrodes, F. Shinsaking and M. Okada, *Matsushita Hiroshi Chubu Kogyo Daigabu Kiyo*, **13-A**, 143–149 (1977) (Japan).

37. Theory and principles of membrane electrodes, R. P. Buck, In: *Ion Selective Electrodes in Analytical Chemistry*, H. Freiser, ed., Vol. 1, Chap. 1, pp. 1–141, Plenum Press, New York (1978).

38. Precipitate-based ion selective electrodes, E. Pungor and K. Toth, In: *Ion Selective Electrodes in Analytical Chemistry*, H. Freiser, ed., Vol. 1, Chap. 2, pp. 143–210, Plenum Press, New York (1978).

39. Ion selective electrodes based on neutral carriers, W. E. Morf and W. Simon, In: *Ion Selective Electrodes in Analytical Chemistry*, H. Freiser, ed., Vol. 1, Chap. 3, pp. 211–286, Plenum Press, New York (1978).

40. Selection of suitable reference electrodes of pH and ion selective measurements, W. C. Clock, *Beckmann* **2**, 22–26 (1978) (Ger).

41. Ion selective electrodes response to anions, K. Masamitsu and T. Kosurhara, *Dogin Nyusu* **8**, 1–5 (1978) (Japan).

42. Functional electrode, N. Katsumi and U. Yoshio, *Bunsenki* **11**, 798–803 (1978) (Japan).

43. Potential pitfalls in the use of ion selective electrodes-reference electrode pairs, R. A. Durst, *Theory Des. Biomed. Appl. Solidstate Chem. Sens. Workshop 1977*, 155–163 (1979).

44. Potential generating processes at interfaces: from electrolyte/metal and electrolyte membrane to electrolyte semiconductor, R. P. Buck, *Theor. Des. Biomed. Appl. Solidstate Chem. Sens. Workshop 1977*, 3–39 (1978) (Eng).

45. The problem of the single ion activity, R. G. Bates, *Denki Kagaku Oyebi Kogyo Butsuri Kagaku* **46** (9), 480–484 (1978) (Eng.).

46. Ion selective electrodes, E. Pungor, K. Toth, and G. Nagy, *Mikrochim Acta* **1** (5–6), 531–545 (1978).

47. Statistical approach for selectivity of ion selective membrane electrodes, C. Liteanu *et al.*, *Anal. Chem.* **50** (8), 1202–1209 (1978).

48. Response time studies on neutral carrier ion-selective membrane electrodes, E. Lindner, K. Toth, E. Pungor, E. Morf Werner, and W. Simon, *Anal. Chem.* **50** (12), 1627–1631 (1978) (Eng).

49. A mixed potential ion selective electrodes theory, K. Camman, *Ion Selective Electrodes Conf. 1978*, 297–306 (1978).

50. Ion selective electrodes in analytical chemistry, H. Freiser, ed., 439 pp., Plenum Press, New York (1978).

51. The present state of art (of ion selective electrodes), E. Pungor, *Ion Selective Electrodes Conf. 1977*, 161–173 (1978).
52. Comparative theoretical aspects of different types of ion selective electrodes, R. P. Buck and P. Richard, W. R. Kenan, Jr., ed., Lab. Chem. Univ. North Carolina, Chapel Hill N.C. *Ion Selective Electrodes Conf. 1977*, 21–55 (1978).
53. Introduction, basic electrode types, classification and selectivity considerations, A. K. Covington, In: *Ion Selective Electrode Methodology*, A. K. Covington, ed., Vol. I, Chap. I, pp. 1–20, CRC Press Inc., Florida (1979).
54. P. Standards and A. K. Covington, In: *Ion Selective Electrode Methodology*, A. K. Covington, ed., Chap. 4, pp. 67–76, CRC Press Inc. Florida (1979).
55. Enzyme electrode, P. Vadgama, In: *Ion Selective Electrode Methodology*, A. K. Covington, ed., Vol. II, Chap. 2, pp. 23–40, CRC Press Inc. Florida (1979).

4.A.2. Instrumentation and Method of Preparation

56. The development of membranes prepared from artificial cation-exchange materials with particular reference to the determination of sodium ion activities, M. R. Wyllie and H. W. Patnode, *J. Phys. Chem.* **54**, 204–227 (1950).
57. Fluoride microanalysis by linear null point potentiometry, R. A. Durst, *Anal. Chem.* **40**, 931–935 (1968).
58. Industrial analysis and control with ion selective electrodes, T. S. Light, In: *Ion Selective Electrodes*, R. A. Durst, ed., Nat. Bur. of Stand. Publ. No. 314, Chap. 10, pp. 349–374 (1969).
59. Ion selective membrane electrodes, II. Nonglass membrane electrode, G. J. Moody, R. B. Oke, and J. D. R. Thomas, *Lab. Pract.* **18**, 1056–1062 (1969).
60. Differential potentiometry with ion selective electrodes. A new instrumental approach, M. J. D. Brand and G. A. Rechnitz, *Anal. Chem.* **42**, 616–622 (1970).
61. Computer approach to ion selective electrode potentiometry by standard addition methods, M. J. D. Brand and G. A. Rechnitz, *Anal. Chem.* **42**, 1172–1177 (1970).
62. Applications of ion selective electrodes in continuous analysis, B. Fleet and A. Y. W. Ho, In: *Ion Selective Electrodes*, E. Pungor, ed., pp. 17–35, Akademiai Kiado, Budapest (1973).
63. Use of sensitive single chemical flow through electrode for automation of potentiometric Cl^- determination in biol. materials according to flow stream principle, U. Tietz, U. Gruenke, P. Hastmann, and E. Keil, *Z. Med. Laboratoriumsdiagn.* **19** (5), 327–332 (1978) (Ger).
64. Automated semi-micro determination with ion selective electrodes, W. A. Lingerak, F. Bakker, and J. Slanina, *Ion Selective electrode Conference 1977*, 453–462 (1978).
65. Application of computers in potentiometric studies of complexation equilibriums, Majer Vladimir, Stulik, and Karel, *Chem. Listy* **72** (8), 785–800 (1978) (Czech).
66. Novel computer evaluation of multiple standard addition with ion selective electrodes, G. Horvai, L. Domokos, and E. Pungor, *Fresenius Z. Anal. Chem.* **292** (2), 132–134 (1978).
67. An evaluation of some pharmaceutical applications of an ion selective electrodes-Auto analyzer II system, A. W. Finnerty, H. Luttrell, Zudeck, and Steven, *Adv. Autom. Anal. Technic Int. Congr.* 7th, **2**, 210–214 (1976) (Pub. 1977) E. C. Barton, ed., Mediad Inc., Tarrytown, N. Y.
68. Evaluation of multiple standard addition data obtained with ion selective electrodes. I. Evaluation using a desk-top computer, G. Horvai, L. Domoskos, and E. Pungor (Altalanos Anal. Kem. Tansz. Budapesti Musz. Fgy, Budapest, Hung). *Magy. Kem. Foly* **84** (11), 481–483 (1978) (Hung.).
69. Electrode system for determining ions in solution, Bocke Jan. Ger Offen. 2726271 (C1 GOIN 27128) 22 Dec. 1977.

70. Method and apparatus for the continuous measurements of gas traces with ion selective electrodes, Fritze, Ulrich, Herschinger and Heinz, Ger Offen. 2723310 (C1 GOIN 27/46) CA 90 (1979) 91782 d.

71. Computer automation of potentiometric analysis with ion selective electrodes, J. Slanina, F. Bakkar, J. J. Moels, J. E. Ordelman, and A. G. M. Bruyen-Hes, *Anal. Chim. Acta* **112** (1), 45–54 (1979).

72. Microcomputer controlled potentiometric analysis system, R. M. Charles and H. Freiser, *Anal. Chem.* **51** (7), 803–807 (1979).

73. Instrumentation for ion selective electrodes, P. R. Burton, In: *Ion Selective Electrode Methodology*, A. K. Covington, ed., Vol. I, Chap. 2, pp. 21–41, CRC Press Inc. Florida (1979).

4.A.3. Applications

74. The use of membrane electrodes in the analysis of ionic concentrations, E. Pungor and E. Hallos-Rockosinyi, *Acta. Chim. Acad. Sci. Hung.* **27**, 63–68 (1961) (German).

75. *Selective ion electrodes in analysis instrumentation*, T. S. Light, Vol. 5, L. Fowler, R. G. Harmon, and D. K. Roe, eds., pp. 73–87, Plenum Press, New York (1968).

76. Activity standards for ion selective electrodes, R. G. Bates and M. Alfenaar, In: *Ion Selective Electrodes*, R. A. Durst, ed., Natl. Bur. of Stand. Publ. No. 314, Chap. 6, pp. 191–214 (1969).

77. Ion selective electrodes in biomedical research, R. N. Khuri, In: *Ion Selective Electrodes*, R. A. Durst, ed., Nat. Bur. of Stand. Publ. No. 314, Chap. 8, pp. 287–312 (1969).

78. Analytical studies on ion selective membrane electrodes, G. A. Rechnitz, In: *Ion Selective Electrodes*, R. A. Durst, ed., Nat. Bur. of Strand. Publ. No. 314, Chap. 9, pp. 313–348 (1969).

79. Analytical techniques and applications of ion selective electrodes, R. A. Durst, In: *Ion Selective Electrodes*, R. A. Durst, ed., Nat. Bur. of Stand. Publ. No. 314, Chap. 11, pp. 375–414 (1969).

80. Application of ion selective electrode systems to industrial processes, R. L. Beals, *Ann. Instrum.* **7**, 74–79 (1969).

81. Ion selective electrodes—Industrial applications, T. S. Light, *Ind. Water Eng.* **6** (9), 33–37 (1969).

82. Measurements of inorganic water pollutants by specific ion electrode, J. M. Riseman, *Am. Lab.* 32–39, July (1969).

83. Ion selective electrodes in science, medicine and technology, R. A. Durst, *Am. Sci.* **59** (3), 354–361 (1971).

84. Ion selective electrode procedure for organophosphate pesticide analysis, G. Baum and F. B. Ward, *Anal. Chem.* **43**, 947–948 (1971).

85. Applications of ion selective electrodes—Application, W. E. Bazzelle, *Anal. Chim. Acta* **54**, 29–39 (1971).

86. Intrinsic end point errors in pipette titrations with ion selective electrodes, P. W. Carr, *Anal. Chem.* **43**, 425–430 (1971).

87. Application of specific ion electrodes to electroplating analyses, M. S. Frant, *Plating* **58**, 686–693 (1971).

88. Ion sensitive electrodes for individual ionic activity measurements, G. Milazzo, In: *Ion Selective Electrodes*, E. Pungor, ed., pp. 115–126, Akademiai Kiado, Budapest (1973).

89. Ion selective electrodes, Crytur and B. Manek, In: *Ion Selective Electrodes*, E. Pungor, ed., pp. 219–224, Akademiai Kiado, Budapest (1973).

90. The biomed. and related roles of ion selective electrodes, G. J. Moody and J. D. R. Thomas, *Prog. Med. Chem.* **14**, 51–104 (1977).

91. Recent analytical applications of ion selective electrodes, G. J. Moody and J. D. R. Thomas, *Lab. Pract. Rev. Curr. Tech. Instrum.* 1–20 (1977).

92. Sources of error in ion selective electrode potentiometry, R. A. Durst, In: *Ion Selective Electrodes in Analytical Chemistry*, H. Freiser, ed., Chap. 5, pp. 311–338, Plenum Press, New York (1978).

93. Applications of ion selective electrodes, G. J. Moody and J. D. R. Thomas, In: *Ion Selective Electrodes in Analytical Chemistry*, H. Freiser, ed., Chap. 6, pp. 339–433, Plenum Press, New York (1978).

94. Applicability of ion selective electrodes in automated systems for clinical analysis, W. Simon, D. Ammann, H. F. Osswald, P. C. Meier, and R. E. Drhneb, *Adv. Autom. Anal. Technicon. Int. Congr. 7th* 1976 (Pub. 1977) **1**, 59–62.

95. Studies on ion selective electrodes. Part II. Calibration of ion selective electrodes with exponential dilution flask, V. Raune, *Kem Kemi* **5** (12), 614–617 (1978).

96. Electrode behaviour of anion selective membranes in solutions of surfactants, S. V. Timofeev, E. A. Materova, L. K. Arkhangelskii, and E. V. Chirkova, *Vestn. Leningr. Univ. Fiz. Khim.* **3**, 139–141 (1978).

97. The use of ion selective electrodes in diary science, E. Tschager, *Milchwirtsch Ber. Bundesanst. Wolfpassing Rotholz* **56**, 189–194 (1978).

98. Evaluation of multiple standard addition data obtained with ion selective electrodes. II. A study on the precision of the multiple standard addition method, G. Horvai, K. Toth, and E. Pungor, *Mogy. Kem. Foly* **84** (11), 483–485 (1978) (Hung.).

99. New achievements in the flow solution analysis with ion selective electrodes, G. Nagy, Z. Feher, K. Toth, and E. Pungor, *Ion Selective Electrode Conf. 1977*, 477–490 (1978).

100. Trends in the standardization of ion selective electrodes, R. G. Bates and R. A. Robinson, *Ion Selective Electrode Conf. 1977* (1978).

101. The use of combination electrodes in automated systems, M. Vandeputte, L. Dryon, and D. L. Massart (Pharm. Inst. Univ. Brussels, St. Genesius Rode. Belg), *Ion Selective Electrodes Conf. 1977*, 583–587 (1978).

102. Analytical techniques involving ion selective electrodes, J. D. R. Thomas, *ISE Conference 1977*, 175–198 (1978).

103. Present status and future applications of coated wire electrodes, H. Freiser, *Theory Des Biomeel Appl. Solidstate Chem. Sensor Workshop 1977*, 177–182 (1978).

104. Survey of flow through analytical systems employing ion selective electrodes as detectors, G. Nagy, Z. Feher, K. Toth, and E. Pungor, *Hung. Sci. Instrum.* **41**, 27–39 (1977).

105. Errors and error propagation using ion selective electrodes evaluation of concentrations, S. Ebel, E. Glaser, and A. Seuring, *Fresenius, Z. Anal. Chem.* **291** (2), 108–112 (1978).

106. Studies on standard addition method in ion selective electrodes—An equation, a plot and a monograph involving ΔE, Chao Tsao-Fang, *K'O Hsueh Tung Pao.* **24** (5), 212–214 (1979).

107. Ion selective electrodes in organic elemental and functional group analysis—a review, W. Selig, Report 1977 (*CA* **90**, 161712 g) (1979).

108. Practical techniques for ion selective electrodes, R. J. Simpson, In: *Ion Selective Electrode Methodology*, A. K. Covington, ed., Vol. I, Chap. 3, pp. 13–66, CRC Press Inc. Florida (1979).

109. Ion Selective Electrodes in Medicine and Medical Research, D. M. Band and T. Treasure, In: *Ion Selective Electrode Methodology*, A. K. Covington, ed., Vol. II, Chap 3, pp. 41–63, CRC Press Inc. Florida (1979).

110. Analytical methods involving ion selective electrodes (including flow methods), K. Toth, G. Nagy, and E. Pungor, In: *Ion Selective Electrode methodology*, A. K. Covington, ed., Vol. II, Chap. 4, pp. 65–122, CRC Press Inc. Florida (1979).

4.B.1. Solid Membrane Electrodes: Theory

1. Electrode for sensing fluoride ion activity in solutions, M. S. Frant and J. W. Ross, Jr., *Science* **154**, 1553–1555 (1966).

2. *Glass electrodes for hydrogen and other cations—Principles and Practice*, G. Eisenman, ed., Marcel Dekker, New York (1967).

3. Theory of potential distribution and response of solidstate membrane electrodes, R. P. Buck, *Anal. Chem.* **40**, 1432–1439 (1968).

4. Alkaline earth and lanthanum ion electrodes of the third kind based on the hydrogen ion responsive glass electrode. Thermodynamic solubility products of long chain normal fatty acids and their alkaline earth and lanthanum salts in water, A. A. A. Al Attar and W. H. Beck, *J. Electroanal. Chem. Interfacial Electrochem.* **27**, 59–67 (1970).

5. Study of the behaviour of membrane and single crystal electrodes with respect to some univalent ions, N. Bottazzini and V. Crespi, *Chim. Ind. (Milan)* **52**, 866–869 (1970).

6. Recent results in the field of precipitate based ion selective electrodes, K. Toth, In: *Ion Selective Electrodes*, E. Pungor, ed., pp. 145–164, Akademiai Kiado, Budapest (1973).

7. The low response of silicone rubber electrodes to low levels of activity, P. L. Bailey and E. Pungor, In: *Ion Selective Electrodes*, E. Pungor, ed., pp. 167–171, Akademiai Kiado, Budapest (1973).

8. Response to Ca^{+2} ions of CaF_2 crystal membrane electrode, G. Malathi *et al.*, *Analyst (Lond)* **103** (1228), 768–770 (1978).

9. Effect of pH on response of a CN^- ISE, M. Gratzl *et al.*, *Anal. Chim. Acta* **102**, 85–90 (1978).

10. Study of mechanism of ion selective electrode using impedance measurements, J. Mertens, P. Vanden Winkel, and J. Verecken, *Bioelectrochem. Broenerg.* **5** (4), 699–712 (1978).

11. Response of Cu(II) ion selective electrode to some complexing agents, I. L. Sekara and F. Josef, *Anal. Lett.* **A11** (5), 415–427 (1978).

12. Dynamic response studies of solidstate Cl^- ISE in presence of Fe(III) with an on line computer, J. W. Bixler, R. Nee, and S. P. Perone, *Anal. Chim. Acta* **99** (2), 225–232 (1978).

13. Transient characteristics of a cadmium ion selective electrode, E. Niki and H. Shirai (Fac. Technol. Univ. Tokyo, Japan), *Elektrokhimiya* **14** (5), 714–718 (Russ) (1978).

14. Grain boundary effects in solidstate ion selective electrodes, R. E. Vander Leest and A. Geven, *J. Electroanal. Chem. Interfacial Electrochem.* **90** (1) 97–104 (1978).

15. Structure and electrical properties of the ion selective electrode membranes based on AgX-Ag_2S (X = Cl, Br, I), Yu. G. Vlasov and S. B. Kocheregin, *Ion Selective Electrode Conf. 1977*, 597–601 (1978).

16. Interpretation of the response mechanism of solidstate ion selective electrodes regarding diffusion process, A. Hulanicki and A. Leuenstam, *ISE Conf. 1977*, 395–402 (1978).

17. The behaviour of solidstate ISE in the presence of complexing agents, W. E. Vander Linden and G. J. M. Heigne, *ISE Conf. 1977*, 445–452 (1978).

18. Polyvinyl chloride matrix membrane ion selective electrodes, G. J. Moody and J. D. R. Thomas, In: *Ion selective Electrode Methodology*, A. K. Covington, ed., Vol. I, Chap. 7, pp. 111–130, CRC Press Inc., Florida (1979).

19. Heterogeneous membrane, carbon support and coated wire ion selective electrodes, R. W. Cattrall, In: *Ion Selective Electrode Methodology*, A. K. Covington, ed., Vol. 1, Chap. 8, pp. 131–173, CRC Press Inc., Florida (1979).

20. Crystalline and pressed powder; Solid membrane electrodes, R. P. Buck, In: *Ion Selective Electrode Methodology*, A. K. Covington, ed., Vol. 1, Chap. 9, pp. 175–250, CRC Press Inc., Florida (1979).

21. Gas-sensing probes, M. Riley, In: *Ion Selective Electrode Methodology*, A. K. Covington, ed., Vol. II, Chap. 1, pp. 1–21, CRC Press Inc., Florida (1979).

4.B.2. Solid Membrane Electrodes: Instrumentation and Methods of Preparation

22. Permselective membrane electrodes, J. S. Parsons, *Anal. Chem.* **30**, 1262–1265 (1958).

23. Electrochemical generation of fluoride ion by solid-state transference, R. A. Durst and J. W. Ross, *Anal. Chem.* **40**, 1343–1344 (1968).

24. Nitrate determinations in soil extracts with the nitrate electrode, A. Oeien and A. R. Selmer-Olsen, *Analyst (Lond)* **94** (1123), 888–894 (1969).

25. Ion selective combination electrode of synthetic materials, D. N. Hoole and G. L. Klein, Ger. Offen. 1918590 (Cl GOIN) 30 Oct. 1969.
26. Plastic electrodes selective for organic ions, T. Higuchi, C. R. Glian, and J. L. Tossounian, *Anal. Chem.* **42**, 1674–1676 (1970).
27. Construction of ion selective glass electrodes by vacuum deposition of metals, J. P. Guignard and S. M. Friedman, *J. Appl. Physiol.* **29**, 254–257 (1970).
28. Copper(I) sulfide ceramic membrane as selective electrodes for copper(II), H. Hirata, K. Higashiyama, and K. Date, *Anal. Chim. Acta* **51**, 209–212 (1970).
29. The p–n–p transistor used exponentially to linearize the voltage out put of the pCO_2 physiological electrode, C. E. W. Hahn, *Rev. Sci. Instrum.* **42**, 1164–1168 (1971).
30. Differential potentiometric determination of parts per billion chloride with ion selective electrodes, T. M. Florence, *J. Electroanal Chem. Interfacial Electrochem.* **31**, 77–86 (1971).
31. Apparatus for producing ion selective electrodes, F. Oehme, Ger. Offen. 2006194 (Cl GOIN) 11 Feb. 1971.
32. Ion specific electrodes, J. Ruzicka, C. G. Lamm, and J. C. Tjell, Ger. Offen. 2034686 (Cl GOIN) 11 Feb. 1971.
33. New type of solid-state ion selective electrodes with insoluble sulfides or halides, J. Ruzicka and C. G. Lamm, *Anal. Chim. Acta* **53** (1), 206–208 (1971).
34. Electrode—the universal ion selective solidstate electrode, I. Halides, J. Ruzicka and C. G. Lamm, *Anal. Chim. Acta* **54** (1), 1–12 (1971).
35. New type of lead(II) ion selective ceramic membrane electrode, H. Hirata and K. Higashiyama, *Anal. Chim. Acta* **54** (3), 415–422 (1971).
36. A further study on a chelate forming selective electrode, P. Gabor-Klatsmanyi, K. Toth, and E. Pungor, In: *Ion Selective Electrodes*, E. Pungor, ed., pp. 183–190, Akademiai Kiado, Budapest (1973).
37. Development of a silicone-rubber potassium membrane electrode, J. Pick, K. Toth, M. Vasale, E. Pungor, and W. Simon, In: *Ion Selective Electrodes*, E. Pungor, ed., pp. 245–252, Akademiai Kiadu, Budapest (1973).
38. Poly(vinyl chloride) matrix membrane ion selective electrodes, G. J. Moody and J. D. R. Thomas, In: *Ion Selective Electrodes in Analytical Chemistry*, H. Freiser, ed., Chap. 4, pp. 287–309, Plenum Press, New York (1978).
39. Electrodes for determination of ion activity, J. Vesely, J. Jindara, and F. Gvegr, Czech. 176359 (Cl GOIN 27130) 15 Jan 1979.
40. Ion selective electrodes process control, J. M. Zimmer Theo, *Mess. Pruef.* **9**, 521–524 (1978).
41. Ion selective electrode, J. Wedzicha, J. Pogorzelski, S. Walizak, Pol. 95511 (Cl GOIN 27/30) 29 Apr 1978.
42. Superhigh sensitivity of a $CuSe$–Ag_2S solid membrane Cu^{2+} ion selective electrode in several metal buffer solutions, Y. Umezawa, *Bull. Chem. Soc. Jpn.* **52** (3), 945–946 (1979).
43. Specific electrode with modular construction, J. Tacussel, Fr. Demonde 2,363,798 (Cl GOIN 27/30) 31 Mar 1978.
44. Ion sensitive inplantable electrodes fabricated by hybrid technology, S. S. Yee and M. A. Afromowitz, *Theory Des Biomed, Appl. Solid State Sensor Workshop 1977*, 81–89 (1978).
45. Potassium film electrodes based on macrocyclic polyesters, Sh. K. Norov *et al.*, *Elektrokhimiya*, **14** (10), 1615 (1978) (Russ).
46. Electrode coatings, S. Grabiec, E. Lubaszha, and I. Janczarski, Pol. 86912 (Cl GOIN 27/30) 30 June 1978.
47. Carbon paste for Ca^{2+} ion selective electrode, G. A. Qureshi, *Libyan J. Sci.* **8A**, 37–41 (1978).
48. Construction and behaviour of micro flow through Cu(II) ion selective electrodes, W. E. Van der Linden and R. Oostervink, *Anal. Chim. Acta* **101** (2), 419–422 (1978).
49. Ion selective electrode, C. J. Battaglia, J. C. Chang, and D. S. Daniel, *Res. Discl.* **176**, 15–16 (1978).
50. Selective ion electrode, M. Lebl, J. Vesely, and J. Jindara, Czech 172449 (Cl GOIN 27/30) 15 May 1978.

51. Measurement of chloride ion concentrations in microsamples, R. K. Popp, J. D. Frantz, G. L. Vogel, and P. E. Hare (Geophys. Lab. Carnegie Inst. Washington, Washington, DC) Year Book—Carnegie Inst. Washington 1977, 913–917 (1978).

52. Use of combination electrodes in automated systems, M. Vandeputte, L. Dryon, and D. L. Massart, *Ion Selective Electrode Conf. 1977*, 583–587 (1978).

53. Construction of Cu^{2+}, Cd^{2+} and Pb^{2+} ion selective electrode, A. Takashi, *Denki Kagaka Oyobi Kogyo Butsuri Kagaku* **46** (5), 259–263 (1978).

54. Polypropylene glycol adducts as sensor for ion selective electrodes, A. M. Y. Jaber, G. J. Moody, and J. D. R. Thomas, *Ion Selective Electrodes Conf. 1977*, 411–417 (1978).

55. A fluoride sensitive electrode based on bismuth fluoride, V. Ebock and C. Neiser, *Z. Chem.* **18** (9), 343–344 (1978) (Ger.).

56. New solid state ion selective electrodes for phosphate and chloride activity measurements, J. Tacussel and J. J. Fombon, *Ion Selective Electrodes Conf. 1977*, 567–575 (1978).

57. Construction of Cl^-, Br^-, I^-, S^-, SCN^- and CN^- ion selective electrodes, A. Takashi, *Denki Kagaku Oyobi Kogyo Butsuri Kagaku* **46** (6), 343–347 (1978).

58. A new type of homogeneous chalcogenide ion selective electrodes, M. Neshkova and H. Sheytanov, *Ion Selective Electrode Conf. 1977*, 503–510 (1978).

59. Technological investigations with ion selective electrodes for the purpose of determination of SO_4^-, W. Misniakiewicz and K. Raszkes, *Ion Selective Electrode Conf. 1977*, 467–475 (1978).

60. The formation of mixed $CuS–Ag_2S$ membranes for Cu^{2+} ion selective electrodes. Part III. Electrode response in presence of complexing agents, G. J. M. Heijne and W. E. Vander Linden, *Anal. Chim. Acta* **96** (1), 13–22 (1978).

61. Applications of chloride electrodes based on mercurous chloride/mercuric sulfide, P. L. Bailey, J. Wilson, S. Karpel, and M. Riley, *Ion Selective Electrodes Conf. 1977*, 201–206 (1978).

62. A new type of homogeneous chalcogenide ion selective electrodes, M. Neshkova and H. Sheytanov, *Ion Selective Electrodes Conf. 1977*, 503–510 (1978).

63. Heterogeneous ion selective electrodes based on epoxy resin. A Cl^- ion selective electrode K. Sykut and E. Nowakowska, *Biul. Lubel Two Nauk Mat. Fiz. Chem.* **19** (1), 89–93 (1977).

64. Cl^- and Br^- ion selective electrodes, J. Siemroth and I. Henning, Ger (East) 127997 (Cl GOIN 27/30) 26 Oct 1977.

65. New solid state ion selective electrodes for phosphate and Cl^- measurement, J. Tacussel and J. J. Fombon, *Ion Selective Electrode Conf. 1977*, 567–575 (1978).

66. K^+ ion selective electrode and digital apparatus for measurements, J. Havas, L. Keskes, and R. Somods, *Orv. Tech.* **15** (2), 41–46 (1977) (Hung.).

67. Ion selective electrode crytur, M. Sember and M. Manek, *Ion Selective Electrode Conf. 1977*, 529–537 (1978).

68. Determination of water miscible organic solvents by ion selective electrodes, G. J. Kakabadse, *Solvents-Neglected Parameter, Solvents Symp. 2nd 1977, CA* **90**, 97119q (1979).

4.B.3. Solid Membrane Electrodes: Applications

69. Analytical study of a sulphide ion selective membrane electrode in alkaline solutions, G. A. Rechnitz and T. M. Hesa, *Anal. Chem.* **40**, 1054–1060 (1968).

70. Activity measurements with a fluoride selective membrane electrode, K. Srinivasan and G. A. Rechnitz, *Anal. Chem.* **40**, 509–512 (1968).

71. Trace fluoride determination with specific ion electrode, E. W. Baumann, *Anal. Chim. Acta* **42** (1), 127–132 (1968).

72. Lead poisoning in children: detection by ion selective electrode, G. A. Rechnitz, *Science* **166** (3904), 532 (1969).

73. Electrometric determination of iodine in organic material, B. Paletta and K. Panzenbeck, *Clin. Chim. Acta* **26**, 11–14 (1969).
74. Analysis of alkaline pulping liquor with sulfide ion selective electrode, J. L. Swartz and T. S. Light, *Tappi* **53**, 90–95 (1970).
75. Potentiometric measurements in aqueous, non-aqueous and biological media using a lead ion selective membrane electrode, G. A. Rechnitz and N. C. Kenny, *Anal. Lett.* **3**, 259–271 (1970).
76. Potentiometric microdetermination of phosphate with an ion selective lead electrode, W. Selig, *Mikrochim. Acta* 564–571 (1970).
77. Semi-micro determination of oxalate with a lead specific electrode, W. Selig, *Microchem. J.* **15**, 452–458 (1970).
78. Determination of the cyanide ion in the atmosphere, drinking water, industrial wastes and biological media by a specific electrode, C. Collombel, J. P. Durand, J. Bureacy, and J. Cotte, *J. Eurip. Toxicol.* **3**, 291–299 (1970).
79. Determination of cyanides with ion selective membrane electrodes, K. Toth and E. Pungor, *Anal. Chim. Acta.* **51**, 221–230 (1970).
80. Determination of fluoride in silicate rocks without separation of aluminium using a specific ion electrodes, B. L. Ingram, *Anal. Chem.* **42**, 1825–1827 (1970).
81. Determination of thiois with a specific ion electrode, L. C. Gruen and B. S. Harrap, *Anal. Biochem.* **42**, 377–381 (1971).
82. Iodide selective electrodes in reacting and in equilibrium systems, J. H. Woodson and H. A. Liebhafsky, *Anal Chem.* **41**, 1894–1897 (1969).
83. Standard addition titration method for the potentiometric determination of fluoride in sea water, T. Anfalt and D. Jagner, *Anal. Chim. Acta* **53**, 13–22 (1970).
84. Potentiometric determination of halide ions in solution, M. Bartusek, J. Senkyr, J. Janosova, and M. Polasek, In: *Ion Selective Electrodes*, E. Pungor, ed., pp. 173–182, Akademiai Kiado, Budapest (1973).
85. Bromide-selective electrode for use in following the Zhabotinsky-type oscillating chemical reaction, E. Koris and M. Bunger, In: *Ion Selective Electrodes*, E. Pungor, ed., pp. 191–203, Akademiai Kiado, Budapest (1973).
86. Potentiometric studies on thioacetamide and thiourea by means of a sulphide ion selective membrane electrode, M. K. Papay, K. Toth, and E. Pungor, In: *Ion Selective Electrodes*, E. Pungor, ed., pp. 225–243, Akademiai Kiado, Budapest (1973).
87. Method for following the formation of aluminium chloride di-isopropylate with a chloride-selective membrane electrode, I. Simonyi and I. Kalman, In: *Ion Selective Electrodes*, E. Pungor, ed., pp. 253–258, Akademiai Kiado, Budapest (1973).
88. *Application of ion selective membrane electrodes in organic analysis. Part II. Applications in organic analysis*, G. E. Baiulescu and V. V. Cosofret (Eng. Trans), Chap. 6–9, 10.3, 11.1.2, 11.2.1, and 15.1 and 15.3.2, Ellis Horwood Ltd. Publ., Chichester (1977).
89. Application of ion selective electrodes in steel industry, M. E. Hofton, *Proc. Chem. Conf. 1975*, **28**, 72–80 (1976).
90. Indirect determination of small amounts of Co in presence of Zn ions and other metallic ions via cyanide ion selective electrodes, A. Duca and M. Florica, *Rev. Chim. (Bucharest)* **28** (12), 1186–1188 (1977).
91. Ion selective electrodes, *Anon (Engl) Res. Disch.* **161**, 29–39 (1977).
92. Behaviour of copper (II) ion selective electrodes in various copper ion buffers, T. Hashizume, *Kitakyushu Kogyo Keto semmon Gakko Kenkyu Hokoku* **11**, 183–188 (1978).
93. Cl⁻ selective liquid membrane electrodes based on lipophilic methyl-tri-N-alkyl ammonium compounds and their applicability to blood serum measurements, K. Hartman, S. Luterotti, H. F. Osswals, M. Oehme, P. C. Meier, D. Ammann, and W. Simon, *Mikrochim. Acta* **2** (3–4), 235–246 (1978).

94. Method for the determination of methanol in binary methanol-water mixtures by use of ion selective electrodes, G. J. Kakabadse, H. A. Maleila *et al., Analyst (Lond)* **103** (1231), 1046–1052 (1978).

95. Application of a copper electrode as a detector for high-performance liquid chromatography, C. R. Loscombe, G. B. Coz, and J. A. W. Dalziel, *J. Chromatogr.* **166** (2), 403–410 (1978).

96. Measurement of critical miscellaneous concentration by ion selective electrodes, J. E. Newberg and V. Smith, *Colloid Polym. Sci.* **256** (5), 494–495 (1978).

97. Measurements of ionic activities in aqueous solution using epoxy based ion exchange membrane electrode, K. M. Joshi and G. M. Ganu, *Ind. J. Technol.* **16** (2), 53–55 (1978).

98. Determination of H_2S in air using ion selective electrodes, Heyessy and Gyorgy, *Munkavedelen,* **24** (1–3), 21–25 (Hung.) (1978).

99. The limits of detection of gas sensing probes—Application to NH_3 sensor, F. Van der Pol, *Anal. Chim. Acta* **97** (2), 245–252 (1978).

100. Solid state ion selective electrode SCN^- ion, Yu. V. Shavnya and Yu. M. Chikin, *Elektrokhimiya* **14** (2), 336 (1978).

101. Determination of tartrate ion activity using a commercial selectrode, L. P. Dorsett and D. E. Mucahy, *Anal. Lett.* **A11** (11), 53–61 (1978).

102. Behaviour of a polycrystalline fluoride-selective membrane electrode in organo-aqueous media, L. I. Manakova, N. V. Bausova, V. E. Moiseev, V. G. Bamburov, and A. P. Sivoplyas, *Zh. Anal. Khim.* **33** (8), 1517–1520 (1978) (Russ.).

103. Determination of concentration of phosphate and arsenate ions, T. Tanaka, K. Hiiro, and A. Kawahara, Japan Kokai 7839193 (Cl GOIN 27/56), 10 April 1978 (*CA* **89** 122625 r-1978).

104. Automated immunoassay with a silver sulfide ion-selective electrode, R. L. Solsky and G. A. Rechnitz, *Anal. Chim. Acta* **99** (2), 241–246 (1978).

105. Determination of F^- in organo-aqueous solutions using a F^- ion selective electrode, S. A. Gava, N. S. Poluckton, and G. N. Koroleva, *Zh. Anal. Khim.* **33** (3), 506–510 (1978).

106. Simple potentiometric method for determination of reducing substances in unine with Cu^{2+} ion selective electrode, D. P. Nikolelis *et al., Anal. Chim. Acta* **98** (2), 227–232 (1978).

107. Determination of sulfates in natural waters using a barium selective electrode, G. Ouzounian and G. Michard, *Anal. Chim. Acta* **96** (2), 405–409 (1978).

108. Applications of ion selective electrodes to metal finishing industries, G. Subramanian, N. Chandra, and G. Prabhakara Rao, *J. Electrochem. Soc. India* **27** (1), 37–42 (1978).

109. Optimisation and control of lab. sulfidization of oxidised copper ores with ion selective electrodes, M. H. Jones and J. T. Woodcock, *Proc. Austr. Inst. Min. Metall.* **266**, 11–19 (1978).

110. Evaluation of ion selective electrodes for control of Na_2S additions during lab. flotation of oxidized ores, M. H. Jones and J. T. Woodcock, *Trans. Inst. Min. Metall Sect. C* **87**, 99–105 (1978).

111. Clinical applications of ion selective electrodes, C. Fuchs, D. Dorn, and C. J. Preusse, *Ion Selective Electrodes Conf. 1977,* 373–378 (1978).

112. Application of chloride ion selective electrodes in the production of synthetic fibers, L. N. Bykova, N. A. Kazaryan, N. S. Chernova, I. D. Shakhova, and N. M. Kvasha, *Ion Selective Electrodes Conf. 1977,* 289–295 (1978).

113. Determination of SO_3^{2-}, SO_2 and HSO_3 by zero current chronopotentiometry, I. Sekera and J. F. Lechner, *Anal. Chim. Acta* **99** (1), 99–104 (1978).

114. Free cyanide analyser, R. A. Purch, *Ion Selective Electrode Conf. 1977,* 359–362 (1978).

115. An investigation of halide ion selective electrodes in mixed solution, L. N. Bykova, N. A. Kzaryn, E. Pungor, and N. S. Chernova, *Ion Selective Electrodes Conf. 1977,* 281–287 (1978).

116. Potentiometric determination of alkylxanthates with ion selective electrodes, G. E. Baillescu, V. V. Cosofret, and M. Blasnic, *Ion Selective Electrode Conf. 1977,* 207–214 (1978).

117. Determination of Pd(II) as PdI_2 by I^- ion selective electrode, V. Vajgand and V. Kalajligeva, *Ion Selective Electrode Conf. 1977,* 577–581 (1978).

118. Analysis of multi-ionic solution with several ion selective electrodes, V. Rauno, *Ion Selective Electrode Conf. 1977*, 589–595 (1978).
119. Experiments with PVC matrix membrane Ca^{2+} ion selective electrodes, B. J. Birch, A. Craggs, G. J. Moody, and J. D. R. Thomas, *J. Chem. Educ.* **55** (11), 740–741 (1978).
120. Testing and characteristics of colbutinol ion selective electrodes, K. Fukamachi and N. Ishibashi, Yakugaku Zasshi, **99** (2), 126–130 (1979) (Japan).
121. Continuous measurement and automatic control of aluminum for paper making systems, F. Deheme, TAPPI Papermakers conf. (Proc) 41–47 (1979).
122. Ion selective electrodes in organic functional group analysis: Microdetermination of hydrazines with the copper electrode, S. S. M. Hassan and M. T. M. Zaki, *Mikrochim. Acta* **1** (1–2), 137–144 (1979).

4.C.1. Liquid Ion Exchange Membrane Electrodes: Theory

1. Liquid ion exchange membranes of extreme selectivity and permeability of anions, K. Sollner and G. M. Shean, *J. Am. Chem. Soc.* **86**, 1901–1902 (1964).
2. Liquid ion exchange membranes of extreme ionic selectivity and high transmissivity, K. Sollner and G. M. Shean, *Protoplasma* **63** (1–3), 174–180 (1967).
3. Calcium selective electrode with liquid ion exchanger, J. W. Ross, Jr., *Science* **156**, 1378–1379 (1967).
4. Electrical phenomena associated with the transport of ions and ion pairs in liquid ion-exchange membranes. I. Zero current properties, J. Sandblom, G. Eisensmair, and J. L. Walker, Jr., *J. Phys. Chem.* **71** (12), 3862–3870 (1967).
5. Liquid ion exchange membranes with weakly ionised groups. I. A theoretical study of their steady state properties, J. Sandblom, *J. Phys. Chem.* **73**, 249–264 (1969).
6. Electrochemical behaviour of liquid anion membranes. Bionic potentials with the $NO_3^--Cl^-$, $NO_3^--Br^-$, Br^--Cl^- couples, G. Scibona, P. R. Danesi, F. Salvemini, and B. Seuppa, *J. Phys. Chem.* **75** (4), 554–561 (1971).
7. Bi-ionic potentials of a liquid membrane electrode selective toward calcium, J. Bagg, O. Nicholoson, and R. Vinen, *J. Phys. Chem.* **75**, 2138–2143 (1971).
8. Selectivity parameter and partition coefficient for a calcium selective electrode, J. Bagg and W. P. Chung, *Aus. J. Chem.* **24**, 1963–1966 (1971).
9. Experience of using chloride electrodes in blood, Z. A. Alagova, G. I. Shumilova, and K. I. Sulka, *Nauchn. Tr. Kofan Gos. Univ.* **232**, 148–152 (1977) (Russ.).
10. Effect of phenol deriv on the selectivity of an organic sulphonate selective electrode, F. Taitiro, O. Satoshi, and H. Hirokazu, *Chem. Lett.* **11**, 1201–1202 (1978).
11. Effect of solvent nature on the selectivity of liquid Br^- ion selective electrodes, G. L. Starobinets, E. M. Rakhmanko, and R. Del Toro Renis, *Vestsi Akad. Nauk BSSR Ser Khim. Nauk* **4**, 75–78 (1978).
12. Dynamic behaviour of a liquid membrane electrode reversible to nicotinate ion, L. Companella, T. Ferri, D. Gozzi, and G. Scorcelletti, *Ion Selective Electrode Conf. 1977*, 307–316 (1978).
13. A mechanistic tool for liquid membrane ion selective electrodes investigation: Polarization of the liquid/liquid interface, J. Koryta *et al.*, *Ion Selective Electrode Conf. 1977*, 441–444 (1978).
14. Exchange kinetics at a K^+ selective liquid membrane electrode, C. Karl, *Anal. Chem.* **50** (7), 936–940 (1978).
15. Relationship between activity and concentration measurements of plasma potassium, D. M. Band, J. Kratochvil, P. A. Poole Wilson, and T. Treame, *Analyst (Lond)* **103** (1224), 246–251 (1978).
16. Recent development in field of neutral carrier based ion selective electrodes, W. E. Morf and W. Simon, *Ion Selective Electrodes Conf. 1977*, 149–159 (1978).

17. Ion transport in free and supported nitrobenzene aliquat nitrate liquid membrane ion selective electrodes. I. Bulk electrical properties including association and dielectric concentration, D. E. Mathis and R. P. Buck, *J. Membr. Sci.* **4** (3), 379–394 (1979).

18. Ion transport in free and supported nitrobenzene aliquat nitrate liquid membrane ion selective electrodes. II. Interfacial kinetics and time dependent phenomena, D. E. Mathis, F. S. Stover, and R. P. Buck, *J. Membr. Sci.* **4** (3), 395–413 (1979).

19. A study of the mechanism of response of liquid ion exchange Ca^{2+} ion selective electrode. Part I. Zero current potential of commercial and modified electrodes, M. Van Nicole and C. Garkch, *J. Electroanal. Chem. Interfacial Electrochem.* **97** (2), 151–161 (1979).

20. Liquid ion exchange types, A. K. Covington and P. Davison, In: *Ion Selective Electrode Methodology*, A. K. Covington, ed., Vol. 1, Chap. 6, pp. 85–110, CRC Press Inc. Florida (1979).

4.C.2. Liquid Ion Exchange Membrane Electrodes: Instrumentation and Method of Preparation

21. A study of Ca–K ion-exchange equilibrium with the help of membrane electrodes, S. K. Bose, *J. Ind. Chem. Soc.* **37**, 465–472 (1960).

22. Liquid anion membrane electrodes sensitive to metal cation concentration, G. Scibona, L. Mantella, and P. R. Danesi, *Anal. Chem.* **42**, 844–848 (1970).

23. A liquid ion exchange electrode specific to iodide ions, A. L. Grekovich, E. A. Materova, and T. I. Pronkina, *Electrokhimiya* **7**, 436–438 (1971); *Sov. Electrochem.* **7**, 421–422 (1971).

24. Anion exchange electrodes, specific for NO_3^- ions, F. A. Belinskaya, E. A. Materova, L. A. Karmanova, and V. K. Kozyreva, *Elektrokhimiya* **7**, 214–217 (1971); *Sov. Electrochem.* **7**, 196 (1971).

25. Cation selectivity of liquid membrane electrodes based upon new ligands, W. Simon and W. E. Morf, In: *Ion Selective Electrodes*, E. Pungor, ed., pp. 127–143, Akademiai Kiado, Budapest (1973).

26. Liquid membrane electrodes with bromide Cl^- functions, Moskvin *et al.*, *Zh. Anal. Khim.* **32** (8), 1559–1563 (1977).

27. Ion selective thiocyanate electrode for monitoring technological solutions at gold-extracting plants, Yu. Shavnya, Yu. M. Chikin, and V. A. Pronin, *Nauchn. Tr. Irkutsk. Gos. Nauchno-Issted Inst. Redk. Tsventn. Met.* **31**, 63–66 (1977) (Russ.).

28. Construction of a new perbromate selective electrode. Kinetic study of the Fe(II) perbromate reaction and microdetermination of iron, L. A. Lazarou and T. P. Hadjioannou, *Anal. Lett.* **A11** (10), 719–795 (1978).

29. Liquid ion selective electrodes based on Zn tetrathiocyanate anion, E. M. Rakhmembo, G. L. Starobinets, V. L. Lomako, and A. Beisis, *Vestsi. Akad. Nauk. BSSR Ser Khim. Nauk* **6**, 68–71 (1978) (Russ.).

30. Calcium ion selective electrodes based on calcium bis sensor and trialkyl phosphate mediators, G. J. Moody, N. S. Nassory, and J. D. R. Thomas, *Analyst (Lond.)* **103** (1222), 68–71 (1978).

31. An ion-extractive liquid membrane anionic surfactant sensitive electrode and its analytical applications, N. Ciocan and D. F. Anghel, *Frtseniuo Z. Anal. Chem.* **290** (3), 237–240 (1978).

32. Ion selective electrodes based on uranyl bis (2-ethylhexyl) phosphate, V. A. Mikhrilov, V. V. Ospov, and N. V. Serebrennikova, *Zh. Anal. Khim.* **33** (6), 1154–1158 (1978).

33. Research and development of selective membrane electrodes. Communication in hydrogen chromate selective electrodes, Yu. I. Urusov, V. V. Sergievskii, A. F. Zhukov, and A. V. Gordievskii, *Zh. Anal. Khim.* **34** (1), 156–160 (1979).

34. Ion exchanger and ion selective electrodes, Y. Shibazaki and N. Hanaoka, Japan Kokai Tokyo Koho 7923092 (C1 BOI J 1/04) 21 Feb. 1979.

35. Substance for indicator membranes of ion selective electrodes for determining potassium, ammonium and ocsium ions, A. V. Gordievskii, Yu. I. Urusov, and V. Syrchenk, *Otkrytiya. Izofret, Prom. Obraztsy Tovarnye Znaki* **6**, 143 (1979).

36. Combination ion selective electrode based on solvent polymeric membranes, R. E. Dohner and W. Simon, *Anal. Lett.* **12** (A3), 205–212 (1979).

4.C.3. Liquid Ion Exchange Membrane Electrodes: Applications

37. Studies with ion exchange calcium electrodes in biological fluids: Some applications in biomedical research and clinical medicine, E. W. Moore, In: *Ion Selective Electrodes*, R. A. Durst, ed., Nat. Bur. of. Stand. Publ. No. 314, Chap. 7, pp. 215–285 (1969).
38. Nitrate determination in soil extracts with the nitrate electrode, A. Oien and A. R. Selmer-Olsen, *Analyst (Lond.)* **94** (1123), 888–894 (1969).
39. Response of calcium selective and divalent cation selective electrodes to cyclohexyl ammonium ion, H. B. Collier, *Anal. Chem.* **42**, 1443 (1970).
40. Determination of nitrogen dioxide and nitric oxide in the parts per million range in flowing gaseous mixtures by means of the nitrate-specific ion electrode, R. DiMartini, *Anal. Chem.* **42**, 1102–1105 (1970).
41. Analysis of plants soils and waters for nitrate by using an ion selective electrode, P. J. Milham, A. S. Awad, R. E. Paull, and J. H. Bull, *Analyst (Lond.)* **95** (1133), 751–757 (1970).
42. Determination of nitrate in waters with the nitrate selective ion electrode, D. R. Keeney, B. H. Byrnes, and J. J. Genson, *Analyst (Lond.)* **95** (1125), 383–386 (1970).
43. Analytical applications of liquid ion exchangers, A. R. Prabhu and S. M. Khopkai, *J. Sci. Ind. Res.* **30**, 16–22 (1971).
44. *Application of ion selective membrane electrodes in organic analysis. Part II. Application in Organic Analysis*, G. E. Baiulescu and V. V. Cosofret (Eng. Trans), Chap. 11, 13, 15, 16.3, Ellis Horwood Ltd. Publ., Chichester (1977).
45. Potentiometric determination of boric acid in elemental boron with a fluoborate specific electrode, R. O. Inlow, Report 1977, NBL 287, 7 pp. (1977).
46. Ion selective electrode for determining anions, K. W. Sykut, J. Dumkiewicz, A. R. Duonkiewicz, Pol. 93,227 (Cl GOIN 27/30) 15 Dec. 1977.
47. A study of surfactant solutions using the liquid membrane electrode selective to alkyl sulphate ions, A. Yamauchi *et al.*, *Bull. Chem. Soc. Jpn.* **51** (10), 2791–2794 (1978).
48. Ion selective membrane electrodes for organic analysis. Acetylcholins selective liquid membrane electrode, E. Hopirtem and M. Miklos, *Rev. Chim (Bucharest)* **29** (12), 1178–1181 (1978) (Rom.).
49. Study of a sodium ion selective electrode, E. A. Materova, Z. S. Alagova, G. I. Shomilova, and L. P. Vatlina, *Vestn. Leningr. Univ. Fiz. Khim.* **4**, 112–115 (1978) (Russ.).
50. Liquid membrane ion selective electrodes for some alkaloids, T. Goina, S. Hobai, and L. Razenberg, *Farmecka (Bucharest)* **26** (3), 141–147 (1978) (Rom.).
51. Application of ion selective electrodes in the purity test for Cl^- and Pb^{2+} in several DAB and DAC substances, Ali Syed Laik, Stock, Wilfried, *Pharm. Ztg.* **123** (41), 1815–1823 (1978).
52. Ion selective electrodes without reference solutions, Sh. K. Norov *et al.*, *Elektrokhimiya* **14** (10), 1613 (1978) (Russ.).
53. Serum Ca^{2+} and K^+ determination by split membrane ion selective electrodes technique, R. W. Cattrall and T. Fong Kwok, *Talanta* **25** (9), 541–543 (1978).
54. Liquid state ion selective electrodes for determination of acidic and basic dyes, A. G. Fog and K. S. Yoo, *Ion Selective Electrode Conf. 1977*, 369–372 (1978).
55. A dual ion selective electrode detector for the simultaneous detection of bromine and chlorine containing compounds in gas chromatography, K. Tsugio *et al.*, *Anal. Chim. Acta* **101** (2), 273–281 (1978).
56. Ion selective electrode responsive to chlorate ion, K. Hitoshi, *Kitakyushu Kogyo Koto semmon Gakko Kenbyu Hokoku* **11**, 159–164 (1978).
57. Determination of gold in cyanide solutions with an ion selective electrode, Yu. V. Shavnya, A. S. Bychkov, O. M. Petrukhim, V. A. Zarinskii, L. V. Bakhtinova, and Yu. A. Zolotov, *Zh. Anal. Khim* **33** (8), 1531–1538 (1978) (Russ.).

58. New chloramine T and picrate ion selective electrode, T. P. Hadjiioannou, M. A. Koupparis, and E. P. Diamandis, *Proc. Anal. Div. Chem. Soc.* **15** (3), 78–80 (1978).

59. Alcohol, lactate and glutamate sensors based on oxidoreductases with regeneration of nicotinamide adenine dinucleotide, A. Malinavskas and J. Kulys, *Anal. Chim. Acta* **98** (1), 31–37 (1978).

60. Measurement of free Ca^{2+} ion in capillary blood and serum, C. F. A. Niel, F. Torben, L. Komarmy, and O. Siggaard-Anderson, *Clin. Chem.* (Winston-Salem, N.C.) **24**(9), 1545–1552 (1978).

61. Determination of Cl^-, F^-, K^+ and Na^+ ion concentration of biological fluids by ion sensitive microcapillary electrode, J. Havas, *et al.*, *Hung. Sci. Instrum.* **43**, 7–14 (1978).

62. Recording intracellularly with K^+ ion selective electrodes from single cortical neurons in awake cats, B. Wong and C. Woody, *Exp. Neurol.* **61** (1), 219–225 (1978).

63. The calibration and use of a calcium ion specific electrode for kinetic studies of microhondrial calcium transport, R. K. Yamazaki, D. L. Mickey, and M. Story, *Anal. Biochem.* **93** (2), 430–441 (1979).

64. Electrochemical sensors for determination of some pharmaceutical products, Luca conslantin, Baloesu Cornel *et al.*, *Lev. Chim.* (*CA* 90-210186) (1979).

4.D.1. Enzyme Electrodes: Theory

1. Mechanism of action of valinomycin on mitochondria, C. Moore and B. C. Pressman, *Biochem. Biophys. Res. Commun.* **15** (6), 562–567 (1964).

2. A highly selective cation electrode system based on in vitro behaviour of macrotetrolides in membranes, Z. Stefanac and W. Simon, *Chimia* **20** (12), 436 (1966) (Ger.).

3. The analytical role of ion selective and gas sensing electrodes in enzymology, C. J. Moody and J. D. R. Thomas, *Analyst* **100**, 609–619 (1975).

4. Standard proton affinity of water-alcohol and water-ketone media determined using G ferricerium electrode, A. A. Pendin, O. L. Kucherova, T. K. Bundina, O. M. Susareva, and B. P. Nikol'skii, *Dokl. Akad. Nauk SSSR* **241** (2), 404–407 (1978) (Phys-Chem) (Russ.).

4.D.2. Enzyme Electrodes: Instrumentation and Method of Preparation

5. Electrodes for measuring urease enzyme activity, J. G. Montalvo, Jr., *Anal. Chem.* **41**, 2093–2094 (1969).

6. A urea-specific enzyme electrode, G. Guilbault and J. G. Montalvo, *J. Amer. Chem. Soc.* **91**, 2164 (1969).

7. An electrode for determination of amino acids, G. G. Guilbault and E. Harbankova, *Anal. Chem.* **42**, 1779–1783 (1970).

8. Enzyme electrode for amygdalin, G. A. Rechnitz and R. Llendada, *Anal. Chem.* **43**, 283 (1971).

9. A novel penicillin enzyme electrode, C. J. Olliff, R. T. Williams, and J. M. Wright, *J. Pharm. Pharmacol.* **30**, 45 pp (1978).

10. An enzyme electrode for acetylcholine, D. Patrick, D. Alain, and T. Daniel, *Biochem. Biophys. Acta* **527** (1), 277–281 (1978).

11. Flexible valinomycin electrodes for on line determination of intravascular and myocardial K^+, J. L. Hill *et al.*, *Am. J. Physiol.* **235** (4), 1455–1459 (1978).

12. Electrochemical studies of NADH in order to produce an enzyme electrode, H. Jaegfeldt and A. Torstensson, *Ion Selective Electrodes Conf. 1977*, 403 (1978).

4.D.3. Enzyme Electrodes: Applications

13. *Application of ion selective membrane electrodes in organic analysis. Part II. Application in organic analysis,* G. Bainlescu and V. V. Cosofret (Eng. Trans), Chap. 10, 12–14, 16.1–16.2, 17, Ellis Horwood Ltd. Publ., Chichester (1977).
14. Determination of glucose in blood by H_2O_2 measurement with an enzyme electrode, F. Scheller, M. Jaenchen, O. Ristau, and I. Seyer, *Fruchdiabetes Pathog Diagn. Pract. Kailsbugu Symp. Dabetes fragen 9th* **2**, 348–351 (1977) (Ger.).
15. *Enzyme electrodes for L. lysine, L. histidine and L. tyrosine and L. arginine,* W. C. White, 173 pp (1977) Avail. From Univ. Microfilms Int. Order No. 7809023.
16. Enzyme electrodes, A. Takasaka, *Riusho Kensa* **22** (5), 556–559 (1978) (Japan).
17. Membrane electrode measurement of lysozyme enzyme using living bacterial cells, P. D'Orazio, M. E. Meyerholf, and G. A. Rechnitz, *Anal. Chem.* **50** (11), 1531–1534 (1978).
18. Lysine specific enzyme electrode for determination of lysine in grains and foodstuff, W. C. White and G. G. Guilbault, *Anal. Chem.* **50** (11), 1481–1486 (1978).
19. An electrode for glucose Part 4. Measurements in diluted and undiluted whole blood samples with a simple arrangement, P. G. Reitnaver, *Z. Med. Laboratoriumsdiagn.* **19** (3), 176–186 (1978) (Ger.).
20. Determination of L-amino acids and alcohols with oxidate enzymes and a tubular iodide-selective electrode, M. Ma Scini and G. Palleschi, *Anal. Chim. Acta* **100**, 215–222 (1978).
21. Enzyme collagen membranes for electrochemical determination of glucose, D. R. Thevenot *et al., Anal. Chem.* **51** (1), 96–100 (1979).
22. Potentiometric enzyme electrode for lactate, T. Shinbo, *et al., Anal. Chim.* **51** (1), 100–104 (1979).
23. Enzyme electrode for exhibitors determination urcate fluoride system, T. M. Canh *et al., Anal. Chem.* **51** (1), 91–95 (1979).

5. Coulometry

The origin of coulometric analysis can be traced to Grower [*Proc. Am. Soc. Testing Materials,* **17**, 129 (1917)] who determined the thickness of tin coatings on copper by measuring the quantity of electricity consumed in the anodic oxidation of tin.

Potentiostatic coulometry has developed as an important analytical tool, after Hickling[1] and Lingane[2] first introduced the potentiostats in electrochemical studies. Lingane has also established[2,4] the fundamental basis for the controlled potential coulometry. The theory of coulometry has been discussed in a number of reviews such as those of Delahay,[7] Meites,[12] Abresch and Classen,[14] and Deford and Miller.[16]

The theory of controlled potential coulometry of electrode reactions which are reversible and not involving kinetic effects, totally irreversible reactions, and those involving coupled chemical reactions have been discussed at length by Bard and Santhanam.[21]

The theory of coulometry has been developed by a number of researchers for employing the coulometric technique for kinetic studies. The important contributions are from Lingane,[33,36] Bard *et al.,*[41,46,48,51,56,57] Meites,[40,49,52,53,62] Rechnitz *et al.,*[43,54,58] Bishop,[69,70] and Kennedy *et al.*[65,68] The instrumentation needed for coulometric studies, such as potentio-stats, current integrators, etc., has been developed and discussed by Ling-ane,[84,88,89,98,116] Meites,[90,91,95,136] Fischer *et al.*,[101,102] Booman,[99,106] Harrar *et al.*,[23,123,124,129] and Schroedder.[121] The special electrodes and cells, for coulometric analysis have also been developed by a number of researchers and they are given in References (114), (115), (126), (128), (132), (133), (137), (138), (140), and (141).

The application of coulometry for analysis of different substances both inorganic and organic up to trace level, determination of thickness of electrodeposits/coatings, and determination of the number of electrons involved in electrode reactions are all discussed in References (145)–(191). The applications of controlled potential coulometry have been extensively discussed by Harrar.[23]

5.1. General Principles, Reviews, and Books

1. Studies in electrode polarisation. Part IV. The automatic control of the potential of a working electrode, A. Hickling, *Trans. Faraday Soc.* **38**, 27–36 (1974).
2. Coulometric analysis, J. J. Lingange, *J. Am. Chem. Soc.* **67**, 1916–1922 (1945).
3. Electroanalysis, S. E. Q. Ashley, *Anal. Chem.* **24**, 91–95 (1952).
4. *Electroanalytical Chemistry*, J. J. Lingane, Interscience Publishers, Inc., New York (1953).
5. Electroanalysis, D. Deford, *Anal. Chem.* **26**, 135–140 (1954).
6. Electroanalysis at controlled potential and related methods, In: *New Instrumental Methods in Electrochemistry*, P. Delahay, Chap. 12, pp. 282–298, Interscience Publishers, Inc., New York (1954).
7. Coulometry at controlled current, In: *New Instrumental Methods in Electrochemistry*, P. Delahay, Chap. 14, pp. 301–315, Interscience Publishers, Inc., New York (1954).
8. Instrumentation in coulometry at controlled current, In: *New Instrumental Methods in Electrochemistry*, P. Delahay, Chap. 19, 402–407, Interscience Publishers, New York (1954).
9. Coulometric methods, W. D. Cooke, In: *Organic Analysis*, J. Mitchell, Jr., I. M. Kolthoff, E. S. Prosakauer, and A. Weissberger, eds., Vol. 2, pp. 169–193, Interscience Publishers, New York (1954).
10. Electroanalysis, D. Deford, *Anal. Chem.* **28**, 660–666 (1956).
11. Electroanalysis and coulometric analysis, D. Deford and R. C. Bowers, *Anal. Chem.* **30**, 613–619 (1958).
12. Controlled potential electrolysis, L. Meites, In: *Physical Methods of Organic Chemistry*, A. Weissberger, ed., Vol. 1, Part IV, Chap. XLIX, pp. 3281–3333, Interscience Publishers, Inc., New York (1959).
13. Electroanalysis and coulometric analysis, D. Deford, *Anal. Chem.* **32**, 31R–37R (1960).
14. *Coulometric analysis*, K. Abresch and I. Classen, (Eng. Trans), General principles, end point detection: Sec. 2–5, pp. 19–55; Electrolysis cells and instrumentation, Sec. 6–12, pp. 56–133; Analytical applications, pp. 137–151, pp. 217–238; Coulometric titrations, pp. 152–216, Chapman and Hall Ltd., London (1961).
15. Electroanalysis and coulometric analysis, A. J. Bard, *Anal. Chem.* **34**, 57R–64R (1962).
16. Coulometric analysis, D. D. Deford and J. W. Miller, In: *Treatise on Analytical Chemistry*, I. M. Kolthoff, P. J. Elving and S. B. Sandell, eds., Part I, Vol. 4, pp. 2475–2531, Interscience Publishers, New York (1963).
17. Electroanalysis and coulometric analysis, A. J. Bard, *Anal. Chem.* **36**, 70R–79R (1964).
18. Controlled potential electroanalysis, coulometry, In: *Polarographic Techniques*, L. Meites, Chap. 10, pp. 511–538, Interscience Publishers, New York (1965).

19. Electroanalysis and coulometric analysis, A. J. Bard, *Anal. Chem.* **38**, 88R–91R (1966).
20. Electroanalysis and coulometric analysis, A. J. Bard, *Anal. Chem.* **40**, 64R–79R (1968).
21. A. J. Bard and K. S. V. Santhanam, In *Electroanalytical Chemistry*, Vol. 4, Theory: pp. 217–251 A. J. Bard, ed.; Application: pp. 252–288; Experimental Methods: pp. 288–309; Marcel Dekker, Inc., New York (1970).
22. Instrumentation for process applications: Coulometry, C. D. Lewis, In: *Treatise on Analytical Chemistry*, I. M. Kolthoff, P. J. Elving, and S. B. Sandell, eds., Part I, Vol. 10, Chap. 105, pp. 6339–6347, Wiley-Interscience, New York (1972).
23. Techniques, apparatus and analytical applications of controlled potential coulometry, J. E. Harrar, In: *Electroanalytical Chemistry*, A. J. Bard, ed., Vol. 8, Principles and scope of coulometric techniques: pp. 4–23; Instrumentation and electrolysis cell design: pp. 23–86; Operational techniques: pp. 86–106; Analytical applications: pp. 106–167; Marcel Dekker, Inc., New York (1975).
24. Continuous analysis based on electrochemistry. III. Coulometry. H. L. Kies, *Rev. Anal. Chem.* **2(3)**, 229–277 (1975).
25. *Coulometry and coulometric titrations*, S. Ebel, (Ger), *Pharm. Unseres. Zeit* **5(5)**, 139–144 (1976).
26. Potential scanning coulometry: Theory and application to the study of chemical reactions associated with the electrode reaction, E. Laviron, L. Roullier, and R. Gavasso, *J. Electroanal. Chem. Interfacial Electrochem.* **75(1)**, 287–300 (1977).
27. *Coulometric analysis*, I. A. Kedrinskii, 32 pp., Sibirskii Tekhnologicheskii Institut., Krasnoyarsk, USSR (1977).
28. Application of controlled potential coulometry to reaction kinetics of species at tracer scale concentration: 'The radio coulometry', K. Samhoun and F. David, Report 1977, IPNO-RC-77-05, 15 pp. (Eng). Avail. INIS from INIS Atomindex 8 (23) Abstr. No. 342851 (1977).
29. Coulometric analyzers at a controlled potential (Review), R. F. Salikhdzhanova, *Zavod. Lab.* **44** (1), 9–11 (1978).
30. Modern electrochemical techniques in chemical, biochemical and pharmaceutical analysis. II. Coulometric techniques, G. T. Patriarche, *Labo-Pharma-Probl. Tech.* **26** (280), 817–826 (Fr.) (1978).
31. Coulometry in non-aqueous media, E. A. M. F. Dahmen and M. Bos, *Proc. Anal. Div. Chem. Soc.* **15** (3), 86–91 (1978).
32. Amperometry, coulometry, conductimetry and potentiometry used in routine biochemistry, F. Paolaggi, *Dimanche, Bid. Lariboisiere* (*C.R.*) **20**, 66–102 (Fr.) (1980).

5.2. Theory and Its Application to Kinetic Studies

33. Controlled potential electroanalysis, J. J. Lingane, *Anal. Chim. Acta* **2**, 591 (1948).
34. Reduction of some nitro-compounds at stirred mercury surface, L. Bergman and J. C. James, *Trans. Faraday Soc.* **50**, 60–65 (1954).
35. Pulse coulometry, M. A. V. Devanathan and Q. Fernando, *Trans. Faraday Soc.* **52**, 1332–1337 (1956).
36. Current efficiency and titration efficiency in coulometric titration with electrogenerated ceric ion. Determination of iodine, J. J. Lingane, C. H. Langford, and F. C. Anson, *Anal. Chim. Acta* **16**, 165–174 (1957).
37. Semi-micro hydrogenation with electrolytically generated hydrogen, J. W. Miller and D. D. DeFord, *Anal. Chem.* **30**, 295–298 (1958).
38. '96493' coulombs, J. J. Lingane, *Anal. Chem.* **30**, 1716–1723 (1958).
39. Constant potential coulometric reduction of organic nitro and halogen compounds, V. B. Ehlers and J. W. Sease, *Anal. Chem.* **31**, 16–22 (1959).
40. Background corrections in controlled potential coulometric analysis, L. Meites and S. Moros, *Anal. Chem.* **31**, 23–28 (1959).

41. Evaluation of the effect of secondary reactions in controlled potential coulometry, D. S. Geske and A. J. Bard, *J. Phys. Chem.* **63**, 1057–1062 (1959).

42. Classification and nomenclature of electroanalytical methods, P. Delahay, G. Charlot, and W. A. Laitinent, *Anal. Chem.* **32**, 1034 (1960).

43. A study of the molybdenum-catalyzed reduction of perchlorate, G. A. Rechnitz and H. A. Laitinen, *Anal. Chem.* **33**, 1473–1477 (1961).

44. Utilisation d'un convertisseur tension-frequence pan l'integration coulometrique, R. Amsmann and J. Desbanes, *J. Electroanal. Chem.* **4**, 121–122 (1962).

45. Precise wide-range direct reading electronic integrators, E. N. Wise, *Anal. Chem.* **34**, 1181 (1962).

46. High speed coulometer based on a voltage to frequency converter, A. J. Bard and E. Solon, *Anal. Chem.* **34**, 1181–1183 (1962).

47. A new method for the study of intermetallic compound formation in mixed amalgams, H. K. Ficker and L. Meites, *Anal. Chim. Acta* **26**, 172–179 (1962).

48. Secondary reaction in controlled potential coulometry. II. Secondary electrode reactions, A. J. Bard and J. S. Mayell, *J. Phys. Chem.* **66**, 2173–2179 (1962).

49. Experimental evaluation of rate constants for dimerization of intermediates formed in controlled potential electrolyses, L. Meites, *J. Electroanal. Chem.* **5**, 270–280 (1963).

50. Physico-chemical hydrodynamics (translated by Scripta Technica Inc. Englewood Cliffs, N.J. Prentice Hall, 1962 700 pp.) V. G. Levich, *Physico Chemical Hydrodynamics*, Prentice Hall (1963).

51. Secondary reactions in controlled potential coulometry. III. Preceding and simultaneous chemical reactions, A. J. Bard and E. Solon, *J. Phys. Chem.* **67**, 2326–2330 (1963).

52. Evaluation of rate constants for consecutive and competing first or pseudo-first order reactions with specific reference to controlled potential electrolysis, R. I. Gelb and L. Meites, *J. Phys. Chem.* **68**, 630–639 (1964).

53. Effect of working electrode potential on the rates and extents of controlled potential electrolysis, L. Meites, *J. Electroanal. Chem.* **7**, 337–342 (1964).

54. Study of kinetics of iridium (III) chlorate reaction by steady-state controlled potential coulometry, G. A. Rechnitz and J. E. McClure, *Anal. Chem.* **36**, 2265–2270 (1964).

55. The value of the Faraday, W. J. Harner and D. N. Craig, *J. Electrochem. Soc.* **111**, 1434 (1964).

56. Coulometric analysis with gas volume measurement, D. M. King and A. J. Bard, *Anal. Chem.* **36**, 2351–2352 (1964).

57. Secondary reactions in controlled potential coulometry. IV. Reversal coulometry and following chemical reactions, A. J. Bard and S. V. Tatwavadi, *J. Phys. Chem.* **68**, 2676–2682 (1964).

58. Application of steady-state controlled potential coulometry to the study of homogeneous solution reactions, G. A. Rechnitz and J. E. McClure, *Talanta* **12** (2), 153–158 (1965).

59. Consecutive electrochemical processes in controlled potential electrolysis. The isolation of intermediates, J. G. Mason, *J. Electroanal. Chem.* **11**, 462–466 (1966).

60. Kinetics and mechanism of controlled potential coulometric oxidation of catechol, G. Sivaramiah and V. R. Krishnan, *Ind. J. Chem.* **4** (12), 541–542 (1966).

61. The application of sub-stoichiometric radiosotopic dilution principles to controlled potential coulometry and solvent extraction, A. R. Lavelsebe, L. T. Mclendon, S. R. DeVoe, P. A. Pella, and W. C. Purdy, *Anal. Chim. Acta* **39**, 151–159 (1967).

62. Current-time curves in controlled potential electrolysis and the rates of reactions of intermediates, S. Karp and L. Meites, *J. Electroanal. Chem. Interfacial Electrochem.* **17**, 253–265 (1968).

63. Undervoltage effects in the determination of silver by scanning and coulometry, R. C. Propst, *J. Electroanal. Chem. Interfacial Electrochem.* **16**, 319–326 (1968).

65. Solid electrolyte coulometry. Silver bromide electrolyte, J. H. Kennedy and F. Chen, *J. Electrochem. Soc.* **115**, (9), 918–924 (1968).

66. Electrochemical reduction of beta-diketones in dimethylsulfoxide, R. C. Buchta and D. H. Evans, *Anal. Chem.* **40**, 2181–2185 (1968).
67. Controlled potential electrolysis and the rate of homogeneous reactions, L. Meites, *Pure Appl. Chem.* **18** (1–2), 35–79 (1969).
68. Solid electrolyte coulometry: Silver sulfide bromide electrolyte, J. H. Kennedy and F. Chen, *J. Electrochem. Soc.* **116** (2), 207–211 (1969).
69. Mass and charge transfer kinetics and coulometric current efficiencies. III. Pattern theory and its application to oxidation reduction electrode processes, R. E. Bishop, *Analyst* **97** (1159), 761–771 (1972).
70. Mass and charge transfer kinetics and coulometric current efficiencies. IV. Application of pattern theory to solvent molecule and solvent ion reactions and the evaluation of current efficiencies, E. Bishop, *Analyst* **97** (1159), 772–782 (1972).
71. Investigation of electrodeposition and electrodissolution of silver at a platinum electrode studied by a galvanostatic method, E. A. Maznichenko, V. V. Sobol, and L. K. Tarasova, *Elektrokhimiya* **8** (12), 1888–1882 (Russ.) (1972).
72. Least squares adjustment of the fundamental constants, E. R. Cohen and B. N. Taylor, *J. Phys. Chem. Ref. Data* **2**, 663–734 (1973).
73. Coupling organic and biological reactions with electrochemical measurements for analytical purposes, P. J. Elving, *Bioelectrochem. Bioenergy* **2** (4), 251–286 (1975).
74. Numerical solution of the disproportionation mechanism in controlled potential coulometry, J. Mock and D. I. Bustin, *J. Electroanal. Chem. Interfacial Electrochem.* **79** (2), 307–317 (1977).
75. Anodic stripping coulometry at a thin film mercury electrode, R. Eggli, *Anal. Chim. Acta* **91** (2), 129–138 (1977).
76. Analogue solution of kinetics of the disproportionation following a first-order chemical reaction, J. Mocak, D. I. Bustin, and V. Pencak, *Chem. Zvesti* **31** (2), 153–164 (1977).
77. Coulometry in non-aqueous media, E. A. M. F. Dahmen, and M. Bos, *Proc. Anal. Div. Chem. Soc.* **15** (3), 86–91 (1978).
78. Measurements of homogeneous reaction rate by concentration-step, Controlled potential electrolysis, S. Unchiyama, G. Muto, and K. Nozaki, *J. Electroanal. Chem. Interfacial Electrochem.* **91** (3), 301–308 (1978).
79. Numerical methods for evaluation of kinetics of coupled chemical reactions in controlled potential coulometry, J. Mock, D. Bustin, and D. Kysela, *Coulom. Anal. Conf.* (Pub. 1979), 269–277 (1978).
80. Advantages of generation of coulometric reagents during anodic dissolution of metals, A. I. Kostromin, A. A. Akhmetov, and I. F. Abdullin, *Zh. Anal. Khim.* **34** (7), 1243–1246 (1979).
81. An electrochemical method for measuring redox potentials of low potential proteins by microcoulometry at controlled potentials, G. D. Watt, *Anal. Biochem.* **99** (2), 399–407 (1979).
82. Radiocoulometry: its application to kinetics and mechanism studies of amalgamation reactions of barium (2^+), calcium (3^+) europium (3^+), samarium (3^+) and californium (3^+) in aqueous media, K. Samhoun and F. David, *J. Electroanal. Chem. Interfacial Electrochem.* **106**, 161–163 (1980).

5.3. Instrumentation, Cells, etc.

83. Apparatus for automatic control of electrodeposition with graded cathode potential, C. W. Caldwell, R. C. Parker, and H. Diehl, *Ind. Eng. Chem. Anal. Ed.* **16**, 532–535 (1944).
84. Electronic trigger circuit for automatic potentiometric and photometric titration, R. H. Muller and J. J. Lingane, *Anal. Chem.* **20**, 795–797 (1948).

85. Apparatus for electrolysis at controlled potential, C. J. Penther and D. J. Pompeo, *Anal. Chem.* **21**, 178–180 (1949).

86. The automatic recording titrator and its application to the continuous measurement of the concentration of organic sulphur compounds in gas streams, R. R. Austin, *Am. Gas. Assoc. Proc. Annual Meeting* **31**, 505–515 (1949).

87. Methods for constant potential control, E. B. Thomas and R. J. Nook, *J. Chem. Educat.* **27**, 217–219 (1950).

88. Improved potentiostat for controlled potential electrolysis, J. J. Lingane and St. L. Jones, *Anal. Chem.* **22**, 1169–1172 (1950).

89. Electromechanical integrator for coulometric analysis, J. J. Lingane and St. L. Jones, *Anal. Chem.* **22**, 1220–1221 (1950).

90. Microcoulometry: I. Construction and operation of an apparatus for the determination of polarographic *n*-values, St. Bogan, L. Meites, E. Peters, and J. M. Sturtevant, *J. Am. Chem. Soc.* **73**, 1584–1587 (1951).

91. Millicoulometric redox-titration, L. Meites, *Anal. Chem.* **24**, 1057–1059 (1952).

92. Controlled potential electroanalyser, D. J. Bode, S. W. Levine, and R. W. Kress, *Anal. Chem.* **25**, 518 (1953).

93. Apparatus for automatic control of cathodic potential in electroanalysis, J. F. Palmer and I. Vogel, *Analyst* **78**, 428–439 (1953).

94. Apparatus and techniques for semi-micro coulometric analyis, G. Packman, *Anal. Chem.* **26**, 784 (1954).

95. Cells, apparatus and methodology for precise analysis by coulometry at controlled potential, L. Meites, *Anal. Chem.* **27**, 116–119 (1955).

96. Automatic limited potential electrolysis using an electronic coulometer, E. L. Martin, *Dissertation Abstracts* **15**, 1323–1324 (1955).

97. The hydrogen ion coulometer, H. W. Hoyer, *J. Phys. Chem.* **60**, 372–373 (1956).

98. A hydrogen-nitrogen gas coulometer, J. A. Page and J. J. Lingane, *Anal. Chim. Acta* **16**, 175–179 (1957).

99. Instrument for controlled potential electrolysis and precision coulometric integration, G. L. Booman, *Anal. Chem.* **29**, 213–218 (1957).

100. Apparatus for automatic controlled potential electrolysis using an electronic coulometer, L. L. Merrit, E. L. Marin, and R. D. Bedi, *Anal. Chem.* **30**, 487–492 (1958).

101. Electronic controlled potential coulometric titrator, D. J. Fischer, M. T. Kelley, and H. C. Jones, *Anal. Chem.* **31**, 488–491 (1959).

102. Electronic controlled potential coulometric titrator, M. T. Kelley, H. C. Jones, and D. J. Fisher, *Anal. Chem.* **31**, 956 (1959).

103. Voltage scanning coulometry for the determination of traces of iron, F. A. Scott, R. M. Peekema, and R. E. Cennally, *Anal. Chem.* **33**, 1024–1027 (1961).

104. A high sensitivity scanning coulometer with automatic background correction and proportional scan rate titration of plutonium and other redox species, R. C. Propst, *Anal. Chem.* **35**, 958–963 (1963).

105. Generalised circuits for electroanalytical instrumentation, W. M. Schwarz and I. Shain, *Anal. Chem.* **35**, 1770–1778 (1963).

106. Simplified use of transfer functions in analysis of operational amplifier electroanalytical instrumentation, I. Shain, J. E. Harrar, and G. L. Booman, *Anal. Chem.* **37**, 1768–1769 (1965).

107. Differential controlled potential coulometry application to the determination of uranium, G. C. Gorde and J. Herrington, *Anal. Chim. Acta* **38**, 369–375 (1967).

108. Differential controlled potential coulometry utilizing substoichiometric radioisotope dilution, P. A. Pella, A. R. Landgrebe et al., *Anal. Chem.* **39**, 1781–1785 (1967).

109. Study on the controlled potential coulometric analyzer, S. Emura and S. Okazaki, *Bunseki Kagaku* **16** (7), 718–720 (1967) (Japan).

110. Lanthanum hexaboride as an electrode material for electrochemical studies, D. J. Curran and K. S. Fletcher, *Anal. Chem.* **40**, 78–82 (1968).

111. An all-transistorized potential-controlled coulometric titrator for analytical purposes, W. Warzanskyj, A. Diez Moreno, and V. Almagro, *J. Electroanal. Chem. Interfacial Electrochem.* **18**, 107–113 (1968).

112. Modular solid-state unit for electrochemical studies, G. Dryhurst, M. Rosen, and P. J. Elving, *Anal. Chim. Acta* **42** (1), 143–152 (1968).

113. Simple low-current potentiostat coulometric analysis, J. T. Stock, *J. Chem. Educ.* **45** (11), 736–738 (1968).

114. Reagent controller for electrolysis at essentially constant current, J. T. Stock, *Microchem. J.* **13** (4), 656–663 (1968).

115. Microcell for coulometric titration, G. D. Christian and F. J. Feldman, *Anal. Chem.* **40**, 1168 (1968).

116. Precise integration of voltage (current) time functions with a fixed-field direct current motor-counter, J. J. Lingane, *Anal. Chim. Acta* **44** (1), 199–203 (1969).

117. Controlled potential coulometers based upon modular electronic units, G. Phillips and G. W. C. Milner, *Analyst* **94**, 833–839 (1969).

118. Controlled potential coulometers based upon modular electronic units. Part II. The determination of ruthenium by controlled potential coulometry, G. Weldrick, G. Phillips, and G. W. C. Milner, *Analyst* **94**, 840–843 (1969).

119. Transistorized power sources for constant current coulometric titration, J. T. Stock, *J. Chem. Educ.* **46** (12), 858–860 (1969).

120. Coulometry of metal ions separated by liquid chromatography, G. Muto, T. Kawaguchi, and Y. Takada, Japan 75, 15, 393 (Cl GOIN 31/08) 04 Jun 1975, Appl. 69147, 604, 18 Jun 1969.

121. Operational amplifier instruments for electrochemistry—circuits for controlled potential electrolysis, coulometry and coulometric titrations, R. R. Schroeder, In: *Electrochemistry, Calculations, Simulations and Instrumentation*, J. S. Mattson, H. B. Mark Jr., and H. C. MacDonald, Jr., eds., Chap. 10, pp. 289–293, Marcel Dekker, New York (1972).

122. Electrolytic gaps between electrodes for the microcoulometry of mercury, J. Grzebalski, J. Maciejewski, and J. Moszczynska, Pol. 84, 467 (C1. GOIN 27/26) 15 Oct 1976, Appl. 160, 680, 10 Feb. 1973, 2 pp.

123. Manual of controlled potential coulometric methods, J. E. Harrar, Report 1970, UCID-15527, 45, PIP, Avail. Dep. NTIS from *Nucl. Sci. Abstr.* **27** (7), 14497 (1973).

124. Determination of palladium by controlled potential coulometry. New platinum working electrode cell for controlled potential coulometry, L. P. Rigdon and J. E. Harrar, *Anal. Chem.* **46**, (6), 696–700 (1974).

125. Versatile, multitime range, integrated circuit, constant current coulometric titrator. Application to determination of microgram and nanogram quantities, L. B. Haycox, 1973, 198 pp. Avail. Univ. Microfilms Ann. Arbor, Mich. Order No. 74,8560. From Diss A6Str. Int. B 34(10), 4858 (1974).

126. Effect of cell geometry on coulometric titrations, T. Damokas, *Magy. Kem. Foly.* **80** (3), 123–126 (1974) (Huong.).

127. Precise set up for coulometric analysis with controlled potential, I. G. Sentyvrin, A. N. Mogilevskii, I. S. Skylarenko, I. A. Trifonova, N. S. Radionova, and I. A. Kazakov, *Zh. Anal. Chim.* **30** (1), 53–58 (1975) (Russ.).

128. Effect of stirring on the reduction current in constant potential coulometry, W. Rutkowski, Sobkowsky, and Aleksandra, *Chem. Anal.* (*Warsaw*) **20** (2), 383–388 (1975) (Pol.).

129. Computer-compatible instrumentation for automated controlled current coulometry, potentiometry and galvanostic measurements, J. E. Harrar and C. L. Pomernacki, *Chem. Instrum.* (*N.Y.*) **7** (4), 229–239 (1976).

130. Methods and instruments for measuring the thickness of layers, W. H. Guehring, *Detektos-kopiya* **4**, 113–125 (1977).

131. Real-time computer optimized scanning potential coulometry for multi-component trace analysis, N. W. Petty, 1977, 127 pp, Avail. Univ. Microfilms Int. Order No. 7731122. From Diss. Abstr. Inst. B38 (3), 3662 (1978).

132. Metal rod and wire electrode holders, E. A. Gaston and B. Edmund, *Anal. Chem.* **49** (12), 1880–1881 (1977).

133. Auxiliary compartment for coulometric titration, G. A. East and E. Bishop, *Anal. Chem.* **49** (12), 1885–1886 (1977).

134. Highly selective coulometric method and equipment for the automated determination of plutonium, D. D. Jackson, R. M. Hollen, F. R. Roensch, and J. E. Rein, *Anal. Chem. Nucl. Fuel.* Re Process Proc. ORNL Conf. 21st 1977, 151–158 (1978).

135. A detailed study of sample injection into flowing streams with potentiometric detection, Z. Feher, G. Nagy, K. Toth, and E. Pungor, *Anal. Chim. Acta* **98** (2), 193–203 (1978).

136. Recent advances in data processing in controlled potential electrolysis, L. Meites, *Coulom. Anal. Conf. 1978* (1979).

137. Applications of thin layer minigrid electrodes. I. Analytical methods for halide and constant current coulometry with logical end point. II. Application for study of biological redox systems, M. L. Meckstoth, 155 pp. (1978) Avail. Univ. Microfilms Inst. Order No. 7904750. From Diss. Abstr. Int. B39 (9), 4311 (1979).

138. Controlled potential coulometry with the flow-through reticulated vitreous carbon electrode, A. N. Strohl and D. J. Curran, *Anal. Chem.* **51** (7), 1050–1053 (1979).

139. Current generator with digital control for coulometry. I. Apparatus for the triangle programmed coulometric titration technique, L. Kovacs, G. Nagy, and M. Kadar, *Magy. Kem. Foly* **85** (7) 331–334 (Hung.) (1979).

140. Flow coulometry with ring-disc electrode, T. Fujinaga, S. Okazaki, and H. Hirai, *Bull. Inst. Chem. Res. Kyoto Univ.* **57** (5–6), 376–380 (1979).

141. Voltammetry and coulometry of iridium in a two-sided thin-layer system, L. V. Eliseeva and O. L. Kabanova, *Zh. Anal. Khim.* **35** (3), 465–470 (1980).

142. Corrections for systematic errors from analog integration in controlled potential coulometry, L. T. Frazzini, M. K. Holland, J. R. Weiss, and E. Pietric, *Anal. Chem.* **52** (13), 2112–2116 (1980).

143. Apparatus for use in rapid and accurate controlled potential coulometric analysis, T. L. Frazzini, M. K. Holland, C. E. Pietri, and J. R. Weiss, U.S. Pat. Appl. 69,152, 24 Oct. 1980, Appl. 23 Aug. 1979, 22 pp. Avail NTIS Order No. PAT Appl. 069152.

144. Determination of reduction potentials and electron transfer stoichiometrics for biological redox species by thin-layer pulse and staircase coulometry C. H. Su and W. R. Heineman, *Anal. Chem.* **53** (4) 594–598 (1981).

5.4. Applications

145. Electrochemical jet test. Determination of local thickness of electrodeposited coatings, A. Ogarew, *J. Appl. Chem.* (*USSR*) **19**, 31188 (1946) (Eng. Trans), *Metal Ind.* (*Lond.*) **70**, 338–40 (1947).

146. Controlled potential electrolytic separation and determination of copper, bismuth, lead, and tin, J. J. Lingane and St. L. Jones, *Anal. Chem.* **23**, 1798–1806 (1951).

147. Galvanic determination of traces of oxygen in gases, P. Hersch, *Nature* **169**, 792–793 (1952).

148. Electrochemical method for oxygen determination in gases, M. G. Jacobson, *Anal. Chem.* **25**, 586–591 (1953).

149. Dual intermediates in coulometric titration equilibria in copper(II) bromide solutions, P. S. Farrington, D. J. Meier, and E. H. Swift, *Anal. Chem.* **25**, 591–595 (1953).

150. Thickness of electrodeposited coatings by the anodic solution method, C. F. Waite, *Plating* **40**, 1245–1248 (1953).

151. A millicoulometer method for the determination of polarographic n-values, Th. De Vries and J. L. Kroon, *J. Am. Chem. Soc.* **75**, 2484–2486 (1953).

152. Standardisation of a titrated solution of hydrochloric acid by constant current coulometry, J. Badoz-Lambling, *Anal. Chim. Acta* **25**, 1574 (1953).

153. Galvanic measurement of dissolved oxygen, A. G. Dowson and I. J. Backland, *Nature* **177**, 712–713 (1956).

154. Coulometric determination of uranium(VI) at controlled potential, G. L. Booman, W. B. Holbook, and J. E. Rein, *Anal. Chem.* **29**, 219–221 (1957).

155. Coulometric titration of aluminium, R. J. Iwamoto, *Anal. Chim. Acta* **19**, 272–276 (1958).

156. Controlled potential coulometric determination of copper and uranium, W. D. Shults and P. F. Thomason, *Anal. Chem.* **31**, 492–494 (1959).

157. Application of coulometric titration to clinical and toxicological analysis, W. C. Purdy, *Fresenius Z. Anal. Chem.* **243**, 17–28 (1968).

158. High sensitivity coulometric analysis in acetonitrile, R. R. Bessette and J. W. Olver, *J. Electroanal. Chem. Interfacial Electrochem.* **17**, 327–334 (1968).

159. Coulometric estimation of phenylthiourea, thioglycolic acid and cysteine, K. S. V. Santhanam and V. R. Krishnan, *Z. Anal. Chim.* **234** (4), 256–260 (1968).

160. The rapid dissolution of plutonium dioxide by a sodium peroxide-sodium hydroxide fusion, followed by determination of the plutonium content by controlled potential coulometry, G. W. C. Milner and D. Crossely, *Analyst* **93**, 429–432 (1968).

161. Coulometric titration in industrial analysis, W. Buetchler, P. Gisske, and J. Meier, *Fresenius Z. Anal. Chim.* **239** (5), 289–294 (1968).

162. Indirect controlled potential coulometry with mercury(II) DPTA complex., T. Kawaguchi and G. Muto, *Bunseki Kagaku* **17** (1), 38–42 (1968) (Japan).

163. Determination of plutonium by controlled current coulometry, J. R. Stokely, Jr. and W. D. Shults, *Anal. Chim. Acta* **45** (3) 528–532 (1969).

164. Controlled potential coulometric determination of americium, J. R. Stokely, Jr. and W. D. Shults, *Anal. Chim. Acta* **35** (3) 417–424 (1969).

165. Controlled potential coulometry and voltammetry of manganese in pyrophosphate medium, J. E. Harrar and L. P. Rigdon, *Anal. Chem.* **41**, 758–765 (1969).

166. Electrochemical methods in ultramicroanalysis. V. Coulometric determination of copper in the big scale, W. F. Helbig, *Z. Anal. Chim.* **246** (6), 353–357 (1969).

167. New coulometric titration method. Application to the determination of uranium, J. J. Lingane, *Anal. Chim. Acta* **50** (1), 1–14 (1970).

168. Electrochemical methods in analytical chemistry. X. Constant current coulometry, F. P. Ijsseling, *Chem. Tech.* **26** (11), 297–230, (12), 325–328 (1971).

169. Coulometric method for determining the thickness of silver electrolyte coatings, T. K. Khamrakulov and P. K. Agasyan, *Zavod. Lab.* **38**, (1), 24–26 (1972).

170. Copper electrode in coulometry, A. I. Kostromin and R. M. Badakshanov, *Zh. Anal. Khim.* **27** (10), 2046–2049 (2972).

171. Simultaneous determination of platinum (II) platinum (IV) by controlled reagent coulometry, O. Ginstrop, *Anal. Chim. Acta* **63** (1), 153–163 (1973).

172. Use of copper electrode in galvanostatic coulometry, A. I. Kostomin and R. M. Badakshanov, *Zh. Anal. Khim.* **29** (9), 1782–1787 (1974).

173. Application of controlled potential coulometry to the automatic recording of liquid chromatography. VI. Liquid chromatographic separation of carboxylic acids with the hydrogen form cation exchange resin, Y. Takata and Y. Arikawa, *Bunseki Kagaku* **23** (12), 1522–1527 (1974).

174. Coulometric microdetermination of organic compounds by controlling the potential. Structure of the apparatus and determination of picric acid as a model substance, J. Senkyr and M. Fialka, *Ser. Fac. Sci-Nat. Univ. Purkynianae Brun.* **5** (5), 19–27 (1975).

175. Controlled potential coulometric determination of some dithiaalkanediols, E. G. Semmertt, F. Rousselet, M. L. Girard, and M. Chemla, *Analysis* **3** (8), 456–459 (1975).

176. Coulometric method for determining the composition of alloys produced by electrolysis, N. Ciureanu and E. Grunichevi, *Rev. Chim. (Bucharest)* **26** (12), 1047–1049 (1975).

177. Application of controlled potential coulometry to the automatic recording of liquid chromatography. VII. Cation exchange chromatography of rare earths, Y. Takta and Y. Arikawa, *Bunseki Kagaku* **24** (12), 762–767 (1975).

178. Use of electrically generated iron(II) in constant current coulometry, A. I. Kostromin, L. L. Makarova, and L. I. Il'ina, *Zh. Anal. Khim.* **31** (2), 240–243 (1976).

179. Precise coulometric determination of actinides: application to trace amounts. I. Uranium, J. Ravenel, C. Soret, and C. Bergey, *Talanta* **23** (8), 569–572 (1976).

180. Coulometry in clinical chemistry, W. C. Purdy, *CRC Crit. Rev. Clin. Lab. Sci.* **7** (3), 227–237 (1977).

181. Recent electroanalytical studies in molten fluorides, D. L. Manning and G. Mamantov, Report 1976, CONF 7608110-1, 20 pp (Eng) Avail. NTIS from Energy Res. Abstr. 2 (24) Abstr. No. 60532 (1977).

182. Cell and method for the coulometric determination of the contents of water soluble components, R. O. Hallberg, C. H. M. Lindstroem, and I. Westerberg, Swed. 407,983 (C1. GOIN 27/42) 30 Ap. 1979, Appl. 77/10,308, 14 Sep 15 7, 9 pp.

183. The detection limit of anodic stripping coulometry at mercury film glassy carbon electrodes, R. Eggli, *Anal. Chim. Acta* **97** (1), 195–198 (1978).

184. Basis for the accuracy and comparability of chemical analysis results. Part II. Use of the Faraday, T. Yoshimoni, *Z. Chem.* **18** (7), 251–254 (1978).

185. Different ranges of fission products by a coulometric method, A. K. Jayak and S. Mukherji, *Nucl. Instrum. Method* **170** (1–3), 193–196 (1980).

186. Controlled current coulometry—high precision assay procedure, J. Bercik, M. Cakrt, and Z. Hlady, *Coulom. Anal. Conf. 1978*, 155–164 (1979).

187. Simultaneous thickness and electrochemical potential determination of individual layers in multilayer nickel deposits, E. P. Harbulak, *Plat. Surf. Finish* **67** (2), 49–54 (1980).

188. Flow through coulometry stripping analysis and the determination of manganese by cathodic stripping voltammetry, A. Trojanek and F. Opekar, *Anal. Chim. Acta* **126**, 15–21 (1981).

189. Precious metal analysis by controlled potential coulometry, J. E. Harrar and M. C. Waggner, *Plat. Surf. Finish* **68** (1), 41–45 (1981).

190. The coulometric response of tubular electrodes applied to flow injection determinations, P. L. Meschi, D. C. Johnson, and G. R. Luecke, *Anal. Chim. Acta* **124** (2), 315–320 (1981).

191. Electrogeneration of coulometric reagents: applications to drug control, J. C. Vire, M. Chateau-Gosselin, and G. J. Patriarche, *Microchim Acta* **1** (3–4), 227–239 (1981).

6. Chronopotentiometry

Although Sand [*Phil. Mag.* **1**, 45 (1901)] and Karaglanoff [*Z. Elektrochem.* **12**, 5 (1906)] laid the theoretical foundation for chronopotentiometry, it was only in the nineteen fifties, that this electroanalytical technique came into prominence through the work of Gierst and Juliard [*Proc. Intern. Comm. Electrochem. Thermodynam. and Kinet.*, Tamburini, Milano (1950), pp. 117–279; *J. Phys. Chem.* **57**, 701 (1953)], and Delahay.[1]

Earlier works pertaining to the theory of chronopotentiometry have been reviewed and discussed by Delahay,[1,3] (published up to 1954), Davis,[4] (published up to 1966), Pannovic,[7] and Galus.[49]

The theory of constant current chronopotentiometry with current reversal has been discussed by Delahay,[1] Nicholson *et al.*,[16] Galus,[49] Bos and Van Dalen,[15] Kizza *et al.*,[37,45] and Pnev and Zakharov.[40] The cyclic chronopotentiometric theory has been suggested by Herman and Bard[12] and later developed by many others.[20,25] Cyclic chronopotentiometry has been briefly but well-discussed by Galus.[49]

Although the usefulness of chronopotentiometry as an analytical tool is limited, because the relationship between concentration and transition time is nonlinear it has gained importance in the studies of kinetics of electrode processes. The latter applications are well discussed in the reviews by Delahay,[1] Davis,[4] Herman and Bard,[50,53] and by a number of other researchers.[59–61] The instrumentation and electrodes for these studies are discussed in References (62)–(70) as well as by Davis.[4] The applications of chronopotentiometry in stripping voltammetric analysis are discussed in References (72), (73), (76), (77), (80)–(83), and (89). The analytical applications of chronopotentiometry are discussed by Davis,[4] and many other researchers.[87–92]

6.1. General Principles and Theory

1. Voltammetry at controlled current, P. Delahay, In: *New Instrumental Methods in Electrochemistry*, Chap. 8, pp. 179–209, Interscience Publishers, Inc., New York (1954).
2. J. J. Lingane, Chronopotentiometry, In: *Electroanalytical Chemistry*, Chap. XXII, pp. 617–638, Interscience Publishers, Inc., New York (1958).
3. Chronopotentiometry, P. Delahay, In: *Treatise on Analytical Chemistry*, I. M. Kolthoff, P. J. Elving, and E. B. Sandell, eds., Part I, Vol. 4, Chap. 44, pp. 2233–2267, Interscience Publishers, New York (1963).
4. Applications of chronopotentiometry to problems in analytical chemistry, D. G. Davis, In: *Electroanalytical Chemistry*, Vol. I, A. J. Bard, ed. pp. 157–196, Marcel Dekker, Inc., New York (1966).
5. Chronopotentiometric measurements of chemical reaction rates. I. Programmed current studies of the ECE mechanism, H. B. Herman and A. J. Bard, *J. Phys. Chem.* **70**, 396–404 (1956).
6. Potentiometry, conductometry, coulometry, electroanalysis, and other electrochemical methods, V. A. Zarinskii, *Zh. Anal. Khim.* **22**, 1669–1678 (1967) (Russ.).
7. Chronopotentiometry, M. Paunovic, *J. Electroanal Chem. Interfacial Electrochem.* **14**, 447–474 (1967).
8. Derivative chronopotentiometry, S. L. Burden, Jr., Diss. Abstr. B27(10), 3424 (1967); Univ. Microfilm, Ann Arbor. Mich., Order No. 67–3650, 147 pp.
9. Rapid chronopotentiometry; effects of double layer charging and adsorption, D. C. Noonan, Diss. Abstr. B.28 (2) 522 (1967), Univ. Microfilm, Ann Arbor. Mich., Order No. 67-9360, 156 pp.
10. Polarography with controlled current density or chronopotentiometry with current density sweep at a dropping mercury electrode, H. L. Kies, *J. Electroanal. Chem. Interfacial Electrochem.* **16**, 279–281 (1968).
11. Chronopotentiometric measurements of chemical reaction rates. II. Kinetics and mechanism of the dehydration of p-hydroxyphenyl hydroxyl amine, H. N. Blount and H. B. Herman, *J. Phys. Chem.* **72**, 3006–3012 (1968).

12. Cyclic chronopotentiometry systems involving kinetic complications, H. B. Herman and A. J. Bard, *J. Electrochem. Soc.* **115**, 1028–1033 (1968).

13. Effect of double layer charging in programmed current chronopotentiometry, W. T. De Vries, *J. Electroanal. Chem. Interfacial Electrochem.* **19** (1/2), 55–60 (1968).

14. Double layer charging in constant current chronopotentiometry at a mercury film electrode, W. T. De Vries, *J. Electroanal. Chem. Interfacial Electrode,* **19** (1/2), 41–53 (1968).

15. Constant-current chronopotentiometry with current reversal at mercury-film electrode, P. Bos and E. Van Dalen, *J. Electroanal. Chem.* **17** (1/2), 21–30 (1968).

16. Theory of chronopotentiometry with current reversal for measuring heterogeneous electron transfer rate constants, F. H. Beyerlein and R. S. Nicholson, *Anal. Chem.* **40**, 286–288 (1968).

17. Chronopotentiometry, E. Tvrzicka and J. Dolezal, *Chem. Listy* **63**, 538–576 (1969) (Czech).

18. Chronopotentiometry and its applications, C. Herdlicka, *Stud. Cercet. Chim.* **17**, 779–799 (1969).

19. Cyclic chronopotentiometry, Determination of types and rates of second order chemical reactions following electron transfer, M. Vukovic and V. Pravdic, *Croat. Chem. Acta,* **42**, 21–32 (1970).

20. New device for measuring electrode reactions by pulse techniques, J. Vondrak and O. Spalek, *Chem. Listy* **64**, 609–614 (1970) (Czech.).

21. Chronopotentiometric measurements of chemical reaction rates. III. Finite difference approach to the ECE mechanism, H. N. Blount and H. B. Herman, *J. Electrochem. Soc.* **117** (4), 504–507 (1970).

22. Amalgam chronopotentiometry in a two side thin film system, V. A. Igolinski and N. M. Igolinskaya, *Electrokhimiya* **7**, 1496–1498 (1971).

23. Chronopotentiometry of mixtures. I. Reduction at a mercury pool and at a mercury film electrode, P. Bos, *J. Electroanal. Chem. Interfacial Electrochem.* **33**, 379–391 (1971).

24. Effect of the diffusion of an adsorbing depolariser on "transient" time at large currents, E. M. Podgaetskii and Yu. V. Filinovskii, *Elektrokhimiya* **7**, 1856–1859 (1971).

25. Cyclic chronopotentiometry on a rotating disk electrode, V. I. Chernenko, K. I. Litovckenko, and Yu. E. Udovenko, *Elektrochimiya* **7**, 1476–1480 (1971).

26. Kinetics of electrode reactions in constant current electrolysis, XVI. Parallel irreversible consecutive reactions of different orders, O. Dracka, *Collect. Czech. Chem. Commun.* **36**, 1889–1897 (1971).

27. Chronopotentiometric measurement adsorption in the case of an isotherm with two plateaus, E. M. Podgaetskii and V. Yu. Filinovskii, *Elektrokhimiya* **7**, 1042–1047 (1971).

28. Kinetics of electrode reactions in constant current electrolysis. XV. Reversible consecutive reactions of the second order, O. Dracka, *Collect. Czech. Chem. Commun.* **36**, 1876–1888 (1971).

29. Current Step; Galvanostatic method or chronopotentiometry, In: *A Guide to the Study of Electrode Kinetics*, H. R. Thirsk and J. A. Harrison, eds., pp. 50–54, Academic Press, London and New York (1972).

30. Transient mass transfer at the rotating disk electrode, L. Nanis and I. Klein, *J. Electrochem. Soc.* **119**, 1683–1687 (1972).

31. Polarographic theory, instrumentation and methodology, R. S. Nicholson, *Anal. Chem.* **44**, 478R–489R (1972).

32. Chronopotentiometry: Theoretical study of metal ion reduction by current inversion at transition time τ. Kinetic parameters of manganese(II), iodate, nickel(II) and manganese(II) oxalate complex, R. Bennes, *J. Electroanal. Chem. Interfacial Electrochem.* **36**, 11–22 (1972) (Fr.).

33. Chronopotentiometry: (I) theoretical principles, A. Kisza, *Wiad. Chem.* **26**, 413–426 (1972) (Pol.).

34. Chronoamperometry, potential sweep chronoamperometry and chronopotentiometry, J. Hladik, In: *Physics of Electrolytes*, J. Hladik, ed., Vol. 2, pp. 867–930, Academic Press, London (1972).

35. The a.c. chronopotentiometry: theoretical study of the reversible deposition of an insoluble substance, N. P. Bansal and H. L. Jindal, *J. Ind. Chem. Soc.* **49**, 957–961 (1972).

36. A.C. chronopotentiometry: Theoretical study of an irreversible electrode processes preceded by a chemical reaction, H. L. Jindal, N. P. Bansal, and K. Bansal, *Rev. Roum. Chim.* **17**, 1969–1975 (1972).

37. Film stripping chronopotentiometric study of reversible electrodissolution of a metal from the surface of an indifferent electrode accompanied by complexing, M. S. Zakharov, V. V. Pnev, and L. A. Moskovskikh, *Tr. Tyumen. Ind. Inst.* 32–34 (1972), *CA* 82:1407077g.

38. Theory of chronopotentiometry with current reversal from a stationary state, A. Kisza and U. Twardoch, *Bull. Acad. Pol. Sci. Ser. Sci. Chem.* **20**, 1063–1067 (1972) (Eng.).

39. Differential and derivative chronopotentiometry, A. J. Engel, 196 pp. (1972), Diss. Abstr. Int. B 1972, **33** (1), 86–87, *CA* 78:23238 h.

40. Stripping chronopotentiometry with periodic current reversal, V. V. Pnev and M. S. Zakharov, *Tr. Tyumen. Ind. Inst.* 57–63 (1972) (Russ.) *CA* 82:1643862.

41. Transient response of a disk electrode, K. Nisancioglu and J. Newman, *J. Electrochem. Soc.* **120**, 1339–1356 (1973).

42. Anodic chronopotentiometry on a film electrode for reversible processes during stirring of solutions. Theory, L. I. Komarova, *Sov. Electrochem.* **9**, 1286–1290 (1973).

43. Current reversal chronopotentiometry of EC processes involving an insoluble product, K. W. Hanck and M. L. Deanhardt, *Anal. Chem.* **45**, 179–182 (1973).

44. Theory of stripping chronopotentiometry for reversible electrode processes involving complexes on a mercury rotating disk microelectrode, V. I. Bakanov, M. S. Zakharov, and N. M. Cheremnykh, *Tr. Tyumen. Industr. Inst.* (22), 46–50 (1974) (Russ.); *CA*:84 51417 h.

45. Theory of chronopotentiometry with current reversal from a stationary state. II. Anodic dissolution of metals with the stepwise reversible electrode process, A. Kisza, *Bull. Acad. Pol. Sci. Ser. Sci. Chim.* **23**, 341–344 (1975).

46. Studies in derivative chronopotentiometry. I. Instrumentation and diffusion controlled systems, P. E. Sturrock, J. L. Hughey, B. Vandrewil, and G. E. O'Brien, *J. Electrochem. Soc.* **122**, 1195–1200 (1975).

47. Stripping chronopotentiometry with a glassy carbon disk electrode. Fundamental factors and comparison with stripping voltammetry, L. Luong and F. Vydra, *Collect. Czech. Chem. Commun.* **40**, 1490–1503 (1975).

48. Studies in derivative chronopotentiometry. IV. Chemical reactions preceding reversible charge transfer, R. H. Gibson and P. E. Sturrock, *J. Electrochem. Soc.* **123**, 1170–1173 (1976).

49. Chronopotentiometry, Z. Galus, In: *Fundamentals of Electrochemical Analysis*, Z. Galus, pp. 43–46, Chap. 5, Chap. 6, pp. 193–196, 208; Chap. 7, pp. 239–245; Chap. 8, pp. 275–278; Chap. 9, pp. 303–305; Chap. 10, pp. 321–322; Chap. 11, pp. 336–337; Chap. 12, pp. 347–351; Chap. 15, pp. 391–393; Chap. 16, pp. 411–413; Chap. 17, pp. 434–440; Chap. 18, pp. 454–456; Ellis Horwood Ltd. Publ., Chichester, Halsted Press, New York (1976).

6.2. Applications

50. Chronopotentiometric measurements of chemical reaction rates. I. Programmed current studies of the ECE mechanism, H. B. Herman and A. J. Bard, *J. Phys. Chem.* **70**, 396–404 (1966).

51. Applications of chronopotentiometry to problems in analytical chemistry, D. G. Davis, In: *Electroanalytical Chemistry*, Vol. I, A. J. Bard, ed., pp. 157–196, Marcel Dekker, Inc., New York (1966).

52. Critique of chronopotentiometry as a tool for study of adsorption, P. J. Lingane, *Anal. Chem.* **39**, 485–494 (1967).

53. Chronopotentiometric measurements of chemical reaction rates. II. Kinetics and mechanism of the dehydration of *p*-hydroxyphenyl hydroxyl-amine, H. N. Blount and H. B. Herman, *J. Phys. Chem.* **72**, 3006–3012 (1968).

54. Aspects of electroanalytical chemistry, J. J. Lingane, *Ind. Chim. Belge* **33**, (Spec. No) 136–139 (1968).

55. Power-of-time current chronopotentiometry: circuit design, construction and evaluation of a new instrumental approach, H. C. Kluge II (Secton Hall Univ., S. Orange N.J.), Diss. Abstr. B28 (7), 2746 (1968) Univ. Microfilms, Ann Arbor, Mich., Order No. 67–16, 351, 129 pp.

56. Chronopotentiometry and its use in analysis and electrochemical studies, V. G. Barikov and O. A. Songina, *Sovrem. Metody Anal. Mater.* 121–134 (1969) (Russ.). (ed) Orient I. M. Otd. 'Metallurfia', Moscow (USSR) (1969).

57. New circuit for current reversal chronopotentiometry, H. B. Herman, E. B. Smith, and B. C. Rudy, *Chem. Instrum.* **2**, 257–265 (1969).

58. Controlled potential and controlled current voltammeter, T. R. Mueller and H. C. Jones, *Chem. Instrum.* **2**, 65–81 (1969).

59. Measuring transition time in chronopotentiometry, R. Yu. Bek, I. I. Burenkov, and A. S. Lifshits, *Izv. Sib. Otd. Akad. Nauk SSSR, Ser. Khim. Nauk.* **2**, 3–6 (1969); *CA* 71:66716 k.

60. Unsteady conditions of electrolysis by an asymmetric rectified current, V. I. Chernenko and K. I. Litovchenko, *Ukr. Khim. Zh.* **35**, 472–477 (1969); *CA* 71:45039 b.

61. Graphical method for correcting the charge of a double layer in chronopotentiometry, D. Deroo, J. Guitton, and J. Besson, *J. Chim. Phys. Physicochim. Biol.* **67**, 1097–1100 (1970).

62. Chronopotentiometry with current reversal switching from a steady state, A. N. Barabc and O. N. Vinogradov Zhalnov, *Tr. Inst. Elektrokhim. Akad. Nauk SSSR, Ural Filia* **15**, 118–125 (1970) (Russ.); *CA* 75:83535a.

63. Electronic instrument for cyclic chronopotentiometry, T. Rabuzin, G. Similjanic, and F. Jovic, *J. Electroannal. Chem. Interfacial Electrochem.* **27**, 397–402 (1970).

64. Apparatus for cyclic chronopotentiometry in non-aqueous solvents, J. H. English, *J. Sci. Instrum.* **3** (I), 69–72 (1970).

65. Apparatus for automatic monitoring of metal cations by stripping chronopotentiometry, G. G. Raneev and I. M. Kogol, *Autom. Proizvod. Processov. Tsvet. Met.* 109–111 (1971) (Russ.).

66. Anodic amalgam voltammetry with a programmed current on a mercury spherical electrode, M. S. Zakharov and V. V. Pnev, *Izv. Tomsk. Politekh. Inst.* 51–54 (1971); *CA* 76:120820 h.

67. Electronic device for measuring transition time in chronopotentiometry, I. I. Burenkov, A. S. Lifshits, and R. Yu. Bek, *Izv. Sib. Otd. Akad. Nauk. SSSR, Ser. Khim. Nauk* **5**, 145–147 (1971).

68. Chronopotentiometry. II. New chronopotentiometric techniques and applications of chronopotentiometry, A. Kisza, *Wiad. Chem.* **26**, 491–502 (1972) (Pol.).

69. Thin layer electrochemistry, R. S. Tyurin, Yu. S. Lyalikov, and S. I. Zhadanov, *Usp. Khim.* **41**, 2272–2299 (1972).

70. Stripping chronopotentiometry with a mercury rotating disk electrode, M. S. Zakharov, V. A. Antipeva, and V. I. Bakanov, *Tr. Tyumen. Ind. Inst.* 3–7 (1972); *CA* 82:147073 c.

71. Release time of a metal from a mercury film electrode in stripping chronopotentio-metry, V. I. Bakanov and M. S. Zakharov, *Tr. Tyumen. Ind. Inst.* 12–14 (1972); *CA* 82:147074 d.

72. Stripping chronopotentiometric study of the kinetics of electrode processes involving the complex ions with a fixed mercury film electrode, M. S. Zakharov, I. V. Shelomentseva, and N. K. Ivanov, *Tr. Tyumen, Ind. Inst.* 23–25 (1972); *CA* 82: 147075 e.

73. Stripping chronopotentiometry. I. Electrodissolution of metal from the surface of an indifferent electrode, M. S. Zakharov, V. V. Pnev, and L. A. Moskovskikh, *Tr. Tyumen. Ind. Inst.* 26–31 (1972); *CA* 82:147076 f.

74. Resistive effects in thin electrochemical cells. Digital simulation of current and potential steps in thin layer electrochemical cells, I. B. Goldberg and A. J. Bard, *J. Electroanal. Chem. Interfacial Electrochem.* **38**, 313–322 (1972).

75. Improvement of the chronopotentiometric method by the use of a potentiostat and a capacity-current addition device, P. Bos and E. Van Dalen, *J. Electroanal. Chem. Interfacial Electrochem.* **45**, 165–179 (1973).

76. Experimental testing of the adsorption theory of chronopotentiometry. I. Electrolytic dissolution of silver from the surface of a graphite electrode, V. V. Pnev, L. A. Moskovskikh, and M. S. Zakharov, *Tr. Tyumen. Ind. Inst.* 60–63 (1973); *CA* 83:67929 s.

77. Inverse voltammetry at graphite electrodes. II. Experimental verification of the absorption. Theory of chronopotentiometry, V. V. Pnev, L. A. Moskovskikh, and M. S. Zakharov, *Usp. Polyargr. Nakopleniem* **157** (1973); *CA* 81:162662 s.

78. A new circuit for cyclic and reverse current chronopotentiometry, M. Declecq and A. Withagen-Declezg, *Anal. Chim. Acta* **63**, 427–433 (1973).

79. Conventional and scanning chronopotentiometry on a rotating disk electrode, V. I. Cherenko and K. I. Litovchenko, *Ukh. Khim. Zh.* **39**, 344–347 (1973) (Russ.).

80. Digital automatic polarograph for stripping chronopotentiometry with amalgam accumulation and trace analysis of elements, *Usp. Polyarogr. Nakopleniem* 220–222 (1973); *CA* 81:145070 e.

81. Automatic instrument for stripping chronopotentiometry with amalgam accumulation, E. V. Galinker, *Usp. Polyarogr. Nakopleniem* 182–184 (1973) (Russ.); *CA* 81:145071 f.

82. Stripping chronopotentiometry on a graphite electrode. Experimental testing of the parametric theory, L. A. Moskovskikh, V. V. Pnev, and M. S. Zakharov, Jr., *Tyumen Ind. Inst.* 73–77 (1973) (Russ.); *CA* 83:1255873.

83. Anodic stripping chronopotentiometry on a glassy carbon disk electrode, F. Vydra and L. Luong, *J. Electroanal. Chem. Interfacial Electrochem.* **54**, 447 (1974).

84. Starting and switching problems and their solution in instruments for fast cyclic chronopotentiometry, F. Jovic and I. Koutusic, *J. Electroanal. Chem. Interfacial Electrochem.* **50**, 269–276 (1974).

85. Measurement of rates of chemical reactions coupled to electron transfer by chronopotentiometry. Disproportionation reaction and the ECE mechanism at a plane electrode, D. Cukman and V. Pravdic, *J. Electroanal. Chem. Interfacial Electrochem.* **49**, 415–419 (1974).

86. Improvements in digital simulation, J. Sandifer and R. P. Buck, *J. Electroanal. Chem. Interfacial Electrochem.* **49**, 161–170 (1974).

87. Multipurpose instruments for chronopotentiometric measurements, A. Dupre and C. Caullet, *Analysis* **3**, 392–395 (1975).

88. Derivative chronopotentiometry. II. Analysis of multicomponent systems, P. E. Sturrock, B. Vandrenil, and R. H. Gibson, *J. Electrochem. Soc.* **122**, 1311–1315 (1975).

89. Stripping chronopotentiometry using a mercury rotating disk electrode. I., V. I. Bakanov, M. S. Zakharov, V. A. Antipeva, and N. M. Cheremnykh, *Elektrokhimiya* **11**, 95–98 (1975).

90. Electroanalytical techniques in trace metal ion analysis, E. E. Brooks and H. B. Mark, *Rev. Anal. Chem.* **3** 1–26 (1975).

91. Chronopotentiometry as an electrochemical method of investigation and analysis, P. K. Agasyan, A. I. Kamanev, and M. I. Luner, *Zh. Anal. Khim.* **31**, 121–142 (1976) (Russ.).

92. Studies in derivative chronopotentiometry. III. Application to submillisecond transition times, P. E. Sturrock and R. H. Gibson, *J. Electrochem. Soc.* **123**, 629–631 (1976).

7. Special Electrode Materials

The special electrode materials employed for electrochemical and electroanalytical studies are of both liquid and solid types. The rotating mercury drop electrode,[90,99,104–107] which was first developed by Kolthoff and co-workers as early as 1956, hanging mercury drop electrode,[102,103,109,114,115] dropping amalgam electrodes,[91,94,110] and streaming mercury electrodes[93,99–101] are the different liquid electrodes employed besides dropping mercury and gallium electrodes. The optically transparent electrodes employed in spectroelectrochemical studies (References 36, 38, 39, 75, 78–80, 87 of Section 7.2), chemically modified electrodes[43,47–49,55–62,70–74] and semiconductor electrodes of different materials[66–69] are all of recent origin and extensively studied by a large number of researchers. The optically transparent electrodes are (1) thin layer of metal deposited on tin oxide film base and (2) minigrids of different metals. The chemically modified electrodes are also prepared by different techniques such as those described by Anson *et al.* and Murray *et al.* Other solid electrodes of special materials employed in electroanalytical studies include pyrolytic graphite, glassy carbon, microelectrodes of Pt, Pd, etc.,[14–34] and metallic electrodes employed as bioelectrodes.[35] The alloy electrodes used in a few electrochemical studies are based on noble metals.[1–13]

7.1. Alloy Electrodes

1. Electrochemical characterisation of the surface composition of heterogeneous platinum–gold alloys, M. W. Breiter, *J. Phys. Chem.* **69**, 901–904 (1965).
2. Hydrogen adsorption on heterogeneous platinum–gold alloys in sulphuric acid solution, M. W. Breiter, *Trans. Faraday Soc.* **61**, 749–754 (1965).
3. Reactivity and surface composition. Anodic methanol oxidation on platinum–gold alloys, M. W. Breiter, *J. Phys. Chem.* **69**, 3377–3383 (1965).
4. Electrolytic codeposited palladium–gold electrodes. Effect of potential cycles on surface properties, R. Woods, *Electrochim. Acta* **14**, 632–635 (1969).
5. Cyclic voltammetry of mixed metal electrodes, J. S. Mayell and W. A. Barber, *J. Electrochem. Soc.* **116**, 1333–1338 (1969).
6. Behaviour of Pt–Rh alloys during electro-oxidation of ethylene, A. A. Michri, A. G. Pshenichnikov, R. Kh. Burshtein, and V. S. Bernard, *Sov. Electrochem.* **5**, 558–560 (1969).
7. Chemisorption of hydrogen and oxygen on platinum–rhodium alloys, K. A. Radyushkina, R. Kh. Burshtein, M. R. Tarasevich, V. V. Kuprina, and L. A. Cheriyaev, *Sov. Electrochem.* **5**, 1309–1312 (1969).
8. Electrolytically codeposited platinum–gold electrodes and their electrocatalytic activity for acetate ion oxidation, R. Woods, *Electrochim. Acta* **14**, 553 (1969).
9. Oxygen reduction on gold alloys in alkaline electrolyte, J. Giner, J. M. Parry, and L. W. Surette, *Adv. Chem. Ser.* **90**, 102–113 (1969).
10. Determination of the surface composition of smooth noble metal alloys by cyclic voltammetry, D. A. J. Rand and R. Woods, *J. Electroanal. Chem. Interfacial Electrochem.* **36**, 57–69 (1972).
11. Adsorption of hydrogen on platinum, osmium and platinum–osmium electrodes, A. A. Sutyagina, I. N. Golyanitskaya, and G. D. Vovchenko, *Sov. Electrochem.* **8**, 884–886 (1972).

12. Electrosorption characteristics of thin layers of noble metal substrates, D. A. J. Rand and R. Woods, *J. Electroanal. Chem. Interfacial Electrochem.* **44**, 83–89 (1973).
13. Chemisorption at electrodes, hydrogen and oxygen on noble metals and their alloys, R. Woods, In: *Electroanalytical chemistry*, A. J. Bard, Vol. 9, pp. 1–162, Marcel Dekker, Inc., New York (1976).

7.2. Solid Electrodes

14. Voltammetry with solid microelectrodes, In: *Polarography*, I. M. Kolthoff and J. J. Lingane, Vol. 1, pp. 399–420, Interscience Publishers, New York (1952).
15. Polarographic oxidation of phenolic compounds (graphite indicator electrode), V. F. Gaylor and P. J. Elving, *Anal. Chem.* **25**, 1078–1082 (1953).
16. Convection controlled limiting currents. I. The platinum wire convection electrode, I. M. Kolthoff and J. Jordon, *J. Am. Chem. Soc.* **76**, 3843 (1954).
17. Rotated and stationary platinum wire electrodes. Residual current voltage curves and dissolution patterns in supporting electrolytes, I. M. Kolthoff and N. Tanaka, *Anal. Chem.* **26**, 632–636 (1954).
18. Polarographic studies with gold, graphite and platinum electrodes, S. S. Lord, Jr. and L. B. Rogers, *Anal. Chem.* **26**, 284–295 (1954).
19. Oxidation of platinum electrodes in poltentiometric redox titrations, J. W. Ross and I. Shain, *Anal. Chem.* **28**, 548–555 (1956).
20. Chemical evidence for oxide films on platinum electrometric electrodes, F. Anson and J. J. Lingane, *J. Am. Chem. Soc.* **79**, 4961–4904 (1957).
21. Behaviour of rotating gold microelectrodes, F. Baumann and I. Shain, *Anal. Chem.* **29**, 303–306 (1957).
22. Studies of gold, platinum and palladium indicating electrodes in strongly oxidising aqueous solutions, J. K. Lee, R. N. Adams, and C. E. Bricker, *J. Am. Chem. Soc. Anal. Chim. Acta* **17**, 321–328 (1957).
23. Anodic chromopotentiometry with platinum and gold electrodes. The iodide–iodine–iodate system, F. C. Anson and J. J. Ligane, *J. Am. Chem. Soc.* **79**, 1015–1020 (1957).
24. Carbon paste electrodes, R. N. Adams, *Anal. Chem.* **30**, 1576 (1959).
25. Voltammetry at inert electrodes. I. Analytical applications of boron carbide electrodes, T. R. Mueller and R. N. Adams, *Anal. Chim. Acta* **23**, 467 (1960).
26. Voltammetry at boron carbide and carbon paste electrodes, T. R. Mueller, C. L. Olson, and R. N. Adams, In: *Advances in Polarography*, I. S. Longmuir, ed., Vol. 1, pp. 198–209, Pergamon Press, London (1960).
27. Voltammetry at electrodes with fixed surfaces, R. N. Adams, In: *Treatise on Analytical Chemistry*, I. M. Kolthoff, P. J. Elving, and E. B. Sandell, eds., Part I. Vol. 4, Chap. 47, pp. 2381–2416, Interscience Publishers, New York (1963).
28. Graphite indicating electrodes: Theory, methodology and applicability, P. J. Elving, I. Fried, and W. R. Turner, In: *Polarography 1964*, G. J. Hills, ed., Vol. 1, pp. 277–297, Macmillan, New York (1965).
29. Behaviour of carbon electrodes in aqueous and non-aqueous systems, R. E. Panzer and P. J. Elving, *J. Electrochem. Soc.* **119**, 864–874 (1972).
30. Electrocapillary phenomena at the stress annealed pyrolytic graphite electrode, I. Morcos, *J. Phys. Chem.* **76**, 2750–2753 (1972).
31. Rapid voltammetric method for the estimation of tocopherols and antioxidants in oils and fats, M. D. McBride and D. H. Evans, *Anal. Chem.* **45**, 446–449 (1973).
32. A study of seven different carbon paste electrodes, J. Lindquist, *J. Electroanal. Chem. Interfacial Electrochem.* **52**, 37 (1974).
33. Etude du comportement de la pate de carbone a compose electroactif incorpore, D. Bauer and M. Ph. Gaillochet, *Electrochim. Acta* **19**, 597–606 (1974).

34. Anodic stripping with collection using thin mercury films, D. Laser and M. Ariel, *J. Electroanal. Chem. Interfacial Electrochem.* **49**, 123 (1974).

35. Metallic electrodes; specialised electrodes, In: *Introduction to Bioelectrodes*, C. D. Ferris, Chap. 1 and 8, Plenum Press, New York and London (1974).

36. Spectroelectrochemistry of optically transparent electrodes, T. Kuwana and N. Winograd, In: *Electroanalytical Chemistry*, A. J. Bard, ed., Vol 7, pp. 29–39, Marcel Dekker, Inc., New York (1974).

37. An improved glassy carbon electrode, S. C. Levy and P. R. Farina, *Anal. Chem.* **47**, 604 (1975).

38. Optically transparent carbon films electrodes for infrared spectroelectrochemistry, J. S. Mattson and C. A. Smith, *Anal. Chem.* **47**, 1122 (1975).

39. Mercury thin film minigrid optically transparent thin layer electrochemical cell, W. R. Heinemann, P. P. Angelis, and J. F. Goelz, *Anal. Chem.* **47**, 1364 (1975).

40. Silver based mercury film electrode. I. General characteristics and stability of the electrode, Z. Stojek and Z. Kublik, *J. Electroanal. Chem. Interfacial Electrochem.* **60**, 349 (1975).

41. Styrene impregnated cobalt-60 irradiated graphite electrodes for anodic stripping analysis, R. G. Clenn and A. F. Sciamannia, *Anal. Chem.* **47**, 276 (1975).

42. The reduction of oxygen at a metallized membrane electrode, C. McCallum and D. Pletcher, *Electrochim. Acta* **20**, 811–814 (1975).

43. Chemically modified tin oxide electrode, P. R. Moses, L. Wier, and R. W. Murray, *Anal. Chem.* **47**, 1882–1886 (1975).

44. A chiral electrode, B. F. Walkins, J. R. Bebiling, E. Kariv, and L. L. Miller, *J. Am. Chem. Soc.* **97**, 3549–3550 (1975).

45. Illustrative electrochemical behaviour reactants irreversibly adsorbed on graphite electrode surfaces, A. P. Brown, C. Koval, and F. C. Anson, *J. Electroanal. Chem. Interfacial Electrochem.* **72**, 379–387 (1976).

46. Graphite epoxy mercury thin film working electrode for anodic stripping voltammetry, J. E. Anderson and D. E. Tallman, *Anal. Chem.* **48**, 209 (1976).

47. Surface electrochemistry of iron porphyrins and iron on tin oxide electrodes, D. G. Davis and R. W. Murray, *Anal. Chem.* **49**, 194–198 (1977).

48. Molecular anchors for the attachment of metal complexes to graphite electrode surfaces, A. P. Brown and F. C. Anson, *J. Electroanal. Chem. Interfacial Electrochem.* **83**, 203–206 (1977).

49. Chemically modified electrodes. VI. Binding and reversible electrochemistry of tetra (aminophenyl) porphyrin on glassy carbon, J. C. Lennox and R. W. Murray, *J. Electroanal. Chem.* **78**, 395–401 (1977).

50. Application of a novel thermistor mercury electrode to the study of changes of activity of an adsorbed enzyme on electrochemical reduction and oxidation, K. S. V. Santhanam, N. Jespersen, and A. J. Bard, *J. Am. Chem. Soc.* **99**, 274–276 (1977).

51. Radiation cured polymer impregnated graphite electrodes for anodic stripping voltammetry, K. G. McLaren and G. E. Batley, *J. Electroanal. Chem. Interfacial Electrochem.* **79**, 169 (1977).

52. Semiconductor electrodes. XIV. Electrochemistry and electroluminescence at n-type TiO_2 in aqueous solutions, R. N. Noufi, P. A. Kohl, S. N. Frank, and A. J. Bard, *J. Electrochem. Soc.* **125**, 246–252 (1978).

53. Semiconductor electrodes. XV. Photoelectrochemical cells with mixed polycrystalline n-type Cds–CdSe electrodes, R. N. Noufii, P. A. Kohl, and A. J. Bard, *J. Electrochem. Soc.* **125**, 375–379 (1978).

54. Electrochemical and solid state studies of phthalocyanine thin film electrodes, H. T. Tachikawa and L. R. Faulkner, *J. Am. Chem. Soc.* **100**, 4379–4385 (1978).

55. Chemically modified electrodes. 10. Electron spectroscopy for chemical analysis and alternating current voltammetry of glassy carbon-bound tetra (aminophenyl) porphyrins, J. C. Lennox and R. W. Murray, *J. Am. Chem. Soc.* **100**, 3710–3714 (1978).

56. Preparation of chemically derivatized platinum and gold electrode surfaces. Synthesis, characterization and surface attachment of trichlorosilyl ferrocene, (1,1' ferrocenediyl) dichlorosilane, and 1,1' bis(triethoxysilyl)ferrocene, M. S. Wrighton, M. C. Palazzoto, A. B. Bocarsly, J. M. Bolts, A. B. Fischer, and L. Nadjo, *J. Am. Chem. Soc.* **100**, 7264–7271 (1978).

57. Introduction of amine functional groups on graphite electrode surfaces and their use in the attachment of ruthenium(II) to the electrode surface, N. Oyama, A. P. Brown, and F. C. Anson, *J. Electroanal. Chem. Interfacial Electrochem.* **87**, 435 (1978).

58. Cyanuric chloride as a general linking agent for modified electrodes: Attachment of redox groups to pyrolytic graphite, A. M. Yacnych and T. Kuwana, *Anal. Chem.* **50**, 640–645 (1978).

59. Electrode surface modification via polymer adsorption, L. C. Miller and M. R. Van De Mark, *J. Am. Chem. Soc.* **100**, 639–640 (1978).

60. Preparation, analysis and use of an electrode surface modified by polymer adsorption, L. L. Miller and M. R. Van De Mark, *J. Electroanal. Chem. Interfacial Electrochem.* **88**, 437–440 (1978).

61. The theory of light induced evolution of hydrogen at semiconductor electrodes, J. O. 'M Bockris and K. Uosaki, *J. Electrochem. Soc.* **125**, 223–227 (1978).

62. Organo-modified metal oxide electrode. IV. Analysis of covalently bound rhodamine B. photoelectrode, M. Fujihira, T. Osa, D. Hursh, and T. Kuwana, *J. Electroanal. Chem. Interfacial. Electrochem.* **88**, 285–288 (1978).

63. Behavior of polymeric sulphur nitride (SN)$_x$ electrodes in aqueous media, R. J. Nowak, W. Kutner, H. B. Mark, and A. G. MacDiamid, *J. Electrochem. Soc.* **125**, 232–240 (1978).

64. The behaviour of an electrochemical detector used in liquid chromatography and continuous flow voltammetry. Part 2. Evaluation of low temperature isotropic carbon for use as an electrode material, B. R. Hepler, S. G. Weber, and W. C. Purdy, *Anal. Chim. Acta* **102**, 41–59 (1978).

65. Evaluation of the basal plane of pyrolytic graphite as an electrochemical detector for liquid chromatography, R. M. Wightman, E. C. Park, S. Borman, and M. A. Dayton, *Anal. Chem.* **50**, 1410–1414 (1978).

66. Charge transfer processes at semiconductor electrodes, R. Memming, In: *Electroanalytical Chemistry*, A. J. Bard, ed., Vol. 11, pp. 1–84, Marcel Dekker, Inc., New York (1979).

67. Semiconductor Electrodes. XVII. Electrochemical behaviour of n- and p-type InP electrodes in acetonitrile solution, P. A. Kohl and A. J. Bard, *J. Electrochem. Soc.* **126**, 598–603 (1979).

68. Semiconductor electrodes. XIX. An investigation of S/Se substitution in single crystal CdSe and CdS, R. N. Noufii, P. A. Kohl, J. W. Rogers, J. M. White, and A. J. Bard, *J. Electrochem. Soc.* **126**, 949–954 (1979).

69. Semiconductor electrodes. XXI. The characterisation and behaviour of n-Type Fe$_2$O$_3$ electrodes in acetonitrile solutions, R. A. Frellein and A. J. Bard, *J. Electrochem. Soc.* **126**, 1892–1898 (1979).

70. Properties of RuO$_2$ working electrodes in nonaqueous solvents, D. R. Rolison, K. Kuo, M. Umana, D. Brundage, and R. W. Murray, *J. Electrochem. Soc.* **126**, 407–414 (1979).

71. Chemically modified carbon electrodes. Part XVII. Metallation of immobilized tetra (aminophenyl) porphyrin with manganese, iron, cobalt, nickel, copper, zinc and electro-chemistry of diprotonated tetraphenyl porphyrin, R. D. Rocklin and R. W. Murray, *J. Electroanal. Chem. Interfacial Electrochem.* **100**, 271–282 (1979).

72. Chemically modified electrodes. VIII. The interaction of aqueous RuCl$_3$ with native and silanized SnO$_2$ electrodes, L. Wier and R. W. Murray, *J. Electrochem. Soc.* **126**, 617–623 (1979).

73. An x-ray photoelectron spectroscopic study of multilayers of an electroactive ferrocene derivative attached to platinum and gold electrodes, A. B. Fischer, M. S. Wrighton, M. Umana, and R. W. Murray, *J. Am. Chem. Soc.* **101**, 3442–3446 (1979).

74. Facile attachment of transition metal complexes to graphite electrodes coated with polymeric

ligands. Observation and control of metal-ligand coordination among reactants confined to electrode surfaces, N. Oyama and F. C. Anson, *J. Am. Chem. Soc.* **101**, 739–741 (1979).

75. Internal reflection spectroscopy at a Hg–Pt optically transparent electrode, J. E. Goelz, A. Y. Yacynych, H. B. Mark, Jr., and W. R. Heineman, *J. Electroanal. Chem. Interfacial Electrochem.* **103**, 277–280 (1979).

76. Two electron oxidations at illuminated n-type semiconducting silicon electrodes. Use of chemically derivatized photoelectrodes, A. B. Bocarsly, E. G. Walton, M. G. Bradley, and M. S. Wrighton, *J. Electroanal. Chem. Interfacial Electrochem.* **100**, 283–306 (1979).

77. Reticulated vitreous carbon flowthrough electrodes, A. N. Strohl and D. I. Curran, *Anal. Chem.* **51**, 353–357 (1979).

78. Optically transparent thin layer electrode techniques for the study of biological redox system, W. R. Heineman, M. L. Meckstroth, B. J. Norris, and C. H. Su, *Bioelectrochem. Bioeng.* **6**, 577–585 (1979).

79. Metallized plastic optically transparent electrodes, R. Cieslinski and N. R. Armstrong, *Anal. Chem.* **51**, 565–568 (1979).

80. Characterisation of Hg–Pt optically transparent electrodes, J. F. Goelz and W. R. Heineman *J. Electroanal. Chem.* **103**, 147–154 (1979).

81. Mercury film nickel minigrid optically transparent thin layer electrochemical cell, W. R. Heineman, T. P. De Angelis, and J. F. Goelz, *Anal. Chem.* **47**, 1364 (1975).

82. A thin layer spectroelectrochemical study of Cob.(I) alamin to Cob.(III) alamin oxidation process, T. M. Kenyhalz and H. B. Mark, Jr., *J. Electroanal. Chem. Interfacial Electrochem.* **23**, 1656–1662 (1976).

83. Optically transparent thin layer electrode for anaerobic measurements on redox enzymes, B. J. Norris, M. L. Meckstroth, and W. R. Heineman, *Anal. Chem.* **48**, 630–632 (1976).

84. Infrared spectrophotometric observations of the absorption of fibrinogen from solution at optically transparent carbon film electrode surfaces, J. S. Mattson and T. T. Jones, *Anal. Chem.* **48**, 2164–2167 (1976).

85. Electrochemical oxidation of 5-6 diaminouracil. An investigation by thin layer spectroelectrochemistry (gold minigrid), J. L. Owens and G. Dryhurst, *J. Electroanal. Chem. Interfacial Electrochem.* **88**, 171–180 (1977).

86. Mercury-gold minigrid optically transparent thin layer electrode, M. L. Meyer, T. P. De Angelis, and W. R. Heineman, *Anal. Chem.* **49**, 602–666 (1977).

87. Carbon and mercury-carbon optically transparent electrodes, T. P. De Angelis, R..W. Hurst, A. M. Yacynycl, H. B. Mark, Jr., W. R. Heineman, and J. S. Mattson, *Anal. Chem.* **49**, 1395–1398 (1977).

88. Optically transparent vitreous carbon electrode, V. E. Norrell and G. Mamantov, *Anal. Chem.* **49**, 1471–1478 (1977).

89. The specific absorption of anions on a Hg–Pt optically transparent electrode by transmission spectroelectrochemistry, W. R. Heineman and J. F. Goelz, *J. Electroanal. Chem. Interfacial Electrodes*, **89**, 437–441 (1978).

7.3. Liquid Electrodes

90. Polarographic current-voltage curves with dropping amalgam electrodes, J. J. Lingane, *J. Am. Chem. Soc.* **61**, 976–977 (1939).

91. Polarographic studies with the dropping mercury electrode. Part XI. The use of dilute amalgams in the dropping electrode, J. Heyrovsky and M. Kalousek, *Collect. Czech. Chem. Commun.* **11**, 464–473 (1939).

92. Oscillographische polarographie, J. Heyrovsky and J. Forejt, *Z. Phys. Chem.* **193**, 77–96 (1943).

93. *Fundamentals and Applications of the Streaming Mercury Electrode*, A. Rius, Academy of Sciences of Madrid, Bermejo, Madrid (1949).
94. A study of the polarographic behaviour of dropping amalgam electrodes, N. H. Furman and W. C. Cooper, *J. Am. Chem. Soc.* **72**, 5667–5676 (1950).
95. Electrochemical phenomena at a rotating mercury electrode. I. Reduction of metal ions, T. S. Lee, *J. Am. Chem. Soc.* **74**, 5001–5008 (1952).
96. Polarographic studies with a stationary mercury-plated platinum electrode, T. L. Marple and L. B. Rogers, *Anal. Chem.* **25**, 1351–1354 (1953).
97. Coulometric determination of submicrogram amounts of cadmium and zinc with stationary mercury plated platinum electrodes, K. W. Gardiner and L. W. Rogers, *Anal. Chem.* **25**, 1393–1397 (1953).
98. The discharge mechanism of simple and complex zinc ions, H. Gerischer, *Z. Physk. Chem.* **202**, 302–317 (1953).
99. Extension of sensitivity of polarographic analysis with rotating amalgam electrodes, W. D. Cooke, *Anal. Chem.* **25**, 215 (1953).
100. Reduction at streaming mercury electrode. I. The limiting current, J. R. Weaver and R. W. Parry, *J. Am. Chem. Soc.* **76**, 6258–6262 (1954).
101. (a) The streaming mercury electrodes (b) The rotating solid electrode (c) The rotating mercury electrode and (d) Vibrating electrode, In: *New Instrumental Methods in Electrochemistry*, P. Delahay, Chap. 9, pp. 240–249, Interscience Publishers, New York (1954).
102. Kinetics of fast electrode reactions (hanging mercury drop electrode), T. Berzins and P. Delahay, *J. Am. Chem. Soc.* **77**, 6448–6453 (1955).
103. Analytical applications of the hanging mercury drop electrode, J. W. Ross, R. D. DeMars, and I. Shain, *Anal. Chem.* **28**, 1768 (1956).
104. The rotated dropping mercury electrode as a new electrode in polarography, W. Stricks and I. M. Kolthoff, *J. Am. Chem. Soc.* **78**, 2085–2094 (1956).
105. Equation for the limiting current at the rotated dropping mercury electrode, Y. Okinaka and I. M. Kolthoff, *J. Am. Chem. Soc.* **79**, 3326–3339 (1957).
106. Factors to be considered in quantitative polarography with the rotated dropping mercury electrode, I. M. Kolthoff and Y. Okinaka, *Anal. Chim. Acta* **18**, 83–96 (1958).
107. A rotating hanging mercury drop electrode, E. Barendrecht, *Nature* **181**, 764–765 (1958).
108. Application de la goutte pendante de mercure a la determination de minimes quantites de differents ions, W. Kemula and Z. Kublik, *Anal. Chim. Acta* **18**, 104–111 (1958).
109. Voltammetry with hanging mercury drop electrode, W. Kemula, In: *Advances in Polarography*, I. S. Longmuir, ed., Vol. 1, pp. 105–143, Pergamon Press, London (1960).
110. The electrodes, In: *Electrochemical Reactions*, G. Charlot, J. Badoz-Lambling, and B. Tremilon, pp. 140–148, Elsevier Publ. Co., Amsterdam, New York (1962).
111. Dropping electrodes, In: *Polarographic Techniques*, L. Meites, pp. 73–83, Interscience Publishers, New York (1965).
112. Cyclic and stripping voltammetry with graphite based thin mercury film electrodes prepared "in situ", Z. Strojek, B. Stepnik, and Z. Kublik, *J. Electroanal. Chem. Interfacial Electrochem.* **74**, 277 (1976).
113. Comparison of spinning dropping mercury electrode response with polarographic and rotating disc electrode theory, H. J. Mortko and R. E. Cover, *Anal. Chem.* **51**, 1144–1149 (1979).
114. Modification of a commercial micrometer hanging mercury drop electrode, J. E. Bonelli, H. E. Taylor, and R. K. Skogerboe, *Anal. Chem.* **51**, 2412–2413 (1979).
115. Static mercury drop electrode, W. M. Peterson, *Am. Lab.* No. 12, 69–78 (1979).
116. High speed device for synchronization of natural drop experiments with a DME, P. D. Tyma, M. J. Weaver, and C. G. Enke, *Anal. Chem.* **51**, 2300–2302 (1979).

8. Special Cells and Electrode Configurations

Thin layer electrochemical cells (TLC) have been widely used in electrode kinetic studies, spectroelectrochemistry,[1-29] monolayer deposition studies, etc. A number of reviews on TLC have also appeared.[1,2,8,15]

Rotating disc electrodes (RDE), rotating ring-disc electrodes (RRDE), rotating double ring electrodes (RDRE), rotating spherical electrodes, rotating ring hemispherical electrodes and rotating conical electrodes are electrodes having special configurations and these are discussed in Ref. (30)–(75), which include a number of detailed reviews.[36,37,39-43] The disc electrode materials employed vary from Pt, gold, and other noble metals to copper, brass, chromium, stainless steel, etc., while for the ring electrode Pt, gold, or other metals can be used depending upon the reaction intermediate to be monitored.

Other electrodes having special configurations and employed in electroanalytical chemistry are ion-selective microelectrodes and microelectrodes used for *in vivo* studies in biological systems.[76-82]

8.1. Thin Layer Cells: Theory, Methodology, and Application

1. Electrochemistry using thin layer cells, C. N. Reilley, *Rev. Pure App. Chem.* **18**, 137 (1968).
2. The theory and practice of electrochemistry with thin layer cells, A. T. Hubbard and F. C. Anson, In: *Electroanalytical Chemistry*, A. J. Bard, ed., Vol. 4, pp. 129–214, Marcel Dekker, Inc., New York (1970).
3. Thin layer electrochemistry, R. S. Tywin and Yu. S. Lyankov, *Zhadonov. S.I. Usp. Khim.* **41**, 2272–2299 (1972).
4. Homogeneous kinetics using electrochemical cells with a large ratio of electrode area to solution volume. Part II. M. Fleishmann, D. Pletcher, and A. Ratinski, *J. Electroanal. Chem. Interfacial Electrochem.* **38**, 329–336 (1972).
5. Spectroelectrochemical study of mixed valence pyrazine-decaamminediruthenium (II, III) complex, V. S. Srinivasan and F. C. Anson, *J. Electrochem. Soc.* **120**, 1359 (1972).
6. Resistive effects in thin electrochemical cells: digital simulations of current and potential steps in thin layer electrochemical cells, I. B. Goldberg and A. J. Bard, *J. Electroanal. Chem. Interfacial Electrochem.* **38**, 313–322 (1972).
7. Resistive effects in thin electrochemical cells: digital simulation of electrochemistry in electron spin resonance cells, I. B. Goldberg, A. J. Bard, and S. W. Feldberg, *J. Phys. Chem.* **76**, 2550 (1972).
8. Electrochemistry in thin layers of solution, A. T. Hubbard, *CRC Crit. Rev. Anal. Chem.* **3**, 201–242 (1973).
9. Construction of an optically transparent thin film electrochemical cell, D. Ratard, P. Belin, and V. Pickton, *Analusis* **2**, 413–419 (1973).
10. Untersuchung der chloridasorption an silber und gold mit hilfe eines zweielektroden-dwins-chichtverfohrens, E. Schmidt and S. Stucki, *J. Electroanal. Chem. Interfacial Electrochem.* **43**, 425–440 (1873).
11. Measurement of enzyme $E^{o\prime}$ values by optically transparent thin layer electrochemical cells, W. R. Heinemann, B. J. Norris, and J. F. Goelz, *Anal. Chem.* **47**, 79 (1975).
12. Mercury film nickel minigrid optically transparent thin layer electrochemical cell, W. R. Heineman, T. P. De Angelis, and J. F. Goelz, *Anal. Chem.* **47**, 1364 (1975).
13. Use of optically transparent thin layer cells for determination of composition and stability constants of metal ion complexes with electrode reaction products, I. Piljack, M. Jkalee, and B. Grabaric, *Anal. Chem.* **47**, 1369 (1975).

14. Differential pulse anodic stripping voltammetry in a thin layer electrochemical cell, T. P. De Angelis and W. R. Heineman, *Anal. Chem.* **48**, 2262–2263 (1976).
15. Thin layer electrochemistry, E. Yeager and J. Kuta, In: *Techniques in Electrochemistry*, E. Yeager and A. J. Salkind, eds., Vol. 1, pp. 167–169 (1976).
17. Theoretical study of a two step reversible electrochemical reaction associated with irreversible chemical reaction in thin layer linear potential sweep voltammetry, V. Plichon and E. Laviron, *J. Electroanal. Chem. Interfacial Electrochem.* **71**, 143–156 (1976).
18. Study of platinum electrodes by means of thin layer electrochemistry and low energy electron diffraction. Part I. Electrode surface structure after exposure to water and aqueous electrolytes, R. M. Ishikawa and A. T. Hubbard, *J. Electroanal. Chem. Interfacial Electrochem.* **69**, 317 (1976).
19. A thin layer spectroelectrochemical study of Cob(I) alamin to Cob(III) alamin oxidation process, T. M. Kenyhelz and H. B. Mark, Jr., *J. Electroanal. Chem. Interfacial Electrochem.* **23**, 1656–1662 (1976).
20. Electrochemical oxidation of 5,6 diaminouracil. An investigation by thin layer spectroelectrochemistry, J. L. Owens and G. Dryhurst, *J. Electroanal. Chem. Interfacial Electrochem.* **88**, 171–180 (1977).
21. A coulometric thin layer flow cell for trace amount detection and its application to adsorption studies, H. Siegenthaler and E. Schmidt, *J. Electroanal. Chem. Interfacial Electrochem.* **80**, 129 (1977).
22. Thin layer differential pulse voltammetry, T. P. DeAngelis, R. E. Bond, E. E. Brooks, and W. R. Heineman, *Anal. Chem.* **49**, 1792 (1977).
23. Small volume high performance cell for nonaqueous spectroelectrochemistry, F. M. Hawkridge, J. E. Pembestin, and H. L. Blount, *Anal. Chem.* **49**, 1646–1647 (1977).
24. An electrochemical thin layer cell for spectroscopic studies of photosynthetic electron transport components, F. M. Hawkridge and B. Ke, *Anal. Biochem.* **78**, 76–85 (1977).
25. Thin layer electrochemical technique for monitoring electrogenerated reactive intermediates, R. L. McCreery, *Anal. Chem.* **49**, 206–209 (1977).
26. Thin layer cell for routine applications, J. Caja, A. Cozerwinskii, and H. B. Mark, Jr., *Anal. Chem.* **51**, 1328–1329 (1979).
27. Thin layer spectroelectrochemistry for monitoring kinetics of electrogenerated species, E. A. Blubaugh, A. M. Yacynych, and W. R. Heineman, *Anal. Chem.* **51**, 561–565 (1979).
28. Studies of the thermodynamics of electron transfer reactions of blue copper proteins, N. Sallauta, F. C. Anson, and H. B. Gray, *J. Am. Chem. Soc.* **101**, 455–458 (1979).
29. A small volume thin layer spectroelectrochemical cell for the study of biological components, C. W. Anderson, H. B. Halsak, and W. R. Heineman, *Anal. Biochem.* **93**, 366–372 (1979).
30. Circulating long optical path thin layer electrochemical cell for spectroelectrochemical characterisation of redox enzymes, J. L. Anderson, *Anal. Chem.* **51**, 2312–2315 (1979).

8.2. Rotating Ring Disc Electrode and Related Configurations

31. Vibrating electrodes in amperometric titrations. Part II. Bromometric determinations of antimony and arsenic, E. D. Harris and A. J. Lindsay, *Analyst* **76**, 647–650 (1951).
32. Hydrogen fluoride solvent system apparatus for polarographic studies. Rotating electrode, J. W. Sargent, A. Clifford, and W. R. Lemmon, *Anal. Chem.* **25**, 1727–1729 (1953).
33. F. W. Stoll and H. BerBack, *Monatsch Chem.* **84**, 1179 (1953).
34. Behaviour of a rotating gold microelectrode, F. Bauman and I. Shain, *Anal. Chem.* **29**, 303 (1957).
35. Diffusion coefficients at rotated microelectrodes, E. R. Nightingale, Jr., *Anal. Chim. Acta* **16**, 493–496 (1957).

36. Die anwendung der rotierenden scheibenelektrode mit einen ringe zur untersuchung von zwischenprodukten elektrochemischer reaktionen, A. N. Frumkin, L. Nekrasov, B. Levich, and Ju. Ivanov, *J. Electroanal. Chem.* **1**, 84 (1959/60).

37. *Physico Chemical Hydrodynamics*, V. G. Levich (Trans. Eng.), Chap. I, III, VI, Prentice Hall, Englewood Cliffs, N.J. (1962).

38. The rotating disk system, A. C. Riddiford, In: *Advances in Electrochemistry and Electrochemical Engineering*, P. Delahay, ed., pp. 47–116, Interscience Pub., New York (1966).

39. Mechanistic analysis of oxygen electrode reactions, A. Damjanovic, In: *Modern Aspects of Electrochemistry*, J. O'M. Bockris and B. E. Conway, eds., Vol. 5, Chap. 5, pp. 409–427, Plenum Press, New York (1969).

40. Transport phenomena in electrochemical kinetics, A. J. Arvia and S. L. Marchiano, In: *Modern Aspects of Electrochemistry*, J. O'M. Bockris and B. E. Conway, ed., vol. 6, Chap. 3, pp. 159–241, Butterworths, London (1971).

41. *Ring-disc electrodes*, W. J. Albery and M. L. Hitchman, Clarendon Press, Oxford (1971).

42. *Rotating disk electrode*, Yu. V. Pleskov and V. Yu. Filinovski, Publ. "Nauka," Moscow (1972) (Russ.).

43. The fundamental principles of current distribution and mass transport in electrochemical cells, J. Newman, In: *Electroanalytical Chemistry*, A. J. Bard, ed., Vol. 6, pp. 187–352, Marcel Dekker, Inc., New York (1973).

44. *Electrochemical systems*, J. S. Newman, Chap. 1, 11–17, and 21, Prentice Hall, Inc., Englewood Cliffs, N.J. (1973).

45. Rotating ring-hemispherical electrode for electroanalytical application, D.-T. Chin, *J. Electroanal. Chem. Interfacial Electrochem.* **120**, 631–635 (1973).

46. Mass transfer on the rotating double ring electrodes. Reactions without gas evolution, V. Yu. Filinovskii, J. V. Kadija, B. Z. Nicolic, and M. B. Nakie, *J. Electroanal. Chem. Interfacial Electrochem.* **54**, 39–46 (1974).

47. Current distribution on a rotating sphere below the limiting current, K. Nisancioglu and J. Newman, *J. Electrochem. Soc.* **121**, 241–246 (1974).

48. Ionic mass transport at horizontal disc electrodes with longitudinal vibration, J. J. Podesta, G. F. Paws, and A. J. Arvia, *Electrochim. Acta* **19**, 583–589 (1974).

49. Submicromolar analysis with rotating and hydrodynamically modulated disc electrodes, B. Miller and S. Bruckenstein, *Anal. Chem.* **46**, 2026–2032 (1974).

50. Rotating disc electrode voltammetry using small sample volumes, B. Miller and S. Bruckenstein, *Anal. Chem.* **46**, 2033 (1974).

51. Etude theorique dune electrode tournante a double anneau. Partie I. Recherche due factem d'efficacite par une methode de simulation numerique, J. Margarit and M. Levy, *J. Electroanal. Chem. Interfacial Electrochem.* **49**, 369–376 (1974).

52. Digital simulation of a rotating photoelectrode: Photolysis of benzophenone in alkaline media, J. R. Lubbers, E. W. Resnick, P. R. Gainer, and D. C. Johnson, *Anal. Chem.* **46**, 865 (1974).

53. Mass transfer to an eccentric rotating disc electrode, C. M. Mohr, Jr. and J. Newman, *J. Electrochem. Soc.* **122**, 928–931 (1975).

54. Mass transfer at a rotating disc with rectangular patch electrodes, A. R. Despic, M. V. Mitrovic, B. Z. Nikovik, and S. D. Cvigoric, *J. Electroanal. Chem. Interfacial electrochem.* **60**, 141–149 (1975).

55. *Rotating disc electrode*, Yu. V. Pleskov and V. Yu. Filinovskii, (Eng. Trans.) Consultant Bureau, New York (1976).

56. The rotating cone electrode, Kirowa-Eisner and E. Gileadi, *J. Electrochem. Soc.* **123**, 22–24 (1976).

57. Glassy carbon rotating ring-disc electrode for molten salt studies, J. Phillips, R. J. Gale, R. G. Wier, and R. A. Osteryoung, *Anal. Chem.* **48**, 126 (1976).

58. Rotating ring-disc enzyme electrode for surface catalysis, F. R. Sluu and G. S. Wilson, *Anal. Chem.* **48**, 1679 (1976).

59. The semitransparent rotating disc electrode, W. J. Albery, M. D. Archer, and R. G. Egdell, *J. Electroanal. Chem. Interfacial Electrochem.* **82**, 199–208 (1977).

60. Unraveling reactions with rotating electrodes, S. Bruckenstein and B. Miller, *Acc. Chem. Rev.* **10**, 54 (1977).

61. Theory of a.c. voltammetry at a rotating disk electrode. Part I. Reversible electrode process, K. Tokuda and H. Matsuda, *J. Electroanal. Chem. Interfacial Electrochem.* **82**, 157–171 (1977).

62. Simultaneous reactions on a rotating disk electrode, R. White and J. Newman, *J. Electroanal. Chem. Interfacial Electrochem.* **82**, 173–186 (1977).

63. Ring disc electrodes. Part 18. Collection efficiency for high frequency a.c., W. J. Albery, R. G. Comptor, and A. R. Hillman, *J. Chem. Soc. Faraday Trans.* **74**, 1007–1019 (1978).

64. Theory of a.c. voltammetry at a rotating disk electrode. Part II. Quasireversible and irreversible redox electrode reactions, K. Tokuda and H. Matsuda, *J. Electroanal. Chem. Interfacial Electrochem.* **90**, 149–163 (1978).

65. Kinetics of oxygen reduction reactions involving catalytic decomposition of hydrogen peroxide applications to porous and rotating ring disk electrode, A. J. Appleby and M. Savy, *J. Electroanal. Chem. Interfacial electrochem.* **92**, 15–30 (1978).

66. Digital simulation of a rotating disc electrode with optically transparent ring, R. Doer and E. W. Grabner, *Ber. Bungenges. Phys. Chem.* **82**, 164–168 (1978).

67. An easy to build rotating dis electrode application to B-diketonates of Co(II) and Co(III), G. Ritzer and M. Gross, *J. Electroanal. Chem. Interfacial Electrochem.* **94**, 209–218 (1978).

68. Construction of a rotating ring disc electrode from irregular electrode materials, P. G. Rowley and J. G. Osteryoung, *Anal. Chem.* **50**, 1015–1016 (1978).

69. Theory of a.c. voltammetry at a rotating disk electrode. Part III. Redox electrode reactions coupled with first order chemical reactions, K. Tokuda and H. Matsuda, *J. Electroanal. Chem. Interfacial Electrochem.* **95**, 147–157 (1979).

70. Application of pulsed current electrolysis to a rotating disk electrode system. II. Electrochemical kinetics, K. Viswanathan and H. Y. Cheh, *J. Electrochem. Soc.* **125**, 1616–1618 (1979).

71. Mass transfer effects of electrolysis by periodic currents, K. Viswanathan and H. Y. Cheh, *J. Electrochem. Soc.* **126**, 398–401 (1979).

72. H. J. Mortko and R. E. Cover, *Anal. Chem.* **51**, 1144–1149 (1979).

73. Mass transfer to a rotating hemisphere in the laminar flow region, J. J. Kim and J. J. Jorne, *J. Electrochem. Soc.* **126**, 1937–1938 (1979).

74. Membrane covered rotated disc electrode, D. A. Gough and J. K. Leypoldt, *Anal. Chem.* **51**, 439–444 (1979).

75. Micrometer voltammetric analysis by ring electrode shielding at a rotating ring-disc electrode, S. Bruckenstein and P. R. Gifford, *Anal. Chem.* **51**, 250–255 (1979).

76. Rotated porous carbon disc electrode, W. J. Blaedel and J. Wang, *Anal. Chem.* **52**, 76–80 (1980).

8.3. Microelectrodes

77. *Ion Selective Microelectrodes*, H. J. Bermann and N. C. Herbert, eds., Plenum Press, New York (1974).

78. Microelectrodes, In: *Introduction to Bio electrodes*, C. D. Ferris, Chap. 4, pp. 57–82, Plenum Press, New York and London (1974).

79. Differential double pulse voltammetry at chemically modified platinum electrodes for in vivo determination of catecholamines, R. F. Lane and A. T. Hubbard, *Anal. Chem.* **48**, 1287 (1976).

80. Brain dopaminergic neurons: in vivo electrochemical information concerning storage, metabolism and release process, R. F. Lane, A. T. Hubbard, and C. D. Blaha, *Bioelectrochem. Bioenerg.* **5**, 504–525 (1978).

81. Methods for electroanalysis "in vivo," J. Koryta, M. Brezina, J. Pradac, and J. Pradacova, In: *Electroanalytical Chemistry*, A. J. Bard, ed., Vol. 11, pp. 85–140, Marcel Dekker Inc., New York (1979).

82. Application of semidifferential electroanalysis to studies of neurotransmitters in the central nervous systems, R. F. Lane, A. T. Hubbard, and C. D. Blaka, *J. Electroanal. Chem. Interfacial Electrochem.* **95**, 117–122 (1979).

83. Normal pulse polarography with carbon fiber electrodes for in vitro and in vivo determination of catecholamines, J. L. Ponchen, R. Cespaglio, F. Gonon, M. Jouve, and J. F. Pujol, *Anal. Chem.* **51**, 1483–1486 (1979).

84. In vivo electrochemistry: Behaviour of microelectrodes in brain tissue, H. Y. Cheng, J. Scheuk, R. Huff, and R. N. Adam, *J. Electroanal. Chem. Interfacial Electrochem.* **100**, 23–31 (1979).

9. Spectroelectroanalytical Chemistry

In the last decade, several spectroscopic techniques have been adopted for electrochemical research; this combination is called spectroelectrochemistry. In spectroelectrochemistry, the spectroscopic methods that have been used are low energy electron diffraction (LEED), auger electron spectroscopy (AES), ultraviolet photoelectron spectroscopy (UPS), X-ray photoelectron spectroscopy (XPS or ESCA), ellipsometric spectroscopy, infrared and ultraviolet visible spectroreflectrometry, Raman spectroscopy, electron spin resonance spectroscopy, etc. While many of these spectroscopic methods employed for surface research can provide considerable information (both qualitative and quantitative) regarding surface composition of electrodes, a few of them will provide considerable information for understanding many aspects of the electrode–electrolyte interface. Baker has discussed[8,12] the principles involved and the method of surface analysis by electron spectroscopy. McIntyre[10] has discussed at length the principles and applications of a number of spectroscopic methods for electrochemical research. The need for employing some of the spectroscopic techniques in understanding many problems in electrocatalysis has been clearly discussed by Yeager.[11]

The basic principles, instrumentation, and some applications of X-ray photoelectron spectroscopy (XPS) and auger electron spectroscopy (AES) have been reviewed by Baker,[8] while the basic principles and applications of XPS and AES to electrochemistry have been discussed in great detail by Augustynski and Balsenc.[14] They have shown how the XPS and AES techniques, which were used originally in surface physics and theoretical chemistry, have come to be used increasingly in electrochemistry, particularly in surface research, as these two techniques provide methods for the identification of surface species and their determination. The applications of XPS for surface analysis (both qualitative and quantitative) of oxide films on gold, platinum, palladium, ruthenium oxide, etc. have been discussed by Augustynski and Balsenc,[14] Kim, Winograd *et al.*,[20,23,25,26,42] and Dickinson.[30] The studies on passivation and corrosion of metals and alloys, particularly stainless steels, employing XPS have been also discussed by many researchers and they are given Ref. (14) and Section 9.2.

The application of AES for studies relating to passivity, surface composition of electrodes at different potentials, under potential deposition, etc., are discussed by Bockris, Hubbard, Staehle, Winograd, Baker et al. in Ref. (48)–(63).

The spectroelectrochemical technique based on LEED has also found application in studying the electrocatalytic nature of the Pt single crystal surfaces for hydrogen electrode reactions by Hubbard[64–67,69,73,74] and Yeager.[68,71]

The Raman spectroscopic technique has found applications in the electrochemical studies pertaining to adsorption, nature of species formed, and other phenomena occurring at the metal/solution interfaces. These studies have been discussed by Fleischmann et al.,[75,76,79,80,87,89,97] Van Duyne et al.,[78,82–84,88,96] Yeager et al.,[86,93,95] Bard,[97] and others.

The reflection spectroscopic techniques are internal reflection spectroscopy (IRS) and specular reflection spectroscopy (SRS), which enable in situ characterization of electrode surfaces (qualitatively as well as quantitatively) and studies of electrode reactions within the compact double layer itself. The principles, instrumentation, and applications of the IRS technique are discussed by Hansen.[2,98–104,107] Laser and Ariel[109,110] and Kuwana and Winograd[7] have reviewed the application of IRS for studies with optically transparent electrodes, while Heineman[113] and Mark et al.[105] have discussed the application of IRS. The applications of the SRS technique in electrochemistry, particularly for studies of hydrogen adsorption and oxide formation on Pt, electrogenerated species, and electrocatalysis by adatoms, have been studied by McIntyre et al.,[117,124] Parsons et al.,[106] Bewick et al.,[118,126,128,129] Pleith et al.,[119,121–122] McCreery et al.,[132,133] and Adzic et al.[134–136]

Ellipsometric spectroscopy has also been shown to be a useful technique for in situ studies of electrode surfaces under different environments. The principles, instrumentation, and applications of ellipsometry in electrochemistry have been reviewed by Muller,[4] Kruger,[5] Conway,[1] and Park.[147] Bockris et al.,[137–140] Veselovskii et al.,[143–146] Parsons et al.,[142,150,152] and Reichman et al.[148,149] have applied ellipsometry to study passivation on nickel and iron, oxide formation on Pt and gold, adsorption at mercury/solution interface, and to study surface processes at noble metal electrodes.

Transmission spectroelectrochemistry (or chronoabsorptometry) (TSE) usually employs optically transparent electrodes OTE and, very often, thin layer electrochemical cells. Optically transparent thin layer electrodes (OTTLE), in which different metallic minigrids serve as OTE, have also been employed in a large number of studies. TSE is used either for determining mechanisms and rates of homogeneous chemical reactions accompanying charge transfer reactions or to study electrode surface phenomena such as the specific adsorption of anions and also for the measurement of spectra and redox potentials of metal complexes and biological redox components. The contributions of Blount,[158,169,192,201,202] Wilson *et al.*,[155,161,177,199,200] Kuwana *et al.*,[154,168,178,196] Hawkridge,[153,179,183,186,202,211] Heineman,[162,163,174,175,184,185,187,194,203,204,206,207,209] Reilley[166,167,191] and Van Duyne,[166,167] and Mark[171,172,185,190] employed TSE (or chronoabsorptometry) for different purposes mentioned above.

Dryhurst,[180,195,205] Mattson *et al.*,[160,176,185] McCreery,[182] and Murray *et al.*[156,173] have employed the TSE for studying the mechanism of organic oxidation reactions, adsorption of organic molecules at carbon electrodes (OTE) by infrared spectrophotometry, studies with TLE to monitor electrogenerated reactive intermediates, and for mechanistic studies employing spectroelectrochemical stopped flow kinetic techniques, respectively. Use of electrochemical concentration methods in spectroscopic analysis, especially atomic absorption spectroscopy (AAS), gives rise to a new spectroelectrochemical technique which enhances the accuracy and sensitivity of AAS as discussed in Section I.[212,215,222–224,229–231] Electron spin resonance spectroscopy has also been applied to solve some electrochemical problems, such as detection of radical intermediates in electrochemical transformations, etc., and these are reviewed by McKineey[236] and by Kastening.[225]

9.1. General Principles and Reviews

1. Special techniques in the study of electrode processes and electrochemical adsorption, B. E. Conway, *Tech. Electrochem.* **1**, 389–568 (1972).
2. Internal reflection spectroscopy in electrochemistry, W. N. Hansen, In: *Advances in Electrochemistry and Electrochemical Engineering*, P. Delahay and C. W. Tobias, eds., Vol. 9, Theory: pp. 1–42; Apparatus: pp. 48–51; Application: pp. 51–60, J. Wiley–Interscience Pub., New York (1973).
3. Specular reflection spectroscopy of the electrode solution interface, J. D. E. McIntyre, In: *Advances in Electrochemistry and Electrochemical Engineering*, P. Delahay and C. W. Tobias, eds., Vol. 9, Theory: pp. 61–102 and 119–136; Instrumentation: pp. 102–119; Application, pp. 136–166, J. Wiley–Interscience Pub., New York (1973).
4. Principles of ellipsometry, R. H. Muller, In: *Advances in Electrochemistry and Electrochemical Engineering*, P. Delahay and C. W. Tobias, eds., Vol. 9, Theory: pp. 167–196; Instrumentation: pp. 197–226, J. Wiley–Interscience Pub., New York (1973).
5. Applications of ellipsometry to electrochemistry, J. Kruger, In: *Advances in Electrochemistry and Electrochemical Engineering*, P. Delahay and C. W. Tobias, eds., Vol. 9, pp. 227–280, J. Wiley–Interscience Pub., New York (1973).
6. Investigation of electrode surfaces by means of combined electrochemical and electron-scattering techniques, A. T. Hubbard, R. M. Ishikawa, and J. A. Schoeffel, *Proc. Symp. Electrocatalysis*, M. W. Breiter, ed., Electrochemical Society, Princeton, N.J., pp. 258–267 (1974).
7. Spectroelectrochemistry at optically transparent electrodes. I. Electrodes under semi-infinite diffusion conditions, T. Kuwana and N. Winograd, In: *Electroanalytical Chemistry*, A. J. Bard, ed., Vol. 7, pp. 1–78, Marcel Dekker, Inc., New York (1974).
8. Surface analysis by electron spectroscopy, B. G. Baker, In: *Modern Aspects of Electrochemistry*, J. O'M. Bockris and B. E. Conway, eds., Vol. 10, pp. 93–160, Plenum Press, New York (1975).

9. Electron desorption of adsorbates, J. H. Leck, *Electron. Fis. Apl.* **17**, 178–182 (1974).
10. New spectroscopic methods for electrochemical research, J. D. E. McIntyre, In: *Trends in Electrochemistry*, J. O'M. Bockris, D. A. J. Rand, and B. J. Welch, eds., pp. 203–231, Plenum Press, New York (1977).
11. Recent advances in the understanding of electrocatalysis and its relation to surface chemistry, E. Yeager, In: *Proceedings of the Symposium on Electrode Materials and Processes for Energy Conversion and Storage*, Proceedings, Vol. 77-6, The Electrochemical Society, Inc., Princeton, N.J. (1977).
12. The application of auger electron spectroscopy to the study of electrode surfaces, B. G. Baker, In: *Trends in Electrochemistry*, J. O'M. Bockris, D. A. J. Rand, and B. J. Welch, eds., pp. 233–239, Plenum Press, New York (1977).
13. Recent advances in some optical experimental methods, R. H. Muller, *Electrochim. Acta* **22**, 951–965 (1977).
14. Application of auger and photoelectron spectroscopy to electrochemical problems, J. Augustynski and L. Balsenc, In: *Modern Aspects of Electrochemistry*, J. O'M. Bockris and B. E. Conway, Eds., Vol. 13, Chap. 4, Auger electron spectroscopy (AES) and X-ray excited photoelectron spectroscopy (XPS), pp. 253–290, AES and XPS for surface analysis, surface analysis applied to electrocatalysis and application of AES and XPS to passivity and corrosion studies, pp. 299–347, Plenum Press, New York (1978).
15. Spectroelectrochemistry: Combination of optical and electrochemical techniques for studies of redox chemistry, W. R. Heineman, *Anal. Chem.* **50**, 390A–402A (1978).
16. Electrocatalysis of fuel cell reactions—Present status and future prospects, W. E. O'Grady and S. Srinivasan, In: *Proceedings of the Workshop on the Electrocatalysis of Fuel Cell Reactions*, W. E. O'Grady, S. Srinivasan, and R. F. Dudley, eds., Proceed. Vol. 79-2, pp. 5–13, The Electrochemical Society Inc., Princeton, N.J. (1979).
17. Electrocatalyst—support interactions, K. Kinoshita, In: *Proceedings of the Workshop on the Electrocatalysis of Fuel Cell Reactions*, W. E. O'Grady, S. Srinivasan, and R. F. Dudley, eds., Proceed. Vol. 79-2, pp. 144–164, The Electrochemical Society, Inc., Princeton, N.J. (1979).
18. Role of surface science and heterogeneous catalysis in electrocatalysis—Present and future status, B. E. Conway and J. A. Joebstl, In: *Proceedings of the Workshop on the Electrocatalysis of Fuel Cell Reactions*, Proceedings Vol. 79-2, W. E. O'Grady, S. Srinivasan, and R. F. Dudley, eds., pp. 212–220, The Electrochemical Society, Inc., Princeton, N.J. (1979).

9.2. X-ray Photoelectron Spectroelectrochemistry

19. Ionisation spectroscopy of contaminated metal surfaces, R. L. Gerlach, *J. Vac. Sci. Technol.* **8**, 599–604 (1971).
20. Electronspectroscopy of platinum oxygen surfaces and application to electrochemical studies, K. S. Kim, N. Winograd, and R. E. Davis, *J. Am. Chem. Soc.* **93**, 6296–6297 (1971).
21. Electron spectroscopy for chemical analysis and Moessbauer investigations of some porous teflon-active carbon–phthalocyanine electrodes, R. Larsson, J. Mrha, and J. Blomqvist, *Acta Chem. Scand.* **26**, 3386–3388 (1972).
22. Trace analysis by ESCA (electron spectroscopy for chemical analysis) electrochemical measurements, J. S. Brinen and J. E. McClure, *Anal. Lett.* **5**, 737–743 (1972).
23. X-ray photoelectron spectra of lead oxides, K. S. Kim, T. J. O'Leary, and N. Winograd, *Anal. Chem.* **45**, 2214 (1973).
24. X-ray photoelectron spectroscopy of adsorbed oxygen and carbonaceous species on platinum electrodes, G. C. Allen, P. M. Tucker, A. Capon, and R. Parsons, *J. Electroanal. Chem. Interfacial Electrochem.* **50**, 335 (1974).

25. X-ray photoelectron spectroscopic studies of palladium oxides and the palladium-oxygen electrode, K. S. Kim, A. F. Grossmann, and N. Winograd, *Anal. Chem.* **46**, 197–200 (1974).
26. X-ray photoelectron spectroscopic studies of ruthenium-oxygen surfaces, K. S. Kim and N. Winograd. *J. Catal.* **35**, 66–72 (1974).
27. ESCA studies of the composition profile of low temperature oxide formed on chromium steels. II. Corrosion in oxygenated water, I. Olefjord and H. Fishmeister, *Corros. Sci.* **15**, 699–707 (1975).
28. Etude potentiostatique et spectroscopique de l' aluminium reconvert par une couche d'oxyde: Effect de differents anions, J. Painot and J. Augustynskii, *Electrochim. Acta* **20**, 747–752 (1975).
29. Electrochemical formation and electron spectroscopic characterization of the platinum sulfide electrode, J. F. Evans, H. N. Blount, and C. R. Ginnard, *J. Electroanal. Chem. Interfacial Electrochem.* **59**, 169 (1975).
30. X-ray photoelectron spectroscopic studies of oxide films on platinum and gold electrodes, Th. Dickinson, A. F. Povey and P. M. A. Sherwood, *J. Chem. Soc. Faraday Trans. I* **71**, 298 (1975).
31. ESCA studies of the passive film on an extremely corrosion resistant amorphous iron alloy, K. Asami, K. Hashimoto, T. Masumoto, and S. Shimodaira, *Corros. Sci.* **16**, 909–914 (1976).
32. XPS study of the interaction between aluminium metal and nitrate ions, J. Augustynski, H. Berthon, and J. Painot, *Chem. Phys. Lett.* **44**, 221–224 (1976).
33. Electrochemical and surface characteristics of tin oxide and indium oxide electrodes, N. R. Armstrong, A. W. C. Lin, M. Fujihira, and T. Kuwana, *Anal. Chem.* **48**(4), 741–750 (1976).
34. The use of X-ray photoelectron spectroscopy in the analysis of passive layers on stainless steel, J. E. Castle and C. R. Clayton, *Corros. Sci.* **17**, 7–26 (1977).
35. XPS determination of compositions of alloy surfaces and surface oxides on mechanically polished iron–chromium alloys, K. Asami, K. Hashimoto, and S. Shimodaira, *Corros. Sci.* **17**, 713–723 (1977).
36. Dissolution and passivation of nickel. An X-ray photoelectron spectroscopic study, Th. Dickenson, A. F. Povey, and P. M. A. Sherwood, *J. Chem. Soc. Faraday I* **73**, 327–343 (1977).
37. X-ray photoelectron spectroscopic studies of tin electrodes after polarization in sodium hydroxide solution, R. O. Ansell, Th. Dickinson, A. F. Povey, and P. M. A. Sherwood, *J. Electrochem. Soc.* **124**, 1360–1364 (1977).
38. On the composition of the passivating film formed on aluminum in chromate solutions, M. Koudelkeva, J. Augustynskii, and H. Berthon, *J. Electrochem. Soc.* **124**, 1165–1168 (1977).
39. X-ray photoelectron and auger spectroscopic study of the underpotential deposition of Ag and Cu on Pt electrodes, J. S. Hammond and N. Winograd, *J. Electrochem. Soc.* **124**, 826–833 (1977).
40. Electrochemical and physicochemical studies of oxygen layers on iridium and ruthenium electrodes, D. A. J. Rand, R. Woods, and D. Michel, In: *Proceed. of the Symposium on Electrode Materials and Processes for Energy Conversion and Storage*, J. D. E. McIntyre, S. Srinivasan, and F. G. Will, eds., Proceed. 77-6, pp. 217–233, The Electrochemical Society Inc., Princeton, N.J. (1977).
41. X-ray photoelectron spectroscopic evidence of distinctive underpotential deposition states of Ag and Cu on Pt substrates, J. S. Hammond and N. Winograd, *J. Electroanal. Chem. Interfacial Electrochem.* **80**, 123–127 (1977).
42. XPS spectroscopic study of potentiostatic and galvanostatic oxidation of Pt electrodes in H_2SO_4 and $HClO_4$, J. S. Hammond and N. Winograd, *J. Electroanal. Chem. Interfacial Electrochem.* **78**, 55–69 (1977).
43. X-ray photoelectron/auger electron spectroscopic studies of tin and iridium metal foils and oxides, A. W. C. Lin, N. R. Armstrong, and T. Kuwana, *Anal. Chem.* **49**(8), 1228–1235 (1977).

44. An XPS study of the passivity of a series of iron–chromium alloys in sulphuric acid, K. Asami, K. Hashimoto, and S. Shimodaira, *Corros. Sci.* **18**, 151–160 (1978).
45. *Passivity of Metals*, J. Augustyskii, R. P. Frankenthal and J. Kruger, eds., The Electrochemical Soc. Inc., Princeton, N.J., p. 989 (1978).
46. A study of ruthenium electrodes by cyclic voltammetry and X-ray emission spectroscopy, D. Michel, D. A. J. Rand, and R. J. Woods, *J. Electroanal. Chem. Interfacial Electrochem.* **89**, 11–27 (1978).
47. X-ray photoelectron spectroscopic studies of RuO_2-based film electrodes, J. Augustynskii, L. Balsenc, and J. Hinden, *J. Electrochem. Soc.* **125**, 1093–1097 (1978).

9.3. Auger Electron Spectroelectrochemistry

48. Chemical analysis of electrodeposited Ni–Ni bonds by auger electron spectroscopy, H. L. Marcus, J. R. Wadrop, F. T. Schuler, and E. F. C. Cain, *J. Electrochem. Soc.* **119**, 1348 (1972).
49. Application of auger electron spectroscopy to the determination of the composition of passive films on type 316 SS, J. B. Lumsden and R. W. Staehle, *Scr. Metall.* **6**, 1205–1208 (1972).
50. Electrochemistry in thin layers of solution, A. T. Hubbard, *CRC Crit. Rev. Anal. Chem.* **3**, 201–242 (1973).
51. Composition of Pt anode surfaces by auger spectroscopy, W. C. Johnson and L. A. Heldt, *J. Electrochem. Soc.* **121**, 34 (1974).
52. The passive film on iron: An application of auger electron spectroscopy, R. W. Revie, B. G. Baker, and J. O'M. Bockris, *J. Electrochem. Soc.* **122**, 1460 (1975).
53. Electron desorption of adsorbates, J. H. Leck, *Electron. Fis. Apl.* **17**, 178–182 (1974).
54. Electrochemical and surface characteristics of tin oxide and indium oxide electrodes, N. R. Armstrong, A. W. C. Lin, M. Fugihira, and T. Kuwana, *Anal. Chem.* **48**(4), 741–750 (1976).
55. X-ray photoelectron and auger spectroscopic study of the underpotential deposition of Ag on Cu and Pt electrodes, J. S. Hammond and N. Winograd, *J. Electrochem. Soc.* **124**, 826–833 (1977).
56. Quantitative elemental analysis of substituted hydrocarbon monolayers on Pt by auger electron spectroscopy with electrochemical calibration, J. A. Schoeffel and A. T. Hubbard, *Anal. Chem.* **49**(14), 2330–2336 (1977).
57. Electrocatalytic properties of single crystal Pt surfaces in aqueous acid electrolytes, P. N. Ross, In: *Proceedings of the Symposium on Electrode Materials and Processes for Energy Conversion and Storage*, J. D. E. McIntyre, S. Srinivasan, and F. G. Will, eds., Proceedings Vol. 77-6, pp. 290–307, The Electrochemical Society Inc., Princeton, N.J. (1977).
58. Auger analysis of the anodic oxide film on iron in neutral solution, M. Seo, M. Sato, J. B. Lumsden, and R. W. Staehle, *Corros. Sci.* **17**, 209–217 (1977).
59. Study by auger spectrometry and cathodic reduction of passive films formed on ferritic stainless steels, M. da cunha Belo, B. Rondot, F. Pons, J. Le Hericy, and J. P. Langeron, *J. Electrochem. Soc.*, **124**, 1317–1324 (1977).
60. The composition of passive films on ferritic stainless steels, A. E. Yaniv, J. B. Lumsden, and R. W. Staehle, *J. Electrochem. Soc.* **124**, 490–496 (1977).
61. A comparison of electrochemical and auger analysis of the surface composition of platinum-rhodium alloys, B. G. Baker, D. A. J. Rand, and R. Woods, *J. Electroanal. Chem. Interfacial Electrochem.* **97**, 189–198 (1979).
62. The use of auger electron spectroscopy for the characterization of the surfaces of spherical gold electrodes, J. Clavilier and J. P. Chauvineau, *J. Electroanal. Chem. Interfacial Electrochem.* **97**, 199–210 (1979).
63. Chemical characterization by auger electron spectroscopy and voltammetry of platinum electrode surfaces prepared in the gas phase, J. Clavilier and J. P. Chauvineau, *J. Electroanal Chem. Interfacial Electrochem.* **100**, 461–472 (1979).

9.4. Low Energy Electron Diffraction Spectroelectrochemistry

64. Electrochemistry in thin layers of solution, A. T. Hubbard, *CRC. Crit. Rev. Anal. Chem.* **3**, 201–242 (1973).

65. Surface characterisation of electrocatalysts by LEED, auguer electron spectroscopy and related techniques, J. A. Joebstl, First Chemical Congress of North American Continent, Mexico City, Nov. 30–Dec. 5, Abs. PHSC-18 (1975).

66. Study of platinum electrodes by means of thin layer electrochemistry and low energy electron diffraction. Part I. Electrode surface structure after exposure to water and aqueous electrolytes, R. M. Ishikawa and A. T. Hubbard, *J. Electroanal. Chem. Interfacial Electrochem.* **69**, 317–338 (1976).

68. Electrode surface studies by LEED-auger, W. E. O'Grady, M. Y. C. Woo, P. L. Hagans, and E. Yeager, *J. Vac. Sci. Technol.* **14**, 365–368 (1977).

69. Ethylene hydrogenation and related reaction on single crystal and polycrystalline Pt electrodes, A. T. Hubbard, J. A. Schoeffel, and H. W. Walter, National Meeting, American Chem. Society, New Orleans, Mar. 1977, Abst. coll. 142.

70. Electrocatalytic properties of single crystal Pt surfaces in aqueous acid electrolytes, P. N. Ross, In: *Proceedings of the Symposium on Electrode Materials and Processes for Energy Conversion and Storage*, J. D. E. McIntyre, S. Srinivasan, and F. G. Will, eds., Vol. 77-6, pp. 290–307, The Electrochemical Society, Inc., Princeton, N.J. (1977).

71. Electrochemical hydrogen adsorption on the Pt(III) and (100) surfaces, W. E. O'Grady, M. Y. C. Woo, P. L. Hagars, and E. Yeager, *Proceed. of the Symp. on Electrode Materials and Processes for Energy Conversion and Storage*, pp. 172–184, Vol. 77-6, The Electrochem. Soc. Inc., Princeton, N.J. (1977).

72. Electrode surface studies by LEED-auger, W. E. O'Grady, M. Y. C. Woo, P. L. Hagans, and E. Yeager, *J. Vac. Sci. Technol.* **14**, 365–368 (1977).

73. Study of platinum electrodes by means of electrochemistry and low energy electron diffraction. Part II. Comparison of the electrochemical activity of Pt(100) and Pt(111) surfaces, A. T. Hubbard, R. M. Ishikawa, and J. Katekaru, *J. Electroanal. Chem. Interfacial Electrochem.* **86**, 271–288 (1978).

74. L.E.E.D. and electrochemistry of iodine on Pt(100) and Pt(111) single crystal surfaces, T. E. Fetter and A. T. Hubbard, *J. Electroanal. Chem. Interfacial Electrochem.* **100**, 473–491 (1979).

9.5. Raman Spectroelectrochemistry

75. Raman spectra from electrode surfaces, M. Fleischmann, P. J. Hendra, and A. J. McQuillan, *J. Chem. Soc. Chem. Commun.* 80–81 (1973).

76. Raman spectra of pyridine adsorbed at a silver electrode, M. Fleischmann, P. J. Hendra, and A. J. McQuillan, *J. Chem. Phys. Lett.* **26**, 163–166 (1974).

77. Laser Raman spectroscopy as a tool for study of diffusion controlled electrochemical processes, J. S. Clarke, A. T. Kuhn, and W. J. Orville-Thomas, *J. Electroanal. Chem. Interfacial Electrochem.* **54**, 253–262 (1974).

78. Resonance Raman spectroelectrochemistry. I. The tetracyano ethylene anion radical, D. L. Jeanmaire, M. R. Suchanaki, and R. P. Van Duyne, *J. Am. Chem. Soc.* **97**, 1699 (1975).

79. Raman spectrosopic investigation of silver electrodes, A. J. McQuillan, P. J. Hendra, and

M. Fleischmann, *J. Electroanal. Chem. Interfacial Electrochem.* **65**, 933–944 (1975).

80. Laser Raman spectroscopy at the surface of a copper electrode, R. L. Paul, A. J. McQuillan, P. J. Hendra, and M. Fleischmann, *J. Electroanal. Chem. Interfacial Electrochem.* **66**, 248–249 (1975).

81. Internal reflection resonance Raman spectroscopy for studies of adsorbed dye layers at electrode-solution interface, M. Fujihire and T. Osa, *J. Am. Chem. Soc.* **98**, 7850–7851 (1976).

82. Resonance Raman spectroelectrochemistry. IV. The oxygen decay chemistry of the tetracyanoquinodimethane dianion, M. R. Suchanski and R. P. Van Duyne, *J. Am. Chem. Soc.* **98**, 250–252 (1976).

83. Resonance Raman spectroelectrochemistry. 2. Scattering spectroscopy accompanying excitation of the lowest $^2B_{1u}$ excited state of the tetracyanoquinodimethane anion radical, D. L. Jeanmaire and R. P. Van Duyne, *J. Am. Chem. Soc.* **98**, 4029–4033 (1976).

84. Resonance Raman spectroelectrochemistry. 3. Tunable dye laser excitation spectroscopy of the lowest $^2B_{1u}$ excited state of the tetracyanoquinodimethane anion radical, D. L. Jeanmaire and R. P. Van Duyne, *J. Am. Chem. Soc.* **98**, 4034–4039 (1976).

85. Resonance Raman spectroscopic detection of electrochemically generated dianion of naphthacene, T. Ikeshoji, T. Mizuno, and T. Sehine, *Chem. Lett.* **11**, 1275–1278 (1976).

86. Light scattering at electrochemical interface, G. Blondeau and E. Yeager, *Prog. Solid State Chem.* **11**, 153 (1976).

87. A Raman spectroscopic study of corrosion of lead electrodes in aqueous chloride media, E. S. Reid, R. P. Cooney, P. J. Hendra, and M. Fleischmann, *J. Electroanal. Chem. Interfacial Electrochem.* **80**, 405–408 (1977).

88. Surface Raman spectroelectrochemistry. Part I. Heterocyclic, aromatic and aliphatic amines adsorbed on the anodized silver electrode, D. L. Jeanmaire and R. P. Van Duyne, *J. Electroanal. Chem. Interfacial Electrochem.* **84**, 1–20 (1977).

89. The Raman spectrum of adsorbed iodine on a platinum electrode surface, R. P. Cooney, E. S. Reid, P. J. Hendra, and M. Fleischmann, *J. Am. Chem. Soc.* **99**, 2002–2003 (1977).

90. Anomalously intense Raman spectra of pyridine at a silver electrode, M. G. Albrecht and J. A. Creighton, *J. Am. Chem. Soc.* **99**, 5215 (1977).

91. Raman spectrum of carbon monoxide on a platinum electrode surface, R. P. Cooney, M. Fleischmann, and P. J. Hendra, *J. Chem. Soc. Chem. Commun.* **7**, 235–237 (1977).

92. Intense Raman spectra at a roughened silver electrode, M. G. Albrecht and J. A. Creighton, *Electrochim. Acta* **23**, 1103–1105 (1978).

93. The Raman spectrum of an adsorbed species on electrode surface, G. Hagen, B. S. Glavski, and E. Yeager, *J. Electroanal. Chem. Interfacial Electrochem.* **88**, 269–275 (1978).

94. Coupled resonance Raman, visible absorption and electrochemistry—applications of a novel circulating cell to multispectral and redox investigations of the haem proteins carboxyhemo-globin and cytochrome C, J. L. Anderson and J. R. Kincaid, *Appl. Spectrosc.* **32**, 356–362 (1978).

96. Resonance Raman spectroelectrochemistry. 6. Ultraviolet laser excitation of the tetracyanoquino dimethane dianion, R. P. Van Duyne, M. R. Suchanski, J. Mlakovits, A. R. Siedle, K. D. Parks, and T. M. Cotton, *J. Am. Chem. Soc.* **101**, 2832–2837 (1979).

97. In situ resonance Raman spectroscopic investigation of the tetrathiafulavalene-tetracyanoquino-dimethane electrode surface, W. L. Wallance, C. D. Jaeger, and A. J. Bard, *J. Am. Chem. Soc.* **101**, 4840–4843 (1979).

9.6. Reflection Spectroelectrochemistry

9.6.1. Internal Reflection Spectroelectrochemistry

 98. Spectrometer cells for single and multiple internal reflection studies in ultraviolet, visible, near infrared and infrared spectral regions, W. N. Hansen and J. A. Horton, *Anal. Chem.* **36**, 783–787 (1964).

 99. Observation of electrode solution interface by means of internal reflection spectrometry, W. N. Hansen, T. Kuwana, and R. A. Osteryoung, *Anal. Chem.* **38**, 1810–1821 (1966).

100. Internal reflection spectroscopic observation of electrode-solution interface, W. N. Hansen, R. A. Osteryoung, and T. Kuwana, *J. Am. Chem. Soc.* **88**, 1062–1063 (1966).

101. Spectroscopic observation of an electrochemical reaction by means of internal reflection, W. N. Hansen, *J. Opt. Soc. Am.* **56**, 380 (1968).

102. Spectroscopic observation of an electrochemical reaction by means of internal reflection, W. N. Hansen, In: *Modern Aspects of Reflectance Spectroscopy*, Proc. Symp. Chicago, 1967, pp. 182–191, W. W. Wendlandt, ed., Plenum Press, New York (1968).

103. Electromodulation of the optical properties of metals, W. N. Hansen, *Surf. Sci.* **16**, 205–216 (1969).

104. Spectroscopic studies of electrochemical reactions of adsorbed dye layers, R. Memming and F. Mollen, *Symp. Faraday Soc.* No. 4, 145–156 (1970).

105. Application of internal reflection spectroscopy to the study of adsorbed layers at interfaces, H. B. Mack Jr. and E. N. Randall, *Symp. Faraday Soc.* **4**, 157–172 (1970).

106. Reflectance studies of the gold/electrolyte interface, B. D. Cahan, J. Horkans, and E. Yeager, *Symp. Faraday Soc.* **4**, pp. 36–44 (1970).

107. Reflection spectroscopy of adsorbed layers, W. N. Hansen, *Symp. Faraday Soc.* No. 4, pp. 27–35 (1970).

108. Optically transparent electrodes, J. W. Strojek, *Chem. Anal. (Warsaw)* **17**, 1023–1029 (1972).

109. Infrared IRS study of electrogenerated species, D. Laser and M. Ariel, *J. Electroanal. Chem. Interfacial Electrochem.* **41**, 381 (1973).

110. Modulated electro-internal reflection spectrometry with a commercial infrared spectrophotometer, D. Laser and M. Ariel, *Anal. Chem.* **45**, 2141 (1973).

111. Spectroelectrochemistry at optically transparent electrodes. I. Electrodes under semi-infinite diffusion conditions, T. Kuwana and N. Winograd, *J. Electroanal. Chem. Interfacial Electrochem.* **7**, 1–78 (1974).

112. Solid state spectroelectrochemistry of cross linked donor bound polymer films, F. B. Kaufman and E. M. Engler, *J. Am. Chem. Soc.* **101**, 547–549 (1979).

113. Internal reflection spectroscopy at a Hg–Pt optically transparent electrode, Harrick rapid scan spectrometer with signal processing model, J. F. Goelz, A. M. Yacynych, H. B. Mark, and W. R. Heineman, *J. Electroanal. Chem. Interfacial Electrochem.* **103**, 277–280 (1979).

9.6.2. Specular Reflection Spectroelectrochemistry

115. Reflectance and ellipsometry of metal/electrolyte interfaces, M. Stedman, *Symp. Faraday Soc.* No. 4, pp. 64–71 (1970).

116. Reflectance studies of adsorption on a platinum electrode, M. A. Barrett and R. Parsons, *Symp. Faraday Soc.* **4**, pp. 72–84 (1970).

117. Specular reflection spectroscopy of electrode surface films, J. D. E. McIntyre and D. M. Kolb, *Symp. Faraday Soc.* **4**, 99–113 (1970).

118. Studies of the cathodic adsorption of hydrogen and the anodic formation of oxide on platinum

in perchloric acid solutions using modulated specular reflectance spectroscopy, A. Bewick and A. M. Tuxford, *Symp. Faraday Soc.* **4**, 114–125 (1970).

119. Studies of adsorbed species at the electrode/electrolyte interface by specular reflection of light, W. J. Pleith, *Symp. Faraday Soc.* **4**, 137–144 (1970).

120. Use of surface reflection in spectroelectrochemistry. Visible spectra of 9,10-diphenyl anthracene radical ions, T. Matsumoto, M. Sato, S. Hirayana, and S. Uemura, *Chem. Lett.* 1077–1080 (1972).

121. Optical methods for studying electrode processes, W. J. Pleith, *Chem. Ing. Tech.* **44**, 221 (1972).

122. Modulation–reflection spectroscopy for studying the kinetics of methylene blue reduction, W. J. Pleith and P. Gruschinske, *Ber. Bunsenges. Phys. Chem.* **76**, 485–491 (1972).

123. Enhanced protein adsorption at the solid-solution interface: Dependence on surface charge, J. S. Mattson and C. A. Smith, *Science* **181**, 1055 (1973).

124. Electrochemical modulation spectroscopy, J. D. E. McIntyre, *Surf. Sci.* **37**, 658–682 (1973).

125. Optical detection of electrochemically generated unstable reduction intermediates of tris (2,2'-bipyridine) chromium(III) by reflectivity measurements of the silver electrode in propylene carbonate, Y. Sato, *Chem. Lett.* 1027–1030 (1973).

126. Studies of the adsorbed hydrogen on platinum cathodes using modulated specular reflectance spectroscopy, A. Bewick and A. M. Tuxford *J. Electroanal. Chem. Interfacial Electrochem.* **47**, 255–264 (1973).

127. The application of reflectance spectroscopy to a study of the anodic oxidation of cuprous sulphide, D. F. A. Koch and R. I. McIntyre, *J. Electroanal. Chem. Interfacial Electrochem.* **71**, 285–296 (1976).

128. Optical studies of the electrode–electrolyte solution interface using reflectance methods. II. The electroreflectance effect at a lead electrode, A. Bewick and J. Robinson, *Surf. Sci.* **55**(1), 349–361 (1976).

129. Optical studies of the electrode–electrolyte solution interface using reflectance methods. Part III. The adsorption of water at a mercury electrode, A. Bewick and J. Robinson, *J. Electroanal. Chem. Interfacial Electrochem.* **71**, 131–141 (1976).

130. Bias potential effects on the anisotropic electroreflectance of single crystal silver, T. E. Furtak and D. W. Lynch, *J. Electroanal. Chem. Interfacial Electrochem.* **79**, 1–17 (1977).

131. On the use of hanging mercury drop and the dropping mercury electrode, situated in a broad homogeneous beam of light as the object in reflectometry and ellipsometry, M. M. J. Pieterse, M. Sluyters-Rehbach, and J. H. Sluyters, *J. Electroanal. Chem. Interfacial Electrochem.* **91**, 55–62 (1978).

132. Observation of electrochemical concentration profiles by absorption spectroelectrochemistry, R. Pruiksma and R. L. McCreery, *Anal. Chem.* **51**, 2253–2257 (1979).

133. Optical monitoring of electrogenerated species via specular reflection at glancing incidence, R. L. McCreery, R. Pruiksma, and R. Fagan, *Anal. Chem.* **51**, 749–752 (1979).

134. Optical and electrochemical study of electrocatalysis by foreign metal adatoms. Oxidation of formic acid on rhodium, R. R. Adzic and A. V. Tripkovic, *J. Electroanal. Chem. Interfacial Electrochem.* **99**, 43–53 (1979).

135. Adsorption of adenine derivative on a gold electrode studied by specular reflectivity measurement, K. Takamura, A. Mori, and F. Watanabe, *J. Electroanal. Chem. Interfacial Electrochem.* **102**, 109–116 (1979).

136. Reflectance study of cation adsorption on oxide layers of gold and platinum electrodes, R. R. Adzic and N. M. Markovic, *J. Electroanal. Chem. Interfacial Electrochem.* **102**, 263–273 (1979).

9.7. Ellipsometry

137. An ellipsometric determination of the mechanism of passivity of nickel, J. O'M. Bockris, A. K. N. Reddy, and B. Rao, *J. Electrochem. Soc.* **113**, 1133–1140 (1966).

138. Ellipsometric studies of oxygen containing film on platinum anodes, A. K. N. Reddy, M. A. Genshaw, and J. O'M. Bockris, *Chem. Phys.* **48**, 671–675 (1968).

139. Electrochemical ellipsometric study of gold, R. S. Sirohi and M. A. Genshaw, *J. Electrochem. Soc.* **116**, 910–914 (1969).

140. Mechanism of film growth and passivation of iron as indicated by transient ellipsometry, J. O'M. Bockris, M. A. Genshaw, and V. Brusic, *Symp. Faraday Soc.* **4**, 177–191 (1970).

141. Ellipsometric techniques and their application to polymer adsorption, R. R. Stromberg, L. E. Smith, and F. L. McCrackin, *Symp. Faraday Soc.* **4**, 192–200 (1970).

142. An ellipsometric investigation of adsorbed layers on platinum electrodes at high anodic potentials, R. Parsons and W. H. M. Visscher, *J. Electroanal. Chem. Interfacial Electrochem.* **36**, 329–336 (1972).

143. Ellipsometric and electrochemical investigation of the Au electrode, Yu. Ya. Vannikov, V. A. Shepelin, and V. I. Veselovskii, *Sov. Electrochem.* **8**, 1201–1204 (1972).

144. Ellipsometric and electrochemical study of the Au electrode. II. Assessment of the thickness and the optical parameters of coatings, Yu. Ya. Vannikov, *Sov. Electrochem.* **8**, 1352–1354 (1972).

145. Ellipsometric and electrochemical investigation of the Pt electrode. I. Formation of a monolayer of oxygen containing species, Yu. Ya. Vannikov, V. A. Shepelin, and V. I. Veselovskii, *Sov. Electrochem.* **9**, 534–536 (1973).

146. An ellipsometric and electrochemical investigation of the Pt electrode. II. Formation and properties of oxide films in the oxygen evolution region, Yu. Ya. Vannikov, V. A. Shepelin, and V. I. Veselovskii, *Sov. Electrochem.* **9**, 624–626 (1973).

147. Ellipsometric optics with special reference to electrochemical systems, W.-K. Park, In: *Electrochemistry*, Physical Chemistry Series One, Vol. 6, Consultant (ed.) A. D. Buckingham, Vol. ed. J. O'M. Bockris, pp. 239–285, Butterworths (London), University Park Press, Baltimore (1973).

148. Combined ellipsometric and reflectometric measurements of surface processes on noble metal electrodes, S. Gottesfeld, M. Babai, and B. Reichman, *Surf. Sci.* **56**(1), 373–393 (1976).

149. The analysis of surface processes on solid electrodes by combined modulated ellipsometric and reflectometric experiments, S. Gottesfeld, M. Babai, and B. Reichman, *Surf. Sci.* **57**(1), 251–265 (1976).

150. Ellipsometry of DNA adsorbed at mercury electrodes—A preliminary study, M. W. Humphreys and R. Parsons, *J. Electroanal. Chem. Interfacial Electrochem.* **75**, 427–436 (1977).

151. Ellipsometric studies of the adsorption of quinolines at the mercury electrode, M. W. Humphreys and R. Parsons, *J. Electroanal. Chem. Interfacial Electrochem.* **82**, 369–390 (1977).

152. An ellipsometric study of the anodic oxidation of nickel in neutral electrolyte, J. L. Ord, J. C. Clayton, and D. J. DeSmet, *J. Electrochem. Soc.* **124**, 1714–1719 (1977).

152. Barbiturates at the mercury/solution interface. An ellipsometric study, K. Kunimatsu and R. Parsons, *J. Electroanal. Chem. Interfacial Electrochem.* **100**, 335–363 (1979).

9.8. Transmission Spectroelectrochemistry

153. On the spectroelectrochemical characterization of the electrocatalytic oxidation of Cu(II) ethylene diamine, D. Meyerstein, F. M. Hawkridge, and T. Kuwana, *J. Electroanal. Chem. Interfacial Electrochem.* **40**, 377 (1972).

154. Electrochemical spectroscopy using the oxide coated optically transparent electrodes, J. W. Strojek and T. Kuwana, *J. Electroanal. Chem. Interfacial Electrochem.* **16**, 471–483 (1968).

155. Theory of potential step transmission chronoabsorptometry, C. Y. Li and G. S. Wilson, *Anal. Chem.* **45**, 2370–2380 (1973).

156. Comparative spectroelectrochemical stopped-flow kinetic and polarographic study of titanium(III) hydroxylamine reaction, M. Petek, T. E. Neal, R. L. McNeely, and R. W. Murray, *Anal. Chem.* **45**, 32 (1973).

157. Automated rapid scan instrument for spectroelectrochemistry in the visible region, E. E. Wells, *Anal. Chem.* **45**, 2022 (1973).
158. Anodic pyridination of 9,10-diphenylanthracene in acetonitrile spectroelectrochemical view, H. N. Blunt, *J. Electronal. Chem. Interfacial Electrochem.* **42**, 271–274 (1973).
159. Application of IR spectroscopy to the study of the platinum electrode at various potentials, Z. A. Markova, A. A. Mikkailova, N. V. Osetrova, and V. S. Bagotski, *Electrokhimiya* **10**, 1794 (1974); *Sov. Electrochem.* **10**, 1701 (1974).
160. Optically transparent carbon film electrodes for infrared spectroelectrochemistry, J. S. Mattson and C. A. Smith, *Anal. Chem.* **47**, 1122 (1975).
161. Some considerations in spectroelectrochemical evaluation of homogeneous electron transfer involving biological molecules, M. D. Rayan and G. S. Wilson, *Anal. Chem.* **47**, 885 (1975).
162. Measurement of enzyme $E^{0\prime}$ values by optically transparent thin layer electrochemical cells, W. R. Heineman, B. J. Norris, and J. F. Goelz, *Anal. Chem.* **47**, 79 (1975).
163. Mercury film nickel minigrid optically transparent thin layer electrochemical cell, W. R. Heineman, T. P. DeAngelis, and J. F. Goelz, *Anal. Chem.* **47**, 1364 (1975).
164. Use of optically transparent thin layer cells for determination of composition and stability constants of metal ion complexes with electrode reaction products, I. Piijac, M. Tkalcce, and B. Grabaric, *Anal. Chem.* **47**, 1369 (1975).
165. Infrared studies of quinone radical anions and dianions generated by flow cell electrolysis, B. R. Clark and D. H. Evans, *J. Electroanal. Chem. Interfacial Electrochem.* **69**, 181–194 (1976).
166. The spectroelectrochemical response for first order E.C. processes with electrode product and reactant adsorption following double potential step excitation, R. P. Van Duyne, T. H. Ridgway, and C. N. Reilley, *J. Electroanal. Chem. Interfacial Electrochem.* **69**, 165–180 (1976).
167. A theoretical investigation of double potential step techniques as applied to the first order, one-half regeneration mechanism at planar electrodes: Chronoamperometry, chronocoulometry and chronoabsorptometry, T. H. Ridgway, R. P. Van Duyne, and C. N. Reilley, *J. Electroanal. Chem. Interfacial Electrochem.* **67**, 1–10 (1976).
168. Spectroelectrochemical kinetic studies of cytochrome-c and cytochrome c oxidase, L. N. Mackey and T. Kuwana, *Biolectrochem. Bioenerget.*, **3**, 596–613 (1976).
169. Reactions of cation radicals of E.E. systems. IV. The kinetics and mechanism of the homogeneous and electrocatalyzed reaction of the cation radical of 9,10-diphenylanthracene with hydrogen sulfide, J. F. Evans and H. N. Blount, *J. Phys. Chem.* **80**, 1011–1017 (1976).
170. The electrochemistry of nitrobenzene and *p*-nitrobenzaldehyde studied by transmission spectroelectrochemical methods in sulfolane, N. R. Armstrong, N. E. Vanderborgh, and R. K. Quinn, *J. Phys. Chem.* **80**, 2740–2745 (1976).
171. The spectroelectrochemical study of the oxidation of 1,2-diaminobenzene: alone and in the presence of Ni(II), A. M. Yacynych and H. B. Mark, Jr., *J. Electrochem. Soc.* **123**, 1346–1351 (1976).
172. A thin layer spectroelectrochemical study of Cob(I) alamin to Cob(III) alamin oxidation process, T. M. Kenyharcz and H. B. Mark, Jr., *J. Electrochem. Soc.* **23**, 1656–1662 (1976).
173. Protonation kinetics and mechanism for 1,8,-dihydroxyanthraquinone and anthraquinone anion radicals in dimethyl formamide solvent, R. M. Wightman, J. R. Cockrell, R. W. Murray, J. N. Burnett, and S. B. Jones, *J. Am. Chem. Soc.* **98**, 2562–2570 (1976).
174. Thin layer spectroelectrochemical study of vitamin B-12 and related cobalamin compounds in aqueous media, T. M. Kenyhercz, T. P. DeAngelis, B. J. Norris, W. R. Heineman, and H. R. Mark, Jr., *J. Am. Chem. Soc.* **98**, 2469–2477 (1976).
175. Optically transparent thin layer electrode for anaerobic measurements on redox enzymes, B. J. Norris, M. L. Meckstroth, and W. R. Heineman, *Anal. Chem.* **48** (3), 630–632 (1976).
176. Infrared spectrophotometric observations of the adsorption of fibrinogen from solution at optically transparent carbon film electrode surfaces, J. S. Mattson and T. T. Jones, *Anal. Chem.* **48** (14), 2164–2167 (1976).

177. Optical pathlength considerations in transmission spectroelectrochemical measurements, F. R. Shu and G. S. Wilson, *Anal. Chem.* **48** (12), 1676–1679 (1976).

178. Spectroelectrochemical investigations of stoichiometry and oxidation reduction potentials of cytochrome c oxidase components in the presence of carbon monoxide: the invisible copper, J. L. Anderson, T. Kuwana, and C. R. Hartzell, *Biochemistry* **15**, 3847–3855 (1976).

179. Redox titration of fluorescence yield of photo system II. B. Ke, F. M. Hawkridge, and S. Sahu, *Proc. Natl. Acad. Sci. USA* **73** (7), 2211–2215 (1976).

180. Electrochemical oxidation of 5,6-diaminouracil. An investigation by thin layer spectroelectrochemistry, J. L. Owens and G. Dryhurst, *J. Electroanal. Chem. Interfacial Electrochem.* **80**, 171–180 (1977).

181. Electrochemistry of vitamin B12. 2. Redox and acid base equilibria in the B12a/B12r system, D. Lexa, J. M. Saveant, and J. Zickler, *J. Am. Chem. Soc.* **99**, 2786–2790 (1977).

182. Thin layer electrochemical technique for monitoring electrogenerated reactive intermediates, R. L. McCreery, *Anal. Chem.* **49** (2), 206–209 (1977).

183. Small volume, high performance cell for non-aqueous spectroelectrochemistry, F. M. Hawkridge, J. E. Pemberton, and H. L. Blount, *Anal. Chem.* **49** (11), 1646–1647 (1977).

184. Mercury–gold minigrid optically transparent thin layer electrode, M. L. Meyer, T. P. DeAngelis, and W. R. Heineman, *Anal. Chem.* **49** (4), 602–606 (1977).

185. Carbon and mercury–carbon optically transparent electrodes, T. P. DeAngelis, R. W. Hurst, A. M. Yacynych, H. B. Mark, Jr., W. R. Heineman, and J. S. Mattson, *Anal. Chem.* **49** (9), 1395–1398 (1977).

186. An electrochemical thin layer cell for spectroscopic studies of photosynthetic electron transport components, F. M. Hawkridge and B. Ke, *Anal. Biochem.* **78**, 76–85 (1977).

187. Thin layer spectroelectrochemical study of tetrakis (4-N-methyl pyridyl) porphinecobalt III), D. F. Rohrbach, E. Deutsch, W. R. Heineman, and R. F. Pasternack, *Inorg. Chem.* **16** (10), 2650–2652 (1977).

188. Circulation cell for electrochemical and simultaneous spectrophotometric measurements, M. Soulard, F. Bloc, and A. Hatter, *Anal. Chim. Acta* **91**, 157 (1977).

189. Spectroelectrochemical studies of olefins. II. A technique for acquisition of the ultraviolet visible spectra of electrogenerated reactive intermediates, E. Steckhan and D. A. Yates, *Ber. Bunsenges. Phys. Chem* **81** (4), 369–374 (1977).

190. Spectroelectrochemical investigation of vitamin B12 and related cobalamins, H. B. Mark, Jr., T. M. Kenyhercz, and P. T. Kissinger, In: Electrochemical Studies of Biological systems, D. T. Sawyer, ed., American Chemical Society, Washington, pp. 1–25 (1977).

191. Analysis of electrochemical mechanisms by finite differences simulation and simplex fitting of double potential step current, charge and absorbance responses, M. K. Hanafey, R. L. Scott, T. H. Ridgeway, and C. N. Reilley, *Anal. Chem.* **50**, 116–137 (1978).

192. Reactions of cation radicals of E.E. systems 7. Mechanistic considerations and relative reactivities of nucleophiles in reaction with the cation radical of 9,10-diphenylanthracene, J. F. Evan and H. N. Blunt, *J. Am. Chem. Soc.* **100**, 4191–4196 (1978).

193. Spectroelectrochemical studies of olefins. 3. The dimerization mechanism of the 4,4′-dimethoxystilbene cation radical in the absence and presence of methanol, E. Steckhan, *J. Am. Chem. Soc.* **100**, 3526–3533 (1978).

194. The specific adsorption of anions on a Hg–Pt optically transparent electrode by transmission spectroelectrochemistry, W. R. Heineman and J. F. Goelz, *J. Electroanal. Chem. Interfacial Electrochem.* **89**, 437–441 (1978).

195. Electrochemical oxidation of uric acid and unanthine—An investigation by cyclic voltammetry, double potential step chronoamperometry and thin layer spectroelectrochemistry, J. L. Owens, H. A. March, Jr., and G. Dryhurst, *J. Electroanal. Chem. Interfacial Electrochem.* **91**, 231–247 (1978).

196. Organo-modified metal oxide electrode. IV. Analysis of covalently bound rhodamine B photoelectrode, M. Fujihira, T. Osa, D. Hursh, and T. Kuwana, *J. Electroanal. Chem. Interfacial Electrochem.* **88**, 285–288 (1978).

197. Spectroelectrochemical studies of some species in fused $PbCl_2 + KCl$ at 440°C, A. De Guibert and V. Plichon, *J. Electroanal. Chem. Interfacial Electrochem.* **90**, 399–411 (1978).
198. Analytical aspects of absorption spectroelectrochemistry at a platinum electrode. I. Study of metal ions, J. F. Tyson and T. S. West, *Talanta* **26**, 117–125 (1979).
199. Analysis of time dependent spectra generated from spectroelectrochemical experiments, D. L. Langhus and G. S. Wilson, *Anal. Chem.* **51**, 1134–1139 (1979).
200. Spectroelectrochemistry and cyclic voltammetry of the E.E. mechanism in a porphyrin diacid reduction, D. L. Langhus and G. S. Wilson, *Anal. Chem.* **51**, 1139–1144 (1979).
201. On the application of open circuit relaxation spectroelectrochemistry to the diagnosis and evaluation of the kinetics of succeeding second order chemical reactions, J. F. Evans and H. N. Blount, *J. Electroanal. Chem. Interfacial Electrochem.* **102**, 289–302 (1979).
202. Spectroelectrochemical determination of heterogeneous electron transfer rate constants, D. E. Albertson, H. N. Blount, and F. M. Hawkridge, *Anal. Chem.* **51**, 556–560 (1979).
203. Transmission spectroelectrochemical study of anion adsorption at a Hg–Pt optically transparent electrode, W. R. Heineman and J. F. Goelz, *J. Electroanal. Chem. Interfacial Electrochem.* **103**, 155–163 (1979).
204. Optically transparent thin layer electrode techniques for the study of biological redox systems, W. R. Heineman, M. L. Meckstroth, B. J. Norris, and C. H. Su, *Bioelectrochem. Bioenerg.* **6**, 577–585 (1979).
205. Enzymatic and electrochemical oxidation of uric acid. A mechanism for the peroxidase catalyzed oxidation of uric acid, H. A. Marsh, Jr. and G. Dryhurst, *J. Electroanal. Chem. Interfacial electrochem.* **95**, 81–90 (1979).
206. Thin layer spectroelectrochemical studies of cobalt and copper schiff base complexes, D. F. Rohrbach, W. R. Heineman, and E. Deutsch, *Inorg. Chem.* **18**, 2536–2542 (1979).
207. A small volume thin layer spectroelectrochemical cell for the study of biological components, C. W. Anderson, H. B. Halsall, and W. R. Heineman, *Anal. Biochem.* **93**, 366–372 (1979),
208. Circulating, long optical path, thin layer electrochemical cell for spectroelectrochemical characterization of redox enzymes, J. L. Anderson, *Anal. Chem.* **51**, 2312–2315 (1979).
209. Characterization of Hg–Pt optically transparent electrodes, J. F. Goelz and W. R. Heineman, *J. Electroanal. Chem. Interfacial Electrochem.* **103**, 147–154 (1979).
210. Oxidation of lead sulphide in molten $PbCl_2 + KCl$; Electrolysis and visible spectrophotometry, A. de Guibert, V. Plichon, and J. Badoz-Lambling, *J. Electroanal. Chem. Interfacial Electrochem.* **105**, 143–148 (1979).
211. On the spectroelectrochemical characterization of the electrocatalytic oxidation of Cu(II) ethylene diamine, D. Meyerstein, F. M. Hawkridge, and T. Kuwana, *J. Electroanal. Chem. Interfacial Electrochem.* **40**, 377 (1972).

9.9. Other Spectroelectrochemical Techniques

212. Use of electrochemical concentration methods in spectroscopic analysis of especially pure substances, V. Z. Krasilshchik and A. F. Yakovleva, *Tr. Vses Nauch-Issled Inst. Khim. Reak. Osobo Chist. Khim. Veshchestv.* No. **33**, 134–142 (1971).
213. Electrohydrodynamic ionization mass spectrometry, A. H. Jones and W. D. France, Jr., *Anal. Chem.* **44**, 1884 (1972).
214. Hanging mercury drop electrodeposition technique for carbon filament flameless atomic absorption analysis. Application to the determination of copper in sea water, C. Fairless and A. J. Bard, *Anal. Chem.* **45**, 2289 (1973).
215. New highly sensitive preconcentrating sampling technique for flameless atomic absorption spectroscopy, M. P. Newton, J. V. Chauvin, and D. G. Davis, *Anal. Lett.* **6**, 89–100 (1973).
216. An isotopic labeling investigation of the mechanism of the electrooxidation of hydrazine at platinum. An electrochemical mass spectrometric study, M. Petek and S. Bruckenstein, *J. Electroanal. Chem. Interfacial Electrochem.* **47**, 329 (1973).

217. X-ray microdetermination of chromium, cobalt, copper, mercury, nickel and zinc in water using electrochemical preconcentration, B. H. Vassos, R. F. Hirsch, and H. Latterman, *Anal. Chem.* **45**, 792 (1973).

218. Lead separation by anodic deposition and isotope ratio mass spectrometry of microgram and smaller samples, I. L. Barnes, T. J. Murphy, J. W. Granolich, and W. R. Shields, *Anal. Chem.* **45**, 1881 (1973).

219. Simultaneous electrochemical–electron spin resonance measurements. II. Kinetic measurements using constant current pulse, J. B. Goldberg and A. J. Bard, *J. Phys. Chem.* **78**, 290 (1974).

220. Simultaneous electrochemical electron spin resonance measurements. III. Determination of rate constants for second order radical anion dimerization, J. B. Goldberg, D. Boyd, R. Hirasawa, and A. J. Bard, *J. Phys. Chem.* **78**, 295 (1974).

221. Application of spin trapping to the detection of radical intermediates in electrochemical transformations, A. J. Bard, J. C. Gilbert, and R. D. Goodwin, *J. Am. Chem. Soc.* **96**, 620 (1974).

222. Electrolytic extraction combined with flame atomic absorption for the determination of metal ions in aqueous solution, J. B. Dawson, D. J. Ellis, T. F. Hartley, M. E. A. Evans, and K. W. Metcalf, *Analyst. (London)* **99**, 602 (1974).

223. Application of electrodeposition techniques to flameless atomic absorption spectrometry. II. Determination of cadmium in sea water, W. Lund and B. V. Larsen, *Anal. Chim. Acta* **72**, 57 (1974).

224. Atomic absorption spectrometric determination of cadmium, lead and zinc in salts or salt solutions by hanging mercury drop electrodeposition and atomization in a graphite furnace, F. O. Jensen, J. Doezal, and F. J. Longmuhr, *Anal. Chim. Acta* **72**, 245–250 (1974).

225. Joint application of electrochemical and ESR techniques, B. Kastening, *Electroanalytical Chemistry*, H. W. Nurnberg, ed., Interscience, New York, pp. 421–494 (1974).

226. Simultaneous electrochemical electron spin resonance measurements with a coaxial microwave cavity, R. D. Allendoerfer, G. A. Martinchek, and S. Bruckenstein, *Anal. Chem.* **47**, 890 (1975).

227. Determination of iron in zirconium by electrolytic dissolution and atomic absorption spectroscopy, M. Mantel and A. Aladjem, *Anal. Lett.* **8**, 415–420 (1975).

228. Flameless atomic absorption spectrometry employing a wire loop atomizer, M. F. Newton and D. G. Davis, *Anal. Chem.* **47**, 2003 (1975).

229. Preconcentration and separation of mercury traces by reduction on metallic copper and determination by flameless atomic absorption spectrometry, S. Dogan and W. Haerdi, *Anal. Chim. Acta* **76**, 345–354 (1975).

230. The application of electrodeposition techniques to flameless atomic absorption spectrometry. Part III. The determination of cadmium in urine, W. Lund, B. V. Larsen, and N. Gundersen, *Anal. Chim. Acta* **81**, 319–324 (1976).

231. The application of electrodeposition techniques to flameless atomic absorption spectrometry. Part IV. Separation and preconcentration on graphite, Y. Thomassen, B. V. Larsen, F. J. Langmuhr, and W. Lund, *Anal. Chim. Acta* **83**, 103–110 (1976).

232. Spectroelectrochemical cell for anaerobic transfer of biological samples for low temperature electron paramagnetic resonance studies, J. L. Anderson, *Anal. Chem.* **48** (6), 921–923 (1976).

233. In situ studies of the passivation and anodic oxidation of cobalt by emission Mossbauer spectroscopy. I. Theoretical background, experimental methods and experimental results for borate solution (pH 8.5), G. W. Simmons, E. Kellerman, and H. Leidheiser, Jr., *J. Electrochem. Soc.* **123**, 1276–1284 (1977).

234. Determination of parts per billion levels of electrodeposited metals by energy dispersive X-ray fluorescence spectrometry, J. A. Boslett, Jr., R. L. R. Towns, R. G. Megargle, K. H. Pearson, and T. C. Furnas, Jr., *Anal. Chem.* **49** (12), 1734–1737 (1977).

235. Determination of heavy metals in sea water by atomic absorption spectrometry after electrodeposition on pyrolytic graphite coated tubes, G. E. Batley and J. P. Matousek, *Anal. Chem.* **49** (13), 2031–2035 (1977).

236. Electron spin resonance and electrochemistry, T. M. McKinney, In: *Electroanalytical Chemistry*, Vol. 10, A. J. Bard, ed., Marcel Dekker Inc., New York, pp. 97–278 (1977).

237. Dynamic X-ray diffraction, R. R. Chianelli, J. C. Scanlon, and B. M. L. Rao, *J. Electrochem. Soc.* **125**, 1563–1566 (1978).

238. Derivatization of surfaces via reaction of strained silicon–carbon bonds. Characterization by photoacoustic spectroscopy, A. B. Fischer, J. B. Kinney, R. H. Staley, and M. S. Wrighton, *J. Am. Chem. Soc.* **101**, 6501–6506 (1979).

239. Observation of semiconductor electrode–dye solution interface by means of fluorescence and laser-induced photoacoustic spectroscopy, T. Iwasaki, T. Sawada, H. Kamada, A. Fujishima, and K. Honda, *J. Phys. Chem.* **83**, 2142–2145 (1979).

240. Characterization and catalytic activity of small platinum particles, R. A. Daka Betta, In: *Proceedings of the Workshop on The Electrocatalysis of Fuel Cell Reactions*, W. E. O'Grady, S. Srinivasan, and R. F. Dudley, eds., Proceed. Vol. 79-2, pp. 203–211, The Electrochemical Society, Inc., Princeton, New Jersey (1979).

Acknowledgment

I wish to thank Dr. H. V. K. Udupa, former Director and Dr. K. S. R. Rajagopalan, present Director, Central Electrochemical Research Institute, Karaikudi, for their encouragement. I also wish to thank Dr. Navin Chandra, Sri R. Chandrasekharan, Sri S. S. Amrit Phale, and Dr. K. Venkateswara Rao of Central Electrochemical Research Institute, Karaikudi, for their kind help in collecting some of the references.

Index